PREFACE

This textbook is an expanded version of *Elementary Linear Algebra*, 6th edition, by Howard Anton, which will soon celebrate its twentieth year as the most widely used textbook in linear algebra. The first ten chapters of this book are identical to the first ten chapters of that text; the eleventh chapter consists of 20 applications of linear algebra drawn from business, economics, engineering, physics, computer science, geometry, approximation theory, ecology, sociology, demography, and genetics. The applications are, for the most part, independent of one another and each comes with a list of mathematical prerequisites. Thus, each instructor has the flexibility to choose those applications that are suitable for his or her students and to incorporate each application anywhere in the course after the mathematical prerequisites have been satisfied.

The changes in this edition are substantial, the most significant being a reorganization to allow for earlier coverage of eigenvalues and eigenvectors. In addition, there are major improvements in exposition, some new text material, changes and additions to the exercises, and new supplementary software and computer-oriented course materials. More details are given later in this preface.

ABOUT THE FIRST TEN CHAPTERS

The objective of the first ten chapters is to present the fundamentals of linear algebra in the clearest possible way. Pedagogy is the main consideration; formalism is secondary. Where possible, basic ideas are studied by means of computational examples and geometrical interpretation.

The treatment of proofs varies. Those proofs that are elementary and have significant pedagogical content are presented precisely, in a style tailored for beginners. A few proofs that are more difficult, but pedagogically valuable, are placed at the ends of the sections and marked "Optional." Still other proofs are omitted completely, with emphasis placed on applying the theorem. Whenever a proof is omitted, the result is usually motivated, often with a discussion about its interpretation in 2-space or 3-space.

It is our experience that Σ-notation is more of a hindrance than a help for beginners in linear algebra. Therefore, we have generally avoided its use.

It is a pedagogical axiom that a teacher should proceed from the familiar to the unfamiliar and from the concrete to the abstract. The ordering of the chapters reflects this tenet.

- **Chapter 1** is concerned with systems of linear equations: how to solve them and some of their properties. It also contains the basic material on matrices and their algebraic properties.

- **Chapter 2** introduces determinants using the classical permutation approach. In our opinion, this approach is less abstract than n-linear alternating forms, gives the student a better intuitive grasp of the subject than an inductive development, and provides the best foundation for a future study of advanced topics in linear algebra.

- **Chapter 3** introduces vectors in 2-space and 3-space as arrows and develops the analytic geometry of lines and planes in 3-space. Depending on the background of the students, this chapter can be omitted without a loss of continuity (see the Guide for the Instructor that follows).

- **Chapter 4** develops the basic results about real finite-dimensional vector spaces; it begins with a study of R^n and proceeds slowly to the general concept of a vector.

- **Chapter 5** is concerned with real inner product spaces. The Gram-Schmidt process is introduced and basic properties about orthonormal bases and orthogonal projections are developed.

- **Chapter 6** is concerned with the eigenvalue problem for matrices and the concept of diagonalization.

- **Chapter 7** is concerned with the basic properties of linear transformations.

- **Chapter 8** gives some applications of linear algebra to problems of approximation, least squares fitting to empirical data, systems of differential equations, Fourier series, and the geometry of plane linear transformations, conic sections, and quadric surfaces.

- **Chapter 9** introduces numerical methods of linear algebra. This chapter gives the reader a basic understanding of the fundamental issues involved in solving practical linear algebra problems. In this chapter some instructors may want to take advantage of the LINEAR-KIT™ software package, which is available as a supplement to this text.

- **Chapter 10** gives a brief introduction to complex numbers and then proceeds to the development of complex vector spaces and inner product spaces. Unitary, Hermitian, and normal matrices are studied, and the chapter concludes with the proof that symmetric matrices have real eigenvalues.

ABOUT CHAPTER 11

This chapter consists of 20 applications of linear algebra. With one clearly marked exception, each application is in its own *independent* section, so that sections can be deleted or permuted freely to fit individual needs and interests. Each topic begins with a list of linear algebra prerequisites so that a reader can tell in advance if he or she has sufficient background to read the section.

Because the topics vary considerably in difficulty, we have included a *subjective* rating of each topic—easy, moderate, more difficult. (See *A Guide for the Instructor* following this preface.) Our evaluation is based more on the intrinsic difficulty of the material rather than the number of prerequisites; thus a topic requiring fewer mathematical prerequisites may be rated harder than one requiring more prerequisites.

Because our primary objective is to present applications of linear algebra, proofs are often omitted. We assume that the reader has met the linear algebra prerequisites and whenever results from other fields are needed, they are stated precisely (with motivation where possible), but usually without proof.

Since there is more material in this book than can be covered in a one-semester or one quarter course, the instructor will have to make a selection of topics. Help in making this selection is provided in the *Guide for the Instructor* below.

NEW FEATURES IN THE SIXTH EDITION

The wide acceptance of the first five editions has been most gratifying, and we are grateful for the many constructive suggestions received from users and reviewers. Substantial portions of the text have been revised for greater clarity, and numerous changes in content and organization have been made in response to reviewer and user suggestions. Here is a summary of the major changes:

- The chapter on eigenvalues, eigenvectors, and diagonalization was moved forward in the text, so it can be covered earlier in the course. This was accomplished by limiting the discussion to matrices and deferring the more general concepts to the later chapter on linear transformations.
- A new application of linear algebra to fractals has been added to Chapter 11.
- The discussion of bases (Section 4.5) has been extensively revised to show more clearly that a basis is the vector-space generalization of a coordinate system.
- The discussions of row space, column space, rank, and nullity (Sections 4.6 and 4.7) have been extensively rewritten.
- The material on inner product spaces has been split off into a separate chapter (Chapter 5).

- As requested by many users, the Gram-Schmidt process has been revised so that the intermediate vectors are orthogonal rather than orthonormal, and the normalization is done at the end of the process.

- The discussion of linear transformations was extensively revised. It now includes the concepts of composition and invertibility, and the parallels between properties of linear transformations and the corresponding properties of their matrix representations are established.

- A section on least squares fitting to empirical data was added to the chapter on applications (Section 8.3).

- Numerous new exercises were added, and many of the drill problems, which had worked their way into departmental and student files over the years, were revised for freshness.

- The material on the geometry of linear operators on R^2, which appeared in the chapter on linear transformations in the previous edition (old Section 5.3), was moved to the chapter on applications (new Section 8.2).

- A number of instructors indicated that they liked to treat transition matrices as matrices of identity operators, so we have added the necessary theorems (Section 7.6).

SUPPLEMENTARY MATERIALS

- *Solutions Manual for Elementary Algebra, Applications Version* by Elizabeth M. Grobe, Charles A. Grobe, Jr., and Chris Rorres contains detailed solutions to most theoretical exercises and many computational exercises.

- *LINEAR-KIT*—A linear algebra software package that can perform most of the basic linear algebra computations using either fractions or decimals. It is available for IBM-compatible computers.

- *LINEAR-KIT Problem Book*—A set of exercises that are intended to be solved using the LINEAR-KIT software.

If any of these supplements is not in your bookstore, ask the bookstore manager to order a copy for you.

A GUIDE FOR
THE INSTRUCTOR

STANDARD COURSE

Chapter 3 can be omitted without loss of continuity if the students have previously studied lines, planes, and geometric vectors in 2-space and 3-space. Depending on the available time and the background of the students, the instructor may wish to add all or part of this chapter to the following suggested core material:

Chapter 1	7 lectures
Chapter 2	5 lectures
Chapter 4	9 lectures
Chapter 5	5 lectures
Chapter 6	4 lectures
Chapter 7	7 lectures

This schedule is rather liberal; it allows a fair amount of classroom time for discussion of homework problems but assumes that little classroom time is devoted to the material marked "optional." The instructor can build on this core as time permits by including lectures on optional material, Chapter 3, Chapter 8, Chapter 9, or Chapter 10.

VARIATIONS OF THE STANDARD COURSE

Several variations of the standard course are easy to implement: Instructors who want to emphasize topics not included in the standard course outline above can reduce the coverage of Chapter 7, as seems appropriate, and use the extra time for numerical methods or other topics. Instructors who prefer to teach linear transformations before eigenvalues can interchange Chapters 6 and 7, deferring the discussion of eigenvalues and eigenvectors in Section 7.6 until after Chapter 6 is covered.

APPLICATIONS-ORIENTED COURSE

Once the necessary core material is covered, the instructor can choose applications from Chapter 8 or Chapter 11. The following table classifies each of the 20 sections in Chapter 11 according to difficulty:

Easy—The average student who has met the stated prerequisites should be able to read the material with no help from the instructor.

Moderate—The average student who has met the stated prerequisites may require a little help from the instructor.

More Difficult—The average student who has met the stated prerequisites will probably need help from the instructor.

	1	2	3	4	5	6	7	8	9	10	11	12	13	14	15	16	17	18	19	20
EASY	●	●																	●	
MODERATE					●	●	●	●	●	●	●				●		●			
MORE DIFFICULT			●									●	●	●		●		●		●

A MICROCOMPUTER-ASSISTED COURSE

It is highly recommended that students with access to IBM-compatible computers be encouraged to use the LINEAR-KIT software that is available through John Wiley & Sons, Inc. This software will save the student many hours of homework time that can be used for other pedagogical purposes. The software is available free to departments adopting this text or can be purchased by individuals through the local bookstore or by using the form in the front of this text. Departmental inquiries should be sent on the university letterhead to:

Order Department
John Wiley & Sons, Inc.
1 Wiley Drive
Somerset, NJ 08873-9969

ACKNOWLEDGMENTS

We express our appreciation for the helpful guidance provided by the following people:

REVIEWERS AND CONTRIBUTORS TO EARLIER EDITIONS

Steven C. Althoen, *University of Michigan–Flint*
C. S. Ballantine, *Oregon State University*
William A. Brown, *University of Maine at Portland–Gorham*
Joseph Buckley, *Western Michigan University*
Hannah Clavner, *Drexel University*
Keith J. Craswell, *Western Washington University*
Dennis DeTurck, *University of Pennsylvania*
Bruce Edwards, *University of Florida*
Harold S. Engelsohn, *Kingsborough Community College*
Garret Etgen, *University of Houston*
Marjorie E. Fitting, *San Jose State University*
David E. Flesner, *Gettysburg College*
Mathew Gould, *Vanderbilt University*
Ralph P. Grimaldi, *Rose–Hulman Institute of Technology*
Collin J. Hightower, *University of Colorado*
Arlene Kleinstein
Lawrence D. Kugler, *University of Michigan*
Roger H. Marty, *Cleveland State University*
Robert M. McConnel, *University of Tennessee*
Harry W. McLaughlin, *Rensselaer Polytechnic Institute*
Douglas McLeod, *Drexel University*
Roberto Mena, *University of Wyoming*
Craig Miller, *University of Pennsylvania*

Donald P. Minassian, *Butler University*
Hal G. Moore, *Brigham Young University*
Robert W. Negus, *Rio Hondo Junior College*
Bart S. Ng, *Purdue University*
Donald R. Sherbert, *University of Illinois*
Charles Shuman, *Drexel University*
Miriam Siegler, *Villanova University*
William F. Trench, *Trinity University*
Oleh J. Tretiak, *Drexel University*
Joseph L. Ullman, *University of Michigan*
James R. Wall, *Auburn University*
Arthur G. Wasserman, *University of Michigan*

REVIEWERS AND CONTRIBUTORS TO THE SIXTH EDITION

Erol Barbut, *University of Idaho*
Risa K. Batterman, *Techsetters, Inc.*
Thomas Cairns, *University of Tulsa*
Douglas E. Cameron, *University of Akron*
Bomshik Chang, *University of British Columbia*
Peter Colwell, *Iowa State University*
Carolyn A. Dean, *University of Michigan–Ann Arbor*
Ken Dunn, *Dalhousie University*
Murray Eisenberg, *University of Massachusetts–Amherst*
Dan Flath, *University of Southern Alabama*
William W. Hager, *University of Florida–Gainesville*
Joseph F. Johnson, *Rutgers University*
Robert L. Kelley, *University of Miami*
Myren Krom, *California State University–Sacramento*
Charles Livingston, *Indiana University*
Nicholas Macri, *Temple University*
Patricia T. McAuley, *SUNY–Binghamton*
Michael R. Meck, *Southern Connecticut State University*
Thomas E. Moore, *Bridgewater State College*
James Osterburg, *University of Cincinnati*
Michael A. Penna, *Indiana University–Purdue University at Indianapolis*
Gerald J. Porter, *University of Pennsylvania*
C. Ray Rosentrater, *Westmont College*
Kenneth Schilling, *University of Michigan–Flint*
Bruce Solomon, *Indiana University*
Mary T. Treanor, *Valparaiso University*
W. Vance Underhill, *East Texas State University*
Evelyn J. Weinstock, *Glassboro State College*
Rugang Ye, *Stanford University*
Frank Zorzitto, *University of Waterloo*
Daniel Zwick, *University of Vermont*

PROBLEM SOLUTIONS

Robin Allred, *Brigham Young University*
Susan Friedman, *Baruch College*
Elena Rakova, *Baruch College*
Mary Karen Solomon, *Brigham Young University*

Thanks are also due to Joan Carrafiello of John Wiley & Sons who did an outstanding job coordinating the work on the answer section.

SUPPLEMENTS

We would like to express our appreciation to the following people for their outstanding work on the supplements.

Marvin L. Brubaker, *Messiah College*
Michael Dagg, *Concentrics, Inc.*
David E. Flesner, *Gettysburg College*
Charles A. Grobe, Jr., *Bowdoin College*
Elizabeth M. Grobe, *Bowdoin College*
Robert S. Smith, *Miami University of Ohio*
Eugene T. Walendziewicz, *Digital Strategies, Inc.*

OTHER CONTRIBUTIONS

Special thanks are due to Dan Flath whose insightful review of the manuscript resulted in hundreds of improvements in the exposition. Thanks are also due to Prof. Gerald J. Porter, who suggested some of the changes in content that appear in this new edition, and to Lilian Brady, whose copyediting skills immensely improved both the style and appearance of the work. Thanks are due to the General Electric Company for graphs and diagrams used in the computed tomography section. We would also like to thank three people who made extraordinary efforts to keep this project on track: Susanne Ingrao and Lucille Buonocore on the Wiley production staff and Ed Burke of Hudson River Studio. Thanks are also due to our editors, Valerie Hunter and Robert Macek for a variety of important contributions. Finally, we are deeply indebted to our assistant, Mary Parker, who contributed to every aspect of this text: coordination, proofreading, word processing, art generation, and most of all, smiles and good cheer at just the right times.

HOWARD ANTON
CHRIS RORRES

CONTENTS

CHAPTER 10 COMPLEX VECTOR SPACES 477

CHAPTER 11 APPLICATIONS OF LINEAR ALGEBRA 527

CHAPTER 1

SYSTEMS OF LINEAR EQUATIONS AND MATRICES

1.1 INTRODUCTION TO SYSTEMS OF LINEAR EQUATIONS

The study of systems of linear equations and their solutions is one of the major topics in linear algebra. In this section we shall introduce some basic terminology and discuss a method for solving such systems.

LINEAR EQUATIONS

A line in the xy-plane can be represented algebraically by an equation of the form

$$a_1 x + a_2 y = b$$

An equation of this kind is called a linear equation in the variables x and y. More generally, we define a ***linear equation*** in the n variables $x_1, x_2, \ldots, x_n$ to be one that can be expressed in the form

$$a_1 x_1 + a_2 x_2 + \cdots + a_n x_n = b$$

where $a_1, a_2, \ldots, a_n$, and b are real constants. The variables in a linear equation are sometimes called the ***unknowns***.

Example 1 The following are linear equations:

$$x + 3y = 7 \qquad x_1 - 2x_2 - 3x_3 + x_4 = 7$$
$$y = \tfrac{1}{2}x + 3z + 1 \qquad x_1 + x_2 + \cdots + x_n = 1$$

Observe that a linear equation does not involve any products or roots of variables. All variables occur only to the first power and do not appear as arguments for trigonometric, logarithmic, or exponential functions. The following are *not* linear equations:

$$x + 3y^2 = 7 \qquad 3x + 2y - z + xz = 4$$
$$y - \sin x = 0 \qquad \sqrt{x_1} + 2x_2 + x_3 = 1 \quad \blacktriangle$$

1

A *solution* of a linear equation $a_1x_1 + a_2x_2 + \cdots + a_nx_n = b$ is a sequence of n numbers $s_1, s_2, \ldots, s_n$ such that the equation is satisfied when we substitute $x_1 = s_1, x_2 = s_2, \ldots, x_n = s_n$. The set of all solutions of the equation is called its *solution set* or sometimes the *general solution* of the equation.

Example 2   Find the solution set of

(a) $4x - 2y = 1$ (b) $x_1 - 4x_2 + 7x_3 = 5$

Solution (a). To find solutions of (a), we can assign an arbitrary value to x and solve for y, or choose an arbitrary value for y and solve for x. If we follow the first approach and assign x an arbitrary value t, we obtain

$$x = t, \qquad y = 2t - \tfrac{1}{2}$$

These formulas describe the solution set in terms of the arbitrary parameter t. Particular numerical solutions can be obtained by substituting specific values for t. For example, $t = 3$ yields the solution $x = 3$, $y = 11/2$; and $t = -1/2$ yields the solution $x = -1/2$, $y = -3/2$.

If we follow the second approach and assign y the arbitrary value t, we obtain

$$x = \tfrac{1}{2}t + \tfrac{1}{4}, \qquad y = t$$

Although these formulas are different from those obtained above, they yield the same solution set as t varies over all possible real numbers. For example, the previous formulas gave the solution $x = 3$, $y = 11/2$ when $t = 3$, while these formulas yield this solution when $t = 11/2$.

Solution (b). To find the solution set of (b) we can assign arbitrary values to any two variables and solve for the third variable. In particular, if we assign arbitrary values s and t to x_2 and x_3, respectively, and solve for x_1, we obtain

$$x_1 = 5 + 4s - 7t, \qquad x_2 = s, \qquad x_3 = t \quad \blacktriangle$$

LINEAR SYSTEMS   A finite set of linear equations in the variables $x_1, x_2, \ldots, x_n$ is called a *system of linear equations* or a *linear system*. A sequence of numbers $s_1, s_2, \ldots, s_n$ is called a *solution* of the system if $x_1 = s_1, x_2 = s_2, \ldots, x_n = s_n$ is a solution of every equation in the system. For example, the system

$$4x_1 - x_2 + 3x_3 = -1$$
$$3x_1 + x_2 + 9x_3 = -4$$

has the solution $x_1 = 1, x_2 = 2, x_3 = -1$ since these values satisfy both equations. However, $x_1 = 1, x_2 = 8, x_3 = 1$ is not a solution since these values satisfy only the first of the two equations in the system.

Not all systems of linear equations have solutions. For example, if we multiply the second equation of the system

$$x + y = 4$$
$$2x + 2y = 6$$

by 1/2, it becomes evident that there are no solutions since the resulting equivalent system

$$x + y = 4$$
$$x + y = 3$$

has contradictory equations

A system of equations that has no solutions is said to be **inconsistent**. If there is at least one solution, it is called **consistent**. To illustrate the possibilities that can occur in solving systems of linear equations, consider a general system of two linear equations in the unknowns x and y:

$$a_1 x + b_1 y = c_1 \qquad (a_1, b_1 \text{ not both zero})$$
$$a_2 x + b_2 y = c_2 \qquad (a_2, b_2 \text{ not both zero})$$

The graphs of these equations are lines; call them l_1 and l_2. Since a point (x, y) lies on a line if and only if the numbers x and y satisfy the equation of the line, the solutions of the system of equations will correspond to points of intersection of l_1 and l_2. There are three possibilities (Figure 1):

(a)   the lines l_1 and l_2 may be parallel, in which case there is no intersection and consequently no solution to the system;

(b)   the lines l_1 and l_2 may intersect at only one point, in which case the system has exactly one solution;

(c)   the lines l_1 and l_2 may coincide, in which case there are infinitely many points of intersection and consequently infinitely many solutions to the system.

Although we have considered only two equations with two unknowns here, we will show later that this same result holds for arbitrary linear systems:

> *Every system of linear equations has either no solutions, exactly one solution, or infinitely many solutions.*

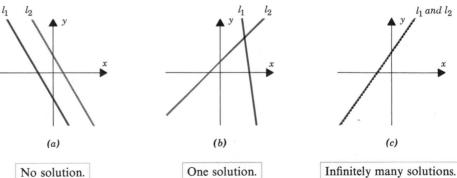

	(a)	(b)	(c)
Figure 1	No solution.	One solution.	Infinitely many solutions.

An arbitrary system of m linear equations in n unknowns will be written

$$a_{11}x_1 + a_{12}x_2 + \cdots + a_{1n}x_n = b_1$$
$$a_{21}x_1 + a_{22}x_2 + \cdots + a_{2n}x_n = b_2$$
$$\vdots \qquad \vdots \qquad \qquad \vdots \qquad \vdots$$
$$a_{m1}x_1 + a_{m2}x_2 + \cdots + a_{mn}x_n = b_m$$

where $x_1, x_2, \ldots, x_n$ are the unknowns and the subscripted a's and b's denote constants.

For example, a general system of three linear equations in four unknowns would be written

$$a_{11}x_1 + a_{12}x_2 + a_{13}x_3 + a_{14}x_4 = b_1$$
$$a_{21}x_1 + a_{22}x_2 + a_{23}x_3 + a_{24}x_4 = b_2$$
$$a_{31}x_1 + a_{32}x_2 + a_{33}x_3 + a_{34}x_4 = b_3$$

The double subscripting on the coefficients of the unknowns is a useful device that is used to specify the location of the coefficient in the system. The first subscript on the coefficient a_{ij} indicates the equation in which the coefficient occurs, and the second subscript indicates which unknown it multiplies. Thus, a_{12} is in the first equation and multiplies unknown x_2.

If we mentally keep track of the location of the $+$'s, the x's, and the $=$'s, a system of m linear equations in n unknowns can be abbreviated by writing only the rectangular array of numbers:

$$\begin{bmatrix} a_{11} & a_{12} & \cdots & a_{1n} & b_1 \\ a_{21} & a_{22} & \cdots & a_{2n} & b_2 \\ \vdots & \vdots & & \vdots & \vdots \\ a_{m1} & a_{m2} & \cdots & a_{mn} & b_m \end{bmatrix}$$

This is called the **augmented matrix** for the system. (The term *matrix* is used in mathematics to denote a rectangular array of numbers. Matrices arise in many contexts; we shall study them in more detail in later sections.) For example, the augmented matrix for the system of equations

$$x_1 + x_2 + 2x_3 = 9$$
$$2x_1 + 4x_2 - 3x_3 = 1$$
$$3x_1 + 6x_2 - 5x_3 = 0$$

is

$$\begin{bmatrix} 1 & 1 & 2 & 9 \\ 2 & 4 & -3 & 1 \\ 3 & 6 & -5 & 0 \end{bmatrix}$$

REMARK. When constructing an augmented matrix, the unknowns must be written in the same order in each equation.

The basic method for solving a system of linear equations is to replace the given system by a new system that has the same solution set but which is easier to solve. This new system is generally obtained in a series of steps by applying the following three types of operations to eliminate unknowns systematically.

1. Multiply an equation through by a nonzero constant.
2. Interchange two equations.
3. Add a multiple of one equation to another.

Since the rows (horizontal lines) of an augmented matrix correspond to the equations in the associated system, these three operations correspond to the following operations on the rows of the augmented matrix.

1. Multiply a row through by a nonzero constant.
2. Interchange two rows.
3. Add a multiple of one row to another row.

ELEMENTARY ROW OPERATIONS

These are called *elementary row operations*. The following example illustrates how these operations can be used to solve systems of linear equations. Since a systematic procedure for finding solutions will be derived in the next section, it is not necessary to worry about how the steps in this example were selected. The main effort at this time should be devoted to understanding the computations and the discussion.

Example 3 In the left column below we solve a system of linear equations by operating on the equations in the system, and in the right column we solve the same system by operating on the rows of the augmented matrix.

$$
\begin{aligned}
x + y + 2z &= 9 \\
2x + 4y - 3z &= 1 \\
3x + 6y - 5z &= 0
\end{aligned}
\qquad
\begin{bmatrix}
1 & 1 & 2 & 9 \\
2 & 4 & -3 & 1 \\
3 & 6 & -5 & 0
\end{bmatrix}
$$

Add -2 times the first equation to the second to obtain

Add -2 times the first row to the second to obtain

$$
\begin{aligned}
x + y + 2z &= 9 \\
2y - 7z &= -17 \\
3x + 6y - 5z &= 0
\end{aligned}
\qquad
\begin{bmatrix}
1 & 1 & 2 & 9 \\
0 & 2 & -7 & -17 \\
3 & 6 & -5 & 0
\end{bmatrix}
$$

Add -3 times the first equation to the third to obtain

Add -3 times the first row to the third to obtain

$$
\begin{aligned}
x + y + 2z &= 9 \\
2y - 7z &= -17 \\
3y - 11z &= -27
\end{aligned}
\qquad
\begin{bmatrix}
1 & 1 & 2 & 9 \\
0 & 2 & -7 & -17 \\
0 & 3 & -11 & -27
\end{bmatrix}
$$

Multiply the second equation by 1/2 to obtain

$$x + y + 2z = 9$$
$$y - \tfrac{7}{2}z = -\tfrac{17}{2}$$
$$3y - 11z = -27$$

Multiply the second row by 1/2 to obtain

$$\begin{bmatrix} 1 & 1 & 2 & 9 \\ 0 & 1 & -\tfrac{7}{2} & -\tfrac{17}{2} \\ 0 & 3 & -11 & -27 \end{bmatrix}$$

Add -3 times the second equation to the third to obtain

$$x + y + 2z = 9$$
$$y - \tfrac{7}{2}z = -\tfrac{17}{2}$$
$$-\tfrac{1}{2}z = -\tfrac{3}{2}$$

Add -3 times the second row to the third to obtain

$$\begin{bmatrix} 1 & 1 & 2 & 9 \\ 0 & 1 & -\tfrac{7}{2} & -\tfrac{17}{2} \\ 0 & 0 & -\tfrac{1}{2} & -\tfrac{3}{2} \end{bmatrix}$$

Multiply the third equation by -2 to obtain

$$x + y + 2z = 9$$
$$y - \tfrac{7}{2}z = -\tfrac{17}{2}$$
$$z = 3$$

Multiply the third row by -2 to obtain

$$\begin{bmatrix} 1 & 1 & 2 & 9 \\ 0 & 1 & -\tfrac{7}{2} & -\tfrac{17}{2} \\ 0 & 0 & 1 & 3 \end{bmatrix}$$

Add -1 times the second equation to the first to obtain

$$x + \tfrac{11}{2}z = \tfrac{35}{2}$$
$$y - \tfrac{7}{2}z = -\tfrac{17}{2}$$
$$z = 3$$

Add -1 times the second row to the first to obtain

$$\begin{bmatrix} 1 & 0 & \tfrac{11}{2} & \tfrac{35}{2} \\ 0 & 1 & -\tfrac{7}{2} & -\tfrac{17}{2} \\ 0 & 0 & 1 & 3 \end{bmatrix}$$

Add $-\tfrac{11}{2}$ times the third equation to the first and $\tfrac{7}{2}$ times the third equation to the second to obtain

$$x = 1$$
$$y = 2$$
$$z = 3$$

Add $-\tfrac{11}{2}$ times the third row to the first and $\tfrac{7}{2}$ times the third row to the second to obtain

$$\begin{bmatrix} 1 & 0 & 0 & 1 \\ 0 & 1 & 0 & 2 \\ 0 & 0 & 1 & 3 \end{bmatrix}$$

The solution

$$x = 1, \quad y = 2, \quad z = 3$$

is now evident. ▲

EXERCISE SET 1.1

1. Which of the following are linear equations in x_1, x_2, and x_3?
   (a) $x_1 + 5x_2 - \sqrt{2}x_3 = 1$ (b) $x_1 - 3x_2 + x_1x_3 = 2$ (c) $x_1 = -7x_2 + 3x_3$
   (d) $x_1^{-2} + x_2 + 8x_3 = 5$ (e) $x_1^{3/5} - 2x_2 + x_3 = 4$ (f) $\pi x_1 - \sqrt{2}x_2 + \tfrac{1}{3}x_3 = 7^{1/3}$

2. Given that k is a constant, which of the following are linear equations?
   (a) $x_1 - x_2 + x_3 = \sin k$ (b) $kx_1 - \dfrac{1}{k}x_2 = 9$ (c) $2^k x_1 + 7x_2 - x_3 = 0$

3. Find the solution set of each of the following linear equations.
   (a) $7x - 5y = 3$
   (b) $3x_1 - 5x_2 + 4x_3 = 7$
   (c) $-8x_1 + 2x_2 - 5x_3 + 6x_4 = 1$
   (d) $3v - 8w + 2x - y + 4z = 0$

4. Find the augmented matrix for each of the following systems of linear equations.

   (a) $3x_1 - 2x_2 = -1$
   $\quad\; 4x_1 + 5x_2 = 3$
   $\quad\; 7x_1 + 3x_2 = 2$

   (b) $2x_1 \qquad\quad + 2x_3 = 1$
   $\quad\; 3x_1 - x_2 + 4x_3 = 7$
   $\quad\; 6x_1 + x_2 - x_3 = 0$

   (c) $x_1 + 2x_2 \qquad - \; x_4 + x_5 = 1$
   $\qquad\quad 3x_2 + x_3 \qquad\; - x_5 = 2$
   $\qquad\qquad\quad\; x_3 + 7x_4 \qquad\; = 1$

   (d) $x_1 \qquad\qquad\qquad = 1$
   $\qquad x_2 \qquad\qquad = 2$
   $\qquad\qquad x_3 = 3$

5. Find a system of linear equations corresponding to each of the following augmented matrices.

   (a) $\begin{bmatrix} 2 & 0 & 0 \\ 3 & -4 & 0 \\ 0 & 1 & 1 \end{bmatrix}$

   (b) $\begin{bmatrix} 3 & 0 & -2 & 5 \\ 7 & 1 & 4 & -3 \\ 0 & -2 & 1 & 7 \end{bmatrix}$

   (c) $\begin{bmatrix} 7 & 2 & 1 & -3 & 5 \\ 1 & 2 & 4 & 0 & 1 \end{bmatrix}$

   (d) $\begin{bmatrix} 1 & 0 & 0 & 0 & 7 \\ 0 & 1 & 0 & 0 & -2 \\ 0 & 0 & 1 & 0 & 3 \\ 0 & 0 & 0 & 1 & 4 \end{bmatrix}$

6. (a) Find a linear equation in the variables x and y that has the general solution $x = 5 + 2t,\; y = t$.

   (b) Show that $x = t,\; y = \frac{1}{2}t - \frac{3}{2}$ is also the general solution of the equation in part (a).

7. The curve $y = ax^2 + bx + c$ shown in Figure 2 passes through the points $(x_1, y_1), (x_2, y_2)$, and (x_3, y_3). Show that the coefficients a, b, and c are a solution of the system of linear equations whose augmented matrix is

   $$\begin{bmatrix} x_1^2 & x_1 & 1 & y_1 \\ x_2^2 & x_2 & 1 & y_2 \\ x_3^2 & x_3 & 1 & y_3 \end{bmatrix}$$

$y = ax^2 + bx + c$

Figure 2

8. For which value(s) of the constant k does the following system of linear equations have no solutions? Exactly one solution? Infinitely many solutions?

   $$x - y = 3$$
   $$2x - 2y = k$$

9. Consider the system of equations

   $$ax + by = k$$
   $$cx + dy = l$$
   $$ex + fy = m$$

   Discuss the relative positions of the lines $ax + by = k$, $cx + dy = l$, and $ex + fy = m$ when
   (a) the system has no solutions
   (b) the system has exactly one solution
   (c) the system has infinitely many solutions

10. Show that if the system of equations in Exercise 9 is consistent, then at least one equation can be discarded from the system without altering the solution set.

11. Let $k = l = m = 0$ in Exercise 9; show that the system must be consistent. What can be said about the point of intersection of the three lines if the system has exactly one solution?

12. Consider the system of equations

$$x + y + 2z = a$$
$$x \quad\ + z = b$$
$$2x + y + 3z = c$$

Show that in order for this system to be consistent, a, b, and c must satisfy $c = a + b$.

13. Prove: If the linear equations $x_1 + kx_2 = c$ and $x_1 + lx_2 = d$ have the same solution set, then the equations are identical.

1.2 GAUSSIAN ELIMINATION

In this section we shall give a systematic procedure for solving systems of linear equations; it is based on the idea of reducing the augmented matrix to a form that is simple enough that the system of equations can be solved by inspection.

REDUCED ROW-ECHELON FORM

In Example 3 of the previous section, we solved the given linear system by reducing the augmented matrix to

$$\begin{bmatrix} 1 & 0 & 0 & 1 \\ 0 & 1 & 0 & 2 \\ 0 & 0 & 1 & 3 \end{bmatrix}$$

from which the solution of the system was evident. This is an example of a matrix that is in *reduced row-echelon form*. To be of this form, a matrix must have the following properties.

1. *If a row does not consist entirely of zeros, then the first nonzero number in the row is a 1. (We call this a **leading** 1.)*
2. *If there are any rows that consist entirely of zeros, then they are grouped together at the bottom of the matrix.*
3. *In any two successive rows that do not consist entirely of zeros, the leading 1 in the lower row occurs farther to the right than the leading 1 in the higher row.*
4. *Each column that contains a leading 1 has zeros everywhere else.*

A matrix having properties **1**, **2**, and **3** (but not necessarily 4) is said to be in *row-echelon form*.

Example 1   The following matrices are in reduced row-echelon form.

$$\begin{bmatrix} 1 & 0 & 0 & 4 \\ 0 & 1 & 0 & 7 \\ 0 & 0 & 1 & -1 \end{bmatrix}, \quad \begin{bmatrix} 1 & 0 & 0 \\ 0 & 1 & 0 \\ 0 & 0 & 1 \end{bmatrix}, \quad \begin{bmatrix} 0 & 1 & -2 & 0 & 1 \\ 0 & 0 & 0 & 1 & 3 \\ 0 & 0 & 0 & 0 & 0 \\ 0 & 0 & 0 & 0 & 0 \end{bmatrix}, \quad \begin{bmatrix} 0 & 0 \\ 0 & 0 \end{bmatrix}$$

The following matrices are in row-echelon form.

$$\begin{bmatrix} 1 & 4 & 3 & 7 \\ 0 & 1 & 6 & 2 \\ 0 & 0 & 1 & 5 \end{bmatrix}, \quad \begin{bmatrix} 1 & 1 & 0 \\ 0 & 1 & 0 \\ 0 & 0 & 0 \end{bmatrix}, \quad \begin{bmatrix} 0 & 1 & 2 & 6 & 0 \\ 0 & 0 & 1 & -1 & 0 \\ 0 & 0 & 0 & 0 & 1 \end{bmatrix}$$

The reader should check to see that each of the above matrices satifies all the necessary requirements. ▲

REMARK.   As the foregoing example illustrates, a matrix in row-echelon form has zeros below each leading 1, whereas a matrix in reduced row-echelon form has zeros both above and below each leading 1.

If, by a sequence of elementary row operations, the augmented matrix for a system of linear equations is put in reduced row-echelon form, then the solution set of the system can be obtained by inspection or, at worst, after a few simple steps. The next example illustrates this point.

Example 2.   Suppose that the augmented matrix for a system of linear equations has been reduced by row operations to the given reduced row-echelon form. Solve the system.

$$\text{(a)} \begin{bmatrix} 1 & 0 & 0 & 5 \\ 0 & 1 & 0 & -2 \\ 0 & 0 & 1 & 4 \end{bmatrix} \qquad \text{(b)} \begin{bmatrix} 1 & 0 & 0 & 4 & -1 \\ 0 & 1 & 0 & 2 & 6 \\ 0 & 0 & 1 & 3 & 2 \end{bmatrix}$$

$$\text{(c)} \begin{bmatrix} 1 & 6 & 0 & 0 & 4 & -2 \\ 0 & 0 & 1 & 0 & 3 & 1 \\ 0 & 0 & 0 & 1 & 5 & 2 \\ 0 & 0 & 0 & 0 & 0 & 0 \end{bmatrix} \qquad \text{(d)} \begin{bmatrix} 1 & 0 & 0 & 0 \\ 0 & 1 & 2 & 0 \\ 0 & 0 & 0 & 1 \end{bmatrix}$$

Solution (a).   The corresponding system of equations is

$$\begin{aligned} x_1 \qquad\qquad &= 5 \\ x_2 \qquad &= -2 \\ x_3 &= 4 \end{aligned}$$

By inspection, $x_1 = 5$, $x_2 = -2$, $x_3 = 4$.

Solution (b). The corresponding system of equations is

$$
\begin{aligned}
x_1 + 4x_4 &= -1 \\
x_2 + 2x_4 &= 6 \\
x_3 + 3x_4 &= 2
\end{aligned}
$$

Since x_1, x_2, and x_3 correspond to leading 1's in the augmented matrix, we call them **leading variables**. The nonleading variables (in this case x_4) are called **free variables**. Solving for the leading variables in terms of the free variable gives

$$
\begin{aligned}
x_1 &= -1 - 4x_4 \\
x_2 &= 6 - 2x_4 \\
x_3 &= 2 - 3x_4
\end{aligned}
$$

Since x_4 can be assigned an arbitrary value, say t, we have infinitely many solutions. The general solution is given by the formulas

$$
x_1 = -1 - 4t, \qquad x_2 = 6 - 2t, \qquad x_3 = 2 - 3t, \qquad x_4 = t
$$

Solution (c). The corresponding system of equations is

$$
\begin{aligned}
x_1 + 6x_2 + 4x_5 &= -2 \\
x_3 + 3x_5 &= 1 \\
x_4 + 5x_5 &= 2
\end{aligned}
$$

Here the leading variables are x_1, x_3, and x_4, and the free variables are x_2 and x_5. Solving for the leading variables in terms of the free variables gives

$$
\begin{aligned}
x_1 &= -2 - 4x_5 - 6x_2 \\
x_3 &= 1 - 3x_5 \\
x_4 &= 2 - 5x_5
\end{aligned}
$$

Since x_5 can be assigned an arbitrary value, t, and x_2 can be assigned an arbitrary value, s, there are infinitely many solutions. The general solution is given by the formulas

$$
x_1 = -2 - 4t - 6s, \qquad x_2 = s, \qquad x_3 = 1 - 3t, \qquad x_4 = 2 - 5t, \qquad x_5 = t
$$

Solution (d). The last equation in the corresponding system of equations is

$$
0x_1 + 0x_2 + 0x_3 = 1
$$

Since this equation can never be satisfied, there is no solution to the system. ▲

GAUSSIAN ELIMINATION

We have just seen how easy it is to solve a system of linear equations, once its augmented matrix is in reduced row-echelon form. Now we shall give a step-by-step procedure that can be used to reduce any matrix to reduced row-echelon form. As we state each step in the procedure, we shall illustrate the idea by reducing the following matrix to reduced row-echelon form.

$$\begin{bmatrix} 0 & 0 & -2 & 0 & 7 & 12 \\ 2 & 4 & -10 & 6 & 12 & 28 \\ 2 & 4 & -5 & 6 & -5 & -1 \end{bmatrix}$$

Step 1. Locate the leftmost column that does not consist entirely of zeros.

$$\begin{bmatrix} 0 & 0 & -2 & 0 & 7 & 12 \\ 2 & 4 & -10 & 6 & 12 & 28 \\ 2 & 4 & -5 & 6 & -5 & -1 \end{bmatrix}$$

$\uparrow$
└─ **Leftmost nonzero column**

Step 2. Interchange the top row with another row, if necessary, to bring a nonzero entry to the top of the column found in Step 1.

$$\begin{bmatrix} 2 & 4 & -10 & 6 & 12 & 28 \\ 0 & 0 & -2 & 0 & 7 & 12 \\ 2 & 4 & -5 & 6 & -5 & -1 \end{bmatrix}$$

> The first and second rows in the previous matrix were interchanged.

Step 3. If the entry that is now at the top of the column found in Step 1 is a, multiply the first row by $1/a$ in order to introduce a leading 1.

$$\begin{bmatrix} 1 & 2 & -5 & 3 & 6 & 14 \\ 0 & 0 & -2 & 0 & 7 & 12 \\ 2 & 4 & -5 & 6 & -5 & -1 \end{bmatrix}$$

> The first row of the previous matrix was multiplied by 1/2.

Step 4. Add suitable multiples of the top row to the rows below so that all entries below leading 1 become zeros.

$$\begin{bmatrix} 1 & 2 & -5 & 3 & 6 & 14 \\ 0 & 0 & -2 & 0 & 7 & 12 \\ 0 & 0 & 5 & 0 & -17 & -29 \end{bmatrix}$$

> -2 times the first row of the previous matrix was added to the third row.

Step 5. Now cover the top row in the matrix and begin again with Step 1 applied to the submatrix that remains. Continue in this way until the *entire* matrix is in row-echelon form.

$$\begin{bmatrix} 1 & 2 & -5 & 3 & 6 & 14 \\ 0 & 0 & -2 & 0 & 7 & 12 \\ 0 & 0 & 5 & 0 & -17 & -29 \end{bmatrix}$$

$\uparrow$
└─ **Leftmost nonzero column in the submatrix**

$$\begin{bmatrix} 1 & 2 & -5 & 3 & 6 & 14 \\ 0 & 0 & 1 & 0 & -\frac{7}{2} & -6 \\ 0 & 0 & 5 & 0 & -17 & -29 \end{bmatrix}$$

The first row in the submatrix was multiplied by $-1/2$ to introduce a leading 1.

$$\begin{bmatrix} 1 & 2 & -5 & 3 & 6 & 14 \\ 0 & 0 & 1 & 0 & -\frac{7}{2} & -6 \\ 0 & 0 & 0 & 0 & \frac{1}{2} & 1 \end{bmatrix}$$

-5 times the first row of the submatrix was added to the second row of the submatrix to introduce a zero below the leading 1.

$$\begin{bmatrix} 1 & 2 & -5 & 3 & 6 & 14 \\ 0 & 0 & 1 & 0 & -\frac{7}{2} & -6 \\ 0 & 0 & 0 & 0 & \frac{1}{2} & 1 \end{bmatrix}$$

The top row in the submatrix was covered, and we returned again to Step 1.

Leftmost nonzero column
in the new submatrix

$$\begin{bmatrix} 1 & 2 & -5 & 3 & 6 & 14 \\ 0 & 0 & 1 & 0 & -\frac{7}{2} & -6 \\ 0 & 0 & 0 & 0 & 1 & 2 \end{bmatrix}$$

The first (and only) row in the new submatrix was multiplied by 2 to introduce a leading 1.

The *entire* matrix is now in row-echelon form. To find the reduced row-echelon form we need the following additional step.

Step 6. Beginning with the last nonzero row and working upward, add suitable multiples of each row to the rows above to introduce zeros above the leading 1's.

$$\begin{bmatrix} 1 & 2 & -5 & 3 & 6 & 14 \\ 0 & 0 & 1 & 0 & 0 & 1 \\ 0 & 0 & 0 & 0 & 1 & 2 \end{bmatrix}$$

$7/2$ times the third row of the previous matrix was added to the second row.

$$\begin{bmatrix} 1 & 2 & -5 & 3 & 0 & 2 \\ 0 & 0 & 1 & 0 & 0 & 1 \\ 0 & 0 & 0 & 0 & 1 & 2 \end{bmatrix}$$

-6 times the third row was added to the first row.

$$\begin{bmatrix} 1 & 2 & 0 & 3 & 0 & 7 \\ 0 & 0 & 1 & 0 & 0 & 1 \\ 0 & 0 & 0 & 0 & 1 & 2 \end{bmatrix}$$

5 times the second row was added to the first row.

The last matrix is in reduced row-echelon form. ▲

The above procedure for reducing a matrix to reduced row-echelon form is called ***Gauss-Jordan elimination***.* If we use only the first five steps, the procedure produces a row-echelon form and is called ***Gaussian elimination***.

Example 3   Solve by Gauss-Jordan elimination.

$$
\begin{aligned}
x_1 + 3x_2 - 2x_3 \qquad\quad + 2x_5 \qquad\qquad &= 0 \\
2x_1 + 6x_2 - 5x_3 - 2x_4 + 4x_5 - 3x_6 &= -1 \\
5x_3 + 10x_4 \qquad\quad + 15x_6 &= 5 \\
2x_1 + 6x_2 \qquad\quad + 8x_4 + 4x_5 + 18x_6 &= 6
\end{aligned}
$$

The augmented matrix for the system is

$$
\begin{bmatrix}
1 & 3 & -2 & 0 & 2 & 0 & 0 \\
2 & 6 & -5 & -2 & 4 & -3 & -1 \\
0 & 0 & 5 & 10 & 0 & 15 & 5 \\
2 & 6 & 0 & 8 & 4 & 18 & 6
\end{bmatrix}
$$

---

* *Karl Friedrich Gauss* (1777–1855)—German mathematician and scientist. Sometimes called the "prince of mathematicians," Gauss ranks with Isaac Newton and Archimedes as one of the three greatest mathematicians who ever lived. In the entire history of mathematics there may never have been a child so precocious as Gauss—by his own account he worked out the rudiments of arithmetic before he could talk. One day, before he was even three years old, his genius became apparent to his parents in a very dramatic way. His father was preparing the weekly payroll for the laborers under his charge while the boy watched quietly from a corner. At the end of the long and tedious calculation, Gauss informed his father that there was an error in the result and stated the answer, which he had worked out in his head. To the astonishment of his parents, a check of the computations showed Gauss to be correct!

In his doctoral dissertation Gauss gave the first complete proof of the fundamental theorem of algebra, which states that every polynomial equation has as many solutions as its degree. At age 19 he solved a problem that baffled Euclid, inscribing a regular polygon of seventeen sides in a circle using straightedge and compass; and in 1801, at age 24, he published his first masterpiece, *Disquisitiones Arithmeticae*, considered by many to be one of the most brilliant achievements in mathematics. In that paper Gauss systematized the study of number theory (properties of the integers) and formulated the basic concepts that form the foundation of that subject.

Among his myriad achievements, Gauss discovered the Gaussian or "bell-shaped" curve that is fundamental in probability, gave the first geometric interpretation of complex numbers and established their fundamental role in mathematics, developed methods of characterizing surfaces intrinsically by means of the curves that they contain, developed the theory of conformal (angle-preserving) maps, and discovered non-Euclidean geometry 30 years before the ideas were published by others. In physics he made major contributions to the theory of lenses and capillary action, and with Wilhelm Weber he did fundamental work in electromagnetism. Gauss invented the heliotrope, bifilar magnetometer, and an electrotelegraph.

Gauss was deeply religious and aristocratic in demeanor. He mastered foreign languages with ease, read extensively, and enjoyed minerology and botany as hobbies. He disliked teaching and was usually cool and discouraging to other mathematicians, possibly because he had already anticipated their work. It has been said that if Gauss had published all of his discoveries, the current state of mathematics would be advanced by 50 years. He was without a doubt the greatest mathematician of the modern era.

Wilhelm Jordan (1842–1899) was a German engineer who specialized in geodesy. His contribution to solving linear systems appeared in his popular book, *Handbuch der Vermessungskunde* (*Handbook of Geodesy*), in 1888.

Adding -2 times the first row to the second and fourth rows gives

$$\begin{bmatrix} 1 & 3 & -2 & 0 & 2 & 0 & 0 \\ 0 & 0 & -1 & -2 & 0 & -3 & -1 \\ 0 & 0 & 5 & 10 & 0 & 15 & 5 \\ 0 & 0 & 4 & 8 & 0 & 18 & 6 \end{bmatrix}$$

Multiplying the second row by -1 and then adding -5 times the second row to the third row and -4 times the second row to the fourth row gives

$$\begin{bmatrix} 1 & 3 & -2 & 0 & 2 & 0 & 0 \\ 0 & 0 & 1 & 2 & 0 & 3 & 1 \\ 0 & 0 & 0 & 0 & 0 & 0 & 0 \\ 0 & 0 & 0 & 0 & 0 & 6 & 2 \end{bmatrix}$$

Interchanging the third and fourth rows and then multiplying the third row of the resulting matrix by $1/6$ gives the row-echelon form

$$\begin{bmatrix} 1 & 3 & -2 & 0 & 2 & 0 & 0 \\ 0 & 0 & 1 & 2 & 0 & 3 & 1 \\ 0 & 0 & 0 & 0 & 0 & 1 & \frac{1}{3} \\ 0 & 0 & 0 & 0 & 0 & 0 & 0 \end{bmatrix}$$

Adding -3 times the third row to the second row and then adding 2 times the second row of the resulting matrix to the first row yields the reduced row-echelon form

$$\begin{bmatrix} 1 & 3 & 0 & 4 & 2 & 0 & 0 \\ 0 & 0 & 1 & 2 & 0 & 0 & 0 \\ 0 & 0 & 0 & 0 & 0 & 1 & \frac{1}{3} \\ 0 & 0 & 0 & 0 & 0 & 0 & 0 \end{bmatrix}$$

The corresponding system of equations is

$$\begin{aligned} x_1 + 3x_2 \quad + 4x_4 + 2x_5 \quad &= 0 \\ x_3 + 2x_4 \quad &= 0 \\ x_6 &= \tfrac{1}{3} \end{aligned}$$

(We have discarded the last equation, $0x_1 + 0x_2 + 0x_3 + 0x_4 + 0x_5 + 0x_6 = 0$, since it will be satisfied automatically by the solutions of the remaining equations.) Solving for the leading variables, we obtain

$$\begin{aligned} x_1 &= -3x_2 - 4x_4 - 2x_5 \\ x_3 &= -2x_4 \\ x_6 &= \tfrac{1}{3} \end{aligned}$$

If we assign the free variables x_2, x_4, and x_5 arbitrary values r, s, and t, respectively, the general solution is given by the formulas

$$x_1 = -3r - 4s - 2t, \qquad x_2 = r, \qquad x_3 = -2s, \qquad x_4 = s, \qquad x_5 = t, \qquad x_6 = \tfrac{1}{3}$$

BACK-SUBSTITUTION

Example 4 It is sometimes preferable to solve a system of linear equations by using Gaussian elimination to bring the augmented matrix into row-echelon form without continuing all the way to the reduced row-echelon form. When this is done, the corresponding system of equations can be solved by a technique called **back-substitution**. We shall illustrate this method using the system of equations in Example 3.

From the computations in Example 3, a row-echelon form of the augmented matrix is

$$\begin{bmatrix} 1 & 3 & -2 & 0 & 2 & 0 & 0 \\ 0 & 0 & 1 & 2 & 0 & 3 & 1 \\ 0 & 0 & 0 & 0 & 0 & 1 & \tfrac{1}{3} \\ 0 & 0 & 0 & 0 & 0 & 0 & 0 \end{bmatrix}$$

To solve the corresponding system of equations

$$\begin{aligned} x_1 + 3x_2 - 2x_3 \quad\quad + 2x_5 \quad\quad &= 0 \\ x_3 + 2x_4 \quad\quad + 3x_6 &= 1 \\ x_6 &= \tfrac{1}{3} \end{aligned}$$

we proceed as follows:

Step 1. Solve the equations for the leading variables.

$$\begin{aligned} x_1 &= -3x_2 + 2x_3 - 2x_5 \\ x_3 &= 1 - 2x_4 - 3x_6 \\ x_6 &= \tfrac{1}{3} \end{aligned}$$

Step 2. Beginning with the bottom equation and working upward, successively substitute each equation into all the equations above it.

Substituting $x_6 = 1/3$ into the second equation yields

$$\begin{aligned} x_1 &= -3x_2 + 2x_3 - 2x_5 \\ x_3 &= -2x_4 \\ x_6 &= \tfrac{1}{3} \end{aligned}$$

Substituting $x_3 = -2x_4$ into the first equation yields

$$\begin{aligned} x_1 &= -3x_2 - 4x_4 - 2x_5 \\ x_3 &= -2x_4 \\ x_6 &= \tfrac{1}{3} \end{aligned}$$

Step 3. Assign arbitrary values to the free variables, if any.

If we assign x_2, x_4, and x_5 the arbitrary values r, s, and t, respectively, the general solution is given by the formulas

$$x_1 = -3r - 4s - 2t, \qquad x_2 = r, \qquad x_3 = -2s, \qquad x_4 = s, \qquad x_5 = t, \qquad x_6 = \tfrac{1}{3}$$

This agrees with the solution obtained in Example 3. ▲

REMARK. The arbitrary values that are assigned to the free variables are often called *parameters*. Although we shall generally use the letters $r, s, t, \ldots$ for the parameters, any letters that do not conflict with the variable names can be used.

Example 5 Solve

$$x + y + 2z = 9$$
$$2x + 4y - 3z = 1$$
$$3x + 6y - 5z = 0$$

by Gaussian elimination and back-substitution.

Solution. This is the system in Example 3 of Section 1.1. In that example we converted the augmented matrix

$$\begin{bmatrix} 1 & 1 & 2 & 9 \\ 2 & 4 & -3 & 1 \\ 3 & 6 & -5 & 0 \end{bmatrix}$$

to the row-echelon form

$$\begin{bmatrix} 1 & 1 & 2 & 9 \\ 0 & 1 & -\frac{7}{2} & -\frac{17}{2} \\ 0 & 0 & 1 & 3 \end{bmatrix}$$

The system corresponding to this matrix is

$$x + y + 2z = 9$$
$$y - \tfrac{7}{2}z = -\tfrac{17}{2}$$
$$z = 3$$

Solving for the leading variables yields

$$x = 9 - y - 2z$$
$$y = -\tfrac{17}{2} + \tfrac{7}{2}z$$
$$z = 3$$

Substituting the bottom equation into those above yields

$$x = 3 - y$$
$$y = 2$$
$$z = 3$$

and substituting the second equation into the top yields

$$x = 1$$
$$y = 2$$
$$z = 3$$

This agrees with the result found by Gauss-Jordan elimination in Example 3 of Section 1.1. ▲

The procedures we have given for reducing a matrix to row-echelon form or reduced row-echelon form are well suited for computer computation because they are systematic. However, these procedures sometimes introduce fractions, which might otherwise be avoided by varying the steps in the right way. Thus, once the basic procedure has been mastered, the reader may wish to vary the steps in specific problems to avoid fractions (see Exercise 15).

REMARK. It can be proved, although we shall not do it, that *every matrix has a unique reduced row-echelon form*; that is, for a given matrix, no matter how the elementary row operations are varied, one will always arrive at the same reduced row-echelon form. In contrast, *a row-echelon form of a matrix is not unique*; by changing the sequence of elementary row operations it is possible to arrive at different row-echelon forms (see Exercise 16). Thus, one talks of *the* reduced row-echelon form of a matrix, but *a* row-echelon form.

REMARK. Most computer algorithms for solving systems of linear equations use Gaussian elimination rather than Gauss-Jordan elimination. Moreover, the basic procedure is often varied in a variety of ways to reduce roundoff error, minimize the required storage space, and obtain maximum speed. For example, many such algorithms do not normalize the leading entry in each row to be a 1. Some of these matters will be discussed in Chapter 9.

EXERCISE SET 1.2

1. Which of the following 3×3 matrices are in reduced row-echelon form?

(a) $\begin{bmatrix} 1 & 0 & 0 \\ 0 & 1 & 0 \\ 0 & 0 & 1 \end{bmatrix}$
(b) $\begin{bmatrix} 1 & 0 & 0 \\ 0 & 1 & 0 \\ 0 & 0 & 0 \end{bmatrix}$
(c) $\begin{bmatrix} 0 & 1 & 0 \\ 0 & 0 & 1 \\ 0 & 0 & 0 \end{bmatrix}$

(d) $\begin{bmatrix} 1 & 0 & 0 \\ 0 & 0 & 1 \\ 0 & 0 & 0 \end{bmatrix}$
(e) $\begin{bmatrix} 1 & 0 & 0 \\ 0 & 0 & 0 \\ 0 & 0 & 1 \end{bmatrix}$
(f) $\begin{bmatrix} 0 & 1 & 0 \\ 1 & 0 & 0 \\ 0 & 0 & 0 \end{bmatrix}$

(g) $\begin{bmatrix} 1 & 1 & 0 \\ 0 & 1 & 0 \\ 0 & 0 & 0 \end{bmatrix}$
(h) $\begin{bmatrix} 1 & 0 & 2 \\ 0 & 1 & 3 \\ 0 & 0 & 0 \end{bmatrix}$
(i) $\begin{bmatrix} 0 & 0 & 1 \\ 0 & 0 & 0 \\ 0 & 0 & 0 \end{bmatrix}$

2. Which of the following 3×3 matrices are in row-echelon form?

(a) $\begin{bmatrix} 1 & 0 & 0 \\ 0 & 1 & 0 \\ 0 & 0 & 1 \end{bmatrix}$ (b) $\begin{bmatrix} 1 & 2 & 0 \\ 0 & 1 & 0 \\ 0 & 0 & 0 \end{bmatrix}$ (c) $\begin{bmatrix} 1 & 0 & 0 \\ 0 & 1 & 0 \\ 0 & 2 & 0 \end{bmatrix}$

(d) $\begin{bmatrix} 1 & 3 & 4 \\ 0 & 0 & 1 \\ 0 & 0 & 0 \end{bmatrix}$ (e) $\begin{bmatrix} 1 & 5 & -3 \\ 0 & 1 & 1 \\ 0 & 0 & 0 \end{bmatrix}$ (f) $\begin{bmatrix} 1 & 2 & 3 \\ 0 & 0 & 0 \\ 0 & 0 & 1 \end{bmatrix}$

3. In each part determine whether the matrix is in row-echelon form, reduced row-echelon form, both, or neither.

(a) $\begin{bmatrix} 1 & 2 & 0 & 3 & 0 \\ 0 & 0 & 1 & 1 & 0 \\ 0 & 0 & 0 & 0 & 1 \\ 0 & 0 & 0 & 0 & 0 \end{bmatrix}$ (b) $\begin{bmatrix} 1 & 0 & 0 & 5 \\ 0 & 0 & 1 & 3 \\ 0 & 1 & 0 & 4 \end{bmatrix}$ (c) $\begin{bmatrix} 1 & 0 & 3 & 1 \\ 0 & 1 & 2 & 4 \end{bmatrix}$

(d) $\begin{bmatrix} 1 & -7 & 5 & 5 \\ 0 & 1 & 3 & 2 \end{bmatrix}$ (e) $\begin{bmatrix} 1 & 3 & 0 & 2 & 0 \\ 1 & 0 & 2 & 2 & 0 \\ 0 & 0 & 0 & 0 & 1 \\ 0 & 0 & 0 & 0 & 0 \end{bmatrix}$ (f) $\begin{bmatrix} 0 & 0 \\ 0 & 0 \\ 0 & 0 \end{bmatrix}$

4. In each part suppose that the augmented matrix for a system of linear equations has been reduced by row operations to the given reduced row-echelon form. Solve the system.

(a) $\begin{bmatrix} 1 & 0 & 0 & -3 \\ 0 & 1 & 0 & 0 \\ 0 & 0 & 1 & 7 \end{bmatrix}$ (b) $\begin{bmatrix} 1 & 0 & 0 & -7 & 8 \\ 0 & 1 & 0 & 3 & 2 \\ 0 & 0 & 1 & 1 & -5 \end{bmatrix}$

(c) $\begin{bmatrix} 1 & -6 & 0 & 0 & 3 & -2 \\ 0 & 0 & 1 & 0 & 4 & 7 \\ 0 & 0 & 0 & 1 & 5 & 8 \\ 0 & 0 & 0 & 0 & 0 & 0 \end{bmatrix}$ (d) $\begin{bmatrix} 1 & -3 & 0 & 0 \\ 0 & 0 & 1 & 0 \\ 0 & 0 & 0 & 1 \end{bmatrix}$

5. In each part suppose that the augmented matrix for a system of linear equations has been reduced by row operations to the given row-echelon form. Solve the system.

(a) $\begin{bmatrix} 1 & -3 & 4 & 7 \\ 0 & 1 & 2 & 2 \\ 0 & 0 & 1 & 5 \end{bmatrix}$ (b) $\begin{bmatrix} 1 & 0 & 8 & -5 & 6 \\ 0 & 1 & 4 & -9 & 3 \\ 0 & 0 & 1 & 1 & 2 \end{bmatrix}$

(c) $\begin{bmatrix} 1 & 7 & -2 & 0 & -8 & -3 \\ 0 & 0 & 1 & 1 & 6 & 5 \\ 0 & 0 & 0 & 1 & 3 & 9 \\ 0 & 0 & 0 & 0 & 0 & 0 \end{bmatrix}$ (d) $\begin{bmatrix} 1 & -3 & 7 & 1 \\ 0 & 1 & 4 & 0 \\ 0 & 0 & 0 & 1 \end{bmatrix}$

6. Solve each of the following systems by Gauss-Jordan elimination.

(a) $\begin{array}{r} x_1 + x_2 + 2x_3 = 8 \\ -x_1 - 2x_2 + 3x_3 = 1 \\ 3x_1 - 7x_2 + 4x_3 = 10 \end{array}$ (b) $\begin{array}{r} 2x_1 + 2x_2 + 2x_3 = 0 \\ -2x_1 + 5x_2 + 2x_3 = 1 \\ 8x_1 + x_2 + 4x_3 = -1 \end{array}$

(c) $\begin{aligned} x - y + 2z - w &= -1 \\ 2x + y - 2z - 2w &= -2 \\ -x + 2y - 4z + w &= 1 \\ 3x \qquad\qquad - 3w &= -3 \end{aligned}$ (d) $\begin{aligned} -2b + 3c &= 1 \\ 3a + 6b - 3c &= -2 \\ 6a + 6b + 3c &= 5 \end{aligned}$

7. Solve each of the systems in Exercise 6 by Gaussian elimination.

8. Solve each of the following systems by Gauss-Jordan elimination.

(a) $\begin{aligned} 2x_1 - 3x_2 &= -2 \\ 2x_1 + x_2 &= 1 \\ 3x_1 + 2x_2 &= 1 \end{aligned}$ (b) $\begin{aligned} 3x_1 + 2x_2 - x_3 &= -15 \\ 5x_1 + 3x_2 + 2x_3 &= 0 \\ 3x_1 + x_2 + 3x_3 &= 11 \\ -6x_1 - 4x_2 + 2x_3 &= 30 \end{aligned}$

(c) $\begin{aligned} 4x_1 - 8x_2 &= 12 \\ 3x_1 - 6x_2 &= 9 \\ -2x_1 + 4x_2 &= -6 \end{aligned}$ (d) $\begin{aligned} 10y - 4z + w &= 1 \\ x + 4y - z + w &= 2 \\ 3x + 2y + z + 2w &= 5 \\ -2x - 8y + 2z - 2w &= -4 \\ x - 6y + 3z &= 1 \end{aligned}$

9. Solve each of the systems in Exercise 8 by Gaussian elimination.

10. Solve each of the following systems by Gauss-Jordan elimination.

(a) $\begin{aligned} 5x_1 - 2x_2 + 6x_3 &= 0 \\ -2x_1 + x_2 + 3x_3 &= 1 \end{aligned}$ (b) $\begin{aligned} x_1 - 2x_2 + x_3 - 4x_4 &= 1 \\ x_1 + 3x_2 + 7x_3 + 2x_4 &= 2 \\ x_1 - 12x_2 - 11x_3 - 16x_4 &= 5 \end{aligned}$ (c) $\begin{aligned} w + 2x - y &= 4 \\ x - y &= 3 \\ w + 3x - 2y &= 7 \\ 2u + 4v + w + 7x &= 7 \end{aligned}$

11. Solve each of the systems in Exercise 10 by Gaussian elimination.

12. Solve the following system by any method.

$$\begin{aligned} 2I_1 - I_2 + 3I_3 + 4I_4 &= 9 \\ I_1 \qquad - 2I_3 + 7I_4 &= 11 \\ 3I_1 - 3I_2 + I_3 + 5I_4 &= 8 \\ 2I_1 + I_2 + 4I_3 + 4I_4 &= 10 \end{aligned}$$

13. Solve the following systems, where a, b, and c are constants.

(a) $\begin{aligned} 2x + y &= a \\ 3x + 6y &= b \end{aligned}$ (b) $\begin{aligned} x_1 + x_2 + x_3 &= a \\ 2x_1 \qquad + 2x_3 &= b \\ 3x_2 + 3x_3 &= c \end{aligned}$

14. For which values of a will the following system have no solutions? Exactly one solution? Infinitely many solutions?

$$\begin{aligned} x + 2y - 3z &= 4 \\ 3x - y + 5z &= 2 \\ 4x + y + (a^2 - 14)z &= a + 2 \end{aligned}$$

15. Reduce

$$\begin{bmatrix} 2 & 1 & 3 \\ 0 & -2 & 7 \\ 3 & 4 & 5 \end{bmatrix}$$

to reduced row-echelon form without introducing any fractions.

16. Find two different row-echelon forms of

$$\begin{bmatrix} 1 & 3 \\ 2 & 7 \end{bmatrix}$$

17. Solve the following system of nonlinear equations for the unknown angles α, β, and γ, where $0 \le \alpha \le 2\pi$, $0 \le \beta \le 2\pi$, and $0 \le \gamma < \pi$.

$$2 \sin \alpha - \cos \beta + 3 \tan \gamma = 3$$
$$4 \sin \alpha + 2 \cos \beta - 2 \tan \gamma = 2$$
$$6 \sin \alpha - 3 \cos \beta + \tan \gamma = 9$$

18. Solve the following system of nonlinear equations for x, y, and z.

$$x^2 + y^2 + z^2 = 6$$
$$x^2 - y^2 + 2z^2 = 2$$
$$2x^2 + y^2 - z^2 = 3$$

19. Figure 1 shows the graph of a cubic equation $y = ax^3 + bx^2 + cx + d$. Find the coefficients a, b, c, and d.

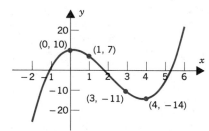

Figure 1

20. Describe the possible reduced row-echelon forms of

$$\begin{bmatrix} a & b & c \\ d & e & f \\ g & h & i \end{bmatrix}$$

21. Show that if $ad - bc \ne 0$, then the reduced row-echelon form of

$$\begin{bmatrix} a & b \\ c & d \end{bmatrix} \quad \text{is} \quad \begin{bmatrix} 1 & 0 \\ 0 & 1 \end{bmatrix}$$

22. Use Exercise 21 to show that if $ad - bc \ne 0$, then the system

$$ax + by = k$$
$$cx + dy = l$$

has exactly one solution.

1.3   HOMOGENEOUS SYSTEMS OF LINEAR EQUATIONS

As we have already pointed out, every system of linear equations has either one solution, infinitely many solutions, or no solutions at all. As we progress, there will be situations in which we will not be interested in finding solutions to a given system, but instead will be concerned with deciding how many solutions the system has. In this section we shall consider various linear systems for which it is possible to reach conclusions about the number of solutions by inspection.

A system of linear equations is said to be *homogeneous* if all the constant terms are zero; that is, the system has the form

$$a_{11}x_1 + a_{12}x_2 + \cdots + a_{1n}x_n = 0$$
$$a_{21}x_1 + a_{22}x_2 + \cdots + a_{2n}x_n = 0$$
$$\vdots \qquad \vdots \qquad \qquad \vdots \qquad \vdots$$
$$a_{m1}x_1 + a_{m2}x_2 + \cdots + a_{mn}x_n = 0$$

Every homogeneous system of linear equations is consistent, since all such systems have $x_1 = 0$, $x_2 = 0$, $\ldots$, $x_n = 0$ as a solution. This solution is called the *trivial solution*; if there are other solutions, they are called *nontrivial solutions*.

Since homogeneous systems of linear equations must be consistent, such systems have one solution or infinitely many solutions. Since one of these solutions is the trivial solution, we can make the following statement.

For a homogeneous system of linear equations, exactly one of the following is true:

1. *The system has only the trivial solution.*
2. *The system has infinitely many nontrivial solutions in addition to the trivial solution.*

There is one case in which a homogeneous system is assured of having nontrivial solutions; namely, whenever the system involves more unknowns than equations. To see why, consider the following example of four equations in five unknowns.

Example 1   Solve the following homogeneous system of linear equations by Gauss-Jordan elimination.

$$\begin{aligned}
2x_1 + 2x_2 - x_3 \qquad\ \ + x_5 &= 0 \\
-x_1 - x_2 + 2x_3 - 3x_4 + x_5 &= 0 \\
x_1 + x_2 - 2x_3 \qquad\ \ - x_5 &= 0 \\
x_3 + x_4 + x_5 &= 0
\end{aligned} \qquad (1)$$

Solution. The augmented matrix for the system is

$$\begin{bmatrix} 2 & 2 & -1 & 0 & 1 & 0 \\ -1 & -1 & 2 & -3 & 1 & 0 \\ 1 & 1 & -2 & 0 & -1 & 0 \\ 0 & 0 & 1 & 1 & 1 & 0 \end{bmatrix}$$

Reducing this matrix to reduced row-echelon form, we obtain

$$\begin{bmatrix} 1 & 1 & 0 & 0 & 1 & 0 \\ 0 & 0 & 1 & 0 & 1 & 0 \\ 0 & 0 & 0 & 1 & 0 & 0 \\ 0 & 0 & 0 & 0 & 0 & 0 \end{bmatrix}$$

The corresponding system of equations is

$$\begin{aligned} x_1 + x_2 \qquad\quad + x_5 &= 0 \\ x_3 \qquad + x_5 &= 0 \\ x_4 \qquad\quad &= 0 \end{aligned} \tag{2}$$

Solving for the leading variables yields

$$\begin{aligned} x_1 &= -x_2 - x_5 \\ x_3 &= -x_5 \\ x_4 &= 0 \end{aligned}$$

Thus, the general solution is

$$x_1 = -s - t, \qquad x_2 = s, \qquad x_3 = -t, \qquad x_4 = 0, \qquad x_5 = t$$

Note that the trivial solution is obtained when $s = t = 0$. ▲

Example 1 illustrates two important points about solving homogeneous systems of linear equations. First, none of the three elementary row operations can alter the final column of zeros in the augmented matrix, so that the system of equations corresponding to the reduced row-echelon form of the augmented matrix must also be a homogeneous system [see system (2)]. Second, depending on whether the reduced row-echelon form of the augmented matrix has any zero rows, the number of equations in the reduced system is the same or less than the number of equations in the original system [compare systems (1) and (2)]. Thus, if the given homogeneous system has m equations in n unknowns with $m < n$, and if there are r nonzero rows in the reduced row-echelon form of the augmented matrix, we will have $r < n$. It follows that the system of equations corresponding to the reduced row-echelon form of the augmented matrix will have the form

$$\begin{aligned} \cdots x_{k_1} \qquad\qquad &+ \Sigma(\ ) = 0 \\ \cdots x_{k_2} \qquad\qquad &+ \Sigma(\ ) = 0 \\ \cdots \qquad\qquad\quad &\quad\ \ \vdots \\ x_{k_r} &+ \Sigma(\ ) = 0 \end{aligned} \tag{3}$$

where $x_{k_1}, x_{k_2}, \ldots, x_{k_r}$ are the leading variables and $\Sigma(\ )$ denotes sums (possibly all different) that involve the $n - r$ free variables. Solving for the leading variables gives

$$
\begin{aligned}
x_{k_1} &= -\Sigma(\ ) \\
x_{k_2} &= -\Sigma(\ ) \\
&\ \ \vdots \\
x_{k_r} &= -\Sigma(\ )
\end{aligned}
$$

As in Example 1, we can assign arbitrary values to the free variables on the right-hand side and thus obtain infinitely many solutions to the system.

In summary, we have the following important theorem.

Theorem 1.3.1. *A homogeneous system of linear equations with more unknowns than equations has infinitely many solutions.*

REMARK. Note that Theorem 1.3.1 applies only to homogeneous systems. A non-homogeneous system with more unknowns than equations need not be consistent (Exercise 12); however, if the system is consistent, it will have infinitely many solutions. This will be proved later.

EXERCISE SET 1.3

1. Without using pencil and paper, determine which of the following homogeneous systems have nontrivial solutions.

(a) $\begin{aligned} 2x_1 - 3x_2 + 4x_3 - x_4 &= 0 \\ 7x_1 + x_2 - 8x_3 + 9x_4 &= 0 \\ 2x_1 + 8x_2 + x_3 - x_4 &= 0 \end{aligned}$
  (b) $\begin{aligned} x_1 + 3x_2 - x_3 &= 0 \\ x_2 - 8x_3 &= 0 \\ 4x_3 &= 0 \end{aligned}$

(c) $\begin{aligned} a_{11}x_1 + a_{12}x_2 + a_{13}x_3 &= 0 \\ a_{21}x_1 + a_{22}x_2 + a_{23}x_3 &= 0 \end{aligned}$
  (d) $\begin{aligned} 3x_1 - 2x_2 &= 0 \\ 6x_1 - 4x_2 &= 0 \end{aligned}$

2. Solve the following homogeneous systems of linear equations by any method.

(a) $\begin{aligned} 2x_1 + x_2 + 3x_3 &= 0 \\ x_1 + 2x_2 &= 0 \\ x_2 + x_3 &= 0 \end{aligned}$
  (b) $\begin{aligned} 3x_1 + x_2 + x_3 + x_4 &= 0 \\ 5x_1 - x_2 + x_3 - x_4 &= 0 \end{aligned}$
  (c) $\begin{aligned} 2x + 2y + 4z &= 0 \\ w - y - 3z &= 0 \\ 2w + 3x + y + z &= 0 \\ -2w + x + 3y - 2z &= 0 \end{aligned}$

3. Solve the following homogeneous systems of linear equations by any method.

(a) $\begin{aligned} 2x - y - 3z &= 0 \\ -x + 2y - 3z &= 0 \\ x + y + 4z &= 0 \end{aligned}$
  (b) $\begin{aligned} v + 3w - 2x &= 0 \\ 2u + v - 4w + 3x &= 0 \\ 2u + 3v + 2w - x &= 0 \\ -4u - 3v + 5w - 4x &= 0 \end{aligned}$
  (c) $\begin{aligned} x_1 + 3x_2 + x_4 &= 0 \\ x_1 + 4x_2 + 2x_3 &= 0 \\ -2x_2 - 2x_3 - x_4 &= 0 \\ 2x_1 - 4x_2 + x_3 + x_4 &= 0 \\ x_1 - 2x_2 - x_3 + x_4 &= 0 \end{aligned}$

4. Solve for Z_1, Z_2, Z_3, Z_4, and Z_5.

$$\begin{aligned}
Z_3 + Z_4 + Z_5 &= 0 \\
-Z_1 - Z_2 + 2Z_3 - 3Z_4 + Z_5 &= 0 \\
Z_1 + Z_2 - 2Z_3 \qquad - Z_5 &= 0 \\
2Z_1 + 2Z_2 - Z_3 \qquad + Z_5 &= 0
\end{aligned}$$

5. Show that the following nonlinear system has eighteen solutions if $0 \le \alpha \le 2\pi$, $0 \le \beta \le 2\pi$, and $0 \le \gamma \le 2\pi$.

$$\begin{aligned}
\sin \alpha + 2 \cos \beta + 3 \tan \gamma &= 0 \\
2 \sin \alpha + 5 \cos \beta + 3 \tan \gamma &= 0 \\
- \sin \alpha - 5 \cos \beta + 5 \tan \gamma &= 0
\end{aligned}$$

6. For which value(s) of γ does the following system of equations have nontrivial solutions?

$$\begin{aligned}
(\lambda - 3)x + \qquad y &= 0 \\
x + (\lambda - 3)y &= 0
\end{aligned}$$

7. Consider the system of equations

$$\begin{aligned}
ax + by &= 0 \\
cx + dy &= 0 \\
ex + fy &= 0
\end{aligned}$$

Discuss the relative positions of the lines $ax + by = 0$, $cx + dy = 0$, and $ex + fy = 0$ when

(a) the system has only the trivial solution   (b) the system has nontrivial solutions

8. Recall from plane geometry that three points, not all lying on a straight line, determine a unique circle. It is shown in analytic geometry that a circle in the xy-plane has an equation of the form

$$ax^2 + ay^2 + bx + cy + d = 0$$

Find an equation of the circle shown in Figure 1.

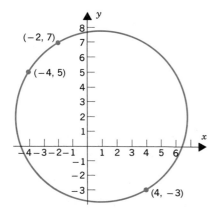

Figure 1

9. Consider the system of equations

$$ax + by = 0$$
$$cx + dy = 0$$

(a) Show that if $x = x_0$, $y = y_0$ is any solution of the system and k is any constant, then $x = kx_0$, $y = ky_0$ is also a solution.
(b) Show that if $x = x_0$, $y = y_0$ and $x = x_1$, $y = y_1$ are any two solutions, then $x = x_0 + x_1$, $y = y_0 + y_1$ is also a solution.

10. Consider the systems of equations

$$\text{(I) } ax + by = k \qquad \text{(II) } ax + by = 0$$
$$cx + dy = l \qquad\qquad\ \ cx + dy = 0$$

(a) Show that if $x = x_1$, $y = y_1$ and $x = x_2$, $y = y_2$ are both solutions of I, then $x = x_1 - x_2$, $y = y_1 - y_2$ is a solution of II.
(b) Show that if $x = x_1$, $y = y_1$ is a solution of I and $x = x_0$, $y = y_0$ is a solution of II, then $x = x_1 + x_0$, $y = y_1 + y_0$ is a solution of I.

11. (a) In the system of equations numbered (3), explain why it would be incorrect to denote the leading variables by $x_1, x_2, \ldots, x_r$ rather than $x_{k_1}, x_{k_2}, \ldots, x_{k_r}$ as we have done.
(b) The system of equations numbered (2) is a specific case of (3). What value does r have in this case? What are $x_{k_1}, x_{k_2}, \ldots, x_{k_r}$ in this case? Write out the sums denoted by $\Sigma(\ )$ in (3).

12. Find an inconsistent linear system that has more unknowns than equations.

1.4   MATRICES AND MATRIX OPERATIONS

Rectangular arrays of real numbers arise in many contexts other than as augmented matrices for systems of linear equations. In this section we shall consider such arrays as objects in their own right and develop some of their properties for use in our later work.

MATRIX NOTATION AND TERMINOLOGY

Definition. A ***matrix*** is a rectangular array of numbers. The numbers in the array are called the ***entries*** in the matrix.

Example 1   The following are matrices.

$$\begin{bmatrix} 1 & 2 \\ 3 & 0 \\ -1 & 4 \end{bmatrix} \qquad \begin{bmatrix} 2 & 1 & 0 & -3 \end{bmatrix} \qquad \begin{bmatrix} -\sqrt{2} & \pi & e \\ 3 & \frac{1}{2} & 0 \\ 0 & 0 & 0 \end{bmatrix} \qquad \begin{bmatrix} 1 \\ 3 \end{bmatrix} \qquad \begin{bmatrix} 4 \end{bmatrix} \ \blacktriangle$$

Matrices vary in size; the *size* of a matrix is described by specifying the number of *rows* (horizontal lines) and *columns* (vertical lines) that occur in the matrix. The first matrix in Example 1 has 3 rows and 2 columns so that its size is 3 by 2 (written 3×2). The first number always indicates the number of rows and the second indicates the number of columns. The remaining matrices in Example 1 thus have sizes 1×4, 3×3, 2×1, and 1×1, respectively.

REMARK. It is common practice to omit the brackets on a 1×1 matrix. Thus, we might write 4 rather than $[4]$. Although this makes it impossible to tell whether 4 denotes the number "four" or the 1×1 matrix whose entry is "four," this rarely causes problems, since it is usually possible to tell which is meant from the context in which the symbol appears.

We shall use capital letters to denote matrices and lowercase letters to denote numerical quantities; thus, we might write

$$A = \begin{bmatrix} 2 & 1 & 7 \\ 3 & 4 & 2 \end{bmatrix} \quad \text{or} \quad C = \begin{bmatrix} a & b & c \\ d & e & f \end{bmatrix}$$

When discussing matrices, it is common to refer to numerical quantities as *scalars*. Until Chapter 10 *all our scalars will be real numbers*.

If A is a matrix, we will use a_{ij} to denote the entry that occurs in row i and column j of A. Thus, a general 3×4 matrix might be written

$$A = \begin{bmatrix} a_{11} & a_{12} & a_{13} & a_{14} \\ a_{21} & a_{22} & a_{23} & a_{24} \\ a_{31} & a_{32} & a_{33} & a_{34} \end{bmatrix}$$

Usually, we shall match the letter used to denote a matrix and the letter used for its entries. Thus, for a matrix B we would use b_{ij} for the entry in row i and column j. A general $m \times n$ matrix might be written

$$B = \begin{bmatrix} b_{11} & b_{12} & \cdots & b_{1n} \\ b_{21} & b_{22} & \cdots & b_{2n} \\ \vdots & \vdots & & \vdots \\ b_{m1} & b_{m2} & \cdots & b_{mn} \end{bmatrix} \quad \text{or} \quad [b_{ij}]_{m \times n}$$

If it is not important to emphasize the size, then we would simply write $[b_{ij}]$ to denote the matrix.

A matrix A with n rows and n columns is called a *square matrix of order n*, and the entries $a_{11}, a_{22}, \ldots, a_{nn}$ are said to be on the *main diagonal* of A (see the shaded entries in Figure 1).

$$\begin{bmatrix} a_{11} & a_{12} & \cdots & a_{1n} \\ a_{21} & a_{22} & \cdots & a_{2n} \\ \vdots & \vdots & & \vdots \\ a_{n1} & a_{n2} & \cdots & a_{nn} \end{bmatrix}$$

Figure 1

A square matrix in which all entries *off* the main diagonal are zero is called a ***diagonal matrix***. Some examples are

$$
\begin{bmatrix} 2 & 0 \\ 0 & -1 \end{bmatrix}
\quad
\begin{bmatrix} 1 & 0 & 0 \\ 0 & -1 & 0 \\ 0 & 0 & 3 \end{bmatrix}
\quad
\begin{bmatrix} 0 & 0 & 0 \\ 0 & 0 & 0 \\ 0 & 0 & 0 \end{bmatrix}
\quad
\begin{bmatrix} 4 & 0 & 0 & 0 \\ 0 & 0 & 0 & 0 \\ 0 & 0 & -\frac{1}{2} & 0 \\ 0 & 0 & 0 & \frac{1}{3} \end{bmatrix}
$$

OPERATIONS ON MATRICES

So far, we have used matrices to abbreviate the work in solving systems of linear equations. For other applications, however, it is desirable to develop an "arithmetic of matrices" in which matrices can be added and multiplied in a useful way. The remainder of this section will be devoted to developing this arithmetic.

Two matrices are said to be ***equal*** if they have the same size and the corresponding entries in the two matrices are equal.

Example 2 Consider the matrices

$$
A = \begin{bmatrix} 2 & 1 \\ 3 & 4 \end{bmatrix}
\quad
B = \begin{bmatrix} 2 & 1 \\ 3 & 5 \end{bmatrix}
\quad
C = \begin{bmatrix} 2 & 1 & 0 \\ 3 & 4 & 0 \end{bmatrix}
$$

Here $A \neq C$ since A and C do not have the same size. For the same reason $B \neq C$. Also, $A \neq B$ since the corresponding entries are not all equal. ▲

Definition. If A and B are any two matrices of the same size, then the ***sum*** $A + B$ is the matrix obtained by adding together the corresponding entries in the two matrices. Matrices of different sizes cannot be added.

Example 3 Consider the matrices

$$
A = \begin{bmatrix} 2 & 1 & 0 & 3 \\ -1 & 0 & 2 & 4 \\ 4 & -2 & 7 & 0 \end{bmatrix}
\quad
B = \begin{bmatrix} -4 & 3 & 5 & 1 \\ 2 & 2 & 0 & -1 \\ 3 & 2 & -4 & 5 \end{bmatrix}
\quad
C = \begin{bmatrix} 1 & 1 \\ 2 & 2 \end{bmatrix}
$$

Then

$$
A + B = \begin{bmatrix} -2 & 4 & 5 & 4 \\ 1 & 2 & 2 & 3 \\ 7 & 0 & 3 & 5 \end{bmatrix}
$$

while $A + C$ and $B + C$ are undefined. ▲

Definition. If A is any matrix and c is any scalar, then the ***product*** cA is the matrix obtained by multiplying each entry of A by c.

Example 4 If A is the matrix

$$A = \begin{bmatrix} 4 & 2 \\ 1 & 3 \\ -1 & 0 \end{bmatrix}$$

then

$$2A = \begin{bmatrix} 8 & 4 \\ 2 & 6 \\ -2 & 0 \end{bmatrix} \quad \text{and} \quad (-1)A = \begin{bmatrix} -4 & -2 \\ -1 & -3 \\ 1 & 0 \end{bmatrix}$$

If B is any matrix, then $-B$ will denote the product $(-1)B$. If A and B are two matrices of the same size, then the matrix $A - B$ is defined to be the sum $A + (-B) = A + (-1)B$. ▲

Example 5 Consider the matrices

$$A = \begin{bmatrix} 2 & 3 & 4 \\ 1 & 2 & 1 \end{bmatrix} \quad \text{and} \quad B = \begin{bmatrix} 0 & 2 & 7 \\ 1 & -3 & 5 \end{bmatrix}$$

From the above definitions

$$-B = \begin{bmatrix} 0 & -2 & -7 \\ -1 & 3 & -5 \end{bmatrix}$$

and

$$A - B = \begin{bmatrix} 2 & 3 & 4 \\ 1 & 2 & 1 \end{bmatrix} + \begin{bmatrix} 0 & -2 & -7 \\ -1 & 3 & -5 \end{bmatrix} = \begin{bmatrix} 2 & 1 & -3 \\ 0 & 5 & -4 \end{bmatrix}$$

Observe that $A - B$ can be obtained directly by subtracting the entries of B from the corresponding entries of A. ▲

Above, we defined the multiplication of a matrix by a scalar, but so far we have not defined the multiplication of two matrices. Since matrices are added by adding corresponding entries and are subtracted by subtracting corresponding entries, it would seem that the most natural definition for multiplying two matrices would be to multiply corresponding entries. However, it turns out that such a definition would not be very useful for most problems. Experience has led mathematicians to the following less natural but more useful definition of matrix multiplication.

Definition. If A is an $m \times r$ matrix and B is an $r \times n$ matrix, then the *product* AB is the $m \times n$ matrix whose entries are determined as follows. To find the entry in row i and column j of AB, single out row i from the matrix A and column j from the matrix B. Multiply the corresponding entries from the row and column together and then add up the resulting products.

Example 6 Consider the matrices

$$A = \begin{bmatrix} 1 & 2 & 4 \\ 2 & 6 & 0 \end{bmatrix} \qquad B = \begin{bmatrix} 4 & 1 & 4 & 3 \\ 0 & -1 & 3 & 1 \\ 2 & 7 & 5 & 2 \end{bmatrix}$$

Since A is a 2×3 matrix and B is a 3×4 matrix, the product AB is a 2×4 matrix. To determine, for example, the entry in row 2 and column 3 of AB, we single out row 2 from A and column 3 from B. Then, as illustrated below, we multiply corresponding entries together and add up these products.

$$\begin{bmatrix} 1 & 2 & 4 \\ 2 & 6 & 0 \end{bmatrix} \begin{bmatrix} 4 & 1 & 4 & 3 \\ 0 & -1 & 3 & 1 \\ 2 & 7 & 5 & 2 \end{bmatrix} = \begin{bmatrix} \square & \square & \square & \square \\ \square & \square & \boxed{26} & \square \end{bmatrix}$$

$$(2 \cdot 4) + (6 \cdot 3) + (0 \cdot 5) = 26$$

The entry in row 1 and column 4 of AB is computed as follows.

$$\begin{bmatrix} 1 & 2 & 4 \\ 2 & 6 & 0 \end{bmatrix} \begin{bmatrix} 4 & 1 & 4 & 3 \\ 0 & -1 & 3 & 1 \\ 2 & 7 & 5 & 2 \end{bmatrix} = \begin{bmatrix} \square & \square & \square & \boxed{13} \\ \square & \square & \square & \square \end{bmatrix}$$

$$(1 \cdot 3) + (2 \cdot 1) + (4 \cdot 2) = 13$$

The computations for the remaining products are

$$\begin{aligned}
(1 \cdot 4) + (2 \cdot 0) + (4 \cdot 2) &= 12 \\
(1 \cdot 1) - (2 \cdot 1) + (4 \cdot 7) &= 27 \\
(1 \cdot 4) + (2 \cdot 3) + (4 \cdot 5) &= 30 \\
(2 \cdot 4) + (6 \cdot 0) + (0 \cdot 2) &= 8 \\
(2 \cdot 1) - (6 \cdot 1) + (0 \cdot 7) &= -4 \\
(2 \cdot 3) + (6 \cdot 1) + (0 \cdot 2) &= 12
\end{aligned} \qquad AB = \begin{bmatrix} 12 & 27 & 30 & 13 \\ 8 & -4 & 26 & 12 \end{bmatrix}$$

The definition of matrix multiplication requires that the number of columns of the first factor A be the same as the number of rows of the second factor B in order to form the product AB. If this condition is not satisfied, the product is undefined. A convenient way to determine whether a product of two matrices is defined is to write down the size of the first factor and, to the right of it, write down the size of the second factor. If, as in Figure 2, the inside numbers are the same, then the product is defined. The outside numbers then give the size of the product.

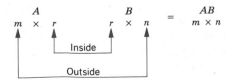

Figure 2

Example 7   Suppose that A is a 3×4 matrix, B is a 4×7 matrix, and C is a 7×3 matrix. Then AB is defined and is a 3×7 matrix; CA is defined and is a 7×4 matrix; BC is defined and is a 4×3 matrix. The products AC, CB, and BA are all undefined. ▲

Example 8   If A is a general $m \times r$ matrix and B is a general $r \times n$ matrix, then, as suggested by the shading below, the entry in row i and column j of AB is given by the formula

$$a_{i1}b_{1j} + a_{i2}b_{2j} + a_{i3}b_{3j} + \cdots + a_{ir}b_{rj}$$

$$AB = \begin{bmatrix} a_{11} & a_{12} & \cdots & a_{1r} \\ a_{21} & a_{22} & \cdots & a_{2r} \\ \vdots & \vdots & & \vdots \\ a_{i1} & a_{i2} & \cdots & a_{ir} \\ \vdots & \vdots & & \vdots \\ a_{m1} & a_{m2} & \cdots & a_{mr} \end{bmatrix} \begin{bmatrix} b_{11} & b_{12} & \cdots & b_{1j} & \cdots & b_{1n} \\ b_{21} & b_{22} & \cdots & b_{2j} & \cdots & b_{2n} \\ \vdots & \vdots & & \vdots & & \vdots \\ b_{r1} & b_{r2} & \cdots & b_{rj} & \cdots & b_{rn} \end{bmatrix}$$

▲

MATRIX FORM OF A LINEAR SYSTEM

Matrix multiplication has an important application to systems of linear equations. Consider any system of m linear equations in n unknowns.

$$\begin{aligned} a_{11}x_1 + a_{12}x_2 + \cdots + a_{1n}x_n &= b_1 \\ a_{21}x_1 + a_{22}x_2 + \cdots + a_{2n}x_n &= b_2 \\ \vdots \qquad\qquad \vdots \qquad\qquad\quad \vdots \quad &\;\; \vdots \\ a_{m1}x_1 + a_{m2}x_2 + \cdots + a_{mn}x_n &= b_m \end{aligned}$$

Since two matrices are equal if and only if their corresponding entries are equal, we can replace the m equations in this system by the single matrix equation

$$\begin{bmatrix} a_{11}x_1 + a_{12}x_2 + \cdots + a_{1n}x_n \\ a_{21}x_1 + a_{22}x_2 + \cdots + a_{2n}x_n \\ \vdots \qquad \vdots \qquad\qquad \vdots \\ a_{m1}x_1 + a_{m2}x_2 + \cdots + a_{mn}x_n \end{bmatrix} = \begin{bmatrix} b_1 \\ b_2 \\ \vdots \\ b_m \end{bmatrix}$$

The $m \times 1$ matrix on the left side of this equation can be written as a product to give

$$\begin{bmatrix} a_{11} & a_{12} & \cdots & a_{1n} \\ a_{21} & a_{22} & \cdots & a_{2n} \\ \vdots & \vdots & & \vdots \\ a_{m1} & a_{m2} & \cdots & a_{mn} \end{bmatrix} \begin{bmatrix} x_1 \\ x_2 \\ \vdots \\ x_n \end{bmatrix} = \begin{bmatrix} b_1 \\ b_2 \\ \vdots \\ b_m \end{bmatrix}$$

If we designate these matrices by A, X, and B, respectively, the original system of m equations in n unknowns has been replaced by the single matrix equation

$$AX = B \tag{1}$$

Some of our later work will be devoted to solving matrix equations such as this for the unknown matrix X. As a consequence of this matrix approach, we will obtain effective new methods for solving systems of linear equations. Matrix A in (1) is called the *coefficient matrix* for the system.

Example 9   At times it is helpful to be able to find a particular row or column in a product AB without computing the entire product. We leave it as an exercise to show that the entries in the jth column of AB are the entries in the product AB_j, where B_j is the matrix formed using only the jth column of B. Thus, if A and B are the matrices in Example 6, the second column of AB can be obtained by the computation

$$\begin{bmatrix} 1 & 2 & 4 \\ 2 & 6 & 0 \end{bmatrix} \begin{bmatrix} 1 \\ -1 \\ 7 \end{bmatrix} = \begin{bmatrix} 27 \\ -4 \end{bmatrix}$$

| **Second column** | **Second column** |
| **of B** | **of AB** |

Similarly, the entries in the ith row of AB are the entries in the product A_iB, where A_i is the matrix formed by using only the ith row of A. Thus the first row in the product AB of Example 6 can be obtained by the computation

$$\begin{bmatrix} 1 & 2 & 4 \end{bmatrix} \begin{bmatrix} 4 & 1 & 4 & 3 \\ 0 & -1 & 3 & 1 \\ 2 & 7 & 5 & 2 \end{bmatrix} = \begin{bmatrix} 12 & 27 & 30 & 13 \end{bmatrix}$$

First row of A   **First row of AB**

We conclude this section by defining two matrix operations that have no analogues in the real numbers.

TRANSPOSE OF A MATRIX

> **Definition.**   If A is any $m \times n$ matrix, then the *transpose of A* is denoted by A^t and is defined to be the $n \times m$ matrix whose first column is the first row of A, whose second column is the second row of A, whose third column is the third row of A, and so on.

Example 10   The following are some examples of matrices and their transposes.

$$A = \begin{bmatrix} a_{11} & a_{12} & a_{13} & a_{14} \\ a_{21} & a_{22} & a_{23} & a_{24} \\ a_{31} & a_{32} & a_{33} & a_{34} \end{bmatrix} \qquad A^t = \begin{bmatrix} a_{11} & a_{21} & a_{31} \\ a_{12} & a_{22} & a_{32} \\ a_{13} & a_{23} & a_{33} \\ a_{14} & a_{24} & a_{34} \end{bmatrix}$$

$$B = \begin{bmatrix} 2 & 3 \\ 1 & 4 \\ 5 & 6 \end{bmatrix} \qquad B^t = \begin{bmatrix} 2 & 1 & 5 \\ 3 & 4 & 6 \end{bmatrix}$$

$$C = \begin{bmatrix} 1 & 3 & 5 \end{bmatrix} \qquad C^t = \begin{bmatrix} 1 \\ 3 \\ 5 \end{bmatrix}$$

$$D = \begin{bmatrix} 3 & 5 & -2 \\ 5 & 4 & 1 \\ -2 & 1 & 7 \end{bmatrix} \qquad D^t = \begin{bmatrix} 3 & 5 & -2 \\ 5 & 4 & 1 \\ -2 & 1 & 7 \end{bmatrix}$$

$$E = \begin{bmatrix} 4 \end{bmatrix} \qquad E^t = \begin{bmatrix} 4 \end{bmatrix} \; \blacktriangle$$

REMARK. Observe that not only are the columns of A^t the rows of A, but the rows of A^t are the columns of A.

Definition. If A is a square matrix, then the *trace of A* is denoted by $\text{tr}(A)$ and is defined to be the sum of the entries on the main diagonal of A.

Example 11 The following are examples of matrices and their traces.

$$A = \begin{bmatrix} a_{11} & a_{12} & a_{13} \\ a_{21} & a_{22} & a_{23} \\ a_{31} & a_{32} & a_{33} \end{bmatrix} \qquad \text{tr}(A) = a_{11} + a_{22} + a_{33}$$

$$B = \begin{bmatrix} -1 & 2 & 7 & 0 \\ 3 & 5 & -8 & 4 \\ 1 & 2 & 7 & -3 \\ 4 & -2 & 1 & 0 \end{bmatrix} \qquad \text{tr}(B) = -1 + 5 + 7 + 0 = 11 \; \blacktriangle$$

EXERCISE SET 1.4

1. Suppose that A, B, C, D, and E are matrices with the following sizes:

A	B	C	D	E
(4×5)	(4×5)	(5×2)	(4×2)	(5×4)

Determine which of the following matrix expressions are defined. For those which are defined, give the size of the resulting matrix.

(a) BA (b) $AC + D$ (c) $AE + B$ (d) $AB + B$
(e) $E(A + B)$ (f) $E(AC)$ (g) $E^t A$ (h) $(A^t + E)D$

2. Solve the following matrix equation for a, b, c, and d.

$$\begin{bmatrix} a - b & b + c \\ 3d + c & 2a - 4d \end{bmatrix} = \begin{bmatrix} 8 & 1 \\ 7 & 6 \end{bmatrix}$$

3. Consider the matrices

$$A = \begin{bmatrix} 3 & 0 \\ -1 & 2 \\ 1 & 1 \end{bmatrix} \quad B = \begin{bmatrix} 4 & -1 \\ 0 & 2 \end{bmatrix} \quad C = \begin{bmatrix} 1 & 4 & 2 \\ 3 & 1 & 5 \end{bmatrix} \quad D = \begin{bmatrix} 1 & 5 & 2 \\ -1 & 0 & 1 \\ 3 & 2 & 4 \end{bmatrix} \quad E = \begin{bmatrix} 6 & 1 & 3 \\ -1 & 1 & 2 \\ 4 & 1 & 3 \end{bmatrix}$$

Compute the following (where possible).
(a) $D + E$ (b) $D - E$ (c) $5A$ (d) $-7C$
(e) $2B - C$ (f) $4E - 2D$ (g) $-3(D + 2E)$ (h) $A - A$
(i) $\text{tr}(D)$ (j) $\text{tr}(D - 3E)$ (k) $4\,\text{tr}(7B)$ (l) $\text{tr}(A)$

4. Using the matrices in Exercise 3, compute the following (where possible).
(a) $2A^t + C$ (b) $D^t - E^t$ (c) $(D - E)^t$ (d) $B^t + 5C^t$
(e) $\frac{1}{2}C^t - \frac{1}{4}A$ (f) $B - B^t$ (g) $2E^t - 3D^t$ (h) $(2E^t - 3D^t)^t$

5. Using the matrices in Exercise 3, compute the following (where possible).
(a) AB (b) BA (c) $(3E)D$ (d) $(AB)C$
(e) $A(BC)$ (f) CC^t (g) $(DA)^t$ (h) $(C^tB)A^t$
(i) $\text{tr}(DD^t)$ (j) $\text{tr}(4E^t - D)$ (k) $\text{tr}(C^tA^t + 2E^t)$

6. Using the matrices in Exercise 3, compute the following (where possible).
(a) $(2D^t - E)A$ (b) $(4B)C + 2B$ (c) $(-AC)^t + 5D^t$
(d) $(BA^t - 2C)^t$ (e) $B^t(CC^t - A^tA)$ (f) $D^tE^t - (ED)^t$

7. Let

$$A = \begin{bmatrix} 3 & -2 & 7 \\ 6 & 5 & 4 \\ 0 & 4 & 9 \end{bmatrix} \quad \text{and} \quad B = \begin{bmatrix} 6 & -2 & 4 \\ 0 & 1 & 3 \\ 7 & 7 & 5 \end{bmatrix}$$

Use the method of Example 9 to find
(a) the first row of AB (b) the third row of AB (c) the second column of AB
(d) the first column of BA (e) the third row of AA (f) the third column of AA

8. Let C, D, and E be the matrices in Exercise 3. Using as few computations as possible, determine the entry in row 2 and column 3 of $C(DE)$.

9. (a) Show that if AB and BA are both defined, then AB and BA are square matrices.
   (b) Show that if A is an $m \times n$ matrix and $A(BA)$ is defined, then B is an $n \times m$ matrix.

10. In each part find matrices A, X, and B which express the given system of linear equations as a single matrix equation $AX = B$.

(a) $2x_1 - 3x_2 + 5x_3 = 7$
    $9x_1 - x_2 + x_3 = -1$
    $x_1 + 5x_2 + 4x_3 = 0$

(b) $4x_1 - 3x_3 + x_4 = 1$
    $5x_1 + x_2 - 8x_4 = 3$
    $2x_1 - 5x_2 + 9x_3 - x_4 = 0$
    $3x_2 - x_3 + 7x_4 = 2$

11. In each part, express the matrix equation as a system of linear equations.

(a) $\begin{bmatrix} 3 & -1 & 2 \\ 4 & 3 & 7 \\ -2 & 1 & 5 \end{bmatrix} \begin{bmatrix} x_1 \\ x_2 \\ x_3 \end{bmatrix} = \begin{bmatrix} 2 \\ -1 \\ 4 \end{bmatrix}$ (b) $\begin{bmatrix} 3 & -2 & 0 & 1 \\ 5 & 0 & 2 & -2 \\ 3 & 1 & 4 & 7 \\ -2 & 5 & 1 & 6 \end{bmatrix} \begin{bmatrix} w \\ x \\ y \\ z \end{bmatrix} = \begin{bmatrix} 0 \\ 0 \\ 0 \\ 0 \end{bmatrix}$

12. Let

$$A = \begin{bmatrix} a_{11} & a_{12} & \cdots & a_{1n} \\ a_{21} & a_{22} & \cdots & a_{2n} \\ \vdots & \vdots & & \vdots \\ a_{m1} & a_{m2} & \cdots & a_{mn} \end{bmatrix}, \quad D = \begin{bmatrix} d_1 & 0 & \cdots & 0 \\ 0 & d_2 & \cdots & 0 \\ \vdots & \vdots & & \vdots \\ 0 & 0 & \cdots & d_m \end{bmatrix}, \quad E = \begin{bmatrix} e_1 & 0 & \cdots & 0 \\ 0 & e_2 & \cdots & 0 \\ \vdots & \vdots & & \vdots \\ 0 & 0 & \cdots & e_n \end{bmatrix}$$

(a) Compute DA and AE.

(b) Examine the rows of DA and the columns of AE; then deduce two simple rules for multiplying a matrix A by a diagonal matrix.

(c) Use the rules you obtained in part (b) to compute AB and BA, where

$$A = \begin{bmatrix} -2 & 1 & 5 \\ 3 & 0 & 2 \\ -7 & 1 & 5 \end{bmatrix} \quad \text{and} \quad B = \begin{bmatrix} 4 & 0 & 0 \\ 0 & 1 & 0 \\ 0 & 0 & -3 \end{bmatrix}$$

13. Show that the product of diagonal matrices is again a diagonal matrix. State a rule for multiplying diagonal matrices.

14. (a) Show that if A has a row of zeros and B is any matrix for which AB is defined, then AB also has a row of zeros.

(b) Find a similar result involving a column of zeros.

15. If a_{ij} is the entry in row i and column j of a matrix A, then in what row and column will a_{ij} appear in A^t?

16. Let A be any $m \times n$ matrix and let 0 be the $m \times n$ matrix each of whose entries is zero. Show that if $kA = 0$, then $k = 0$ or $A = 0$.

17. Let I be the $n \times n$ matrix whose entry in row i and column j is

$$\begin{cases} 1 & \text{if} \quad i = j \\ 0 & \text{if} \quad i \neq j \end{cases}$$

Show that $AI = IA = A$ for every $n \times n$ matrix A.

18. In each part find a 6×6 matrix $[a_{ij}]$ that satisfies the stated condition. Make your answers as general as possible by using letters rather than specific numbers for the nonzero entries.

(a) $a_{ij} = 0$ if $i \neq j$ (b) $a_{ij} = 0$ if $i > j$ (c) $a_{ij} = 0$ if $i < j$ (d) $a_{ij} = 0$ if $|i - j| > 1$

19. (a) Show that the entries in the jth column of AB are the entries in the product AB_j, where B_j is the matrix formed from the jth column of B.

(b) Show that the entries in the ith row of AB are the entries in the product A_iB, where A_i is the matrix formed from the ith row of A.

1.5 RULES OF MATRIX ARITHMETIC

In this section we shall discuss some properties of the arithmetic operations on matrices. We shall see that many of the basic rules of arithmetic for real numbers also hold for matrices but a few do not.

PROPERTIES OF MATRIX OPERATIONS

For real numbers a and b, we always have $ab = ba$, which is called the *commutative law for multiplication*. For matrices, however, AB and BA need not be equal. Equality can fail to hold for three reasons. It can happen, for example, that AB is defined but BA is undefined. This is the case if A is a 2×3 matrix and B is a 3×4 matrix. Also, it can happen that AB and BA are both defined but have different sizes. This is the situation if A is a 2×3 matrix and B is a 3×2 matrix. Finally, as Example 1 shows, it is possible to have $AB \neq BA$ even if both AB and BA are defined and have the same size.

Example 1 Consider the matrices

$$A = \begin{bmatrix} -1 & 0 \\ 2 & 3 \end{bmatrix} \qquad B = \begin{bmatrix} 1 & 2 \\ 3 & 0 \end{bmatrix}$$

Multiplying gives

$$AB = \begin{bmatrix} -1 & -2 \\ 11 & 4 \end{bmatrix} \qquad BA = \begin{bmatrix} 3 & 6 \\ -3 & 0 \end{bmatrix}$$

Thus, $AB \neq BA$. ▲

Although the commutative law for multiplication is not valid in matrix arithmetic, many familiar laws of arithmetic are valid for matrices. Some of the most important ones and their names are summarized in the following theorem.

Theorem 1.5.1. *Assuming that the sizes of the matrices are such that the indicated operations can be performed, the following rules of matrix arithmetic are valid.*

(a) $A + B = B + A$	*(Commutative law for addition)*
(b) $A + (B + C) = (A + B) + C$	*(Associative law for addition)*
(c) $A(BC) = (AB)C$	*(Associative law for multiplication)*
(d) $A(B + C) = AB + AC$	*(Distributive law)*
(e) $(B + C)A = BA + CA$	*(Distributive law)*
(f) $A(B - C) = AB - AC$	
(g) $(B - C)A = BA - CA$	
(h) $a(B + C) = aB + aC$	
(i) $a(B - C) = aB - aC$	
(j) $(a + b)C = aC + bC$	
(k) $(a - b)C = aC - bC$	
(l) $a(bC) = (ab)C$	
(m) $a(BC) = (aB)C = B(aC)$	

Each of the equations in this theorem asserts an equality between matrices. To prove one of these equalities, it is necessary to show that the matrix on the left side has the same size as the matrix on the right side, and that corresponding entries on the two sides are equal. As an illustration, we shall prove (*h*). Some of the remaining proofs are given as exercises.

Proof (*h*). Since the left side involves the operation $B + C$, B and C must have the same size, say $m \times n$. It follows that $a(B + C)$ and $aB + aC$ are also $m \times n$ matrices, and thus have the same size.

Let l_{ij} be any entry on the left side, and let r_{ij} be the corresponding entry on the right side. To complete the proof, we must show $l_{ij} = r_{ij}$. If we let b_{ij} and c_{ij} be the entries in the *i*th row and *j*th column of B and C, respectively, then from the definitions of the matrix operations we have

$$l_{ij} = a(b_{ij} + c_{ij}) \qquad \text{and} \qquad r_{ij} = ab_{ij} + ac_{ij}$$

Since $a(b_{ij} + c_{ij}) = ab_{ij} + ac_{ij}$, we have $l_{ij} = r_{ij}$, which completes the proof. ∎

Although the operations of matrix addition and matrix multiplication were defined for pairs of matrices, associative laws (*b*) and (*c*) enable us to denote sums and products of three matrices as $A + B + C$ and ABC without inserting any parentheses. This is justified by the fact that no matter how parentheses are inserted, the associative laws guarantee that the same end result will be obtained. Without going into details we observe that similar results are valid for sums or products involving four or more matrices. In general, *given any sum or any product of matrices, pairs of parentheses can be inserted or deleted anywhere within the expression without affecting the end result.*

Example 2 As an illustration of the associative law for matrix multiplication, consider

$$A = \begin{bmatrix} 1 & 2 \\ 3 & 4 \\ 0 & 1 \end{bmatrix} \qquad B = \begin{bmatrix} 4 & 3 \\ 2 & 1 \end{bmatrix} \qquad C = \begin{bmatrix} 1 & 0 \\ 2 & 3 \end{bmatrix}$$

Then

$$AB = \begin{bmatrix} 1 & 2 \\ 3 & 4 \\ 0 & 1 \end{bmatrix} \begin{bmatrix} 4 & 3 \\ 2 & 1 \end{bmatrix} = \begin{bmatrix} 8 & 5 \\ 20 & 13 \\ 2 & 1 \end{bmatrix}$$

so that

$$(AB)C = \begin{bmatrix} 8 & 5 \\ 20 & 13 \\ 2 & 1 \end{bmatrix} \begin{bmatrix} 1 & 0 \\ 2 & 3 \end{bmatrix} = \begin{bmatrix} 18 & 15 \\ 46 & 39 \\ 4 & 3 \end{bmatrix}$$

On the other hand

$$BC = \begin{bmatrix} 4 & 3 \\ 2 & 1 \end{bmatrix} \begin{bmatrix} 1 & 0 \\ 2 & 3 \end{bmatrix} = \begin{bmatrix} 10 & 9 \\ 4 & 3 \end{bmatrix}$$

so that

$$A(BC) = \begin{bmatrix} 1 & 2 \\ 3 & 4 \\ 0 & 1 \end{bmatrix} \begin{bmatrix} 10 & 9 \\ 4 & 3 \end{bmatrix} = \begin{bmatrix} 18 & 15 \\ 46 & 39 \\ 4 & 3 \end{bmatrix}$$

Thus, $(AB)C = A(BC)$, as guaranteed by Theorem 1.5.1c. ▲

ZERO MATRICES   A matrix, all of whose entries are zero, such as

$$\begin{bmatrix} 0 & 0 \\ 0 & 0 \end{bmatrix} \qquad \begin{bmatrix} 0 & 0 & 0 \\ 0 & 0 & 0 \\ 0 & 0 & 0 \end{bmatrix} \qquad \begin{bmatrix} 0 & 0 & 0 & 0 \\ 0 & 0 & 0 & 0 \end{bmatrix} \qquad \begin{bmatrix} 0 \\ 0 \\ 0 \\ 0 \end{bmatrix} \qquad [0]$$

is called a *zero matrix.* A zero matrix will be denoted by *0*; if it is important to emphasize the size, we shall write $0_{m \times n}$ for the $m \times n$ zero matrix.

If A is any matrix and *0* is the zero matrix with the same size, it is obvious that $A + 0 = 0 + A = A$. The matrix *0* plays much the same role in these matrix equations as the number 0 plays in the numerical equations $a + 0 = 0 + a = a$.

Since we already know that some of the rules of arithmetic for real numbers do not carry over to matrix arithmetic, it would be foolhardy to assume that all the properties of the real number zero carry over to zero matrices. For example, consider the following two standard results in the arithmetic of real numbers.

(i) If $ab = ac$ and $a \neq 0$, then $b = c$. (This is called the *cancellation law.*)
(ii) If $ad = 0$, then at least one of the factors on the left is 0.

As the next example shows, the corresponding results are not generally true in matrix arithmetic.

Example 3   Consider the matrices

$$A = \begin{bmatrix} 0 & 1 \\ 0 & 2 \end{bmatrix} \qquad B = \begin{bmatrix} 1 & 1 \\ 3 & 4 \end{bmatrix} \qquad C = \begin{bmatrix} 2 & 5 \\ 3 & 4 \end{bmatrix} \qquad D = \begin{bmatrix} 3 & 7 \\ 0 & 0 \end{bmatrix}$$

Here

$$AB = AC = \begin{bmatrix} 3 & 4 \\ 6 & 8 \end{bmatrix}$$

Although $A \neq 0$, it is *incorrect* to cancel the A from both sides of the equation $AB = AC$ and write $B = C$. Thus, the cancellation law fails to hold for matrices.

Also, $AD = 0$, yet $A \neq 0$ and $D \neq 0$, so that result (ii) listed above does not carry over to matrix arithmetic. ▲

In spite of the above examples, there are a number of familiar properties of the real number 0 that do carry over to zero matrices. Some of the more important ones are summarized in the next theorem. The proofs are left as exercises.

Theorem 1.5.2. *Assuming that the sizes of the matrices are such that the indicated operations can be performed, the following rules of matrix arithmetic are valid.*

(a) $A + 0 = 0 + A = A$
(b) $A - A = 0$
(c) $0 - A = -A$
(d) $A0 = 0; \quad 0A = 0$

As an application of our results on matrix arithmetic, we shall prove the following theorem, which was anticipated earlier in the text.

Theorem 1.5.3. *Every system of linear equations has either no solutions, exactly one solution, or infinitely many solutions.*

Proof. If $AX = B$ is a system of linear equations, exactly one of the following is true: (a) the system has no solutions, (b) the system has exactly one solution, or (c) the system has more than one solution. The proof will be complete if we can show that the system has infinitely many solutions in case (c).

Assume that $AX = B$ has more than one solution, and let $X_0 = X_1 - X_2$, where X_1 and X_2 are any two distinct solutions. Because X_1 and X_2 are distinct, the matrix X_0 is nonzero; moreover

$$AX_0 = A(X_1 - X_2) = AX_1 - AX_2 = B - B = 0$$

If we now let k be any scalar, then

$$
\begin{aligned}
A(X_1 + kX_0) &= AX_1 + A(kX_0) \\
&= AX_1 + k(AX_0) \\
&= B + k0 \\
&= B + 0 \\
&= B
\end{aligned}
$$

But this says that $X_1 + kX_0$ is a solution of $AX = B$. Since X_0 is nonzero and there are infinitely many choices for k, $AX = B$ has infinitely many solutions. ▌

IDENTITY MATRICES

Of special interest are square matrices with 1's on the main diagonal and 0's off the main diagonal, such as

$$
\begin{bmatrix} 1 & 0 \\ 0 & 1 \end{bmatrix}, \quad
\begin{bmatrix} 1 & 0 & 0 \\ 0 & 1 & 0 \\ 0 & 0 & 1 \end{bmatrix}, \quad
\begin{bmatrix} 1 & 0 & 0 & 0 \\ 0 & 1 & 0 & 0 \\ 0 & 0 & 1 & 0 \\ 0 & 0 & 0 & 1 \end{bmatrix}, \quad \text{and so on.}
$$

A matrix of this form is called an ***identity matrix*** and is denoted by I. If it is important to emphasize the size, we shall write I_n for the $n \times n$ identity matrix.

If A is an $m \times n$ matrix, then, as illustrated in the next example

$$AI_n = A \quad \text{and} \quad I_m A = A$$

Thus, an identity matrix plays much the same role in matrix arithmetic as the number 1 plays in the numerical relationships $a \cdot 1 = 1 \cdot a = a$.

Example 4 Consider the matrix

$$A = \begin{bmatrix} a_{11} & a_{12} & a_{13} \\ a_{21} & a_{22} & a_{23} \end{bmatrix}$$

Then

$$I_2 A = \begin{bmatrix} 1 & 0 \\ 0 & 1 \end{bmatrix} \begin{bmatrix} a_{11} & a_{12} & a_{13} \\ a_{21} & a_{22} & a_{23} \end{bmatrix} = \begin{bmatrix} a_{11} & a_{12} & a_{13} \\ a_{21} & a_{22} & a_{23} \end{bmatrix} = A$$

and

$$AI_3 = \begin{bmatrix} a_{11} & a_{12} & a_{13} \\ a_{21} & a_{22} & a_{23} \end{bmatrix} \begin{bmatrix} 1 & 0 & 0 \\ 0 & 1 & 0 \\ 0 & 0 & 1 \end{bmatrix} = \begin{bmatrix} a_{11} & a_{12} & a_{13} \\ a_{21} & a_{22} & a_{23} \end{bmatrix} = A \quad \blacktriangle$$

INVERSE OF A MATRIX

Definition. If A is a square matrix, and if a matrix B can be found such that $AB = BA = I$, then A is said to be **invertible** and B is called an **inverse** of A.

Example 5 The matrix

$$B = \begin{bmatrix} 3 & 5 \\ 1 & 2 \end{bmatrix} \quad \text{is an inverse of} \quad A = \begin{bmatrix} 2 & -5 \\ -1 & 3 \end{bmatrix}$$

since

$$AB = \begin{bmatrix} 2 & -5 \\ -1 & 3 \end{bmatrix} \begin{bmatrix} 3 & 5 \\ 1 & 2 \end{bmatrix} = \begin{bmatrix} 1 & 0 \\ 0 & 1 \end{bmatrix} = I$$

and

$$BA = \begin{bmatrix} 3 & 5 \\ 1 & 2 \end{bmatrix} \begin{bmatrix} 2 & -5 \\ -1 & 3 \end{bmatrix} = \begin{bmatrix} 1 & 0 \\ 0 & 1 \end{bmatrix} = I$$

Example 6 The matrix

$$A = \begin{bmatrix} 1 & 4 & 0 \\ 2 & 5 & 0 \\ 3 & 6 & 0 \end{bmatrix}$$

is not invertible. To see why, let

$$B = \begin{bmatrix} b_{11} & b_{12} & b_{13} \\ b_{21} & b_{22} & b_{23} \\ b_{31} & b_{32} & b_{33} \end{bmatrix}$$

be any 3×3 matrix. From Example 9 of Section 1.4, the third column of BA is

$$\begin{bmatrix} b_{11} & b_{12} & b_{13} \\ b_{21} & b_{22} & b_{23} \\ b_{31} & b_{32} & b_{33} \end{bmatrix} \begin{bmatrix} 0 \\ 0 \\ 0 \end{bmatrix} = \begin{bmatrix} 0 \\ 0 \\ 0 \end{bmatrix}$$

Thus,

$$BA \neq I = \begin{bmatrix} 1 & 0 & 0 \\ 0 & 1 & 0 \\ 0 & 0 & 1 \end{bmatrix} \quad \blacktriangle$$

PROPERTIES OF INVERSES

It is reasonable to ask whether an invertible matrix can have more than one inverse. The next theorem shows the answer is no—an invertible matrix has exactly one inverse.

Theorem 1.5.4. *If B and C are both inverses of the matrix A, then B = C.*

Proof. Since B is an inverse of A, $BA = I$. Multiplying both sides on the right by C gives $(BA)C = IC = C$. But $(BA)C = B(AC) = BI = B$, so that $C = B$. ∎

As a consequence of this important result, we can now speak of "the" inverse of an invertible matrix. If A is invertible, then its inverse will be denoted by the symbol A^{-1}. Thus,

$$AA^{-1} = I \qquad \text{and} \qquad A^{-1}A = I$$

The inverse of A plays much the same role in matrix arithmetic that reciprocal a^{-1} plays in the numerical relationships $aa^{-1} = 1$ and $a^{-1}a = 1$.

Example 7 Consider the 2×2 matrix

$$A = \begin{bmatrix} a & b \\ c & d \end{bmatrix}$$

If $ad - bc \neq 0$, then

$$A^{-1} = \frac{1}{ad - bc} \begin{bmatrix} d & -b \\ -c & a \end{bmatrix} = \begin{bmatrix} \dfrac{d}{ad - bc} & -\dfrac{b}{ad - bc} \\ -\dfrac{c}{ad - bc} & \dfrac{a}{ad - bc} \end{bmatrix} \tag{1}$$

since $AA^{-1} = I_2$ and $A^{-1}A = I_2$ (verify). In the next section we shall show how to find inverses of invertible matrices whose sizes are greater than 2×2. ▲

Theorem 1.5.5. *If A and B are invertible matrices of the same size, then*

(a) AB is invertible
(b) $(AB)^{-1} = B^{-1}A^{-1}$

Proof. If we can show that $(AB)(B^{-1}A^{-1}) = (B^{-1}A^{-1})(AB) = I$, then we will have simultaneously established that the matrix AB is invertible and that $(AB)^{-1} = B^{-1}A^{-1}$. But $(AB)(B^{-1}A^{-1}) = A(BB^{-1})A^{-1} = AIA^{-1} = AA^{-1} = I$. Similarly, $(B^{-1}A^{-1})(AB) = I$. ∎

Although we will not prove it, this result can be extended to include three or more factors. Thus, we can state the following general rule.

A product of invertible matrices is always invertible, and the inverse of the product is the product of the inverses in the reverse order.

Example 8 Consider the matrices

$$A = \begin{bmatrix} 1 & 2 \\ 1 & 3 \end{bmatrix} \quad B = \begin{bmatrix} 3 & 2 \\ 2 & 2 \end{bmatrix} \quad AB = \begin{bmatrix} 7 & 6 \\ 9 & 8 \end{bmatrix}$$

Applying Formula (1), we obtain

$$A^{-1} = \begin{bmatrix} 3 & -2 \\ -1 & 1 \end{bmatrix} \quad B^{-1} = \begin{bmatrix} 1 & -1 \\ -1 & \frac{3}{2} \end{bmatrix} \quad (AB)^{-1} = \begin{bmatrix} 4 & -3 \\ -\frac{9}{2} & \frac{7}{2} \end{bmatrix}$$

Also

$$B^{-1}A^{-1} = \begin{bmatrix} 1 & -1 \\ -1 & \frac{3}{2} \end{bmatrix}\begin{bmatrix} 3 & -2 \\ -1 & 1 \end{bmatrix} = \begin{bmatrix} 4 & -3 \\ -\frac{9}{2} & \frac{7}{2} \end{bmatrix}$$

Therefore, $(AB)^{-1} = B^{-1}A^{-1}$ as guaranteed by Theorem 1.5.5. ▲

Next, we shall define powers of a square matrix and discuss their properties.

POWERS OF A MATRIX

Definition. If A is a square matrix, then we define the nonnegative integer powers of A to be

$$A^0 = I \qquad A^n = \underbrace{AA \cdots A}_{n \text{ factors}} \qquad (n > 0)$$

Moreover, if A is invertible, then we define the negative integer powers to be

$$A^{-n} = (A^{-1})^n = \underbrace{A^{-1}A^{-1} \cdots A^{-1}}_{n \text{ factors}}$$

Example 9 If

$$A = \begin{bmatrix} 1 & 2 \\ 1 & 3 \end{bmatrix}$$

then from Example 8

$$A^{-1} = \begin{bmatrix} 3 & -2 \\ -1 & 1 \end{bmatrix}$$

so

$$A^3 = \begin{bmatrix} 1 & 2 \\ 1 & 3 \end{bmatrix} \begin{bmatrix} 1 & 2 \\ 1 & 3 \end{bmatrix} \begin{bmatrix} 1 & 2 \\ 1 & 3 \end{bmatrix} = \begin{bmatrix} 11 & 30 \\ 15 & 41 \end{bmatrix}$$

$$A^{-3} = (A^{-1})^3 = \begin{bmatrix} 3 & -2 \\ -1 & 1 \end{bmatrix} \begin{bmatrix} 3 & -2 \\ -1 & 1 \end{bmatrix} \begin{bmatrix} 3 & -2 \\ -1 & 1 \end{bmatrix} = \begin{bmatrix} 41 & -30 \\ -15 & 11 \end{bmatrix}$$ ▲

Computing high powers of large matrices can involve an enormous amount of computation. However, the situation is greatly simplified for diagonal matrices. For example, the kth power of a diagonal matrix

$$D = \begin{bmatrix} d_1 & 0 & \cdots & 0 \\ 0 & d_2 & \cdots & 0 \\ \vdots & \vdots & & \vdots \\ 0 & 0 & \cdots & d_n \end{bmatrix}$$

is

$$D^k = \begin{bmatrix} d_1{}^k & 0 & \cdots & 0 \\ 0 & d_2{}^k & \cdots & 0 \\ \vdots & \vdots & & \vdots \\ 0 & 0 & \cdots & d_n{}^k \end{bmatrix} \tag{2}$$

Thus, to raise a diagonal matrix to the kth power, we need only raise each diagonal entry to the kth power (verify).

The following theorem, which we state without proof, shows that the familiar laws of exponents are valid for powers of matrices.

Theorem 1.5.6. *If A is a square matrix and r and s are integers, then*

$$A^r A^s = A^{r+s} \qquad (A^r)^s = A^{rs}$$

The next theorem provides some additional properties of matrix exponents.

Theorem 1.5.7. *If A is an invertible matrix, then:*

(a) A^{-1} *is invertible and* $(A^{-1})^{-1} = A$.

(b) A^n *is invertible and* $(A^n)^{-1} = (A^{-1})^n$ *for* $n = 0, 1, 2, \ldots$

(c) *For any nonzero scalar* k, kA *is invertible and* $(kA)^{-1} = \dfrac{1}{k} A^{-1}$.

Proof

(a) Since $AA^{-1} = A^{-1}A = I$, A^{-1} is invertible and $(A^{-1})^{-1} = A$.

(b) This part is left as an exercise.

(c) If k is any nonzero scalar, results (l) and (m) of Theorem 1.5.1 enable us to write

$$(kA)\left(\frac{1}{k}A^{-1}\right) = \frac{1}{k}(kA)A^{-1} = \left(\frac{1}{k}k\right)AA^{-1} = (1)I = I$$

Similarly $\left(\dfrac{1}{k}A^{-1}\right)(kA) = I$ so that kA is invertible and $(kA)^{-1} = \dfrac{1}{k}A^{-1}$. ∎

PROPERTIES OF THE TRANSPOSE

We conclude this section with a theorem that lists the main properties of the transpose operation. The proofs are discussed in the exercises.

Theorem 1.5.8. *If the sizes of the matrices are such that the stated operations can be performed, then*

(a) $(A^t)^t = A$

(b) $(A + B)^t = A^t + B^t$

(c) $(kA)^t = kA^t$, *where* k *is any scalar*

(d) $(AB)^t = B^t A^t$

Although we shall not prove it, part (d) of this theorem can be extended to include three or more factors. More precisely, we can state the following general rule.

The transpose of a product of matrices is equal to the product of their transposes in the reverse order.

REMARK. Note the similarity between this result and the result following Theorem 1.5.5 about the inverse of a product of matrices.

EXERCISE SET 1.5

1. Let

$$A = \begin{bmatrix} 2 & -1 & 3 \\ 0 & 4 & 5 \\ -2 & 1 & 4 \end{bmatrix} \quad B = \begin{bmatrix} 8 & -3 & -5 \\ 0 & 1 & 2 \\ 4 & -7 & 6 \end{bmatrix} \quad C = \begin{bmatrix} 0 & -2 & 3 \\ 1 & 7 & 4 \\ 3 & 5 & 9 \end{bmatrix} \quad a = 4 \quad b = -7$$

Show
(a) $A + (B + C) = (A + B) + C$ (b) $(AB)C = A(BC)$ (c) $(a + b)C = aC + bC$ (d) $a(B - C) = aB - aC$

2. Using the matrices and scalars in Exercise 1, show
(a) $a(BC) = (aB)C = B(aC)$ (b) $A(B - C) = AB - AC$ (c) $(B + C)A = BA + CA$ (d) $a(bC) = (ab)C$

3. Using the matrices and scalars in Exercise 1, show
(a) $(A^t)^t = A$ (b) $(A + B)^t = A^t + B^t$ (c) $(aC)^t = aC^t$ (d) $(AB)^t = B^t A^t$

4. Use the formula given in Example 7 to compute the inverses of the following matrices.

$$A = \begin{bmatrix} 3 & 1 \\ 5 & 2 \end{bmatrix} \quad B = \begin{bmatrix} 2 & -3 \\ 4 & 4 \end{bmatrix} \quad C = \begin{bmatrix} 2 & 0 \\ 0 & 3 \end{bmatrix}$$

5. Verify that the three matrices A, B, and C in Exercise 4 satisfy the relationships $(AB)^{-1} = B^{-1}A^{-1}$ and $(ABC)^{-1} = C^{-1}B^{-1}A^{-1}$.

6. Let A and B be square matrices of the same size. Is $(AB)^2 = A^2 B^2$ a valid matrix identity? Justify your answer.

7. Let A be an invertible matrix whose inverse is

$$\begin{bmatrix} 2 & -1 \\ 3 & 5 \end{bmatrix}$$

Find the matrix A.

8. Let A be an invertible matrix, and suppose that the inverse of $7A$ is

$$\begin{bmatrix} -3 & 7 \\ 1 & -2 \end{bmatrix}$$

Find the matrix A.

9. Let A be the matrix

$$\begin{bmatrix} 2 & 0 \\ 4 & 1 \end{bmatrix}$$

Compute A^3, A^{-3}, and $A^2 - 2A + I$.

10. Let A be the matrix

$$\begin{bmatrix} 1 & 0 & 1 \\ 1 & 1 & 0 \\ 0 & 1 & 1 \end{bmatrix}$$

Determine whether A is invertible, and if so, find its inverse. [**Hint.** Solve $AX = I$ by equating corresponding entries on the two sides.]

11. Find the inverse of

$$\begin{bmatrix} \cos\theta & \sin\theta \\ -\sin\theta & \cos\theta \end{bmatrix}$$

12. (a) Find 2×2 matrices A and B such that $(A + B)^2 \neq A^2 + 2AB + B^2$.

(b) Show that, if A and B are square matrices such that $AB = BA$, then

$$(A + B)^2 = A^2 + 2AB + B^2$$

(c) Find an expansion of $(A + B)^2$ that is valid for all square matrices A and B having the same size.

13. Consider the matrix

$$A = \begin{bmatrix} a_{11} & 0 & 0 & \cdots & 0 \\ 0 & a_{22} & 0 & \cdots & 0 \\ \vdots & \vdots & \vdots & & \vdots \\ 0 & 0 & 0 & \cdots & a_{nn} \end{bmatrix}$$

where $a_{11}a_{22} \cdots a_{nn} \neq 0$. Show that A is invertible and find its inverse.

14. Show that if a square matrix A satisfies $A^2 - 3A + I = 0$, then $A^{-1} = 3I - A$.

15. (a) Show that a matrix with a row of zeros cannot have an inverse.

(b) Show that a matrix with a column of zeros cannot have an inverse.

16. Is the sum of two invertible matrices necessarily invertible?

17. Let A and B be square matrices such that $AB = 0$. Show that if A is invertible then $B = 0$.

18. In Theorem 1.5.2 why didn't we write part (d) as $A0 = 0 = 0A$?

19. The real equation $a^2 = 1$ has exactly two solutions. Find at least eight different 3×3 matrices that satisfy the matrix equation $A^2 = I_3$. [**Hint.** Look for solutions in which all the entries off the main diagonal are zero.]

20. Let $AX = B$ be any consistent system of linear equations, and let X_1 be a fixed solution. Show that every solution to the system can be written in the form $X = X_1 + X_0$ where X_0 is a solution to $AX = 0$. Show also that every matrix of this form is a solution.

21. Let

$$A = \begin{bmatrix} a_{11} & a_{12} & a_{13} \\ a_{21} & a_{22} & a_{23} \\ a_{31} & a_{32} & a_{33} \end{bmatrix} \quad \text{and} \quad B = \begin{bmatrix} b_{11} & b_{12} & b_{13} \\ b_{21} & b_{22} & b_{23} \\ b_{31} & b_{32} & b_{33} \end{bmatrix}$$

Show that

(a) $(A^t)^t = A$ (b) $(A + B)^t = A^t + B^t$ (c) $(AB)^t = B^t A^t$ (d) $(kA)^t = kA^t$

22. (a) Verify that if the circled entries are interchanged, as indicated in the following figure, then the resulting matrix is the transpose of the original matrix.

(b) Use the result in (a) to help find a nonzero 3×3 matrix A such that $A^t = A$.

(c) Use the result in (a) to help find a nonzero 3×3 matrix A such that $A^t = -A$.

(d) In general, if A is a square matrix, then A^t results when the entries that are symmetrically placed on the two sides of the main diagonal are interchanged. Use this fact to help find a 4×4 nonzero matrix such that $A = A^t$.

23. A square matrix A is called **symmetric** if $A^t = A$ and **skew–symmetric** if $A^t = -A$. Show that if B is a square matrix, then
    (a) BB^t and $B + B^t$ are symmetric   (b) $B - B^t$ is skew-symmetric

24. If A is a square matrix and n is a positive integer, is it true that $(A^n)^t = (A^t)^n$? Justify your answer.

25. Apply parts (d) and (m) of Theorem 1.5.1 to the matrices A, B, and $(-1)C$ to derive the result in part (f).

26. Use parts (b) and (c) of Theorem 1.5.8 to prove that $(A - B)^t = A^t - B^t$.

27. Prove: If A is invertible, then A^t is invertible and $(A^t)^{-1} = (A^{-1})^t$.

28. Prove part (b) of Theorem 1.5.1.

29. Prove Theorem 1.5.2.

30. Prove part (c) of Theorem 1.5.1.

31. Prove part (b) of Theorem 1.5.7.

32. Consider the laws of exponents $A^r A^s = A^{r+s}$ and $(A^r)^s = A^{rs}$.
    (a) Show that if A is any square matrix, these laws are valid for all nonnegative integer values of r and s.
    (b) Show that if A is invertible, these laws hold for all negative integer values of r and s.

33. Show that if A is invertible and k is any nonzero scalar, then $(kA)^n = k^n A^n$ for all integer values of n.

34. (a) Show that if A is invertible and $AB = AC$, then $B = C$.
    (b) Explain why part (a) and Example 3 do not contradict one another.

35. Prove: If R is a square matrix in reduced row-echelon form, and if R has no zero rows, then $R = I$.

1.6 ELEMENTARY MATRICES AND A METHOD FOR FINDING A^{-1}

In this section we shall develop an algorithm for finding the inverse of an invertible matrix, and we shall discuss some basic properties of invertible matrices.

ELEMENTARY MATRICES

Definition. An $n \times n$ matrix is called an **elementary matrix** if it can be obtained from the $n \times n$ identity matrix I_n by performing a single elementary row operation.

Example 1 Listed below are four elementary matrices and the operations that produce them.

$$\begin{bmatrix} 1 & 0 \\ 0 & -3 \end{bmatrix} \qquad \begin{bmatrix} 1 & 0 & 0 & 0 \\ 0 & 0 & 0 & 1 \\ 0 & 0 & 1 & 0 \\ 0 & 1 & 0 & 0 \end{bmatrix} \qquad \begin{bmatrix} 1 & 0 & 3 \\ 0 & 1 & 0 \\ 0 & 0 & 1 \end{bmatrix} \qquad \begin{bmatrix} 1 & 0 & 0 \\ 0 & 1 & 0 \\ 0 & 0 & 1 \end{bmatrix}$$

Multiply the second row of I_2 by -3.	Interchange the second and fourth rows of I_4.	Add 3 times the third row of I_3 to the first row.	Multiply the first row of I_3 by 1.

When a matrix A is multiplied on the *left* by an elementary matrix E, the effect is to perform an elementary row operation on A. This is the content of the following theorem, which we state without proof.

> **Theorem 1.6.1.** *If the elementary matrix E results from performing a certain row operation on* I_m *and if A is an m × n matrix, then the product EA is the matrix that results when this same row operation is performed on A.*

Example 2 Consider the matrix

$$A = \begin{bmatrix} 1 & 0 & 2 & 3 \\ 2 & -1 & 3 & 6 \\ 1 & 4 & 4 & 0 \end{bmatrix}$$

and consider the elementary matrix

$$E = \begin{bmatrix} 1 & 0 & 0 \\ 0 & 1 & 0 \\ 3 & 0 & 1 \end{bmatrix}$$

which results from adding 3 times the first row of I_3 to the third row. The product EA is

$$EA = \begin{bmatrix} 1 & 0 & 2 & 3 \\ 2 & -1 & 3 & 6 \\ 4 & 4 & 10 & 9 \end{bmatrix}$$

which is precisely the same matrix that results when we add 3 times the first row of A to the third row. ▲

REMARK. Theorem 1.6.1 is primarily of theoretical interest and will be used for developing some results about matrices and systems of linear equations. Computationally, it is preferable to perform row operations directly rather than multiply on the left by an elementary matrix.

If an elementary row operation is applied to an identity matrix I to produce an elementary matrix E, then there is a second row operation that, when applied to E, produces I back again. For example, if E is obtained by multiplying the ith row of I by a nonzero constant c, then I can be recovered if the ith row of E is multiplied by $1/c$. The various possibilities are listed in Table 1.

TABLE 1

Row Operation on I that Produces E	Row Operation on E that Reproduces I
Multiply row i by $c \neq 0$	Multiply row i by $1/c$
Interchange rows i and j	Interchange rows i and j
Add c times row i to row j	Add $-c$ times row i to row j

The operations on the right side of this table are called the ***inverse operations*** of the corresponding operations on the left.

Example 3 In each of the following, an elementary row operation is applied to the 2×2 identity matrix to obtain an elementary matrix E, then E is restored to the identity matrix by applying the inverse row operation.

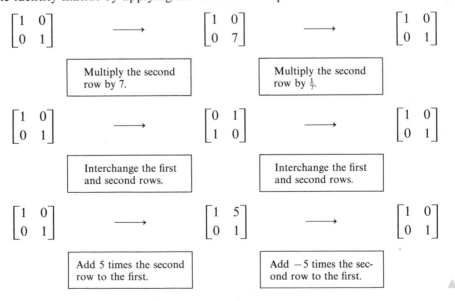

The next theorem gives an important property of elementary matrices.

> **Theorem 1.6.2.**   *Every elementary matrix is invertible, and the inverse is also an elementary matrix.*

Proof. If E is an elementary matrix, then E results from performing some row operation on I. Let E_0 be the matrix that results when the inverse of this operation is performed by I. Applying Theorem 1.6.1 and using the fact that inverse row operations cancel the effect of each other, it follows that

$$E_0 E = I \qquad \text{and} \qquad E E_0 = I$$

Thus, the elementary matrix E_0 is the inverse of E. ∎

ROW EQUIVALENCE

If a matrix B can be obtained from a matrix A by performing a finite sequence of elementary row operations, then obviously we can get from B back to A by performing the inverses of these elementary row operations in reverse order. Matrices that can be obtained from one another by a finite sequence of elementary row operations are said to be ***row equivalent***.

The next theorem establishes some fundamental relationships between $n \times n$ matrices and systems of n linear equations in n unknowns. These results are extremely important and will be used many times in later sections.

> **Theorem 1.6.3.**   *If A is an $n \times n$ matrix, then the following statements are equivalent, that is, all true or all false.*
>
> *(a) A is invertible.*
> *(b) $AX = 0$ has only the trivial solution.*
> *(c) A is row equivalent to I_n.*

Proof. We shall prove the equivalence by establishing the chain of implications: $(a) \Rightarrow (b) \Rightarrow (c) \Rightarrow (a)$.

$(a) \Rightarrow (b)$: Assume A is invertible and let X_0 be any solution of $AX = 0$; thus, $AX_0 = 0$. Multiplying both sides of this equation by A^{-1} gives $A^{-1}(AX_0) = A^{-1}0$, or $(A^{-1}A)X_0 = 0$, or $IX_0 = 0$, or $X_0 = 0$. Thus, $AX = 0$ has only the trivial solution.

$(b) \Rightarrow (c)$: Let $AX = 0$ be the matrix form of the system

$$
\begin{aligned}
a_{11}x_1 + a_{12}x_2 + \cdots + a_{1n}x_n &= 0 \\
a_{21}x_1 + a_{22}x_2 + \cdots + a_{2n}x_n &= 0 \\
\vdots \qquad \vdots \qquad\qquad \vdots \quad\;\; \vdots \\
a_{n1}x_1 + a_{n2}x_2 + \cdots + a_{nn}x_n &= 0
\end{aligned}
\tag{1}
$$

and assume that the system has only the trivial solution. If we solve by Gauss-

Jordan elimination, then the system of equations corresponding to the reduced row-echelon form of the augmented matrix will be

$$
\begin{aligned}
x_1 \qquad\qquad &= 0 \\
x_2 \qquad\quad &= 0 \\
\ddots \qquad & \\
x_n &= 0
\end{aligned}
\tag{2}
$$

Thus, the augmented matrix

$$
\begin{bmatrix}
a_{11} & a_{12} & \cdots & a_{1n} & 0 \\
a_{21} & a_{22} & \cdots & a_{2n} & 0 \\
\vdots & \vdots & & \vdots & \vdots \\
a_{n1} & a_{n2} & \cdots & a_{nn} & 0
\end{bmatrix}
$$

for (1) can be reduced to the augmented matrix

$$
\begin{bmatrix}
1 & 0 & 0 & \cdots & 0 & 0 \\
0 & 1 & 0 & \cdots & 0 & 0 \\
0 & 0 & 1 & \cdots & 0 & 0 \\
\vdots & \vdots & \vdots & & \vdots & \vdots \\
0 & 0 & 0 & \cdots & 1 & 0
\end{bmatrix}
$$

for (2) by a sequence of elementary row operations. If we disregard the last column (of zeros) in each of these matrices, we can conclude that A can be reduced to I_n by a sequence of elementary row operations; that is, A is a row equivalent to I_n.

$(c) \Rightarrow (a)$: Assume A is row equivalent to I_n, so that A can be reduced to I_n by a finite sequence of elementary row operations. By Theorem 1.6.1 each of these operations can be accomplished by multiplying on the left by an appropriate elementary matrix. Thus, we can find elementary matrices $E_1, E_2, \ldots, E_k$ such that

$$
E_k \cdots E_2 E_1 A = I_n
\tag{3}
$$

By Theorem 1.6.2, $E_1, E_2, \ldots, E_k$ are invertible. Multiplying both sides of Equation (3) on the left successively by $E_k^{-1}, \ldots, E_2^{-1}, E_1^{-1}$ we obtain

$$
A = E_1^{-1} E_2^{-1} \cdots E_k^{-1} I_n = E_1^{-1} E_2^{-1} \cdots E_k^{-1}
\tag{4}
$$

Since (4) expresses A as a product of invertible matrices, we can conclude that A is invertible. ∎

REMARK. Part (c) of Theorem 1.6.3 is equivalent to stating that I_n is the reduced row-echelon form of A. This is because the reduced row-echelon form of A is unique, and I_n is in reduced row-echelon form; so I_n must be the reduced row-echelon form of A.

**A METHOD FOR
INVERTING
MATRICES**

As our first application of Theorem 1.6.3, we shall establish a method for determining the inverse of an invertible matrix. Inverting the left and right sides of (4) yields $A^{-1} = E_k \cdots E_2 E_1$, or equivalently

$$A^{-1} = E_k \cdots E_2 E_1 I_n \tag{5}$$

which tells us that A^{-1} can be obtained by multiplying I_n successively on the left by the elementary matrices $E_1, E_2, \ldots, E_k$. Since each multiplication on the left by one of these elementary matrices performs a row operation, it follows, by comparing Equations (3) and (5), that *the sequence of row operations that reduces A to I_n will reduce I_n to A^{-1}.* Thus, we have the following result:

> *To find the inverse of an invertible matrix A, we must find a sequence of elementary row operations that reduces A to the identity and then perform this same sequence of operations on I_n to obtain A^{-1}.*

A simple method for carrying out this procedure is given in the following example.

Example 4   Find the inverse of

$$A = \begin{bmatrix} 1 & 2 & 3 \\ 2 & 5 & 3 \\ 1 & 0 & 8 \end{bmatrix}$$

Solution. We want to reduce A to the identity matrix by row operations and simultaneously apply these operations to I to produce A^{-1}. To accomplish this we shall adjoin the identity matrix to the right side of A, thereby producing a matrix of the form

$$[A | I]$$

then we shall apply row operations to this matrix until the left side is reduced to I; these operations will convert the right side to A^{-1}, so that the final matrix will have the form

$$[I | A^{-1}]$$

The computations are as follows:

$$\begin{bmatrix} 1 & 2 & 3 & \vdots & 1 & 0 & 0 \\ 2 & 5 & 3 & \vdots & 0 & 1 & 0 \\ 1 & 0 & 8 & \vdots & 0 & 0 & 1 \end{bmatrix}$$

$$\begin{bmatrix} 1 & 2 & 3 & \vdots & 1 & 0 & 0 \\ 0 & 1 & -3 & \vdots & -2 & 1 & 0 \\ 0 & -2 & 5 & \vdots & -1 & 0 & 1 \end{bmatrix}$$

We added -2 times the first row to the second and -1 times the first row to the third.

$$\begin{bmatrix} 1 & 2 & 3 & \vdots & 1 & 0 & 0 \\ 0 & 1 & -3 & \vdots & -2 & 1 & 0 \\ 0 & 0 & -1 & \vdots & -5 & 2 & 1 \end{bmatrix}$$

We added 2 times the second row to the third.

$$\begin{bmatrix} 1 & 2 & 3 & \vdots & 1 & 0 & 0 \\ 0 & 1 & -3 & \vdots & -2 & 1 & 0 \\ 0 & 0 & 1 & \vdots & 5 & -2 & -1 \end{bmatrix}$$ We multiplied the third row by -1.

$$\begin{bmatrix} 1 & 2 & 0 & \vdots & -14 & 6 & 3 \\ 0 & 1 & 0 & \vdots & 13 & -5 & -3 \\ 0 & 0 & 1 & \vdots & 5 & -2 & -1 \end{bmatrix}$$ We added 3 times the third row to the second and -3 times the third row to the first.

$$\begin{bmatrix} 1 & 0 & 0 & \vdots & -40 & 16 & 9 \\ 0 & 1 & 0 & \vdots & 13 & -5 & -3 \\ 0 & 0 & 1 & \vdots & 5 & -2 & -1 \end{bmatrix}$$ We added -2 times the second row to the first.

Thus,

$$A^{-1} = \begin{bmatrix} -40 & 16 & 9 \\ 13 & -5 & -3 \\ 5 & -2 & -1 \end{bmatrix} \quad \blacktriangle$$

Often it will not be known in advance whether a given matrix is invertible. If an $n \times n$ matrix A is not invertible, then it cannot be reduced to I_n by elementary row operations [part (c) of Theorem 1.6.3]. Stated another way, the reduced row-echelon form of A has at least one row of zeros. Thus, if the procedure in the last example is attempted on a matrix that is not invertible, then at some point in the computations a row of zeros will occur on the *left side*. It can then be concluded that the given matrix is not invertible, and the computations can be stopped.

Example 5   Consider the matrix

$$A = \begin{bmatrix} 1 & 6 & 4 \\ 2 & 4 & -1 \\ -1 & 2 & 5 \end{bmatrix}$$

Applying the procedure of Example 4 yields

$$\begin{bmatrix} 1 & 6 & 4 & \vdots & 1 & 0 & 0 \\ 2 & 4 & -1 & \vdots & 0 & 1 & 0 \\ -1 & 2 & 5 & \vdots & 0 & 0 & 1 \end{bmatrix}$$

$$\begin{bmatrix} 1 & 6 & 4 & \vdots & 1 & 0 & 0 \\ 0 & -8 & -9 & \vdots & -2 & 1 & 0 \\ 0 & 8 & 9 & \vdots & 1 & 0 & 1 \end{bmatrix}$$ We added -2 times the first row to the second and added the first row to the third.

$$\begin{bmatrix} 1 & 6 & 4 & \vdots & 1 & 0 & 0 \\ 0 & -8 & -9 & \vdots & -2 & 1 & 0 \\ 0 & 0 & 0 & \vdots & -1 & 1 & 1 \end{bmatrix}$$ We added the second row to the third.

Since we have obtained a row of zeros on the left side, A is not invertible.

Example 6   In Example 4 we showed that

$$A = \begin{bmatrix} 1 & 2 & 3 \\ 2 & 5 & 3 \\ 1 & 0 & 8 \end{bmatrix}$$

is an invertible matrix. From Theorem 1.6.3 it follows that the system of equations

$$
\begin{aligned}
x_1 + 2x_2 + 3x_3 &= 0 \\
2x_1 + 5x_2 + 3x_3 &= 0 \\
x_1 \qquad\quad + 8x_3 &= 0
\end{aligned}
$$

has only the trivial solution.   ▲

EXERCISE SET 1.6

1. Which of the following are elementary matrices?

(a) $\begin{bmatrix} 1 & 0 \\ -5 & 1 \end{bmatrix}$ (b) $\begin{bmatrix} -5 & 1 \\ 1 & 0 \end{bmatrix}$ (c) $\begin{bmatrix} 1 & 0 \\ 0 & \sqrt{3} \end{bmatrix}$ (d) $\begin{bmatrix} 0 & 0 & 1 \\ 0 & 1 & 0 \\ 1 & 0 & 0 \end{bmatrix}$

(e) $\begin{bmatrix} 1 & 1 & 0 \\ 0 & 0 & 1 \\ 0 & 0 & 0 \end{bmatrix}$ (f) $\begin{bmatrix} 1 & 0 & 0 \\ 0 & 1 & 9 \\ 0 & 0 & 1 \end{bmatrix}$ (g) $\begin{bmatrix} 2 & 0 & 0 & 2 \\ 0 & 1 & 0 & 0 \\ 0 & 0 & 1 & 0 \\ 0 & 0 & 0 & 1 \end{bmatrix}$

2. Find a row operation that will restore the given elementary matrix to an identity matrix.

(a) $\begin{bmatrix} 1 & 0 \\ -3 & 1 \end{bmatrix}$ (b) $\begin{bmatrix} 1 & 0 & 0 \\ 0 & 1 & 0 \\ 0 & 0 & 3 \end{bmatrix}$ (c) $\begin{bmatrix} 0 & 0 & 0 & 1 \\ 0 & 1 & 0 & 0 \\ 0 & 0 & 1 & 0 \\ 1 & 0 & 0 & 0 \end{bmatrix}$ (d) $\begin{bmatrix} 1 & 0 & -\frac{1}{7} & 0 \\ 0 & 1 & 0 & 0 \\ 0 & 0 & 1 & 0 \\ 0 & 0 & 0 & 1 \end{bmatrix}$

3. Consider the matrices

$$A = \begin{bmatrix} 3 & 4 & 1 \\ 2 & -7 & -1 \\ 8 & 1 & 5 \end{bmatrix}, \quad B = \begin{bmatrix} 8 & 1 & 5 \\ 2 & -7 & -1 \\ 3 & 4 & 1 \end{bmatrix}, \quad C = \begin{bmatrix} 3 & 4 & 1 \\ 2 & -7 & -1 \\ 2 & -7 & 3 \end{bmatrix}$$

Find elementary matrices E_1, E_2, E_3, and E_4 such that
(a) $E_1 A = B$ (b) $E_2 B = A$ (c) $E_3 A = C$ (d) $E_4 C = A$

4. In Exercise 3 is it possible to find an elementary matrix E such that $EB = C$? Justify your answer.

In Exercises 5–7 use the method shown in Examples 4 and 5 to find the inverse of the given matrix if the matrix is invertible.

5. (a) $\begin{bmatrix} 1 & 4 \\ 2 & 7 \end{bmatrix}$ (b) $\begin{bmatrix} -3 & 6 \\ 4 & 5 \end{bmatrix}$ (c) $\begin{bmatrix} 6 & -4 \\ -3 & 2 \end{bmatrix}$

6. (a) $\begin{bmatrix} 3 & 4 & -1 \\ 1 & 0 & 3 \\ 2 & 5 & -4 \end{bmatrix}$ **(b)** $\begin{bmatrix} -1 & 3 & -4 \\ 2 & 4 & 1 \\ -4 & 2 & -9 \end{bmatrix}$ **(c)** $\begin{bmatrix} 1 & 0 & 1 \\ 0 & 1 & 1 \\ 1 & 1 & 0 \end{bmatrix}$ **(d)** $\begin{bmatrix} 2 & 6 & 6 \\ 2 & 7 & 6 \\ 2 & 7 & 7 \end{bmatrix}$ **(e)** $\begin{bmatrix} 1 & 0 & 1 \\ -1 & 1 & 1 \\ 0 & 1 & 0 \end{bmatrix}$

7. (a) $\begin{bmatrix} \frac{1}{5} & \frac{1}{5} & -\frac{2}{5} \\ \frac{1}{5} & \frac{1}{5} & \frac{1}{10} \\ \frac{1}{5} & -\frac{4}{5} & \frac{1}{10} \end{bmatrix}$ **(b)** $\begin{bmatrix} \sqrt{2} & 3\sqrt{2} & 0 \\ -4\sqrt{2} & \sqrt{2} & 0 \\ 0 & 0 & 1 \end{bmatrix}$ **(c)** $\begin{bmatrix} 1 & 0 & 0 & 0 \\ 1 & 3 & 0 & 0 \\ 1 & 3 & 5 & 0 \\ 1 & 3 & 5 & 7 \end{bmatrix}$

(d) $\begin{bmatrix} -8 & 17 & 2 & \frac{1}{3} \\ 4 & 0 & \frac{2}{5} & -9 \\ 0 & 0 & 0 & 0 \\ -1 & 13 & 4 & 2 \end{bmatrix}$ **(e)** $\begin{bmatrix} 0 & 0 & 2 & 0 \\ 1 & 0 & 0 & 1 \\ 0 & -1 & 3 & 0 \\ 2 & 1 & 5 & -3 \end{bmatrix}$

8. Find the inverse of each of the following 4×4 matrices, where $k_1, k_2, k_3, k_4,$ and k are all nonzero.

(a) $\begin{bmatrix} k_1 & 0 & 0 & 0 \\ 0 & k_2 & 0 & 0 \\ 0 & 0 & k_3 & 0 \\ 0 & 0 & 0 & k_4 \end{bmatrix}$ **(b)** $\begin{bmatrix} 0 & 0 & 0 & k_1 \\ 0 & 0 & k_2 & 0 \\ 0 & k_3 & 0 & 0 \\ k_4 & 0 & 0 & 0 \end{bmatrix}$ **(c)** $\begin{bmatrix} k & 0 & 0 & 0 \\ 1 & k & 0 & 0 \\ 0 & 1 & k & 0 \\ 0 & 0 & 1 & k \end{bmatrix}$

9. Consider the matrix

$$A = \begin{bmatrix} 1 & 0 \\ -5 & 2 \end{bmatrix}$$

(a) Find elementary matrices E_1 and E_2 such that $E_2 E_1 A = I$.
(b) Write A^{-1} as a product of two elementary matrices.
(c) Write A as a product of two elementary matrices.

10. In each part perform the stated row operation on

$$\begin{bmatrix} 2 & -1 & 0 \\ 4 & 5 & -3 \\ 1 & -4 & 7 \end{bmatrix}$$

by multiplying A on the left by a suitable elementary matrix. Check your answer in each case by performing the row operation directly on A.
(a) Interchange the first and third rows.
(b) Multiply the second row by $1/3$.
(c) Add twice the second row to the first row.

11. Express the matrix

$$A = \begin{bmatrix} 0 & 1 & 7 & 8 \\ 1 & 3 & 3 & 8 \\ -2 & -5 & 1 & -8 \end{bmatrix}$$

in the form $A = EFGR$, where E, F, and G are elementary matrices, and R is in row-echelon form.

12. Show that if

$$A = \begin{bmatrix} 1 & 0 & 0 \\ 0 & 1 & 0 \\ a & b & c \end{bmatrix}$$

is an elementary matrix, then at least one entry in the third row must be a zero.

13. Show that

$$A = \begin{bmatrix} 0 & a & 0 & 0 & 0 \\ b & 0 & c & 0 & 0 \\ 0 & d & 0 & e & 0 \\ 0 & 0 & f & 0 & g \\ 0 & 0 & 0 & h & 0 \end{bmatrix}$$

is not invertible for any values of the entries.

14. Prove that if A is an $m \times n$ matrix, there is an invertible matrix C such that CA is in reduced row-echelon form.

15. Prove that if A is an invertible matrix and B is row equivalent to A, then B is also invertible.

1.7   FURTHER RESULTS ON SYSTEMS OF EQUATIONS AND INVERTIBILITY

In this section we shall establish more results about systems of linear equations and invertibility of matrices. Our work will lead to a totally new method for solving n equations in n unknowns.

SOLVING LINEAR SYSTEMS BY MATRIX INVERSION

We begin with a theorem that provides a new method for solving certain systems of linear equations.

Theorem 1.7.1. *If A is an invertible $n \times n$ matrix, then for each $n \times 1$ matrix B, the system of equations $AX = B$ has exactly one solution, namely, $X = A^{-1}B$.*

Proof.   Since $A(A^{-1}B) = B$, it follows that $X = A^{-1}B$ is a solution of $AX = B$. To show that this is the only solution, we will assume that X_0 is an arbitrary solution and then show that X_0 must be the solution $A^{-1}B$.

If X_0 is any solution, then $AX_0 = B$. Multiplying both sides by A^{-1}, we obtain $X_0 = A^{-1}B$. ▮

Example 1   Consider the system of linear equations

$$x_1 + 2x_2 + 3x_3 = 5$$
$$2x_1 + 5x_2 + 3x_3 = 3$$
$$x_1 \qquad + 8x_3 = 17$$

In matrix form this system can be written as $AX = B$, where

$$A = \begin{bmatrix} 1 & 2 & 3 \\ 2 & 5 & 3 \\ 1 & 0 & 8 \end{bmatrix} \qquad X = \begin{bmatrix} x_1 \\ x_2 \\ x_3 \end{bmatrix} \qquad B = \begin{bmatrix} 5 \\ 3 \\ 17 \end{bmatrix}$$

In Example 4 of the previous section we showed that A is invertible and

$$A^{-1} = \begin{bmatrix} -40 & 16 & 9 \\ 13 & -5 & -3 \\ 5 & -2 & -1 \end{bmatrix}$$

By Theorem 1.7.1 the solution of the system is

$$X = A^{-1}B = \begin{bmatrix} -40 & 16 & 9 \\ 13 & -5 & -3 \\ 5 & -2 & -1 \end{bmatrix} \begin{bmatrix} 5 \\ 3 \\ 17 \end{bmatrix} = \begin{bmatrix} 1 \\ -1 \\ 2 \end{bmatrix}$$

or $x_1 = 1$, $x_2 = -1$, $x_3 = 2$. ▲

REMARK. Note that the method of Example 1 applies only when the system has as many equations as unknowns and the coefficient matrix is invertible.

Frequently, one is concerned with solving a sequence of systems

$$AX = B_1, \quad AX = B_2, \quad AX = B_3, \dots, \quad AX = B_k$$

each of which has the same square coefficient matrix A. If A is invertible, then the solutions

$$X_1 = A^{-1}B_1, \quad X_2 = A^{-1}B_2, \quad X_3 = A^{-1}B_3, \dots, \quad X_k = A^{-1}B_k$$

can be obtained with one matrix inversion and k matrix multiplications. However, a more efficient method is to form the matrix

$$[A \mid B_1 \mid B_2 \mid \cdots \mid B_k] \tag{1}$$

in which the coefficient matrix A is "augmented" by all k of the matrices B_1, $B_2, \dots, B_k$. By reducing (1) to reduced row-echelon form we can solve all k systems at once by Gauss-Jordan elimination. This method has the added advantage that it applies even when A is not invertible.

Example 2   Solve the systems

(a) $x_1 + 2x_2 + 3x_3 = 4$ (b) $x_1 + 2x_2 + 3x_3 = 1$
  $2x_1 + 5x_2 + 3x_3 = 5$ $2x_1 + 5x_2 + 3x_3 = 6$
  $x_1 \qquad + 8x_3 = 9$ $x_1 \qquad + 8x_3 = -6$

Solution. The two systems have the same coefficient matrix. If we augment this coefficient matrix with the columns of constants on the right sides of these systems, we obtain

$$\begin{bmatrix} 1 & 2 & 3 & | & 4 & | & 1 \\ 2 & 5 & 3 & | & 5 & | & 6 \\ 1 & 0 & 8 & | & 9 & | & -6 \end{bmatrix}$$

Reducing this matrix to reduced row-echelon form yields (verify)

$$\begin{bmatrix} 1 & 0 & 0 & | & 1 & | & 2 \\ 0 & 1 & 0 & | & 0 & | & 1 \\ 0 & 0 & 1 & | & 1 & | & -1 \end{bmatrix}$$

It follows from the last two columns that the solution of system (a) is $x_1 = 1$, $x_2 = 0$, $x_3 = 1$ and of system (b) is $x_1 = 2$, $x_2 = 1$, $x_3 = -1$. ▲

Let us digress for a moment to illustrate a type of application that leads to a set of linear systems, all of which have the same coefficient matrix. In various applied problems physical systems arise that are described as *black boxes*. This term is intended to convey the idea that the internal workings of the system are unknown or unimportant for the problem—one simply imagines that if an input, such as electrical current, force, or heat, is applied to the black box, then a certain output will result (Figure 1).

Often, the input and output can be described mathematically by matrices with one column. For example, if the black box is a microprocessor, then the input might be an $n \times 1$ matrix whose entries are n voltages read across certain input terminals, and the output might be an $m \times 1$ matrix whose entries are the resulting currents in m output wires. Mathematically speaking, such a system does nothing more than transform an $n \times 1$ input matrix into an $m \times 1$ output matrix.

For many black-box systems the input matrix M_{IN} and the output matrix M_{OUT} are related by a matrix equation

$$AM_{\mathrm{IN}} = M_{\mathrm{OUT}}$$

where the entries in A are physical parameters determined by the system. Such systems are called **linear physical systems**.

In many applications it is important to know what input must be applied to the system to achieve a certain desired output. This amounts to solving the equation $AX = M_{\mathrm{OUT}}$ for the unknown input X, given the desired output M_{OUT}. Thus, if we have a succession of different output matrices $B_1, \ldots, B_k$, and we want to determine the input matrices that produce these given outputs, we must successively solve the k systems of linear equations

$$AX = B_1, \quad AX = B_2, \ldots, \quad AX = B_k$$

each of which has the same coefficient matrix A.

PROPERTIES OF INVERTIBLE MATRICES

Up to now, to show that an $n \times n$ matrix A is invertible, it has been necessary to find an $n \times n$ matrix B such that

$$AB = I \quad \text{and} \quad BA = I$$

The next theorem shows that if we produce an $n \times n$ matrix B satisfying *either* condition, then the other condition holds automatically.

Theorem 1.7.2. *Let A be a square matrix.*

(a) If B is a square matrix satisfying $BA = I$, then $B = A^{-1}$.
(b) If B is a square matrix satisfying $AB = I$, then $B = A^{-1}$.

We shall prove part (*a*) and leave part (*b*) as an exercise.

Proof (a). Assume $BA = I$. If we can show that A is invertible, the proof can be completed by multiplying $BA = I$ on both sides by A^{-1} to obtain

$$BAA^{-1} = IA^{-1} \quad \text{or} \quad BI = IA^{-1} \quad \text{or} \quad B = A^{-1}$$

To show that A is invertible, it suffices to show that the system $AX = 0$ has only the trivial solution (see Theorem 1.6.3). Let X_0 be any solution of this system. If we multiply both sides of $AX_0 = 0$ on the left by B, we obtain $BAX_0 = B0$ or $IX_0 = 0$ or $X_0 = 0$. Thus, the system of equations $AX = 0$ has only the trivial solution. ▊

We are now in a position to add a fourth statement equivalent to the three given in Theorem 1.6.3.

Theorem 1.7.3. *If A is an $n \times n$ matrix, then the following statements are equivalent.*

(a) A is invertible.
(b) $AX = 0$ has only the trivial solution.
(c) A is row equivalent to I_n.
(d) $AX = B$ is consistent for every $n \times 1$ matrix B.

Proof. Since we proved in Theorem 1.6.3 that (*a*), (*b*), and (*c*) are equivalent, it will be sufficient to prove (*a*) $\Rightarrow$ (*d*) and (*d*) $\Rightarrow$ (*a*).

(*a*) $\Rightarrow$ (*d*): If A is invertible and B is any $n \times 1$ matrix, then $X = A^{-1}B$ is a solution of $AX = B$ by Theorem 1.7.1. Thus, $AX = B$ is consistent.

(*d*) $\Rightarrow$ (*a*): If the system $AX = B$ is consistent for every $n \times 1$ matrix B, then in particular, the systems

$$AX = \begin{bmatrix} 1 \\ 0 \\ 0 \\ \vdots \\ 0 \end{bmatrix}, \quad AX = \begin{bmatrix} 0 \\ 1 \\ 0 \\ \vdots \\ 0 \end{bmatrix}, \quad \ldots, \quad AX = \begin{bmatrix} 0 \\ 0 \\ 0 \\ \vdots \\ 1 \end{bmatrix}$$

are consistent. Let $X_1, X_2, \ldots, X_n$ be solutions of the respective systems, and let us form an $n \times n$ matrix C having these solutions as columns. Thus, C has the form

$$C = [X_1 | X_2 | \cdots | X_n]$$

As discussed in Example 9 of Section 1.4, the successive columns of the product AC will be

$$AX_1, AX_2, \ldots, AX_n$$

Thus,

$$AC = [AX_1 | AX_2 | \cdots | AX_n] = \begin{bmatrix} 1 & 0 & \cdots & 0 \\ 0 & 1 & \cdots & 0 \\ 0 & 0 & \cdots & 0 \\ \vdots & \vdots & & \vdots \\ 0 & 0 & \cdots & 1 \end{bmatrix} = I$$

By part (*b*) of Theorem 1.7.2 it follows that $C = A^{-1}$. Thus, A is invertible. ▮

In our later work the following fundamental problem will occur frequently in various contexts.

> **A Fundamental Problem.** Let A be a fixed $m \times n$ matrix. Find all $m \times 1$ matrices B such that the system of equations $AX = B$ is consistent.

If A is an invertible matrix, Theorem 1.7.1 completely solves this problem by asserting that for *every* $m \times 1$ matrix B, $AX = B$ has the unique solution $X = A^{-1}B$. If A is not square, or if A is square but not invertible, then Theorem 1.7.1 does not apply. In these cases the matrix B must satisfy certain conditions in order for $AX = B$ to be consistent. The following example illustrates how Gaussian elimination can be used to determine such conditions.

Example 3   What conditions must b_1, b_2, and b_3 satisfy in order for the system of equations

$$\begin{aligned} x_1 + x_2 + 2x_3 &= b_1 \\ x_1 \qquad\;\; + x_3 &= b_2 \\ 2x_1 + x_2 + 3x_3 &= b_3 \end{aligned}$$

to be consistent?

Solution. The augmented matrix is

$$\begin{bmatrix} 1 & 1 & 2 & b_1 \\ 1 & 0 & 1 & b_2 \\ 2 & 1 & 3 & b_3 \end{bmatrix}$$

which can be reduced to row-echelon form as follows.

$$\begin{bmatrix} 1 & 1 & 2 & b_1 \\ 0 & -1 & -1 & b_2 - b_1 \\ 0 & -1 & -1 & b_3 - 2b_1 \end{bmatrix}$$

$\boxed{\text{$-1$ times the first row was added to the second and -2 times the first row was added to the third.}}$

$$\begin{bmatrix} 1 & 1 & 2 & b_1 \\ 0 & 1 & 1 & b_1 - b_2 \\ 0 & -1 & -1 & b_3 - 2b_1 \end{bmatrix}$$

$\boxed{\text{The second row was multiplied by -1.}}$

$$\begin{bmatrix} 1 & 1 & 1 & b_2 \\ 0 & 1 & 1 & b_1 - b_2 \\ 0 & 0 & 0 & b_3 - b_2 - b_1 \end{bmatrix}$$

$\boxed{\text{The second row was added to the third.}}$

It is now evident from the third row in the matrix that the system has a solution if and only if b_1, b_2, and b_3 satisfy the condition

$$b_3 - b_2 - b_1 = 0 \qquad \text{or} \qquad b_3 = b_1 + b_2$$

To express this condition another way, $AX = B$ is consistent if and only if B is a matrix of the form

$$B = \begin{bmatrix} b_1 \\ b_2 \\ b_1 + b_2 \end{bmatrix}$$

where b_1 and b_2 are arbitrary.

Example 4   What conditions must b_1, b_2, and b_3 satisfy in order for the system of equations

$$\begin{aligned} x_1 + 2x_2 + 3x_3 &= b_1 \\ 2x_1 + 5x_2 + 3x_3 &= b_2 \\ x_1 \qquad\quad + 8x_3 &= b_3 \end{aligned}$$

to be consistent?

Solution. The augmented matrix is

$$\begin{bmatrix} 1 & 2 & 3 & b_1 \\ 2 & 5 & 3 & b_2 \\ 1 & 0 & 8 & b_3 \end{bmatrix}$$

Reducing this to reduced row-echelon form yields (verify)

$$\begin{bmatrix} 1 & 0 & 0 & -40b_1 + 16b_2 + 9b_3 \\ 0 & 1 & 0 & 13b_1 - 5b_2 - 3b_3 \\ 0 & 0 & 1 & 5b_1 - 2b_2 - b_3 \end{bmatrix} \qquad (2)$$

In this case there are no restrictions on b_1, b_2, and b_3; that is, the given system $AX = B$ has the unique solution

$$x_1 = -40b_1 + 16b_2 + 9b_3, \quad x_2 = 13b_1 - 5b_2 - 3b_3, \quad x_3 = 5b_1 - 2b_2 - b_3 \quad (3)$$

for all B. ▲

REMARK. Because the system $AX = B$ in the foregoing example is consistent for all B, it follows from Theorem 1.7.3 that A is invertible. We leave it for the reader to verify that the formulas in (2) can also be obtained by calculating $X = A^{-1}B$.

EXERCISE SET 1.7

In Exercises 1–8 solve the system using the method of Example 1.

1. $x_1 + x_2 = 2$
$5x_1 + 6x_2 = 9$

2. $4x_1 - 3x_2 = -3$
$2x_1 - 5x_2 = 9$

3. $x_1 + 3x_2 + x_3 = 4$
$2x_1 + 2x_2 + x_3 = -1$
$2x_1 + 3x_2 + x_3 = 3$

4. $5x_1 + 3x_2 + 2x_3 = 4$
$3x_1 + 3x_2 + 2x_3 = 2$
$x_2 + x_3 = 5$

5. $x + y + z = 5$
$x + y - 4z = 10$
$-4x + y + z = 0$

6. $-x - 2y - 3z = 0$
$w + x + 4y + 4z = 7$
$w + 3x + 7y + 9z = 4$
$-w - 2x - 4y - 6z = 6$

7. $3x_1 + 5x_2 = b_1$
$x_1 + 2x_2 = b_2$

8. $x_1 + 2x_2 + 3x_3 = b_1$
$2x_1 + 5x_2 + 5x_3 = b_2$
$3x_1 + 5x_2 + 8x_3 = b_3$

9. Solve the following general system using the method of Example 1.

$$x_1 + 2x_2 + x_3 = b_1$$
$$x_1 - x_2 + x_3 = b_2$$
$$x_1 + x_2 = b_3$$

Use the resulting formulas to find the solution if
(a) $b_1 = -1$, $b_2 = 3$, $b_3 = 4$ (b) $b_1 = 5$, $b_2 = 0$, $b_3 = 0$ (c) $b_1 = -1$, $b_2 = -1$, $b_3 = 3$

10. Solve the three systems in Exercise 9 using the method of Example 2.

In Exercises 11–14 use the method of Example 2 to solve the systems in all the parts simultaneously.

11. $x_1 - 5x_2 = b_1$
$3x_1 + 2x_2 = b_2$

(a) $b_1 = 1, \quad b_2 = 4$
(b) $b_1 = -2, \quad b_2 = 5$

12. $-x_1 + 4x_2 + x_3 = b_1$
$x_1 + 9x_2 - 2x_3 = b_2$
$6x_1 + 4x_2 - 8x_3 = b_3$

(a) $b_1 = 0, \quad b_2 = 1, \quad b_3 = 0$
(b) $b_1 = -3, \quad b_2 = 4, \quad b_3 = -5$

13. $4x_1 - 7x_2 = b_1$
$x_1 + 2x_2 = b_2$

(a) $b_1 = 0, \quad b_2 = 1$
(b) $b_1 = -4, \quad b_2 = 6$
(c) $b_1 = -1, \quad b_2 = 3$
(d) $b_1 = -5, \quad b_2 = 1$

14. $x_1 + 3x_2 + 5x_3 = b_1$
$-x_1 - 2x_2 \quad\quad = b_2$
$2x_1 + 5x_2 + 4x_3 = b_3$

(a) $b_1 = 1, \quad b_2 = 0, \quad b_3 = -1$
(b) $b_1 = 0, \quad b_2 = 1, \quad b_3 = 1$
(c) $b_1 = -1, \quad b_2 = -1, \quad b_3 = 0$

15. The method of Example 2 can be used for linear systems with infinitely many solutions. Use that method to solve the systems in both parts simultaneously.

(a) $x_1 - 2x_2 + x_3 = -2$
$2x_1 - 5x_2 + x_3 = 1$
$3x_1 - 7x_2 + 2x_3 = -1$

(b) $x_1 - 2x_2 + x_3 = 1$
$2x_1 - 5x_2 + x_3 = -1$
$3x_1 - 7x_2 + 2x_3 = 0$

In Exercises 16–19 find the conditions that b's must satisfy for the system to be consistent.

16. $6x_1 - 4x_2 = b_1$
$3x_1 - 2x_2 = b_2$

17. $x_1 - 2x_2 + 5x_3 = b_1$
$4x_1 - 5x_2 + 8x_3 = b_2$
$-3x_1 + 3x_2 - 3x_3 = b_3$

18. $x_1 - 2x_2 - x_3 = b_1$
$-4x_1 + 5x_2 + 2x_3 = b_2$
$-4x_1 + 7x_2 + 4x_3 = b_3$

19. $x_1 - x_2 + 3x_3 + 2x_1 = b_1$
$-2x_1 + x_2 + 5x_3 + x_1 = b_2$
$-3x_1 + 2x_2 + 2x_3 - x_1 = b_3$
$4x_1 - 3x_2 + x_3 + 3x_1 = b_4$

20. Consider the matrices

$$A = \begin{bmatrix} 2 & 1 & 2 \\ 2 & 2 & -2 \\ 3 & 1 & 1 \end{bmatrix} \quad \text{and} \quad X = \begin{bmatrix} x_1 \\ x_2 \\ x_3 \end{bmatrix}$$

(a) Show that the equation $AX = X$ can be rewritten as $(A - I)X = 0$ and use this result to solve $AX = X$ for X.
(b) Solve $AX = 4X$.

21. Solve the following matrix equation for X.

$$\begin{bmatrix} 1 & -1 & 1 \\ 2 & 3 & 0 \\ 0 & 2 & -1 \end{bmatrix} X = \begin{bmatrix} 2 & -1 & 5 & 7 & 8 \\ 4 & 0 & -3 & 0 & 1 \\ 3 & 5 & -7 & 2 & 1 \end{bmatrix}$$

22. In each part determine whether the homogeneous system has a nontrivial solution (without using pencil and paper); then state whether the given matrix is invertible.

(a) $2x_1 + x_2 - 3x_3 + x_4 = 0$
$5x_2 + 4x_3 + 3x_4 = 0$
$x_3 + 2x_4 = 0$
$3x_4 = 0$

$$\begin{bmatrix} 2 & 1 & -3 & 1 \\ 0 & 5 & 4 & 3 \\ 0 & 0 & 1 & 2 \\ 0 & 0 & 0 & 3 \end{bmatrix}$$

(b) $5x_1 + x_2 + 4x_3 + x_4 = 0$
$2x_3 - x_4 = 0$
$x_3 + x_4 = 0$
$7x_4 = 0$

$$\begin{bmatrix} 5 & 1 & 4 & 1 \\ 0 & 0 & 2 & -1 \\ 0 & 0 & 1 & 1 \\ 0 & 0 & 0 & 7 \end{bmatrix}$$

23. Let $AX = 0$ be a homogeneous system of n linear equations in n unknowns that has only the trivial solution. Show that if k is any positive integer, then the system $A^kX = 0$ also has only the trivial solution.

24. Let $AX = 0$ be a homogeneous system of n linear equations in n unknowns, and let Q be an invertible matrix. Show that $AX = 0$ has just the trivial solution if and only if $(QA)X = 0$ has just the trivial solution.

25. Show that an $n \times n$ matrix A is invertible if and only if it can be written as a product of elementary matrices.

26. Use part (a) of Theorem 1.7.2 to prove part (b).

SUPPLEMENTARY EXERCISES

1. Use Gauss-Jordan elimination to solve for x' and y' in terms of x and y.

$$x = \tfrac{3}{5}x' - \tfrac{4}{5}y'$$
$$y = \tfrac{4}{5}x' + \tfrac{3}{5}y'$$

2. Use Gauss-Jordan elimination to solve for x' and y' in terms of x and y.

$$x = x' \cos\theta - y' \sin\theta$$
$$y = x' \sin\theta + y' \cos\theta$$

3. Find a homogeneous linear system with two equations that are not multiples of one another and such that

$$x_1 = 1, \quad x_2 = -1, \quad x_3 = 1, \quad x_4 = 2$$

and

$$x_1 = 2, \quad x_2 = 0, \quad x_3 = 3, \quad x_4 = -1$$

are solutions of the system.

4. A box containing pennies, nickels, and dimes has 13 coins with a total value of 83 cents. How many coins of each type are in the box?

5. Find positive integers that satisfy

$$x + y + z = 9$$
$$x + 5y + 10z = 44$$

6. For which value(s) of a does the following system have zero, one, infinitely many solutions?

$$x_1 + x_2 + x_3 = 4$$
$$x_3 = 2$$
$$(a^2 - 4)x_3 = a - 2$$

7. Let

$$\begin{bmatrix} a & 0 & b & 2 \\ a & a & 4 & 4 \\ 0 & a & 2 & b \end{bmatrix}$$

be the augmented matrix for a linear system. For what values of a and b does the system have

(a) a unique solution,   (b) a one-parameter solution,

(c) a two-parameter solution,   (d) no solution?

8. Solve for x, y, and z.

$$xy - 2\sqrt{y} + 3zy = 8$$
$$2xy - 3\sqrt{y} + 2zy = 7$$
$$-xy + \sqrt{y} + 2zy = 4$$

9. Find a matrix K such that $AKB = C$ given that

$$A = \begin{bmatrix} 1 & 4 \\ -2 & 3 \\ 1 & -2 \end{bmatrix} \quad B = \begin{bmatrix} 2 & 0 & 0 \\ 0 & 1 & -1 \end{bmatrix} \quad C = \begin{bmatrix} 8 & 6 & -6 \\ 6 & -1 & 1 \\ -4 & 0 & 0 \end{bmatrix}$$

10. How should the coefficients a, b, and c be chosen so that the system

$$ax + by - 3z = -3$$
$$-2x - by + cz = -1$$
$$ax + 3y - cz = -3$$

has the solution $x = 1$, $y = -1$, and $z = 2$?

11. In each part solve the matrix equation for X.

(a) $X \begin{bmatrix} -1 & 0 & 1 \\ 1 & 1 & 0 \\ 3 & 1 & -1 \end{bmatrix} = \begin{bmatrix} 1 & 2 & 0 \\ -3 & 1 & 5 \end{bmatrix}$ (b) $X \begin{bmatrix} 1 & -1 & 2 \\ 3 & 0 & 1 \end{bmatrix} = \begin{bmatrix} -5 & -1 & 0 \\ 6 & -3 & 7 \end{bmatrix}$

(c) $\begin{bmatrix} 3 & 1 \\ -1 & 2 \end{bmatrix} X - X \begin{bmatrix} 1 & 4 \\ 2 & 0 \end{bmatrix} = \begin{bmatrix} 2 & -2 \\ 5 & 4 \end{bmatrix}$

12. (a) Express the equations

$$\begin{aligned} y_1 &= x_1 - x_2 + x_3 \\ y_2 &= 3x_1 + x_2 - 4x_3 \\ y_3 &= -2x_1 - 2x_2 + 3x_3 \end{aligned} \quad \text{and} \quad \begin{aligned} z_1 &= 4y_1 - y_2 + y_3 \\ z_2 &= -3y_1 + 5y_2 - y_3 \end{aligned}$$

in the matrix forms $Y = AX$ and $Z = BY$. Then use these to obtain a direct relationship $Z = CX$ between Z and X.

(b) Use the equation $Z = CX$ obtained in (a) to express z_1 and z_2 in terms of x_1, x_2, and x_3.

(c) Check the result in (b) by directly substituting the equations for y_1, y_2, and y_3 into the equations for z_1 and z_2 and then simplifying.

13. If A is $m \times n$ and B is $n \times p$, how many multiplication operations and how many addition operations are needed to calculate the matrix product AB?

14. Let A be a square matrix.

(a) Show that $(I - A)^{-1} = I + A + A^2 + A^3$ if $A^4 = 0$.

(b) Show that $(I - A)^{-1} = I + A + A^2 + \cdots + A^n$ if $A^{n+1} = 0$.

15. Find values of a, b, and c so that the graph of the polynomial $p(x) = ax^2 + bx + c$ passes through the points $(1, 2)$, $(-1, 6)$, and $(2, 3)$.

16. (For readers who have studied calculus.) Find values of a, b, and c so that the graph of the polynomial $p(x) = ax^2 + bx + c$ passes through the point $(-1, 0)$ and has a horizontal tangent at $(2, -9)$.

17. Let J_n be the $n \times n$ matrix each of whose entries is 1. Show that

$$(I - J_n)^{-1} = I - \frac{1}{n-1} J_n$$

18. Show that if a square matrix A satisfies $A^3 + 4A^2 - 2A + 7I = 0$, then so does A^t.

19. Prove: If B is invertible, then $AB^{-1} = B^{-1}A$ if and only if $AB = BA$.

20. Prove: If A is invertible, then $A + B$ and $I + BA^{-1}$ are both invertible or both not invertible.

21. Prove that if A and B are $n \times n$ matrices, then

(a) $\text{tr}(A + B) = \text{tr}(A) + \text{tr}(B)$ **(b)** $\text{tr}(kA) = k \, \text{tr}(A)$ **(c)** $\text{tr}(A^t) = \text{tr}(A)$ **(d)** $\text{tr}(AB) = \text{tr}(BA)$

22. Use Exercise 21 to show that there are no square matrices A and B such that

$$AB - BA = I$$

23. Prove: If A is an $m \times n$ matrix and B is the $n \times 1$ matrix each of whose entries is $1/n$, then

$$AB = \begin{bmatrix} \bar{r}_1 \\ \bar{r}_2 \\ \vdots \\ \bar{r}_m \end{bmatrix}$$

where $\bar{r}_i$ is the average of the entries in the ith row of A.

24. **(For readers who have studied calculus.)** If the entries of the matrix

$$C = \begin{bmatrix} c_{11}(x) & c_{12}(x) & \cdots & c_{1n}(x) \\ c_{21}(x) & c_{22}(x) & \cdots & c_{2n}(x) \\ \vdots & \vdots & & \vdots \\ c_{m1}(x) & c_{m2}(x) & \cdots & c_{mn}(x) \end{bmatrix}$$

are differentiable functions of x, then we define

$$\frac{dC}{dx} = \begin{bmatrix} c'_{11}(x) & c'_{12}(x) & \cdots & c'_{1n}(x) \\ c'_{21}(x) & c'_{22}(x) & \cdots & c'_{2n}(x) \\ \vdots & \vdots & & \vdots \\ c'_{m1}(x) & c'_{m2}(x) & \cdots & c'_{mn}(x) \end{bmatrix}$$

Show that if the entries in A and B are differentiable functions of x and the sizes of the matrices are such that the stated operations can be performed, then

(a) $\dfrac{d}{dx}(kA) = k\dfrac{dA}{dx}$ (b) $\dfrac{d}{dx}(A + B) = \dfrac{dA}{dx} + \dfrac{dB}{dx}$ (c) $\dfrac{d}{dx}(AB) = \dfrac{dA}{dx}B + A\dfrac{dB}{dx}$

25. **(For readers who have studied calculus.)** Use part (c) of Exercise 24 to show that

$$\frac{dA^{-1}}{dx} = -A^{-1}\frac{dA}{dx}A^{-1}$$

State all the assumptions you make in obtaining this formula.

26. Find the values of A, B, and C that will make the equation

$$\frac{x^2 + x - 2}{(3x - 1)(x^2 + 1)} = \frac{A}{3x - 1} + \frac{Bx + C}{x^2 + 1}$$

an identity. [**Hint.** Multiply through by $(3x - 1)(x^2 + 1)$ and equate the corresponding coefficients of the polynomials on each side of the resulting equation.]

27. If P is an $n \times 1$ matrix such that $P^tP = 1$, then $H = I - 2PP^t$ is called the corresponding **Householder matrix** (named after the American mathematician A. S. Householder).
    (a) Verify that $P^tP = 1$ if $P^t = \begin{bmatrix} \frac{3}{4} & \frac{1}{6} & \frac{1}{4} & \frac{5}{12} & \frac{5}{12} \end{bmatrix}$ and compute the corresponding Householder matrix.
    (b) Prove that if H is any Householder matrix, then $H = H^t$ and $H^tH = I$.
    (c) Verify that the Householder matrix found in part (a) satisfies the conditions proved in part (b).

28. Assuming that the stated inverses exist, prove the following equalities.
    (a) $(C^{-1} + D^{-1})^{-1} = C(C + D)^{-1}D$ (b) $(I + CD)^{-1}C = C(I + DC)^{-1}$
    (c) $(C + DD^t)^{-1}D = C^{-1}D(I + D^tC^{-1}D)^{-1}$

CHAPTER 2

DETERMINANTS

2.1 THE DETERMINANT FUNCTION

We are all familiar with functions such as $f(x) = \sin x$ and $f(x) = x^2$, which associate a real number $f(x)$ with a real value of the variable x. Since both x and $f(x)$ assume only real values, such functions are described as "real-valued functions of a real variable." In this section we shall study the **determinant function**, *which is a "real-valued function of a matrix variable" in the sense that it associates a real number $f(X)$ with a matrix X. Our work on determinant functions will have important applications to the theory of systems of linear equations and will also lead us to an explicit formula for the inverse of an invertible matrix.*

PERMUTATIONS   Before we can define the determinant function, it will be necessary to establish some results concerning permutations.

> **Definition.** A **permutation** of the set of integers $\{1, 2, \ldots, n\}$ is an arrangement of these integers in some order without omissions or repetitions.

Example 1   There are six different permutations of the set of integers $\{1, 2, 3\}$. These are

$$(1, 2, 3) \quad (2, 1, 3) \quad (3, 1, 2)$$
$$(1, 3, 2) \quad (2, 3, 1) \quad (3, 2, 1)$$

3!

One convenient method of systematically listing permutations is to use a *permutation tree*. This method is illustrated in our next example.

Example 2 List all permutations of the set of integers {1, 2, 3, 4}.

Solution. Consider Figure 1. The four dots labeled 1, 2, 3, 4 at the top of the figure represent the possible choices for the first number in the permutation. The three branches emanating from these dots represent the possible choices for the second position in the permutation. Thus, if the permutation begins $(2, -, -, -)$, the three possibilities for the second position are 1, 3, and 4. The two branches emanating from each dot in the second position represent the possible choices for the third position. Thus, if the permutation begins $(2, 3, -, -)$, the two possible choices for the third position are 1 and 4. Finally, the single branch emanating from each dot in the third position represents the only possible choice for the fourth position. Thus, if the permutation begins $(2, 3, 4, -)$, the only choice for the fourth position is 1. The different permutations can now be listed by tracing out all the possible paths through the "tree" from the first position to the last position. We obtain the following list by this process.

(1, 2, 3, 4)	(2, 1, 3, 4)	(3, 1, 2, 4)	(4, 1, 2, 3)
(1, 2, 4, 3)	(2, 1, 4, 3)	(3, 1, 4, 2)	(4, 1, 3, 2)
(1, 3, 2, 4)	(2, 3, 1, 4)	(3, 2, 1, 4)	(4, 2, 1, 3)
(1, 3, 4, 2)	(2, 3, 4, 1)	(3, 2, 4, 1)	(4, 2, 3, 1)
(1, 4, 2, 3)	(2, 4, 1, 3)	(3, 4, 1, 2)	(4, 3, 1, 2)
(1, 4, 3, 2)	(2, 4, 3, 1)	(3, 4, 2, 1)	(4, 3, 2, 1)

From this example we see that there are 24 permutations of {1, 2, 3, 4}. This result could have been anticipated without actually listing the permutations by arguing as follows. Since the first position can be filled in four ways and then the second position in three ways, there are $4 \cdot 3$ ways of filling the first two positions. Since the third position can then be filled in two ways, there are $4 \cdot 3 \cdot 2$ ways of filling the first three positions. Finally, since the last position can then be filled in only one way, there are $4 \cdot 3 \cdot 2 \cdot 1 = 24$ ways of filling all four positions. In general, the set {1, 2, . . . , n} will have $n(n-1)(n-2) \cdots 2 \cdot 1 = n!$ different permutations.

To denote a general permutation of the set {1, 2, . . . , n}, we shall write $(j_1, j_2, \ldots, j_n)$. Here, j_1 is the first integer in the permutation, j_2 is the second, and so on. An *inversion* is said to occur in a permutation $(j_1, j_2, \ldots, j_n)$ whenever a larger integer precedes a smaller one. The total number of inversions occurring in a permutation can be obtained as follows: (1) find the number of integers that are less than j_1 and that follow j_1 in the permutation; (2) find the number of integers

Figure 1

that are less than j_2 and that follow j_2 in the permutation. Continue this counting process for $j_3, \ldots, j_{n-1}$. The sum of these numbers will be the total number of inversions in the permutation.

Example 3   Determine the number of inversions in the following permutations:

$$\text{(a) } (6, 1, 3, 4, 5, 2) \qquad \text{(b) } (2, 4, 1, 3) \qquad \text{(c) } (1, 2, 3, 4)$$

Solution.

(a)   The number of inversions is $5 + 0 + 1 + 1 + 1 = 8$.
(b)   The number of inversions is $1 + 2 + 0 = 3$.
(c)   There are no inversions in this permutation.   ▲

Definition. A permutation is called *even* if the total number of inversions is an even integer and is called *odd* if the total number of inversions is an odd integer.

Example 4   The following table classifies the various permutations of $\{1, 2, 3\}$ as even or odd.

Permutation	Number of Inversions	Classification
(1, 2, 3)	0	even
(1, 3, 2)	1	odd
(2, 1, 3)	1	odd
(2, 3, 1)	2	even
(3, 1, 2)	2	even
(3, 2, 1)	3	odd

Consider the $n \times n$ matrix

$$A = \begin{bmatrix} a_{11} & a_{12} & \cdots & a_{1n} \\ a_{21} & a_{22} & \cdots & a_{2n} \\ \vdots & \vdots & & \vdots \\ a_{n1} & a_{n2} & \cdots & a_{nn} \end{bmatrix}$$

DEFINITION OF A DETERMINANT

By an *elementary product from A* we shall mean any product of n entries from A, no two of which come from the same row or same column.

Example 5   List all elementary products from the matrices

$$\text{(a) } \begin{bmatrix} a_{11} & a_{12} \\ a_{21} & a_{22} \end{bmatrix} \qquad \text{(b) } \begin{bmatrix} a_{11} & a_{12} & a_{13} \\ a_{21} & a_{22} & a_{23} \\ a_{31} & a_{32} & a_{33} \end{bmatrix}$$

Solution (*a*). Since each elementary product has two factors, and since each factor comes from a different row, an elementary product can be written in the form

$$a_{1\_}a_{2\_}$$

where the blanks designate column numbers. Since no two factors in the product come from the same column, the column numbers must be 1 2 or 2 1. Thus, the only elementary products are $a_{11}a_{22}$ and $a_{12}a_{21}$.

Solution (*b*). Since each elementary product has three factors, each of which comes from a different row, an elementary product can be written in the form

$$a_{1\_}a_{2\_}a_{3\_}$$

Since no two factors in the product come from the same column, the column numbers have no repetitions; consequently, they must form a permutation of the set $\{1, 2, 3\}$. These $3! = 6$ permutations yield the following list of elementary products.

$$a_{11}a_{22}a_{33} \qquad a_{12}a_{21}a_{33} \qquad a_{13}a_{21}a_{32}$$

$$a_{11}a_{23}a_{32} \qquad a_{12}a_{23}a_{31} \qquad a_{13}a_{22}a_{31}$$

As this example points out, an $n \times n$ matrix A has $n!$ elementary products. They are the products of the form $a_{1j_1}a_{2j_2} \dots a_{nj_n}$, where $(j_1, j_2, \dots, j_n)$ is a permutation of the set $\{1, 2, \dots, n\}$. By a ***signed elementary product from*** A we shall mean an elementary product $a_{1j_1}a_{2j_2} \dots a_{nj_n}$ multiplied by $+1$ or -1. We use the $+$ if $(j_1, j_2, \dots, j_n)$ is an even permutation and the $-$ if $(j_1, j_2, \dots, j_n)$ is an odd permutation.

Example 6 List all signed elementary products from the matrices.

$$(a) \begin{bmatrix} a_{11} & a_{12} \\ a_{21} & a_{22} \end{bmatrix} \qquad (b) \begin{bmatrix} a_{11} & a_{12} & a_{13} \\ a_{21} & a_{22} & a_{23} \\ a_{31} & a_{32} & a_{33} \end{bmatrix}$$

Solution.

(a)

Elementary Product	Associated Permutation	Even or Odd	Signed Elementary Product
$a_{11}a_{22}$	(1, 2)	even	$a_{11}a_{22}$
$a_{12}a_{21}$	(2, 1)	odd	$-a_{12}a_{21}$

(b)

Elementary Product	Associated Permutation	Even or Odd	Signed Elementary Product
$a_{11}a_{22}a_{33}$	(1, 2, 3)	even	$a_{11}a_{22}a_{33}$
$a_{11}a_{23}a_{32}$	(1, 3, 2)	odd	$-a_{11}a_{23}a_{32}$
$a_{12}a_{21}a_{33}$	(2, 1, 3)	odd	$-a_{12}a_{21}a_{33}$
$a_{12}a_{23}a_{31}$	(2, 3, 1)	even	$a_{12}a_{23}a_{31}$
$a_{13}a_{21}a_{32}$	(3, 1, 2)	even	$a_{13}a_{21}a_{32}$
$a_{13}a_{22}a_{31}$	(3, 2, 1)	odd	$-a_{13}a_{22}a_{31}$

We are now in a position to define the determinant function.

> **Definition.** Let A be a square matrix. The ***determinant function*** is denoted by ***det***, and we define $\det(A)$ to be the sum of all signed elementary products from A. The number $\det(A)$ is called the ***determinant of*** A.

EVALUATING 2×2 AND 3×3 DETERMINANTS

Example 7   Referring to Example 6, we obtain

(a) $\det\begin{bmatrix} a_{11} & a_{12} \\ a_{21} & a_{22} \end{bmatrix} = a_{11}a_{22} - a_{12}a_{21}$

(b) $\det\begin{bmatrix} a_{11} & a_{12} & a_{13} \\ a_{21} & a_{22} & a_{23} \\ a_{31} & a_{32} & a_{33} \end{bmatrix} = a_{11}a_{22}a_{33} + a_{12}a_{23}a_{31} + a_{13}a_{21}a_{32}$
$$- a_{13}a_{22}a_{31} - a_{12}a_{21}a_{33} - a_{11}a_{23}a_{32}$$

It is useful to have the two formulas in this example available for ready reference. To avoid memorizing these unwieldy expressions, however, we suggest using the mnemonic devices described in Figure 2. The first formula in Example 7 is obtained from Figure 2*a* by multiplying the entries on the rightward arrow and subtracting the product of the entries on the leftward arrow. The second formula in Example 7 is obtained by recopying the first and second columns as shown in Figure 2*b*. The determinant is then computed by summing the products on the rightward arrows and subtracting the products on the leftward arrows.

Figure 2   (*a*)   (*b*)

Example 8 Evaluate the determinants of

$$A = \begin{bmatrix} 3 & 1 \\ 4 & -2 \end{bmatrix} \quad \text{and} \quad B = \begin{bmatrix} 1 & 2 & 3 \\ -4 & 5 & 6 \\ 7 & -8 & 9 \end{bmatrix}$$

Solution. Using the method of Figure 2a gives

$$\det(A) = (3)(-2) - (1)(4) = -10$$

Using the method of Figure 2b gives

$$\det(B) = (45) + (84) + (96) - (105) - (-48) - (-72) = 240$$

$$\begin{bmatrix} 1 & 2 & 3 \\ -4 & 5 & 6 \\ 7 & -8 & 9 \end{bmatrix} \begin{matrix} 1 & 2 \\ -4 & 5 \\ 7 & -8 \end{matrix} \blacktriangle$$

Warning. We emphasize that the methods described in Figure 2 do not work for determinants of 4 × 4 matrices or higher.

Evaluating determinants directly from the definition leads to computational difficulties. Indeed, evaluating a 4 × 4 determinant directly would involve computing 4! = 24 signed elementary products, and a 10 × 10 determinant would involve 10! = 3,628,800 signed elementary products. Even the fastest of digital computers cannot handle the computation of a 25 × 25 determinant by this method in a practical amount of time. Much of the remainder of this chapter is devoted, therefore, to developing properties of determinants that will simplify their evaluation.

We conclude this section with some comments about terminology and notation. First, we note that the symbol $|A|$ is an alternative notation for $\det(A)$. For example, the determinant of a 3 × 3 matrix can be written as

$$\det \begin{bmatrix} a_{11} & a_{12} & a_{13} \\ a_{21} & a_{22} & a_{23} \\ a_{31} & a_{32} & a_{33} \end{bmatrix} \quad \text{or} \quad \begin{vmatrix} a_{11} & a_{12} & a_{13} \\ a_{21} & a_{22} & a_{23} \\ a_{31} & a_{32} & a_{33} \end{vmatrix}$$

In the latter notation the determinant of the matrix A in Example 8 would be written as

$$\begin{vmatrix} 3 & 1 \\ 4 & -2 \end{vmatrix} = -10$$

REMARK. Strictly speaking, the determinant of a matrix is a number. However, it is common practice to "abuse" the terminology slightly and use the term "determinant" to refer to the *matrix* whose determinant is being computed. Thus, we might refer to

$$\begin{vmatrix} 3 & 1 \\ 4 & -2 \end{vmatrix}$$

as a 2 × 2 determinant and call 3 the entry in the first row and first column of the determinant.

Finally, we note that the determinant of A is often written symbolically as

$$\det(A) = \sum \pm a_{1j_1} a_{2j_2} \dots a_{nj_n}$$

where $\sum$ indicates that the terms are to be summed over all permutations $(j_1, j_2, \dots, j_n)$ and the $+$ or $-$ is selected in each term according to whether the permutation is even or odd. This notation is useful when the definition of a determinant needs to be emphasized.

EXERCISE SET 2.1

1. Find the number of inversions in each of the following permutations of $\{1, 2, 3, 4, 5\}$.
   (a) $(4\ 1\ 3\ 5\ 2)$ (b) $(5\ 3\ 4\ 2\ 1)$ (c) $(3\ 2\ 5\ 4\ 1)$ (d) $(5\ 4\ 3\ 2\ 1)$ (e) $(1\ 2\ 3\ 4\ 5)$ (f) $(1\ 4\ 2\ 3\ 5)$

2. Classify each of the permutations in Exercise 1 as even or odd.

In Exercises 3–12 evaluate the determinant.

3. $\begin{vmatrix} 3 & 5 \\ -2 & 4 \end{vmatrix}$ 4. $\begin{vmatrix} 4 & 1 \\ 8 & 2 \end{vmatrix}$ 5. $\begin{vmatrix} -5 & 6 \\ -7 & -2 \end{vmatrix}$ 6. $\begin{vmatrix} \sqrt{2} & \sqrt{6} \\ 4 & \sqrt{3} \end{vmatrix}$ 7. $\begin{vmatrix} a-3 & 5 \\ -3 & a-2 \end{vmatrix}$ 8. $\begin{vmatrix} -2 & 7 & 6 \\ 5 & 1 & -2 \\ 3 & 8 & 4 \end{vmatrix}$

9. $\begin{vmatrix} -2 & 1 & 4 \\ 3 & 5 & -7 \\ 1 & 6 & 2 \end{vmatrix}$ 10. $\begin{vmatrix} -1 & 1 & 2 \\ 3 & 0 & -5 \\ 1 & 7 & 2 \end{vmatrix}$ 11. $\begin{vmatrix} 3 & 0 & 0 \\ 2 & -1 & 5 \\ 1 & 9 & -4 \end{vmatrix}$ 12. $\begin{vmatrix} c & -4 & 3 \\ 2 & 1 & c^2 \\ 4 & c-1 & 2 \end{vmatrix}$

13. Find all values of λ for which $\det(A) = 0$.

   (a) $\begin{bmatrix} \lambda-2 & 1 \\ -5 & \lambda+4 \end{bmatrix}$ (b) $\begin{bmatrix} \lambda-4 & 0 & 0 \\ 0 & \lambda & 2 \\ 0 & 3 & \lambda-1 \end{bmatrix}$

14. Classify each permutation of $\{1, 2, 3, 4\}$ as even or odd.

15. Use the results in Exercise 14 to construct a formula for the determinant of a 4×4 matrix.

16. Use the formula obtained in Exercise 15 to evaluate

   $$\begin{vmatrix} 4 & -9 & 9 & 2 \\ -2 & 5 & 6 & 4 \\ 1 & 2 & -5 & -3 \\ 1 & -2 & 0 & -2 \end{vmatrix}$$

17. Use the determinant definition to evaluate

   (a) $\begin{vmatrix} 0 & 0 & 0 & 0 & -3 \\ 0 & 0 & 0 & -4 & 0 \\ 0 & 0 & -1 & 0 & 0 \\ 0 & 2 & 0 & 0 & 0 \\ 5 & 0 & 0 & 0 & 0 \end{vmatrix}$ (b) $\begin{vmatrix} 5 & 0 & 0 & 0 & 0 \\ 0 & 0 & 0 & 0 & -4 \\ 0 & 0 & 3 & 0 & 0 \\ 0 & 0 & 0 & 1 & 0 \\ 0 & -2 & 0 & 0 & 0 \end{vmatrix}$

18. Solve for x.

$$\begin{vmatrix} x & -1 \\ 3 & 1-x \end{vmatrix} = \begin{vmatrix} 1 & 0 & -3 \\ 2 & x & -6 \\ 1 & 3 & x-5 \end{vmatrix}$$

19. Show that the value of the determinant

$$\begin{vmatrix} \sin\theta & \cos\theta & 0 \\ -\cos\theta & \sin\theta & 0 \\ \sin\theta - \cos\theta & \sin\theta + \cos\theta & 1 \end{vmatrix}$$

does not depend on θ.

20. Prove that if a square matrix A has a column of zeros, then $\det(A) = 0$.

2.2 EVALUATING DETERMINANTS BY ROW REDUCTION

In this section we shall show that the determinant of a matrix can be evaluated by reducing the matrix to row-echelon form. This method is important since it avoids the lengthy computations involved in directly applying the determinant definition.

We first consider two classes of matrices whose determinants can be easily evaluated, regardless of the size of the matrix.

> **Theorem 2.2.1.** *If A is any square matrix that contains a row of zeros, then $det(A) = 0$.*

Proof. Since a signed elementary product from A contains one factor from each row of A, every signed elementary product contains a factor from the row of zeros and consequently has value zero. Since $\det(A)$ is the sum of all signed elementary products, we obtain $\det(A) = 0$. ∎

DETERMINANTS OF TRIANGULAR MATRICES

A square matrix is called ***upper triangular*** if all the entries below the main diagonal are zeros. Similarly, a square matrix is called ***lower triangular*** if all the entries above the main diagonal are zeros. A matrix that is either upper or lower triangular is called ***triangular***.

Example 1   A general 4×4 upper triangular matrix has the form

$$\begin{bmatrix} a_{11} & a_{12} & a_{13} & a_{14} \\ 0 & a_{22} & a_{23} & a_{24} \\ 0 & 0 & a_{33} & a_{34} \\ 0 & 0 & 0 & a_{44} \end{bmatrix}$$

A general 4×4 lower triangular matrix has the form

$$\begin{bmatrix} a_{11} & 0 & 0 & 0 \\ a_{21} & a_{22} & 0 & 0 \\ a_{31} & a_{32} & a_{33} & 0 \\ a_{41} & a_{42} & a_{43} & a_{44} \end{bmatrix}$$ ▲

REMARK. A diagonal matrix has zeros above and below the main diagonal, so it is both upper and lower triangular.

Example 2 Evaluate $\det(A)$, where

$$A = \begin{bmatrix} a_{11} & 0 & 0 & 0 \\ a_{21} & a_{22} & 0 & 0 \\ a_{31} & a_{32} & a_{33} & 0 \\ a_{41} & a_{42} & a_{43} & a_{44} \end{bmatrix}$$

Solution. The only elementary product from A that can be nonzero is $a_{11}a_{22}a_{33}a_{44}$. To see that this is so, consider a typical elementary product $a_{1j_1}a_{2j_2}a_{3j_3}a_{4j_4}$. Since $a_{12} = a_{13} = a_{14} = 0$, we must have $j_1 = 1$ in order to have a nonzero elementary product. If $j_1 = 1$, we must have $j_2 \neq 1$, since no two factors come from the same column. Further, since $a_{23} = a_{24} = 0$, we must have $j_2 = 2$ in order to have a nonzero product. Continuing in this way, we obtain $j_3 - 3$ and $j_4 - 4$. Since $a_{11}a_{22}a_{33}a_{44}$ is multiplied by $+1$ in forming the signed elementary product, we obtain

$$\det(A) = a_{11}a_{22}a_{33}a_{44}$$ ▲

An argument similar to the one just presented can be applied to any triangular matrix to yield the following general result.

Theorem 2.2.2. *If A is an $n \times n$ triangular matrix (upper triangular, lower triangular, or diagonal), then $\det(A)$ is the product of the entries on the main diagonal; that is, $\det(A) = a_{11}a_{22} \cdots a_{nn}$.*

Example 3

$$\begin{vmatrix} 2 & 7 & -3 & 8 & 3 \\ 0 & -3 & 7 & 5 & 1 \\ 0 & 0 & 6 & 7 & 6 \\ 0 & 0 & 0 & 9 & 8 \\ 0 & 0 & 0 & 0 & 4 \end{vmatrix} = (2)(-3)(6)(9)(4) = -1296$$ ▲

EFFECT OF ELEMENTARY ROW OPERATIONS ON A DETERMINANT

The next theorem shows how an elementary row operation on a matrix affects the value of its determinant.

Theorem 2.2.3. *Let A be any $n \times n$ matrix.*

(a) *If A' is the matrix that results when a single row of A is multiplied by a constant k, then $\det(A') = k \det(A)$.*

(b) *If A' is the matrix that results when two rows of A are interchanged, then $\det(A') = -\det(A)$.*

(c) *If A' is the matrix that results when a multiple of one row of A is added to another row, then $\det(A') = \det(A)$.*

We shall omit the general proof (see Exercise 15).

Example 4 Consider the matrices

$$A = \begin{bmatrix} 1 & 2 & 3 \\ 0 & 1 & 4 \\ 1 & 2 & 1 \end{bmatrix} \qquad A_1 = \begin{bmatrix} 4 & 8 & 12 \\ 0 & 1 & 4 \\ 1 & 2 & 1 \end{bmatrix} \qquad A_2 = \begin{bmatrix} 0 & 1 & 4 \\ 1 & 2 & 3 \\ 1 & 2 & 1 \end{bmatrix}$$

$$A_3 = \begin{bmatrix} 1 & 2 & 3 \\ -2 & -3 & 2 \\ 1 & 2 & 1 \end{bmatrix}$$

If we evaluate the determinants of these matrices by the method in Example 8 of Section 2.1 we obtain

$$\det(A) = -2, \qquad \det(A_1) = -8, \qquad \det(A_2) = 2, \qquad \det(A_3) = -2$$

Observe that A_1 is obtained by multiplying the first row of A by 4; A_2 by interchanging the first two rows; and A_3 by adding -2 times the third row of A to the second. As predicted by Theorem 2.2.3, we have the relationships

$$\det(A_1) = 4 \det(A) \qquad \det(A_2) = -\det(A) \qquad \text{and} \qquad \det(A_3) = \det(A) \quad \blacktriangle$$

Example 5 Statement (*a*) in Theorem 2.2.3 has an alternative interpretation that is sometimes useful. This result allows us to bring a "common factor" from any row of a square matrix through the determinant sign. For example, consider the matrices

$$A = \begin{bmatrix} a_{11} & a_{12} & a_{13} \\ a_{21} & a_{22} & a_{33} \\ a_{31} & a_{32} & a_{33} \end{bmatrix} \qquad B = \begin{bmatrix} a_{11} & a_{12} & a_{13} \\ ka_{21} & ka_{22} & ka_{23} \\ a_{31} & a_{32} & a_{33} \end{bmatrix}$$

where the second row of B has a common factor of k. Since B is the matrix that results when the second row of A is multiplied by k, statement (*a*) in Theorem 2.2.3 asserts that $\det(B) = k \det(A)$; that is,

$$\begin{vmatrix} a_{11} & a_{12} & a_{13} \\ ka_{21} & ka_{22} & ka_{23} \\ a_{31} & a_{32} & a_{33} \end{vmatrix} = k \begin{vmatrix} a_{11} & a_{12} & a_{13} \\ a_{21} & a_{22} & a_{23} \\ a_{31} & a_{32} & a_{33} \end{vmatrix} \quad \blacktriangle$$

**EVALUATING
DETERMINANTS
BY ROW
REDUCTION**

We shall now formulate an alternative method for evaluating determinants that will avoid the large amount of computation involved in directly applying the determinant definition. The basic idea of this method is to apply elementary row operations to reduce the given matrix A to a matrix R that is in row-echelon form. Since a row-echelon form of a square matrix is always upper triangular (Exercise 14), $\det(R)$ can be evaluated easily using Theorem 2.2.2. The value of $\det(A)$ can then be obtained by using Theorem 2.2.3 to relate the unknown value of $\det(A)$ to the known value of $\det(R)$. The following example illustrates this method.

Example 6   Evaluate $\det(A)$ where

$$A = \begin{bmatrix} 0 & 1 & 5 \\ 3 & -6 & 9 \\ 2 & 6 & 1 \end{bmatrix}$$

Solution.   Reducing A to row-echelon form and applying Theorem 2.2.3, we obtain

$$\det(A) = \begin{vmatrix} 0 & 1 & 5 \\ 3 & -6 & 9 \\ 2 & 6 & 1 \end{vmatrix} = - \begin{vmatrix} 3 & -6 & 9 \\ 0 & 1 & 5 \\ 2 & 6 & 1 \end{vmatrix}$$

> The first and second rows of A were interchanged.

$$= -3 \begin{vmatrix} 1 & -2 & 3 \\ 0 & 1 & 5 \\ 2 & 6 & 1 \end{vmatrix}$$

> A common factor of 3 from the first row of the preceding matrix was taken through the det sign (see Example 5).

$$= -3 \begin{vmatrix} 1 & -2 & 3 \\ 0 & 1 & 5 \\ 0 & 10 & -5 \end{vmatrix}$$

> 2 times the first row of the preceding matrix was added to the third row.

$$= -3 \begin{vmatrix} 1 & -2 & 3 \\ 0 & 1 & 5 \\ 0 & 0 & -55 \end{vmatrix}$$

> -10 times the second row of the preceding matrix was added to the third row.

$$= (-3)(-55) \begin{vmatrix} 1 & -2 & 3 \\ 0 & 1 & 5 \\ 0 & 0 & 1 \end{vmatrix}$$

> A common factor of -55 from the last row of the preceding matrix was taken through the det sign.

$$= (-3)(-55)(1) = 165 \quad \blacktriangle$$

REMARK.   The method of row reduction is well suited for computer evaluation of determinants because it is systematic and easily programmed. However, in subsequent sections we will develop methods that are often easier for hand computation.

Example 7   Evaluate det(A), where

$$A = \begin{bmatrix} 1 & 3 & -2 & 4 \\ 2 & 6 & -4 & 8 \\ 3 & 9 & 1 & 5 \\ 1 & 1 & 4 & 8 \end{bmatrix}$$

$$\det(A) = \begin{vmatrix} 1 & 3 & -2 & 4 \\ 0 & 0 & 0 & 0 \\ 3 & 9 & 1 & 5 \\ 1 & 1 & 4 & 8 \end{vmatrix}$$

$\boxed{\begin{array}{l} -2 \text{ times the first} \\ \text{row of } A \text{ was added} \\ \text{to the second row.} \end{array}}$

There is no further reduction required since it follows from Theorem 2.2.1 that det(A) = 0. ▲

It should be evident from this example that whenever a square matrix has two proportional rows (like the first and second rows of A), it is possible to introduce a row of zeros by adding a suitable multiple of one of these rows to the other. Thus, *if a square matrix has two proportional rows, its determinant is zero.* In particular, a matrix with two identical rows has determinant zero.

Example 8   Each of the following matrices has two proportional rows; thus, by inspection, each has a zero determinant.

$$\begin{bmatrix} -1 & 4 \\ -2 & 8 \end{bmatrix} \qquad \begin{bmatrix} 2 & 7 & 8 \\ 3 & 2 & 4 \\ 2 & 7 & 8 \end{bmatrix} \qquad \begin{bmatrix} 3 & -1 & 4 & -5 \\ 6 & -2 & 5 & 2 \\ 5 & 8 & 1 & 4 \\ -9 & 3 & -12 & 15 \end{bmatrix} \quad ▲$$

EXERCISE SET 2.2

1. Evaluate the following determinants by inspection.

(a) $\begin{vmatrix} 3 & -17 & 4 \\ 0 & 5 & 1 \\ 0 & 0 & -2 \end{vmatrix}$
(b) $\begin{vmatrix} \sqrt{2} & 0 & 0 & 0 \\ -8 & \sqrt{2} & 0 & 0 \\ 7 & 0 & -1 & 0 \\ 9 & 5 & 6 & 1 \end{vmatrix}$
(c) $\begin{vmatrix} -2 & 1 & 3 \\ 1 & -7 & 4 \\ -2 & 1 & 3 \end{vmatrix}$
(d) $\begin{vmatrix} 1 & -2 & 3 \\ 2 & -4 & 6 \\ 5 & -8 & 1 \end{vmatrix}$

In Exercises 2–9 evaluate the determinant of the given matrix by reducing the matrix to row-echelon form.

2. $\begin{bmatrix} 3 & 6 & -9 \\ 0 & 0 & -2 \\ -2 & 1 & 5 \end{bmatrix}$
3. $\begin{bmatrix} 0 & 3 & 1 \\ 1 & 1 & 2 \\ 3 & 2 & 4 \end{bmatrix}$
4. $\begin{bmatrix} 1 & -3 & 0 \\ -2 & 4 & 1 \\ 5 & -2 & 2 \end{bmatrix}$
5. $\begin{bmatrix} 3 & -6 & 9 \\ -2 & 7 & -2 \\ 0 & 1 & 5 \end{bmatrix}$

6. $\begin{bmatrix} 1 & -2 & 3 & 1 \\ 5 & -9 & 6 & 3 \\ -1 & 2 & -6 & -2 \\ 2 & 8 & 6 & 1 \end{bmatrix}$
 7. $\begin{bmatrix} 2 & 1 & 3 & 1 \\ 1 & 0 & 1 & 1 \\ 0 & 2 & 1 & 0 \\ 0 & 1 & 2 & 3 \end{bmatrix}$
 8. $\begin{bmatrix} 0 & 1 & 1 & 1 \\ \frac{1}{2} & \frac{1}{2} & 1 & \frac{1}{2} \\ \frac{2}{3} & \frac{1}{3} & \frac{1}{3} & 0 \\ -\frac{1}{3} & \frac{2}{3} & 0 & 0 \end{bmatrix}$
 9. $\begin{bmatrix} 1 & 3 & 1 & 5 & 3 \\ -2 & -7 & 0 & -4 & 2 \\ 0 & 0 & 1 & 0 & 1 \\ 0 & 0 & 2 & 1 & 1 \\ 0 & 0 & 0 & 1 & 1 \end{bmatrix}$

10. Given that $\begin{vmatrix} a & b & c \\ d & e & f \\ g & h & i \end{vmatrix} = -6$, find

(a) $\begin{vmatrix} d & e & f \\ g & h & i \\ a & b & c \end{vmatrix}$
 (b) $\begin{vmatrix} 3a & 3b & 3c \\ -d & -e & -f \\ 4g & 4h & 4i \end{vmatrix}$
 (c) $\begin{vmatrix} a+g & b+h & c+i \\ d & e & f \\ g & h & i \end{vmatrix}$
 (d) $\begin{vmatrix} -3a & -3b & -3c \\ d & e & f \\ g-4d & h-4e & i-4f \end{vmatrix}$

11. Use row reduction to show that

$$\begin{vmatrix} 1 & 1 & 1 \\ a & b & c \\ a^2 & b^2 & c^2 \end{vmatrix} = (b-a)(c-a)(c-b)$$

12. Use an argument like that given in Example 2 to show

(a) $\det \begin{bmatrix} 0 & 0 & a_{13} \\ 0 & a_{22} & a_{23} \\ a_{31} & a_{32} & a_{33} \end{bmatrix} = -a_{13}a_{22}a_{31}$
 (b) $\det \begin{bmatrix} 0 & 0 & 0 & a_{14} \\ 0 & 0 & a_{23} & a_{24} \\ 0 & a_{32} & a_{33} & a_{34} \\ a_{41} & a_{42} & a_{43} & a_{44} \end{bmatrix} = a_{14}a_{23}a_{32}a_{41}$

13. Prove that Theorem 2.2.1 is true when the word "row" is replaced by "column."

14. Prove that a row-echelon form of a square matrix is always upper triangular.

15. Prove the following special cases of Theorem 2.2.3.

(a) $\begin{vmatrix} ka_{11} & ka_{12} & ka_{13} \\ a_{21} & a_{22} & a_{23} \\ a_{31} & a_{32} & a_{33} \end{vmatrix} = k \begin{vmatrix} a_{11} & a_{12} & a_{13} \\ a_{21} & a_{22} & a_{23} \\ a_{31} & a_{32} & a_{33} \end{vmatrix}$
 (b) $\begin{vmatrix} a_{21} & a_{22} & a_{23} \\ a_{11} & a_{12} & a_{13} \\ a_{31} & a_{32} & a_{33} \end{vmatrix} = - \begin{vmatrix} a_{11} & a_{12} & a_{13} \\ a_{21} & a_{22} & a_{23} \\ a_{31} & a_{32} & a_{33} \end{vmatrix}$

(c) $\begin{vmatrix} a_{11}+ka_{21} & a_{12}+ka_{22} & a_{13}+ka_{23} \\ a_{21} & a_{22} & a_{23} \\ a_{31} & a_{32} & a_{33} \end{vmatrix} = \begin{vmatrix} a_{11} & a_{12} & a_{13} \\ a_{21} & a_{22} & a_{23} \\ a_{31} & a_{32} & a_{33} \end{vmatrix}$

2.3   PROPERTIES OF THE DETERMINANT FUNCTION

In this section we shall develop some of the fundamental properties of the determinant function. Our work here will give us some further insight into the relationship between a square matrix and its determinant. One of the immediate consequences of this material will be an important determinant test for the invertibility of a matrix.

BASIC PROPERTIES OF DETERMINANTS

Recall that the determinant of an $n \times n$ matrix A is defined to be the sum of all signed elementary products from A. Since an elementary product has one factor from each row and one factor from each column, it is obvious that A and A^t have precisely the same set of elementary products. Although we shall omit the details, it can be shown that A and A^t actually have the same *signed* elementary products; this yields the following theorem.

Theorem 2.3.1. *If A is any square matrix, then $det(A) = det(A^t)$.*

REMARK. Because of this result, nearly every theorem about determinants that contains the word "row" in its statement is also true when the word "column" is substituted for "row." To prove a column statement one need only transpose the matrix in question to convert the column statement to a row statement, and then apply the corresponding known result for rows.

Example 1 Use Theorem 2.3.1 and the fact that interchanging two rows of a square matrix A changes the sign of its determinant to prove that interchanging two columns of A changes the sign of its determinant.

Solution. Let A' be the matrix that results when column r and column s of A are interchanged. Thus, $(A')^t$ is the matrix that results when *row* r and *row* s of A^t are interchanged. Therefore,

$$\det(A') = \det(A'^t) \qquad \text{[Theorem 2.3.1]}$$
$$= -\det(A^t) \qquad \text{[Theorem 2.2.3}b\text{]}$$
$$= -\det(A) \qquad \text{[Theorem 2.3.1]}$$

which proves the result. ▲

The next two examples use column properties of determinants.

Example 2 By inspection, the matrix

$$\begin{bmatrix} 1 & -2 & 7 \\ -4 & 8 & 5 \\ 2 & -4 & 3 \end{bmatrix}$$

has a zero determinant since the first and second columns are proportional. ▲

Example 3 Compute the determinant of

$$A = \begin{bmatrix} 1 & 0 & 0 & 3 \\ 2 & 7 & 0 & 6 \\ 0 & 6 & 3 & 0 \\ 7 & 3 & 1 & -5 \end{bmatrix}$$

Solution. This determinant could be computed as before by using elementary row operations to reduce A to row-echelon form. In contrast, we can put A in lower triangular form in one step by adding -3 times the first column to the fourth to obtain

$$\det(A) = \det \begin{bmatrix} 1 & 0 & 0 & 0 \\ 2 & 7 & 0 & 0 \\ 0 & 6 & 3 & 0 \\ 7 & 3 & 1 & -26 \end{bmatrix} = (1)(7)(3)(-26) = -546$$

This example points out that it is always wise to keep an eye open for column operations that can shorten computations. ▲

Suppose that A and B are $n \times n$ matrices and k is any scalar. We now consider possible relationships between $\det(A)$, $\det(B)$, and

$$\det(kA), \qquad \det(A + B), \qquad \text{and} \qquad \det(AB)$$

Since a common factor of any row of a matrix can be moved through the det sign, and since each of the n rows in kA has a common factor of k, we obtain

$$\det(kA) = k^n \det(A) \tag{1}$$

Example 4 Consider the matrices

$$A = \begin{bmatrix} 3 & 1 \\ 2 & 2 \end{bmatrix} \qquad \text{and} \qquad 5A = \begin{bmatrix} 15 & 5 \\ 10 & 10 \end{bmatrix}$$

By direct calculation, $\det(A) = 4$ and $\det(5A) = 100$. This agrees with relationship (1), which asserts that $\det(5A) = 5^2 \det(A)$. ▲

Unfortunately, no simple relationship exists between $\det(A)$, $\det(B)$, and $\det(A + B)$ in general. In particular, we emphasize that $\det(A + B)$ is usually *not* equal to $\det(A) + \det(B)$. The following example illustrates this fact.

Example 5 Consider

$$A = \begin{bmatrix} 1 & 2 \\ 2 & 5 \end{bmatrix} \qquad B = \begin{bmatrix} 3 & 1 \\ 1 & 3 \end{bmatrix} \qquad A + B = \begin{bmatrix} 4 & 3 \\ 3 & 8 \end{bmatrix}$$

We have $\det(A) = 1$, $\det(B) = 8$, and $\det(A + B) = 23$; thus $\det(A + B) \neq \det(A) + \det(B)$.

In spite of this negative result, there is one important relationship concerning sums of determinants that is often useful. To obtain it, consider two 2×2 matrices that differ only in the second row:

$$A = \begin{bmatrix} a_{11} & a_{12} \\ a_{21} & a_{22} \end{bmatrix} \qquad \text{and} \qquad A' = \begin{bmatrix} a_{11} & a_{12} \\ a'_{21} & a'_{22} \end{bmatrix}$$

From the formula in Example 7 of Section 2.1, we obtain

$$\det(A) + \det(A') = (a_{11}a_{22} - a_{12}a_{21}) + (a_{11}a'_{22} - a_{12}a'_{21})$$
$$= a_{11}(a_{22} + a'_{22}) - a_{12}(a_{21} + a'_{21})$$
$$= \det \begin{bmatrix} a_{11} & a_{12} \\ a_{21} + a'_{21} & a_{22} + a'_{22} \end{bmatrix}$$

Thus,

$$\det \begin{bmatrix} a_{11} & a_{12} \\ a_{21} & a_{22} \end{bmatrix} + \det \begin{bmatrix} a_{11} & a_{12} \\ a'_{21} & a'_{22} \end{bmatrix} = \det \begin{bmatrix} a_{11} & a_{12} \\ a_{21} + a'_{21} & a_{22} + a'_{22} \end{bmatrix} \quad \blacktriangle$$

This example is a special case of the following general result.

> **Theorem 2.3.2.** *Let A, A', and A'' be n × n matrices that differ only in a single row, say the rth, and assume that the rth row of A'' can be obtained by adding corresponding entries in the rth rows of A and A'. Then*
>
> $$det(A'') = det(A) + det(A')$$
>
> *The same result holds for columns.*

Example 6 By evaluating the determinants, the reader can check that

$$\det \begin{bmatrix} 1 & 7 & 5 \\ 2 & 0 & 3 \\ 1+0 & 4+1 & 7+(-1) \end{bmatrix} = \det \begin{bmatrix} 1 & 7 & 5 \\ 2 & 0 & 3 \\ 1 & 4 & 7 \end{bmatrix} + \det \begin{bmatrix} 1 & 7 & 5 \\ 2 & 0 & 3 \\ 0 & 1 & -1 \end{bmatrix} \quad \blacktriangle$$

DETERMINANT OF A MATRIX PRODUCT

The next theorem relates the determinant of a product of matrices and the determinants of the factors.

> **Theorem 2.3.3.** *If A and B are square matrices of the same size, then*
>
> $$det(AB) = det(A)det(B)$$

When one considers the complexity of the definitions of matrix multiplication and determinants, the elegant simplicity of this result is surprising. The proof is quite complicated and will be omitted.

Example 7 Consider the matrices

$$A = \begin{bmatrix} 3 & 1 \\ 2 & 1 \end{bmatrix} \qquad B = \begin{bmatrix} -1 & 3 \\ 5 & 8 \end{bmatrix} \qquad AB = \begin{bmatrix} 2 & 17 \\ 3 & 14 \end{bmatrix}$$

We leave it for the reader to verify that

$$\det(A) = 1 \qquad \det(B) = -23 \qquad \text{and} \qquad \det(AB) = -23$$

Thus, $\det(AB) = \det(A)\det(B)$ as predicted by Theorem 2.3.3. $\blacktriangle$

In Theorem 1.7.3, we listed three important statements that are equivalent to the invertibility of a matrix. The next example will help us to add another result to that list.

Example 8   The purpose of this example is to show that if the reduced row-echelon form R of a *square* matrix has no rows consisting entirely of zeros, then R must be the identity matrix. This can be illustrated by considering the following $n \times n$ matrix:

$$R = \begin{bmatrix} r_{11} & r_{12} & \cdots & r_{1n} \\ r_{21} & r_{22} & \cdots & r_{2n} \\ \vdots & \vdots & & \vdots \\ r_{n1} & r_{n2} & \cdots & r_{nn} \end{bmatrix}$$

Either the last row in this matrix consists entirely of zeros or it does not. If not, the matrix contains no zero rows, and consequently each of the n rows has a leading entry of 1. Since these leading 1's occur progressively further to the right as we move down the matrix, each of these 1's must occur on the main diagonal. Since the other entries in the same column as one of these 1's are zero, R must be I. Thus, either R has a row of zeros or $R = I$. ▲

DETERMINANT TEST FOR INVERTIBILITY

Theorem 2.3.4.   *A square matrix A is invertible if and only if $\det(A) \neq 0$.*

Proof.   If A is invertible, then $I = AA^{-1}$ so that $1 = \det(I) = \det(A)\det(A^{-1})$. Thus, $\det(A) \neq 0$. Conversely, assume that $\det(A) \neq 0$. We shall show that A is row equivalent to I, and thus conclude from Theorem 1.6.3 that A is invertible. Let R be the reduced row-echelon form of A. Since R can be obtained from A by a finite sequence of elementary row operations, we can find a sequence of elementary matrices $E_1, E_2, \ldots, E_k$ such that $E_k \cdots E_2 E_1 A = R$. Thus,

$$\det(R) = \det(E_k) \cdots \det(E_2)\det(E_1)\det(A)$$

Since we are assuming that $\det(A) \neq 0$, it follows from this equation that $\det(R$` Therefore, R does not have any zero rows, so that $R = I$ (see Example 8). ▌

Corollary 2.3.5.   If A is invertible, then

$$\det(A^{-1}) - \frac{1}{\det(A)}$$

Proof.   Since $A^{-1}A = I$, it follows that $\det(A^{-1}A) = \det(I)$. Therefore, we have $\det(A^{-1})\det(A) = 1$. Since $\det(A) \neq 0$, the proof can be completed by dividing through by $\det(A)$. ▌

Example 9 Since the first and third rows of

$$A = \begin{bmatrix} 1 & 2 & 3 \\ 1 & 0 & 1 \\ 2 & 4 & 6 \end{bmatrix}$$

are proportional, $\det(A) = 0$. Thus, A is not invertible. ▲

EXERCISE SET 2.3

1. Verify that $\det(A) = \det(A^t)$ for

(a) $A = \begin{bmatrix} -2 & 3 \\ 1 & 4 \end{bmatrix}$ (b) $\begin{bmatrix} 2 & -1 & 3 \\ 1 & 2 & 4 \\ 5 & -3 & 6 \end{bmatrix}$

2. Verify that $\det(AB) = \det(A)\det(B)$ for

$$A = \begin{bmatrix} 2 & 1 & 0 \\ 3 & 4 & 0 \\ 0 & 0 & 2 \end{bmatrix} \quad \text{and} \quad B = \begin{bmatrix} 1 & -1 & 3 \\ 7 & 1 & 2 \\ 5 & 0 & 1 \end{bmatrix}$$

3. By inspection, explain why $\det(A) = 0$.

$$A = \begin{bmatrix} -2 & 8 & 1 & 4 \\ 3 & 2 & 5 & 1 \\ 1 & 0 & 7 & 0 \\ 4 & -6 & 4 & -3 \end{bmatrix}$$

4. Use Theorem 2.3.4 to determine which of the following matrices are invertible.

(a) $\begin{bmatrix} 1 & 0 & -1 \\ 9 & -1 & 4 \\ 8 & 9 & -1 \end{bmatrix}$ (b) $\begin{bmatrix} 4 & 2 & 8 \\ -2 & 1 & -4 \\ 3 & 1 & 6 \end{bmatrix}$ (c) $\begin{bmatrix} \sqrt{2} & -\sqrt{7} & 0 \\ 3\sqrt{2} & -3\sqrt{7} & 0 \\ 5 & -9 & 0 \end{bmatrix}$ (d) $\begin{bmatrix} -3 & 0 & 1 \\ 5 & 0 & 6 \\ 8 & 0 & 3 \end{bmatrix}$

5. Let

$$A = \begin{bmatrix} a & b & c \\ d & e & f \\ g & h & i \end{bmatrix}$$

Assuming that $\det(A) = -7$, find

(a) $\det(3A)$ (b) $\det(2A^{-1})$ (c) $\det((2A)^{-1})$ (d) $\det \begin{bmatrix} a & g & d \\ b & h & e \\ c & i & f \end{bmatrix}$

6. Without directly evaluating, show that $x = 0$ and $x = 2$ satisfy

$$\begin{vmatrix} x^2 & x & 2 \\ 2 & 1 & 1 \\ 0 & 0 & -5 \end{vmatrix} = 0$$

7. Without directly evaluating, show that

$$\det \begin{bmatrix} b+c & c+a & b+a \\ a & b & c \\ 1 & 1 & 1 \end{bmatrix} = 0$$

In Exercises 8–11 prove the identity without evaluating the determinants.

8. $\begin{vmatrix} a_1 & b_1 & a_1+b_1+c_1 \\ a_2 & b_2 & a_2+b_2+c_2 \\ a_3 & b_3 & a_3+b_3+c_3 \end{vmatrix} = \begin{vmatrix} a_1 & b_1 & c_1 \\ a_2 & b_2 & c_2 \\ a_3 & b_3 & c_3 \end{vmatrix}$

9. $\begin{vmatrix} a_1+b_1 & a_1-b_1 & c_1 \\ a_2+b_2 & a_2-b_2 & c_2 \\ a_3+b_3 & a_3-b_3 & c_3 \end{vmatrix} = -2\begin{vmatrix} a_1 & b_1 & c_1 \\ a_2 & b_2 & c_2 \\ a_3 & b_3 & c_3 \end{vmatrix}$

10. $\begin{vmatrix} a_1+b_1t & a_2+b_2t & a_3+b_3t \\ a_1t+b_1 & a_2t+b_2 & a_3t+b_3 \\ c_1 & c_2 & c_3 \end{vmatrix} = (1-t^2)\begin{vmatrix} a_1 & a_2 & a_3 \\ b_1 & b_2 & b_3 \\ c_1 & c_2 & c_3 \end{vmatrix}$

11. $\begin{vmatrix} a_1 & b_1+ta_1 & c_1+rb_1+sa_1 \\ a_2 & b_2+ta_2 & c_2+rb_2+sa_2 \\ a_3 & b_3+ta_3 & c_3+rb_3+sa_3 \end{vmatrix} = \begin{vmatrix} a_1 & a_2 & a_3 \\ b_1 & b_2 & b_3 \\ c_1 & c_2 & c_3 \end{vmatrix}$

12. For which value(s) of k does A fail to be invertible?

(a) $A = \begin{bmatrix} k-3 & -2 \\ -2 & k-2 \end{bmatrix}$ (b) $A = \begin{bmatrix} 1 & 2 & 4 \\ 3 & 1 & 6 \\ k & 3 & 2 \end{bmatrix}$

13. Use Theorem 2.3.4 to show that

$$\begin{bmatrix} \sin^2 \alpha & \sin^2 \beta & \sin^2 \gamma \\ \cos^2 \alpha & \cos^2 \beta & \cos^2 \gamma \\ 1 & 1 & 1 \end{bmatrix}$$

is not invertible for any values of α, β, and γ.

14. Let A and B be $n \times n$ matrices. Show that if A is invertible, then $\det(B) = \det(A^{-1}BA)$.

15. (a) Express

$$\begin{vmatrix} a_1+b_1 & c_1+d_1 \\ a_2+b_2 & c_2+d_2 \end{vmatrix}$$

as a sum of four determinants whose entries contain no sums.

(b) Express

$$\begin{vmatrix} a_1+b_1 & c_1+d_1 & e_1+f_1 \\ a_2+b_2 & c_2+d_2 & e_2+f_2 \\ a_3+b_3 & c_3+d_3 & e_3+f_3 \end{vmatrix}$$

as a sum of eight determinants whose entries contain no sums.

2.4 COFACTOR EXPANSION; CRAMER'S RULE

In this section we shall consider a method for evaluating determinants that is useful for hand computations and important theoretically. As a consequence of our work here, we will obtain a formula for the inverse of an invertible matrix as well as a formula for the solution to certain systems of linear equations in terms of determinants.

MINORS AND COFACTORS

Definition. If A is a square matrix, then the **minor of entry a_{ij}** is denoted by M_{ij} and is defined to be the determinant of the submatrix that remains after the ith row and jth column are deleted from A. The number $(-1)^{i+j}M_{ij}$ is denoted by C_{ij} and is called the **cofactor of entry a_{ij}.**

Example 1   Let

$$A = \begin{bmatrix} 3 & 1 & -4 \\ 2 & 5 & 6 \\ 1 & 4 & 8 \end{bmatrix}$$

The minor of entry a_{11} is

$$M_{11} = \begin{vmatrix} 3 & 1 & -4 \\ 2 & 5 & 6 \\ 1 & 4 & 8 \end{vmatrix} = \begin{vmatrix} 5 & 6 \\ 4 & 8 \end{vmatrix} = 16$$

The cofactor of a_{11} is

$$C_{11} = (-1)^{1+1}M_{11} = M_{11} = 16$$

Similarly, the minor of entry a_{32} is

$$M_{32} = \begin{vmatrix} 3 & 1 & -4 \\ 2 & 5 & 6 \\ 1 & 4 & 8 \end{vmatrix} = \begin{vmatrix} 3 & -4 \\ 2 & 6 \end{vmatrix} = 26$$

The cofactor of a_{32} is

$$C_{32} = (-1)^{3+2}M_{32} = -M_{32} = -26 \quad \blacktriangle$$

Notice that the cofactor and the minor of an element a_{ij} differ only in sign, that is, $C_{ij} = \pm M_{ij}$. A quick way for determining whether to use the $+$ or $-$ is to use the fact that the sign relating C_{ij} and M_{ij} is in the ith row and jth column of the "checkerboard" array

$$\begin{bmatrix} + & - & + & - & + & \cdots \\ - & + & - & + & - & \cdots \\ + & - & + & - & + & \cdots \\ - & + & - & + & - & \cdots \\ \vdots & \vdots & \vdots & \vdots & \vdots & \end{bmatrix}$$

For example, $C_{11} = M_{11}$, $C_{21} = -M_{21}$, $C_{12} = -M_{12}$, $C_{22} = M_{22}$, and so on.

COFACTOR EXPANSIONS

Consider the general 3×3 matrix

$$A = \begin{bmatrix} a_{11} & a_{12} & a_{13} \\ a_{21} & a_{22} & a_{23} \\ a_{31} & a_{32} & a_{33} \end{bmatrix}$$

In Example 7 of Section 2.1 we showed

$$\det(A) = a_{11}a_{22}a_{33} + a_{12}a_{23}a_{31} + a_{13}a_{21}a_{32}$$
$$- a_{13}a_{22}a_{31} - a_{12}a_{21}a_{33} - a_{11}a_{23}a_{32} \quad (1)$$

which can be rewritten as

$$\det(A) = a_{11}(a_{22}a_{33} - a_{23}a_{32}) + a_{21}(a_{13}a_{32} - a_{12}a_{33}) + a_{31}(a_{12}a_{23} - a_{13}a_{22})$$

Since the expressions in parentheses are just the cofactors C_{11}, C_{21}, and C_{31} (verify), we have

$$\det(A) = a_{11}C_{11} + a_{21}C_{21} + a_{31}C_{31} \quad (2)$$

Equation (2) shows that the determinant of A can be computed by multiplying the entries in the first column of A by their cofactors and adding the resulting products. This method of evaluating $\det(A)$ is called **cofactor expansion** along the first column of A.

Example 2 Let

$$A = \begin{bmatrix} 3 & 1 & 0 \\ -2 & -4 & 3 \\ 5 & 4 & -2 \end{bmatrix}$$

Evaluate $\det(A)$ by cofactor expansion along the first column of A.

Solution. From (2)

$$\det(A) = 3 \begin{vmatrix} -4 & 3 \\ 4 & -2 \end{vmatrix} - (-2) \begin{vmatrix} 1 & 0 \\ 4 & -2 \end{vmatrix} + 5 \begin{vmatrix} 1 & 0 \\ -4 & 3 \end{vmatrix}$$

$$= 3(-4) - (-2)(-2) + 5(3) = -1 \quad \blacktriangle$$

By rearranging the terms in (1) in various ways, it is possible to obtain other formulas like (2). There should be no trouble checking that all of the following are correct (see Exercise 28):

$$\begin{aligned} \det(A) &= a_{11}C_{11} + a_{12}C_{12} + a_{13}C_{13} \\ &= a_{11}C_{11} + a_{21}C_{21} + a_{31}C_{31} \\ &= a_{21}C_{21} + a_{22}C_{22} + a_{23}C_{23} \\ &= a_{12}C_{12} + a_{22}C_{22} + a_{32}C_{32} \\ &= a_{31}C_{31} + a_{32}C_{32} + a_{33}C_{33} \\ &= a_{13}C_{13} + a_{23}C_{23} + a_{33}C_{33} \end{aligned} \quad (3)$$

Notice that in each equation the entries and cofactors all come from the same row or column. These equations are called the **cofactor expansions** of $\det(A)$.

The results we have just given for 3×3 matrices form a special case of the following general theorem, which we state without proof.

Theorem 2.4.1. *The determinant of an $n \times n$ matrix A can be computed by multiplying the entries in any row (or column) by their cofactors and adding the resulting products; that is, for each $1 \leq i \leq n$ and $1 \leq j \leq n$,*

$$det(A) = a_{1j}C_{1j} + a_{2j}C_{2j} + \cdots + a_{nj}C_{nj}$$

(cofactor expansion along the jth column)

and

$$det(A) = a_{i1}C_{i1} + a_{i2}C_{i2} + \cdots + a_{in}C_{in}$$

(cofactor expansion along the ith row)

Example 3 Let A be the matrix in Example 2. Evaluate $det(A)$ by cofactor expansion along the first row.

Solution.

$$det(A) = \begin{vmatrix} 3 & 1 & 0 \\ -2 & -4 & 3 \\ 5 & 4 & -2 \end{vmatrix} = 3 \begin{vmatrix} -4 & 3 \\ 4 & -2 \end{vmatrix} - (1) \begin{vmatrix} -2 & 3 \\ 5 & -2 \end{vmatrix} + 0 \begin{vmatrix} -2 & -4 \\ 5 & 4 \end{vmatrix}$$

$$= 3(-4) - (1)(-11) + 0 = -1$$

This agrees with the result obtained in Example 2. ▲

REMARK. In this example it was unnecessary to compute the last cofactor, since it was multiplied by zero. In general, the best strategy for evaluating a determinant by cofactor expansion is to expand along a row or column having the largest number of zeros.

Cofactor expansion and row or column operations can sometimes be used in combination to provide an effective method for evaluating determinants. The following example illustrates this idea.

Example 4 Evaluate $det(A)$ where

$$A = \begin{bmatrix} 3 & 5 & -2 & 6 \\ 1 & 2 & -1 & 1 \\ 2 & 4 & 1 & 5 \\ 3 & 7 & 5 & 3 \end{bmatrix}$$

Solution. By adding suitable multiples of the second row to the remaining rows, we obtain

$$\det(A) = \begin{vmatrix} 0 & -1 & 1 & 3 \\ 1 & 2 & -1 & 1 \\ 0 & 0 & 3 & 3 \\ 0 & 1 & 8 & 0 \end{vmatrix}$$

$$= - \begin{vmatrix} -1 & 1 & 3 \\ 0 & 3 & 3 \\ 1 & 8 & 0 \end{vmatrix}$$

> Cofactor expansion along the first column.

$$= - \begin{vmatrix} -1 & 1 & 3 \\ 0 & 3 & 3 \\ 0 & 9 & 3 \end{vmatrix}$$

> We added the first row to the third row.

$$= -(-1) \begin{vmatrix} 3 & 3 \\ 9 & 3 \end{vmatrix}$$

> Cofactor expansion along the first column.

$$= -18 \quad \blacktriangle$$

ADJOINT OF A MATRIX

In a cofactor expansion we compute $\det(A)$ by multiplying the entries in a row or column by their cofactors and adding the resulting products. It turns out that if one multiplies the entries in any row by the corresponding cofactors from a *different* row, the sum of these products is always zero. (This result also holds for columns.) Although we omit the general proof, the next example illustrates the idea of the proof in a special case.

Example 5   Let

$$A = \begin{bmatrix} a_{11} & a_{12} & a_{13} \\ a_{21} & a_{22} & a_{23} \\ a_{31} & a_{32} & a_{33} \end{bmatrix}$$

Consider the quantity

$$a_{11}C_{31} + a_{12}C_{32} + a_{13}C_{33}$$

that is formed by multiplying the entries in the first row by the cofactors of the corresponding entries in the third row and adding the resulting products. We now show that this quantity is equal to zero by the following trick. Construct a new matrix A' by replacing the third row of A with another copy of the first row. Thus,

$$A' = \begin{bmatrix} a_{11} & a_{12} & a_{13} \\ a_{21} & a_{22} & a_{23} \\ a_{11} & a_{12} & a_{13} \end{bmatrix}$$

Let C'_{31}, C'_{32}, C'_{33} be the cofactors of the entires in the third row of A'. Since the first two rows of A and A' are the same, and since the computation of C_{31}, C_{32}, C_{33}, C'_{31}, C'_{32}, and C'_{33} involve only entries from the first two rows of A and A', it follows that

$$C_{31} = C'_{31} \qquad C_{32} = C'_{32}, \qquad C_{33} = C'_{33}$$

Since A' has two identical rows,

$$\det(A') = 0 \tag{4}$$

On the other hand, evaluating $\det(A')$ by cofactor expansion along the third row gives

$$\det(A') = a_{11}C'_{31} + a_{12}C'_{32} + a_{13}C'_{33}$$
$$= a_{11}C_{31} + a_{12}C_{32} + a_{13}C_{33} \tag{5}$$

From (4) and (5) we obtain

$$a_{11}C_{31} + a_{12}C_{32} + a_{13}C_{33} = 0 \quad \blacktriangle$$

Definition. If A is any $n \times n$ matrix and C_{ij} is the cofactor of a_{ij}, then the matrix

$$\begin{bmatrix} C_{11} & C_{12} & \cdots & C_{1n} \\ C_{21} & C_{22} & \cdots & C_{2n} \\ \vdots & \vdots & & \vdots \\ C_{n1} & C_{n2} & \cdots & C_{nn} \end{bmatrix}$$

is called the *matrix of cofactors from* A. The transpose of this matrix is called the *adjoint of* A and is denoted by $\mathrm{adj}(A)$.

Example 6 Let

$$A = \begin{bmatrix} 3 & 2 & -1 \\ 1 & 6 & 3 \\ 2 & -4 & 0 \end{bmatrix}$$

The cofactors of A are

$$C_{11} = 12 \qquad C_{12} = 6 \qquad C_{13} = -16$$
$$C_{21} = 4 \qquad C_{22} = 2 \qquad C_{23} = 16$$
$$C_{31} = 12 \qquad C_{32} = -10 \qquad C_{33} = 16$$

so that the matrix of cofactors is

$$\begin{bmatrix} 12 & 6 & -16 \\ 4 & 2 & 16 \\ 12 & -10 & 16 \end{bmatrix}$$

and the adjoint of A is

$$\text{adj}(A) = \begin{bmatrix} 12 & 4 & 12 \\ 6 & 2 & -10 \\ -16 & 16 & 16 \end{bmatrix}$$

We are now in a position to derive a formula for the inverse of an invertible matrix.

FORMULA FOR THE INVERSE OF A MATRIX

Theorem 2.4.2. *If A is an invertible matrix, then*

$$A^{-1} = \frac{1}{det(A)} \, adj(A) \tag{6}$$

Proof. We show first that

$$A \, \text{adj}(A) = \det(A)I$$

Consider the product

$$A \, \text{adj}(A) = \begin{bmatrix} a_{11} & a_{12} & \cdots & a_{1n} \\ a_{21} & a_{22} & \cdots & a_{2n} \\ \vdots & \vdots & & \vdots \\ a_{i1} & a_{i2} & \cdots & a_{in} \\ \vdots & \vdots & & \vdots \\ a_{n1} & a_{n2} & \cdots & a_{nn} \end{bmatrix} \begin{bmatrix} C_{11} & C_{21} & \cdots & C_{j1} & \cdots & C_{n1} \\ C_{12} & C_{22} & \cdots & C_{j2} & \cdots & C_{n2} \\ \vdots & \vdots & & \vdots & & \vdots \\ C_{1n} & C_{2n} & \cdots & C_{jn} & \cdots & C_{nn} \end{bmatrix}$$

The entry in the ith row and jth column of the product $A \, \text{adj}(A)$ is

$$a_{i1}C_{j1} + a_{i2}C_{j2} + \cdots + a_{in}C_{jn} \tag{7}$$

(see the shaded lines above).

If $i = j$ then (7) is the cofactor expansion of $\det(A)$ along the ith row of A (Theorem 2.4.1), and if $i \neq j$ then the a's and the cofactors come from different rows of A, so the value of (7) is zero. Therefore,

$$A \, \text{adj}(A) = \begin{bmatrix} \det(A) & 0 & \cdots & 0 \\ 0 & \det(A) & \cdots & 0 \\ \vdots & \vdots & & \vdots \\ 0 & 0 & \cdots & \det(A) \end{bmatrix} = \det(A)I \tag{8}$$

Since A is invertible, $\det(A) \neq 0$. Therefore, Equation (8) can be rewritten as

$$\frac{1}{\det(A)} \left[A \, \text{adj}(A) \right] = I$$

or

$$A\left[\frac{1}{\det(A)}\,\text{adj}(A)\right] = I$$

Multiplying both sides on the left by A^{-1} yields

$$A^{-1} = \frac{1}{\det(A)}\,\text{adj}(A)\quad\blacksquare$$

Example 7 Use (6) to find the inverse of the matrix A in Example 6.

Solution. The reader can check that $\det(A) = 64$. Thus,

$$A^{-1} = \frac{1}{\det(A)}\,\text{adj}(A) = \frac{1}{64}\begin{bmatrix} 12 & 4 & 12 \\ 6 & 2 & -10 \\ -16 & 16 & 16 \end{bmatrix}$$

$$= \begin{bmatrix} \frac{12}{64} & \frac{4}{64} & \frac{12}{64} \\ \frac{6}{64} & \frac{2}{64} & -\frac{10}{64} \\ -\frac{16}{64} & \frac{16}{64} & \frac{16}{64} \end{bmatrix}\quad\blacktriangle$$

CRAMER'S RULE Although the method in the foregoing example is reasonable for inverting 3×3 matrices by hand, the inversion algorithm discussed in Section 1.6 is more efficient for larger matrices. It should be kept in mind, however, that the method of Section 1.6 is just a computational procedure, whereas Formula (6) is an actual formula for the inverse that can be used to study its properties without computing it (Exercise 25, for example). In a similar vein, it is often useful to have a formula for the solution of a system of equations that can be used to study properties of the solution without solving the system. The next theorem establishes such a formula for certain systems of n equations in n unknowns. The formula is known as **Cramer's Rule**.*

---

* *Gabriel Cramer* (1704–1752), Swiss mathematician. Although Cramer does not rank with the great mathematicians of his time, his contributions as a disseminator of mathematical ideas have earned him a well-deserved place in the history of mathematics. Cramer traveled extensively and met many of the leading mathematicians of his day. These contacts and friendships led to extensive correspondence in which information about new mathematical discoveries was transmitted.

Cramer's most widely known work, *Introduction à l'analyse des lignes courbes algébriques* (1750), was a study and classification of algebraic curves; Cramer's Rule appeared in the appendix. Although the rule bears his name, variations of the basic idea were formulated earlier by various mathematicians. However, Cramer's superior notation helped clarify and popularize the technique.

Overwork combined with a fall from a carriage eventually led to his death in 1752. Cramer was apparently a good-natured and pleasant person, though he never married. His interests were broad. He wrote on philosophy of law and government and the history of mathematics. He served in public office, participated in artillery and fortifications activities for the government, instructed workers on techniques of cathedral repair, and undertook excavations of cathedral archives. Cramer received numerous honors for his activities.

Theorem 2.4.3 (*Cramer's Rule*). *If $AX = B$ is a system of n linear equations in n unknowns such that $\det(A) \neq 0$, then the system has a unique solution. This solution is*

$$x_1 = \frac{\det(A_1)}{\det(A)}, \quad x_2 = \frac{\det(A_2)}{\det(A)}, \ldots, \quad x_n = \frac{\det(A_n)}{\det(A)}$$

where A_j is the matrix obtained by replacing the entries in the jth column of A by the entries in the matrix

$$B = \begin{bmatrix} b_1 \\ b_2 \\ \vdots \\ b_n \end{bmatrix}$$

Proof. If $\det(A) \neq 0$, then A is invertible and, by Theorem 1.7.1, $X = A^{-1}B$ is the unique solution of $AX = B$. Therefore, by Theorem 2.4.2 we have

$$X = A^{-1}B = \frac{1}{\det(A)} \operatorname{adj}(A)B = \frac{1}{\det(A)} \begin{bmatrix} C_{11} & C_{21} & \cdots & C_{n1} \\ C_{12} & C_{22} & \cdots & C_{n2} \\ \vdots & \vdots & & \vdots \\ C_{1n} & C_{2n} & \cdots & C_{nn} \end{bmatrix} \begin{bmatrix} b_1 \\ b_2 \\ \vdots \\ b_n \end{bmatrix}$$

Multiplying the matrices out gives

$$X = \frac{1}{\det(A)} \begin{bmatrix} b_1 C_{11} + b_2 C_{21} + \cdots + b_n C_{n1} \\ b_1 C_{12} + b_2 C_{22} + \cdots + b_n C_{n2} \\ \vdots \quad\quad \vdots \quad\quad\quad\quad \vdots \\ b_1 C_{1n} + b_2 C_{2n} + \cdots + b_n C_{nn} \end{bmatrix}$$

The entry in the jth row of X is therefore

$$x_j = \frac{b_1 C_{1j} + b_2 C_{2j} + \cdots + b_n C_{nj}}{\det(A)} \tag{9}$$

Now let

$$A_j = \begin{bmatrix} a_{11} & a_{12} & \cdots & a_{1j-1} & b_1 & a_{1j+1} & \cdots & a_{1n} \\ a_{21} & a_{22} & \cdots & a_{2j-1} & b_2 & a_{2j+1} & \cdots & a_{2n} \\ \vdots & \vdots & & \vdots & \vdots & \vdots & & \vdots \\ a_{n1} & a_{n2} & \cdots & a_{nj-1} & b_n & a_{nj+1} & \cdots & a_{nn} \end{bmatrix}$$

Since A_j differs from A only in the jth column, it follows that the cofactors of entries $b_1, b_2, \ldots, b_n$ in A_j are the same as the cofactors of the corresponding entries in the jth column of A. The cofactor expansion of $\det(A_j)$ along the jth column is therefore

$$\det(A_j) = b_1 C_{1j} + b_2 C_{2j} + \cdots + b_n C_{nj}$$

Substituting this result in (9) gives

$$x_j = \frac{\det(A_j)}{\det(A)} \quad \blacksquare$$

Example 8 Use Cramer's Rule to solve

$$
\begin{aligned}
x_1 + + 2x_3 &= 6 \\
-3x_1 + 4x_2 + 6x_3 &= 30 \\
-x_1 - 2x_2 + 3x_3 &= 8
\end{aligned}
$$

Solution.

$$
A = \begin{bmatrix} 1 & 0 & 2 \\ -3 & 4 & 6 \\ -1 & -2 & 3 \end{bmatrix} \qquad A_1 = \begin{bmatrix} 6 & 0 & 2 \\ 30 & 4 & 6 \\ 8 & -2 & 3 \end{bmatrix}
$$

$$
A_2 = \begin{bmatrix} 1 & 6 & 2 \\ -3 & 30 & 6 \\ -1 & 8 & 3 \end{bmatrix} \qquad A_3 = \begin{bmatrix} 1 & 0 & 6 \\ -3 & 4 & 30 \\ -1 & -2 & 8 \end{bmatrix}
$$

Therefore,

$$x_1 = \frac{\det(A_1)}{\det(A)} = \frac{-40}{44} = \frac{-10}{11} \qquad x_2 = \frac{\det(A_2)}{\det(A)} = \frac{72}{44} = \frac{18}{11}$$

$$x_3 = \frac{\det(A_3)}{\det(A)} = \frac{152}{44} = \frac{38}{11} \quad \blacktriangle$$

To solve a system of n equations in n unknowns by Cramer's Rule, it is necessary to evaluate $n + 1$ determinants of $n \times n$ matrices. For systems with more than three equations, Gaussian elimination is more efficient, since it is only necessary to reduce one $n \times (n + 1)$ augmented matrix. However, Cramer's Rule does give a formula for the solution.

EXERCISE SET 2.4

1. Let

$$A = \begin{bmatrix} 1 & -2 & 3 \\ 6 & 7 & -1 \\ -3 & 1 & 4 \end{bmatrix}$$

(a) Find all the minors of A.   (b) Find all the cofactors.

2. Let

$$A = \begin{bmatrix} 4 & -1 & 1 & 6 \\ 0 & 0 & -3 & 3 \\ 4 & 1 & 0 & 14 \\ 4 & 1 & 3 & 2 \end{bmatrix}$$

Find

(a) M_{13} and C_{13} (b) M_{23} and C_{23} (c) M_{22} and C_{22} (d) M_{21} and C_{21}

3. Evaluate the determinant of the matrix in Exercise 1 by a cofactor expansion along

(a) the first row   (b) the first column   (c) the second row
(d) the second column   (e) the third row   (f) the third column

4. For the matrix in Exercise 1, find

(a) adj(A)   (b) A^{-1} using Theorem 2.4.2

In Exercises 5–10 evaluate det(A) by a cofactor expansion along a row or column of your choice.

5. $A = \begin{bmatrix} -3 & 0 & 7 \\ 2 & 5 & 1 \\ -1 & 0 & 5 \end{bmatrix}$ **6.** $A = \begin{bmatrix} 3 & 3 & 1 \\ 1 & 0 & -1 \\ 1 & -3 & 5 \end{bmatrix}$ **7.** $A = \begin{bmatrix} 1 & k & k^2 \\ 1 & k & k^2 \\ 1 & k & k^2 \end{bmatrix}$

8. $A = \begin{bmatrix} k+1 & k-1 & 7 \\ 2 & k-3 & 4 \\ 5 & k+1 & k \end{bmatrix}$ **9.** $A = \begin{bmatrix} 3 & 3 & 0 & 5 \\ 2 & 2 & 0 & -2 \\ 4 & 1 & -3 & 0 \\ 2 & 10 & 3 & 2 \end{bmatrix}$ **10.** $A = \begin{bmatrix} 4 & 0 & 0 & 1 & 0 \\ 3 & 3 & 3 & -1 & 0 \\ 1 & 2 & 4 & 2 & 3 \\ 9 & 4 & 6 & 2 & 3 \\ 2 & 2 & 4 & 2 & 3 \end{bmatrix}$

In Exercises 11–14 find A^{-1} using Theorem 2.4.2.

11. $A = \begin{bmatrix} 2 & 5 & 5 \\ -1 & -1 & 0 \\ 2 & 4 & 3 \end{bmatrix}$ **12.** $A = \begin{bmatrix} 2 & 0 & 3 \\ 0 & 3 & 2 \\ -2 & 0 & -4 \end{bmatrix}$

13. $A = \begin{bmatrix} 2 & -3 & 5 \\ 0 & 1 & -3 \\ 0 & 0 & 2 \end{bmatrix}$ **14.** $A = \begin{bmatrix} 2 & 0 & 0 \\ 8 & 1 & 0 \\ -5 & 3 & 6 \end{bmatrix}$

15. Let

$$A = \begin{bmatrix} 1 & 3 & 1 & 1 \\ 2 & 5 & 2 & 2 \\ 1 & 3 & 8 & 9 \\ 1 & 3 & 2 & 2 \end{bmatrix}$$

(a) Evaluate A^{-1} using Theorem 2.4.2.
(b) Evaluate A^{-1} using the method of Example 4 in Section 1.6.
(c) Which method involves less computation?

In Exercises 16–21 solve by Cramer's Rule, where it applies.

16. $7x_1 - 2x_2 = 3$
$3x_1 + x_2 = 5$

17. $4x + 5y = 2$
$11x + y + 2z = 3$
$x + 5y + 2z = 1$

18. $x - 4y + z = 6$
$4x - y + 2z = -1$
$2x + 2y - 3z = -20$

19. $x_1 - 3x_2 + x_3 = 4$
$2x_1 - x_2 = -2$
$4x_1 - 3x_3 = 0$

20. $-x_1 - 4x_2 + 2x_3 + x_4 = -32$
$2x_1 - x_2 + 7x_3 + 9x_4 = 14$
$-x_1 + x_2 + 3x_3 + x_4 = 11$
$x_1 - 2x_2 + x_3 - 4x_4 = -4$

21. $3x_1 - x_2 + x_3 = 4$
$-x_1 + 7x_2 - 2x_3 = 1$
$2x_1 + 6x_2 - x_3 = 5$

22. Show that the matrix

$$A = \begin{bmatrix} \cos\theta & \sin\theta & 0 \\ -\sin\theta & \cos\theta & 0 \\ 0 & 0 & 1 \end{bmatrix}$$

is invertible for all values of θ; then find A^{-1} using Theorem 2.4.2.

23. Use Cramer's Rule to solve for y without solving for x, z, and w.

$$4x + y + z + w = 6$$
$$3x + 7y - z + w = 1$$
$$7x + 3y - 5z + 8w = -3$$
$$x + y + z + 2w = 3$$

24. Let $AX = B$ be the system in Exercise 23.
   (a) Solve by Cramer's Rule.   (b) Solve by Gauss-Jordan elimination.
   (c) Which method involves fewer computations?

25. Prove that if $\det(A) = 1$ and all the entries in A are integers, then all the entries in A^{-1} are all integers.

26. Let $AX = B$ be a system of n linear equations in n unknowns with integer coefficients and integer constants. Prove that if $\det(A) = 1$, the solution X has integer entries.

27. Prove that if A is an invertible upper triangular matrix, then A^{-1} is upper triangular.

28. Derive the first and last cofactor expansions listed in Formula (3).

29. Prove: The equation of the line through the distinct points (a_1, b_1) and (a_2, b_2) can be written as

$$\begin{vmatrix} x & y & 1 \\ a_1 & b_1 & 1 \\ a_2 & b_2 & 1 \end{vmatrix} = 0$$

30. Prove: (x_1, y_1), (x_2, y_2), and (x_3, y_3) are collinear points if and only if

$$\begin{vmatrix} x_1 & y_1 & 1 \\ x_2 & y_2 & 1 \\ x_3 & y_3 & 1 \end{vmatrix} = 0$$

31. Prove: The equation of the plane through the noncollinear points (a_1, b_1, c_1), (a_2, b_2, c_2), and (a_3, b_3, c_3) can be written as

$$\begin{vmatrix} x & y & z & 1 \\ a_1 & b_1 & c_1 & 1 \\ a_2 & b_2 & c_2 & 1 \\ a_3 & b_3 & c_3 & 1 \end{vmatrix} = 0$$

SUPPLEMENTARY EXERCISES

1. Use Cramer's Rule to solve for x' and y' in terms of x and y.

$$x = \tfrac{3}{5}x' - \tfrac{4}{5}y'$$
$$y = \tfrac{4}{5}x' + \tfrac{3}{5}y'$$

2. Use Cramer's Rule to solve for x' and y' in terms of x and y.

$$x = x' \cos \theta - y' \sin \theta$$
$$y = x' \sin \theta + y' \cos \theta$$

3. By examining the determinant of the coefficient matrix, show that the following system has a nontrivial solution if and only if $\alpha = \beta$.

$$\begin{aligned} x + \;\; y + \alpha z &= 0 \\ x + \;\; y + \beta z &= 0 \\ \alpha x + \beta y + \;\; z &= 0 \end{aligned}$$

4. Let A be a 3×3 matrix, each of whose entries is 1 or 0. What is the largest possible value for $|A|$?

5. (a) For the triangle in Figure 1 below, use trigonometry to show

$$\begin{aligned} b \cos \gamma + c \cos \beta &= a \\ c \cos \alpha + a \cos \gamma &= b \\ a \cos \beta + b \cos \alpha &= c \end{aligned}$$

and then apply Cramer's Rule to show

$$\cos \alpha = \frac{b^2 + c^2 - a^2}{2bc}$$

(b) Use Cramer's Rule to obtain similar formulas for $\cos \beta$ and $\cos \gamma$.

Figure 1

6. Use determinants to show that for all real values of λ the only solution of

$$x - 2y = \lambda x$$
$$x - y = \lambda y$$

is $x = 0$, $y = 0$.

7. Prove: If A is invertible, then adj(A) is invertible and

$$[\text{adj}(A)]^{-1} = \frac{1}{\det(A)} A = \text{adj}(A^{-1})$$

8. Prove: If A is an $n \times n$ matrix, then $\det[\text{adj}(A)] = [\det(A)]^{n-1}$.

9. (For readers who have studied calculus.) Show that if $f_1(x)$, $f_2(x)$, $g_1(x)$, and $g_2(x)$ are differentiable functions, and if

$$W = \begin{vmatrix} f_1(x) & f_2(x) \\ g_1(x) & g_2(x) \end{vmatrix} \quad \text{then} \quad \frac{dW}{dx} = \begin{vmatrix} f'_1(x) & f'_2(x) \\ g_1(x) & g_2(x) \end{vmatrix} + \begin{vmatrix} f_1(x) & f_2(x) \\ g'_1(x) & g'_2(x) \end{vmatrix}$$

10. (a) In Figure 2 below, the area of triangle ABC is expressible as

$$\text{area } ABC = \text{area } ADEC + \text{area } CEFB - \text{area } ADFB$$

Use this and the fact that the area of a trapezoid equals $\frac{1}{2}$ the altitude times the sum of the parallel sides to show

$$\text{area } ABC = \frac{1}{2} \begin{vmatrix} x_1 & y_1 & 1 \\ x_2 & y_2 & 1 \\ x_3 & y_3 & 1 \end{vmatrix}$$

[*Note.* In the derivation of this formula, the vertices are labeled so the triangle is traced counterclockwise proceeding from (x_1, y_1) to (x_2, y_2) to (x_3, y_3). For a clockwise orientation, the determinant above yields the *negative* of the area]

(b) Use the result in (a) to find the area of the triangle with vertices $(3, 3)$, $(4, 0)$, $(-2, -1)$.

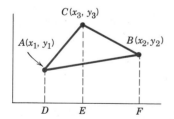

Figure 2

11. Prove: If the entries in each row of an $n \times n$ matrix A add up to zero, then $\det(A) = 0$. [**Hint.** Consider the product AX, where X is the $n \times 1$ matrix, each of whose entries is one.]

12. Let A be an $n \times n$ matrix and B the matrix that results when the rows of A are written in reverse order. How are $\det(A)$ and $\det(B)$ related?

13. How will A^{-1} be affected if
(a) the ith and jth rows of A are interchanged;
(b) the ith row of A is multiplied by a nonzero scalar, c;
(c) c times the ith row of A is added to the jth row?

14. Let A be an $n \times n$ matrix. Suppose that B_1 is obtained by adding the same number t to each entry in the ith row of A and B_2 is obtained by subtracting t from each entry in the ith row of A. Show that $\det(A) = \frac{1}{2}[\det(B_1) + \det(B_2)]$.

15. Let

$$A = \begin{bmatrix} a_{11} & a_{12} & a_{13} \\ a_{21} & a_{22} & a_{23} \\ a_{31} & a_{32} & a_{33} \end{bmatrix}$$

(a) Express $\det(\lambda I - A)$ as a polynomial $p(\lambda) = \lambda^3 + b\lambda^2 + c\lambda + d$.
(b) Express the coefficients b and d in terms of determinants and traces.

16. Without directly evaluating the determinant, show that

$$\begin{vmatrix} \sin \alpha & \cos \alpha & \sin(\alpha + \delta) \\ \sin \beta & \cos \beta & \sin(\beta + \delta) \\ \sin \gamma & \cos \gamma & \sin(\gamma + \delta) \end{vmatrix} = 0$$

17. Use the fact that 21375, 38798, 34162, 40223, and 79154 are all divisible by 19 to show that

$$\begin{vmatrix} 2 & 1 & 3 & 7 & 5 \\ 3 & 8 & 7 & 9 & 8 \\ 3 & 4 & 1 & 6 & 2 \\ 4 & 0 & 2 & 2 & 3 \\ 7 & 9 & 1 & 5 & 4 \end{vmatrix}$$

is divisible by 19 without directly evaluating the determinant.

CHAPTER 3

VECTORS IN 2-SPACE AND 3-SPACE

Readers familiar with the contents of this chapter can go to Chapter 4 with no loss of continuity.

3.1 INTRODUCTION TO VECTORS (GEOMETRIC)

*Many physical quantities, such as area, length, mass, and temperature, are completely described once the magnitude of the quantity is given. Such quantities are called **scalars**. Other physical quantities, called **vectors**, are not completely determined until both a magnitude and a direction are specified. For example, wind movement is usually described by giving the speed and the direction, say 20 mph northeast. The wind speed and wind direction together form a vector quantity called the wind **velocity**. Other examples of vectors are force and displacement. In this section vectors in 2-space and 3-space will be introduced geometrically, arithmetic operations on vectors will be defined, and some basic properties of these operations will be established.*

GEOMETRIC VECTORS

Vectors can be represented geometrically as directed line segments or arrows in 2-space or 3-space; the direction of the arrow specifies the direction of the vector, and the length of the arrow describes its magnitude. The tail of the arrow is called the ***initial point*** of the vector, and the tip of the arrow the ***terminal point***. We shall denote vectors in lowercase boldface type (for instance, **a**, **k**, **v**, **w**, and **x**). When discussing vectors, we shall refer to numbers as ***scalars***. All our scalars will be real numbers and will be denoted in ordinary lowercase type (for instance, a, k, v, w, and x).

If, as in Figure 1a, the initial point of a vector **v** is A and the terminal point is B, we write

$$\mathbf{v} = \overrightarrow{AB}$$

(a) (b)

Figure 1 | The vector $\overrightarrow{AB}$. | | Equivalent vectors. |

Vectors with the same length and same direction, such as those in Figure 1*b*, are called *equivalent*. Since we want a vector to be determined solely by its length and direction, equivalent vectors are regarded as *equal* even though they may be located in different positions. If **v** and **w** are equivalent, we write

$$\mathbf{v} = \mathbf{w}$$

> **Definition.** If **v** and **w** are any two vectors, then the *sum* **v** + **w** is the vector determined as follows: Position the vector **w** so that its initial point coincides with the terminal point of **v**. The vector **v** + **w** is represented by the arrow from the initial point of **v** to the terminal point of **w** (Figure 2*a*).

In Figure 2*b* we have constructed two sums, **v** + **w** (blue arrows) and **w** + **v** (white arrows). It is evident that

$$\mathbf{v} + \mathbf{w} = \mathbf{w} + \mathbf{v}$$

and that the sum coincides with the diagonal of the parallelogram determined by **v** and **w** when these vectors are positioned so they have the same initial point.

The vector of length zero is called the *zero vector* and is denoted by **0**. We define

$$\mathbf{0} + \mathbf{v} = \mathbf{v} + \mathbf{0} = \mathbf{v}$$

for every vector **v**. Since there is no natural direction for the zero vector, we shall agree that it can be assigned any direction that is convenient for the problem being considered. If **v** is any nonzero vector, then −**v**, the *negative* of **v**, is defined to be

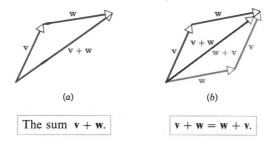

(a) (b)

Figure 2 | The sum **v** + **w**. | | **v** + **w** = **w** + **v**. |

the vector having the same magnitude as **v**, but oppositely directed (Figure 3).

Figure 3 | The negative of **v** has the same length as **v**, but is oppositely directed. |

This vector has the property

$$\mathbf{v} + (-\mathbf{v}) = \mathbf{0}$$

(Why?) In addition, we define $-\mathbf{0} = \mathbf{0}$. Subtraction of vectors is defined as follows.

> **Definition.** If **v** and **w** are any two vectors, then *difference* of **w** from **v** is defined by
>
> $$\mathbf{v} - \mathbf{w} = \mathbf{v} + (-\mathbf{w})$$
>
> (Figure 4a).

To obtain the difference $\mathbf{v} - \mathbf{w}$ without constructing $-\mathbf{w}$, position **v** and **w** so their initial points coincide; the vector from the terminal point of **w** to the terminal point of **v** is then the vector $\mathbf{v} - \mathbf{w}$ (Figure 4b).

Figure 4   (a)   (b)

> **Definition.** If **v** is a nonzero vector and k is a nonzero real number (scalar), then the *product* $k\mathbf{v}$ is defined to be the vector whose length is $|k|$ times the length of **v** and whose direction is the same as that of **v** if $k > 0$ and opposite to that of **v** if $k < 0$. We define $k\mathbf{v} = \mathbf{0}$ if $k = 0$ or $\mathbf{v} = \mathbf{0}$.

Figure 5 illustrates the relation between a vector **v** and the vectors $\frac{1}{2}\mathbf{v}$, $(-1)\mathbf{v}$, $2\mathbf{v}$, and $(-3)\mathbf{v}$. Note that the vector $(-1)\mathbf{v}$ has the same length as **v**, but is oppositely directed. Thus, $(-1)\mathbf{v}$ is just the negative of **v**; that is,

$$(-1)\mathbf{v} = -\mathbf{v}$$

Figure 5

A vector of the form $k\mathbf{v}$ is called a *scalar multiple* of $\mathbf{v}$. As evidenced by Figure 5, vectors that are scalar multiples of each other are parallel. Conversely, it can be shown that nonzero parallel vectors are scalar multiples of each other. We omit the proof.

VECTORS IN COORDINATE SYSTEMS

Problems involving vectors can often be simplified by introducing a rectangular coordinate system. For the moment we shall restrict the discussion to vectors in 2-space (the plane). Let $\mathbf{v}$ be any vector in the plane, and assume, as in Figure 6, that $\mathbf{v}$ has been positioned so its initial point is at the origin of a rectangular coordinate system. The coordinates (v_1, v_2) of the terminal point of $\mathbf{v}$ are called the *components of* $\mathbf{v}$, and we write

$$\mathbf{v} = (v_1, v_2)$$

If equivalent vectors, $\mathbf{v}$ and $\mathbf{w}$, are located so their initial points fall at the origin, then it is obvious that their terminal points must coincide (since the vectors have the same length and direction); thus, the vectors have the same components. Conversely, vectors with the same components are equivalent since they have the same length and same direction. In summary, two vectors

$$\mathbf{v} = (v_1, v_2) \quad \text{and} \quad \mathbf{w} = (w_1, w_2)$$

are equivalent if and only if

$$v_1 = w_1 \quad \text{and} \quad v_2 = w_2$$

Figure 6 | v_1 and v_2 are the components of $\mathbf{v}$.

The operations of vector addition and multiplication by scalars are very easy to carry out in terms of components. As illustrated in Figure 7, if

$$\mathbf{v} = (v_1, v_2) \quad \text{and} \quad \mathbf{w} = (w_1, w_2)$$

then

$$\mathbf{v} + \mathbf{w} = (v_1 + w_1, v_2 + w_2) \tag{1}$$

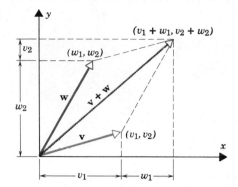

Figure 7

If $\mathbf{v} = (v_1, v_2)$ and k is any scalar, then by using a geometric argument involving similar triangles, it can be shown (Exercise 15) that

$$kv = (kv_1, kv_2) \tag{2}$$

(Figure 8). Thus, for example, if $\mathbf{v} = (1, -2)$ and $\mathbf{w} = (7, 6)$, then

$$\mathbf{v} + \mathbf{w} = (1, -2) + (7, 6) = (1 + 7, -2 + 6) = (8, 4)$$

and

$$4\mathbf{v} = 4(1, -2) = (4(1), 4(-2)) = (4, -8)$$

Since $\mathbf{v} - \mathbf{w} = \mathbf{v} + (-1)\mathbf{w}$, it follows from Formulas (1) and (2) that

$$\mathbf{v} - \mathbf{w} = (v_1 - w_1, v_2 - w_2)$$

(Verify.)

Figure 8

VECTORS IN 3-SPACE

Just as vectors in the plane can be described by pairs of real numbers, vectors in 3-space can be described by triples of real numbers by introducing a *rectangular coordinate* system. To construct such a coordinate system, select a point O, called the *origin*, and choose three mutually perpendicular lines, called *coordinate axes*, passing through the origin. Label these axes x, y, and z, and select a positive direction for each coordinate axis as well as a unit of length for measuring distances (Figure 9a). Each pair of coordinate axes determines a plane called a *coordinate plane*. These are referred to as the *xy-plane*, the *xz-plane*, and the *yz-plane*. To

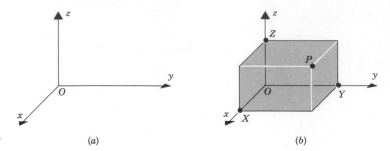

Figure 9   (a)   (b)

each point P in 3-space we assign a triple of numbers (x, y, z), called the ***coordinates of P***, as follows: Pass three planes through P parallel to the coordinate planes, and denote the points of intersections of these planes with the three coordinate axes by X, Y, and Z (Figure 9b). The coordinates of P are defined to be the signed lengths

$$x = OX \qquad y = OY \qquad z = OZ$$

In Figure 10 we have constructed the points whose coordinates ae $(4, 5, 6)$ and $(-3, 2, -4)$.

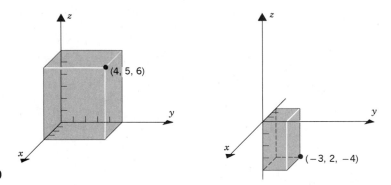

Figure 10

Rectangular coordinate systems in 3-space fall into two categories, ***left-handed*** and ***right-handed***. A right-handed system has the property that an ordinary screw pointed in the positive direction on the z-axis would be advanced if the positive x-axis is rotated $90°$ toward the positive y-axis (Figure 11a); the system is left-

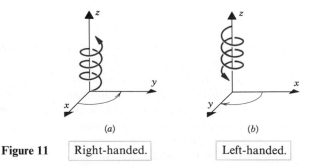

| (a) | (b) |

Figure 11 | Right-handed. | | Left-handed. |

handed if the screw would be retracted (Figure 11*b*).

REMARK. In this book we shall use only right-handed coordinate systems.

If, as in Figure 12, a vector **v** in 3-space is positioned so its initial point is at the origin of a rectangular coordinate system, then the coordinates of the terminal point are called the ***components*** of **v**, and we write

$$\mathbf{v} = (v_1, v_2, v_3)$$

If $\mathbf{v} = (v_1, v_2, v_3)$ and $\mathbf{w} = (w_1, w_2, w_3)$ are two vectors in 3-space, then arguments similar to those used for vectors in a plane can be used to establish the following results.

> **v** and **w** are equivalent if and only if $v_1 = w_1$, $v_2 = w_2$, and $v_3 = w_3$
> $\mathbf{v} + \mathbf{w} = (v_1 + w_1, v_2 + w_2, v_3 + w_3)$
> $k\mathbf{v} = (kv_1, kv_2, kv_3)$, where k is any scalar

Figure 12

Example 1   If $\mathbf{v} = (1, -3, 2)$ and $\mathbf{w} = (4, 2, 1)$, then

$$\mathbf{v} + \mathbf{w} = (5, -1, 3), \qquad 2\mathbf{v} = (2, -6, 4), \qquad -\mathbf{w} = (-4, -2, -1),$$
$$\mathbf{v} - \mathbf{w} = \mathbf{v} + (-\mathbf{w}) = (-3, -5, 1) \quad \blacktriangle$$

Sometimes a vector is positioned so that its initial point is not at the origin. If the vector $\overrightarrow{P_1 P_2}$ has initial point $P_1(x_1, y_1, z_1)$ and terminal point $P_2(x_2, y_2, z_2)$, then

$$\overrightarrow{P_1 P_2} = (x_2 - x_1, y_2 - y_1, z_2 - z_1)$$

That is, the components of $\overrightarrow{P_1 P_2}$ are obtained by subtracting the coordinates of the initial point from the coordinates of the terminal point. This may be seen using Figure 13: The vector $\overrightarrow{P_1 P_2}$ is the difference of vectors $\overrightarrow{OP_2}$ and $\overrightarrow{OP_1}$, so

$$\overrightarrow{P_1 P_2} = \overrightarrow{OP_2} - \overrightarrow{OP_1} = (x_2, y_2, z_2) - (x_1, y_1, z_1) = (x_2 - x_1, y_2 - y_1, z_2 - z_1)$$

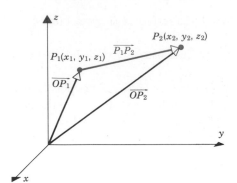

Figure 13

Example 2 The components of the vector $\mathbf{v} = \overrightarrow{P_1P_2}$ with initial point $P_1(2, -1, 4)$ and terminal point $P_2(7, 5, -8)$ are

$$\mathbf{v} = (7 - 2, 5 - (-1), (-8) - 4) = (5, 6, -12) \quad \blacktriangle$$

In 2-space the vector with initial point $P_1(x_1, y_1)$ and terminal point $P_2(x_2, y_2)$ is

$$\overrightarrow{P_1P_2} = (x_2 - x_1, y_2 - y_1)$$

TRANSLATION OF AXES

The solutions to many problems can be simplified by translating the coordinate axes to obtain new axes parallel to the original ones.

In Figure 14*a* we have translated the xy-coordinate axes to obtain an $x'y'$-coordinate system whose origin O' is at the point $(x, y) = (k, l)$. A point P in 2-space now has both (x, y) coordinates and (x', y') coordinates. To see how the two are related, consider the vector $\overrightarrow{O'P}$ (Figure 14*b*). In the xy-system its initial point is at (k, l) and its terminal point is at (x, y), so $\overrightarrow{O'P} = (x - k, y - l)$. In the $x'y'$-system its initial point is at $(0, 0)$ and its terminal point is at (x', y'), so $\overrightarrow{O'P} = (x', y')$. Therefore

$$x' = x - k \qquad y' = y - l$$

These formulas are called the *translation equations*.

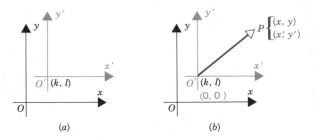

Figure 14   (*a*)   (*b*)

Example 3 Suppose that an xy-coordinate system is translated to obtain an $x'y'$-coordinate system whose origin has xy-coordinates $(k, l) = (4, 1)$.

(a) Find the $x'y'$-coordinates of the point with the xy-coordinates $P(2, 0)$.
(b) Find the xy-coordinates of the point with $x'y'$-coordinates $Q(-1, 5)$.

Solution (a). The translation equations are

$$x' = x - 4 \qquad y' = y - 1$$

so the $x'y'$-coordinates of $P(2, 0)$ are $x' = 2 - 4 = -2$ and $y' = 0 - 1 = -1$.

Solution (b). The translation equations in (a) can be rewritten as

$$x - x' + 4 \qquad y - y' + 1$$

so the xy-coordinates of Q are $x = -1 + 4 = 3$ and $y = 5 + 1 = 6$. ▲

In 3-space the translation equations are

$$x' = x - k \qquad y' = y - l \qquad z' = z - m$$

where (k, l, m) are the xyz-coordinates of the $x'y'z'$-origin.

EXERCISE SET 3.1

1. Draw a right-handed coordinate system and locate the points whose coordinates are
   (a) $(3, 4, 5)$ (b) $(-3, 4, 5)$ (c) $(3, -4, 5)$ (d) $(3, 4, -5)$
   (e) $(-3, -4, 5)$ (f) $(-3, 4, -5)$ (g) $(3, -4, -5)$ (h) $(-3, -4, -5)$
   (i) $(-3, 0, 0)$ (j) $(3, 0, 3)$ (k) $(0, 0, -3)$ (l) $(0, 3, 0)$

2. Sketch the following vectors with the initial points located at the origin:
   (a) $\mathbf{v}_1 = (3, 6)$ (b) $\mathbf{v}_2 = (-4, -8)$ (c) $\mathbf{v}_3 = (-4, -3)$ (d) $\mathbf{v}_4 = (5, -4)$ (e) $\mathbf{v}_5 = (3, 0)$
   (f) $\mathbf{v}_6 = (0, -7)$ (g) $\mathbf{v}_7 = (3, 4, 5)$ (h) $\mathbf{v}_8 = (3, 3, 0)$ (i) $\mathbf{v}_9 = (0, 0, -3)$

3. Find the components of the vector having initial point P_1 and terminal point P_2.
   (a) $P_1(4, 8)$, $P_2(3, 7)$ (b) $P_1(3, -5)$, $P_2(-4, -7)$ (c) $P_1(-5, 0)$, $P_2(-3, 1)$
   (d) $P_1(0, 0)$, $P_2(a, b)$ (e) $P_1(3, -7, 2)$, $P_2(-2, 5, -4)$ (f) $P_1(-1, 0, 2)$, $P_2(0, -1, 0)$
   (g) $P_1(a, b, c)$, $P_2(0, 0, 0)$ (h) $P_1(0, 0, 0)$, $P_2(a, b, c)$

4. Find a vector $\mathbf{u}$ with initial point $P(-1, 3, -5)$ such that
   (a) $\mathbf{u}$ has the same direction as $\mathbf{v} = (6, 7, -3)$ (b) $\mathbf{u}$ is oppositely directed to $\mathbf{v} = (6, 7, -3)$

5. Find a vector $\mathbf{u}$ with terminal point $Q(3, 0, -5)$ such that
   (a) $\mathbf{u}$ has the same direction as $\mathbf{v} = (4, -2, -1)$ (b) $\mathbf{u}$ is oppositely directed to $\mathbf{v} = (4, -2, -1)$

6. Let $\mathbf{u} = (-3, 1, 2)$, $\mathbf{v} = (4, 0, -8)$, and $\mathbf{w} = (6, -1, -4)$. Find the components of
   (a) $\mathbf{v} - \mathbf{w}$ (b) $6\mathbf{u} + 2\mathbf{v}$ (c) $-\mathbf{v} + \mathbf{u}$ (d) $5(\mathbf{v} - 4\mathbf{u})$ (e) $-3(\mathbf{v} - 8\mathbf{w})$ (f) $(2\mathbf{u} - 7\mathbf{w}) - (8\mathbf{v} + \mathbf{u})$

7. Let $\mathbf{u}$, $\mathbf{v}$, and $\mathbf{w}$ be the vectors in Exercise 6. Find the components of the vector $\mathbf{x}$ that satisfies $2\mathbf{u} - \mathbf{v} + \mathbf{x} = 7\mathbf{x} + \mathbf{w}$.

8. Let **u**, **v**, and **w** be the vectors in Exercise 6. Find scalars c_1, c_2, and c_3 such that

$$c_1\mathbf{u} + c_2\mathbf{v} + c_3\mathbf{w} = (2, 0, 4)$$

9. Show that there do not exist scalars c_1, c_2, and c_3 such that

$$c_1(-2, 9, 6) + c_2(-3, 2, 1) + c_3(1, 7, 5) = (0, 5, 4)$$

10. Find all scalars c_1, c_2, and c_3 such that

$$c_1(1, 2, 0) + c_2(2, 1, 1) + c_3(0, 3, 1) = (0, 0, 0)$$

11. Let P be the point $(2, 3, -2)$ and Q the point $(7, -4, 1)$.
    (a) Find the midpoint of the line segment connecting P and Q.
    (b) Find the point on the line segment connecting P and Q that is 3/4 of the way from P to Q.

12. Suppose an xy-coordinate system is translated to obtain an $x'y'$-coordinate system whose origin O' has xy-coordinates $(2, -3)$.
    (a) Find the $x'y'$-coordinates of the point P whose xy-coordinates are $(7, 5)$.
    (b) Find the xy-coordinates of the point Q whose $x'y'$-coordinates are $(-3, 6)$.
    (c) Draw the xy and $x'y'$-coordinate axes and locate the points P and Q.

13. Suppose an xyz-coordinate system is translated to obtain an $x'y'z'$-coordinate system. Let **v** be a vector whose components are $\mathbf{v} = (v_1, v_2, v_3)$ in the xyz-system. Show that **v** has the same components in the $x'y'z'$-system.

14. Find the components of **u**, **v**, **u** + **v**, and **u** − **v** for the vectors shown in Figure 15.

Figure 15

15. Prove geometrically that if $\mathbf{v} = (v_1, v_2)$, then $k\mathbf{v} = (kv_1, kv_2)$. (Restrict the proof to the case $k > 0$ illustrated in Figure 8. The complete proof would involve various cases that depend on the sign of k and the quadrant in which the vector falls.)

3.2 NORM OF A VECTOR; VECTOR ARITHMETIC

In this section we shall establish the basic rules of vector arithmetic.

PROPERTIES OF VECTOR OPERATIONS

The following theorem lists the most important properties of vectors in 2-space and 3-space.

> **Theorem 3.2.1.** *If* **u**, **v**, *and* **w** *are vectors in 2- or 3-space and k and l are scalars, then the following relationships hold.*
>
> (a) $\mathbf{u} + \mathbf{v} = \mathbf{v} + \mathbf{u}$ (e) $k(l\mathbf{u}) = (kl)\mathbf{u}$
> (b) $(\mathbf{u} + \mathbf{v}) + \mathbf{w} = \mathbf{u} + (\mathbf{v} + \mathbf{w})$ (f) $k(\mathbf{u} + \mathbf{v}) = k\mathbf{u} + k\mathbf{v}$
> (c) $\mathbf{u} + \mathbf{0} = \mathbf{0} + \mathbf{u} = \mathbf{u}$ (g) $(k + l)\mathbf{u} = k\mathbf{u} + l\mathbf{u}$
> (d) $\mathbf{u} + (-\mathbf{u}) = \mathbf{0}$ (h) $1\mathbf{u} = \mathbf{u}$

Before discussing the proof, we note that we have developed two approaches to vectors: *geometric*, in which vectors are represented by arrows or directed line segments, and *analytic*, in which vectors are represented by pairs or triples of numbers called components. As a consequence, the equations in Theorem 3.2.1 can be proved either geometrically or analytically. To illustrate, we shall prove part (*b*) both ways. The remaining proofs are left as exercises.

Proof of part (b) (analytic). We shall give the proof for vectors in 3-space; the proof for 2-space is similar. If $\mathbf{u} = (u_1, u_2, u_3)$, $\mathbf{v} = (v_1, v_2, v_3)$, and $\mathbf{w} = (w_1, w_2, w_3)$, then

$$
\begin{aligned}
(\mathbf{u} + \mathbf{v}) + \mathbf{w} &= [(u_1, u_2, u_3) + (v_1, v_2, v_3)] + (w_1, w_2, w_3) \\
&= (u_1 + v_1, u_2 + v_2, u_3 + v_3) + (w_1, w_2, w_3) \\
&= ([u_1 + v_1] + w_1, [u_2 + v_2] + w_2, [u_3 + v_3] + w_3) \\
&= (u_1 + [v_1 + w_1], u_2 + [v_2 + w_2], u_3 + [v_3 + w_3]) \\
&= (u_1, u_2, u_3) + (v_1 + w_1, v_2 + w_2, v_3 + w_3) \\
&= \mathbf{u} + (\mathbf{v} + \mathbf{w})
\end{aligned}
$$

Proof of part (b) (geometric). Let **u**, **v**, and **w** be represented by $\overrightarrow{PQ}$, $\overrightarrow{QR}$, and $\overrightarrow{RS}$ as shown in Figure 1. Then

$$\mathbf{v} + \mathbf{w} = \overrightarrow{QS} \qquad \text{and} \qquad \mathbf{u} + (\mathbf{v} + \mathbf{w}) = \overrightarrow{PS}$$

Also,

$$\mathbf{u} + \mathbf{v} = \overrightarrow{PR} \qquad \text{and} \qquad (\mathbf{u} + \mathbf{v}) + \mathbf{w} = \overrightarrow{PS}$$

Therefore,

$$\mathbf{u} + (\mathbf{v} + \mathbf{w}) = (\mathbf{u} + \mathbf{v}) + \mathbf{w} \quad \blacksquare$$

REMARK. In light of part (*b*) of this theorem, the symbol $\mathbf{u} + \mathbf{v} + \mathbf{w}$ is unambiguous since the same sum is obtained no matter where parentheses are inserted. Moreover, if the vectors **u**, **v**, and **w** are placed "tip to tail," then the sum $\mathbf{u} + \mathbf{v} + \mathbf{w}$ is the vector from the initial point of **u** to the terminal point of **w** (Figure 1).

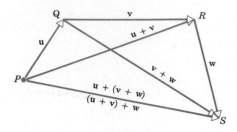

Figure 1 | The vectors $\mathbf{u} + (\mathbf{v} + \mathbf{w})$ and $(\mathbf{u} + \mathbf{v}) + \mathbf{w}$ are equal. |

NORM OF A VECTOR

The *length* of a vector $\mathbf{u}$ is often called the *norm* of $\mathbf{u}$ and is denoted by $\|\mathbf{u}\|$. It follows from the theorem of Pythagoras that the norm of a vector $\mathbf{u} = (u_1, u_2)$ in 2-space is

$$\|\mathbf{u}\| = \sqrt{u_1^2 + u_2^2} \tag{1}$$

(Figure 2*a*). Let $\|\mathbf{u}\| = (u_1, u_2, u_3)$ be a vector in 3-space. Using Figure 2*b* and two applications of the theorem of Pythagoras, we obtain

$$\begin{aligned}
\|\mathbf{u}\|^2 &= (OR)^2 + (RP)^2 \\
&= (OQ)^2 + (OS)^2 + (RP)^2 \\
&= u_1^2 + u_2^2 + u_3^2
\end{aligned}$$

Thus,

$$\|\mathbf{u}\| = \sqrt{u_1^2 + u_2^2 + u_3^2} \tag{2}$$

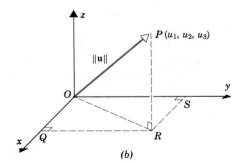

Figure 2   *(a)*   *(b)*

A vector of norm 1 is called a *unit vector*.

If $P_1(x_1, y_1, z_1)$ and $P_2(x_2, y_2, z_2)$ are two points in 3-space, then the *distance* d between them is the norm of the vector $\overrightarrow{P_1P_2}$ (Figure 3). Since

$$\overrightarrow{P_1P_2} = (x_2 - x_1, y_2 - y_1, z_2 - z_1)$$

it follows from (2) that

$$d = \sqrt{(x_2 - x_1)^2 + (y_2 - y_1)^2 + (z_2 - z_1)^2} \tag{3}$$

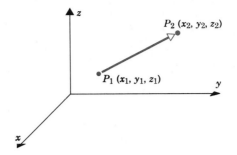

Figure 3   The distance between P_1 and P_2 is the norm of the vector $\overrightarrow{P_1P_2}$.

Similarly, if $P_1(x_1, y_1)$ and $P_2(x_2, y_2)$ are points in 2-space, then the distance between them is given by

$$d = \sqrt{(x_2 - x_1)^2 + (y_2 - y_1)^2} \tag{4}$$

Example 1   The norm of the vector $\mathbf{u} = (-3, 2, 1)$ is

$$\|\mathbf{u}\| = \sqrt{(-3)^2 + (2)^2 + (1)^2} = \sqrt{14}$$

The distance d between the points $P_1(2, -1, -5)$ and $P_2(4, -3, 1)$ is

$$d = \sqrt{(4 - 2)^2 + (-3 + 1)^2 + (1 + 5)^2} = \sqrt{44} = 2\sqrt{11} \quad \triangle$$

From the definition of the product $k\mathbf{u}$, the length of the vector $k\mathbf{u}$ is $|k|$ times the length of $\mathbf{u}$. Expressed as an equation, this statement says that

$$\|k\mathbf{u}\| = |k|\,\|\mathbf{u}\| \tag{5}$$

This useful formula is applicable in both 2-space and 3-space.

EXERCISE SET 3.2

1. Find the norm of **v**.
   (a) $\mathbf{v} = (4, -3)$ (b) $\mathbf{v} = (2, 3)$ (c) $\mathbf{v} = (-5, 0)$
   (d) $\mathbf{v} = (2, 2, 2)$ (e) $\mathbf{v} = (-7, 2, -1)$ (f) $\mathbf{v} = (0, 6, 0)$

2. Find the distance between P_1 and P_2.
   (a) $P_1(3, 4)$, $P_2(5, 7)$ (b) $P_1(-3, 6)$, $P_2(-1, -4)$
   (c) $P_1(7, -5, 1)$, $P_2(-7, -2, -1)$ (d) $P_1(3, 3, 3)$, $P_2(6, 0, 3)$

3. Let $\mathbf{u} = (2, -2, 3)$, $\mathbf{v} = (1, -3, 4)$, $\mathbf{w} = (3, 6, -4)$. In each part evaluate the expression.
   (a) $\|\mathbf{u} + \mathbf{v}\|$ (b) $\|\mathbf{u}\| + \|\mathbf{v}\|$ (c) $\|-2\mathbf{u}\| + 2\|\mathbf{u}\|$

   (d) $\|3\mathbf{u} - 5\mathbf{v} + \mathbf{w}\|$ (e) $\dfrac{1}{\|\mathbf{w}\|}\,\mathbf{w}$ (f) $\left\|\dfrac{1}{\|\mathbf{w}\|}\,\mathbf{w}\right\|$

4. Let $\mathbf{v} = (-1, 2, 5)$. Find all scalars k such that $\|k\mathbf{v}\| = 4$.

5. Let $\mathbf{u} = (7, -3, 1)$, $\mathbf{v} = (9, 6, 6)$, $\mathbf{w} = (2, 1, -8)$, $k = -2$, and $l = 5$. Verify that these vectors and scalars satisfy the stated equalities from Theorem 3.2.1.
   (a) part (*b*)   (b) part (*e*)   (c) part (*f*)   (d) part (*g*)

6. (a) Show that if $\mathbf{v}$ is any nonzero vector, then

$$\frac{1}{\|\mathbf{v}\|}\,\mathbf{v}$$

   is a unit vector.
   (b) Use the result in part (a) to find a unit vector that has the same direction as the vector $\mathbf{v} = (3, 4)$.
   (c) Use the result in part (a) to find a unit vector that is oppositely directed to the vector $\mathbf{v} = (-2, 3, -6)$.

7. (a) Show that the components of the vector $\mathbf{v} = (v_1, v_2)$ in Figure 4 are $v_1 = \|\mathbf{v}\| \cos \theta$ and $v_2 = \|\mathbf{v}\| \sin \theta$.
   (b) Let $\mathbf{u}$ and $\mathbf{v}$ be the vectors in Figure 5. Use the result in part (a) to find the components of $4\mathbf{u} - 5\mathbf{v}$.

(a)

Figure 4

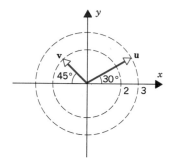

(b)

Figure 5

8. Let $\mathbf{p}_0 = (x_0, y_0, z_0)$ and $\mathbf{p} = (x, y, z)$. Describe the set of all points (x, y, z) for which $\|\mathbf{p} - \mathbf{p}_0\| = 1$.

9. Prove geometrically that if $\mathbf{u}$ and $\mathbf{v}$ are vectors in 2- or 3-space, then $\|\mathbf{u} + \mathbf{v}\| \le \|\mathbf{u}\| + \|\mathbf{v}\|$.

10. Prove parts (*a*), (*c*), and (*e*) of Theorem 3.2.1 analytically.

11. Prove parts (*d*), (*g*), and (*h*) of Theorem 3.2.1 analytically.

12. Prove part (*f*) of Theorem 3.2.1 geometrically.

3.3 DOT PRODUCT; PROJECTIONS

In this section we shall discuss a way of multiplying vectors in 2-space or 3-space, and we shall give some applications of this multiplication to geometry.

DOT PRODUCT OF VECTORS

Let $\mathbf{u}$ and $\mathbf{v}$ be two nonzero vectors in 2-space or 3-space, and assume these vectors have been positioned so their initial points coincide. By the *angle between* $\mathbf{u}$ *and* $\mathbf{v}$, we shall mean the angle θ determined by $\mathbf{u}$ and $\mathbf{v}$ that satisfies $0 \le \theta \le \pi$ (Figure 1).

Figure 1

The angle θ between **u** and **v** satisfies $0° \leq \theta \leq \pi$.

Definition. If **u** and **v** are vectors in 2-space or 3-space and θ is the angle between **u** and **v**, then the ***dot product*** or ***Euclidean inner product*** **u · v** is defined by

$$\mathbf{u} \cdot \mathbf{v} = \begin{cases} \|\mathbf{u}\|\,\|\mathbf{v}\|\cos\theta & \textit{if } \mathbf{u} \neq \mathbf{0} \textit{ and } \mathbf{v} \neq \mathbf{0} \\ 0 & \textit{if } \mathbf{u} = \mathbf{0} \textit{ or } \mathbf{v} = \mathbf{0} \end{cases} \tag{1}$$

Example 1   As shown in Figure 2, the angle between the vectors $\mathbf{u} = (0, 0, 1)$ and $\mathbf{v} = (0, 2, 2)$ is 45°. Thus,

$$\mathbf{u} \cdot \mathbf{v} = \|\mathbf{u}\|\,\|\mathbf{v}\|\cos\theta = \left(\sqrt{0^2 + 0^2 + 1^2}\right)\left(\sqrt{0^2 + 2^2 + 2^2}\right)\left(\frac{1}{\sqrt{2}}\right) = 2 \quad \blacktriangle$$

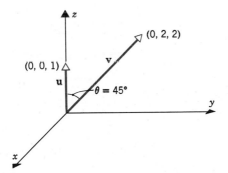

Figure 2

COMPONENT FORMULA FOR THE DOT PRODUCT

For purposes of computation, it is desirable to have a formula that expresses the dot product of two vectors in terms of the components of the vectors. We will derive such a formula for vectors in 3-space; the derivation for vectors in 2-space is similar.

Let $\mathbf{u} = (u_1, u_2, u_3)$ and $\mathbf{v} = (v_1, v_2, v_3)$ be two nonzero vectors. If, as in Figure 3, θ is the angle between **u** and **v**, then the law of cosines yields

$$\|\overrightarrow{PQ}\|^2 = \|\mathbf{u}\|^2 + \|\mathbf{v}\|^2 - 2\|\mathbf{u}\|\,\|\mathbf{v}\|\cos\theta \tag{2}$$

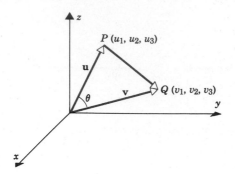

Figure 3

Since $\overrightarrow{PQ} = \mathbf{v} - \mathbf{u}$, we can rewrite (2) as

$$\|\mathbf{u}\| \, \|\mathbf{v}\| \cos \theta = \tfrac{1}{2}(\|\mathbf{u}\|^2 + \|\mathbf{v}\|^2 - \|\mathbf{v} - \mathbf{u}\|^2)$$

or

$$\mathbf{u} \cdot \mathbf{v} = \tfrac{1}{2}(\|\mathbf{u}\|^2 + \|\mathbf{v}\|^2 - \|\mathbf{v} - \mathbf{u}\|^2)$$

Substituting

$$\|\mathbf{u}\|^2 = u_1^2 + u_2^2 + u_3^2 \qquad \|\mathbf{v}\|^2 = v_1^2 + v_2^2 + v_3^2$$

and

$$\|\mathbf{v} - \mathbf{u}\|^2 = (v_1 - u_1)^2 + (v_2 - u_2)^2 + (v_3 - u_3)^2$$

we obtain after simplifying

$$\mathbf{u} \cdot \mathbf{v} = u_1 v_1 + u_2 v_2 + u_3 v_3 \tag{3}$$

If $\mathbf{u} = (u_1, u_2)$ and $\mathbf{v} = (v_1, v_2)$ are two vectors in 2-space, then the coresponding formula is

$$\mathbf{u} \cdot \mathbf{v} = u_1 v_1 + u_2 v_2 \tag{4}$$

FINDING THE ANGLE BETWEEN VECTORS

If $\mathbf{u}$ and $\mathbf{v}$ are nonzero vectors, then Formula (1) can be written as

$$\cos \theta = \frac{\mathbf{u} \cdot \mathbf{v}}{\|\mathbf{u}\| \, \|\mathbf{v}\|} \tag{5}$$

Example 2 Consider the vectors

$$\mathbf{u} = (2, -1, 1) \qquad \text{and} \qquad \mathbf{v} = (1, 1, 2)$$

Find $\mathbf{u} \cdot \mathbf{v}$ and determine the angle θ between $\mathbf{u}$ and $\mathbf{v}$.

Solution

$$\mathbf{u} \cdot \mathbf{v} = u_1v_1 + u_2v_2 + u_3v_3 = (2)(1) + (-1)(1) + (1)(2) = 3$$

For the given vectors we have $\|\mathbf{u}\| = \|\mathbf{v}\| = \sqrt{6}$, so that from (5)

$$\cos\theta = \frac{\mathbf{u} \cdot \mathbf{v}}{\|\mathbf{u}\|\,\|\mathbf{v}\|} = \frac{3}{\sqrt{6}\sqrt{6}} = \frac{1}{2}$$

Thus, $\theta = 60°$.

Example 3   Find the angle between a diagonal of a cube and one of its edges.

Solution. Let k be the length of an edge and introduce a coordinate system as shown in Figure 4.

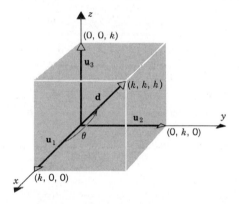

If we let $\mathbf{u}_1 = (k, 0, 0)$, $\mathbf{u}_2 = (0, k, 0)$, and $\mathbf{u}_3 = (0, 0, k)$, then the vector

$$\mathbf{d} = (k, k, k) = \mathbf{u}_1 + \mathbf{u}_2 + \mathbf{u}_3$$

is a diagonal of the cube. The angle θ between $\mathbf{d}$ and the edge $\mathbf{u}_1$ satisfies

$$\cos\theta = \frac{\mathbf{u}_1 \cdot \mathbf{d}}{\|\mathbf{u}_1\|\,\|\mathbf{d}\|} = \frac{k^2}{(k)(\sqrt{3k^2})} = \frac{1}{\sqrt{3}}$$

Thus,

$$\theta = \cos^{-1}\left(\frac{1}{\sqrt{3}}\right) \approx 54°44'$$

The following theorem shows how the dot product can be used to obtain information about the angle between two vectors; it also establishes an important relationship between the norm and the dot product.

Theorem 3.3.1. *Let* **u** *and* **v** *be vectors in 2- or 3-space.*

(a) $\mathbf{v} \cdot \mathbf{v} = \|\mathbf{v}\|^2$; *that is,* $\|\mathbf{v}\| = (\mathbf{v} \cdot \mathbf{v})^{1/2}$

(b) *If the vectors* **u** *and* **v** *are nonzero and* θ *is the angle between them, then*

θ *is acute*	*if and only if*	$\mathbf{u} \cdot \mathbf{v} > 0$
θ *is obtuse*	*if and only if*	$\mathbf{u} \cdot \mathbf{v} < 0$
$\theta = \pi/2$	*if and only if*	$\mathbf{u} \cdot \mathbf{v} = 0$

Proof (a). Since the angle θ between **v** and **v** is 0, we have

$$\mathbf{v} \cdot \mathbf{v} = \|\mathbf{v}\|\, \|\mathbf{v}\| \cos \theta = \|\mathbf{v}\|^2 \cos 0 = \|\mathbf{v}\|^2$$

Proof (b). Since θ satisfies $0 \le \theta \le \pi$, it follows that: θ is acute if and only if $\cos \theta > 0$; θ is obtuse if and only if $\cos \theta < 0$; and $\theta = \pi/2$ if and only if $\cos \theta = 0$. But $\cos \theta$ has the same sign as $\mathbf{u} \cdot \mathbf{v}$ since $\mathbf{u} \cdot \mathbf{v} = \|\mathbf{u}\|\, \|\mathbf{v}\| \cos \theta$, $\|\mathbf{u}\| > 0$, and $\|\mathbf{v}\| > 0$. Thus, the result follows. ▮

Example 4 If $\mathbf{u} = (1, -2, 3)$, $\mathbf{v} = (-3, 4, 2)$, and $\mathbf{w} = (3, 6, 3)$, then

$$\mathbf{u} \cdot \mathbf{v} = (1)(-3) + (-2)(4) + (3)(2) = -5$$
$$\mathbf{v} \cdot \mathbf{w} = (-3)(3) + (4)(6) + (2)(3) = 21$$
$$\mathbf{u} \cdot \mathbf{w} = (1)(3) + (-2)(6) + (3)(3) = 0$$

Therefore, **u** and **v** make an obtuse angle, **v** and **w** make an acute angle, and **u** and **w** are perpendicular. ▲

ORTHOGONAL VECTORS

Perpendicular vectors are also called *orthogonal* vectors. In light of Theorem 3.3.1*b*, two *nonzero* vectors are orthogonal if and only if their dot product is zero. If we agree to consider **u** and **v** to be perpendicular when either or both of these vectors is **0**, then we can state without exception that *two vectors* **u** *and* **v** *are orthogonal (perpendicular) if and only if* $\mathbf{u} \cdot \mathbf{v} = 0$. To indicate that **u** and **v** are orthogonal vectors we write $\mathbf{u} \perp \mathbf{v}$.

Example 5 Show that in 2-space the nonzero vector $\mathbf{n} = (a, b)$ is perpendicular to the line $ax + by + c = 0$.

Solution. Let $P_1(x_1, y_1)$ and $P_2(x_2, y_2)$ be distinct points on the line, so that

$$\begin{aligned} ax_1 + by_1 + c &= 0 \\ ax_2 + by_2 + c &= 0 \end{aligned} \tag{6}$$

Since the vector $\overrightarrow{P_1 P_2} = (x_2 - x_1, y_2 - y_1)$ runs along the line (Figure 5), we need only show that **n** and $\overrightarrow{P_1 P_2}$ are perpendicular. But on subtracting the equations in (6) we obtain

$$a(x_2 - x_1) + b(y_2 - y_1) = 0$$

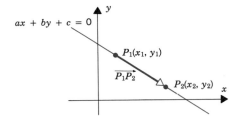

Figure 5

which can be expressed in the form

$$(a, b) \cdot (x_2 - x_1, y_2 - y_1) = 0$$

or

$$\mathbf{n} \cdot \overrightarrow{P_1 P_2} = 0$$

Thus, $\mathbf{n}$ and $\overrightarrow{P_1 P_2}$ are perpendicular. ▲

The following theorem lists the most important properties of the dot product. They are useful in calculations involving vectors.

Theorem 3.3.2. *If* $\mathbf{u}$, $\mathbf{v}$, *and* $\mathbf{w}$ *are vectors in 2- or 3-space and k is a scalar, then*

(a) $\mathbf{u} \cdot \mathbf{v} = \mathbf{v} \cdot \mathbf{u}$
(b) $\mathbf{u} \cdot (\mathbf{v} + \mathbf{w}) = \mathbf{u} \cdot \mathbf{v} + \mathbf{u} \cdot \mathbf{w}$
(c) $k(\mathbf{u} \cdot \mathbf{v}) = (k\mathbf{u}) \cdot \mathbf{v} = \mathbf{u} \cdot (k\mathbf{v})$
(d) $\mathbf{v} \cdot \mathbf{v} > 0$ *if* $\mathbf{v} \neq \mathbf{0}$ *and* $\mathbf{v} \cdot \mathbf{v} = 0$ *if* $\mathbf{v} = \mathbf{0}$

Proof. We shall prove (c) for vectors in 3-space and leave the remaining proofs as exercises. Let $\mathbf{u} = (u_1, u_2, u_3)$ and $\mathbf{v} = (v_1, v_2, v_3)$; then

$$k(\mathbf{u} \cdot \mathbf{v}) = k(u_1 v_1 + u_2 v_2 + u_3 v_3)$$
$$= (k u_1)v_1 + (k u_2)v_2 + (k u_3)v_3$$
$$= (k\mathbf{u}) \cdot \mathbf{v}$$

Similarly,

$$k(\mathbf{u} \cdot \mathbf{v}) = \mathbf{u} \cdot (k\mathbf{v})$$ ▌

ORTHOGONAL PROJECTIONS

In many applications it is of interest to "decompose" a vector $\mathbf{u}$ into a sum of two terms, one parallel to a specified nonzero vector $\mathbf{a}$ and the other perpendicular to $\mathbf{a}$. If $\mathbf{u}$ and $\mathbf{a}$ are positioned so their initial points coincide at a point Q, we can decompose the vector $\mathbf{u}$ as follows (Figure 6): Drop a perpendicular from the tip of $\mathbf{u}$ to the line through $\mathbf{a}$, and construct the vector $\mathbf{w}_1$ from Q to the foot of this perpendicular. Next form the difference

$$\mathbf{w}_2 = \mathbf{u} - \mathbf{w}_1$$

Figure 6   The vector **u** is the sum of $\mathbf{w}_1$ and $\mathbf{w}_2$, where $\mathbf{w}_1$ is parallel to **a** and $\mathbf{w}_2$ is perpendicular to **a**.

As indicated in Figure 6, the vector $\mathbf{w}_1$ is parallel to **a**, the vector $\mathbf{w}_2$ is perpendicular to **a**, and

$$\mathbf{w}_1 + \mathbf{w}_2 = \mathbf{w}_1 + (\mathbf{u} - \mathbf{w}_1) = \mathbf{u}$$

The vector $\mathbf{w}_1$ is called the ***orthogonal projection of*** **u** ***on*** **a** or sometimes the ***vector component of*** **u** ***along*** **a**. It is denoted by

$$\text{proj}_\mathbf{a}\, \mathbf{u} \qquad\qquad (7)$$

The vector $\mathbf{w}_2$ is called the ***vector component of*** **u** ***orthogonal to*** **a**. Since $\mathbf{w}_2 = \mathbf{u} - \mathbf{w}_1$ this vector can be written in notation (7) as

$$\mathbf{w}_2 = \mathbf{u} - \text{proj}_\mathbf{a}\, \mathbf{u}$$

The following theorem gives formulas for calculating the vectors $\text{proj}_\mathbf{a}\, \mathbf{u}$ and $\mathbf{u} - \text{proj}_\mathbf{a}\, \mathbf{u}$.

Theorem 3.3.3. *If* **u** *and* **a** *are vectors in 2-space or 3-space and if* $\mathbf{a} \neq \mathbf{0}$, *then*

$$\text{proj}_\mathbf{a}\, \mathbf{u} = \frac{\mathbf{u} \cdot \mathbf{a}}{\|\mathbf{a}\|^2}\, \mathbf{a} \qquad \textit{(vector component of } \mathbf{u} \textit{ along } \mathbf{a})$$

$$\mathbf{u} - \text{proj}_\mathbf{a}\, \mathbf{u} = \mathbf{u} - \frac{\mathbf{u} \cdot \mathbf{a}}{\|\mathbf{a}\|^2}\, \mathbf{a} \qquad \textit{(vector component of } \mathbf{u} \textit{ orthogonal to } \mathbf{a})$$

Proof. Let $\mathbf{w}_1 = \text{proj}_\mathbf{a}\, \mathbf{u}$ and $\mathbf{w}_2 = \mathbf{u} - \text{proj}_\mathbf{a}\, \mathbf{u}$. Since $\mathbf{w}_1$ is parallel to **a**, it must be a scalar multiple of **a**, so it can be written in the form $\mathbf{w}_1 = k\mathbf{a}$. Thus

$$\mathbf{u} = \mathbf{w}_1 + \mathbf{w}_2 = k\mathbf{a} + \mathbf{w}_2 \qquad\qquad (8)$$

Taking the dot product of both sides of (8) with **a** and using Theorems 3.3.1*a* and 3.3.2 yields

$$\mathbf{u} \cdot \mathbf{a} = (k\mathbf{a} + \mathbf{w}_2) \cdot \mathbf{a} = k\|\mathbf{a}\|^2 + \mathbf{w}_2 \cdot \mathbf{a} \qquad\qquad (9)$$

But $\mathbf{w}_2 \cdot \mathbf{a} = 0$ since $\mathbf{w}_2$ is perpendicular to $\mathbf{a}$; so (9) yields

$$k = \frac{\mathbf{u} \cdot \mathbf{a}}{\|\mathbf{a}\|^2}$$

Since $\text{proj}_\mathbf{a}\, \mathbf{u} = \mathbf{w}_1 = k\mathbf{a}$, we obtain

$$\text{proj}_\mathbf{a}\, \mathbf{u} = \frac{\mathbf{u} \cdot \mathbf{a}}{\|\mathbf{a}\|^2}\, \mathbf{a} \quad \blacksquare$$

Example 6   Let $\mathbf{u} = (2, -1, 3)$ and $\mathbf{a} = (4, -1, 2)$. Find the vector component of $\mathbf{u}$ along $\mathbf{a}$ and the vector component of $\mathbf{u}$ orthogonal to $\mathbf{a}$.

Solution.

$$\mathbf{u} \cdot \mathbf{a} = (2)(4) + (-1)(-1) + (3)(2) = 15$$
$$\|\mathbf{a}\|^2 = 4^2 + (-1)^2 + 2^2 = 21$$

Thus, the vector component of $\mathbf{u}$ along $\mathbf{a}$ is

$$\text{proj}_\mathbf{a}\, \mathbf{u} = \frac{\mathbf{u} \cdot \mathbf{a}}{\|\mathbf{a}\|^2}\, \mathbf{a} = \frac{15}{21}(4, -1, 2) = \left(\frac{20}{7}, -\frac{5}{7}, \frac{10}{7}\right)$$

and the vector component of $\mathbf{u}$ orthogonal to $\mathbf{a}$ is

$$\mathbf{u} - \text{proj}_\mathbf{a}\, \mathbf{u} = (2, -1, 3) - \left(\frac{20}{7}, -\frac{5}{7}, \frac{10}{7}\right) = \left(-\frac{6}{7}, -\frac{2}{7}, \frac{11}{7}\right)$$

As a check, the reader may wish to verify that the vectors $\mathbf{u} - \text{proj}_\mathbf{a}\, \mathbf{u}$ and $\mathbf{a}$ are perpendicular by showing that their dot product is zero. ▲

A formula for the length of the vector component of $\mathbf{u}$ along $\mathbf{a}$ can be obtained by writing

$$\|\text{proj}_\mathbf{a}\, \mathbf{u}\| = \left\| \frac{\mathbf{u} \cdot \mathbf{a}}{\|\mathbf{a}\|^2}\, \mathbf{a} \right\|$$

$$= \left| \frac{\mathbf{u} \cdot \mathbf{a}}{\|\mathbf{a}\|^2} \right| \|\mathbf{a}\| \qquad \boxed{\begin{array}{l}\text{Formula (5) of}\\ \text{Section 3.2.}\end{array}}$$

$$= \frac{|\mathbf{u} \cdot \mathbf{a}|}{\|\mathbf{a}\|^2} \|\mathbf{a}\| \qquad \boxed{\text{Since } \|\mathbf{a}\|^2 > 0.}$$

which yields

$$\|\text{proj}_\mathbf{a}\, \mathbf{u}\| = \frac{|\mathbf{u} \cdot \mathbf{a}|}{\|\mathbf{a}\|} \qquad (10)$$

If θ denotes the angle between $\mathbf{u}$ and $\mathbf{a}$, then $\mathbf{u} \cdot \mathbf{a} = \|\mathbf{u}\|\, \|\mathbf{a}\| \cos \theta$, so that (10) can also be written as

$$\|\text{proj}_\mathbf{a}\, \mathbf{u}\| = \|\mathbf{u}\|\, |\cos \theta| \qquad (11)$$

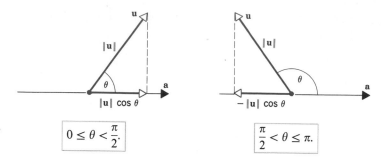

Figure 7

$$0 \leq \theta < \frac{\pi}{2}.$$

$$\frac{\pi}{2} < \theta \leq \pi.$$

(Verify.) A geometric interpretation of this result is given in Figure 7.

As an example, we will use vector methods to derive a formula for the distance from a point in the plane to a line.

Example 7 Find a formula for the distance D between the point $P_0(x_0, y_0)$ and the line $ax + by + c = 0$.

Solution. Let $Q(x_1, y_1)$ be any point on the line and position the vector

$$\mathbf{n} = (a, b)$$

so that its initial point is at Q.

By virtue of Example 5, the vector $\mathbf{n}$ is perpendicular to the line (Figure 8). As indicated in the figure, the distance D is equal to the length of the orthogonal projection of $\overrightarrow{QP_0}$ on $\mathbf{n}$; thus, from (10),

$$D = \|\text{proj}_{\mathbf{n}} \overrightarrow{QP_0}\| = \frac{|\overrightarrow{QP_0} \cdot \mathbf{n}|}{\|\mathbf{n}\|}$$

But

$$\overrightarrow{QP_0} = (x_0 - x_1, y_0 - y_1)$$

$$\overrightarrow{QP_0} \cdot \mathbf{n} = a(x_0 - x_1) + b(y_0 - y_1)$$

$$\|\mathbf{n}\| = \sqrt{a^2 + b^2}$$

so that

$$D = \frac{|a(x_0 - x_1) + b(y_0 - y_1)|}{\sqrt{a^2 + b^2}} \tag{12}$$

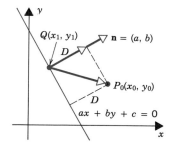

Figure 8

Since the point $Q(x_1, y_1)$ lies on the line, its coordinates satisfy the equation of the line, so

$$ax_1 + by_1 + c = 0$$

or

$$c = -ax_1 - by_1$$

Substituting this expression in (12) yields the formula

$$D = \frac{|ax_0 + by_0 + c|}{\sqrt{a^2 + b^2}} \quad \blacktriangle \tag{13}$$

Example 8 It follows from Formula (13) that the distance D from the point $(1, -2)$ to the line $3x + 4y - 6 = 0$ is

$$D = \frac{|(3)(1) + 4(-2) - 6|}{\sqrt{3^2 + 4^2}} = \frac{|-11|}{\sqrt{25}} = \frac{11}{5} \quad \blacktriangle$$

EXERCISE SET 3.3

1. Find $\mathbf{u} \cdot \mathbf{v}$.
   (a) $\mathbf{u} = (2, 3)$, $\mathbf{v} = (5, -7)$
   (b) $\mathbf{u} = (-6, -2)$, $\mathbf{v} = (4, 0)$
   (c) $\mathbf{u} = (1, -5, 4)$, $\mathbf{v} = (3, 3, 3)$
   (d) $\mathbf{u} = (-2, 2, 3)$, $\mathbf{v} = (1, 7, -4)$

2. In each part of Exercise 1, find the cosine of the angle θ between $\mathbf{u}$ and $\mathbf{v}$.

3. Determine whether $\mathbf{u}$ and $\mathbf{v}$ make an acute angle, make an obtuse angle, or are orthogonal.
   (a) $\mathbf{u} = (6, 1, 4)$, $\mathbf{v} = (2, 0, -3)$
   (b) $\mathbf{u} = (0, 0, -1)$, $\mathbf{v} = (1, 1, 1)$
   (c) $\mathbf{u} = (-6, 0, 4)$, $\mathbf{v} = (3, 1, 6)$
   (d) $\mathbf{u} = (2, 4, -8)$, $\mathbf{v} = (5, 3, 7)$

4. Find the orthogonal projection of $\mathbf{u}$ on $\mathbf{a}$.
   (a) $\mathbf{u} = (6, 2)$, $\mathbf{a} = (3, -9)$
   (b) $\mathbf{u} = (-1, -2)$, $\mathbf{a} = (-2, 3)$
   (c) $\mathbf{u} = (3, 1, -7)$, $\mathbf{a} = (1, 0, 5)$
   (d) $\mathbf{u} = (1, 0, 0)$, $\mathbf{a} = (4, 3, 8)$

5. In each part of Exercise 4, find the vector component of $\mathbf{u}$ orthogonal to $\mathbf{a}$.

6. In each part find $\|\text{proj}_\mathbf{a}\, \mathbf{u}\|$.
   (a) $\mathbf{u} = (1, -2)$, $\mathbf{a} = (-4, -3)$
   (b) $\mathbf{u} = (5, 6)$, $\mathbf{a} = (2, -1)$
   (c) $\mathbf{u} = (3, 0, 4)$ $\mathbf{a} = (2, 3, 3)$
   (d) $\mathbf{u} = (3, -2, 6)$, $\mathbf{a} = (1, 2, -7)$

7. Let $\mathbf{u} = (5, -2, 1)$, $\mathbf{v} = (1, 6, 3)$, and $k = -4$. Verify Theorem 3.3.2 for these quantities.

8. (a) Show that $\mathbf{v} = (a, b)$ and $\mathbf{w} = (-b, a)$ are orthogonal vectors.
   (b) Use the result in part (a) to find two vectors that are orthogonal to $\mathbf{v} = (2, -3)$.
   (c) Find two unit vectors that are orthogonal to $(-3, 4)$.

9. Let $\mathbf{u} = (3, 4)$, $\mathbf{v} = (5, -1)$, and $\mathbf{w} = (7, 1)$. Evaluate the expression.
   (a) $\mathbf{u} \cdot (7\mathbf{v} + \mathbf{w})$
   (b) $\|(\mathbf{u} \cdot \mathbf{w})\mathbf{w}\|$
   (c) $\|\mathbf{u}\|(\mathbf{v} \cdot \mathbf{w})$
   (d) $(\|\mathbf{u}\|\mathbf{v}) \cdot \mathbf{w}$

10. Explain why each of the following expressions makes no sense.
    (a) $\mathbf{u} \cdot (\mathbf{v} \cdot \mathbf{w})$
    (b) $(\mathbf{u} \cdot \mathbf{v}) + \mathbf{w}$
    (c) $\|\mathbf{u} \cdot \mathbf{v}\|$
    (d) $k \cdot (\mathbf{u} + \mathbf{v})$

11. Use vectors to find the cosines of the interior angles of the triangle with vertices $(0, -1)$, $(1, -2)$, and $(4, 1)$.

12. Show that $A(3, 0, 2)$, $B(4, 3, 0)$, and $C(8, 1, -1)$ are vertices of a right triangle. At which vertex is the right angle?

13. Suppose $\mathbf{a} \cdot \mathbf{b} = \mathbf{a} \cdot \mathbf{c}$ and $\mathbf{a} \neq \mathbf{0}$. Does it follow that $\mathbf{b} = \mathbf{c}$? Explain.

14. Let $\mathbf{p} = (2, k)$ and $\mathbf{q} = (3, 5)$. Find k such that
    (a) $\mathbf{p}$ and $\mathbf{q}$ are parallel
    (b) $\mathbf{p}$ and $\mathbf{q}$ are orthogonal
    (c) the angle between $\mathbf{p}$ and $\mathbf{q}$ is $\pi/3$
    (d) the angle between $\mathbf{p}$ and $\mathbf{q}$ is $\pi/4$.

15. Use Formula (13) to calculate the distance between the point and the line.
    (a) $4x + 3y + 4 = 0$; $(-3, 1)$
    (b) $y = -4x + 2$; $(2, -5)$
    (c) $3x + y = 5$; $(1, 8)$

16. Establish the identity $\|\mathbf{u} + \mathbf{v}\|^2 + \|\mathbf{u} - \mathbf{v}\|^2 = 2\|\mathbf{u}\|^2 + 2\|\mathbf{v}\|^2$.

17. Establish the identity $\mathbf{u} \cdot \mathbf{v} = \frac{1}{4}\|\mathbf{u} + \mathbf{v}\|^2 - \frac{1}{4}\|\mathbf{u} - \mathbf{v}\|^2$.

18. Find the angle between a diagonal of a cube and one of its faces.

19. Let $\mathbf{i}$, $\mathbf{j}$, and $\mathbf{k}$ be unit vectors along the positive x, y, and z axes of a rectangular coordinate system in 3-space. If $\mathbf{v} = (a, b, c)$ is a vector, then the angles α, β, and γ between $\mathbf{v}$ and the vectors $\mathbf{i}$, $\mathbf{j}$, and $\mathbf{k}$, respectively, are called the *direction angles* of $\mathbf{v}$ (Figure 9), and the numbers $\cos \alpha$, $\cos \beta$, and $\cos \gamma$ are called the *direction cosines* of $\mathbf{v}$.
    (a) Show that $\cos \alpha = a/\|\mathbf{v}\|$.
    (b) Find $\cos \beta$ and $\cos \gamma$.
    (c) Show that $\mathbf{v}/\|\mathbf{v}\| = (\cos \alpha, \cos \beta, \cos \gamma)$.
    (d) Show that $\cos^2 \alpha + \cos^2 \beta + \cos^2 \gamma = 1$.

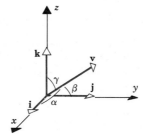

Figure 9

20. Use the result in Exercise 19 to estimate, to the nearest degree, the angles that a diagonal of a box with dimensions 10 cm × 15 cm × 25 cm makes with edges of the box. [*Note:* A calculator or trigonometric table is needed.]

21. Referring to Exercise 19, show that $\mathbf{v}_1$ and $\mathbf{v}_2$ are perpendicular vectors in 3-space if and only if their direction cosines satisfy

$$\cos \alpha_1 \cos \alpha_2 + \cos \beta_1 \cos \beta_2 + \cos \gamma_1 \cos \gamma_2 = 0$$

22. Show that if $\mathbf{v}$ is orthogonal to both $\mathbf{w}_1$ and $\mathbf{w}_2$, then $\mathbf{v}$ is orthogonal to $k_1\mathbf{w}_1 + k_2\mathbf{w}_2$ for all scalars k_1 and k_2.

23. Let $\mathbf{u}$ and $\mathbf{v}$ be nonzero vectors in 2- or 3-space, and let $k = \|\mathbf{u}\|$ and $l = \|\mathbf{v}\|$. Show that the vector $\mathbf{w} = l\mathbf{u} + k\mathbf{v}$ bisects the angle between $\mathbf{u}$ and $\mathbf{v}$.

3.4   CROSS PRODUCT

In many applications of vectors to problems in geometry, physics, and engineering, it is of interest to construct a vector in 3-space that is perpendicular to two given vectors. In this section we shall introduce a type of vector multiplication that produces such a vector.

CROSS PRODUCT OF VECTORS

Definition. If $\mathbf{u} = (u_1, u_2, u_3)$ and $\mathbf{v} = (v_1, v_2, v_3)$ are vectors in 3-space, then the *cross product* $\mathbf{u} \times \mathbf{v}$ is the vector defined by

$$\mathbf{u} \times \mathbf{v} = (u_2 v_3 - u_3 v_2, u_3 v_1 - u_1 v_3, u_1 v_2 - u_2 v_1)$$

or in determinant notation

$$\mathbf{u} \times \mathbf{v} = \left(\begin{vmatrix} u_2 & u_3 \\ v_2 & v_3 \end{vmatrix}, -\begin{vmatrix} u_1 & u_3 \\ v_1 & v_3 \end{vmatrix}, \begin{vmatrix} u_1 & u_2 \\ v_1 & v_2 \end{vmatrix} \right) \tag{1}$$

REMARK. Instead of memorizing (1), you can obtain the components of $\mathbf{u} \times \mathbf{v}$ as follows:

• Form the 2 × 3 matrix

$$\begin{bmatrix} u_1 & u_2 & u_3 \\ v_1 & v_2 & v_3 \end{bmatrix}$$

whose first row contains the components of $\mathbf{u}$ and whose second row contains the components of $\mathbf{v}$.

• To find the first component of $\mathbf{u} \times \mathbf{v}$, delete the first column and take the determinant; to find the second component, delete the second column and take the negative of the determinant; and to find the third component, delete the third column and take the determinant.

Example 1   Find $\mathbf{u} \times \mathbf{v}$, where $\mathbf{u} = (1, 2, -2)$ and $\mathbf{v} = (3, 0, 1)$.

Solution.

$$\begin{bmatrix} 1 & 2 & -2 \\ 3 & 0 & 1 \end{bmatrix}$$

$$\mathbf{u} \times \mathbf{v} = \left(\begin{vmatrix} 2 & -2 \\ 0 & 1 \end{vmatrix}, -\begin{vmatrix} 1 & -2 \\ 3 & 1 \end{vmatrix}, \begin{vmatrix} 1 & 2 \\ 3 & 0 \end{vmatrix} \right)$$

$$= (2, -7, -6) \ \blacktriangle$$

There is an important difference between the dot product and cross product of two vectors—the dot product is a scalar and the cross product is a vector. The following theorem gives an important relationship between the dot product and cross product and also shows that $\mathbf{u} \times \mathbf{v}$ is orthogonal to both $\mathbf{u}$ and $\mathbf{v}$.

Theorem 3.4.1. *If* **u** *and* **v** *are vectors in 3-space, then*

(a) $\mathbf{u} \cdot (\mathbf{u} \times \mathbf{v}) = 0$ (*$\mathbf{u} \times \mathbf{v}$ is orthogonal to* **u**)

(b) $\mathbf{v} \cdot (\mathbf{u} \times \mathbf{v}) = 0$ (*$\mathbf{u} \times \mathbf{v}$ is orthogonal to* **v**)

(c) $\|\mathbf{u} \times \mathbf{v}\|^2 = \|\mathbf{u}\|^2 \|\mathbf{v}\|^2 - (\mathbf{u} \cdot \mathbf{v})^2$ (*Lagrange's* identity*)

Proof (a). Let $\mathbf{u} = (u_1, u_2, u_3)$ and $\mathbf{v} = (v_1, v_2, v_3)$. Then

$$\mathbf{u} \cdot (\mathbf{u} \times \mathbf{v}) = (u_1, u_2, u_3) \cdot (u_2 v_3 - u_3 v_2, u_3 v_1 - u_1 v_3, u_1 v_2 - u_2 v_1)$$
$$= u_1(u_2 v_3 - u_3 v_2) + u_2(u_3 v_1 - u_1 v_3) + u_3(u_1 v_2 - u_2 v_1)$$
$$= 0$$

Proof (b). Similar to (a).

Proof (c). Since

$$\|\mathbf{u} \times \mathbf{v}\|^2 = (u_2 v_3 - u_3 v_2)^2 + (u_3 v_1 - u_1 v_3)^2 + (u_1 v_2 - u_2 v_1)^2 \tag{2}$$

and

$$\|\mathbf{u}\|^2 \|\mathbf{v}\|^2 - (\mathbf{u} \cdot \mathbf{v})^2 = (u_1^2 + u_2^2 + u_3^2)(v_1^2 + v_2^2 + v_3^2) - (u_1 v_1 + u_2 v_2 + u_3 v_3)^2 \tag{3}$$

Lagrange's identity can be established by "multiplying out" the right sides of (2) and (3) and verifying their equality. ▊

Example 2 Consider the vectors

$$\mathbf{u} = (1, 2, -2) \quad \text{and} \quad \mathbf{v} = (3, 0, 1)$$

In Example 1 we showed that

$$\mathbf{u} \times \mathbf{v} = (2, -7, -6)$$

Since

$$\mathbf{u} \cdot (\mathbf{u} \times \mathbf{v}) = (1)(2) + (2)(-7) + (-2)(-6) = 0$$

---

* *Joseph Louis Lagrange* (1736–1813). French–Italian mathematican and astronomer. Lagrange, the son of a public official, was born in Turin, Italy. (Baptismal records list his name as Giuseppe Lodovico Lagrangia.) Although his father wanted him to be a lawyer, Lagrange was attracted to mathematics and astronomy after reading a memoir by the astronomer Halley. At age 16 he began to study mathematics on his own and by age 19 was appointed to a professorship at the Royal Artillery School in Turin. The following year he solved some famous problems using new methods that eventually blossomed into a branch of mathematics called calculus of variations. These methods and Lagrange's applications of them to problems in celestial mechanics were so monumental that by age 25 he was regarded by many of his contemporaries as the greatest living mathematician. One of Lagrange's most famous works is a memoir, *Mécanique Analytique*, in which he reduced the theory of mechanics to a few general formulas from which all other necessary equations could be derived.

It is an interesting historical fact that Lagrange's father speculated unsuccessfully in several financial ventures, so his family was forced to live quite modestly. Lagrange himself stated that if his family had money, he would not have made mathematics his vocation.

Napoleon was a great admirer of Lagrange and showered him with honors—count, senator, and Legion of Honor. In spite of his fame, Lagrange was always a shy and modest man. On his death, he was buried with honor in Pantheon.

and

$$\mathbf{v} \cdot (\mathbf{u} \times \mathbf{v}) = (3)(2) + (0)(-7) + (1)(-6) = 0$$

$\mathbf{u} \times \mathbf{v}$ is orthogonal to both $\mathbf{u}$ and $\mathbf{v}$ as guaranteed by Theorem 3.4.1. ▲

The main arithmetic properties of the cross product are listed in the next theorem.

Theorem 3.4.2. *If* $\mathbf{u}$, $\mathbf{v}$, *and* $\mathbf{w}$ *are any vectors in 3-space and k is any scalar, then*

(a) $\mathbf{u} \times \mathbf{v} = -(\mathbf{v} \times \mathbf{u})$

(b) $\mathbf{u} \times (\mathbf{v} + \mathbf{w}) = (\mathbf{u} \times \mathbf{v}) + (\mathbf{u} \times \mathbf{w})$

(c) $(\mathbf{u} + \mathbf{v}) \times \mathbf{w} = (\mathbf{u} \times \mathbf{w}) + (\mathbf{v} \times \mathbf{w})$

(d) $k(\mathbf{u} \times \mathbf{v}) = (k\mathbf{u}) \times \mathbf{v} = \mathbf{u} \times (k\mathbf{v})$

(e) $\mathbf{u} \times \mathbf{0} = \mathbf{0} \times \mathbf{u} = \mathbf{0}$

(f) $\mathbf{u} \times \mathbf{u} = \mathbf{0}$

The proofs follow immediately from Formula (1) and properties of determinants; for example, (a) can be proved as follows:

Proof (a). Interchanging $\mathbf{u}$ and $\mathbf{v}$ in (1) interchanges the rows of the three determinants on the right side of (1) and hence changes the sign of each component in the cross product. Thus, $\mathbf{u} \times \mathbf{v} = -(\mathbf{v} \times \mathbf{u})$. ▌

The proofs of the remaining parts are left as exercises.

Example 3 Consider the vectors

$$\mathbf{i} = (1, 0, 0) \qquad \mathbf{j} = (0, 1, 0) \qquad \mathbf{k} = (0, 0, 1)$$

These vectors each have length 1 and lie along the coordinate axes (Figure 1). They are called the ***standard unit vectors*** in 3-space. Every vector $\mathbf{v} = (v_1, v_2, v_3)$ in 3-space is expressible in terms of $\mathbf{i}$, $\mathbf{j}$, and $\mathbf{k}$ since we can write

$$\mathbf{v} = (v_1, v_2, v_3) = v_1(1, 0, 0) + v_2(0, 1, 0) + v_3(0, 0, 1) = v_1\mathbf{i} + v_2\mathbf{j} + v_3\mathbf{k}$$

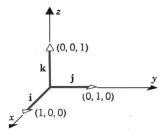

Figure 1 The standard unit vectors.

For example,

$$(2, -3, 4) = 2\mathbf{i} - 3\mathbf{j} + 4\mathbf{k}$$

From (1) we obtain

$$\mathbf{i} \times \mathbf{j} = \left(\begin{vmatrix} 0 & 0 \\ 1 & 0 \end{vmatrix}, \; -\begin{vmatrix} 1 & 0 \\ 0 & 0 \end{vmatrix}, \; \begin{vmatrix} 1 & 0 \\ 0 & 1 \end{vmatrix} \right) = (0, 0, 1) = \mathbf{k} \quad \blacktriangle$$

The reader should have no trouble obtaining the following results:

$$\mathbf{i} \times \mathbf{i} = \mathbf{j} \times \mathbf{j} = \mathbf{k} \times \mathbf{k} = \mathbf{0}$$
$$\mathbf{i} \times \mathbf{j} = \mathbf{k}, \quad \mathbf{j} \times \mathbf{k} = \mathbf{i}, \quad \mathbf{k} \times \mathbf{i} = \mathbf{j}$$
$$\mathbf{j} \times \mathbf{i} = -\mathbf{k}, \quad \mathbf{k} \times \mathbf{j} = -\mathbf{i}, \quad \mathbf{i} \times \mathbf{k} = -\mathbf{j}$$

Figure 2 is helpful for remembering these results.

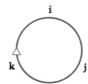

Figure 2

Referring to this diagram, the cross product of two consecutive vectors going clockwise is the next vector around, and the cross product of two consecutive vectors going counterclockwise is the negative of the next vector around.

DETERMINANT FORMULA FOR CROSS PRODUCT

It is also worth noting that a cross product can be represented symbolically in the form of a 3 × 3 determinant:

$$\mathbf{u} \times \mathbf{v} = \begin{vmatrix} \mathbf{i} & \mathbf{j} & \mathbf{k} \\ u_1 & u_2 & u_3 \\ v_1 & v_2 & v_3 \end{vmatrix} = \begin{vmatrix} u_2 & u_3 \\ v_2 & v_3 \end{vmatrix} \mathbf{i} - \begin{vmatrix} u_1 & u_3 \\ v_1 & v_3 \end{vmatrix} \mathbf{j} + \begin{vmatrix} u_1 & u_2 \\ v_1 & v_2 \end{vmatrix} \mathbf{k} \qquad (4)$$

For example, if $\mathbf{u} = (1, 2, -2)$ and $\mathbf{v} = (3, 0, 1)$, then

$$\mathbf{u} \times \mathbf{v} = \begin{vmatrix} \mathbf{i} & \mathbf{j} & \mathbf{k} \\ 1 & 2 & -2 \\ 3 & 0 & 1 \end{vmatrix} = 2\mathbf{i} - 7\mathbf{j} - 6\mathbf{k}$$

which agrees with the result obtained in Example 1.

Warning. It is not true in general that $\mathbf{u} \times (\mathbf{v} \times \mathbf{w}) = (\mathbf{u} \times \mathbf{v}) \times \mathbf{w}$. For example,

$$\mathbf{i} \times (\mathbf{j} \times \mathbf{j}) = \mathbf{i} \times \mathbf{0} = \mathbf{0}$$

and

$$(\mathbf{i} \times \mathbf{j}) \times \mathbf{j} = \mathbf{k} \times \mathbf{j} = -(\mathbf{j} \times \mathbf{k}) = -\mathbf{i}$$

so that

$$\mathbf{i} \times (\mathbf{j} \times \mathbf{j}) \neq (\mathbf{i} \times \mathbf{j}) \times \mathbf{j}$$

We know from Theorem 3.4.1 that $\mathbf{u} \times \mathbf{v}$ is orthogonal to both $\mathbf{u}$ and $\mathbf{v}$. If $\mathbf{u}$ and $\mathbf{v}$ are nonzero vectors, it can be shown that the direction of $\mathbf{u} \times \mathbf{v}$ can be determined using the following "right-hand rule"* (Figure 3): Let θ be the angle between $\mathbf{u}$ and $\mathbf{v}$, and suppose $\mathbf{u}$ is rotated through the angle θ until it coincides with $\mathbf{v}$. If the fingers of the right hand are cupped so they point in the direction of rotation, then the thumb indicates (roughly) the direction of $\mathbf{u} \times \mathbf{v}$.

Figure 3

The reader may find it instructive to practice this rule with the products

$$\mathbf{i} \times \mathbf{j} = \mathbf{k} \qquad \mathbf{j} \times \mathbf{k} = \mathbf{i} \qquad \mathbf{k} \times \mathbf{i} = \mathbf{j}$$

GEOMETRIC APPLICATIONS OF THE CROSS PRODUCT

If $\mathbf{u}$ and $\mathbf{v}$ are nonzero vectors in 3-space, then the norm of $\mathbf{u} \times \mathbf{v}$ has a useful geometric interpretation. Lagrange's identity, given in Theorem 3.4.1, states that

$$\|\mathbf{u} \times \mathbf{v}\|^2 = \|\mathbf{u}\|^2 \|\mathbf{v}\|^2 - (\mathbf{u} \cdot \mathbf{v})^2 \tag{5}$$

If θ denotes the angle between $\mathbf{u}$ and $\mathbf{v}$, then $\mathbf{u} \cdot \mathbf{v} = \|\mathbf{u}\| \|\mathbf{v}\| \cos \theta$, so that (5) can be rewritten as

$$\begin{aligned}
\|\mathbf{u} \times \mathbf{v}\|^2 &= \|\mathbf{u}\|^2 \|\mathbf{v}\|^2 - \|\mathbf{u}\|^2 \|\mathbf{v}\|^2 \cos^2 \theta \\
&= \|\mathbf{u}\|^2 \|\mathbf{v}\|^2 (1 - \cos^2 \theta) \\
&= \|\mathbf{u}\|^2 \|\mathbf{v}\|^2 \sin^2 \theta
\end{aligned}$$

Thus,

$$\|\mathbf{u} \times \mathbf{v}\| = \|\mathbf{u}\| \|\mathbf{v}\| \sin \theta \tag{6}$$

But $\|\mathbf{v}\| \sin \theta$ is the altitude of the parallelogram determined by $\mathbf{u}$ and $\mathbf{v}$ (Figure 4). Thus, from (6), the area A of this parallelogram is given by

$$A = (\text{base})(\text{altitude}) = \|\mathbf{u}\| \|\mathbf{v}\| \sin \theta = \|\mathbf{u} \times \mathbf{v}\|$$

In other words, *the norm of* $\mathbf{u} \times \mathbf{v}$ *is equal to the area of the parallelogram determined by* $\mathbf{u}$ *and* $\mathbf{v}$.

---

* Recall that we agreed to consider only right-handed coordinate systems in this text. Had we used left-handed systems instead, a "left-hand rule" would apply here.

Figure 4

Example 4 Find the area of the triangle determined by the points $P_1(2, 2, 0)$, $P_2(-1, 0, 2)$, and $P_3(0, 4, 3)$.

Solution. The area A of the triangle is 1/2 the area of the parallelogram determined by the vectors $\overrightarrow{P_1P_2}$ and $\overrightarrow{P_1P_3}$ (Figure 5). Using the method discussed in Example 2 of Section 3.1, $\overrightarrow{P_1P_2} = (-3, -2, 2)$ and $\overrightarrow{P_1P_3} = (-2, 2, 3)$. It follows that

$$\overrightarrow{P_1P_2} \times \overrightarrow{P_1P_3} = (-10, 5, -10)$$

and consequently

$$A = \tfrac{1}{2}\|\overrightarrow{P_1P_2} \times \overrightarrow{P_1P_3}\| = \tfrac{1}{2}(15) = \tfrac{15}{2} \quad \blacktriangle$$

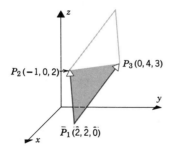

Figure 5

SCALAR TRIPLE PRODUCT

> **Definition.** If **u**, **v**, and **w** are vectors in 3-space, then
>
> $$\mathbf{u} \cdot (\mathbf{v} \times \mathbf{w})$$
>
> is called the *scalar triple product* of **u**, **v**, and **w**.

The scalar triple product of $\mathbf{u} = (u_1, u_2, u_3)$, $\mathbf{v} = (v_1, v_2, v_3)$, and $\mathbf{w} = (w_1, w_2, w_3)$ can be calculated from the formula

$$\mathbf{u} \cdot (\mathbf{v} \times \mathbf{w}) = \begin{vmatrix} u_1 & u_2 & u_3 \\ v_1 & v_2 & v_3 \\ w_1 & w_2 & w_3 \end{vmatrix} \tag{7}$$

This follows from Formula (4) since

$$\mathbf{u} \cdot (\mathbf{v} \times \mathbf{w}) = \mathbf{u} \cdot \left(\begin{vmatrix} v_2 & v_3 \\ w_2 & w_3 \end{vmatrix} \mathbf{i} - \begin{vmatrix} v_1 & v_3 \\ w_1 & w_3 \end{vmatrix} \mathbf{j} + \begin{vmatrix} v_1 & v_2 \\ w_1 & w_2 \end{vmatrix} \mathbf{k} \right)$$

$$= \begin{vmatrix} v_2 & v_3 \\ w_2 & w_3 \end{vmatrix} u_1 - \begin{vmatrix} v_1 & v_3 \\ w_1 & w_3 \end{vmatrix} u_2 + \begin{vmatrix} v_1 & v_2 \\ w_1 & w_2 \end{vmatrix} u_3$$

$$= \begin{vmatrix} u_1 & u_2 & u_3 \\ v_1 & v_2 & v_3 \\ w_1 & w_2 & w_3 \end{vmatrix}$$

Example 5 Calculate the scalar triple product $\mathbf{u} \cdot (\mathbf{v} \times \mathbf{w})$ of the vectors

$$\mathbf{u} = 3\mathbf{i} - 2\mathbf{j} - 5\mathbf{k}, \qquad \mathbf{v} = \mathbf{i} + 4\mathbf{j} - 4\mathbf{k}, \qquad \mathbf{w} = 3\mathbf{j} + 2\mathbf{k}$$

Solution. From (7)

$$\mathbf{u} \cdot (\mathbf{v} \times \mathbf{w}) = \begin{vmatrix} 3 & -2 & -5 \\ 1 & 4 & -4 \\ 0 & 3 & 2 \end{vmatrix}$$

$$= 3 \begin{vmatrix} 4 & -4 \\ 3 & 2 \end{vmatrix} - (-2) \begin{vmatrix} 1 & -4 \\ 0 & 2 \end{vmatrix} + (-5) \begin{vmatrix} 1 & 4 \\ 0 & 3 \end{vmatrix}$$

$$= 60 + 4 - 15 = 49 \; \blacktriangle$$

REMARK. The symbol $(\mathbf{u} \cdot \mathbf{v}) \times \mathbf{w}$ makes no sense since we cannot form the cross product of a scalar and a vector. Thus, no ambiguity arises if we write $\mathbf{u} \cdot \mathbf{v} \times \mathbf{w}$ rather than $\mathbf{u} \cdot (\mathbf{v} \times \mathbf{w})$. However, for clarity we shall usually keep the parentheses.
It follows from (7) that

$$\mathbf{u} \cdot (\mathbf{v} \times \mathbf{w}) = \mathbf{w} \cdot (\mathbf{u} \times \mathbf{v}) = \mathbf{v} \cdot (\mathbf{w} \times \mathbf{u})$$

since the 3×3 determinants that represent these products can be obtained from one another by *two* row interchanges. (Verify.) These relationships can be remembered by moving the vectors $\mathbf{u}$, $\mathbf{v}$, and $\mathbf{w}$ clockwise around the vertices of the triangle in Figure 6.

Figure 6

The scalar triple product $\mathbf{u} \cdot (\mathbf{v} \times \mathbf{w})$ has a useful geometric interpretation. If we assume, for the moment, that the vectors $\mathbf{u}$, $\mathbf{v}$, $\mathbf{w}$ do not all lie in the same plane, then the three vectors form adjacent sides of a parallelepiped (Figure 7) when they

Figure 7 $h = \|\text{proj}_{\mathbf{v} \times \mathbf{w}} \mathbf{u}\|$

are positioned with a common initial point. If the parallelogram determined by **v** and **w** is regarded as the base of the parallelepiped, then the area of the base is $\|\mathbf{v} \times \mathbf{w}\|$, and the height h is the length of the orthogonal projection of **u** on $\mathbf{v} \times \mathbf{w}$ (Figure 7). Therefore, by Formula (10) of Section 3.3

$$h = \|\text{proj}_{\mathbf{v} \times \mathbf{w}} \mathbf{u}\| = \frac{|\mathbf{u} \cdot (\mathbf{v} \times \mathbf{w})|}{\|\mathbf{v} \times \mathbf{w}\|}$$

It follows that the volume V of the parallelepiped is

$$V = (\text{area of base}) \cdot \text{height} = \|\mathbf{v} \times \mathbf{w}\| \frac{|\mathbf{u} \cdot (\mathbf{v} \times \mathbf{w})|}{\|\mathbf{v} \times \mathbf{w}\|}$$

or, on simplifying,

$$V = \begin{bmatrix} \text{volume of parallelepiped with} \\ \text{adjacent sides } \mathbf{u}, \mathbf{v}, \text{ and } \mathbf{w} \end{bmatrix} = |\mathbf{u} \cdot (\mathbf{v} \times \mathbf{w})| \qquad (8)$$

REMARK. It follows from this formula that

$$\mathbf{u} \cdot (\mathbf{v} \times \mathbf{w}) = \pm V$$

where the $+$ or $-$ results depending on whether **u** makes an acute or obtuse angle with $(\mathbf{v} \times \mathbf{w})$ (Theorem 3.3.1).

Formula (8) leads to a useful test for ascertaining whether three given vectors lie in the same plane. Since three vectors not in the same plane determine a parallelepiped of positive volume, it follows from (8) that $|\mathbf{u} \cdot (\mathbf{v} \times \mathbf{w})| = 0$ if and only if the vectors **u**, **v**, and **w** lie in the same plane. Thus, we have the following result.

Theorem 3.4.3. *If the vectors* $\mathbf{u} = (u_1, u_2, u_3)$, $\mathbf{v} = (v_1, v_2, v_3)$, *and* $\mathbf{w} = (w_1, w_2, w_3)$ *have the same initial point, then they lie in the same plane if and only if*

$$\mathbf{u} \cdot (\mathbf{v} \times \mathbf{w}) = \begin{vmatrix} u_1 & u_2 & u_3 \\ v_1 & v_2 & v_3 \\ w_1 & w_2 & w_3 \end{vmatrix} = 0$$

Initially, we defined a vector to be a directed line segment or arrow in 2-space or 3-space; coordinate systems and components were introduced later in order to simplify computations with vectors. Thus, a vector has a "mathematical existence" regardless of whether a coordinate system has been introduced. Further, the components of a vector are not determined by the vector alone; they depend as well on the coordinate system chosen. For example, in Figure 8 we have indicated a fixed vector **v** in the plane and two different coordinate systems. In the xy-coordinate system the components of **v** are (1, 1), and in the $x'y'$-system they are ($\sqrt{2}$, 0).

This raises an important question about our definition of cross product. Since we defined the cross product $\mathbf{u} \times \mathbf{v}$ in terms of the components of **u** and **v**, and since these components depend on the coordinate system chosen, it seems possible that two *fixed* vectors **u** and **v** might have different cross products in different coordinate systems. Fortunately, this is not the case. To see that this is so, we need only recall that

(i) $\mathbf{u} \times \mathbf{v}$ is perpendicular to both **u** and **v**.
(ii) The orientation of $\mathbf{u} \times \mathbf{v}$ is determined by the right-hand rule.
(iii) $\|\mathbf{u} \times \mathbf{v}\| = \|\mathbf{u}\| \|\mathbf{v}\| \sin \theta$.

These three properties completely determine the vector $\mathbf{u} \times \mathbf{v}$: properties (i) and (ii) determine the direction, and property (iii) determines the length. Since these properties of $\mathbf{u} \times \mathbf{v}$ depend only on the lengths and relative positions of **u** and **v** and not on the particular right-hand coordinate system being used, the vector $\mathbf{u} \times \mathbf{v}$ will remain unchanged if a different right-hand coordinate system is introduced. Thus, we say that the definition of $\mathbf{u} \times \mathbf{v}$ is *coordinate free*. This result is of importance to physicists and engineers who often work with many coordinate systems in the same problem.

Figure 8

Example 6 Consider two perpendicular vectors **u** and **v**, each of length 1 (as shown in Figure 9*a*). If we introduce an xyz-coordinate system as shown in Figure 9*b*, then

$$\mathbf{u} = (1, 0, 0) = \mathbf{i} \quad \text{and} \quad \mathbf{v} = (0, 1, 0) = \mathbf{j}$$

so that

$$\mathbf{u} \times \mathbf{v} = \mathbf{i} \times \mathbf{j} = \mathbf{k} = (0, 0, 1)$$

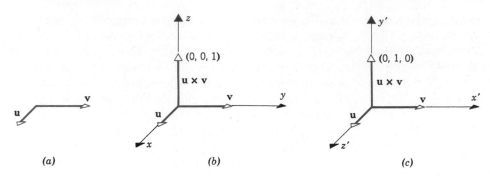

Figure 9   *(a)*   *(b)*   *(c)*

However, if we introduce an $x'y'z'$-coordinate system as shown in Figure 9c, then

$$\mathbf{u} = (0, 0, 1) = \mathbf{k} \quad \text{and} \quad \mathbf{v} = (1, 0, 0) = \mathbf{i}$$

so that

$$\mathbf{u} \times \mathbf{v} = \mathbf{k} \times \mathbf{i} = \mathbf{j} = (0, 1, 0)$$

But it is clear from Figures 9b and 9c that the vector $(0, 0, 1)$ in the xyz-system is the same as the vector $(0, 1, 0)$ in the $x'y'z'$-system. Thus, we obtain the same vector $\mathbf{u} \times \mathbf{v}$ whether we compute with coordinates from the xyz-system or with coordinates from the $x'y'z'$-system. ▲

EXERCISE SET 3.4

1. Let $\mathbf{u} = (3, 2, -1)$, $\mathbf{v} = (0, 2, -3)$, and $\mathbf{w} = (2, 6, 7)$. Compute
   (a) $\mathbf{v} \times \mathbf{w}$ (b) $\mathbf{u} \times (\mathbf{v} \times \mathbf{w})$ (c) $(\mathbf{u} \times \mathbf{v}) \times \mathbf{w}$
   (d) $(\mathbf{u} \times \mathbf{v}) \times (\mathbf{v} \times \mathbf{w})$ (e) $\mathbf{u} \times (\mathbf{v} - 2\mathbf{w})$ (f) $(\mathbf{u} \times \mathbf{v}) - 2\mathbf{w}$

2. Find a vector that is orthogonal to both $\mathbf{u}$ and $\mathbf{v}$.
   (a) $\mathbf{u} = (-6, 4, 2)$, $\mathbf{v} = (3, 1, 5)$ (b) $\mathbf{u} = (-2, 1, 5)$, $\mathbf{v} = (3, 0, -3)$

3. Find the area of the triangle having vertices P, Q, and R.
   (a) $P(2, 6, -1)$, $Q(1, 1, 1)$, $R(4, 6, 2)$ (b) $P(1, -1, 2)$, $Q(0, 3, 4)$, $R(6, 1, 8)$

4. Find the area of the parallelogram determined by $\mathbf{u}$ and $\mathbf{v}$.
   (a) $\mathbf{u} = (1, -1, 2)$, $\mathbf{v} = (0, 3, 1)$ (b) $\mathbf{u} = (2, 3, 0)$, $\mathbf{v} = (-1, 2, -2)$

5. Verify Theorem 3.4.1 for the vectors $\mathbf{u} = (4, 2, 1)$ and $\mathbf{v} = (-3, 2, 7)$.

6. Verify Theorem 3.4.2 for $\mathbf{u} = (5, -1, 2)$, $\mathbf{v} = (6, 0, -2)$, $\mathbf{w} = (1, 2, -1)$, and $k = -5$.

7. What is wrong with the expression $\mathbf{u} \times \mathbf{v} \times \mathbf{w}$?

8. Find the scalar triple product $\mathbf{u} \cdot (\mathbf{v} \times \mathbf{w})$.
   (a) $\mathbf{u} = (-1, 2, 4)$, $\mathbf{v} = (3, 4, -2)$, $\mathbf{w} = (-1, 2, 5)$
   (b) $\mathbf{u} = (3, -1, 6)$, $\mathbf{v} = (2, 4, 3)$, $\mathbf{w} = (5, -1, 2)$

9. Suppose $\mathbf{u} \cdot (\mathbf{v} \times \mathbf{w}) = 3$. Find
   (a) $\mathbf{u} \cdot (\mathbf{w} \times \mathbf{v})$ (b) $(\mathbf{v} \times \mathbf{w}) \cdot \mathbf{u}$ (c) $\mathbf{w} \cdot (\mathbf{u} \times \mathbf{v})$ (d) $\mathbf{v} \cdot (\mathbf{u} \times \mathbf{w})$ (e) $(\mathbf{u} \times \mathbf{w}) \cdot \mathbf{v}$ (f) $\mathbf{v} \cdot (\mathbf{w} \times \mathbf{w})$

10. Find the volume of the parallelepiped with sides $\mathbf{u}$, $\mathbf{v}$, and $\mathbf{w}$.
    (a) $\mathbf{u} = (2, -6, 2)$, $\mathbf{v} = (0, 4, -2)$, $\mathbf{w} = (2, 2, -4)$ (b) $\mathbf{u} = (3, 1, 2)$, $\mathbf{v} = (4, 5, 1)$, $\mathbf{w} = (1, 2, 4)$

11. Determine whether **u**, **v**, and **w** lie in the same plane when positioned so that their initial points coincide.

(a) $\mathbf{u} = (-1, -2, 1)$, $\mathbf{v} = (3, 0, -2)$, $\mathbf{w} = (5, -4, 0)$
(b) $\mathbf{u} = (5, -2, 1)$, $\mathbf{v} = (4, -1, 1)$, $\mathbf{w} = (1, -1, 0)$
(c) $\mathbf{u} = (4, -8, 1)$, $\mathbf{v} = (2, 1, -2)$, $\mathbf{w} = (3, -4, 12)$

12. Find all unit vectors parallel to the *yz*-plane that are perpendicular to the vector $(3, -1, 2)$.

13. Find all unit vectors in the plane determined by $\mathbf{u} = (3, 0, 1)$ and $\mathbf{v} = (1, -1, 1)$ that are perpendicular to the vector $\mathbf{w} = (1, 2, 0)$.

14. Let $\mathbf{a} = (a_1, a_2, a_3)$, $\mathbf{b} = (b_1, b_2, b_3)$, $\mathbf{c} = (c_1, c_2, c_3)$, and $\mathbf{d} = (d_1, d_2, d_3)$. Show that

$$(\mathbf{a} + \mathbf{d}) \cdot (\mathbf{b} \times \mathbf{c}) = \mathbf{a} \cdot (\mathbf{b} \times \mathbf{c}) + \mathbf{d} \cdot (\mathbf{b} \times \mathbf{c})$$

15. Simplify $(\mathbf{u} + \mathbf{v}) \times (\mathbf{u} - \mathbf{v})$.

16. Use the cross product to find the sine of the angle between the vectors $\mathbf{u} = (2, 3, -6)$ and $\mathbf{v} = (2, 3, 6)$.

17. (a) Find the area of the triangle having vertices $A(1, 0, 1)$, $B(0, 2, 3)$, and $C(2, 1, 0)$.
(b) Use the result of part (a) to find the length of the altitude from vertex C to side AB.

18. Show that if **u** is a vector from any point on a line to a point P not on the line, and **v** is a vector parallel to the line, then the distance between P and the line is given by $\|\mathbf{u} \times \mathbf{v}\|/\|\mathbf{v}\|$.

19. Use the result of Exercise 18 to find the distance between the point P and the line through the points A and B:

(a) $P(-3, 1, 2)$, $A(1, 1, 0)$, $B(-2, 3, -4)$ (b) $P(4, 3, 0)$, $A(2, 1, -3)$, $B(0, 2, -1)$

20. Prove: If θ is the angle between **u** and **v** and $\mathbf{u} \cdot \mathbf{v} \neq 0$, then $\tan \theta = \|\mathbf{u} \times \mathbf{v}\|/(\mathbf{u} \cdot \mathbf{v})$.

21. Consider the parallelepiped with sides $\mathbf{u} = (3, 2, 1)$, $\mathbf{v} = (1, 1, 2)$, and $\mathbf{w} = (1, 3, 3)$.
(a) Find the area of the face determined by **u** and **w**.
(b) Find the angle between **u** and the plane containing the face determined by **v** and **w**.

22. Find a vector **n** perpendicular to the plane determined by the points $A(0, -2, 1)$, $B(1, -1, -2)$, and $C(-1, 1, 0)$.

23. Let **m** and **n** be vectors whose components in the *xyz*-system of Figure 9 are $\mathbf{m} = (0, 0, 1)$ and $\mathbf{n} = (0, 1, 0)$.
(a) Find the components of **m** and **n** in the $x'y'z'$-system of Figure 9.
(b) Compute $\mathbf{m} \times \mathbf{n}$ using the components in the *xyz*-system.
(c) Compute $\mathbf{m} \times \mathbf{n}$ using the components in the $x'y'z'$-system.
(d) Show that the vectors obtained in (b) and (c) are the same.

24. Prove the following identities.
(a) $(\mathbf{u} + k\mathbf{v}) \times \mathbf{v} = \mathbf{u} \times \mathbf{v}$ (b) $\mathbf{u} \cdot (\mathbf{v} \times \mathbf{z}) = -(\mathbf{u} \times \mathbf{z}) \cdot \mathbf{v}$

25. Let **u**, **v**, and **w** be nonzero vectors in 3-space with the same initial point, but such that no two of them are collinear. Show that
(a) $\mathbf{u} \times (\mathbf{v} \times \mathbf{w})$ lies in the plane determined by **v** and **w**
(b) $(\mathbf{u} \times \mathbf{v}) \times \mathbf{w}$ lies in the plane determined by **u** and **v**

26. Prove that $\mathbf{x} \times (\mathbf{y} \times \mathbf{z}) = (\mathbf{x} \cdot \mathbf{z})\mathbf{y} - (\mathbf{x} \cdot \mathbf{y})\mathbf{z}$. [*Hint.* First prove the result in the case where $\mathbf{z} = \mathbf{i} = (1, 0, 0)$, then when $\mathbf{z} = \mathbf{j} = (0, 1, 0)$, and then when $\mathbf{z} = \mathbf{k} = (0, 0, 1)$. Finally prove it for an arbitrary vector $\mathbf{z} = (z_1, z_2, z_3)$ by writing $\mathbf{z} = z_1\mathbf{i} + z_2\mathbf{j} + z_3\mathbf{k}$.]

27. Let $\mathbf{x} = (1, 3, -1)$, $\mathbf{y} = (1, 1, 2)$, and $\mathbf{z} = (3, -1, 2)$. Calculate $\mathbf{x} \times (\mathbf{y} \times \mathbf{z})$ using Exercise 26; then check your result by calculating directly.

28. Prove: If $\mathbf{a}$, $\mathbf{b}$, $\mathbf{c}$, and $\mathbf{d}$ lie in the same plane, then $(\mathbf{a} \times \mathbf{b}) \times (\mathbf{c} \times \mathbf{d}) = \mathbf{0}$.

29. It is a theorem of solid geometry that the volume of a tetrahedron is $\frac{1}{3}$(area of base) · (height). Use this result to prove that the volume of a tetrahedron whose sides are the vectors $\mathbf{a}$, $\mathbf{b}$, and $\mathbf{c}$ is $\frac{1}{6}|\mathbf{a} \cdot (\mathbf{b} \times \mathbf{c})|$ (Figure 10).

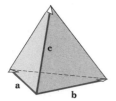

Figure 10

30. Use the result of Exercise 29 to find the volume of the tetrahedron with vertices P, Q, R, and S.
(a) $P(-1, 2, 0)$, $Q(2, 1, -3)$, $R(1, 0, 1)$, $S(3, -2, 3)$
(b) $P(0, 0, 0)$, $Q(1, 2, -1)$, $R(3, 4, 0)$, $S(-1, -3, 4)$.

31. Prove parts (*a*) and (*b*) of Theorem 3.4.2.

32. Prove parts (*c*) and (*d*) of Theorem 3.4.2.

33. Prove parts (*e*) and (*f*) of Theorem 3.4.2.

3.5 LINES AND PLANES IN 3-SPACE

In this section we shall use vectors to derive equations of lines and planes in 3-space, and we shall use these equations to solve some basic geometric problems.

PLANES IN 3-SPACE

In plane analytic geometry a line can be specified by giving its slope and one of its points. Similarly, a plane in 3-space can be specified by giving its inclination and specifying one of its points. A convenient method for describing the inclination is to specify a nonzero vector (called a ***normal***) that is perpendicular to the plane.

Suppose we want the equation of the plane passing through the point $P_0(x_0, y_0, z_0)$ and having the nonzero vector $\mathbf{n} = (a, b, c)$ as a normal. It is evident from Figure 1 that the plane consists precisely of those points $P(x, y, z)$ for which the vector $\overrightarrow{P_0P}$ is orthogonal to $\mathbf{n}$, that is, for which

$$\mathbf{n} \cdot \overrightarrow{P_0P} = 0 \tag{1}$$

Since $\overrightarrow{P_0P} = (x - x_0, y - y_0, z - z_0)$, Equation (1) can be written as

$$a(x - x_0) + b(y - y_0) + c(z - z_0) = 0 \tag{2}$$

We call this the ***point-normal*** form of the equation of a plane.

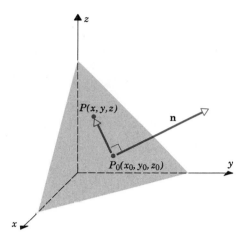

Figure 1

Example 1 Find an equation of the plane passing through the point $(3, -1, 7)$ and perpendicular to the vector $\mathbf{n} = (4, 2, -5)$.

Solution. From (2) a point-normal form is

$$4(x - 3) + 2(y + 1) - 5(z - 7) = 0$$

By multiplying out and collecting terms, (2) can be rewritten in the form

$$ax + by + cz + d = 0$$

where a, b, c, and d are constants, and a, b, and c are not all zero. For example, the equation in Example 1 can be rewritten as

$$4x + 2y - 5z + 25 = 0$$

As the next theorem shows, every equation of the form $ax + by + cz + d = 0$ represents a plane in 3-space.

Theorem 3.5.1. *If a, b, c, and d are constants and a, b, and c are not all zero, then the graph of the equation*

$$ax + by + cz + d = 0 \tag{3}$$

is a plane having the vector $\mathbf{n} = (a, b, c)$ as a normal.

Equation (3) is a linear equation in x, y, and z; it is called the **general form** of the equation of a plane.

Proof. By hypothesis, the coefficients a, b, and c are not all zero. Assume, for the moment, that $a \neq 0$. Then the equation $ax + by + cz + d = 0$ can be rewritten as $a\,x + (d/a)) + by + cz = 0$. But this is a point-normal form of the plane passing through the point $(-d/a, 0, 0)$ and having $\mathbf{n} = (a, b, c)$ as a normal.

If $a = 0$, then either $b \neq 0$ or $c \neq 0$. A straightforward modification of the above argument will handle these other cases. ▮

Just as the solutions of a system of linear equations

$$ax + by = k_1$$
$$cx + dy = k_2$$

correspond to points of intersection of the lines $ax + by = k_1$ and $cx + dy = k_2$ in the xy-plane, so the solutions of a system

$$ax + by + cz = k_1$$
$$dx + ey + fz = k_2 \tag{4}$$
$$gx + hy + iz = k_3$$

correspond to the points of intersection of the three planes $ax + by + cz = k_1$, $dx + ey + fz = k_2$, and $gx + hy + iz = k_3$.

In Figure 2 we have illustrated some of the geometric possibilities that occur when (4) has zero, one, or infinitely many solutions.

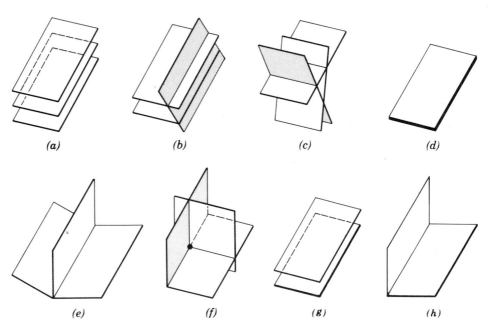

(a)   *(b)*   *(c)*   *(d)*

(e)   *(f)*   *(g)*   *(h)*

Figure 2

(a) No solutions (3 parallel planes).   (b) No solutions (2 planes parallel). (c) No solutions (3 planes with no common intersection).   (d) Infinitely many solutions (3 coincident planes).   (e) Infinitely many solutions (3 planes intersecting in a line).   (f) One solution (3 planes intersecting at a point).   (g) No solutions (2 coincident planes parallel to a third plane). (h) Infinitely many solutions (2 coincident planes intersecting a third plane).

Example 2   Find the equation of the plane passing through the points $P_1(1, 2, -1)$, $P_2(2, 3, 1)$, and $P_3(3, -1, 2)$.

Solution. Since the three points lie in the plane, their coordinates must satisfy the general equation $ax + by + cz + d = 0$ of the plane. Thus,

$$a + 2b - c + d = 0$$
$$2a + 3b + c + d = 0$$
$$3a - b + 2c + d = 0$$

Solving this system gives

$$a = -\tfrac{9}{16}t \qquad b = -\tfrac{1}{16}t \qquad c = \tfrac{5}{16}t \qquad d = t$$

Letting $t = -16$, for example, yields the desired equation

$$9x + y - 5z - 16 = 0$$

We note that any other choice of t gives a multiple of this equation, so that any value of $t \neq 0$ would also give a valid equation of the plane.

Alternative solution. Since $P_1(1, 2, -1)$, $P_2(2, 3, 1)$, and $P_3(3, -1, 2)$ lie in the plane, the vectors $\overrightarrow{P_1P_2} = (1, 1, 2)$ and $\overrightarrow{P_1P_3} = (2, -3, 3)$ are parallel to the plane. Therefore, $\overrightarrow{P_1P_2} \times \overrightarrow{P_1P_3} = (9, 1, -5)$ is normal to the plane, since it is perpendicular to both $\overrightarrow{P_1P_2}$ and $\overrightarrow{P_1P_3}$. From this and the fact that P_1 lies in the plane, a point-normal form for the equation of the plane is

$$9(x - 1) + (y - 2) - 5(z + 1) = 0$$

or

$$9x + y - 5z - 16 = 0 \quad \blacktriangle$$

LINES IN 3-SPACE   We shall now show how to obtain equations for lines in 3-space. Suppose l is the line in 3-space through the point $P_0(x_0, y_0, z_0)$ and parallel to the nonzero vector $\mathbf{v} = (a, b, c)$. It is clear (Figure 3) that l consists precisely of those points $P(x, y, z)$ for which the vector $\overrightarrow{P_0P}$ is parallel to $\mathbf{v}$, that is, for which there is a scalar t such that

$$\overrightarrow{P_0P} = t\mathbf{v} \tag{5}$$

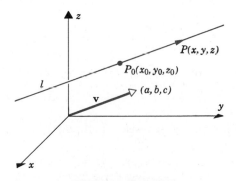

Figure 3 $\overrightarrow{P_0P}$ is parallel to $\mathbf{v}$.

In terms of components, (5) can be written as

$$(x - x_0, y - y_0, z - z_0) = (ta, tb, tc)$$

from which it follows that

$$\begin{aligned} x &= x_0 + ta \\ y &= y_0 + tb \qquad \text{where } -\infty < t < +\infty \\ z &= z_0 + tc \end{aligned}$$

These equations are called **parametric equations** for l since the line l is traced out by $P(x, y, z)$ as the parameter t varies from $-\infty$ to $+\infty$.

REMARK. For brevity, we shall omit the phrase "where $-\infty < t < +\infty$."

Example 3 The line through the point $(1, 2, -3)$ and parallel to the vector $\mathbf{v} = (4, 5, -7)$ has parametric equations

$$\begin{aligned} x &= 1 + 4t \\ y &= 2 + 5t \qquad \text{where } -\infty < t < +\infty \\ z &= -3 - 7t \end{aligned}$$

Example 4

(a) Find parametric equations for the line l passing through the points $P_1(2, 4, -1)$ and $P_2(5, 0, 7)$.
(b) Where does the line intersect the xy-plane?

Solution (a). Since the vector $\overrightarrow{P_1 P_2} = (3, -4, 8)$ is parallel to l and $P_1(2, 4, -1)$ lies on l, the line l is given by

$$\begin{aligned} x &= 2 + 3t \\ y &= 4 - 4t \qquad \text{where } -\infty < t < +\infty \\ z &= -1 + 8t \end{aligned}$$

Solution (b). The line intersects the xy-plane at the point where $z = -1 + 8t = 0$, that is, where $t = 1/8$. Substituting this value of t in the parametric equations for l yields as the point of intersection

$$(x, y, z) = (\tfrac{19}{8}, \tfrac{7}{2}, 0)$$

Example 5 Find parametric equations for the line of intersection of the planes

$$3x + 2y - 4z - 6 = 0 \qquad \text{and} \qquad x - 3y - 2z - 4 = 0$$

Solution. The line of intersection consists of all points (x, y, z) that satisfy the two equations in the system

$$\begin{aligned} 3x + 2y - 4z &= 6 \\ x - 3y - 2z &= 4 \end{aligned}$$

Solving this system gives

$$x = \tfrac{26}{11} + \tfrac{16}{11}t, \qquad y = -\tfrac{6}{11} - \tfrac{2}{11}t, \qquad z = t$$

The parametric equations for l are therefore

$$
\begin{aligned}
x &= \tfrac{26}{11} + \tfrac{16}{11}t \\
y &= -\tfrac{6}{11} - \tfrac{2}{11}t \qquad \text{where } -\infty < t < +\infty \\
z &= t
\end{aligned}
$$

In some problems, a line

$$
\begin{aligned}
x &= x_0 + at \\
y &= y_0 + bt \qquad \text{where } -\infty < t < +\infty \\
z &= z_0 + ct
\end{aligned}
\tag{6}
$$

is given, and it is of interest to find two planes whose intersection is the given line. Since there are infinitely many planes through the line, there are always infinitely many such pairs of planes. To find two such planes when a, b, and c are all different from zero, we can rewrite each equation in (6) in the form

$$\frac{x - x_0}{a} = t \qquad \frac{y - y_0}{b} = t \qquad \frac{z - z_0}{c} = t$$

Eliminating the parameter t shows that the line consists of all points (x, y, z) that satisfy the equations

$$\frac{x - x_0}{a} = \frac{y - y_0}{b} = \frac{z - z_0}{c}$$

These are called **symmetric equations** for the line. It follows that the line can be viewed as the intersection of the planes

$$\frac{x - x_0}{a} = \frac{y - y_0}{b} \qquad \text{and} \qquad \frac{y - y_0}{b} = \frac{z - z_0}{c}$$

or as the intersection of the planes

$$\frac{x - x_0}{a} = \frac{z - z_0}{c} \qquad \text{and} \qquad \frac{y - y_0}{b} = \frac{z - z_0}{c}$$

and so forth.

Example 6 Find two planes whose intersection is the line

$$
\begin{aligned}
x &= 3 + 2t \\
y &= -4 + 7t \qquad \text{where } -\infty < t < +\infty \\
z &= 1 + 3t
\end{aligned}
$$

Solution. Since the symmetric equations for this line are

$$\frac{x - 3}{2} = \frac{y + 4}{7} = \frac{z - 1}{3} \tag{7}$$

the line is the intersection of the planes

$$\frac{x - 3}{2} = \frac{y + 4}{7} \qquad \text{and} \qquad \frac{y + 4}{7} = \frac{z - 1}{3}$$

or equivalently

$$7x - 2y - 29 = 0 \qquad \text{and} \qquad 3y - 7z + 19 = 0$$

Other solutions can be obtained by choosing different pairs of equations from (7). ▲

SOME PROBLEMS INVOLVING DISTANCE

We conclude this section by discussing two basic "distance problems" in 3-space:

> **Problems.**
>
> (a) Find the distance between a point and a plane.
> (b) Find the distance between two parallel planes.

The two problems are related. If we can find the distance between a point and a plane, then we can find the distance between parallel planes by computing the distance between either one of the planes and an arbitrary point P_0 in the other (Figure 4).

Figure 4 | The distance between the parallel planes V and W is equal to the distance between P_0 and W.

> **Theorem 3.5.2.** *The distance D between a point $P_0(x_0, y_0, z_0)$ and the plane $ax + by + cz + d = 0$ is*
>
> $$D = \frac{|ax_0 + by_0 + cz_0 + d|}{\sqrt{a^2 + b^2 + c^2}} \tag{8}$$

Proof. Let $Q(x_1, y_1, z_1)$ be any point in the plane. Position the normal $\mathbf{n} = (a, b, c)$ so that its initial point is at Q. As illustrated in Figure 5, the distance D is equal to the length of the orthogonal projection of $\overrightarrow{QP_0}$ on $\mathbf{n}$. Thus, from (10) of Section 3.3,

$$D = \|\text{proj}_{\mathbf{n}} \overrightarrow{QP_0}\| = \frac{|\overrightarrow{QP_0} \cdot \mathbf{n}|}{\|\mathbf{n}\|}$$

But

$$\overrightarrow{QP_0} = (x_0 - x_1, y_0 - y_1, z_0 - z_1)$$
$$\overrightarrow{QP_0} \cdot \mathbf{n} = a(x_0 - x_1) + b(y_0 - y_1) + c(z_0 - z_1)$$
$$\|\mathbf{n}\| = \sqrt{a^2 + b^2 + c^2}$$

Thus,

$$D = \frac{|a(x_0 - x_1) + b(y_0 - y_1) + c(z_0 - z_1)|}{\sqrt{a^2 + b^2 + c^2}} \qquad (9)$$

Since the point $Q(x_1, y_1, z_1)$ lies in the plane, its coordinates satisfy the equation of the plane; thus

$$ax_1 + by_1 + cz_1 + d = 0$$

or

$$d = -ax_1 - by_1 - cz_1$$

Substituting this expression in (9) yields (8). ▮

Figure 5

REMARK. Note the similarity between (8) and the formula for the distance between a point and a line in 2-space [(13) of Section 3.3].

Example 7 Find the distance D between the point $(1, -4, -3)$ and the plane $2x - 3y + 6z = -1$.

Solution. To apply (8), we first rewrite the equation of the plane in the form

$$2x - 3y + 6z + 1 = 0$$

Then

$$D = \frac{|(2)(1) + (-3)(-4) + 6(-3) + 1|}{\sqrt{2^2 + (-3)^2 + 6^2}} = \frac{|-3|}{7} = \frac{3}{7} \quad \blacktriangle$$

Given two planes, either they intersect, in which case we can ask for their line of intersection, as in Example 5, or they are parallel, in which case we can ask for the distance between them. The following example illustrates the latter problem.

Example 8 The planes

$$x + 2y - 2z = 3 \quad \text{and} \quad 2x + 4y - 4z = 7$$

are parallel since their normals, $(1, 2, -2)$ and $(2, 4, -4)$, are parallel vectors. Find the distance between these planes.

Solution. To find the distance D between the planes, we may select an arbitrary point in one of the planes and compute its distance to the other plane. By setting $y = z = 0$ in the equation $x + 2y - 2z = 3$, we obtain the point $P_0(3, 0, 0)$ in this plane. From (8), the distance between P_0 and the plane $2x + 4y - 4z = 7$ is

$$D = \frac{|(2)(3) + 4(0) + (-4)(0) - 7|}{\sqrt{2^2 + 4^2 + (-4)^2}} = \frac{1}{6} \quad \blacktriangle$$

EXERCISE SET 3.5

1. Find a point-normal form of the equation of the plane passing through P and having $\mathbf{n}$ as a normal.
   (a) $P(-1, 3, -2)$; $\mathbf{n} = (-2, 1, -1)$ (b) $P(1, 1, 4)$; $\mathbf{n} = (1, 9, 8)$
   (c) $P(2, 0, 0)$; $\mathbf{n} = (0, 0, 2)$ (d) $P(0, 0, 0)$; $\mathbf{n} = (1, 2, 3)$

2. Write the equations of the planes in Exercise 1 in general form.

3. Find a point-normal form.
   (a) $-3x + 7y + 2z = 10$ (b) $x - 4z = 0$

4. Find an equation for the plane passing through the given points.
   (a) $P(-4, -1, -1)$, $Q(-2, 0, 1)$, $R(-1, -2, -3)$ (b) $P(5, 4, 3)$, $Q(4, 3, 1)$, $R(1, 5, 4)$

5. Determine whether the planes are parallel.
   (a) $4x - y + 2z = 5$ and $7x - 3y + 4z = 8$
   (b) $x - 4y - 3z - 2 = 0$ and $3x - 12y - 9z - 7 = 0$
   (c) $2y = 8x - 4z + 5$ and $x = \frac{1}{2}z + \frac{1}{4}y$

6. Determine whether the line and plane are parallel.
   (a) $x = -5 - 4t$, $y = 1 - t$, $z = 3 + 2t$; $x + 2y + 3z - 9 = 0$
   (b) $x = 3t$, $y = 1 + 2t$, $z = 2 - t$; $4x - y + 2z = 1$

7. Determine whether the planes are perpendicular.

    (a) $3x - y + z - 4 = 0$, $x + 2z = -1$ (b) $x - 2y + 3z = 4$, $-2x + 5y + 4z = -1$

8. Determine whether the line and plane are perpendicular.

    (a) $x = -2 - 4t$, $y = 3 - 2t$, $z = 1 + 2t$; $2x + y - z = 5$

    (b) $x = 2 + t$, $y = 1 - t$, $z = 5 + 3t$; $6x + 6y - 7 = 0$

9. Find parametric equations for the line passing through P and parallel to **n**.

    (a) $P(3, -1, 2)$, $\mathbf{n} = (2, 1, 3)$ (b) $P(-2, 3, -3)$; $\mathbf{n} = (6, -6, -2)$

    (c) $P(2, 2, 6)$; $\mathbf{n} = (0, 1, 0)$ (d) $P(0, 0, 0)$; $\mathbf{n} = (1, -2, 3)$

10. Find symmetric equations for the lines in parts (a) and (b) of Exercise 9.

11. Find parametric equations for the line passing through the given points.

    (a) $(5, -2, 4)$, $(7, 2, -4)$ (b) $(0, 0, 0)$, $(2, -1, -3)$

12. Find parametric equations for the line of intersection of the given planes.

    (a) $7x - 2y + 3z = -2$ and $-3x + y + 2z + 5 = 0$ (b) $2x + 3y - 5z = 0$ and $y = 0$

13. In each part find equations for two planes whose intersection is the given line.

    (a) $x = 7 - 4t$ (b) $x = 4t$

        $y = -5 - 2t$ $y = 2t$

        $z = 5 + t$ $z = 7t$

14. Show that the line

$$x = 0$$
$$y = t \qquad -\infty < t < +\infty$$
$$z = t$$

    (a) lies in the plane $6x + 4y - 4z = 0$

    (b) is parallel to and below the plane $5x - 3y + 3z = 1$

    (c) is parallel to and above the plane $6x + 2y - 2z = 3$

15. Find an equation for the plane through $(-2, 1, 7)$ that is perpendicular to the line $x - 4 = 2t$, $y + 2 = 3t$, $z = -5t$.

16. Find an equation of

    (a) the xy-plane   (b) the xz-plane   (c) the yz-plane

17. Find an equation of the plane that contains the point (x_0, y_0, z_0) and is

    (a) parallel to the xy-plane   (b) parallel to the yz-plane   (c) parallel to the xz-plane

18. Find an equation for the plane that passes through the origin and is parallel to the plane $7x + 4y - 2z + 3 = 0$.

19. Find an equation for the plane that passes through the point $(3, -6, 7)$ and is parallel to the plane $5x - 2y + z - 5 = 0$.

20. Find the point of intersection of the line

$$x - 9 = -5t$$
$$y + 1 = -t$$
$$z - 3 = t$$

    and the plane $2x - 3y + 4z + 7 = 0$.

21. Find an equation for the plane that contains the line $x = -1 + 3t$, $y = 5 + 2t$, $z = 2 - t$ and is perpendicular to the plane $2x - 4y + 2z = 9$.

22. Find an equation for the plane that passes through $(2, 4, -1)$ and contains the line of intersection of the planes $x - y - 4z = 2$ and $-2x + y + 2z = 3$.

23. Show that the points $(-1, -2, -3)$, $(-2, 0, 1)$, $(-4, -1, -1)$, and $(2, 0, 1)$ lie in the same plane.

24. Find parametric equations for the line through $(-2, 5, 0)$ that is parallel to the planes $2x + y - 4z = 0$ and $-x + 2y + 3z + 1 = 0$.

25. Find an equation for the plane through $(-2, 1, 5)$ that is perpendicular to the planes $4x - 2y + 2z = -1$ and $3x + 3y - 6z = 5$.

26. Find an equation for the plane through $(2, -1, 4)$ that is perpendicular to the line of intersection of the planes $4x + 2y + 2z = -1$ and $3x + 6y + 3z = 7$.

27. Find an equation for the plane that is perpendicular to the plane $8x - 2y + 6z = 1$ and passes through the points $P_1(-1, 2, 5)$ and $P_2(2, 1, 4)$.

28. Show that the lines

$$x = 3 - 2t \qquad x = 5 + 2t$$
$$y = 4 + \ t \quad \text{and} \quad y = 1 - \ t$$
$$z = 1 - \ t \qquad z = 7 + \ t$$

are parallel, and find an equation for the plane they determine.

29. Find an equation for the plane that contains the point $(1, -1, 2)$ and the line $x = t$, $y = t + 1$, $z = -3 + 2t$.

30. Find an equation for the plane that contains the line $x = 1 + t$, $y = 3t$, $z = 2t$ and is parallel to the line of intersection of the planes $-x + 2y + z = 0$ and $x + z + 1 = 0$.

31. Find an equation for the plane, each of whose points is equidistant from $(-1, -4, -2)$ and $(0, -2, 2)$.

32. Show that the line

$$x - 5 = \ -t$$
$$y + 3 = \quad 2t$$
$$z + 1 = -5t$$

is parallel to the plane $-3x + y + z - 9 = 0$.

33. Show that the lines

$$x - 3 = 4t \qquad x + 1 = 12t$$
$$y - 4 = \ t \quad \text{and} \quad y - 7 = \ 6t$$
$$z - 1 = 0 \qquad z - 5 = \ 3t$$

intersect, and find the point of intersection.

34. Find an equation for the plane containing the lines in Exercise 33.

35. Find parametric equations for the line of intersection of the planes
(a) $-3x + 2y + z = -5$ and $7x + 3y - 2z = -2$
(b) $5x - 7y + 2z = 0$ and $y = 0$

36. Show that the plane whose intercepts with the coordinate axes are $x = a$, $y = b$, and $z = c$ has equation

$$\frac{x}{a} + \frac{y}{b} + \frac{z}{c} = 1$$

provided a, b, and c are nonzero.

37. Find the distance between the point and the plane.
(a) $(3, 1, -2)$; $x + 2y - 2z = 4$
(b) $(-1, 2, 1)$; $2x + 3y - 4z = 1$
(c) $(0, 3, -2)$; $x - y - z = 3$

38. Find the distance between the given parallel planes.
(a) $3x - 4y + z = 1$ and $6x - 8y + 2z = 3$
(b) $-4x + y - 3z = 0$ and $8x - 2y + 6z = 0$
(c) $2x - y + z = 1$ and $2x - y + z = -1$

39. Two intersecting planes in 3-space determine two angles of intersection, an acute angle $(0 \leq \theta \leq 90°)$ and its supplement $180° - \theta$ (Figure 6a). If $\mathbf{n}_1$ and $\mathbf{n}_2$ are nonzero normals to the planes, then the angle between $\mathbf{n}_1$ and $\mathbf{n}_2$ is θ or $180° - \theta$, depending on the directions of the normals (Figure 6b). In each part find the acute angle of intersection of the planes to the nearest degree.
(a) $x = 0$ and $2x - y + z - 4 = 0$
(b) $x + 2y - 2z = 5$ and $6x - 3y + 2z = 8$
[*Note.* A calculator or trigonometric tables are needed.]

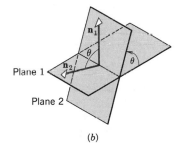

(a)   (b)   **Figure 6**

40. Find the acute angle of intersection between the plane $x - y - 3z = 5$ and the line $x = 2 - t$, $y = 2t$, $z = 3t - 1$ to the nearest degree. [*Hint.* See Exercise 39.]

CHAPTER 4

VECTOR SPACES

4.1 EUCLIDEAN *n*-SPACE

The idea of using pairs of numbers to locate points in the plane and triples of numbers to locate points in 3-space was first clearly spelled out in the mid-seventeenth century. By the latter part of the nineteenth century mathematicians and physicists began to realize that there was no need to stop with triples. It was recognized that quadruples of numbers (a_1, a_2, a_3, a_4) could be regarded as points in "4-dimensional" space, quintuples $(a_1, a_2, \ldots, a_5)$ as points in "5-dimensional" space, and so on. Although our geometric visualization does not extend beyond 3-space, it is nevertheless possible to extend many familiar ideas beyond 3-space by working with analytic or numerical properties of points and vectors rather than the geometric properties. In this section we shall make these ideas more precise.

VECTORS IN
***n*-SPACE**

Definition. If n is a positive integer, then an ***ordered n-tuple*** is a sequence of n real numbers $(a_1, a_2, \ldots, a_n)$. The set of all ordered n-tuples is called ***n-space*** and is denoted by R^n.

When $n = 2$ or 3, it is usual to use the terms ***ordered pair*** and ***ordered triple***, respectively, rather than ordered 2-tuple and 3-tuple. When $n = 1$, each ordered n-tuple consists of one real number, and so R^1 may be viewed as the set of real numbers. It is usual to write R rather than R^1 for this set.

It might have occurred to the reader in the study of 3-space that the symbol (a_1, a_2, a_3) has two different geometric interpretations: it can be interpreted as a point, in which case a_1, a_2, and a_3 are the coordinates (Figure 1*a*), or it can be

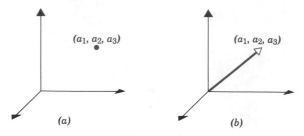

(a)   (b)

Figure 1 | The ordered triple (a_1, a_2, a_3) can be interpreted geometrically as a point or a vector.

interpreted as a vector, in which case a_1, a_2, and a_3 are the components (Figure 1*b*). It follows, therefore, that an ordered *n*-tuple $(a_1, a_2, \ldots, a_n)$ can be viewed either as a "generalized point" or a "generalized vector"—the distinction is mathematically unimportant. Thus, we can describe the 5-tuple $(-2, 4, 0, 1, 6)$ either as a point in R^5 or a vector in R^5.

Definition. Two vectors $\mathbf{u} = (u_1, u_2, \ldots, u_n)$ and $\mathbf{v} = (v_1, v_2, \ldots, v_n)$ in R^n are called *equal* if

$$u_1 = v_1, u_2 = v_2, \ldots, u_n = v_n$$

The *sum* $\mathbf{u} + \mathbf{v}$ is defined by

$$\mathbf{u} + \mathbf{v} = (u_1 + v_1, u_2 + v_2, \ldots, u_n + v_n)$$

and if k is any scalar, the *scalar multiple* $k\mathbf{u}$ is defined by

$$k\mathbf{u} = (ku_1, ku_2, \ldots, ku_n)$$

The operations of addition and scalar multiplication in this definition are called the *standard operations* on R^n.

The *zero vector* in R^n is denoted by $\mathbf{0}$ and is defined to be the vector

$$\mathbf{0} = (0, 0, \ldots, 0)$$

If $\mathbf{u} = (u_1, u_2, \ldots, u_n)$ is any vector in R^n, then the *negative* (or *additive inverse*) of $\mathbf{u}$ is denoted by $-\mathbf{u}$ and is defined by

$$-\mathbf{u} = (-u_1, -u_2, \ldots, -u_n)$$

The *difference* of vectors in R^n is defined by

$$\mathbf{v} - \mathbf{u} = \mathbf{v} + (-\mathbf{u})$$

or in terms of components

$$\mathbf{v} - \mathbf{u} = (v_1 - u_1, v_2 - u_2, \ldots, v_n - u_n)$$

**PROPERTIES
OF VECTOR
OPERATIONS IN
n-SPACE**

The most important arithmetic properties of addition and scalar multiplication of vectors in R^n are listed in the following theorem. The proofs are all easy and are left as exercises.

Theorem 4.1.1. *If* $\mathbf{u} = (u_1, u_2, \ldots, u_n)$, $\mathbf{v} = (v_1, v_2, \ldots, v_n)$, *and* $\mathbf{w} = (w_1, w_2, \ldots, w_n)$ *are vectors in* R^n *and* k *and* l *are scalars, then*

(a) $\mathbf{u} + \mathbf{v} = \mathbf{v} + \mathbf{u}$
(b) $\mathbf{u} + (\mathbf{v} + \mathbf{w}) = (\mathbf{u} + \mathbf{v}) + \mathbf{w}$
(c) $\mathbf{u} + \mathbf{0} = \mathbf{0} + \mathbf{u} = \mathbf{u}$
(d) $\mathbf{u} + (-\mathbf{u}) = \mathbf{0}$, *that is,* $\mathbf{u} - \mathbf{u} = \mathbf{0}$
(e) $k(l\mathbf{u}) = (kl)\mathbf{u}$
(f) $k(\mathbf{u} + \mathbf{v}) = k\mathbf{u} + k\mathbf{v}$
(g) $(k + l)\mathbf{u} = k\mathbf{u} + l\mathbf{u}$
(h) $1\mathbf{u} = \mathbf{u}$

This theorem enables us to manipulate vectors in R^n without expressing the vectors in terms of components. For example, to solve the vector equation $\mathbf{x} + \mathbf{u} = \mathbf{v}$ for $\mathbf{x}$, we can add $-\mathbf{u}$ to both sides and proceed as follows.

$$(\mathbf{x} + \mathbf{u}) + (-\mathbf{u}) = \mathbf{v} + (-\mathbf{u})$$
$$\mathbf{x} + (\mathbf{u} - \mathbf{u}) = \mathbf{v} - \mathbf{u}$$
$$\mathbf{x} + \mathbf{0} = \mathbf{v} - \mathbf{u}$$
$$\mathbf{x} = \mathbf{v} - \mathbf{u}$$

The reader will find it instructive to name the parts of Theorem 4.1.1 that justify the last three steps in this computation.

**EUCLIDEAN
n-SPACE**

To extend the notions of distance, norm, and angle to R^n, we begin with the following generalization of the dot product on R^2 and R^3 [Formula (3) of Section 3.3].

Definition. If $\mathbf{u} = (u_1, u_2, \ldots, u_n)$ and $\mathbf{v} = (v_1, v_2, \ldots, v_n)$ are any vectors in R^n, then the ***Euclidean inner product*** $\mathbf{u} \cdot \mathbf{v}$ is defined by

$$\mathbf{u} \cdot \mathbf{v} = u_1 v_1 + u_2 v_2 + \cdots + u_n v_n$$

Observe that when $n = 2$ or 3, the Euclidean inner product is the ordinary dot product [Formulas (3) and (4) of Section 3.3].

Example 1 The Euclidean inner product of the vectors

$$\mathbf{u} = (-1, 3, 5, 7) \quad \text{and} \quad \mathbf{v} = (5, -4, 7, 0)$$

in R^4 is

$$\mathbf{u} \cdot \mathbf{v} = (-1)(5) + (3)(-4) + (5)(7) + (7)(0) = 18 \quad \blacktriangle$$

Since so many of the familiar ideas from 2-space and 3-space carry over, it is common to refer to R^n with the operations of addition, scalar multiplication, and the Euclidean inner product as ***Euclidean n-space***.

The four main arithmetic properties of the Euclidean inner product are listed in the next theorem.

> **Theorem 4.1.2.** *If* **u**, **v**, *and* **w** *are vectors in* R^n *and* k *is any scalar, then*
>
> (a) $\mathbf{u} \cdot \mathbf{v} = \mathbf{v} \cdot \mathbf{u}$
> (b) $(\mathbf{u} + \mathbf{v}) \cdot \mathbf{w} = \mathbf{u} \cdot \mathbf{w} + \mathbf{v} \cdot \mathbf{w}$
> (c) $(k\mathbf{u}) \cdot \mathbf{v} = k(\mathbf{u} \cdot \mathbf{v})$
> (d) $\mathbf{v} \cdot \mathbf{v} \geq \mathbf{0}$. *Further,* $\mathbf{v} \cdot \mathbf{v} = 0$ *if and only if* $\mathbf{v} = \mathbf{0}$.

We shall prove parts (b) and (d) and leave proofs of the rest as exercises.

Proof (b). Let $\mathbf{u} = (u_1, u_2, \ldots, u_n)$, $\mathbf{v} = (v_1, v_2, \ldots, v_n)$, and $\mathbf{w} = (w_1, w_2, \ldots, w_n)$. Then

$$\begin{aligned}
(\mathbf{u} + \mathbf{v}) \cdot \mathbf{w} &= (u_1 + v_1, u_2 + v_2, \ldots, u_n + v_n) \cdot (w_1, w_2, \ldots, w_n) \\
&= (u_1 + v_1)w_1 + (u_2 + v_2)w_2 + \cdots + (u_n + v_n)w_n \\
&= (u_1 w_1 + u_2 w_2 + \cdots + u_n w_n) + (v_1 w_1 + v_2 w_2 + \cdots + v_n w_n) \\
&= \mathbf{u} \cdot \mathbf{w} + \mathbf{v} \cdot \mathbf{w}
\end{aligned}$$

Proof (d). $\mathbf{v} \cdot \mathbf{v} = v_1^2 + v_2^2 + \cdots + v_n^2 \geq 0$. Further, equality holds if and only if $v_1 = v_2 = \cdots = v_n = 0$, that is, if and only if $\mathbf{v} = \mathbf{0}$. ∎

Example 2 Theorem 4.1.2 allows us to perform computations with Euclidean inner products in much the same way that we perform them with ordinary arithmetic products. For example,

$$\begin{aligned}
(3\mathbf{u} + 2\mathbf{v}) \cdot (4\mathbf{u} + \mathbf{v}) &= (3\mathbf{u}) \cdot (4\mathbf{u} + \mathbf{v}) + (2\mathbf{v}) \cdot (4\mathbf{u} + \mathbf{v}) \\
&= (3\mathbf{u}) \cdot (4\mathbf{u}) + (3\mathbf{u}) \cdot \mathbf{v} + (2\mathbf{v}) \cdot (4\mathbf{u}) + (2\mathbf{v}) \cdot \mathbf{v} \\
&= 12(\mathbf{u} \cdot \mathbf{u}) + 3(\mathbf{u} \cdot \mathbf{v}) + 8(\mathbf{v} \cdot \mathbf{u}) + 2(\mathbf{v} \cdot \mathbf{v}) \\
&= 12(\mathbf{u} \cdot \mathbf{u}) + 11(\mathbf{u} \cdot \mathbf{v}) + 2(\mathbf{v} \cdot \mathbf{v})
\end{aligned}$$

The reader should determine which parts of Theorem 4.1.2 were used in each step. ▲

NORM AND DISTANCE IN EUCLIDEAN n-SPACE

By analogy with the familiar formulas in R^2 and R^3, we define the ***Euclidean norm*** (or ***Euclidean length***) of a vector $\mathbf{u} = (u_1, u_2, \ldots, u_n)$ in R^n by

$$\|\mathbf{u}\| = (\mathbf{u} \cdot \mathbf{u})^{1/2} = \sqrt{u_1^2 + u_2^2 + \cdots + u_n^2} \tag{1}$$

[Compare this formula to Formulas (1) and (2) in Section 3.2.]

Similarly, the **Euclidean distance** between the points $\mathbf{u} = (u_1, u_2, \ldots, u_n)$ and $\mathbf{v} = (v_1, v_2, \ldots, v_n)$ in R^n is defined by

$$d(\mathbf{u}, \mathbf{v}) = \|\mathbf{u} - \mathbf{v}\| = \sqrt{(u_1 - v_1)^2 + (u_2 - v_2)^2 + \cdots + (u_n - v_n)^2} \qquad (2)$$

[See Formulas (3) and (4) of Section 3.2.]

Example 3   If $\mathbf{u} = (1, 3, -2, 7)$ and $\mathbf{v} = (0, 7, 2, 2)$, then

$$\|\mathbf{u}\| = \sqrt{(1)^2 + (3)^2 + (-2)^2 + (7)^2} = \sqrt{63} = 3\sqrt{7}$$

and

$$d(\mathbf{u}, \mathbf{v}) = \sqrt{(1 - 0)^2 + (3 - 7)^2 + (-2 - 2)^2 + (7 - 2)^2} = \sqrt{58} \quad \blacktriangle$$

ALTERNATIVE NOTATIONS FOR VECTORS IN R^n

We conclude this section by noting that it is possible to use the matrix notation

$$\mathbf{u} = \begin{bmatrix} u_1 \\ u_2 \\ \vdots \\ u_n \end{bmatrix}$$

rather than the horizontal notation $\mathbf{u} = (u_1, u_2, \ldots u_n)$ to denote vectors in R^n. This is justified because the matrix operations

$$\mathbf{u} + \mathbf{v} = \begin{bmatrix} u_1 \\ u_2 \\ \vdots \\ u_n \end{bmatrix} + \begin{bmatrix} v_1 \\ v_2 \\ \vdots \\ v_n \end{bmatrix} = \begin{bmatrix} u_1 + v_1 \\ u_2 + v_2 \\ \vdots \\ u_n + v_n \end{bmatrix}$$

$$k\mathbf{u} = k\begin{bmatrix} u_1 \\ u_2 \\ \vdots \\ u_n \end{bmatrix} = \begin{bmatrix} ku_1 \\ ku_2 \\ \vdots \\ ku_n \end{bmatrix}$$

produce the same results as the vector operations

$$\mathbf{u} + \mathbf{v} = (u_1, u_2, \ldots, u_n) + (v_1, v_2, \ldots, v_n) = (u_1 + v_1, u_2 + v_2, \ldots, u_n + v_n)$$

$$k\mathbf{u} = k(u_1, u_2, \ldots, u_n) = (ku_1, ku_2, \ldots, ku_n)$$

The only difference is that results are displayed vertically in one case and horizontally in the other. We will use both notations at various times. However, from here on, we shall denote $n \times 1$ matrices by lowercase boldface letters. Thus, a system of linear equations will be written as

$$A\mathbf{x} = \mathbf{b}$$

rather than $AX = B$ as before.

REMARK. Some authors use the notation R_n for Euclidean *n*-space when vectors are written in vertical notation and R^n for Euclidean *n*-space when vectors are written in horizontal notation. We will use R^n in both cases. This will not cause any problems, since the notation being used will always be clear from the context.

A MATRIX FORMULA FOR THE DOT PRODUCT

If we use matrix notation for the vectors

$$\mathbf{u} = \begin{bmatrix} u_1 \\ u_2 \\ \vdots \\ u_n \end{bmatrix} \quad \text{and} \quad \mathbf{v} = \begin{bmatrix} v_1 \\ v_2 \\ \vdots \\ v_n \end{bmatrix}$$

and omit the brackets on 1×1 matrices, then it follows that

$$\mathbf{v}^t\mathbf{u} = \begin{bmatrix} v_1 & v_2 & \cdots & v_n \end{bmatrix} \begin{bmatrix} u_1 \\ u_2 \\ \vdots \\ u_n \end{bmatrix} = \begin{bmatrix} u_1v_1 + u_2v_2 + \cdots + u_nv_n \end{bmatrix} = \begin{bmatrix} \mathbf{u} \cdot \mathbf{v} \end{bmatrix} = \mathbf{u} \cdot \mathbf{v}$$

Thus, for vectors in vertical notation we have the following matrix formula for the Euclidean inner product:

$$\mathbf{v}^t\mathbf{u} = \mathbf{u} \cdot \mathbf{v} \tag{3}$$

For example, if

$$\mathbf{u} = \begin{bmatrix} -1 \\ 3 \\ 5 \\ 7 \end{bmatrix} \quad \text{and} \quad \mathbf{v} = \begin{bmatrix} 5 \\ -4 \\ 7 \\ 0 \end{bmatrix}$$

then

$$\mathbf{u} \cdot \mathbf{v} = \mathbf{v}^t\mathbf{u} = \begin{bmatrix} 5 & -4 & 7 & 0 \end{bmatrix} \begin{bmatrix} -1 \\ 3 \\ 5 \\ 7 \end{bmatrix} = [18] = 18$$

EXERCISE SET 4.1

1. Let $\mathbf{u} = (-3, 2, 1, 0)$, $\mathbf{v} = (4, 7, -3, 2)$, and $\mathbf{w} = (5, -2, 8, 1)$. Find
   (a) $\mathbf{v} - \mathbf{w}$ (b) $2\mathbf{u} + 7\mathbf{v}$ (c) $-\mathbf{u} + (\mathbf{v} - 4\mathbf{w})$
   (d) $6(\mathbf{u} - 3\mathbf{v})$ (e) $-\mathbf{v} - \mathbf{w}$ (f) $(6\mathbf{v} - \mathbf{w}) - (4\mathbf{u} + \mathbf{v})$

2. Let $\mathbf{u}$, $\mathbf{v}$, and $\mathbf{w}$ be the vectors in Exercise 1. Find the vector $\mathbf{x}$ that satisfies
   $5\mathbf{x} - 2\mathbf{v} = 2(\mathbf{w} - 5\mathbf{x})$.

3. Let $u_1 = (-1, 3, 2, 0)$, $u_2 = (2, 0, 4, -1)$, $u_3 = (7, 1, 1, 4)$, and $u_4 = (6, 3, 1, 2)$. Find scalars c_1, c_2, c_3, and c_4 such that $c_1u_1 + c_2u_2 + c_3u_3 + c_4u_4 = (0, 5, 6, -3)$.

4. Show that there do not exist scalars c_1, c_2, and c_3 such that

   det = 8

   $$c_1(1, 0, 1, 0) + c_2(1, 0, -2, 1) + c_3(2, 0, 1, 2) = (1, -2, 2, 3)$$

5. In each part compute the Euclidean norm of the vector.

   $\|u\| = \sqrt{u_1^2 + u_2^2}$ etc

   (a) $(-2, 5)$ (b) $(1, 2, -2)$ (c) $(3, 4, 0, -12)$ (d) $(-2, 1, 1, -3, 4)$

6. Let $u = (4, 1, 2, 3)$, $v = (0, 3, 8, -2)$, and $w = (3, 1, 2, 2)$. Evaluate the expression.

   (a) $\|u + v\|$ (b) $\|u\| + \|v\|$ (c) $\|-2u\| + 2\|u\|$

   (d) $\|3u - 5v + w\|$ (e) $\dfrac{1}{\|w\|} w$ (f) $\left\| \dfrac{1}{\|w\|} w \right\|$

7. Show that if v is a nonzero vector in R^n, then $(1/\|v\|)v$ has Euclidean norm 1.

8. Let $v = (-2, 3, 0, 6)$. Find all scalars k such that $\|kv\| = 5$.

9. Find the Euclidean inner product $u \cdot v$.

   (a) $u = (2, 5)$, $v = (-4, 3)$ (b) $u = (4, 8, 2)$, $v = (0, 1, 3)$

   (c) $u = (3, 1, 4, -5)$, $v = (2, 2, -4, -3)$ (d) $u = (-1, 1, 0, 4, -3)$, $v = (-2, -2, 0, 2, -1)$

10. (a) Find two vectors in R^2 with Euclidean norm 1 whose Euclidean inner product with $(3, -1)$ is zero.

    (b) Show that there are infinitely many vectors in R^3 with Euclidean norm 1 whose Euclidean inner product with $(1, -3, 5)$ is zero.

11. Find the Euclidean distance between u and v.   *$\|u - v\| = \sqrt{(diff)^2 + (diff)^2}$ etc*

    (a) $u = (1, -2)$, $v = (2, 1)$ (b) $u = (2, -2, 2)$, $v = (0, 4, -2)$

    (c) $u = (0, -2, -1, 1)$, $v = (-3, 2, 4, 4)$ (d) $u = (3, -3, -2, 0, -3)$, $v = (-4, 1, -1, 5, 0)$

12. Verify parts (b), (e), (f), and (g) of Theorem 4.1.1 for $u = (2, 0, -3, 1)$, $v = (4, 0, 3, 5)$, $w = (1, 6, 2, -1)$, $k = 5$, and $l = -3$.

13. Verify parts (b) and (c) of Theorem 4.1.2 for the values of u, v, w, and k in Exercise 12.

14. Prove the identity

    $$\|u + v\|^2 + \|u - v\|^2 = 2\|u\|^2 + 2\|v\|^2$$

    for vectors in R^n. Interpret this result geometrically in R^2.

15. Prove the identity

    $$u \cdot v = \tfrac{1}{4}\|u + v\|^2 - \tfrac{1}{4}\|u - v\|^2$$

    for vectors in R^n.

16. Prove: If u, v, and w are vectors in R^n and k is any scalar, then

    (a) $u \cdot (kv) = k(u \cdot v)$ (b) $u \cdot (v + w) = u \cdot v + u \cdot w$

17. Prove (a) through (d) of Theorem 4.1.1.

18. Prove (e) through (h) of Theorem 4.1.1.

19. Prove (a) and (c) of Theorem 4.1.2.

20. Suppose that $a_1, a_2, \ldots, a_n$ are positive real numbers. In R^2, the vectors $\mathbf{v}_1 = (a_1, 0)$ and $\mathbf{v}_2 = (0, a_2)$ determine a rectangle of area $A = a_1 a_2$ (Figure 2a), and in R^3 the vectors $\mathbf{v}_1 = (a_1, 0, 0)$, $\mathbf{v}_2 = (0, a_2, 0)$, and $\mathbf{v}_3 = (0, 0, a_3)$ determine a box of volume $V = a_1 a_2 a_3$ (Figure 2b). The area A and the volume V are sometimes called the **Euclidean measure** of the rectangle and box, respectively.

(a) How would you define the Euclidean measure of the "box" in R^n that is determined by the vectors

$$\mathbf{v}_1 = (a_1, 0, 0, \ldots, 0), \mathbf{v}_2 = (0, a_2, 0, \ldots, 0), \ldots, \mathbf{v}_n = (0, 0, 0, \ldots, a_n)?$$

(b) How would you define the Euclidean length of the diagonal of the box in part (a)?

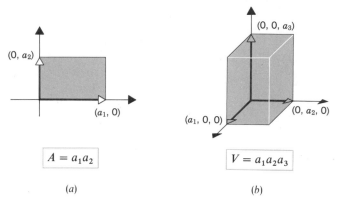

(a) (b) **Figure 2**

4.2 GENERAL VECTOR SPACES

In this section we shall generalize the concept of a vector still further. We shall state a set of axioms which, if satisfied by a class of objects, will entitle those objects to be called "vectors." The axioms will be chosen by abstracting the most important properties of vectors in R^n; as a consequence, vectors in R^n will automatically satisfy these axioms. Thus, our new concept of a vector will include our old vectors and many new kinds of vectors as well. These new types of vectors will include, among other things, various kinds of matrices and functions. Our work in this section is not an idle exercise in theoretical mathematics; it will provide a powerful tool for extending our geometric visualization to a wide variety of important mathematical problems where geometric intuition would not otherwise be available. Briefly stated, the idea is this: We can visualize vectors in R^2 and R^3 geometrically as arrows, which enables us to draw physical or mental pictures to help solve problems. Because the axioms that we will be using to create our new kinds of vectors will be based on properties of vectors in R^2 and R^3, these new vectors will have many of the familiar properties of vectors in R^2 and R^3. Consequently, when we want to solve a problem involving our new kinds of vectors, say matrices or functions, we may be able to get a foothold on the problem by visualizing geometrically what the corresponding problem would be like in R^2 and R^3.

VECTOR SPACE AXIOMS

Definition. Let V be an arbitrary nonempty set of objects on which two operations are defined, addition and multiplication by scalars (numbers). By *addition* we mean a rule for associating with each pair of objects **u** and **v** in V an object $\mathbf{u} + \mathbf{v}$, called the *sum* of **u** and **v**; by *scalar multiplication* we mean a rule for associating with each scalar k and each object **u** in V an object $k\mathbf{u}$, called the *scalar multiple* of **u** by k. If the following axioms are satisfied by all objects **u**, **v**, **w** in V and all scalars k and l, then we call V a *vector space* and we call the objects in V *vectors*:

(1) If **u** and **v** are objects in V, then $\mathbf{u} + \mathbf{v}$ is in V.

(2) $\mathbf{u} + \mathbf{v} = \mathbf{v} + \mathbf{u}$

(3) $\mathbf{u} + (\mathbf{v} + \mathbf{w}) = (\mathbf{u} + \mathbf{v}) + \mathbf{w}$

(4) There is an object **0** in V, called a *zero vector* for V, such that $\mathbf{0} + \mathbf{u} = \mathbf{u} + \mathbf{0} = \mathbf{u}$ for all **u** in V.

(5) For each **u** in V, there is an object $-\mathbf{u}$ in V, called a *negative* of **u**, such that $\mathbf{u} + (-\mathbf{u}) = (-\mathbf{u}) + \mathbf{u} = \mathbf{0}$.

(6) If k is any scalar and **u** is any object in V, then $k\mathbf{u}$ is in V.

(7) $k(\mathbf{u} + \mathbf{v}) = k\mathbf{u} + k\mathbf{v}$

(8) $(k + l)\mathbf{u} = k\mathbf{u} + l\mathbf{u}$

(9) $k(l\mathbf{u}) = (kl)(\mathbf{u})$

(10) $1\mathbf{u} = \mathbf{u}$

REMARK. Depending on the application, scalars may be real numbers or complex numbers. Vector spaces in which the scalars are complex numbers are called *complex vector spaces*, and those in which the scalars must be real are called *real vector spaces*. In Chapter 10 we shall discuss complex vector spaces; until then, *all of our scalars will be real numbers*.

The reader should keep in mind that, in the definition of a vector space, neither the nature of the vectors nor the operations is specified. Any kind of object can be a vector, and the operations of addition and scalar multiplication may not have any relationship or similarity to the standard vector operations on R^n. The only requirement is that the ten vector space axioms be satisfied. Some authors use the notations $\oplus$ and $\odot$ for vector addition and scalar multiplication to distinguish these operations from addition and multiplication of real numbers; we will not use this convention, however.

EXAMPLES OF VECTOR SPACES

The following examples will illustrate the variety of possible vector spaces. In each example we will specify a nonempty set V and two operations: addition and scalar multiplication; then we shall verify that the ten vector space axioms are satisfied, thereby entitling V, with the specified operations, to be called a vector space.

Example 1 The set $V = R^n$ with the standard operations of addition and scalar multiplication defined in the previous section is a vector space. Axioms 1 and 6 follow from the definitions of the standard operations on R^n; the remaining axioms follow from Theorem 4.1.1. ▲

The three most important special cases of Example 1 are R (the real numbers), R^2 (the vectors in the plane), and R^3 (the vectors in 3-space).

Example 2 Show that the set V of all 2×2 matrices with real entries is a vector space if vector addition is defined to be matrix addition and vector scalar multiplication is defined to be matrix scalar multiplication.

Solution. In this example we will find it convenient to verify the axioms in the following order: 1, 6, 2, 3, 7, 8, 9, 4, 5, and 10. Let

$$\mathbf{u} = \begin{bmatrix} u_{11} & u_{12} \\ u_{21} & u_{22} \end{bmatrix} \quad \text{and} \quad \mathbf{v} = \begin{bmatrix} v_{11} & v_{12} \\ v_{21} & v_{22} \end{bmatrix}$$

To prove Axiom 1, we must show that $\mathbf{u} + \mathbf{v}$ is an object in V; that is, we must show that $\mathbf{u} + \mathbf{v}$ is a 2×2 matrix. But this follows from the definition of matrix addition, since

$$\mathbf{u} + \mathbf{v} = \begin{bmatrix} u_{11} & u_{12} \\ u_{21} & u_{22} \end{bmatrix} + \begin{bmatrix} v_{11} & v_{12} \\ v_{21} & v_{22} \end{bmatrix} = \begin{bmatrix} u_{11} + v_{11} & u_{12} + v_{12} \\ u_{21} + v_{21} & u_{22} + v_{22} \end{bmatrix}$$

Similarly, Axiom 6 holds because for any real number k we have

$$k\mathbf{u} = k\begin{bmatrix} u_{11} & u_{12} \\ u_{21} & u_{22} \end{bmatrix} = \begin{bmatrix} ku_{11} & ku_{12} \\ ku_{21} & ku_{22} \end{bmatrix}$$

so that $k\mathbf{u}$ is a 2×2 matrix and consequently is an object in V.

Axiom 2 follows from Theorem 1.5.1a since

$$\mathbf{u} + \mathbf{v} = \begin{bmatrix} u_{11} & u_{12} \\ u_{21} & u_{22} \end{bmatrix} + \begin{bmatrix} v_{11} & v_{12} \\ v_{21} & v_{22} \end{bmatrix} = \begin{bmatrix} v_{11} & v_{12} \\ v_{21} & v_{22} \end{bmatrix} + \begin{bmatrix} u_{11} & u_{12} \\ u_{21} & u_{22} \end{bmatrix} = \mathbf{v} + \mathbf{u}$$

Similarly, Axiom 3 follows from part (*b*) of that theorem; and Axioms 7, 8, and 9 follow from parts (*h*), (*j*), and (*l*), respectively.

To prove Axiom 4, we must find an object $\mathbf{0}$ in V such that $\mathbf{0} + \mathbf{u} = \mathbf{u} + \mathbf{0} = \mathbf{u}$ for all $\mathbf{u}$ in V. This can be done by defining $\mathbf{0}$ to be

$$\mathbf{0} = \begin{bmatrix} 0 & 0 \\ 0 & 0 \end{bmatrix}$$

With this definition

$$\mathbf{0} + \mathbf{u} = \begin{bmatrix} 0 & 0 \\ 0 & 0 \end{bmatrix} + \begin{bmatrix} u_{11} & u_{12} \\ u_{21} & u_{22} \end{bmatrix} = \begin{bmatrix} u_{11} & u_{12} \\ u_{21} & u_{22} \end{bmatrix} = \mathbf{u}$$

and similarly $\mathbf{u} + \mathbf{0} = \mathbf{u}$. To prove Axiom 5 we must show that each object $\mathbf{u}$ in V has a negative $-\mathbf{u}$ such that $\mathbf{u} + (-\mathbf{u}) = \mathbf{0}$ and $(-\mathbf{u}) + \mathbf{u} = \mathbf{0}$. This can be done by defining the negative of **u** to be

$$-\mathbf{u} = \begin{bmatrix} -u_{11} & -u_{12} \\ -u_{21} & -u_{22} \end{bmatrix}$$

With this definition

$$\mathbf{u} + (-\mathbf{u}) = \begin{bmatrix} u_{11} & u_{12} \\ u_{21} & u_{22} \end{bmatrix} + \begin{bmatrix} -u_{11} & -u_{12} \\ -u_{21} & -u_{22} \end{bmatrix} = \begin{bmatrix} 0 & 0 \\ 0 & 0 \end{bmatrix} = \mathbf{0}$$

and similarly $(-\mathbf{u}) + \mathbf{u} = \mathbf{0}$. Finally, Axiom 10 is a simple computation:

$$1\mathbf{u} = 1\begin{bmatrix} u_{11} & u_{12} \\ u_{21} & u_{22} \end{bmatrix} = \begin{bmatrix} u_{11} & u_{12} \\ u_{21} & u_{22} \end{bmatrix} = \mathbf{u} \quad \blacktriangle$$

Example 3   Example 2 is a special case of a more general class of vector spaces. The arguments in that example can be adapted to show that the set V of all $m \times n$ matrices with real entries, together with the operations of matrix addition and scalar multiplication, is a vector space. The $m \times n$ zero matrix is the zero vector **0**, and if **u** is the $m \times n$ matrix U, then the matrix $-U$ is the negative $-\mathbf{u}$ of the vector **u**. We shall denote this vector space by the symbol M_{mn}. $\blacktriangle$

Example 4   Let V be the set of real-valued functions defined on the entire real line. If $\mathbf{f} = f(x)$ and $\mathbf{g} = g(x)$ are two such functions and k is any real number, define the sum function $\mathbf{f} + \mathbf{g}$ and the scalar multiple $k\mathbf{f}$ by

$$(\mathbf{f} + \mathbf{g})(x) = f(x) + g(x)$$
$$(k\mathbf{f})(x) = kf(x)$$

In other words, the value of the function $\mathbf{f} + \mathbf{g}$ at x is obtained by adding together the value of **f** and **g** at x (Figure 1*a*). Similarly, the value of $k\mathbf{f}$ at x is k times the value of **f** at x (Figure 1*b*). In the exercises we shall ask you to show that V is a vector space with respect to these operations. The zero vector in this space is the constant function that is identically zero for all values of x, and the negative of a function **f** is the function $-\mathbf{f}$ whose value at any point x is $-f(x)$. Geometrically, the graph of $-\mathbf{f}$ is the reflection of the graph of **f** across the x-axis (Figure 1*c*). $\blacktriangle$

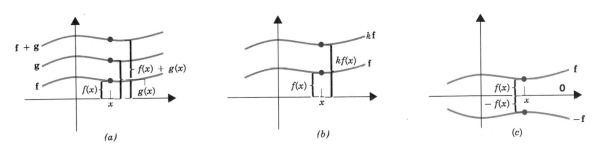

(a)   (b)   (c)

Figure 1

Example 5 Let $V = R^2$ and define addition and scalar multiplication operations as follows: If $\mathbf{u} = (u_1, u_2)$ and $\mathbf{v} = (v_1, v_2)$, then define

$$\mathbf{u} + \mathbf{v} = (u_1 + v_1, u_2 + v_2)$$

and if k is any real number, then define

$$k\mathbf{u} = (ku_1, 0)$$

For example, if $\mathbf{u} = (2, 4)$, $\mathbf{v} = (-3, 5)$, and $k = 7$, then

$$\mathbf{u} + \mathbf{v} = (2 + (-3), 4 + 5) = (-1, 9)$$
$$k\mathbf{u} = 7\mathbf{u} = (7 \cdot 2, 0) = (14, 0)$$

The addition operation is the standard addition operation on R^2, but the scalar multiplication operation is not the standard scalar multiplication. In the exercises we will ask you to show that the first nine vector space axioms are satisfied; however, there are values of $\mathbf{u}$ for which Axiom 10 fails to hold. For example, if $\mathbf{u} = (u_1, u_2)$ is such that $u_2 \neq 0$, then

$$1\mathbf{u} = 1(u_1, u_2) = (1 \cdot u_1, 0) = (u_1, 0) \neq \mathbf{u}$$

Thus, V is not a vector space with the stated operations. ▲

Example 6 Let V be any plane through the origin in R^3. We shall show that the points in V form a vector space under the standard addition and scalar multiplication operations for vectors in R^3. From Example 1, we know that R^3 itself is a vector space under these operations. Thus, Axioms 2, 3, 7, 8, 9, and 10 hold for all points in R^3 and consequently for all points in the plane V. We therefore need only show that Axioms 1, 4, 5, and 6 are satisfied.

Since the plane V passes through the origin, it has an equation of the form

$$ax + by + cz = 0 \tag{1}$$

(Theorem 3.5.1). Thus, if $\mathbf{u} = (u_1, u_2, u_3)$ and $\mathbf{v} = (v_1, v_2, v_3)$ are points in V, then $au_1 + bu_2 + cu_3 = 0$ and $av_1 + bv_2 + cv_3 = 0$. Adding these equations gives

$$a(u_1 + v_1) + b(u_2 + v_2) + c(u_3 + v_3) = 0$$

This equality tells us that the coordinates of the point

$$\mathbf{u} + \mathbf{v} = (u_1 + v_1, u_2 + v_2, u_3 + v_3)$$

satisfy (1); thus, $\mathbf{u} + \mathbf{v}$ lies in the plane V. This proves that Axiom 1 is satisfied. Multiplying $au_1 + bu_2 + cu_3 = 0$ through by -1 gives

$$a(-u_1) + b(-u_2) + c(-u_3) = 0$$

Thus, $-\mathbf{u} = (-u_1, -u_2, -u_3)$ lies in V. This establishes Axiom 5. The verifications of Axioms 4 and 6 are left as exercises. ▲

Example 7 Let V consist of a single object, which we denote by $\mathbf{0}$, and define

$$\mathbf{0} + \mathbf{0} = \mathbf{0}$$
$$k\mathbf{0} = \mathbf{0}$$

for all scalars k. It is easy to check that all the vector-space axioms are satisfied. We call this the *zero vector space*. ▲

SOME PROPERTIES OF VECTORS

As we progress, we shall add more examples of vector spaces to our list. We conclude this section with a theorem that gives a useful list of vector properties.

Theorem 4.2.1. *Let V be a vector space, $\mathbf{u}$ a vector in V, and k a scalar; then:*

(*a*) $0\mathbf{u} = \mathbf{0}$
(*b*) $k\mathbf{0} = \mathbf{0}$
(*c*) $(-1)\mathbf{u} = -\mathbf{u}$
(*d*) *If $k\mathbf{u} = \mathbf{0}$, then $k = 0$ or $\mathbf{u} = \mathbf{0}$.*

We shall prove parts (*a*) and (*c*) and leave proofs of the remaining parts as exercises.

Proof (*a*). We can write

$$0\mathbf{u} + 0\mathbf{u} = (0 + 0)\mathbf{u} \qquad\qquad \text{[Axiom 8]}$$
$$= 0\mathbf{u} \qquad\qquad \text{[Property of the number 0]}$$

By Axiom 5 the vector $0\mathbf{u}$ has a negative, $-0\mathbf{u}$. Adding this negative to both sides above yields

$$[0\mathbf{u} + 0\mathbf{u}] + (-0\mathbf{u}) = 0\mathbf{u} + (-0\mathbf{u})$$

or

$$0\mathbf{u} + [0\mathbf{u} + (-0\mathbf{u})] = 0\mathbf{u} + (-0\mathbf{u}) \qquad\qquad \text{[Axiom 3]}$$
$$0\mathbf{u} + \mathbf{0} = \mathbf{0} \qquad\qquad \text{[Axiom 5]}$$
$$0\mathbf{u} = \mathbf{0} \qquad\qquad \text{[Axiom 4]}$$

Proof (*c*). To show $(-1)\mathbf{u} = -\mathbf{u}$, we must demonstrate that $\mathbf{u} + (-1)\mathbf{u} = \mathbf{0}$. To see this, observe that

$$\mathbf{u} + (-1)\mathbf{u} = 1\mathbf{u} + (-1)\mathbf{u} \qquad\qquad \text{[Axiom 10]}$$
$$= (1 + (-1))\mathbf{u} \qquad\qquad \text{[Axiom 8]}$$
$$= 0\mathbf{u} \qquad\qquad \text{[Property of numbers]}$$
$$= \mathbf{0} \qquad\qquad \text{[Part (*a*) above]}$$

EXERCISE SET 4.2

In Exercises 1–14 a set of objects is given together with operations of addition and scalar multiplication. Determine which sets are vector spaces under the given operations. For those that are not, list all axioms that fail to hold.

1. The set of all triples of real numbers (x, y, z) with the operations

$$(x, y, z) + (x', y', z') = (x + x', y + y', z + z') \quad \text{and} \quad k(x, y, z) = (kx, y, z)$$

2. The set of all triples of real numbers (x, y, z) with the operations

$$(x, y, z) + (x', y', z') = (x + x', y + y', z + z') \quad \text{and} \quad k(x, y, z) = (0, 0, 0)$$

3. The set of all pairs of real numbers (x, y) with the operations

$$(x, y) + (x', y') = (x + x', y + y') \quad \text{and} \quad k(x, y) = (2kx, 2ky)$$

4. The set of all real numbers x with the standard operations of addition and multiplication.

5. The set of all pairs of real numbers of the form $(x, 0)$ with the standard operations on R^2.

6. The set of all pairs of real numbers of the form (x, y), where $x \geq 0$, with the standard operations on R^2.

7. The set of all n-tuples of real numbers of the form $(x, x, \ldots, x)$ with the standard operations on R^n.

8. The set of all pairs of real numbers (x, y) with the operations

$$(x, y) + (x', y') = (x + x' + 1, y + y' + 1) \quad \text{and} \quad k(x, y) = (kx, ky)$$

9. The set of all positive real numbers x with the operations $x + x' = xx'$ and $kx = x^k$.

10. The set of all 2×2 matrices of the form

$$\begin{bmatrix} a & 1 \\ 1 & b \end{bmatrix}$$

with matrix addition and scalar multiplication.

11. The set of all 2×2 matrices of the form

$$\begin{bmatrix} a & 0 \\ 0 & b \end{bmatrix}$$

with matrix addition and scalar multiplication.

12. The set of all real-valued functions f defined everywhere on the real line and such that $f(1) = 0$, with the operations defined in Example 4.

13. The set of all 2×2 matrices of the form

$$\begin{bmatrix} a & a+b \\ a+b & b \end{bmatrix}$$

with matrix addition and scalar multiplication.

14. The set whose only element is the moon. The operations are moon + moon = moon and k(moon) = moon, where k is a real number.

15. Prove that a line passing through the origin in R^3 is a vector space under the standard operations on R^3.

16. Complete the unfinished details in Example 4.

17. Complete the unfinished details in Example 6.

18. Prove part (*b*) of Theorem 4.2.1.

19. Prove part (*d*) of Theorem 4.2.1.

20. Prove that a vector space cannot have more than one zero vector.

21. Prove that a vector has exactly one negative.

22. Show that the first nine vector space axioms are satisfied if $V = R^2$ has the addition and scalar multiplication operations defined in Example 5.

4.3 SUBSPACES

It is possible for one vector space to be contained within a larger vector space. For example, planes through the origin are vector spaces that are contained within the larger vector space R^3 (Example 6 of the previous section). In this section we shall study this important concept in more detail.

DEFINITION OF A SUBSPACE

Definition. A subset W of a vector space V is called a ***subspace*** of V if W is itself a vector space under the addition and scalar multiplication defined on V.

In general, one must verify the ten vector space axioms to show that a set W with addition and scalar multiplication forms a vector space. However, if W is part of a larger set V that is already known to be a vector space, then certain axioms need not be verified for W because they are "inherited" from V. For example, there is no need to check that $\mathbf{u} + \mathbf{v} = \mathbf{v} + \mathbf{u}$ (Axiom 2) for W because this holds for all vectors in V and consequently for all vectors in W. Other axioms inherited by W from V are 3, 7, 8, 9, and 10. Thus, to show that a set W is a subspace of a vector space V, we need only verify Axioms 1, 4, 5, and 6. The following theorem shows that even Axioms 4 and 5 can be dispensed with.

Theorem 4.3.1. *If W is a set of one or more vectors from a vector space V, then W is a subspace of V if and only if the following conditions hold.*

(a) If $\mathbf{u}$ and $\mathbf{v}$ are vectors in W, then $\mathbf{u} + \mathbf{v}$ is in W.
(b) If k is any scalar and $\mathbf{u}$ is any vector in W, then $k\mathbf{u}$ is in W.

Proof. If W is a subspace of V, then all the vector-space axioms are satisfied; in particular, Axioms 1 and 6 hold. But these are precisely conditions (*a*) and (*b*).

Conversely, assume conditions (*a*) and (*b*) hold. Since these conditions are vector-space Axioms 1 and 6, we need only show that W satisfies the remaining eight axioms. Axioms 2, 3, 7, 8, 9, and 10 are automatically satisfied by the vectors in W since they are satisfied by all vectors in V. Therefore, to complete the proof, we need only verify that Axioms 4 and 5 are satisfied by vectors in W.

Let $\mathbf{u}$ be any vector in W. By condition (b), $k\mathbf{u}$ is in W for every scalar k. Setting $k = 0$, it follows from Theorem 4.2.1 that $0\mathbf{u} = \mathbf{0}$ is in W, and setting $k = -1$, it follows that $(-1)\mathbf{u} = -\mathbf{u}$ is in W. ▮

REMARK. A set W of one or more vectors from a vector space V is said to be ***closed under addition*** if condition (a) in Theorem 4.3.1 holds and ***closed under scalar multiplication*** if condition (b) holds. Thus, Theorem 4.3.1 states that W is a subspace of V if and only if W is closed under addition and closed under scalar multiplication.

EXAMPLES OF SUBSPACES

Example 1   In Example 6 of the previous section we verified the ten vector space axioms to show that the points in a plane through the origin of R^3 form a subspace of R^3. In light of Theorem 4.3.1 we can see that much of that work was unnecessary; it would have been sufficient to verify that the plane is closed under addition and scalar multiplication (Axioms 1 and 6). In the previous section we verified those two axioms algebraically; however, they can also be proved geometrically as follows: Let W be any plane through the origin and let $\mathbf{u}$ and $\mathbf{v}$ be any vectors in W. Then $\mathbf{u} + \mathbf{v}$ must lie in W because it is the diagonal of the parallelogram determined by $\mathbf{u}$ and $\mathbf{v}$ (Figure 1), and $k\mathbf{u}$ must lie in W for any scalar k because $k\mathbf{u}$ lies on a line through $\mathbf{u}$. Thus, W is closed under addition and scalar multiplication, so it is a subspace of R^3. ▲

Figure 1 | The vectors $\mathbf{u} + \mathbf{v}$ and $k\mathbf{u}$ both lie in the same plane as $\mathbf{u}$ and $\mathbf{v}$.

Example 2   Show that a line through the origin of R^3 is a subspace of R^3.

Solution. Let W be a line through the origin of R^3. It is evident geometrically that the sum of two vectors on this line also lies on the line and that a scalar multiple of a vector on the line is on the line as well (Figure 2). Thus, W is closed under addition and scalar multiplication, so it is a subspace of R^3. In the exercises we will ask you to prove this result algebraically using parametric equations for the line. ▲

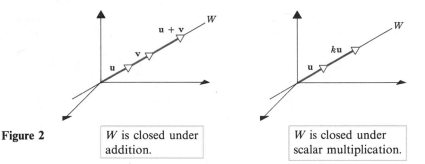

Figure 2   W is closed under addition.   W is closed under scalar multiplication.

Example 3   Let W be the set of all points (x, y) in R^2 that lie in the first quadrant, that is, such that $x \geq 0$ and $y \geq 0$. The set W is *not* a subspace of R^2 since it is not closed under scalar multiplication. For example, $\mathbf{v} = (1, 1)$ lies in W, but $(-1)\mathbf{v} = -\mathbf{v} = (-1, -1)$ does not (Figure 3). ▲

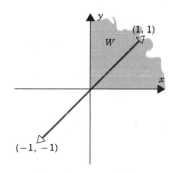

Figure 3   W is not closed under scalar multiplication.

Every nonzero vector space V has at least two subspaces: V itself is a subspace, and the set $\{\mathbf{0}\}$ consisting of just the zero vector in V is a subspace called the ***zero subspace***. Combining this with Examples 1 and 2, we obtain the following list of subspaces of R^3:

- $\{\mathbf{0}\}$
- Lines through the origin
- Planes through the origin
- R^3

Similarly, the subspaces of R^2 are

- $\{\mathbf{0}\}$
- Lines through the origin
- R^2

Later, we will show that these are the only subspaces of R^2 and R^3.

Example 4 Show that the set W of all 2×2 matrices having zeros on the main diagonal is a subspace of the vector space M_{22} of all 2×2 matrices.

Solution. Let

$$A = \begin{bmatrix} 0 & a_{12} \\ a_{21} & 0 \end{bmatrix} \quad \text{and} \quad B = \begin{bmatrix} 0 & b_{12} \\ b_{21} & 0 \end{bmatrix}$$

be any two matrices in W and k any scalar. Then

$$kA = \begin{bmatrix} 0 & ka_{12} \\ ka_{21} & 0 \end{bmatrix} \quad \text{and} \quad A + B = \begin{bmatrix} 0 & a_{12} + b_{12} \\ a_{21} + b_{21} & 0 \end{bmatrix}$$

Since kA and $A + B$ have zeros on the main diagonal, they lie in W. Thus, W is a subspace of M_{22}. ▲

Example 5 Let n be a positive integer, and let W consist of all functions expressible in the form

$$p(x) = a_0 + a_1 x + \cdots + a_n x^n \tag{1}$$

where $a_0, \ldots, a_n$ are real numbers. Thus, W consists of the zero function together with all real polynomials of degree $\leq n$. The set W is a subspace of the vector space of all real-valued functions discussed in Example 4 of the previous section. To see this, let $\mathbf{p}$ and $\mathbf{q}$ be the polynomials

$$p(x) = a_0 + a_1 x + \cdots + a_n x^n$$

and

$$q(x) = b_0 + b_1 x + \cdots + b_n x^n$$

Then

$$(\mathbf{p} + \mathbf{q})(x) = p(x) + q(x) = (a_0 + b_0) + (a_1 + b_1)x + \cdots + (a_n + b_n)x^n$$

and

$$(k\mathbf{p})(x) = kp(x) = (ka_0) + (ka_1)x + \cdots + (ka_n)x^n$$

These functions have the form given in (1), so $\mathbf{p} + \mathbf{q}$ and $k\mathbf{p}$ lie in W. We shall denote the vector space W in this example by the symbol P_n. ▲

Example 6 **(For readers who have studied calculus.)** Recall from calculus that if $\mathbf{f}$ and $\mathbf{g}$ are continuous functions and k is a constant, then $\mathbf{f} + \mathbf{g}$ and $k\mathbf{f}$ are also continuous functions. It follows that the set of all continuous functions is a subspace of the vector space of all real-valued functions. This space is denoted by $C(-\infty, +\infty)$. A closely related example is the vector space of all functions that are continuous on a closed interval $a \leq x \leq b$. This space is denoted by $C[a, b]$. ▲

SOLUTION SPACES OF HOMOGENEOUS SYSTEMS

Example 7 Consider a system of m linear equations in n unknowns

$$a_{11}x_1 + a_{12}x_2 + \cdots + a_{1n}x_n = b_1$$
$$a_{21}x_1 + a_{22}x_2 + \cdots + a_{2n}x_n = b_2$$
$$\vdots \qquad \vdots \qquad \qquad \vdots \qquad \vdots$$
$$a_{m1}x_1 + a_{m2}x_2 + \cdots + a_{mn}x_n = b_m$$

or, in matrix notation, $A\mathbf{x} = \mathbf{b}$. A vector*

$$\mathbf{s} = \begin{bmatrix} s_1 \\ s_2 \\ \vdots \\ s_n \end{bmatrix}$$

in R^n is called a ***solution vector*** of the system if $x_1 = s_1, x_2 = s_2, \ldots, x_n = s_n$ is a solution of the system. In this example we shall show that the set of solution vectors of a *homogeneous* system is a subspace of R^n.

Let $A\mathbf{x} = \mathbf{0}$ be a homogeneous system of m linear equations in n unknowns, let W be the set of solution vectors, and let $\mathbf{s}$ and $\mathbf{s}'$ be vectors in W. To show that W is closed under addition and scalar multiplication, we must demonstrate that $\mathbf{s} + \mathbf{s}'$ and $k\mathbf{s}$ are also solution vectors, where k is any scalar. Since $\mathbf{s}$ and $\mathbf{s}'$ are solution vectors we have

$$A\mathbf{s} = \mathbf{0} \qquad \text{and} \qquad A\mathbf{s}' = \mathbf{0}$$

Therefore,

$$A(\mathbf{s} + \mathbf{s}') = A\mathbf{s} + A\mathbf{s}' = \mathbf{0} + \mathbf{0} = \mathbf{0}$$

and

$$A(k\mathbf{s}) = k(A\mathbf{s}) = k\mathbf{0} = \mathbf{0}$$

These equations show that $\mathbf{x} = \mathbf{s} + \mathbf{s}'$ and $\mathbf{x} = k\mathbf{s}$ satisfy the equation $A\mathbf{x} = \mathbf{0}$. Thus, $\mathbf{s} + \mathbf{s}'$ and $k\mathbf{s}$ are solution vectors. ▲

The subspace W in the foregoing example is called the ***solution space*** of the system $A\mathbf{x} = \mathbf{0}$.

The following definition is one of the most fundamental concepts in the study of vectors.

LINEAR COMBINATIONS OF VECTORS

> **Definition.** A vector $\mathbf{w}$ is called a ***linear combination*** of the vectors $\mathbf{v}_1, \mathbf{v}_2, \ldots, \mathbf{v}_r$ if it can be expressed in the form
>
> $$\mathbf{w} = k_1\mathbf{v}_1 + k_2\mathbf{v}_2 + \cdots + k_r\mathbf{v}_r$$
>
> where $k_1, k_2, \ldots, k_r$ are scalars.

---

* In this example we are using the matrix notation for vectors in R^n.

Example 8 Consider the vectors $\mathbf{u} = (1, 2, -1)$ and $\mathbf{v} = (6, 4, 2)$ in R^3. Show that $\mathbf{w} = (9, 2, 7)$ is a linear combination of $\mathbf{u}$ and $\mathbf{v}$ and that $\mathbf{w}' = (4, -1, 8)$ is *not* a linear combination of $\mathbf{u}$ and $\mathbf{v}$.

Solution. In order for $\mathbf{w}$ to be a linear combination of $\mathbf{u}$ and $\mathbf{v}$, there must be scalars k_1 and k_2 such that $\mathbf{w} = k_1\mathbf{u} + k_2\mathbf{v}$; that is,

$$(9, 2, 7) = k_1(1, 2, -1) + k_2(6, 4, 2)$$

or

$$(9, 2, 7) = (k_1 + 6k_2, 2k_1 + 4k_2, -k_1 + 2k_2)$$

Equating corresponding components gives

$$k_1 + 6k_2 = 9$$
$$2k_1 + 4k_2 = 2$$
$$-k_1 + 2k_2 = 7$$

Solving this system yields $k_1 = -3$, $k_2 = 2$ so that

$$\mathbf{w} = -3\mathbf{u} + 2\mathbf{v}$$

Similarly, for $\mathbf{w}'$ to be a linear combination of $\mathbf{u}$ and $\mathbf{v}$, there must be scalars k_1 and k_2 such that $\mathbf{w}' = k_1\mathbf{u} + k_2\mathbf{v}$; that is,

$$(4, -1, 8) = k_1(1, 2, -1) + k_2(6, 4, 2)$$

or

$$(4, -1, 8) = (k_1 + 6k_2, 2k_1 + 4k_2, -k_1 + 2k_2)$$

Equating corresponding components gives

$$k_1 + 6k_2 = \quad 4$$
$$2k_1 + 4k_2 = -1$$
$$-k_1 + 2k_2 = \quad 8$$

This system of equations is inconsistent (verify), so that no such scalars k_1 and k_2 exist. Consequently, $\mathbf{w}'$ is not a linear combination of $\mathbf{u}$ and $\mathbf{v}$. ▲

SPANNING

> **Definition.** If $\mathbf{v}_1, \mathbf{v}_2, \ldots, \mathbf{v}_r$ are vectors in a vector space V and if every vector in V is expressible as a linear combination of these vectors, then we say that $\mathbf{v}_1, \mathbf{v}_2, \ldots, \mathbf{v}_r$ *span* V.

Example 9 The vectors $\mathbf{i} = (1, 0, 0)$, $\mathbf{j} = (0, 1, 0)$, and $\mathbf{k} = (0, 0, 1)$ span R^3 because every vector (a, b, c) in R^3 can be written as

$$(a, b, c) = a\mathbf{i} = b\mathbf{j} + c\mathbf{k}$$

which is a linear combination of $\mathbf{i}$, $\mathbf{j}$, and $\mathbf{k}$. ▲

Example 10 The polynomials $1, x, x^2, \ldots, x^n$ span the vector space P_n defined in Example 5 since each polynomial $\mathbf{p}$ in P_n can be written as

$$\mathbf{p} = a_0 + a_1 x + \cdots + a_n x^n$$

which is a linear combination of $1, x, x^2, \ldots, x^n$. ▲

Example 11 Determine whether $\mathbf{v}_1 = (1, 1, 2)$, $\mathbf{v}_2 = (1, 0, 1)$, and $\mathbf{v}_3 = (2, 1, 3)$ span the vector space R^3.

Solution. We must determine whether an arbitrary vector $\mathbf{b} = (b_1, b_2, b_3)$ in R^3 can be expressed as a linear combination

$$\mathbf{b} = k_1 \mathbf{v}_1 + k_2 \mathbf{v}_2 + k_3 \mathbf{v}_3$$

of the vectors $\mathbf{v}_1, \mathbf{v}_2$, and $\mathbf{v}_3$. Expressing this equation in terms of components gives

$$(b_1, b_2, b_3) = k_1(1, 1, 2) + k_2(1, 0, 1) + k_3(2, 1, 3)$$

or

$$(b_1, b_2, b_3) = (k_1 + k_2 + 2k_3, \; k_1 + k_3, \; 2k_1 + k_2 + 3k_3)$$

or

$$
\begin{aligned}
k_1 + k_2 + 2k_3 &= b_1 \\
k_1 \qquad + k_3 &= b_2 \\
2k_1 + k_2 + 3k_3 &= b_3
\end{aligned}
$$

The problem thus reduces to determining whether this system is consistent for all values of b_1, b_2, and b_3. By parts (a) and (d) of Theorem 1.7.3, this system will be consistent for all b_1, b_2, and b_3 if and only if the coefficient matrix

$$
A = \begin{bmatrix} 1 & 1 & 2 \\ 1 & 0 & 1 \\ 2 & 1 & 3 \end{bmatrix}
$$

is invertible. But $\det(A) = 0$ (verify), so that A is not invertible; consequently, $\mathbf{v}_1$, $\mathbf{v}_2$, and $\mathbf{v}_3$ do not span R^3. ▲

In general, a given set of vectors $\{\mathbf{v}_1, \mathbf{v}_2, \ldots, \mathbf{v}_r\}$ in a vector space V may or may not span V. If they span, then every vector in V is expressible as a linear combination of $\mathbf{v}_1, \mathbf{v}_2, \ldots, \mathbf{v}_r$, and if they do not span, then some vectors are so expressible while others are not. The following theorem shows that if we group together all vectors in V that are expressible as linear combinations of the vectors $\mathbf{v}_1, \mathbf{v}_2, \ldots, \mathbf{v}_r$, then we obtain a subspace of V. This subspace is called the *linear space spanned* by $\mathbf{v}_1, \mathbf{v}_2, \ldots, \mathbf{v}_r$, or more simply, the *space spanned* by $\mathbf{v}_1, \mathbf{v}_2, \ldots, \mathbf{v}_r$.

Theorem 4.3.2. *If* $\mathbf{v}_1, \mathbf{v}_2, \ldots, \mathbf{v}_r$ *are vectors in a vector space V, then*

(a) *the set W of all linear combinations of* $\mathbf{v}_1, \mathbf{v}_2, \ldots, \mathbf{v}_r$ *is a subspace of V;*

(b) *W is the smallest subspace of V that contains* $\mathbf{v}_1, \mathbf{v}_2, \ldots, \mathbf{v}_r$ *in the sense that every other subspace of V that contains* $\mathbf{v}_1, \mathbf{v}_2, \ldots, \mathbf{v}_r$ *must contain W.*

Proof (*a*). To show that W is a subspace of V, we must prove it is closed under addition and scalar multiplication. If $\mathbf{u}$ and $\mathbf{v}$ are vectors in W, then

$$\mathbf{u} = c_1\mathbf{v}_1 + c_2\mathbf{v}_2 + \cdots + c_r\mathbf{v}_r$$

and

$$\mathbf{v} = k_1\mathbf{v}_1 + k_2\mathbf{v}_2 + \cdots + k_r\mathbf{v}_r$$

where $c_1, c_2, \ldots, c_r, k_1, k_2, \ldots, k_r$ are scalars. Therefore,

$$\mathbf{u} + \mathbf{v} = (c_1 + k_1)\mathbf{v}_1 + (c_2 + k_2)\mathbf{v}_2 + \cdots + (c_r + k_r)\mathbf{v}_r$$

and, for any scalar k,

$$k\mathbf{u} = (kc_1)\mathbf{v}_1 + (kc_2)\mathbf{v}_2 + \cdots + (kc_r)\mathbf{v}_r$$

Thus, $\mathbf{u} + \mathbf{v}$ and $k\mathbf{u}$ are linear combinations of $\mathbf{v}_1, \mathbf{v}_2, \ldots, \mathbf{v}_r$ and consequently lie in W. Therefore, W is closed under addition and scalar multiplication.

Proof (*b*). Each vector $\mathbf{v}_i$ is a linear combination of $\mathbf{v}_1, \mathbf{v}_2, \ldots, \mathbf{v}_r$ since we can write

$$\mathbf{v}_i = 0\mathbf{v}_1 + 0\mathbf{v}_2 + \cdots + 1\mathbf{v}_i + \cdots + 0\mathbf{v}_r$$

Therefore, the subspace W contains each of the vectors $\mathbf{v}_1, \mathbf{v}_2, \ldots, \mathbf{v}_r$. Let W' be any other subspace that contains $\mathbf{v}_1, \mathbf{v}_2, \ldots, \mathbf{v}_r$. Since W' is closed under addition and scalar multiplication, it must contain all linear combinations of $\mathbf{v}_1, \mathbf{v}_2, \ldots, \mathbf{v}_r$. Thus, W' contains each vector of W. ▌

REMARK. There is a subtlety in terminology that is important to keep in mind: To say that $\mathbf{v}_1, \mathbf{v}_2, \ldots, \mathbf{v}_r$ *span* W means that every vector in W is a linear combination of these vectors with the possibility that vectors outside of W may also be linear combinations of $\mathbf{v}_1, \mathbf{v}_2, \ldots, \mathbf{v}_r$ not precluded. However, to say that W is the *space spanned* by $\mathbf{v}_1, \mathbf{v}_2, \ldots, \mathbf{v}_r$ means that every vector in W is a linear combination of $\mathbf{v}_1, \mathbf{v}_2, \ldots, \mathbf{v}_r$ and there are no vectors outside of W with this property.

We introduce the following notation.

Definition. The linear space W spanned by a set of vectors $S = \{\mathbf{v}_1, \mathbf{v}_2, \ldots, \mathbf{v}_r\}$ will be denoted by

$$\text{lin}(S) \qquad \text{or} \qquad \text{lin}\{\mathbf{v}_1, \mathbf{v}_2, \ldots, \mathbf{v}_r\}$$

Example 12 If $\mathbf{v}_1$ and $\mathbf{v}_2$ are noncollinear vectors in R^3 with initial points at the origin, then $\text{lin}\{\mathbf{v}_1, \mathbf{v}_2\}$, which consists of all linear combinations $k_1\mathbf{v}_1 + k_2\mathbf{v}_2$, is the plane determined by $\mathbf{v}_1$ and $\mathbf{v}_2$ (Figure 4*a*). Similarly, if $\mathbf{v}$ is a nonzero vector in R^2 or R^3, then $\text{lin}\{\mathbf{v}\}$, which is the set of all scalar multiples $k\mathbf{v}$, is the line determined by $\mathbf{v}$ (Figure 4*b*). ▲

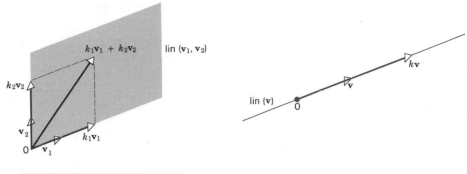

Figure 4

lin$\{v_1, v_2\}$ consists of all linear combinations of v_1 and v_2

lin$\{v\}$ consists of all scalar multiples of v

EXERCISE SET 4.3

1. Use Theorem 4.3.1 to determine which of the following are subspaces of R^3.
  (a) all vectors of the form $(a, 0, 0)$
  (b) all vectors of the form $(a, 1, 1)$
  (c) all vectors of the form (a, b, c), where $b = a + c$
  (d) all vectors of the form (a, b, c), where $b = a + c + 1$

2. Use Theorem 4.3.1 to determine which of the following are subspaces of M_{22}.
  (a) all matrices of the form
  $$\begin{bmatrix} a & b \\ c & d \end{bmatrix}$$
  where a, b, c, and d are integers
  (b) all matrices of the form
  $$\begin{bmatrix} a & b \\ c & d \end{bmatrix}$$
  where $a + d = 0$
  (c) all 2×2 matrices A such that $A = A^t$
  (d) all 2×2 matrices A such that $\det(A) = 0$

3. Use Theorem 4.3.1 to determine which of the following are subspaces of P_3.
  (a) all polynomials $a_0 + a_1x + a_2x^2 + a_3x^3$ for which $a_0 = 0$
  (b) all polynomials $a_0 + a_1x + a_2x^2 + a_3x^3$ for which $a_0 + a_1 + a_2 + a_3 = 0$
  (c) all polynomials $a_0 + a_1x + a_2x^2 + a_3x^3$ for which a_0, a_1, a_2, and a_3 are integers
  (d) all polynomials of the form $a_0 + a_1x$, where a_0 and a_1 are real numbers

4. Use Theorem 4.3.1 to determine which of the following are subspaces of the space of all real-valued functions f defined on the entire real line.
  (a) all f such that $f(x) \leq 0$ for all x (b) all f such that $f(0) = 0$
  (c) all f such that $f(0) = 2$ (d) all constant functions
  (e) all f of the form $k_1 + k_2 \sin x$, where k_1 and k_2 are real numbers

5. Which of the following are linear combinations of $\mathbf{u} = (0, -2, 2)$ and $\mathbf{v} = (1, 3, -1)$?
   (a) $(2, 2, 2)$ (b) $(3, 1, 5)$ (c) $(0, 4, 5)$ (d) $(0, 0, 0)$

6. Express the following as linear combinations of $\mathbf{u} = (2, 1, 4)$, $\mathbf{v} = (1, -1, 3)$, and $\mathbf{w} = (3, 2, 5)$.
   (a) $(-9, -7, -15)$ (b) $(6, 11, 6)$ (c) $(0, 0, 0)$ (d) $(7, 8, 9)$

7. Express the following as linear combinations of $\mathbf{p}_1 = 2 + x + 4x^2$, $\mathbf{p}_2 = 1 - x + 3x^2$, and $\mathbf{p}_3 = 3 + 2x + 5x^2$.
   (a) $-9 - 7x - 15x^2$ (b) $6 + 11x + 6x^2$ (c) 0 (d) $7 + 8x + 9x^2$

8. Which of the following are linear combinations of

   $$A = \begin{bmatrix} 4 & 0 \\ -2 & -2 \end{bmatrix}, \quad B = \begin{bmatrix} 1 & -1 \\ 2 & 3 \end{bmatrix}, \quad C = \begin{bmatrix} 0 & 2 \\ 1 & 4 \end{bmatrix}?$$

   (a) $\begin{bmatrix} 6 & -8 \\ -1 & -8 \end{bmatrix}$ (b) $\begin{bmatrix} 0 & 0 \\ 0 & 0 \end{bmatrix}$ (c) $\begin{bmatrix} 6 & 0 \\ 3 & 8 \end{bmatrix}$ (d) $\begin{bmatrix} -1 & 5 \\ 7 & 1 \end{bmatrix}$

9. In each part determine whether the given vectors span R^3.
   (a) $\mathbf{v}_1 = (2, 2, 2)$, $\mathbf{v}_2 = (0, 0, 3)$, $\mathbf{v}_3 = (0, 1, 1)$
   (b) $\mathbf{v}_1 = (2, -1, 3)$, $\mathbf{v}_2 = (4, 1, 2)$, $\mathbf{v}_3 = (8, -1, 8)$
   (c) $\mathbf{v}_1 = (3, 1, 4)$, $\mathbf{v}_2 = (2, -3, 5)$, $\mathbf{v}_3 = (5, -2, 9)$, $\mathbf{v}_4 = (1, 4, -1)$
   (d) $\mathbf{v}_1 = (1, 2, 6)$, $\mathbf{v}_2 = (3, 4, 1)$, $\mathbf{v}_3 = (4, 3, 1)$, $\mathbf{v}_4 = (3, 3, 1)$

10. Let $\mathbf{f} = \cos^2 x$ and $\mathbf{g} = \sin^2 x$. Which of the following lie in the space spanned by $\mathbf{f}$ and $\mathbf{g}$?
    (a) $\cos 2x$ (b) $3 + x^2$ (c) 1 (d) $\sin x$ (e) 0

11. Determine whether the following polynomials span P_2.

    $$\mathbf{p}_1 = 1 - x + 2x^2 \qquad \mathbf{p}_2 = 3 + x$$
    $$\mathbf{p}_3 = 5 - x + 4x^2 \qquad \mathbf{p}_4 = -2 - 2x + 2x^2$$

12. Let $\mathbf{v}_1 = (2, 1, 0, 3)$, $\mathbf{v}_2 = (3, -1, 5, 2)$, and $\mathbf{v}_3 = (-1, 0, 2, 1)$. Which of the following vectors are in $\text{lin}\{\mathbf{v}_1, \mathbf{v}_2, \mathbf{v}_3\}$?
    (a) $(2, 3, -7, 3)$ (b) $(0, 0, 0, 0)$ (c) $(1, 1, 1, 1)$ (d) $(-4, 6, -13, 4)$

13. Find an equation for the plane spanned by the vectors $\mathbf{u} = (-1, 1, 1)$ and $\mathbf{v} = (3, 4, 4)$.

14. Find parametric equations for the line spanned by the vector $\mathbf{u} = (3, -2, 5)$.

15. Show that the solution vectors of a consistent nonhomogeneous system of m linear equations in n unknowns do not form a subspace of R^n.

16. A line L through the origin in R^3 can be represented by parametric equations of the form $x = at$, $y = bt$, and $z = ct$. Use these equations to show that L is a subspace of R^3; that is, if $\mathbf{v}_1 = (x_1, y_1, z_1)$ and $\mathbf{v}_2 = (x_2, y_2, z_2)$ are points on L and k is any real number, then $k\mathbf{v}_1$ and $\mathbf{v}_1 + \mathbf{v}_2$ are also points on L.

17. (**For readers who have studied calculus.**) Show that the following sets of functions are subspaces of the vector space of real-valued functions defined on the entire real line.
    (a) all everywhere continuous functions   (b) all everywhere differentiable functions
    (c) all everywhere differentiable functions that satisfy $\mathbf{f}' + 2\mathbf{f} = 0$

4.4   LINEAR INDEPENDENCE

In the previous section we learned that a set of vectors $S = \{v_1, v_2, \ldots, v_r\}$ spans a given vector space V if every vector in V is expressible as a linear combination of the vectors in S. In general, there may be more than one way to express a vector in V as a linear combination of a spanning set. In this section we shall study conditions under which each vector in V is expressible as a linear combination of the spanning vectors in exactly one way. Spanning sets with this property play a fundamental role in the study of vector spaces.

DEFINITION OF LINEAR INDEPENDENCE

Definition. If $S = \{v_1, v_2, \ldots, v_r\}$ is a nonempty set of vectors, then the vector equation

$$k_1 v_1 + k_2 v_2 + \cdots + k_r v_r = 0$$

has at least one solution, namely

$$k_1 = 0, \quad k_2 = 0, \quad \ldots, \quad k_r = 0$$

If this is the only solution, then S is called a ***linearly independent*** set. If there are other solutions, then S is called a ***linearly dependent*** set.

Example 1   If $v_1 = (2, -1, 0, 3)$, $v_2 = (1, 2, 5, -1)$, and $v_3 = (7, -1, 5, 8)$, then the set of vectors $S = \{v_1, v_2, v_3\}$ is linearly dependent, since $3v_1 + v_2 - v_3 = 0$.   ▲

Example 2   The polynomials

$$p_1 = 1 - x, \quad p_2 = 5 + 3x - 2x^2, \quad \text{and} \quad p_3 = 1 + 3x - x^2$$

form a linearly dependent set in P_2 since $3p_1 - p_2 + 2p_3 = 0$.   ▲

Example 3   Consider the vectors $i = (1, 0, 0)$, $j = (0, 1, 0)$, and $k = (0, 0, 1)$ in R^3. In terms of components the vector equation

$$k_1 i + k_2 j + k_3 k = 0$$

becomes

$$k_1(1, 0, 0) + k_2(0, 1, 0) + k_3(0, 0, 1) = (0, 0, 0)$$

or equivalently

$$(k_1, k_2, k_3) = (0, 0, 0)$$

This implies that $k_1 = 0$, $k_2 = 0$, and $k_3 = 0$, so the set $S = \{i, j, k\}$ is linearly independent. A similar argument can be used to show that the vectors

$$e_1 = (1, 0, 0, \ldots, 0), \quad e_2 = (0, 1, 0, \ldots, 0), \quad \ldots, \quad e_n = (0, 0, 0, \ldots, 1)$$

form a linearly independent set in R^n.   ▲

Example 4 Determine whether the vectors

$$\mathbf{v}_1 = (1, -2, 3) \qquad \mathbf{v}_2 = (5, 6, -1) \qquad \mathbf{v}_3 = (3, 2, 1)$$

form a linearly dependent set or a linearly independent set.

Solution. In terms of components the vector equation

$$k_1\mathbf{v}_1 + k_2\mathbf{v}_2 + k_3\mathbf{v}_3 = \mathbf{0}$$

becomes

$$k_1(1, -2, 3) + k_2(5, 6, -1) + k_3(3, 2, 1) = (0, 0, 0)$$

or equivalently

$$(k_1 + 5k_2 + 3k_3, \ -2k_1 + 6k_2 + 2k_3, \ 3k_1 - k_2 + k_3) = (0, 0, 0)$$

Equating corresponding components gives

$$\begin{aligned} k_1 + 5k_2 + 3k_3 &= 0 \\ -2k_1 + 6k_2 + 2k_3 &= 0 \\ 3k_1 - k_2 + k_3 &= 0 \end{aligned}$$

Thus, $\mathbf{v}_1$, $\mathbf{v}_2$, and $\mathbf{v}_3$ form a linearly dependent set if this system has a nontrivial solution, or a linearly independent set if it has only the trivial solution. Solving this system yields

$$k_1 = -\tfrac{1}{2}t, \qquad k_2 = -\tfrac{1}{2}t, \qquad k_3 = t$$

Thus, the system has nontrivial solutions and $\mathbf{v}_1$, $\mathbf{v}_2$, and $\mathbf{v}_3$ form a linearly dependent set. Alternatively, we could show the existence of nontrivial solutions without solving the system by showing that the coefficient matrix has determinant zero and consequently is not invertible (verify). ▲

The term "linearly dependent" suggests that the vectors "depend" on each other in some way. The following theorem shows that this is in fact the case.

Theorem 4.4.1. *A set S with two or more vectors is*

(a) *linearly dependent if and only if at least one of the vectors in S is expressible as a linear combination of the other vectors in S;*

(b) *linearly independent if and only if no vector in S is expressible as a linear combination of the other vectors in S.*

We shall prove part (*a*) and leave the proof of part (*b*) as an exercise.

Proof (*a*). Let $S = \{\mathbf{v}_1, \mathbf{v}_2, \dots, \mathbf{v}_r\}$ be a set with two or more vectors. If we assume that S is linearly dependent, then there are scalars $k_1, k_2, \dots, k_r$, not all zero, such that

$$k_1\mathbf{v}_1 + k_2\mathbf{v}_2 + \cdots + k_r\mathbf{v}_r = \mathbf{0} \tag{1}$$

To be specific, suppose that $k_1 \neq 0$. Then (1) can be rewritten as

$$\mathbf{v}_1 = \left(-\frac{k_2}{k_1}\right)\mathbf{v}_2 + \cdots + \left(-\frac{k_r}{k_1}\right)\mathbf{v}_r$$

which expresses $\mathbf{v}_1$ as a linear combination of the other vectors in S. Similarly, if $k_j \neq 0$ in (1) for some $j = 2, 3, \ldots, r$, then $\mathbf{v}_j$ is expressible as a linear combination of the other vectors in S.

Conversely, let us assume that at least one of the vectors in S is expressible as a linear combination of the other vectors. To be specific, suppose that

$$\mathbf{v}_1 = c_2\mathbf{v}_2 + c_3\mathbf{v}_3 + \cdots + c_r\mathbf{v}_r$$

so

$$\mathbf{v}_1 - c_2\mathbf{v}_2 - c_3\mathbf{v}_3 - \cdots - c_r\mathbf{v}_r = \mathbf{0}$$

It follows that S is linearly dependent since the equation

$$k_1\mathbf{v}_1 + k_2\mathbf{v}_2 + \cdots + k_r\mathbf{v}_r = \mathbf{0}$$

is satisfied by

$$k_1 = 1, \quad k_2 = -c_2, \quad \ldots, \quad k_r = -c_r$$

which are not all zero. The proof in the case where some vector other than $\mathbf{v}_1$ is expressible as a linear combination of the other vectors in S is similar. ∎

Example 5   In Example 1 we saw that the vectors

$$\mathbf{v}_1 = (2, -1, 0, 3), \qquad \mathbf{v}_2 = (1, 2, 5, -1), \qquad \text{and} \qquad \mathbf{v}_3 = (7, -1, 5, 8)$$

form a linearly dependent set. It follows from Theorem 4.4.1 that at least one of these vectors is expressible as a linear combination of the other two. In this example each vector is expressible as a linear combination of the other two since it follows from the equation $3\mathbf{v}_1 + \mathbf{v}_2 - \mathbf{v}_3 = \mathbf{0}$ (see Example 1) that

$$\mathbf{v}_1 = -\tfrac{1}{3}\mathbf{v}_2 + \tfrac{1}{3}\mathbf{v}_3, \qquad \mathbf{v}_2 = -3\mathbf{v}_1 + \mathbf{v}_3, \qquad \text{and} \qquad \mathbf{v}_3 = 3\mathbf{v}_1 + \mathbf{v}_2 \quad ▲$$

Example 6   In Example 3 we saw that the vectors $\mathbf{i} = (1, 0, 0)$, $\mathbf{j} = (0, 1, 0)$, and $\mathbf{k} = (0, 0, 1)$ form a linearly independent set. Thus, it follows from Theorem 4.4.1 that none of these vectors is expressible as a linear combination of the other two. To see directly that this is so, suppose that $\mathbf{k}$ is expressible as

$$\mathbf{k} = k_1\mathbf{i} + k_2\mathbf{j}$$

Then, in terms of components,

$$(0, 0, 1) = k_1(1, 0, 0) + k_2(0, 1, 0)$$

or

$$(0, 0, 1) = (k_1, k_2, 0)$$

But this equation is not satisfied by any values of k_1 and k_2, so $\mathbf{k}$ cannot be expressed as a linear combination of $\mathbf{i}$ and $\mathbf{j}$. Similarly, $\mathbf{i}$ is not expressible as a linear combination of $\mathbf{j}$ and $\mathbf{k}$, and $\mathbf{j}$ is not expressible as a linear combination of $\mathbf{i}$ and $\mathbf{k}$. ▲

The following theorem gives two simple facts about linear independence that are important to know. The proof is left as an exercise.

Theorem 4.4.2
(*a*) *A set that contains the zero vector is linearly dependent.*
(*b*) *A set with exactly two vectors is linearly independent if and only if neither vector is a scalar multiple of the other.*

Linear independence has some useful geometric interpretations in R^2 and R^3:

· In R^2 or R^3, a set of two vectors is linearly independent if and only if the vectors do not lie on the same line when they are placed with their initial points at the origin (Figure 1).

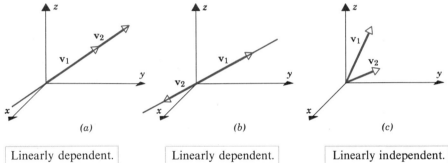

| (a) | (b) | (c) |

Figure 1   Linearly dependent.   Linearly dependent.   Linearly independent.

· In R^3, a set of three vectors is linearly independent if and only if the vectors do not lie in the same plane when they are placed with their initial points at the origin (Figure 2).

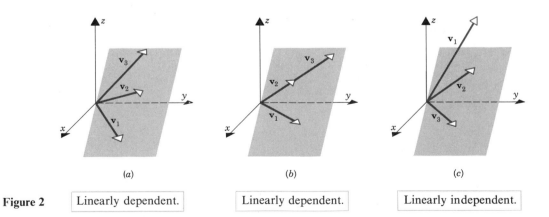

| (a) | (b) | (c) |

Figure 2   Linearly dependent.   Linearly dependent.   Linearly independent.

To see why the first result is true, suppose that two vectors in R^2 or R^3 are positioned with their initial points at the origin. By Theorem 4.4.2b, the two vectors are linearly independent if and only if they are not scalar multiples of one another. Geometrically, this is equivalent to stating that the vectors do not lie on the same line.

To see why the second result is true, suppose that three vectors in R^3 are positioned with their initial points at the origin. By Theorem 4.4.1b, the vectors are linearly independent if and only if none of the vectors is a linear combination of the other two, or equivalently, if and only if none of the vectors is in the space spanned by the other two. But the space spanned by two vectors in R^3 is either a plane or a subset of a plane (a line through the origin or the origin itself). (See Exercise 17.) Thus, to assert that no vector is a linear combination of the other two is equivalent to asserting that no vector lies in the plane of the other two, which is the same as asserting that the three vectors do not lie in the same plane.

We conclude this section with a theorem that shows that a linearly independent set in R^n can contain at most n vectors.

Theorem 4.4.3. *Let* $S = \{v_1, v_2, \ldots, v_r\}$ *be a set of vectors in* R^n. *If* $r > n$, *then* S *is linearly dependent.*

Proof. Suppose

$$v_1 = (v_{11}, v_{12}, \ldots, v_{1n})$$
$$v_2 = (v_{21}, v_{22}, \ldots, v_{2n})$$
$$\vdots$$
$$v_r = (v_{r1}, v_{r2}, \ldots, v_{rn})$$

Consider the equation

$$k_1 v_1 + k_2 v_2 + \cdots + k_r v_r = 0$$

If, as illustrated in Example 4, we express both sides of this equation in terms of components and then equate corresponding components, we obtain the system

$$v_{11} k_1 + v_{21} k_2 + \cdots + v_{r1} k_r = 0$$
$$v_{12} k_1 + v_{22} k_2 + \cdots + v_{r2} k_r = 0$$
$$\vdots \qquad \vdots \qquad\qquad \vdots \qquad \vdots$$
$$v_{1n} k_1 + v_{2n} k_2 + \cdots + v_{rn} k_r = 0$$

This is a homogeneous system of n equations in the r unknowns $k_1, \ldots, k_r$. Since $r > n$, it follows from Theorem 1.3.1 that the system has nontrivial solutions. Therefore, $S = \{v_1, v_2, \ldots, v_r\}$ is a linearly dependent set. ∎

In particular, the foregoing theorem tells us that a set in R^2 with more than two vectors is linearly dependent, and a set in R^3 with more than three vectors is linearly dependent.

EXERCISE SET 4.4

1. Explain why the following are linearly dependent sets of vectors. (Solve this problem by inspection.)
   (a) $u_1 = (-1, 2, 4)$ and $u_2 = (5, -10, -20)$ in R^3 　　(b) $u_1 = (3, -1)$, $u_2 = (4, 5)$, $u_3 = (-4, 7)$ in R^2

   (c) $p_1 = 3 - 2x + x^2$ and $p_2 = 6 - 4x + 2x^2$ in P_2 　　(d) $A = \begin{bmatrix} -3 & 4 \\ 2 & 0 \end{bmatrix}$ and $B = \begin{bmatrix} 3 & -4 \\ -2 & 0 \end{bmatrix}$ in M_{22}

2. Which of the following sets of vectors in R^3 are linearly dependent?
   (a) $(4, -1, 2)$, $(-4, 10, 2)$ 　　(b) $(-3, 0, 4)$, $(5, -1, 2)$, $(1, 1, 3)$
   (c) $(8, -1, 3)$, $(4, 0, 1)$ 　　(d) $(-2, 0, 1)$, $(3, 2, 5)$, $(6, -1, 1)$, $(7, 0, -2)$

3. Which of the following sets of vectors in R^4 are linearly dependent?
   (a) $(3, 8, 7, -3)$, $(1, 5, 3, -1)$, $(2, -1, 2, 6)$, $(1, 4, 0, 3)$
   (b) $(0, 0, 2, 2)$, $(3, 3, 0, 0)$, $(1, 1, 0, -1)$
   (c) $(0, 3, -3, -6)$, $(-2, 0, 0, -6)$, $(0, -4, -2, -2)$, $(0, -8, 4, -4)$
   (d) $(3, 0, -3, 6)$, $(0, 2, 3, 1)$, $(0, -2, -2, 0)$, $(-2, 1, 2, 1)$

4. Which of the following sets of vectors in P_2 are linearly dependent?
   (a) $2 - x + 4x^2$, $3 + 6x + 2x^2$, $2 + 10x - 4x^2$ 　　(b) $3 + x + x^2$, $2 - x + 5x^2$, $4 - 3x^2$
   (c) $6 - x^2$, $1 + x + 4x^2$ 　　(d) $1 + 3x + 3x^2$, $x + 4x^2$, $5 + 6x + 3x^2$, $7 + 2x - x^2$

5. Let V be the vector space of all real-valued functions defined on the entire real line. Which of the following sets of vectors in V are linearly dependent?
   (a) 6, $3\sin^2 x$, $2\cos^2 x$ 　　(b) x, $\cos x$ 　　(c) 1, $\sin x$, $\sin 2x$
   (d) $\cos 2x$, $\sin^2 x$, $\cos^2 x$ 　　(e) $(3 - x)^2$, $x^2 - 6x$, 5 　　(f) 0, $\cos^3 \pi x$, $\sin^5 3\pi x$

6. Assume that v_1, v_2, and v_3 are vectors in R^3 that have their initial points at the origin. In each part determine whether the three vectors lie in a plane.
   (a) $v_1 = (2, -2, 0)$, $v_2 = (6, 1, 4)$, $v_3 = (2, 0, -4)$ 　　(b) $v_1 = (-6, 7, 2)$, $v_2 = (3, 2, 4)$, $v_3 = (4, -1, 2)$

7. Assume that v_1, v_2, and v_3 are vectors in R^3 that have their initial points at the origin. In each part determine whether the three vectors lie on the same line.
   (a) $v_1 = (-1, 2, 3)$, $v_2 = (2, -4, -6)$, $v_3 = (-3, 6, 0)$ 　　(b) $v_1 = (2, -1, 4)$, $v_2 = (4, 2, 3)$, $v_3 = (2, 7, -6)$
   (c) $v_1 = (4, 6, 8)$, $v_2 = (2, 3, 4)$, $v_3 = (-2, -3, -4)$

8. (a) Show that the vectors $v_1 = (0, 3, 1, -1)$, $v_2 = (6, 0, 5, 1)$, and $v_3 = (4, -7, 1, 3)$ form a linearly dependent set in R^4.
   (b) Express each vector as a linear combination of the other two.

9. For which real values of λ do the following vectors form a linearly dependent set in R^3?
   $$v_1 = (\lambda, -\tfrac{1}{2}, -\tfrac{1}{2}), \quad v_2 = (-\tfrac{1}{2}, \lambda, -\tfrac{1}{2}), \quad v_3 = (-\tfrac{1}{2}, -\tfrac{1}{2}, \lambda)$$

10. Show that if $\{v_1, v_2, v_3\}$ is a linearly independent set of vectors, then so are $\{v_1, v_2\}$, $\{v_1, v_3\}$, $\{v_2, v_3\}$, $\{v_1\}$, $\{v_2\}$, and $\{v_3\}$.

11. Show that if $S = \{v_1, v_2, \ldots, v_r\}$ is a linearly independent set of vectors, then so is every nonempty subset of S.

12. Show that if $\{v_1, v_2, v_3\}$ is a linearly dependent set of vectors in a vector space V, and v_4 is any vector in V, then $\{v_1, v_2, v_3, v_4\}$ is also linearly dependent.

13. Show that if $\{v_1, v_2, \ldots, v_r\}$ is a linearly dependent set of vectors in a vector space V, and if $v_{r+1}, \ldots, v_n$ are any vectors in V, then $\{v_1, v_2, \ldots, v_{r+1}, \ldots, v_n\}$ is also linearly dependent.

14. Show that every set with more than three vectors from P_2 is linearly dependent.

15. Show that if $\{v_1, v_2\}$ is linearly independent and v_3 does not lie in $\text{lin}\{v_1, v_2\}$, then $\{v_1, v_2, v_3\}$ is linearly independent.

16. Prove: For any vectors u, v, and w, the vectors $u - v$, $v - w$, and $w - u$ form a linearly dependent set.

17. Prove: The space spanned by two vectors in R^3 is either a line through the origin, a plane through the origin, or the origin itself.

18. **(For readers who have studied calculus.)** Let V be the vector space of real-valued functions defined on the entire real line. If f, g, and h are vectors in V that are twice differentiable, then the function $w = w(x)$ defined by

$$w(x) = \begin{vmatrix} f(x) & g(x) & h(x) \\ f'(x) & g'(x) & h'(x) \\ f''(x) & g''(x) & h''(x) \end{vmatrix}$$

is called the **Wronskian** of f, g, and h. Prove that f, g, and h form a linearly independent set if the Wronskian is not the zero vector in V [i.e., $w(x)$ is not identically zero].

19. **(For readers who have studied calculus.)** Use the Wronskian (Exercise 18) to show that the following sets of vectors are linearly independent.

    (a) $1,\ x,\ e^x$ (b) $\sin x,\ \cos x,\ x \sin x$ (c) $e^x,\ xe^x,\ x^2e^x$ (d) $1,\ x,\ x^2$

20. Under what conditions is a set with one vector linearly independent?

21. Are the vectors v_1, v_2, and v_3 in Figure 3a linearly independent? What about those in Figure 3b? Explain.

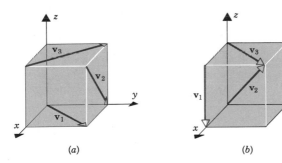

     (a)   (b)   **Figure 3**

22. Use part (a) of Theorem 4.4.1 to prove part (b).

23. Prove part (a) of Theorem 4.4.2.

24. Prove part (b) of Theorem 4.4.2.

4.5 BASIS AND DIMENSION

We usually think of a line as being one-dimensional, a plane as two-dimensional, and space around us as three-dimensional. It is the primary purpose of this section to make this intuitive notion of dimension precise.

NON-RECTANGULAR COORDINATE SYSTEMS

In plane analytic geometry we learned to associate a pair of coordinates (a, b) with a point P in the plane by projecting P onto a pair of perpendicular coordinate axes (Figure 1a). By this process, each point in the plane is assigned a unique set of coordinates and, conversely, each pair of coordinates is associated with a unique point in the plane. We describe this by saying that the coordinate system establishes a **one-to-one correspondence** between points in the plane and ordered pairs of real numbers. Although perpendicular coordinate axes are the most common, any two nonparallel lines can be used to define a coordinate system in the plane. For example, in Figure 1b, we attached a pair of coordinates (a, b) to the point P by projecting P parallel to the nonperpendicular coordinate axes. Similarly, in 3-space any three noncoplanar coordinate axes can be used to define a coordinate system (Figure 1c).

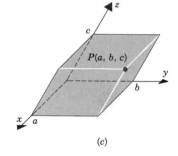

Figure 1 (a)   (b)   (c)

Coordinates of P in a rectangular coordinate system in 2-space.	Coordinates of P in a nonrectangular coordinate system in 2-space.	Coordinates of P in a nonrectangular coordinate system in 3-space.

Our first objective in this section is to extend the concept of a coordinate system to general vector spaces. As a start, it will be helpful to reformulate the notion of a coordinate system in 2-space or 3-space using vectors rather than coordinate axes to specify the coordinate system. This can be done by replacing each coordinate axis with a vector of length 1 that points in the positive direction of the axis. In Figure 2a, for example, $\mathbf{v}_1$ and $\mathbf{v}_2$ are such vectors. As illustrated in that figure, if P is any point in the plane, the vector $\overrightarrow{OP}$ can be written as a linear combination of $\mathbf{v}_1$ and $\mathbf{v}_2$ by projecting P parallel to $\mathbf{v}_1$ and $\mathbf{v}_2$ to make $\overrightarrow{OP}$ the diagonal of a parallelogram determined by vectors $a\mathbf{v}_1$ and $b\mathbf{v}_2$:

$$\overrightarrow{OP} = a\mathbf{v}_1 + b\mathbf{v}_2$$

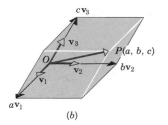

Figure 2   (a)   (b)

It is evident that the numbers a and b in this vector formula are precisely the coordinates of P in the coordinate system of Figure 1b. Similarly, the coordinates (a, b, c) of the point P in Figure 1c can be obtained by expressing $\overrightarrow{OP}$ as a linear combination of the vectors shown in Figure 2b.

Informally stated, vectors that specify a coordinate system are called "basis vectors" for that system. Although we used basis vectors of length 1 in the foregoing discussion, we shall see in a moment that this is not essential—nonzero vectors of any length will suffice.

The scales of measurement along the coordinate axes are essential ingredients of any coordinate system. Usually, one tries to use the same scale on each axis and have the integer points on the axes spaced one unit of distance apart. However, this is not always practical or appropriate: Unequal scales, or scales in which the integer points are more or less than one unit apart, may be required to fit a particular graph on a printed page or to represent physical quantities with diverse units in the same coordinate system (time in seconds on one axis and temperature in hundreds of degrees on another, for example). When a coordinate system is specified by a set of basis vectors, then the lengths of those vectors correspond to the distances between successive integer points on the coordinate axes (Figure 3). Thus, it is the directions of the basis vectors that define the positive directions of the coordinate axes and the lengths of the basis vectors that establish the scales of measurement.

The following key definition will make the foregoing ideas more precise and enable us to extend the concept of a coordinate system to general vector spaces.

BASIS FOR A VECTOR SPACE

Definition. If V is any vector space and $S = \{v_1, v_2, \ldots, v_n\}$ is a set of vectors in V, then S is called a ***basis*** for V if the following two conditions hold:

(a) S is linearly independent;   *one solution k, = 0 etc*
(b) S spans V.   *(express bb as line- combinations)*

A basis is the vector space generalization of a coordinate system in 2-space and 3-space. The following theorem will help us to see why this is so.

Theorem 4.5.1. *If $S = \{v_1, v_2, \ldots, v_n\}$ is a basis for a vector space V, then every vector v in V can be expressed in the form $v = c_1 v_1 + c_2 v_2 + \cdots + c_n v_n$ in exactly one way.*

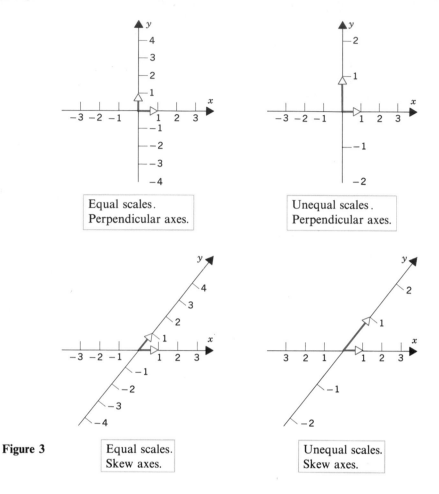

Equal scales.
Perpendicular axes.

Unequal scales.
Perpendicular axes.

Figure 3

Equal scales.
Skew axes.

Unequal scales.
Skew axes.

Proof. Since S spans V, it follows from the definition of a spanning set that every vector in V is expressible as a linear combination of the vectors in S. To see that there is only *one* way to express a vector as a linear combination of the vectors in S, suppose that some vector $\mathbf{v}$ can be written as

$$\mathbf{v} = c_1\mathbf{v}_1 + c_2\mathbf{v}_2 + \cdots + c_n\mathbf{v}_n$$

and also as

$$\mathbf{v} = k_1\mathbf{v}_1 + k_2\mathbf{v}_2 + \cdots + k_n\mathbf{v}_n$$

Subtracting the second equation from the first gives

$$\mathbf{0} = (c_1 - k_1)\mathbf{v}_1 + (c_2 - k_2)\mathbf{v}_2 + \cdots + (c_n - k_n)\mathbf{v}_n$$

Since the right side of this equation is a linear combination of vectors in S, the linear independence of S implies that

$$c_1 - k_1 = 0, \quad c_2 - k_2 = 0, \quad \ldots, \quad c_n - k_n = 0$$

that is,

$$c_1 = k_1, \quad c_2 = k_2, \quad \ldots, \quad c_n = k_n$$

Thus, the two expressions for $\mathbf{v}$ are the same. ▮

COORDINATES RELATIVE TO A BASIS

If $S = \{v_1, v_2, \ldots, v_n\}$ is a basis for a vector space V, and

$$v = c_1 v_1 + c_2 v_2 + \cdots + c_n v_n$$

is the expression for v in terms of the basis S, then the scalars $c_1, c_2, \ldots, c_n$ are called the **coordinates** of v relative to the basis S. The vector $(c_1, c_2, \ldots, c_n)$ constructed from these coordinates is called the **coordinate vector of v relative to** S; it is denoted by

$$(v)_S = (c_1, c_2, \ldots, c_n)$$

It should be noted that coordinate vectors depend not only on the basis S but also on the order in which the basis vectors are written; a change in the order of the basis vectors results in a corresponding change of order for the entries in the coordinate vectors.

Example 1   In Example 3 of the previous section we showed that if

$$i = (1, 0, 0), \qquad j = (0, 1, 0), \qquad \text{and} \qquad k = (0, 0, 1)$$

then $S = \{i, j, k\}$ is a linearly independent set in R^3. This set also spans R^3 since any vector $v = (a, b, c)$ in R^3 can be written as

$$v = (a, b, c) = a(1, 0, 0) + b(0, 1, 0) + c(0, 0, 1) = ai + bj + ck \qquad (1)$$

Thus, S is a basis for R^3; it is called the **standard basis** for R^3. Looking at the coefficients of i, j, and k in (1), it follows that the coordinates of v relative to the standard basis are a, b, and c, so

$$(v)_S = (a, b, c)$$

Comparing this result to (1) we see that

$$v = (v)_S$$

This equation states that the components of a vector v relative to a rectangular *xyz*-coordinate system and the coordinates of v relative to the standard basis are the same; thus, the coordinate system and the basis produce precisely the same one-to-one correspondence between points in 3-space and ordered triples of real numbers (Figure 4). ▲

The results in the foregoing example are a special case of those in the next example.

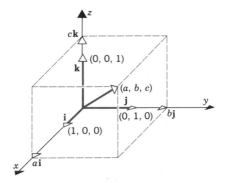

Example 2   In Example 3 of the previous section we showed that if

$$\mathbf{e}_1 = (1, 0, 0, \ldots, 0), \quad \mathbf{e}_2 = (0, 1, 0, \ldots, 0), \quad \ldots, \quad \mathbf{e}_n = (0, 0, 0, \ldots, 1)$$

then

$$S = \{\mathbf{e}_1, \mathbf{e}_2, \ldots, \mathbf{e}_n\}$$

is a linearly independent set in R^n. This set also spans R^n since any vector $\mathbf{v} = (v_1, v_2, \ldots, v_n)$ in R^n can be written as

$$\mathbf{v} = v_1\mathbf{e}_1 + v_2\mathbf{e}_2 + \cdots + v_n\mathbf{e}_n \tag{2}$$

STANDARD BASIS FOR R^n

Thus, S is a basis for R^n; it is called the ***standard basis for R^n***. It follows from (2) that the coordinates of $\mathbf{v} = (v_1, v_2, \ldots, v_n)$ relative to the standard basis are v_1, $v_2, \ldots, v_n$, so

$$(\mathbf{v})_S = (v_1, v_2, \ldots, v_n)$$

As in the previous example, we have

$$\mathbf{v} = (\mathbf{v})_S$$

so a vector $\mathbf{v}$ and its coordinate vector relative to the standard basis for R^n are the same.   ▲

REMARK.   It should be emphasized that in general a vector and its coordinate vector will not be the same; the equality that we observed in the two foregoing examples is a special situation that occurs only with the standard basis for R^n.

Example 3   Let $\mathbf{v}_1 = (1, 2, 1)$, $\mathbf{v}_2 = (2, 9, 0)$, and $\mathbf{v}_3 = (3, 3, 4)$. Show that the set $S = \{\mathbf{v}_1, \mathbf{v}_2, \mathbf{v}_3\}$ is a basis for R^3.

Solution.   To show that the set S spans R^3, we must show that an arbitrary vector $\mathbf{b} = (b_1, b_2, b_3)$ can be expressed as a linear combination

$$\mathbf{b} = c_1\mathbf{v}_1 + c_2\mathbf{v}_2 + c_3\mathbf{v}_3$$

of the vectors in S. Expressing this equation in terms of components gives

$$(b_1, b_2, b_3) = c_1(1, 2, 1) + c_2(2, 9, 0) + c_3(3, 3, 4)$$

or

$$(b_1, b_2, b_3) = (c_1 + 2c_2 + 3c_3, 2c_1 + 9c_2 + 3c_3, c_1 + 4c_3)$$

or on equating corresponding components

$$\begin{aligned} c_1 + 2c_2 + 3c_3 &= b_1 \\ 2c_1 + 9c_2 + 3c_3 &= b_2 \\ c_1 \qquad\quad + 4c_3 &= b_3 \end{aligned} \tag{3}$$

Thus, to show that S spans R^3, we must demonstrate that system (3) has a solution for all choices of $\mathbf{b} = (b_1, b_2, b_3)$.

To prove that S is linearly independent, we must show that the only solution of

$$c_1\mathbf{v}_1 + c_2\mathbf{v}_2 + c_3\mathbf{v}_3 = \mathbf{0} \tag{4}$$

is $c_1 = c_2 = c_3 = 0$. As above, if (4) is expressed in terms of components, the verification of independence reduces to showing that the homogeneous system

$$c_1 + 2c_2 + 3c_3 = 0$$
$$2c_1 + 9c_2 + 3c_3 = 0 \qquad (5)$$
$$c_1 \qquad\quad + 4c_3 = 0$$

has only the trivial solution. Observe that systems (3) and (5) have the same coefficient matrix. Thus, by parts (*a*), (*b*), and (*d*) of Theorem 1.7.3, we can simultaneously prove that S is linearly independent and spans R^3 by demonstrating that in systems (3) and (5) the matrix of coefficients

$$A = \begin{bmatrix} 1 & 2 & 3 \\ 2 & 9 & 3 \\ 1 & 0 & 4 \end{bmatrix}$$

is invertible. Since

$$\det(A) = \begin{vmatrix} 1 & 2 & 3 \\ 2 & 9 & 3 \\ 1 & 0 & 4 \end{vmatrix} = -1$$

(verify), it follows from Theorem 2.3.4 that A is invertible. Thus, S is a basis for R^3. ▲

Example 4 Let $S - \{v_1, v_2, v_3\}$ be the basis for R^3 in the previous example.

(a) Find the coordinate vector of $v = (5, -1, 9)$ with respect to S.
(b) Find the vector v in R^3 whose coordinate vector with respect to the basis S is $(v)_S = (-1, 3, 2)$.

Solution (*a*). We must find scalars c_1, c_2, c_3 such that

$$v = c_1 v_1 + c_2 v_2 + c_3 v_3$$

or, in terms of components,

$$(5, -1, 9) = c_1(1, 2, 1) + c_2(2, 9, 0) + c_3(3, 3, 4)$$

Equating corresponding components gives

$$c_1 + 2c_2 + 3c_3 = \quad 5$$
$$2c_1 + 9c_2 + 3c_3 = -1$$
$$c_1 \qquad\quad + 4c_3 = \quad 9$$

Solving this system, we obtain $c_1 = 1$, $c_2 = -1$, $c_3 = 2$ (verify). Therefore,

$$(v)_S = (1, -1, 2)$$

Solution (*b*). Using the definition of the coordinate vector $(v)_S$, we obtain

$$v = (-1)v_1 + 3v_2 + 2v_3 = (-1)(1, 2, 1) + 3(2, 9, 0) + 2(3, 3, 4) = (11, 31, 7) \quad ▲$$

Example 5

(a) Show that $S = \{1, x, x^2, \ldots, x^n\}$ is a basis for the vector space P_n of polynomials of the form $a_0 + a_1 x + \cdots + a_n x^n$.

(b) Find the coordinate vector of the polynomial $\mathbf{p} = a_0 + a_1 x + a_2 x^2$ relative to the basis $S = \{1, x, x^2\}$ for P_2.

Solution (a). We showed that S spans P_n in Example 10 of Section 4.3. To see that S is linearly independent, assume that some linear combination of vectors in S is the zero vector; that is,

$$a_0 + a_1 x + \cdots + a_n x^n = 0 \qquad \text{for all } x \tag{6}$$

We must show that $a_0 = a_1 = \cdots = a_n = 0$. From algebra, a polynomial of degree n with coefficients not all zero has at most n distinct roots. Since (6) holds for all values of x, every value of x is a root of the left-hand side. This implies that $a_0 = a_1 = \cdots = a_n = 0$; otherwise, $a_0 + a_1 x + \cdots + a_n x^n$ could have at most n roots. The set S is therefore linearly independent. The basis S is called the **standard basis for P_n**.

Solution (b). The coordinates of $\mathbf{p} = a_0 + a_1 x + a_2 x^2$ are the scalar coefficients of the basis vectors 1, x, and x^2, so $(\mathbf{p})_S = (a_0, a_1, a_2)$. ▲

Example 6 Let

$$M_1 = \begin{bmatrix} 1 & 0 \\ 0 & 0 \end{bmatrix} \qquad M_2 = \begin{bmatrix} 0 & 1 \\ 0 & 0 \end{bmatrix} \qquad M_3 = \begin{bmatrix} 0 & 0 \\ 1 & 0 \end{bmatrix} \qquad M_4 = \begin{bmatrix} 0 & 0 \\ 0 & 1 \end{bmatrix}$$

The set $S = \{M_1, M_2, M_3, M_4\}$ is a basis for the vector space M_{22} of 2×2 matrices. To see that S spans M_{22}, note that a typical vector (matrix)

$$\begin{bmatrix} a & b \\ c & d \end{bmatrix}$$

can be written as

$$\begin{bmatrix} a & b \\ c & d \end{bmatrix} = a \begin{bmatrix} 1 & 0 \\ 0 & 0 \end{bmatrix} + b \begin{bmatrix} 0 & 1 \\ 0 & 0 \end{bmatrix} + c \begin{bmatrix} 0 & 0 \\ 1 & 0 \end{bmatrix} + d \begin{bmatrix} 0 & 0 \\ 0 & 1 \end{bmatrix}$$

$$= a M_1 + b M_2 + c M_3 + d M_4$$

To see that S is linearly independent, assume that

$$a M_1 + b M_2 + c M_3 + d M_4 = 0$$

That is,

$$a \begin{bmatrix} 1 & 0 \\ 0 & 0 \end{bmatrix} + b \begin{bmatrix} 0 & 1 \\ 0 & 0 \end{bmatrix} + c \begin{bmatrix} 0 & 0 \\ 1 & 0 \end{bmatrix} + d \begin{bmatrix} 0 & 0 \\ 0 & 1 \end{bmatrix} = \begin{bmatrix} 0 & 0 \\ 0 & 0 \end{bmatrix}$$

It follows that

$$\begin{bmatrix} a & b \\ c & d \end{bmatrix} = \begin{bmatrix} 0 & 0 \\ 0 & 0 \end{bmatrix}$$

Thus, $a = b = c = d = 0$ so that S is linearly independent. The basis S in this example is called the *standard basis for* M_{22}. More generally, the **standard basis for** M_{mn} consists of the mn different matrices with a single 1 and zeros for the remaining entries.

Example 7   If $S = \{v_1, v_2, \ldots, v_r\}$ is a *linearly independent* set in a vector space V, then S is a basis for the subspace lin(S) since the set S spans lin(S) by definition of lin(S). ▲

DIMENSION

> **Definition.** A nonzero vector space V is called **finite-dimensional** if it contains a finite set of vectors $\{v_1, v_2, \ldots, v_n\}$ that forms a basis. If no such set exists, V is called **infinite dimensional**. In addition, we shall regard the zero vector space to be finite-dimensional even though it has no linearly independent sets and consequently no basis.

Example 8   By Examples 2, 5, and 6, R^n, P_n, and M_{mn} are finite-dimensional vector spaces. Some examples of infinite-dimensional vector spaces are given in the exercises. ▲

The next theorem provides the key to the concept of dimension for vector spaces. From it we shall obtain some of the most important results in linear algebra.

> **Theorem 4.5.2.** *If $S = \{v_1, v_2, \ldots, v_n\}$ is a basis for a vector space V, then every set with more than n vectors from V is linearly dependent.*

Proof. Let $S' = \{w_1, w_2, \ldots, w_m\}$ be any set of m vectors in V where $m > n$. We want to show that S' is linearly dependent. Since $S = \{v_1, v_2, \ldots, v_n\}$ is a basis, each w_i can be expressed as a linear combination of the vectors in S, say,

$$\begin{aligned}
w_1 &= a_{11}v_1 + a_{21}v_2 + \cdots + a_{n1}v_n \\
w_2 &= a_{12}v_1 + a_{22}v_2 + \cdots + a_{n2}v_n \\
&\ \ \vdots \qquad \vdots \qquad \vdots \qquad \vdots \\
w_m &= a_{1m}v_1 + a_{2m}v_2 + \cdots + a_{nm}v_n
\end{aligned} \tag{7}$$

To show that S' is linearly dependent, we must find scalars $k_1, k_2, \ldots, k_m$, not all zero, such that

$$k_1 w_1 + k_2 w_2 + \cdots + k_m w_m = 0 \tag{8}$$

Using the equations in (7), we can rewrite (8) as

$$(k_1 a_{11} + k_2 a_{12} + \cdots + k_m a_{1m})\mathbf{v}_1$$

$$+ (k_1 a_{21} + k_2 a_{22} + \cdots + k_m a_{2m})\mathbf{v}_2$$

$$\ddots$$

$$+ (k_1 a_{n1} + k_2 a_{n2} + \cdots + k_m a_{nm})\mathbf{v}_n = \mathbf{0}$$

The problem of proving that S' is a linearly dependent set thus reduces to showing there are scalars $k_1, k_2, \ldots, k_m$, not all zero, which satisfy

$$\begin{aligned}
a_{11}k_1 + a_{12}k_2 + \cdots + a_{1m}k_m &= 0 \\
a_{21}k_1 + a_{22}k_2 + \cdots + a_{2m}k_m &= 0 \\
\vdots \qquad \vdots \qquad\qquad \vdots \qquad \vdots \\
a_{n1}k_1 + a_{n2}k_2 + \cdots + a_{nm}k_m &= 0
\end{aligned} \qquad (9)$$

Since (9) has more unknowns than equations, the proof is complete, since Theorem 1.3.1 guarantees the existence of nontrivial solutions. ▮

As a consequence, we obtain the following result.

Theorem 4.5.3. *Any two bases for a finite-dimensional vector space have the same number of vectors.*

Proof. Let $S = \{\mathbf{v}_1, \mathbf{v}_2, \ldots, \mathbf{v}_n\}$ and $S' = \{\mathbf{w}_1, \mathbf{w}_2, \ldots, \mathbf{w}_m\}$ be two bases for a finite-dimensional vector space V. Since S is a basis and S' is a linearly independent set, Theorem 4.5.2 implies that $m \leq n$. Similarly, since S' is a basis and S is linearly independent, we also have $n \leq m$. Therefore, $m = n$. ▮

Example 9 The standard basis for R^n contains n vectors (Example 2). Therefore, every basis for R^n contains n vectors. ▲

Example 10 The standard basis for P_n (Example 5) contains $n + 1$ vectors; thus, every basis for P_n contains $n + 1$ vectors. ▲

The number of vectors in a basis for a finite-dimensional vector space is an extremely important quantity. By Example 9, every basis for R^2 has two vectors and every basis for R^3 has three vectors. Since R^2 (the plane) is intuitively two-dimensional and R^3 is intuitively three-dimensional, the dimension of these spaces is the same as the number of vectors that occur in their bases. This suggests the following definition.

Definition. The **dimension** of a finite-dimensional vector space V is defined to be the number of vectors in a basis for V. In addition, we define the zero vector space to have dimension zero.

From Examples 9 and 10, R^n is an n-dimensional vector space and P_n is an $n + 1$-dimensional vector space.

Example 11 Determine a basis for and the dimension of the solution space of the homogeneous system

$$
\begin{aligned}
2x_1 + 2x_2 - \;x_3 \qquad\quad + x_5 &= 0 \\
-x_1 - \;x_2 + 2x_3 - 3x_4 + x_5 &= 0 \\
x_1 + \;x_2 - 2x_3 \qquad\quad - x_5 &= 0 \\
x_3 + \;x_4 + x_5 &= 0
\end{aligned}
$$

Solution. In Example 1 of Section 1.3 it was shown that the general solution of the given system is

$$
x_1 = -s - t, \qquad x_2 = s, \qquad x_3 = -t, \qquad x_4 = 0, \qquad x_5 = t
$$

Therefore, the solution vectors can be written as

$$
\begin{bmatrix} x_1 \\ x_2 \\ x_3 \\ x_4 \\ x_5 \end{bmatrix}
=
\begin{bmatrix} -s-t \\ s \\ -t \\ 0 \\ t \end{bmatrix}
=
\begin{bmatrix} -s \\ s \\ 0 \\ 0 \\ 0 \end{bmatrix}
+
\begin{bmatrix} -t \\ 0 \\ -t \\ 0 \\ t \end{bmatrix}
= s \begin{bmatrix} -1 \\ 1 \\ 0 \\ 0 \\ 0 \end{bmatrix}
+ t \begin{bmatrix} -1 \\ 0 \\ -1 \\ 0 \\ 1 \end{bmatrix}
$$

which shows that the vectors

$$
\mathbf{v}_1 = \begin{bmatrix} -1 \\ 1 \\ 0 \\ 0 \\ 0 \end{bmatrix}
\qquad \text{and} \qquad
\mathbf{v}_2 = \begin{bmatrix} -1 \\ 0 \\ -1 \\ 0 \\ 1 \end{bmatrix}
$$

span the solution space. Since they are also linearly independent (verify), $\{\mathbf{v}_1, \mathbf{v}_2\}$ is a basis, and the solution space is two-dimensional. ▲

In general, to show that a set of vectors $\{\mathbf{v}_1, \mathbf{v}_2, \dots, \mathbf{v}_n\}$ is a basis for a vector space V, we must show that the vectors are linearly independent and span V. However, if we happen to know that V has dimension n (so that $\{\mathbf{v}_1, \mathbf{v}_2, \dots, \mathbf{v}_n\}$ contains the right number of vectors for a basis), then it suffices to check *either* linear independence *or* spanning—the remaining condition will hold automatically. This is the content of the following theorem, whose proof is left as an exercise.

Theorem 4.5.4

(*a*) *If* $S = \{\mathbf{v}_1, \mathbf{v}_2, \dots, \mathbf{v}_n\}$ *is a set of n linearly independent vectors in an n-dimensional space V, then S is a basis for V.*

(*b*) *If* $S = \{\mathbf{v}_1, \mathbf{v}_2, \dots, \mathbf{v}_n\}$ *is a set of n vectors that spans an n-dimensional space V, then S is a basis for V.*

Example 12 Show that $v_1 = (-3, 7)$ and $v_2 = (5, 5)$ form a basis for R^2.

Solution. Since neither vector is a scalar multiple of the other, $S = \{v_1, v_2\}$ is linearly independent. Since R^2 is two-dimensional, S is a basis for R^2 by part (a) of the foregoing theorem. ▲

We conclude this section with another important theorem that involves the concept of dimension. The proofs are left as exercises.

Theorem 4.5.5. *Let V be an n-dimensional vector space.*

(a) *If $S = \{v_1, v_2, \ldots, v_r\}$ is a linearly independent set of vectors in V, and if $r < n$, then S can be enlarged to a basis for V; that is, there are vectors $v_{r+1}, \ldots, v_n$ such that $\{v_1, v_2, \ldots, v_r, v_{r+1}, \ldots, v_n\}$ is a basis for V.*

(b) *If W is a subspace of V, then $dim(W) \leq dim(V)$; moreover, $dim(W) = dim(V)$; if and only if $W = V$.*

Part (b) of the foregoing theorem states that for any finite-dimensional vector space, the dimension of a subspace cannot exceed the dimension of the vector space itself and that the only way a subspace can have the same dimension as the vector space itself is if the subspace is the entire vector space. Figure 5 illustrates this idea in R^3. In that figure observe that successively larger subspaces increase in dimension.

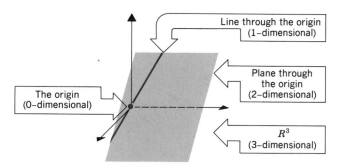

Figure 5

EXERCISE SET 4.5

1. Explain why the following sets of vectors are *not* bases for the indicated vector spaces. (Solve this problem by inspection.)
   (a) $u_1 = (1, 2)$, $u_2 = (0, 3)$, $u_3 = (2, 7)$ for R^2
   (b) $u_1 = (-1, 3, 2)$, $u_2 = (6, 1, 1)$ for R^3
   (c) $p_1 = 1 + x + x^2$, $p_2 = x - 1$ for P_2

   (d) $A = \begin{bmatrix} 1 & 1 \\ 2 & 3 \end{bmatrix}$, $B = \begin{bmatrix} 6 & 0 \\ -1 & 4 \end{bmatrix}$, $C = \begin{bmatrix} 3 & 0 \\ 1 & 7 \end{bmatrix}$, $D = \begin{bmatrix} 5 & 1 \\ 4 & 2 \end{bmatrix}$, $E = \begin{bmatrix} 7 & 1 \\ 2 & 9 \end{bmatrix}$ for M_{22}

2. Which of the following sets of vectors are bases for R^2?
   (a) $(2, 1)$, $(3, 0)$ (b) $(4, 1)$, $(-7, -8)$ (c) $(0, 0)$, $(1, 3)$ (d) $(3, 9)$, $(-4, -12)$

3. Which of the following sets of vectors are bases for R^3?
   (a) (1, 0, 0), (2, 2, 0), (3, 3, 3) (b) (3, 1, −4), (2, 5, 6), (1, 4, 8)
   (c) (2, −3, 1), (4, 1, 1), (0, −7, 1) (d) (1, 6, 4), (2, 4, −1), (−1, 2, 5)

4. Which of the following sets of vectors are bases for P_2?
   (a) $1 − 3x + 2x^2$, $1 + x + 4x^2$, $1 − 7x$ (b) $4 + 6x + x^2$, $−1 + 4x + 2x^2$, $5 + 2x − x^2$
   (c) $1 + x + x^2$, $x + x^2$, x^2 (d) $−4 + x + 3x^2$, $6 + 5x + 2x^2$, $8 + 4x + x^2$

5. Show that the following set of vectors is a basis for M_{22}.

$$\begin{bmatrix} 3 & 6 \\ 3 & -6 \end{bmatrix} \quad \begin{bmatrix} 0 & -1 \\ -1 & 0 \end{bmatrix} \quad \begin{bmatrix} 0 & -8 \\ -12 & -4 \end{bmatrix} \quad \begin{bmatrix} 1 & 0 \\ -1 & 2 \end{bmatrix}$$

6. Let V be the space spanned by $v_1 = \cos^2 x$, $v_2 = \sin^2 x$, $v_3 = \cos 2x$.
   (a) Show that $S = \{v_1, v_2, v_3\}$ is not a basis for V. (b) Find a basis for V.

7. Find the coordinate vector of **w** relative to the basis $S = \{u_1, u_2\}$ for R^2.
   (a) $u_1 = (1, 0)$, $u_2 = (0, 1)$; $w = (3, −7)$ (b) $u_1 = (2, −4)$, $u_2 = (3, 8)$; $w = (1, 1)$
   (c) $u_1 = (1, 1)$, $u_2 = (0, 2)$; $w = (a, b)$

8. Find the coordinate vector of **v** relative to the basis $S = \{v_1, v_2, v_3\}$.
   (a) $v = (2, −1, 3)$; $v_1 = (1, 0, 0)$, $v_2 = (2, 2, 0)$, $v_3 = (3, 3, 3)$
   (b) $v = (5, −12, 3)$; $v_1 = (1, 2, 3)$, $v_2 = (−4, 5, 6)$, $v_3 = (7, −8, 9)$

9. Find the coordinate vector of **p** relative to the basis $S = \{p_1, p_2, p_3\}$.
   (a) $p = 4 − 3x + x^2$; $p_1 = 1$, $p_2 = x$, $p_3 = x^2$
   (b) $p = 2 − x + x^2$; $p_1 = 1 + x$, $p_2 = 1 + x^2$, $p_3 = x + x^2$

10. Find the coordinate vector of A relative to the basis $S = \{A_1, A_2, A_3, A_4\}$.

$$A = \begin{bmatrix} 2 & 0 \\ -1 & 3 \end{bmatrix} \quad A_1 = \begin{bmatrix} -1 & 1 \\ 0 & 0 \end{bmatrix} \quad A_2 = \begin{bmatrix} 1 & 1 \\ 0 & 0 \end{bmatrix} \quad A_3 = \begin{bmatrix} 0 & 0 \\ 1 & 0 \end{bmatrix} \quad A_4 = \begin{bmatrix} 0 & 0 \\ 0 & 1 \end{bmatrix}$$

In Exercises 11–16 determine the dimension of and a basis for the solution space of the system.

11. $\begin{aligned} x_1 + x_2 − x_3 &= 0 \\ −2x_1 − x_2 + 2x_3 &= 0 \\ −x_1 \quad\quad + x_3 &= 0 \end{aligned}$
   12. $\begin{aligned} 3x_1 + x_2 + x_3 + x_4 &= 0 \\ 5x_1 − x_2 + x_3 − x_4 &= 0 \end{aligned}$
   13. $\begin{aligned} x_1 − 4x_2 + 3x_3 − x_4 &= 0 \\ 2x_1 − 8x_2 + 6x_3 − 2x_4 &= 0 \end{aligned}$

14. $\begin{aligned} x_1 − 3x_2 + x_3 &= 0 \\ 2x_1 − 6x_2 + 2x_3 &= 0 \\ 3x_1 − 9x_2 + 3x_3 &= 0 \end{aligned}$
   15. $\begin{aligned} 2x_1 + x_2 + 3x_3 &= 0 \\ x_1 \quad\quad + 5x_3 &= 0 \\ x_2 + x_3 &= 0 \end{aligned}$
   16. $\begin{aligned} x + y + z &= 0 \\ 3x + 2y − 2z &= 0 \\ 4x + 3y − z &= 0 \\ 6x + 5y + z &= 0 \end{aligned}$

17. Determine bases for the following subspaces of R^3.
   (a) the plane $3x − 2y + 5z = 0$ (b) the plane $x − y = 0$
   (c) the line $x = 2t$, $y = −t$, $z = 4t$ (d) all vectors of the form (a, b, c) where $b = a + c$

18. Determine the dimensions of the following subspaces of R^4.
   (a) all vectors of the form $(a, b, c, 0)$
   (b) all vectors of the form (a, b, c, d), where $d = a + b$ and $c = a − b$
   (c) all vectors of the form (a, b, c, d), where $a = b = c = d$

19. Determine the dimension of the subspace of P_3 consisting of all polynomials
   $a_0 + a_1x + a_2x^2 + a_3x^3$ for which $a_0 = 0$.

20. Let $\{v_1, v_2, v_3\}$ be a basis for a vector space V. Show that $\{u_1, u_2, u_3\}$ is also a basis, where $u_1 = v_1$, $u_2 = v_1 + v_2$, and $u_3 = v_1 + v_2 + v_3$.

21. Show that the vector space of all real-valued functions defined on the entire real line is infinite-dimensional. [**Hint.** Assume it is finite-dimensional with dimension n, and obtain a contradiction by producing $n + 1$ linearly independent vectors.]

22. Show that every subspace of a finite-dimensional vector space is finite-dimensional.

23. Let V be a subspace of a finite-dimensional vector space W. Show that $\dim(V) \leq \dim(W)$. [**Hint.** V is finite dimensional by Exercise 22.]

24. Show that the only subspaces of R^3 are lines through the origin, planes through the origin, the zero subspace, and R^3 itself. [**Hint.** By Exercise 23 the subspaces of R^3 must be 0-dimensional, 1-dimensional, 2-dimensional, or 3-dimensional.]

25. Let S be a basis for an n-dimensional vector space V. Show that if $v_1, v_2, \ldots, v_r$ form a linearly independent set of vectors in V, then the coordinate vectors $(v_1)_S, (v_2)_S, \ldots, (v_r)_S$ form a linearly independent set in R^n, and conversely.

26. Using the notation from Exercise 25, show that if $v_1, v_2, \ldots, v_r$ span V, then the coordinate vectors $(v_1)_S, (v_2)_S, \ldots, (v_r)_S$ span R^n, and conversely.

27. Find a basis for the subspace of P_2 spanned by the given vectors.
    (a) $-1 + x - 2x^2$, $3 + 3x + 6x^2$, 9 (b) $1 + x$, x^2, $-2 + 2x^2$, $-3x$
    (c) $1 + x - 3x^2$, $2 + 2x - 6x^2$, $3 + 3x - 9x^2$
    [**Hint.** Let S be the standard basis for P_2 and work with the coordinate vectors relative to S; note Exercises 25, 26.]

28. Figure 6 shows a rectangular xy-coordinate system and an $x'y'$-coordinate system with skewed axes. Assuming that 1-unit scales are used on all the axes, find the $x'y'$-coordinates of the points whose xy-coordinates are given.
    (a) $(1, 1)$ (b) $(1, 0)$ (c) $(0, 1)$ (d) (a, b)

29. Figure 7 shows a rectangular xy-coordinate system determined by the unit basis vectors **i** and **j** and an $x'y'$-coordinate system determined by unit basis vectors u_1 and u_2. Find the $x'y'$-coordinates of the points whose xy-coordinates are given.
    (a) $(\sqrt{3}, 1)$ (b) $(1, 0)$ (c) $(0, 1)$ (d) (a, b)

Figure 6

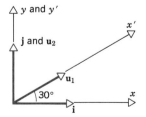

Figure 7

30. Prove part (a) of Theorem 4.5.4.

31. Prove part (b) of Theorem 4.5.4.

32. Prove part (a) of Theorem 4.5.5.

33. Prove part (b) of Theorem 4.5.5.

4.6 ROW SPACE, COLUMN SPACE, AND NULLSPACE

In this section we shall study three important vector spaces associated with matrices. Our results will provide us with a deeper understanding of the relationships between the solutions of a linear system and properties of its coefficient matrix. As a by-product of our work, we will obtain some computational algorithms for finding bases of vector spaces.

We begin with some definitions.

ROW AND COLUMN VECTORS

Definition. For an $m \times n$ matrix

$$A = \begin{bmatrix} a_{11} & a_{12} & \cdots & a_{1n} \\ a_{21} & a_{22} & \cdots & a_{2n} \\ \vdots & \vdots & & \vdots \\ a_{m1} & a_{m2} & \cdots & a_{mn} \end{bmatrix}$$

the vectors

$$\mathbf{r}_1 = (a_{11}, a_{12}, \ldots, a_{1n})$$
$$\mathbf{r}_2 = (a_{21}, a_{22}, \ldots, a_{2n})$$
$$\vdots \qquad \vdots$$
$$\mathbf{r}_m = (a_{m1}, a_{m2}, \ldots, a_{mn})$$

formed from the rows of A are called the **row vectors** of A, and the vectors

$$\mathbf{c}_1 = \begin{bmatrix} a_{11} \\ a_{21} \\ \vdots \\ a_{m1} \end{bmatrix}, \quad \mathbf{c}_2 = \begin{bmatrix} a_{12} \\ a_{22} \\ \vdots \\ a_{m2} \end{bmatrix}, \quad \ldots, \quad \mathbf{c}_n = \begin{bmatrix} a_{1n} \\ a_{2n} \\ \vdots \\ a_{mn} \end{bmatrix}$$

formed from the columns of A are called the **column vectors** of A.

Example 1 Let

$$A = \begin{bmatrix} 2 & 1 & 0 \\ 3 & -1 & 4 \end{bmatrix}$$

The row vectors of A are

$$\mathbf{r}_1 = (2, 1, 0) \quad \text{and} \quad \mathbf{r}_2 = (3, -1, 4)$$

and the column vectors of A are

$$\mathbf{c}_1 = \begin{bmatrix} 2 \\ 3 \end{bmatrix} \quad \mathbf{c}_2 = \begin{bmatrix} 1 \\ -1 \end{bmatrix} \quad \text{and} \quad \mathbf{c}_3 = \begin{bmatrix} 0 \\ 4 \end{bmatrix} \quad \blacktriangle$$

An $m \times n$ matrix A with column vectors $\mathbf{c}_1, \mathbf{c}_2, \ldots, \mathbf{c}_n$ is sometimes denoted by

$$A = [\mathbf{c}_1 | \mathbf{c}_2 | \cdots | \mathbf{c}_n]$$

and an $m \times n$ matrix A with row vectors $\mathbf{r}_1, \mathbf{r}_2, \ldots, \mathbf{r}_m$ by

$$A = \begin{bmatrix} \mathbf{r}_1 \\ \hline \mathbf{r}_2 \\ \hline \vdots \\ \hline \mathbf{r}_m \end{bmatrix}$$

Column vectors provide a useful way of thinking about matrix multiplication. For example, suppose that

$$A = \begin{bmatrix} a_{11} & a_{12} & \cdots & a_{1n} \\ a_{21} & a_{22} & \cdots & a_{2n} \\ \vdots & \vdots & & \vdots \\ a_{m1} & a_{m2} & \cdots & a_{mn} \end{bmatrix} \quad \text{and} \quad \mathbf{x} = \begin{bmatrix} x_1 \\ x_2 \\ \vdots \\ x_n \end{bmatrix}$$

Then

$$A\mathbf{x} = \begin{bmatrix} a_{11}x_1 + a_{12}x_2 + \cdots + a_{1n}x_n \\ a_{21}x_1 + a_{22}x_2 + \cdots + a_{2n}x_n \\ \vdots & \vdots & & \vdots \\ a_{m1}x_1 + a_{m2}x_2 + \cdots + a_{mn}x_n \end{bmatrix} = x_1 \begin{bmatrix} a_{11} \\ a_{21} \\ \vdots \\ a_{m1} \end{bmatrix} + x_2 \begin{bmatrix} a_{12} \\ a_{22} \\ \vdots \\ a_{m2} \end{bmatrix} + \cdots + x_n \begin{bmatrix} a_{1n} \\ a_{2n} \\ \vdots \\ a_{mn} \end{bmatrix}$$

or equivalently

$$A\mathbf{x} = x_1\mathbf{c}_1 + x_2\mathbf{c}_2 + \cdots + x_n\mathbf{c}_n \tag{1}$$

where $\mathbf{c}_1, \mathbf{c}_2, \ldots, \mathbf{c}_n$ are the column vectors of A. In words, (1) tells us that the product $A\mathbf{x}$ is a linear combination of the column vectors of A and that the scalar coefficients of the linear combination are the entries of $\mathbf{x}$. ▲

Example 2 The matrix product

$$\begin{bmatrix} -1 & 3 & 2 \\ 1 & 2 & -3 \\ 2 & 1 & -2 \end{bmatrix} \begin{bmatrix} 2 \\ -1 \\ 3 \end{bmatrix} = \begin{bmatrix} 1 \\ -9 \\ -3 \end{bmatrix} \tag{2}$$

can be written as the linear combination

$$2\begin{bmatrix} -1 \\ 1 \\ 2 \end{bmatrix} - 1\begin{bmatrix} 3 \\ 2 \\ 1 \end{bmatrix} + 3\begin{bmatrix} 2 \\ -3 \\ -2 \end{bmatrix} = \begin{bmatrix} 1 \\ -9 \\ -3 \end{bmatrix}$$

by applying (1) to the left side of (2). ▲

The following definition defines three important vector spaces associated with a matrix.

ROW SPACE,
COLUMN SPACE,
AND NULLSPACE

Definition. If A is an $m \times n$ matrix, then the subspace of R^n spanned by the row vectors of A is called the **row space** of A, and the subspace of R^m spanned by the column vectors is called the **column space** of A. The solution space of the homogeneous system of equations $A\mathbf{x} = \mathbf{0}$, which is a subspace of R^n, is called the **nullspace** of A.

In this section and the next we shall be concerned with the following general questions:

- What relationships exist between the row space, column space, and nullspace of a matrix?
- What relationships exist between the solutions of a linear system $A\mathbf{x} = \mathbf{b}$ and the row space, column space, and nullspace of the coefficient matrix A.

To answer these questions we shall need a way to find bases for the row space, column space, and nullspace of a matrix.

Example 3 Find a basis for the nullspace of

$$A = \begin{bmatrix} 2 & 2 & -1 & 0 & 1 \\ -1 & -1 & 2 & -3 & 1 \\ 1 & 1 & -2 & 0 & -1 \\ 0 & 0 & 1 & 1 & 1 \end{bmatrix}$$

Solution. The nullspace of A is the solution space of the homogeneous system

$$\begin{aligned} 2x_1 + 2x_2 - x_3 \quad\quad + x_5 &= 0 \\ -x_1 - x_2 + 2x_3 - 3x_4 + x_5 &= 0 \\ x_1 + x_2 - 2x_3 \quad\quad - x_5 &= 0 \\ x_3 + x_4 + x_5 &= 0 \end{aligned}$$

In Example 11 of Section 4.5 we showed that the vectors

$$\mathbf{v}_1 = \begin{bmatrix} -1 \\ 1 \\ 0 \\ 0 \\ 0 \end{bmatrix} \quad \text{and} \quad \mathbf{v}_2 = \begin{bmatrix} -1 \\ 0 \\ -1 \\ 0 \\ 1 \end{bmatrix}$$

form a basis for this space. ▲

The following theorem, whose proof is deferred to the end of this section, will lead to techniques for computing bases for the row and column spaces of a matrix.

> **Theorem 4.6.1.** *Elementary row operations do not change the row space or the nullspace of a matrix.*

It follows from this theorem that a matrix and all of its row-echelon forms have the same row space and the same nullspace. In the exercises we shall ask you to show that the nonzero row vectors of a matrix in row-echelon form are linearly independent and consequently form a basis for the row space (Exercise 10). This fact and Theorem 4.6.1 yield the following result.

> **Theorem 4.6.2.** *The nonzero row vectors in any row-echelon form of a matrix A form a basis for the row space of A.*

Example 4 Find a basis for the space spanned by the vectors

$$\mathbf{v}_1 = (1, -2, 0, 0, 3) \qquad \mathbf{v}_2 = (2, -5, -3, -2, 6) \qquad \mathbf{v}_3 = (0, 5, 15, 10, 0)$$
$$\mathbf{v}_4 = (2, 6, 18, 8, 6)$$

Solution. The space spanned by these vectors is the row space of the matrix

$$\begin{bmatrix} 1 & -2 & 0 & 0 & 3 \\ 2 & -5 & -3 & -2 & 6 \\ 0 & 5 & 15 & 10 & 0 \\ 2 & 6 & 18 & 8 & 6 \end{bmatrix}$$

Reducing this matrix to row-echelon form we obtain

$$\begin{bmatrix} 1 & -2 & 0 & 0 & 3 \\ 0 & 1 & 3 & 2 & 0 \\ 0 & 0 & 1 & 1 & 0 \\ 0 & 0 & 0 & 0 & 0 \end{bmatrix}$$

The nonzero row vectors in this matrix are

$$\mathbf{w}_1 = (1, -2, 0, 0, 3) \qquad \mathbf{w}_2 = (0, 1, 3, 2, 0) \qquad \mathbf{w}_3 = (0, 0, 1, 1, 0)$$

These vectors form a basis for the row space and consequently form a basis for the space spanned by $\mathbf{v}_1$, $\mathbf{v}_2$, $\mathbf{v}_3$, and $\mathbf{v}_4$. ▲

REMARK. Observe that the basis vectors produced by the method in the foregoing example are not all row vectors in the original matrix. Later in this section, we shall show how to obtain a basis for the row space of a matrix that consists entirely of row vectors of that matrix.

Although elementary row operations do not change the row space and null-space of a matrix, they usually change the column space. However, we shall show that even though the column space of the matrix may change, whatever relationships of linear independence or linear dependence that exist among the column vectors prior to a row operation will also hold for the corresponding columns of the matrix that results from that operation. To make this more precise, suppose that a matrix B results by performing an elementary row operation on an $m \times n$ matrix A. By Theorem 4.6.1 the two homogeneous linear systems

$$A\mathbf{x} = \mathbf{0} \qquad \text{and} \qquad B\mathbf{x} = \mathbf{0}$$

have the same solution set. Thus, the first system has a nontrivial solution if and only if the same is true of the second. But if the column vectors of A and B, respectively, are

$$\mathbf{c}_1, \mathbf{c}_2, \ldots, \mathbf{c}_n \qquad \text{and} \qquad \mathbf{c}'_1, \mathbf{c}'_2, \ldots, \mathbf{c}'_n$$

then from (1) the two systems can be rewritten as

$$x_1\mathbf{c}_1 + x_2\mathbf{c}_2 + \cdots + x_n\mathbf{c}_n = \mathbf{0} \tag{3}$$

and

$$x_1\mathbf{c}'_1 + x_2\mathbf{c}'_2 + \cdots + x_n\mathbf{c}'_n = \mathbf{0} \tag{4}$$

Thus, (3) has a nontrivial solution for $x_1, x_2, \ldots, x_n$ if and only if the same is true of (4). This implies that the column vectors of A are linearly independent if and only if the same is true of B. Although we shall omit the proof, this conclusion also applies to any subset of the column vectors. Thus, we have the following result.

Theorem 4.6.3. *If A and B are row equivalent matrices, then*

(a) a given set of column vectors of A is linearly independent if and only if the corresponding column vectors of B are linearly independent;

(b) a given set of column vectors of A forms a basis for the column space of A if and only if the corresponding column vectors of B form a basis for the column space of B.

Example 5 The reduced row-echelon form of

$$A = \begin{bmatrix} 1 & 2 & 0 & 2 & 5 \\ -2 & -5 & 1 & -1 & -8 \\ 0 & -3 & 3 & 4 & 1 \\ 3 & 6 & 0 & -7 & 2 \end{bmatrix}$$

is

$$R = \begin{bmatrix} 1 & 0 & 2 & 0 & 1 \\ 0 & 1 & -1 & 0 & 1 \\ 0 & 0 & 0 & 1 & 1 \\ 0 & 0 & 0 & 0 & 0 \end{bmatrix}$$

We leave it for the reader to verify that the column vectors

$$\mathbf{c}_1' = \begin{bmatrix} 1 \\ 0 \\ 0 \\ 0 \end{bmatrix}, \quad \mathbf{c}_2' = \begin{bmatrix} 0 \\ 1 \\ 0 \\ 0 \end{bmatrix}, \quad \mathbf{c}_4' = \begin{bmatrix} 0 \\ 0 \\ 1 \\ 0 \end{bmatrix}$$

form a basis for the column space of R; thus the corresponding column vectors of A, namely,

$$\mathbf{c}_1 = \begin{bmatrix} 1 \\ -2 \\ 0 \\ 3 \end{bmatrix}, \quad \mathbf{c}_2 = \begin{bmatrix} 2 \\ -5 \\ -3 \\ 6 \end{bmatrix}, \quad \mathbf{c}_4 = \begin{bmatrix} 2 \\ -1 \\ 4 \\ -7 \end{bmatrix}$$

form a basis for the column space of A. ▲

The following result (Exercise 11) makes it easy to find a basis for the column space of a matrix by reducing it to any row-echelon form.

Theorem 4.6.4. *If a matrix is in row-echelon form, then the column vectors that contain the leading 1's form a basis for the column space of that matrix.*

Example 6 Find a basis for the column space of the matrix

$$A = \begin{bmatrix} 1 & -2 & 0 & 0 & 3 \\ 2 & -5 & -3 & -2 & 6 \\ 0 & 5 & 15 & 10 & 0 \\ 2 & 6 & 18 & 8 & 6 \end{bmatrix}$$

Solution. Reducing this matrix to row-echelon form we obtain (see Example 4)

$$R = \begin{bmatrix} 1 & -2 & 0 & 0 & 3 \\ 0 & 1 & 3 & 2 & 0 \\ 0 & 0 & 1 & 1 & 0 \\ 0 & 0 & 0 & 0 & 0 \end{bmatrix}$$

The first three columns of R contain the leading 1's, so these column vectors form a basis for the column space of R. The corresponding column vectors in the matrix A form a basis for the column space of A; these are

$$\mathbf{c}_1 = \begin{bmatrix} 1 \\ 2 \\ 0 \\ 2 \end{bmatrix}, \quad \mathbf{c}_2 = \begin{bmatrix} -2 \\ -5 \\ 5 \\ 6 \end{bmatrix}, \quad \mathbf{c}_3 = \begin{bmatrix} 0 \\ -3 \\ 15 \\ 18 \end{bmatrix} \quad ▲$$

Observe that the procedure in the foregoing example produces a basis for the column space of A consisting entirely of column vectors of A. In the remark following Example 4 we promised to give a procedure for finding a basis for the row space of a matrix consisting entirely of row vectors. The following example shows how this can be done.

Example 7   Find a basis for the row space of

$$A = \begin{bmatrix} 1 & -2 & 0 & 0 & 3 \\ 2 & -5 & -3 & -2 & 6 \\ 0 & 5 & 15 & 10 & 0 \\ 2 & 6 & 18 & 8 & 6 \end{bmatrix}$$

consisting entirely of row vectors from A.

Solution. We will transpose A, thereby converting the row space of A into the column space of A^t; then we will use the method of the previous example to find a basis for the column space of A^t; and then we will transpose again to convert column vectors back to row vectors. Transposing A yields

$$A^t = \begin{bmatrix} 1 & 2 & 0 & 2 \\ -2 & -5 & 5 & 6 \\ 0 & -3 & 15 & 18 \\ 0 & -2 & 10 & 8 \\ 3 & 6 & 0 & 6 \end{bmatrix}$$

Reducing this matrix to row-echelon form yields

$$\begin{bmatrix} 1 & 2 & 0 & 2 \\ 0 & 1 & -5 & -10 \\ 0 & 0 & 0 & 1 \\ 0 & 0 & 0 & 0 \\ 0 & 0 & 0 & 0 \end{bmatrix}$$

The first, second, and fourth columns contain the leading 1's, so the corresponding column vectors in A^t from a basis for the column space of A^t; these are

$$\mathbf{c}_1 = \begin{bmatrix} 1 \\ -2 \\ 0 \\ 0 \\ 3 \end{bmatrix} \quad \mathbf{c}_2 = \begin{bmatrix} 2 \\ -5 \\ -3 \\ -2 \\ 6 \end{bmatrix} \quad \text{and} \quad \mathbf{c}_4 = \begin{bmatrix} 2 \\ 6 \\ 18 \\ 8 \\ 6 \end{bmatrix}$$

Transposing again and adjusting the notation appropriately yields the basis vectors

$$\mathbf{r}_1 = \begin{bmatrix} 1 & -2 & 0 & 0 & 3 \end{bmatrix} \qquad \mathbf{r}_2 = \begin{bmatrix} 2 & -5 & -3 & -2 & 6 \end{bmatrix}$$

and

$$\mathbf{r}_4 = \begin{bmatrix} 2 & 6 & 18 & 8 & 6 \end{bmatrix}$$

for the row space of A.   ▲

We know from Theorem 4.6.3 that elementary row operations do not alter relationships of linear independence and linear dependence among the column vectors; however, Formulas (3) and (4) imply an even deeper result. Because these formulas actually have *the same scalar coefficients* $x_1, x_2, \ldots, x_n$, it follows that elementary row operations do not alter the *formulas* (linear combinations) that relate linearly dependent column vectors. We omit the formal proof.

Example 8

(a) Find a subset of the vectors

$$\mathbf{v}_1 = (1, -2, 0, 3), \qquad \mathbf{v}_2 = (2, -5, -3, 6)$$
$$\mathbf{v}_3 = (0, 1, 3, 0), \qquad \mathbf{v}_4 = (2, -1, 4, -7), \qquad \mathbf{v}_5 = (5, -8, 1, 2)$$

that forms a basis for the space spanned by these vectors.

(b) Express the vectors not in the basis as a linear combination of the basis vectors.

Solution (a). We begin by constructing a matrix that has $\mathbf{v}_1, \mathbf{v}_2, \ldots, \mathbf{v}_5$ as its column vectors:

$$
\begin{bmatrix}
1 & 2 & 0 & 2 & 5 \\
-2 & -5 & 1 & -1 & -8 \\
0 & -3 & 3 & 4 & 1 \\
3 & 6 & 0 & -7 & 2
\end{bmatrix} \tag{5}
$$
$$
\begin{array}{ccccc}
\uparrow & \uparrow & \uparrow & \uparrow & \uparrow \\
\mathbf{v}_1 & \mathbf{v}_2 & \mathbf{v}_3 & \mathbf{v}_4 & \mathbf{v}_5
\end{array}
$$

The first part of our problem can be solved by finding a basis for the column space of this matrix. Reducing the matrix to *reduced* row-echelon form and denoting the column vectors of the resulting matrix by $\mathbf{w}_1, \mathbf{w}_2, \mathbf{w}_3, \mathbf{w}_4$, and $\mathbf{w}_5$ yields

$$
\begin{bmatrix}
1 & 0 & 2 & 0 & 1 \\
0 & 1 & -1 & 0 & 1 \\
0 & 0 & 0 & 1 & 1 \\
0 & 0 & 0 & 0 & 0
\end{bmatrix} \tag{6}
$$
$$
\begin{array}{ccccc}
\uparrow & \uparrow & \uparrow & \uparrow & \uparrow \\
\mathbf{w}_1 & \mathbf{w}_2 & \mathbf{w}_3 & \mathbf{w}_4 & \mathbf{w}_5
\end{array}
$$

The leading 1's occur in columns 1, 2, and 4, so that by Theorem 4.6.4

$$\{\mathbf{w}_1, \mathbf{w}_2, \mathbf{w}_4\}$$

is a basis for the column space of (6) and consequently

$$\{\mathbf{v}_1, \mathbf{v}_2, \mathbf{v}_4\}$$

is a basis for the column space of (5).

Solution (b). We shall start by expressing $\mathbf{w}_3$ and $\mathbf{w}_5$ as linear combinations of the basis vectors $\mathbf{w}_1, \mathbf{w}_2, \mathbf{w}_4$. The simplest way of doing this is to express $\mathbf{w}_3$ and $\mathbf{w}_5$ in terms of basis vectors with smaller subscripts. Thus, we shall express $\mathbf{w}_3$ as a linear combination of $\mathbf{w}_1$ and $\mathbf{w}_2$, and we shall express $\mathbf{w}_5$ as a linear combination of $\mathbf{w}_1$, $\mathbf{w}_2$, and $\mathbf{w}_4$. By inspection of (6), these linear combinations are

$$\mathbf{w}_3 = 2\mathbf{w}_1 - \mathbf{w}_2$$
$$\mathbf{w}_5 = \mathbf{w}_1 + \mathbf{w}_2 + \mathbf{w}_4$$

We call these the ***dependency equations***. The corresponding relationships in (5) are

$$\mathbf{v}_3 = 2\mathbf{v}_1 - \mathbf{v}_2$$
$$\mathbf{v}_5 = \mathbf{v}_1 + \mathbf{v}_2 + \mathbf{v}_4$$

The procedure illustrated in the foregoing example is sufficiently important that we shall summarize the steps.

Given a set of vectors $S = \{\mathbf{v}_1, \mathbf{v}_2, \ldots, \mathbf{v}_k\}$ in R^n, the following procedure produces a subset of those vectors that is a basis for lin(S) and expresses those vectors of S that are not in the basis as linear combinations of the basis vectors.

Step 1. Form the matrix A having $\mathbf{v}_1, \mathbf{v}_2, \ldots, \mathbf{v}_k$ as its column vectors.

Step 2. Reduce the matrix A to its reduced row-echelon form R, and let $\mathbf{w}_1, \mathbf{w}_2, \ldots, \mathbf{w}_k$ be the column vectors of R.

Step 3. Identify the columns that contain the leading 1's in R. The corresponding column vectors of A are the basis vectors for lin(S).

Step 4. Express each column vector of R that does *not* contain a leading 1 as a linear combination of preceding column vectors that do contain leading 1's. (You will be able to do this by inspection.) This yields a set of dependency equations involving the column vectors of R. The corresponding equations for the column vectors of A express the vectors not in the basis as linear combinations of the basis vectors.

OPTIONAL

Proof of Theorem 4.6.1. Suppose that the row vectors of a matrix A are $\mathbf{r}_1, \mathbf{r}_2, \ldots,$ $\mathbf{r}_m$, and let B be obtained from A by performing an elementary row operation. We shall show that every vector in the row space of B is also in the row space of A, and conversely that every vector in the row space of A is in the row space of B. We can then conclude that A and B have the same row space.

Consider the possibilities: If the row operation is a row interchange, then B and A have the same row vectors and consequently have the same row space. If the row operation is multiplication of a row by a nonzero scalar or addition of a multiple of one row to another, then the row vectors $\mathbf{r}'_1, \mathbf{r}'_2, \ldots, \mathbf{r}'_m$ of B are linear combinations of $\mathbf{r}_1, \mathbf{r}_2, \ldots, \mathbf{r}_m$; thus, they lie in the row space of A. Since a vector space is closed under addition and scalar multiplication, all linear combinations of $\mathbf{r}'_1, \mathbf{r}'_2, \ldots, \mathbf{r}'_m$ will also lie in the row space of A. Therefore, each vector in the row space of B is in the row space of A.

Since B is obtained from A by performing a row operation, A can be obtained from B by performing the inverse operation (Section 1.6). Thus, the argument above shows that the row space of A is contained in the row space of B. ∎

EXERCISE SET 4.6

1. List the row vectors and column vectors of the matrix

$$\begin{bmatrix} 2 & -1 & 0 & 1 \\ 3 & 5 & 7 & -1 \\ 1 & 4 & 2 & 7 \end{bmatrix}$$

2. As illustrated in Example 2 express the given matrix product as a linear combination of the column vectors of the left factor.

(a) $\begin{bmatrix} 2 & 3 \\ -1 & 4 \end{bmatrix}\begin{bmatrix} 1 \\ 2 \end{bmatrix}$

(b) $\begin{bmatrix} 4 & 0 & -1 \\ 3 & 6 & 2 \\ 0 & -1 & 4 \end{bmatrix}\begin{bmatrix} -2 \\ 3 \\ 5 \end{bmatrix}$

(c) $\begin{bmatrix} -3 & 6 & 2 \\ 5 & -4 & 0 \\ 2 & 3 & -1 \\ 1 & 8 & 3 \end{bmatrix}\begin{bmatrix} -1 \\ 2 \\ 5 \end{bmatrix}$

(d) $\begin{bmatrix} 2 & 1 & 5 \\ 6 & 3 & -8 \end{bmatrix}\begin{bmatrix} 3 \\ 0 \\ -5 \end{bmatrix}$

3. Find a basis for the nullspace of A.

(a) $A = \begin{bmatrix} 1 & -1 & 3 \\ 5 & -4 & -4 \\ 7 & -6 & 2 \end{bmatrix}$

(b) $A = \begin{bmatrix} 2 & 0 & -1 \\ 4 & 0 & -2 \\ 0 & 0 & 0 \end{bmatrix}$

(c) $A = \begin{bmatrix} 1 & 4 & 5 & 2 \\ 2 & 1 & 3 & 0 \\ -1 & 3 & 2 & 2 \end{bmatrix}$

(d) $A = \begin{bmatrix} 1 & 4 & 5 & 6 & 9 \\ 3 & -2 & 1 & 4 & -1 \\ -1 & 0 & -1 & -2 & -1 \\ 2 & 3 & 5 & 7 & 8 \end{bmatrix}$

(e) $A = \begin{bmatrix} 1 & -3 & 2 & 2 & 1 \\ 0 & 3 & 6 & 0 & -3 \\ 2 & -3 & -2 & 4 & 4 \\ 3 & -6 & 0 & 6 & 5 \\ -2 & 9 & 2 & -4 & -5 \end{bmatrix}$

4. For the matrices in Exercise 3, find a basis for the row space of A by reducing the matrix to row-echelon form.

5. For the matrices in Exercise 3, find a basis for the column space of A.

6. For the matrices in Exercise 3, find a basis for the row space of A consisting entirely of row vectors of A.

7. Find a basis for the subspace of R^4 spanned by the given vectors.
   (a) $(1, 1, -4, -3)$, $(2, 0, 2, -2)$, $(2, -1, 3, 2)$ (b) $(-1, 1, -2, 0)$, $(3, 3, 6, 0)$, $(9, 0, 0, 3)$
   (c) $(1, 1, 0, 0)$, $(0, 0, 1, 1)$, $(-2, 0, 2, 2)$, $(0, -3, 0, 3)$

8. Find a subset of the vectors that forms a basis for the space spanned by the vectors; then express each vector that is not in the basis as a linear combination of the basis vectors.
   (a) $\mathbf{v}_1 = (1, 0, 1, 1)$, $\mathbf{v}_2 = (-3, 3, 7, 1)$, $\mathbf{v}_3 = (-1, 3, 9, 3)$, $\mathbf{v}_4 = (-5, 3, 5, -1)$
   (b) $\mathbf{v}_1 = (1, -2, 0, 3)$, $\mathbf{v}_2 = (2, -4, 0, 6)$, $\mathbf{v}_3 = (-1, 1, 2, 0)$, $\mathbf{v}_4 = (0, -1, 2, 3)$
   (c) $\mathbf{v}_1 = (1, -1, 5, 2)$, $\mathbf{v}_2 = (-2, 3, 1, 0)$, $\mathbf{v}_3 = (4, -5, 9, 4)$, $\mathbf{v}_4 = (0, 4, 2, -3)$, $\mathbf{v}_5 = (-7, 18, 2, -8)$

9. (a) Let

$$A = \begin{bmatrix} 0 & 1 & 0 \\ 1 & 0 & 0 \\ 0 & 0 & 0 \end{bmatrix}$$

and consider a rectangular xyz-coordinate system in 3-space. Show that the nullspace of A consists of all points on the z-axis, and the column space consists of all points in the xy-plane (Figure 1).

(b) Find a 3×3 matrix whose nullspace is the x-axis and whose column space is the yz-plane.

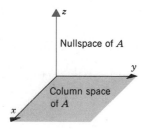

Figure 1

10. Prove that the nonzero row vectors of a matrix in row-echelon form are linearly independent.

11. Prove Theorem 4.6.4.

4.7   RANK AND NULLITY

In this section we shall study relationships between the dimensions of the row space, column space, and nullspace of a matrix. Our work will reveal some new and important facts about systems of linear equations.

ROW SPACE AND COLUMN SPACE HAVE EQUAL DIMENSIONS

In Examples 6 and 7 of the previous section, we found that the row space and column space of the matrix

$$A = \begin{bmatrix} 1 & -2 & 0 & 0 & 3 \\ 2 & -5 & -3 & -2 & 6 \\ 0 & 5 & 15 & 10 & 0 \\ 2 & 6 & 18 & 8 & 6 \end{bmatrix}$$

each have three basis vectors; that is, both are three-dimensional. It is not accidental that these dimensions are the same; it is a consequence of the following general result.

Theorem 4.7.1. *If A is any matrix, then the row space and column space of A have the same dimension.*

Proof. Let R be the reduced row-echelon form of A. It follows from Theorem 4.6.2 that

$$\dim(\text{row space of } A) = \dim(\text{row space of } R)$$

and it follows from Theorem 4.6.3*b* that

$$\dim(\text{column space of } A) = \dim(\text{column space of } R)$$

Thus, the proof will be complete if we can show that the row space and column space of R have the same dimension. But the dimension of the row space of R is the number of nonzero rows (Theorem 4.6.2) and the dimension of the column space of R is the number of columns that contain leading 1's (Theorem 4.6.4). However, the leading 1's occur in the nonzero rows, so the number of leading 1's and the number of nonzero rows is the same. This shows that row space and column space of R have the same dimension. ▌

The dimensions of the row space, column space, and nullspace of a matrix are such important numbers that there is some notation and terminology associated with them.

RANK AND NULLITY

Definition. The common dimension of the row space and column space of a matrix A is called the **rank** of A and is denoted by $\text{rank}(A)$; the dimension of the nullspace of A is called the **nullity** of A and is denoted by $\text{nullity}(A)$.

Example 1 Find the rank and nullity of the matrix

$$A = \begin{bmatrix} -1 & 2 & 0 & 4 & 5 & -3 \\ 3 & -7 & 2 & 0 & 1 & 4 \\ 2 & -5 & 2 & 4 & 6 & 1 \\ 4 & -9 & 2 & -4 & -4 & 7 \end{bmatrix}$$

Solution. The reduced row-echelon form of A is

$$\begin{bmatrix} 1 & 0 & -4 & -28 & -37 & 13 \\ 0 & 1 & -2 & -12 & -16 & 5 \\ 0 & 0 & 0 & 0 & 0 & 0 \\ 0 & 0 & 0 & 0 & 0 & 0 \end{bmatrix} \tag{1}$$

(verify). Since there are two nonzero rows (or equivalently, two leading 1's), the row space and column space are both two-dimensional, so rank(A) = 2. To find the nullity of A, we must find the dimension of the solution space of the linear system $A\mathbf{x} = \mathbf{0}$. This system can be solved by reducing the augmented matrix to reduced row-echelon form. The resulting matrix will be identical to (1), except with an additional last column of zeros, and the corresponding system of equations will be

$$x_1 - 4x_3 - 28x_4 - 37x_5 + 13x_6 = 0$$
$$x_2 - 2x_3 - 12x_4 - 16x_5 + 5x_6 = 0$$

or, on solving for the leading variables,

$$x_1 = 4x_3 + 28x_4 + 37x_5 - 13x_6$$
$$x_2 = 2x_3 + 12x_4 + 16x_5 - 5x_6$$

(2)

It follows that the general solution of the system is

$$x_1 = 4r + 28s + 37t - 13u$$
$$x_2 = 2r + 12s + 16t - 5u$$
$$x_3 = r$$
$$x_4 = s$$
$$x_5 = t$$
$$x_6 = u$$

or equivalently,

$$
\begin{bmatrix} x_1 \\ x_2 \\ x_3 \\ x_4 \\ x_5 \\ x_6 \end{bmatrix}
= r \begin{bmatrix} 4 \\ 2 \\ 1 \\ 0 \\ 0 \\ 0 \end{bmatrix}
+ s \begin{bmatrix} 28 \\ 12 \\ 0 \\ 1 \\ 0 \\ 0 \end{bmatrix}
+ t \begin{bmatrix} 37 \\ 16 \\ 0 \\ 0 \\ 1 \\ 0 \end{bmatrix}
+ u \begin{bmatrix} -13 \\ -5 \\ 0 \\ 0 \\ 0 \\ 1 \end{bmatrix}
$$

(3)

The four vectors on the right side of (3) form a basis for the solution space, so nullity(A) = 4. ▲

The following theorem establishes an important relationship between the rank and nullity of a matrix.

DIMENSION THEOREM

> **Theorem 4.7.2 (*Dimension Theorem for Matrices*).** *If A is a matrix with n columns, then*
>
> $$rank(A) + nullity(A) = n$$
>
> (4)

Proof. Since A has n columns, the homogeneous linear system $A\mathbf{x} = \mathbf{0}$ has n unknowns (variables). These fall into two categories: the leading variables and the free variables. Thus,

$$\begin{bmatrix} \text{number of leading} \\ \text{variables} \end{bmatrix} + \begin{bmatrix} \text{number of free} \\ \text{variables} \end{bmatrix} = n$$

But the number of leading variables is the same as the number of leading 1's in the reduced row-echelon form of A, and this is the rank of A. Thus,

$$\text{rank}(A) + \begin{bmatrix} \text{number of free} \\ \text{variables} \end{bmatrix} = n$$

The number of free variables is equal to the nullity of A. This is so because the nullity of A is the dimension of the solution space of $A\mathbf{x} = \mathbf{0}$, which is the same as the number of parameters in the general solution [see (3), for example], which is the same as the number of free variables. Thus

$$\text{rank}(A) + \text{nullity}(A) = n \quad \blacksquare$$

REMARK. The proof of the foregoing theorem used two facts worth noting:

· $\text{rank}(A) =$ number of leading variables that occur in solving $A\mathbf{x} = \mathbf{0}$
· $\text{nullity}(A) =$ number of parameters in the general solution of $A\mathbf{x} = \mathbf{0}$

Example 2   The matrix

$$A = \begin{bmatrix} -1 & 2 & 0 & 4 & 5 & -3 \\ 3 & -7 & 2 & 0 & 1 & 4 \\ 2 & -5 & 2 & 4 & 6 & 1 \\ 4 & -9 & 2 & -4 & -4 & 7 \end{bmatrix}$$

has 6 columns, so

$$\text{rank}(A) + \text{nullity}(A) = 6$$

This is consistent with Example 1, where we showed that $\text{rank}(A) = 2$ and nullity$(A) = 4$. ▲

Example 3   Find the number of parameters in the solution set of $A\mathbf{x} = \mathbf{0}$ if A is a 5×7 matrix of rank 3.

Solution. From (4)

$$\text{nullity}(A) = n - \text{rank}(A) = 7 - 3 = 4$$

Thus, there are four parameters. ▲

THEOREMS ON RANK AND NULLITY

We shall now explore relationships between the rank of a matrix A and the solutions, if any, of a *nonhomogeneous* linear system $A\mathbf{x} = \mathbf{b}$. We begin with a theorem that ties together all the major topics we have studied so far: matrices, systems of equations, determinants, vector spaces, rank, and nullity; this theorem builds on Theorem 1.7.3 and Theorem 2.3.4.

Theorem 4.7.3. *If A is an n × n matrix, then the following statements are equivalent:*

(a) *A is invertible.*
(b) *A**x** = **0** has only the trivial solution.*
(c) *A is row equivalent to* I_n.
(d) *A**x** = **b** is consistent for every n × 1 matrix **b**.*
(e) *det(A) ≠ 0.*
(f) *A has nullity 0.*
(g) *A has rank n.*
(h) *The row vectors of A are linearly independent.*
(i) *The column vectors of A are linearly independent.*

Proof. We already know that (a), (b), (c), (d), and (e) are equivalent. We shall complete the proof by showing that (f), (g), (h), and (i) are all equivalent to (b). We will do this by proving the sequence of implications (b) ⇒ (f) ⇒ (g) ⇒ (h) ⇒ (i) ⇒ (b).

(b) ⇒ (f). Assume that $A\mathbf{x} = \mathbf{0}$ has only the trivial solution; then the nullspace of A is the zero subspace of R^n, which has dimension zero. Thus, nullity(A) = 0.

(f) ⇒ (g). Assume that nullity(A) = 0; then by the dimension theorem

$$\text{rank}(A) = n - \text{nullity}(A) = n - 0 = n$$

(g) ⇒ (h). Assume that rank(A) − n; then the row space of A is n-dimensional. Since the n row vectors of A span the row space of A, it follows from Theorem 4.5.4b that the row vectors of A are linearly independent.

(h) ⇒ (i). Assume that the row vectors of A are linearly independent; then the row space of A is n-dimensional. By Theorem 4.7.1 the column space of A is also n-dimensional. Since the column vectors of A span the column space, the column vectors of A are linearly independent by Theorem 4.5.4b.

(i) ⇒ (b). Let $\mathbf{c}_1, \mathbf{c}_2, \ldots, \mathbf{c}_n$ denote the column vectors of A, and assume that these vectors are linearly independent; then the only scalars $x_1, x_2, \ldots, x_n$ that satisfy the vector equation

$$x_1\mathbf{c}_1 + x_2\mathbf{c}_2 + \cdots + x_n\mathbf{c}_n = \mathbf{0}$$

are $x_1 = 0, x_2 = 0, \ldots, x_n = 0$. But this vector equation is an alternative form of the linear system $A\mathbf{x} = \mathbf{0}$ [see (1) in Section 4.6], so that this system has only the trivial solution. ▌

If $A\mathbf{x} = \mathbf{b}$ is a linear system of m equations in the n unknowns $x_1, x_2, \ldots, x_n$ and if $\mathbf{c}_1, \mathbf{c}_2, \ldots, \mathbf{c}_n$ are the column vectors of A, then this system can be written as

$$x_1\mathbf{c}_1 + x_2\mathbf{c}_2 + \cdots + x_n\mathbf{c}_n = \mathbf{b} \tag{5}$$

Since the left side of this equation is a linear combination of the column vectors of A, it follows that the system $A\mathbf{x} = \mathbf{b}$ is consistent if and only if **b** is a linear combination of the column vectors of A. Thus, we have the following useful theorem.

Theorem 4.7.4. *A system of linear equations* $A\mathbf{x} = \mathbf{b}$ *is consistent if and only if* $\mathbf{b}$ *is in the column space of* A.

Example 4 Let $A\mathbf{x} = \mathbf{b}$ be the linear system

$$\begin{bmatrix} -1 & 3 & 2 \\ 1 & 2 & -3 \\ 2 & 1 & -2 \end{bmatrix} \begin{bmatrix} x_1 \\ x_2 \\ x_3 \end{bmatrix} = \begin{bmatrix} 1 \\ -9 \\ -3 \end{bmatrix}$$

Show that $\mathbf{b}$ is in the column space of A, and express $\mathbf{b}$ as a linear combination of the column vectors of A.

Solution. Solving the system by Gaussian elimination yields (verify)

$$x_1 = 2, \qquad x_2 = -1, \qquad x_3 = 3$$

Since the system is consistent, $\mathbf{b}$ is in the column space of A. Moreover, from (5) and the solution obtained, it follows that

$$2\begin{bmatrix} -1 \\ 1 \\ 2 \end{bmatrix} - \begin{bmatrix} 3 \\ 2 \\ 1 \end{bmatrix} + 3\begin{bmatrix} 2 \\ -3 \\ -2 \end{bmatrix} = \begin{bmatrix} 1 \\ -9 \\ -3 \end{bmatrix} \quad \blacktriangle$$

The following theorem is a direct consequence of Theorem 4.7.4.

Theorem 4.7.5. *A system of linear equations* $A\mathbf{x} = \mathbf{b}$ *is consistent if and only if the rank of the coefficient matrix* A *is the same as the rank of the augmented matrix* $[A\,|\,\mathbf{b}]$.

Although we shall leave the proof as an exercise, the foregoing result is fairly obvious if one remembers that the rank of a matrix is equal to the number of nonzero rows in its reduced row-echelon form. For example, the augmented matrix for the system

$$\begin{aligned}
x_1 - 2x_2 - 3x_3 + 2x_4 &= -4 \\
-3x_1 + 7x_2 - x_3 + x_4 &= -3 \\
2x_1 - 5x_2 + 4x_3 - 3x_4 &= 7 \\
-3x_1 + 6x_2 + 9x_3 - 6x_4 &= -1
\end{aligned}$$

is

$$\begin{bmatrix} 1 & -2 & -3 & 2 & | & -4 \\ -3 & 7 & -1 & 1 & | & -3 \\ 2 & -5 & 4 & -3 & | & 7 \\ -3 & 6 & 9 & -6 & | & -1 \end{bmatrix}$$

which has the following reduced row-echelon form (verify):

$$\begin{bmatrix} 1 & 0 & -23 & 16 & | & 0 \\ 0 & 1 & -10 & 7 & | & 0 \\ 0 & 0 & 0 & 0 & | & 1 \\ 0 & 0 & 0 & 0 & | & 0 \end{bmatrix}$$

Because of the row

$$0 \quad 0 \quad 0 \quad 0 \quad 1$$

we see that the system is inconsistent. However, it is also because of this row that the reduced row-echelon form of the augmented matrix has fewer zero rows than the reduced row-echelon form of the coefficient matrix. Thus, the coefficient matrix and the augmented matrix for the system have different ranks.

RELATIONSHIP BETWEEN HOMOGENEOUS AND NONHOMOGENEOUS LINEAR SYSTEMS

The next theorem establishes a fundamental relationship between the solutions of a nonhomogeneous linear system $A\mathbf{x} = \mathbf{b}$ and those of the corresponding homogeneous linear system $A\mathbf{x} = \mathbf{0}$ with the same coefficient matrix.

Theorem 4.7.6. *If $\mathbf{x}_0$ denotes any single solution of a consistent nonhomogeneous linear system $A\mathbf{x} = \mathbf{b}$, and if $\mathbf{v}_1, \mathbf{v}_2, \ldots, \mathbf{v}_k$ form a basis for the solution space of the homogeneous system $A\mathbf{x} = \mathbf{0}$, then every solution of $A\mathbf{x} = \mathbf{b}$ can be expressed in the form*

$$\mathbf{x} = \mathbf{x}_0 + c_1\mathbf{v}_1 + c_2\mathbf{v}_2 + \cdots + c_k\mathbf{v}_k \tag{6}$$

and, conversely, for all choices of $c_1, c_2, \ldots, c_k$, the vector $\mathbf{x}$ in this formula is a solution of $A\mathbf{x} = \mathbf{b}$.

The proof is deferred to the end of this section.

REMARK. There is some terminology associated with Formula (6). The vector $\mathbf{x}_0$ is called a *particular solution* of $A\mathbf{x} = \mathbf{b}$. The expression $\mathbf{x}_0 + c_1\mathbf{v}_1 + c_2\mathbf{v}_2 + \cdots + c_k\mathbf{v}_k$ is called the *general solution* of $A\mathbf{x} = \mathbf{b}$, and the expression $c_1\mathbf{v}_1 + c_2\mathbf{v}_2 + \cdots + c_k\mathbf{v}_k$ is called the *general solution* of $A\mathbf{x} = \mathbf{0}$. With this terminology, Formula (6) states, *the general solution of $A\mathbf{x} = \mathbf{b}$ is the sum of any particular solution of $A\mathbf{x} = \mathbf{b}$ and the general solution of $A\mathbf{x} = \mathbf{0}$.*

Example 5 In Example 3 of Section 1.2 we solved the nonhomogeneous linear system

$$\begin{array}{rcrcrcrcrcrcr} x_1 & + & 3x_2 & - & 2x_3 & & & & + & 2x_5 & & & = & 0 \\ 2x_1 & + & 6x_2 & - & 5x_3 & - & 2x_4 & + & 4x_5 & - & 3x_6 & = & -1 \\ & & & & 5x_3 & + & 10x_4 & & & + & 15x_6 & = & 5 \\ 2x_1 & + & 6x_2 & & & + & 8x_4 & + & 4x_5 & + & 18x_6 & = & 6 \end{array} \tag{7}$$

and obtained

$$x_1 = -3r - 4s - 2t, \quad x_2 = r, \quad x_3 = -2s, \quad x_4 = s, \quad x_5 = t, \quad x_6 = \tfrac{1}{3}$$

This result can be written in vector form as

$$\begin{bmatrix} x_1 \\ x_2 \\ x_3 \\ x_4 \\ x_5 \\ x_6 \end{bmatrix} = \begin{bmatrix} -3r - 4s - 2t \\ r \\ -2s \\ s \\ t \\ \frac{1}{3} \end{bmatrix} = \begin{bmatrix} 0 \\ 0 \\ 0 \\ 0 \\ 0 \\ \frac{1}{3} \end{bmatrix} + r \begin{bmatrix} -3 \\ 1 \\ 0 \\ 0 \\ 0 \\ 0 \end{bmatrix} + s \begin{bmatrix} -4 \\ 0 \\ -2 \\ 1 \\ 0 \\ 0 \end{bmatrix} + t \begin{bmatrix} -2 \\ 0 \\ 0 \\ 0 \\ 1 \\ 0 \end{bmatrix}$$

which is the general solution of (7). Comparing this to (6), the vector

$$\mathbf{x}_0 = \begin{bmatrix} 0 \\ 0 \\ 0 \\ 0 \\ 0 \\ 0 \\ \frac{1}{3} \end{bmatrix}$$

is a particular solution of (7) and

$$\mathbf{x} = r \begin{bmatrix} -3 \\ 1 \\ 0 \\ 0 \\ 0 \\ 0 \end{bmatrix} + s \begin{bmatrix} -4 \\ 0 \\ -2 \\ 1 \\ 0 \\ 0 \end{bmatrix} + t \begin{bmatrix} -2 \\ 0 \\ 0 \\ 0 \\ 1 \\ 0 \end{bmatrix}$$

is the general solution of the homogeneous system

$$\begin{array}{rcl} x_1 + 3x_2 - 2x_3 \quad\quad + 2x_5 \quad\quad &=& 0 \\ 2x_1 + 6x_2 - 5x_3 - 2x_4 + 4x_5 - 3x_6 &=& 0 \\ 5x_3 + 10x_4 \quad\quad + 15x_6 &=& 0 \\ 2x_1 + 6x_2 \quad\quad + 8x_4 + 4x_5 + 18x_6 &=& 0 \end{array}$$

(verify). ▲

In Formula (6), the scalars $c_1, c_2, \ldots, c_k$ are the arbitrary parameters in the general solution of $A\mathbf{x} = \mathbf{b}$ and of $A\mathbf{x} = \mathbf{0}$. Thus, the two systems have the same number of parameters in their general solutions. As noted in the remark following Theorem 4.7.2, the number of such parameters is the nullity of A. This fact and Theorem 4.7.2 yield the following result.

Theorem 4.7.7. *If $A\mathbf{x} = \mathbf{b}$ is a consistent linear system of m equations in n unknowns, and if A has rank r, then the general solution of the system contains $n - r$ parameters.*

Example 6 If A is a 5×7 matrix with rank 4, and if $Ax = b$ is a consistent linear system, then the general solution of the system contains $7 - 4 = 3$ parameters. ▲

OPTIONAL

Proof of Theorem 4.7.6. Assume that x_0 is any fixed solution of $Ax = b$ and that x is an arbitrary solution. Then

$$Ax_0 = b \qquad \text{and} \qquad Ax = b$$

Subtracting these equations yields

$$Ax - Ax_0 = 0$$

or

$$A(x - x_0) = 0$$

which shows that $x - x_0$ is a solution of the homogeneous system $Ax = 0$. Since $v_1, v_2, \ldots, v_k$ is a basis for the solution space of this system, we can express $x - x_0$ as a linear combination of these vectors, say

$$x - x_0 = c_1 v_1 + c_2 v_2 + \cdots + c_k v_k$$

Thus,

$$x = x_0 + c_1 v_1 + c_2 v_2 + \cdots + c_k v_k$$

which proves the first part of the theorem. Conversely, for all choices of the scalars $c_1, c_2, \ldots, c_k$ in (6) we have

$$Ax = A(x_0 + c_1 v_1 + c_2 v_2 + \cdots + c_k v_k)$$

or

$$Ax = Ax_0 + c_1(Av_1) + c_2(Av_2) + \cdots + c_k(Av_k)$$

But x_0 is a solution of the nonhomogeneous system and $v_1, v_2, \ldots, v_k$ are solutions of the homogeneous system, so the last equation implies that

$$Ax = b + 0 + 0 + \cdots + 0 = b$$

which shows that x is a solution of $Ax = b$. ∎

EXERCISE SET 4.7

1. Find the rank and nullity of the matrix; then verify that the values obtained satisfy Formula (4) of the dimension theorem.

(a) $A = \begin{bmatrix} 1 & -1 & 3 \\ 5 & -4 & -4 \\ 7 & -6 & 2 \end{bmatrix}$

(b) $A = \begin{bmatrix} 2 & 0 & -1 \\ 4 & 0 & -2 \\ 0 & 0 & 0 \end{bmatrix}$

(c) $A = \begin{bmatrix} 1 & 4 & 5 & 2 \\ 2 & 1 & 3 & 0 \\ -1 & 3 & 2 & 2 \end{bmatrix}$

(d) $A = \begin{bmatrix} 1 & 4 & 5 & 6 & 9 \\ 3 & -2 & 1 & 4 & -1 \\ -1 & 0 & -1 & -2 & -1 \\ 2 & 3 & 5 & 7 & 8 \end{bmatrix}$

(e) $A = \begin{bmatrix} 1 & -3 & 2 & 2 & 1 \\ 0 & 3 & 6 & 0 & -3 \\ 2 & -3 & -2 & 4 & 4 \\ 3 & -6 & 0 & 6 & 5 \\ -2 & 9 & 2 & -4 & -5 \end{bmatrix}$

2. Determine whether **b** is in the column space of A, and if so, express **b** as a linear combination of the column vectors of A.

(a) $A = \begin{bmatrix} 1 & 3 \\ 4 & -6 \end{bmatrix}$; $\mathbf{b} = \begin{bmatrix} -2 \\ 10 \end{bmatrix}$

(b) $A = \begin{bmatrix} 1 & 1 & 2 \\ 1 & 0 & 1 \\ 2 & 1 & 3 \end{bmatrix}$; $\mathbf{b} = \begin{bmatrix} -1 \\ 0 \\ 2 \end{bmatrix}$

(c) $A = \begin{bmatrix} 1 & -1 & 1 \\ 9 & 3 & 1 \\ 1 & 1 & 1 \end{bmatrix}$; $\mathbf{b} = \begin{bmatrix} 5 \\ 1 \\ -1 \end{bmatrix}$

(d) $A = \begin{bmatrix} 1 & -1 & 1 \\ 1 & 1 & -1 \\ -1 & -1 & 1 \end{bmatrix}$; $\mathbf{b} = \begin{bmatrix} 2 \\ 0 \\ 0 \end{bmatrix}$

(e) $A = \begin{bmatrix} 1 & 2 & 0 & 1 \\ 0 & 1 & 2 & 1 \\ 1 & 2 & 1 & 3 \\ 0 & 1 & 2 & 2 \end{bmatrix}$; $\mathbf{b} = \begin{bmatrix} 4 \\ 3 \\ 5 \\ 7 \end{bmatrix}$

3. Suppose that $x_1 = -1$, $x_2 = 2$, $x_3 = 4$, $x_4 = -3$ is a solution of a nonhomogeneous linear system $A\mathbf{x} = \mathbf{b}$ and that the solution set of the homogeneous system $A\mathbf{x} = \mathbf{0}$ is given by the formulas

$$x_1 = -3r + 4s, \qquad x_2 = r - s, \qquad x_3 = r, \qquad x_4 = s$$

(a) Find the vector form of the general solution of $A\mathbf{x} = \mathbf{0}$.
(b) Find the vector form of the general solution of $A\mathbf{x} = \mathbf{b}$.

4. Find the vector form of the general solution of the given linear system $A\mathbf{x} = \mathbf{b}$; then use that result to find the vector form of the general solution of $A\mathbf{x} = \mathbf{0}$.

(a) $\begin{aligned} x_1 - 3x_2 &= 1 \\ 2x_1 - 6x_2 &= 2 \end{aligned}$

(b) $\begin{aligned} x_1 + x_2 + 2x_3 &= 5 \\ x_1 \quad\;\; + x_3 &= -2 \\ 2x_1 + x_2 + 3x_3 &= 3 \end{aligned}$

(c) $\begin{aligned} x_1 - 2x_2 + x_3 + 2x_4 &= -1 \\ 2x_1 - 4x_2 + 2x_3 + 4x_4 &= -2 \\ -x_1 + 2x_2 - x_3 - 2x_4 &= 1 \\ 3x_1 - 6x_2 + 3x_3 + 6x_4 &= -3 \end{aligned}$

(d) $\begin{aligned} x_1 + 2x_2 - 3x_3 + x_4 &= 4 \\ -2x_1 + x_2 + 2x_3 + x_4 &= -1 \\ -x_1 + 3x_2 - x_3 + 2x_4 &= 3 \\ 4x_1 - 7x_2 \quad\;\; - 5x_4 &= -5 \end{aligned}$

5. In each part, find the largest possible value for the rank of A and the smallest possible value for the nullity of A.
(a) A is 4×4 (b) A is 3×5 (c) A is 5×3

6. If A is an $m \times n$ matrix, what is the largest possible value for its rank and the smallest possible value for its nullity? [**Hint.** See Exercise 5.]

7. In each part, use the information in the table to determine whether the linear system $A\mathbf{x} = \mathbf{b}$ is consistent. If so, state the number of parameters in its general solution.

	Size of A	Rank(A)	Rank[A ⦙ b]
(a)	3×3	3	3
(b)	3×3	2	3
(c)	3×3	1	1
(d)	5×9	2	2
(e)	5×9	2	3
(f)	4×4	0	0
(g)	6×2	2	2

8. For each of the matrices in Exercise 7, find the nullity of A, and determine the number of parameters in the general solution of the homogeneous linear system $A\mathbf{x} = \mathbf{0}$.

9. Let

$$A = \begin{bmatrix} a_{11} & a_{12} & a_{13} \\ a_{21} & a_{22} & a_{23} \end{bmatrix}$$

Show that A has rank 2 if and only if one or more of the determinants

$$\begin{vmatrix} a_{11} & a_{12} \\ a_{21} & a_{22} \end{vmatrix}, \quad \begin{vmatrix} a_{11} & a_{13} \\ a_{21} & a_{23} \end{vmatrix}, \quad \begin{vmatrix} a_{12} & a_{13} \\ a_{22} & a_{23} \end{vmatrix}$$

is nonzero.

10. Suppose that A is a 3×3 matrix whose nullspace is a line through the origin in 3-space. Can the row or column space of A also be a line through the origin? Explain.

11. Suppose that A is a 3×3 matrix whose column space is a plane through the origin in 3-space. Can the nullspace be a plane through the origin? Can the row space? Explain.

12. (a) Prove: If A is a 3×5 matrix, then the column vectors of A are linearly dependent.
(b) Prove: If A is a 5×3 matrix, then the row vectors of A are linearly dependent.

13. Prove: If A is a matrix that is not square, then either the row vectors of A or the column vectors of A are linearly dependent. [**Hint.** See Exercise 12.]

14. Prove that the row vectors of an $n \times n$ invertible matrix A form a basis for R^n.

15. Prove: If $A\mathbf{x} = \mathbf{0}$ has only the trivial solution, then $A\mathbf{x} = \mathbf{b}$ has at most one solution for any $\mathbf{b}$.

16. Prove Theorem 4.7.5.

SUPPLEMENTARY EXERCISES

1. In each part, the solution space is a subspace of R^3 and so must be a line through the origin, a plane through the origin, all of R^3, or the origin only. For each system, determine which is the case. If the subspace is a plane, find an equation for it, and if it is a line, find parametric equations.

(a) $0x + 0y + 0z = 0$

(b) $\begin{aligned} 2x - 3y + z &= 0 \\ 6x - 9y + 3z &= 0 \\ -4x + 6y - 2z &= 0 \end{aligned}$

(c) $\begin{aligned} x - 2y + 7z &= 0 \\ -4x + 8y + 5z &= 0 \\ 2x - 4y + 3z &= 0 \end{aligned}$

(d) $\begin{aligned} x + 4y + 8z &= 0 \\ 2x + 5y + 6z &= 0 \\ 3x + y - 4z &= 0 \end{aligned}$

2. (a) Express $(4a, a - b, a + 2b)$ as a linear combination of $(4, 1, 1)$ and $(0, -1, 2)$.
(b) Express $(3a + b + 3c, -a + 4b - c, 2a + b + 2c)$ as a linear combination of $(3, -1, 2)$ and $(1, 4, 1)$.
(c) Express $(2a - b + 4c, 3a - c, 4b + c)$ as a linear combination of three nonzero vectors.

3. Let W be the space spanned by $\mathbf{f} = \sin x$ and $\mathbf{g} = \cos x$.
(a) Show that for any value of θ, $\mathbf{f}_1 = \sin(x + \theta)$ and $\mathbf{g}_1 = \cos(x + \theta)$ are vectors in W.
(b) Show that $\mathbf{f}_1$ and $\mathbf{g}_1$ form a basis for W.

4. (a) Express $\mathbf{v} = (1, 1)$ as a linear combination of $\mathbf{v}_1 = (1, -1)$, $\mathbf{v}_2 = (3, 0)$, $\mathbf{v}_3 = (2, 1)$ in two different ways.
(b) Show that this does not violate Theorem 4.5.1.

5. Let A be an $n \times n$ matrix, and let $v_1, v_2, \ldots, v_n$ be linearly independent vectors in R^n expressed as $n \times 1$ matrices. What must be true about A for $Av_1, Av_2, \ldots, Av_n$ to be linearly independent?

6. Must a basis for P_n contain a polynomial of degree k for each $k = 0, 1, 2, \ldots, n$? Justify your answer.

7. For purposes of this problem, let us define a "checkerboard matrix" to be a square matrix $A = [a_{ij}]$ such that

$$a_{ij} = \begin{cases} 1 & \text{if } i + j \text{ is even} \\ 0 & \text{if } i + j \text{ is odd} \end{cases}$$

Find the rank and nullity of the following checkerboard matrices:
(a) the 3×3 checkerboard matrix   (b) the 4×4 checkerboard matrix   (c) the $n \times n$ checkerboard matrix

8. For purposes of this exercise, let us define an "*X*-matrix" to be a square matrix with an odd number of rows and columns that has 0's everywhere except on the two diagonals, where it has 1's. Find the rank and nullity of the following *X*-matrices:

(a) $\begin{bmatrix} 1 & 0 & 1 \\ 0 & 1 & 0 \\ 1 & 0 & 1 \end{bmatrix}$ (b) $\begin{bmatrix} 1 & 0 & 0 & 0 & 1 \\ 0 & 1 & 0 & 1 & 0 \\ 0 & 0 & 1 & 0 & 0 \\ 0 & 1 & 0 & 1 & 0 \\ 1 & 0 & 0 & 0 & 1 \end{bmatrix}$ (c) the *X*-matrix of size $(2n + 1) \times (2n + 1)$

9. In each part, show that the set of polynomials is a subspace of P_n and find a basis for it.
(a) all polynomials in P_n such that $p(-x) = p(x)$ (b) all polynomials in P_n such that $p(0) = 0$

10. **(For readers who have studied calculus.)** Show that the set of all polynomials in P_n that have a horizontal tangent at $x = 0$ is a subspace of P_n. Find a basis for this subspace.

11. In advanced linear algebra one proves the following determinant criterion for rank: *The rank of a matrix A is r if and only if A has some r × r submatrix with a nonzero determinant, and all square submatrices of larger size have determinant zero.* (A submatrix of A is any matrix obtained by deleting rows or columns of A. The matrix A itself is also considered to be a submatrix of A.) In each part use this criterion to find the rank of the matrix.

(a) $\begin{bmatrix} 1 & 2 & 0 \\ 2 & 4 & -1 \end{bmatrix}$ (b) $\begin{bmatrix} 1 & 2 & 3 \\ 2 & 4 & 6 \end{bmatrix}$ (c) $\begin{bmatrix} 1 & 0 & 1 \\ 2 & -1 & 3 \\ 3 & -1 & 4 \end{bmatrix}$ (d) $\begin{bmatrix} 1 & -1 & 2 & 0 \\ 3 & 1 & 0 & 0 \\ -1 & 2 & 4 & 0 \end{bmatrix}$

12. Use the result in Exercise 11 to find the possible ranks for matrices of the form

$$\begin{bmatrix} 0 & 0 & 0 & 0 & 0 & a_{16} \\ 0 & 0 & 0 & 0 & 0 & a_{26} \\ 0 & 0 & 0 & 0 & 0 & a_{36} \\ 0 & 0 & 0 & 0 & 0 & a_{46} \\ a_{51} & a_{52} & a_{53} & a_{54} & a_{55} & a_{56} \end{bmatrix}$$

13. Prove: If S is a basis for a vector space V, then for any vectors u and v in V and any scalar k the following relationships hold:
(a) $(u + v)_S = (u)_S + (v)_S$ (b) $(ku)_S = k(u)_S$

CHAPTER 5

INNER PRODUCT SPACES

5.1 INNER PRODUCTS

In Section 4.1 we defined the Euclidean inner product on R^n and used it to extend the concepts of length and distance to Euclidean n-space. In this section we shall use the most important properties of the Euclidean inner product as axioms for defining a general concept of an inner product; then we will show how inner products can be used to define notions of length and distance in vector spaces other than R^n.

GENERAL INNER PRODUCTS

In Theorem 4.1.2, we collected the most important properties of the Euclidean inner product. Except for a difference in notation, the properties listed in that theorem are precisely the axioms in the following definition.

Definition. An ***inner product*** on a real vector space V is a function that associates a real number $\langle \mathbf{u}, \mathbf{v} \rangle$ with each pair of vectors $\mathbf{u}$ and $\mathbf{v}$ in V in such a way that the following axioms are satisfied for all vectors $\mathbf{u}$, $\mathbf{v}$, and $\mathbf{w}$ in V and all scalars k.

(1) $\langle \mathbf{u}, \mathbf{v} \rangle = \langle \mathbf{v}, \mathbf{u} \rangle$ [symmetry axiom]
(2) $\langle \mathbf{u} + \mathbf{v}, \mathbf{w} \rangle = \langle \mathbf{u}, \mathbf{w} \rangle + \langle \mathbf{v}, \mathbf{w} \rangle$ [additivity axiom]
(3) $\langle k\mathbf{u}, \mathbf{v} \rangle = k\langle \mathbf{u}, \mathbf{v} \rangle$ [homogeneity axiom]
(4) $\langle \mathbf{v}, \mathbf{v} \rangle \geq 0$ [positivity axiom]
    where $\langle \mathbf{v}, \mathbf{v} \rangle = 0$
    if and only if $\mathbf{v} = \mathbf{0}$

A real vector space with an inner product is called a ***real inner product space***.

REMARK. In Chapter 10 we shall study complex inner products, that is, inner products whose values are complex numbers. Until that time we shall use the term "inner product space" to mean "real inner product space."

Because the inner product axioms are based on properties of the Euclidean inner product, the Euclidean inner product automatically satisfies these axioms; this is the content of the following example.

Example 1   If $\mathbf{u} = (u_1, u_2, \ldots, u_n)$ and $\mathbf{v} = (v_1, v_2, \ldots, v_n)$ are vectors in R^n, then the formula

$$\langle \mathbf{u}, \mathbf{v} \rangle = \mathbf{u} \cdot \mathbf{v} = u_1 v_1 + u_2 v_2 + \cdots + u_n v_n$$

defines $\langle \mathbf{u}, \mathbf{v} \rangle$ to be the Euclidean inner product on R^n. The four inner product axioms hold by Theorem 4.1.2. ▲

The Euclidean inner product is the most important inner product on R^n. However, there are various applications in which it is desirable to modify the Euclidean inner product by *weighting* its terms differently. More precisely, if

$$w_1, w_2, \ldots, w_n$$

are *positive* real numbers, which we shall call **weights**, and if $\mathbf{u} = (u_1, u_2, \ldots, u_n)$ and $\mathbf{v} = (v_1, v_2, \ldots, v_n)$ are vectors in R^n, then it can be shown (Exercise 26) that the formula

$$\langle \mathbf{u}, \mathbf{v} \rangle = w_1 u_1 v_1 + w_2 u_2 v_2 + \cdots + w_n u_n v_n \tag{1}$$

defines an inner product on R^n; it is called the **weighted Euclidean inner product with weights $w_1, w_2, \ldots, w_n$**.

Example 2   Let $\mathbf{u} = (u_1, u_2)$ and $\mathbf{v} = (v_1, v_2)$ be vectors in R^2. Verify that the weighted Euclidean inner product

$$\langle \mathbf{u}, \mathbf{v} \rangle = 3u_1 v_1 + 2u_2 v_2$$

satisfies the four inner product axioms.

Solution.   Note first that if $\mathbf{u}$ and $\mathbf{v}$ are interchanged in this equation, the right side remains the same. Therefore,

$$\langle \mathbf{u}, \mathbf{v} \rangle = \langle \mathbf{v}, \mathbf{u} \rangle$$

If $\mathbf{w} = (w_1, w_2)$, then

$$\langle \mathbf{u} + \mathbf{v}, \mathbf{w} \rangle = 3(u_1 + v_1)w_1 + 2(u_2 + v_2)w_2$$
$$= (3u_1 w_1 + 2u_2 w_2) + (3v_1 w_1 + 2v_2 w_2)$$
$$= \langle \mathbf{u}, \mathbf{w} \rangle + \langle \mathbf{v}, \mathbf{w} \rangle$$

which establishes the second axiom.

Next,

$$\langle k\mathbf{u}, \mathbf{v}\rangle = 3(ku_1)v_1 + 2(ku_2)v_2$$
$$= k(3u_1v_1 + 2u_2v_2)$$
$$= k\langle \mathbf{u}, \mathbf{v}\rangle$$

which establishes the third axiom.

Finally,

$$\langle \mathbf{v}, \mathbf{v}\rangle = 3v_1v_1 + 2v_2v_2 = 3v_1^2 + 2v_2^2$$

Obviously, $\langle \mathbf{v}, \mathbf{v}\rangle = 3v_1^2 + 2v_2^2 \geq 0$. Further, $\langle \mathbf{v}, \mathbf{v}\rangle = 3v_1^2 + 2v_2^2 = 0$ if and only if $v_1 = v_2 = 0$, that is, if and only if $\mathbf{v} = (v_1, v_2) = \mathbf{0}$. Thus, the fourth axiom is satisfied. ▲

LENGTH AND DISTANCE IN INNER PRODUCT SPACES

Before discussing more examples of inner products, we shall pause to explain how inner products are used to introduce notions of length and distance in inner product spaces. Recall that in Euclidean n-space the Euclidean length of a vector $\mathbf{u} = (u_1, u_2, \ldots, u_n)$ can be expressed in terms of the Euclidean inner product as

$$\|\mathbf{u}\| = (\mathbf{u} \cdot \mathbf{u})^{1/2}$$

and the Euclidean distance between two arbitrary points $\mathbf{u} = (u_1, u_2, \ldots, u_n)$ and $\mathbf{v} = (v_1, v_2, \ldots, v_n)$ can be expressed as

$$d(\mathbf{u}, \mathbf{v}) = \|\mathbf{u} - \mathbf{v}\| = [(\mathbf{u} - \mathbf{v}) \cdot (\mathbf{u} - \mathbf{v})]^{1/2}$$

[see Formulas (1) and (2) of Section 4.1]. Motivated by these formulas, we make the following definition.

> **Definition.** If V is an inner product space, then the *norm* (or *length*) of a vector $\mathbf{u}$ in V is denoted by $\|\mathbf{u}\|$ and is defined by
>
> $$\|\mathbf{u}\| = \langle \mathbf{u}, \mathbf{u}\rangle^{1/2}$$
>
> The *distance* between two points (vectors) $\mathbf{u}$ and $\mathbf{v}$ is denoted by $d(\mathbf{u}, \mathbf{v})$ and is defined by
>
> $$d(\mathbf{u}, \mathbf{v}) = \|\mathbf{u} - \mathbf{v}\|$$

Example 3 If $\mathbf{u} = (u_1, u_2, \ldots, u_n)$ and $\mathbf{v} = (v_1, v_2, \ldots, v_n)$ are vectors in R^n with the Euclidean inner product, then

$$\|\mathbf{u}\| = \langle \mathbf{u}, \mathbf{u}\rangle^{1/2} = (\mathbf{u} \cdot \mathbf{u})^{1/2} = \sqrt{u_1^2 + u_2^2 + \cdots + u_n^2}$$

and

$$d(\mathbf{u}, \mathbf{v}) = \|\mathbf{u} - \mathbf{v}\| = \langle \mathbf{u} - \mathbf{v}, \mathbf{u} - \mathbf{v}\rangle^{1/2} = [(\mathbf{u} - \mathbf{v}) \cdot (\mathbf{u} - \mathbf{v})]^{1/2}$$
$$= \sqrt{(u_1 - v_1)^2 + (u_2 - v_2)^2 + \cdots + (u_n - v_n)^2}$$

Observe that these are simply the standard formulas for the Euclidean norm and distance discussed in Section 4.1 [see Formulas (1) and (2) in that section]. ▲

Example 4   It is important to keep in mind that norm and distance depend on the inner product being used. If the inner product is changed, then the norms and distances between vectors also change. For example, for the vectors $\mathbf{u} = (1, 0)$ and $\mathbf{v} = (0, 1)$ in R^2 with the Euclidean inner product, we have

$$\|\mathbf{u}\| = \sqrt{1^2 + 0^2} = 1$$

and

$$d(\mathbf{u}, \mathbf{v}) = \|\mathbf{u} - \mathbf{v}\| = \|(1, -1)\| = \sqrt{1^2 + (-1)^2} = \sqrt{2}$$

However, if we change to the weighted Euclidean inner product

$$\langle \mathbf{u}, \mathbf{v} \rangle = 3u_1v_1 + 2u_2v_2$$

then we obtain

$$\|\mathbf{u}\| = \langle \mathbf{u}, \mathbf{u} \rangle^{1/2} = [3(1)(1) + 2(0)(0)]^{1/2} = \sqrt{3}$$

and

$$d(\mathbf{u}, \mathbf{v}) = \|\mathbf{u} - \mathbf{v}\| = \langle (1, -1), (1, -1) \rangle^{1/2}$$
$$= [3(1)(1) + 2(-1)(-1)]^{1/2} = \sqrt{5} \quad \blacktriangle$$

UNIT CIRCLES AND SPHERES IN INNER PRODUCT SPACES

If V is an inner product space, then the set of points in V that satisfy

$$\|\mathbf{u}\| = 1$$

is called the **unit sphere** or sometimes the **unit circle** in V. In R^2 and R^3 these are the points that lie one unit away from the origin.

Example 5

(a)   Sketch the unit circle in an xy-coordinate system in R^2 using the Euclidean inner product $\langle \mathbf{u}, \mathbf{v} \rangle = u_1v_1 + u_2v_2$.
(b)   Sketch the unit circle in an xy-coordinate system in R^2 using the weighted Euclidean inner product $\langle \mathbf{u}, \mathbf{v} \rangle = \frac{1}{9}u_1v_1 + \frac{1}{4}u_2v_2$.

Solution (a). If $\mathbf{u} = (x, y)$, then $\|\mathbf{u}\| = \langle \mathbf{u}, \mathbf{u} \rangle^{1/2} = \sqrt{x^2 + y^2}$, so the equation of the unit circle is $\sqrt{x^2 + y^2} = 1$, or on squaring both sides

$$x^2 + y^2 = 1$$

As expected, the graph of this equation is a circle of radius 1 centered at the origin (Figure 1*a*).

Solution (b). If $\mathbf{u} = (x, y)$, then $\|\mathbf{u}\| = \langle \mathbf{u}, \mathbf{u} \rangle^{1/2} = \sqrt{\frac{1}{9}x^2 + \frac{1}{4}y^2}$, so the equation of the unit circle is $\sqrt{\frac{1}{9}x^2 + \frac{1}{4}y^2} = 1$, or on squaring both sides

$$\frac{x^2}{9} + \frac{y^2}{4} = 1$$

The graph of this equation is the ellipse shown in Figure 1*b*. $\blacktriangle$

(a)   (b)

Figure 1 | The unit circle with the Euclidean norm $\|\mathbf{u}\| = \sqrt{x^2 + y^2}$. | The unit circle with the norm $\|\mathbf{u}\| = \sqrt{\frac{1}{9}x^2 + \frac{1}{4}y^2}$.

It would not be unreasonable for you to feel uncomfortable with the results in the last example. For even though our definitions of length and distance reduce to the standard definitions when applied to R^2 with the Euclidean inner product, it does require a stretch of the imagination to think of the unit "circle" as having an elliptical shape. However, even though nonstandard inner products distort familiar spaces and lead to strange values for lengths and distances, many of the basic theorems of Euclidean geometry continue to apply in these unusual spaces. For example, it is a basic fact in Euclidean geometry that the sum of the lengths of two sides of a triangle is at least as large as the length of the third side (Figure 2*a*). We shall see later that this familiar result holds in all inner product spaces, regardless of how unusual the inner product might be. As another example, recall the theorem from Euclidean geometry which states that the sum of the squares of the diagonals of a parallelogram is equal to the sum of the squares of the four sides (Figure 2*b*). This result also holds in all inner product spaces, regardless of the inner product (Exercise 20).

(a)   (b)

Figure 2 | $\|\mathbf{u} + \mathbf{v}\| \leq \|\mathbf{u}\| + \|\mathbf{v}\|$ | $\|\mathbf{u} + \mathbf{v}\|^2 + \|\mathbf{u} - \mathbf{v}\|^2 = 2(\|\mathbf{u}\|^2 + \|\mathbf{v}\|^2)$

INNER PRODUCTS GENERATED BY MATRICES

The Euclidean inner product and the weighted Euclidean inner products are special cases of a general class of inner products on R^n, which we shall now describe. Let

$$\mathbf{u} = \begin{bmatrix} u_1 \\ u_2 \\ \vdots \\ u_n \end{bmatrix} \quad \text{and} \quad \mathbf{v} = \begin{bmatrix} v_1 \\ v_2 \\ \vdots \\ v_n \end{bmatrix}$$

be vectors in R^n (expressed as $n \times 1$ matrices), and let A be an *invertible* $n \times n$ matrix. It can be shown (Exercise 30) that if $\mathbf{u} \cdot \mathbf{v}$ is the Euclidean inner product on R^n, then the formula

$$\langle \mathbf{u}, \mathbf{v} \rangle = A\mathbf{u} \cdot A\mathbf{v} \tag{2}$$

defines an inner product; it is called the ***inner product on R^n generated by A***.

Recalling that the Euclidean inner product $\mathbf{u} \cdot \mathbf{v}$ can be written as the matrix product $\mathbf{v}^t\mathbf{u}$ [see (3) in Section 4.1], it follows that (2) can be written in the alternative form

$$\langle \mathbf{u}, \mathbf{v} \rangle = (A\mathbf{v})^t A\mathbf{u}$$

or equivalently

$$\langle \mathbf{u}, \mathbf{v} \rangle = \mathbf{v}^t A^t A\mathbf{u} \tag{3}$$

Example 6 The inner product on R^n generated by the $n \times n$ identity matrix is the Euclidean inner product, since substituting $A = I$ in (2) yields

$$\langle \mathbf{u}, \mathbf{v} \rangle = I\mathbf{u} \cdot I\mathbf{v} = \mathbf{u} \cdot \mathbf{v}$$

The weighted Euclidean inner product $\langle \mathbf{u}, \mathbf{v} \rangle = 3u_1 v_1 + 2u_2 v_2$ discussed in Example 2 is the inner product on R^2 generated by

$$A = \begin{bmatrix} \sqrt{3} & 0 \\ 0 & \sqrt{2} \end{bmatrix}$$

because substituting this matrix in (3) yields

$$\langle \mathbf{u}, \mathbf{v} \rangle = \begin{bmatrix} v_1 & v_2 \end{bmatrix} \begin{bmatrix} \sqrt{3} & 0 \\ 0 & \sqrt{2} \end{bmatrix} \begin{bmatrix} \sqrt{3} & 0 \\ 0 & \sqrt{2} \end{bmatrix} \begin{bmatrix} u_1 \\ u_2 \end{bmatrix}$$

$$= \begin{bmatrix} v_1 & v_2 \end{bmatrix} \begin{bmatrix} 3 & 0 \\ 0 & 2 \end{bmatrix} \begin{bmatrix} u_1 \\ u_2 \end{bmatrix}$$

$$= 3u_1 v_1 + 2u_2 v_2$$

In general, the weighted Euclidean inner product

$$\langle \mathbf{u}, \mathbf{v} \rangle = w_1 u_1 v_1 + w_2 u_2 v_2 + \cdots + w_n u_n v_n$$

is the inner product on R^n generated by

$$A = \begin{bmatrix} \sqrt{w_1} & 0 & 0 & \cdots & 0 \\ 0 & \sqrt{w_2} & 0 & \cdots & 0 \\ \vdots & \vdots & \vdots & & \vdots \\ 0 & 0 & 0 & \cdots & \sqrt{w_n} \end{bmatrix} \tag{4}$$

(verify). ▲

In the following examples we shall describe some inner products on vector spaces other than R^n.

Example 7   If

$$U = \begin{bmatrix} u_1 & u_2 \\ u_3 & u_4 \end{bmatrix} \quad \text{and} \quad V = \begin{bmatrix} v_1 & v_2 \\ v_3 & v_4 \end{bmatrix}$$

are any two 2×2 matrices, then the following formula defines an inner product on M_{22} (verify):

$$\langle U, V \rangle = u_1 v_1 + u_2 v_2 + u_3 v_3 + u_4 v_4$$

For example, if

$$U = \begin{bmatrix} 1 & 2 \\ 3 & 4 \end{bmatrix} \quad \text{and} \quad V = \begin{bmatrix} -1 & 0 \\ 3 & 2 \end{bmatrix}$$

then

$$\langle U, V \rangle = 1(-1) + 2(0) + 3(3) + 4(2) = 16$$

The norm of a matrix U relative to this inner product is

$$\|U\| = \langle U, U \rangle^{1/2} = \sqrt{u_1^2 + u_2^2 + u_3^2 + u_4^2}$$

and the unit sphere in this space consists of all 2×2 matrices U whose entries satisfy the equation $\|U\| = 1$, which on squaring yields

$$u_1^2 + u_2^2 + u_3^2 + u_4^2 = 1 \quad \triangle$$

Example 8   If

$$\mathbf{p} = a_0 + a_1 x + a_2 x^2 \quad \text{and} \quad \mathbf{q} = b_0 + b_1 x + b_2 x^2$$

are any two vectors in P_2, then the following formula defines an inner product on P_2 (verify):

$$\langle \mathbf{p}, \mathbf{q} \rangle = a_0 b_0 + a_1 b_1 + a_2 b_2$$

The norm of the polynomial $\mathbf{p}$ relative to this inner product is

$$\|\mathbf{p}\| = \langle \mathbf{p}, \mathbf{p} \rangle^{1/2} = \sqrt{a_0^2 + a_1^2 + a_2^2}$$

and the unit sphere in this space consists of all polynomials $\mathbf{p}$ in P_2 whose coefficients satisfy the equation $\|\mathbf{p}\| = 1$, which on squaring yields

$$a_0^2 + a_1^2 + a_2^2 = 1 \quad \triangle$$

Example 9   (For readers who have studied calculus.)   Let $\mathbf{f} = f(x)$ and $\mathbf{g} = g(x)$ be two continuous functions in $C[a, b]$ and define

$$\langle \mathbf{f}, \mathbf{g} \rangle = \int_a^b f(x)g(x)\, dx \tag{5}$$

We shall show that this formula defines an inner product on $C[a, b]$ by verifying the four inner product axioms for functions $\mathbf{f} = f(x)$, $\mathbf{g} = g(x)$, and $\mathbf{s} = s(x)$ in $C[a, b]$:

(1) $\langle \mathbf{f}, \mathbf{g} \rangle = \int_a^b f(x)g(x) \, dx = \int_a^b g(x)f(x) \, dx = \langle \mathbf{g}, \mathbf{f} \rangle$

which proves that Axiom 1 holds.

(2) $\langle \mathbf{f} + \mathbf{g}, \mathbf{s} \rangle = \int_a^b (f(x) + g(x))s(x) \, dx$

$$= \int_a^b f(x)s(x) \, dx + \int_a^b g(x)s(x) \, dx$$

$$= \langle \mathbf{f}, \mathbf{s} \rangle + \langle \mathbf{g}, \mathbf{s} \rangle$$

which proves that Axiom 2 holds.

(3) $\langle k\mathbf{f}, \mathbf{g} \rangle = \int_a^b kf(x)g(x) \, dx = k \int_a^b f(x)g(x) \, dx = k\langle \mathbf{f}, \mathbf{g} \rangle$

which proves that Axiom 3 holds.

(4) If $\mathbf{f} = f(x)$ is any function in $C[a, b]$, then $f^2(x) \geq 0$ for all x in $[a, b]$; therefore

$$\langle \mathbf{f}, \mathbf{f} \rangle = \int_a^b f^2(x) \, dx \geq 0$$

Further, because $f^2(x) \geq 0$ and $\mathbf{f} = f(x)$ is continuous on $[a, b]$, it follows that $\int_a^b f^2(x) \, dx = 0$ if and only if $f(x) = 0$ for all x in $[a, b]$. Therefore, $\langle \mathbf{f}, \mathbf{f} \rangle = \int_a^b f^2(x) \, dx = 0$ if and only if $\mathbf{f} = \mathbf{0}$. This proves that Axiom 4 holds. ▲

Example 10 **(For readers who have studied calculus.)** If $C[a, b]$ has the inner product defined in the preceding example, then the norm of a function $\mathbf{f} = f(x)$ relative to this inner product is

$$\|\mathbf{f}\| = \langle \mathbf{f}, \mathbf{f} \rangle^{1/2} = \sqrt{\int_a^b f^2(x) \, dx} \tag{6}$$

and the unit sphere in this space consists of all functions $\mathbf{f}$ in $C[a, b]$ which satisfy the equation $\|\mathbf{f}\| = 1$, which on squaring yields

$$\int_a^b f^2(x) \, dx = 1 \quad \blacktriangle$$

REMARK **(For readers who have studied calculus).** Since polynomials are continuous functions on $(-\infty, \infty)$, they are continuous on any closed interval $[a, b]$. Thus, for all such intervals the vector space P_n is a subspace of $C[a, b]$, and Formula (5) defines an inner product on P_n.

REMARK **(For readers who have studied calculus).** Recall from calculus that the arc length of a curve $y = f(x)$ over an interval $[a, b]$ is given by the formula

$$L = \int_a^b \sqrt{1 + [f'(x)]^2} \, dx \tag{7}$$

Do not confuse this concept of arc length with $\|\mathbf{f}\|$, which is the length (norm) of $\mathbf{f}$ when $\mathbf{f}$ is viewed as a vector in $C[a, b]$. Formulas (6) and (7) are quite different.

**SOME
PROPERTIES OF
INNER
PRODUCTS**

The following theorem lists some basic algebraic properties of inner products.

> **Theorem 5.1.1.** *If* $\mathbf{u}$, $\mathbf{v}$, *and* $\mathbf{w}$ *are vectors in a real inner product space, and* k *is any scalar, then*
>
> (a) $\langle \mathbf{0}, \mathbf{v} \rangle = \langle \mathbf{v}, \mathbf{0} \rangle = 0$
> (b) $\langle \mathbf{u}, \mathbf{v} + \mathbf{w} \rangle = \langle \mathbf{u}, \mathbf{v} \rangle + \langle \mathbf{u}, \mathbf{w} \rangle$
> (c) $\langle \mathbf{u}, k\mathbf{v} \rangle = k \langle \mathbf{u}, \mathbf{v} \rangle$
> (d) $\langle \mathbf{u} - \mathbf{v}, \mathbf{w} \rangle = \langle \mathbf{u}, \mathbf{w} \rangle - \langle \mathbf{v}, \mathbf{w} \rangle$
> (e) $\langle \mathbf{u}, \mathbf{v} - \mathbf{w} \rangle = \langle \mathbf{u}, \mathbf{v} \rangle - \langle \mathbf{u}, \mathbf{w} \rangle$

Proof. We shall prove part (*b*) and leave the proofs of the remaining parts as exercises.

$$
\begin{aligned}
\langle \mathbf{u}, \mathbf{v} + \mathbf{w} \rangle &= \langle \mathbf{v} + \mathbf{w}, \mathbf{u} \rangle && \text{[by symmetry]} \\
&= \langle \mathbf{v}, \mathbf{u} \rangle + \langle \mathbf{w}, \mathbf{u} \rangle && \text{[by additivity]} \\
&= \langle \mathbf{u}, \mathbf{v} \rangle + \langle \mathbf{u}, \mathbf{w} \rangle && \text{[by symmetry]} \quad \blacksquare
\end{aligned}
$$

Example 11 Since Theorem 5.1.1 is a general result, it is guaranteed to hold for all real inner product spaces. However, it will be instructive to verify these properties for the inner product (3). We shall verify property (*b*) and leave the others as exercises.

$$
\begin{aligned}
\langle \mathbf{u}, \mathbf{v} + \mathbf{w} \rangle &= (\mathbf{v} + \mathbf{w})^t A^t A \mathbf{u} \\
&= (\mathbf{v}^t + \mathbf{w}^t) A^t A \mathbf{u} && \text{[property of transpose]} \\
&= (\mathbf{v}^t A^t A \mathbf{u}) + (\mathbf{w}^t A^t A \mathbf{u}) && \text{[property of matrix multiplication]} \\
&= \langle \mathbf{u}, \mathbf{v} \rangle + \langle \mathbf{u}, \mathbf{w} \rangle && \blacktriangle
\end{aligned}
$$

The following example illustrates how Theorem 5.1.1 and the defining properties of inner products can be used to perform algebraic computations with inner products. As you read through the example, you will find it instructive to justify the steps.

Example 12

$$
\begin{aligned}
\langle \mathbf{u} - 2\mathbf{v}, 3\mathbf{u} + 4\mathbf{v} \rangle &= \langle \mathbf{u}, 3\mathbf{u} + 4\mathbf{v} \rangle - \langle 2\mathbf{v}, 3\mathbf{u} + 4\mathbf{v} \rangle \\
&= \langle \mathbf{u}, 3\mathbf{u} \rangle + \langle \mathbf{u}, 4\mathbf{v} \rangle - \langle 2\mathbf{v}, 3\mathbf{u} \rangle - \langle 2\mathbf{v}, 4\mathbf{v} \rangle \\
&= 3\langle \mathbf{u}, \mathbf{u} \rangle + 4\langle \mathbf{u}, \mathbf{v} \rangle - 6\langle \mathbf{v}, \mathbf{u} \rangle - 8\langle \mathbf{v}, \mathbf{v} \rangle \\
&= 3\|\mathbf{u}\|^2 + 4\langle \mathbf{u}, \mathbf{v} \rangle - 6\langle \mathbf{u}, \mathbf{v} \rangle - 8\|\mathbf{v}\|^2 \\
&= 3\|\mathbf{u}\|^2 - 2\langle \mathbf{u}, \mathbf{v} \rangle - 8\|\mathbf{v}\|^2 \quad \blacktriangle
\end{aligned}
$$

EXERCISE SET 5.1

1. Let $\langle \mathbf{u}, \mathbf{v} \rangle$ be the Euclidean inner product on R^2, and let $\mathbf{u} = (3, -2)$, $\mathbf{v} = (4, 5)$, $\mathbf{w} = (-1, 6)$, and $k = -4$. Find
   (a) $\langle \mathbf{u}, \mathbf{v} \rangle = \langle \mathbf{v}, \mathbf{u} \rangle$ (b) $\langle \mathbf{u} + \mathbf{v}, \mathbf{w} \rangle = \langle \mathbf{u}, \mathbf{w} \rangle + \langle \mathbf{v}, \mathbf{w} \rangle$ (c) $\langle \mathbf{u}, \mathbf{v} + \mathbf{w} \rangle = \langle \mathbf{u}, \mathbf{v} \rangle + \langle \mathbf{u}, \mathbf{w} \rangle$
   (d) $\langle k\mathbf{u}, \mathbf{v} \rangle = k\langle \mathbf{u}, \mathbf{v} \rangle = \langle \mathbf{u}, k\mathbf{v} \rangle$ (e) $\langle \mathbf{0}, \mathbf{v} \rangle = \langle \mathbf{v}, \mathbf{0} \rangle = 0$

2. Repeat Exercise 1 for the weighted Euclidean inner product $\langle \mathbf{u}, \mathbf{v} \rangle = 4u_1v_1 + 5u_2v_2$.

3. Compute $\langle \mathbf{u}, \mathbf{v} \rangle$ using the inner product in Example 7.
   (a) $\mathbf{u} = \begin{bmatrix} 3 & -2 \\ 4 & 8 \end{bmatrix}$, $\mathbf{v} = \begin{bmatrix} -1 & 3 \\ 1 & 1 \end{bmatrix}$ (b) $\mathbf{u} = \begin{bmatrix} 1 & 2 \\ -3 & 5 \end{bmatrix}$, $\mathbf{v} = \begin{bmatrix} 4 & 6 \\ 0 & 8 \end{bmatrix}$

4. Compute $\langle \mathbf{p}, \mathbf{q} \rangle$ using the inner product in Example 8.
   (a) $\mathbf{p} = -2 + x + 3x^2$, $\mathbf{q} = 4 - 7x^2$ (b) $\mathbf{p} = -5 + 2x + x^2$, $\mathbf{q} = 3 + 2x - 4x^2$

5. (a) Use Formula (2) to show that $\langle \mathbf{u}, \mathbf{v} \rangle = 9u_1v_1 + 4u_2v_2$ is the inner product on R^2 generated by

   $$A = \begin{bmatrix} 3 & 0 \\ 0 & 2 \end{bmatrix}$$

   (b) Use the inner product in (a) to compute $\langle \mathbf{u}, \mathbf{v} \rangle$ if $\mathbf{u} = (-3, 2)$ and $\mathbf{v} = (1, 7)$.

6. (a) Use Formula (2) to show that $\langle \mathbf{u}, \mathbf{v} \rangle = 5u_1v_1 - u_1v_2 - u_2v_1 + 10u_2v_2$ is the inner product on R^2 generated by

   $$A = \begin{bmatrix} 2 & 1 \\ -1 & 3 \end{bmatrix}$$

   (b) Use the inner product in (a) to compute $\langle \mathbf{u}, \mathbf{v} \rangle$ if $\mathbf{u} = (0, -3)$ and $\mathbf{v} = (6, 2)$.

7. Let $\mathbf{u} = (u_1, u_2)$ and $\mathbf{v} = (v_1, v_2)$. In each part, the given expression is an inner product on R^2. Find a matrix that generates it.
   (a) $\langle \mathbf{u}, \mathbf{v} \rangle = 3u_1v_1 + 5u_2v_2$ (b) $\langle \mathbf{u}, \mathbf{v} \rangle = 4u_1v_1 + 6u_2v_2$

8. Let $\mathbf{u} = (u_1, u_2)$ and $\mathbf{v} = (v_1, v_2)$. Show that the following are inner products on R^2 by verifying that the inner product axioms hold.
   (a) $\langle \mathbf{u}, \mathbf{v} \rangle = 3u_1v_1 + 5u_2v_2$ (b) $\langle \mathbf{u}, \mathbf{v} \rangle = 4u_1v_1 + u_2v_1 + u_1v_2 + 4u_2v_2$

9. Let $\mathbf{u} = (u_1, u_2, u_3)$ and $\mathbf{v} = (v_1, v_2, v_3)$. Determine which of the following are inner products on R^3. For those that are not, list the axioms that do not hold.
   (a) $\langle \mathbf{u}, \mathbf{v} \rangle = u_1v_1 + u_3v_3$ (b) $\langle \mathbf{u}, \mathbf{v} \rangle = u_1^2v_1^2 + u_2^2v_2^2 + u_3^2v_3^2$
   (c) $\langle \mathbf{u}, \mathbf{v} \rangle = 2u_1v_1 + u_2v_2 + 4u_3v_3$ (d) $\langle \mathbf{u}, \mathbf{v} \rangle = u_1v_1 - u_2v_2 + u_3v_3$

10. In each part use the given inner product on R^2 to find $\|\mathbf{w}\|$, where $\mathbf{w} = (-1, 3)$.
    (a) the Euclidean inner product
    (b) the weighted Euclidean inner product $\langle \mathbf{u}, \mathbf{v} \rangle = 3u_1v_1 + 2u_2v_2$, where $\mathbf{u} = (u_1, u_2)$ and $\mathbf{v} = (v_1, v_2)$
    (c) the inner product generated by the matrix

    $$A = \begin{bmatrix} 1 & 2 \\ -1 & 3 \end{bmatrix}$$

11. Use the inner products in Exercise 10 to find $d(\mathbf{u}, \mathbf{v})$ for $\mathbf{u} = (-1, 2)$ and $\mathbf{v} = (2, 5)$.

12. Let P_2 have the inner product in Example 8. In each part find $\|\mathbf{p}\|$.
    (a) $\mathbf{p} = -2 + 3x + 2x^2$ (b) $\mathbf{p} = 4 - 3x^2$

13. Let M_{22} have the inner product in Example 7. In each part find $\|A\|$.

(a) $A = \begin{bmatrix} -2 & 5 \\ 3 & 6 \end{bmatrix}$ (b) $A = \begin{bmatrix} 0 & 0 \\ 0 & 0 \end{bmatrix}$

14. Let P_2 have the inner product in Example 8. Find $d(\mathbf{p}, \mathbf{q})$.

$$\mathbf{p} = 3 - x + x^2, \qquad \mathbf{q} = 2 + 5x^2$$

15. Let M_{22} have the inner product in Example 7. Find $d(A, B)$.

(a) $A = \begin{bmatrix} 2 & 6 \\ 9 & 4 \end{bmatrix}$, $B = \begin{bmatrix} -4 & 7 \\ 1 & 6 \end{bmatrix}$ (b) $A = \begin{bmatrix} -2 & 4 \\ 1 & 0 \end{bmatrix}$, $B = \begin{bmatrix} -5 & 1 \\ 6 & 2 \end{bmatrix}$

16. Suppose that $\mathbf{u}$, $\mathbf{v}$, and $\mathbf{w}$ are vectors such that

$$\langle \mathbf{u}, \mathbf{v} \rangle = 2, \quad \langle \mathbf{v}, \mathbf{w} \rangle = -3, \quad \langle \mathbf{u}, \mathbf{w} \rangle = 5, \quad \|\mathbf{u}\| = 1, \quad \|\mathbf{v}\| = 2, \quad \|\mathbf{w}\| = 3$$

Evaluate the given expression.
(a) $\langle \mathbf{u} + \mathbf{v}, \mathbf{v} + \mathbf{w} \rangle$ (b) $\langle 2\mathbf{v} - \mathbf{w}, 3\mathbf{u} + 2\mathbf{w} \rangle$ (c) $\langle \mathbf{u} - \mathbf{v} - 2\mathbf{w}, 4\mathbf{u} + \mathbf{v} \rangle$
(d) $\|\mathbf{u} + \mathbf{v}\|$ (e) $\|2\mathbf{w} - \mathbf{v}\|$ (f) $\|\mathbf{u} - 2\mathbf{v} + 4\mathbf{w}\|$

17. **(For readers who have studied calculus.)** Let the vector space P_2 have the inner product

$$\langle \mathbf{p}, \mathbf{q} \rangle = \int_{-1}^{1} p(x)q(x)\,dx$$

(a) Find $\|\mathbf{p}\|$ for $\mathbf{p} = 1$, $\mathbf{p} = x$, and $\mathbf{p} = x^2$.   (b) Find $d(\mathbf{p}, \mathbf{q})$ if $\mathbf{p} = 1$ and $\mathbf{q} = x$.

18. Sketch the unit circle in R^2 using the given inner product.
(a) $\langle \mathbf{u}, \mathbf{v} \rangle = \tfrac{1}{4}u_1 v_1 + \tfrac{1}{16}u_2 v_2$ (b) $\langle \mathbf{u}, \mathbf{v} \rangle = 2u_1 v_1 + u_2 v_2$

19. Find a weighted Euclidean inner product on R^2 for which the unit circle is the ellipse shown in Figure 3.

Figure 3

20. Show that the following identity holds for vectors in any inner product space.

$$\|\mathbf{u} + \mathbf{v}\|^2 + \|\mathbf{u} - \mathbf{v}\|^2 = 2\|\mathbf{u}\|^2 + 2\|\mathbf{v}\|^2$$

21. Show that the following identity holds for vectors in any inner product space.

$$\langle \mathbf{u}, \mathbf{v} \rangle = \tfrac{1}{4}\|\mathbf{u} + \mathbf{v}\|^2 - \tfrac{1}{4}\|\mathbf{u} - \mathbf{v}\|^2$$

22. Let $U = \begin{bmatrix} u_1 & u_2 \\ u_3 & u_4 \end{bmatrix}$ and $V = \begin{bmatrix} v_1 & v_2 \\ v_3 & v_4 \end{bmatrix}$.

Determine whether $\langle U, V \rangle = u_1 v_1 + u_2 v_3 + u_3 v_2 + u_4 v_4$ is an inner product on M_{22}.

23. Let $\mathbf{p} = p(x)$ and $\mathbf{q} = q(x)$ be polynomials in P_2. Show that

$$\langle \mathbf{p}, \mathbf{q} \rangle = p(0)q(0) + p(1/2)q(1/2) + p(1)q(1)$$

is an inner product on P_2.

24. Prove: If $\langle \mathbf{u}, \mathbf{v} \rangle$ is the Euclidean inner product on R^n, and if A is an $n \times n$ matrix, then
$$\langle \mathbf{u}, A\mathbf{v} \rangle = \langle A^t\mathbf{u}, \mathbf{v} \rangle$$
[*Hint.* Use the fact that $\langle \mathbf{u}, \mathbf{v} \rangle = \mathbf{u} \cdot \mathbf{v} = \mathbf{v}^t\mathbf{u}$.]

25. Verify the result in Exercise 24 for the Euclidean inner product on R^3 and
$$\mathbf{u} = \begin{bmatrix} -1 \\ 2 \\ 4 \end{bmatrix} \quad \mathbf{v} = \begin{bmatrix} 3 \\ 0 \\ -2 \end{bmatrix} \quad A = \begin{bmatrix} 1 & -2 & 1 \\ 3 & 4 & 0 \\ 5 & -1 & 2 \end{bmatrix}$$

26. Let $\mathbf{u} = (u_1, u_2, \ldots, u_n)$ and $\mathbf{v} = (v_1, v_2, \ldots, v_n)$. Show that
$$\langle \mathbf{u}, \mathbf{v} \rangle = w_1 u_1 v_1 + w_2 u_2 v_2 + \cdots + w_n u_n v_n$$
is an inner product on R^n if $w_1, w_2, \ldots, w_n$ are positive real numbers.

27. (For readers who have studied calculus.) Use the inner product
$$\langle \mathbf{p}, \mathbf{q} \rangle = \int_{-1}^{1} p(x)q(x)\,dx$$
to compute $\langle \mathbf{p}, \mathbf{q} \rangle$ for the vectors $\mathbf{p} = p(x)$ and $\mathbf{q} = q(x)$ in P_3.
(a) $\mathbf{p} = 1 - x + x^2 + 5x^3 \qquad \mathbf{q} = x - 3x^2$
(b) $\mathbf{p} = x - 5x^3 \qquad \mathbf{q} = 2 + 8x^2$

28. (For readers who have studied calculus.) In each part use the inner product
$$\langle \mathbf{f}, \mathbf{g} \rangle = \int_0^1 f(x)g(x)\,dx$$
to compute $\langle \mathbf{f}, \mathbf{g} \rangle$ for the vectors $\mathbf{f} = f(x)$ and $\mathbf{g} = g(x)$ in $C[0, 1]$.

(a) $\mathbf{f} = \cos 2\pi x, \quad \mathbf{g} = \sin 2\pi x$ \qquad (b) $\mathbf{f} = x, \quad \mathbf{g} = e^x$ \qquad (c) $\mathbf{f} = \tan \dfrac{\pi}{4} x, \quad \mathbf{g} = 1$

29. The *trace* of a square matrix A is denoted by $\text{tr}(A)$ and is defined to be the sum of the entries on the main diagonal of A. Show that the inner product in Example 7 can be written as $\langle U, V \rangle = \text{tr}(U^t V)$.

30. Prove that Formula (2) defines an inner product on R^n. [*Hint.* Use the alternative version of Formula (2) given by (3).]

31. Show that matrix (4) generates the weighted Euclidean inner product
$\langle \mathbf{u}, \mathbf{v} \rangle = w_1 u_1 v_1 + w_2 u_2 v_2 + \cdots + w_n u_n v_n$ on R^n.

32. Prove part (*a*) of Theorem 5.1.1.

33. Prove part (*c*) of Theorem 5.1.1.

5.2 ANGLE AND ORTHOGONALITY IN INNER PRODUCT SPACES

In this section we shall define the notion of an angle between two vectors in an inner product space, and we shall use this concept to obtain some important properties of vectors in inner product spaces.

CAUCHY-SCHWARZ INEQUALITY

Recall from Formula (1) of Section 3.3 that if **u** and **v** are nonzero vectors in R^2 or R^3 and θ is the angle between them, then

$$\mathbf{u} \cdot \mathbf{v} = \|\mathbf{u}\| \, \|\mathbf{v}\| \cos \theta \tag{1}$$

or alternatively

$$\cos \theta = \frac{\mathbf{u} \cdot \mathbf{v}}{\|\mathbf{u}\| \, \|\mathbf{v}\|} \tag{2}$$

If we square both sides of (1) and use the relationships $\|\mathbf{u}\|^2 = \mathbf{u} \cdot \mathbf{u}$, $\|\mathbf{v}\|^2 = \mathbf{v} \cdot \mathbf{v}$, and $\cos^2 \theta \le 1$, we obtain the inequality

$$(\mathbf{u} \cdot \mathbf{v})^2 \le (\mathbf{u} \cdot \mathbf{u})(\mathbf{v} \cdot \mathbf{v})$$

The following theorem shows that this inequality can be generalized to any real inner product space. The resulting inequality, called the *Cauchy-Schwarz* * *inequality*, will enable us to define angles in general real inner product spaces.

Theorem 5.2.1 (*Cauchy-Schwarz Inequality*). *If* **u** *and* **v** *are vectors in a real inner product space, then*

$$\langle \mathbf{u}, \mathbf{v} \rangle^2 \le \langle \mathbf{u}, \mathbf{u} \rangle \langle \mathbf{v}, \mathbf{v} \rangle \tag{3}$$

Proof. We warn the reader in advance that the proof presented here depends on a clever trick that is not easy to motivate. If $\mathbf{u} = \mathbf{0}$, then $\langle \mathbf{u}, \mathbf{v} \rangle = \langle \mathbf{u}, \mathbf{u} \rangle = 0$,

---

* *Augustin Louis (Baron de) Cauchy* (1789–1857), French mathematician. Cauchy's early education was acquired from his father, a barrister and master of the classics. Cauchy entered L'Ecole Polytechnique in 1805 to study engineering, but because of poor health, was advised to concentrate on mathematics. His major mathematical work began in 1811 with a series of brilliant solutions to some difficult outstanding problems.

Cauchy's mathematical contributions for the next 35 years were brilliant and staggering in quantity, over 700 papers filling 26 modern volumes. Cauchy's work initiated the era of modern analysis; he brought to mathematics standards of precision and rigor undreamed of by earlier mathematicians.

Cauchy's life was inextricably tied to the political upheavals of the time. A strong partisan of the Bourbons, he left his wife and children in 1830 to follow the Bourbon king Charles X into exile. For his loyalty he was made a baron by the ex-king. Cauchy eventually returned to France, but refused to accept a university position until the government waived its requirement that he take a loyalty oath.

It is difficult to get a clear picture of the man. Devoutly Catholic, he sponsored charitable work for unwed mothers and criminals, and relief for Ireland. Yet other aspects of his life cast him in an unfavorable light. The Norwegian mathematician Abel described him as "mad, infinitely Catholic, and bigoted." Some writers praise his teaching, yet others say he rambled incoherently and, according to a report of the day, he once devoted an entire lecture to extracting the square root of seventeen to ten decimal places by a method well known to his students. In any event, Cauchy is undeniably one of the greatest minds in the history of science.

Herman Amandus Schwarz (1843–1921), German mathematician. Schwarz was the leading mathematician in Berlin in the first third of the twentieth century. Because of a devotion to his teaching duties at the University of Berlin and a propensity for treating both important and trivial facts with equal thoroughness, he did not publish in great volume. He tended to focus on narrow concrete problems, but his techniques were often extremely clever and influenced the work of other mathematicians. A version of the inequality that bears his name appeared in a paper about surfaces of minimal area published in 1885.

so that the two sides of (3) are equal. Assume now that $\mathbf{u} \neq \mathbf{0}$. Let $a = \langle \mathbf{u}, \mathbf{u} \rangle$, $b = 2\langle \mathbf{u}, \mathbf{v} \rangle$, $c = \langle \mathbf{v}, \mathbf{v} \rangle$, and let t be any real number. By the positivity axiom, the inner product of any vector with itself is always nonnegative. Therefore,

$$0 \leq \langle (t\mathbf{u} + \mathbf{v}), (t\mathbf{u} + \mathbf{v}) \rangle = \langle \mathbf{u}, \mathbf{u} \rangle t^2 + 2\langle \mathbf{u}, \mathbf{v} \rangle t + \langle \mathbf{v}, \mathbf{v} \rangle$$
$$= at^2 + bt + c$$

This inequality implies that the quadratic polynomial $at^2 + bt + c$ has either no real roots or a repeated real root. Therefore, its discriminant must satisfy $b^2 - 4ac \leq 0$. Expressing the coefficients a, b, and c in terms of the vectors $\mathbf{u}$ and $\mathbf{v}$ gives $4\langle \mathbf{u}, \mathbf{v} \rangle^2 - 4\langle \mathbf{u}, \mathbf{u} \rangle \langle \mathbf{v}, \mathbf{v} \rangle \leq 0$, or equivalently, $\langle \mathbf{u}, \mathbf{v} \rangle^2 \leq \langle \mathbf{u}, \mathbf{u} \rangle \langle \mathbf{v}, \mathbf{v} \rangle$. ∎

We note that the Cauchy-Schwarz inequality can be written in two useful alternative forms. Since $\|\mathbf{u}\|^2 = \langle \mathbf{u}, \mathbf{u} \rangle$ and $\|\mathbf{v}\|^2 = \langle \mathbf{v}, \mathbf{v} \rangle$, it follows from (3) that

$$\langle \mathbf{u}, \mathbf{v} \rangle^2 \leq \|\mathbf{u}\|^2 \|\mathbf{v}\|^2 \tag{4}$$

or, upon taking square roots, that

$$|\langle \mathbf{u}, \mathbf{v} \rangle| \leq \|\mathbf{u}\| \|\mathbf{v}\| \tag{5}$$

Example 1 If $\mathbf{u} = (u_1, u_2, \ldots, u_n)$ and $\mathbf{v} = (v_1, v_2, \ldots, v_n)$ are any two vectors in R^n with the Euclidean inner product, then form (5) of the Cauchy-Schwarz inequality applied to $\mathbf{u}$ and $\mathbf{v}$ yields

$$|u_1 v_1 + u_2 v_2 + \cdots + u_n v_n| \leq (u_1^2 + u_2^2 + \cdots + u_n^2)^{1/2}(v_1^2 + v_2^2 + \cdots + v_n^2)^{1/2}$$

This is called ***Cauchy's inequality***. ▲

PROPERTIES OF LENGTH AND DISTANCE IN INNER PRODUCT SPACES

Through the years mathematicians have isolated what are considered to be the most important properties of Euclidean length and distance in R^2 and R^3. They are listed in the following table.

TABLE 1

Basic Properties of Length	Basic Properties of Distance
$L1.$ $\|\mathbf{u}\| \geq 0$	$D1.$ $d(\mathbf{u}, \mathbf{v}) \geq 0$
$L2.$ $\|\mathbf{u}\| = 0$ if and only if $\mathbf{u} = \mathbf{0}$	$D2.$ $d(\mathbf{u}, \mathbf{v}) = 0$ if and only if $\mathbf{u} = \mathbf{v}$
$L3.$ $\|k\mathbf{u}\| = |k| \|\mathbf{u}\|$	$D3.$ $d(\mathbf{u}, \mathbf{v}) = d(\mathbf{v}, \mathbf{u})$
$L4.$ $\|\mathbf{u} + \mathbf{v}\| \leq \|\mathbf{u}\| + \|\mathbf{v}\|$ [triangle inequality]	$D4.$ $d(\mathbf{u}, \mathbf{v}) \leq d(\mathbf{u}, \mathbf{w}) + d(\mathbf{w}, \mathbf{v})$ [triangle inequality]

The next theorem is a strong justification for our definitions of norm and distance in inner product spaces.

> **Theorem 5.2.2.** *If V is an inner product space, then the norm $\|\mathbf{u}\| = \langle \mathbf{u}, \mathbf{u} \rangle^{1/2}$ and the distance $d(\mathbf{u}, \mathbf{v}) = \|\mathbf{u} - \mathbf{v}\|$ satisfy all the properties listed in Table 1.*

We shall prove property $L4$ and leave the proofs of the remaining parts as exercises.

Proof of Property L4. By definition

$$
\begin{aligned}
\|\mathbf{u} + \mathbf{v}\|^2 &= \langle \mathbf{u} + \mathbf{v}, \mathbf{u} + \mathbf{v} \rangle \\
&= \langle \mathbf{u}, \mathbf{u} \rangle + 2\langle \mathbf{u}, \mathbf{v} \rangle + \langle \mathbf{v}, \mathbf{v} \rangle \\
&\le \langle \mathbf{u}, \mathbf{u} \rangle + 2|\langle \mathbf{u}, \mathbf{v} \rangle| + \langle \mathbf{v}, \mathbf{v} \rangle \qquad \text{[property of absolute value]} \\
&\le \langle \mathbf{u}, \mathbf{u} \rangle + 2\|\mathbf{u}\| \|\mathbf{v}\| + \langle \mathbf{v}, \mathbf{v} \rangle \qquad \text{[by (5)]} \\
&= \|\mathbf{u}\|^2 + 2\|\mathbf{u}\| \|\mathbf{v}\| + \|\mathbf{v}\|^2 \\
&= (\|\mathbf{u}\| + \|\mathbf{v}\|)^2
\end{aligned}
$$

Taking square roots gives

$$\|\mathbf{u} + \mathbf{v}\| \le \|\mathbf{u}\| + \|\mathbf{v}\| \quad\blacksquare$$

ANGLE BETWEEN VECTORS

We shall now show how the Cauchy-Schwarz inequality can be used to define angles in general inner product spaces. Suppose that $\mathbf{u}$ and $\mathbf{v}$ are nonzero vectors in an inner product space V. If we divide both sides of Formula (4) by $\|\mathbf{u}\|^2 \|\mathbf{v}\|^2$, we obtain

$$\left[\frac{\langle \mathbf{u}, \mathbf{v} \rangle}{\|\mathbf{u}\| \|\mathbf{v}\|} \right]^2 \le 1$$

or equivalently

$$-1 \le \frac{\langle \mathbf{u}, \mathbf{v} \rangle}{\|\mathbf{u}\| \|\mathbf{v}\|} \le 1 \tag{6}$$

Now if θ is an angle whose radian measure varies from 0 to π, then $\cos \theta$ assumes every value between -1 and 1 inclusive exactly once (Figure 1).

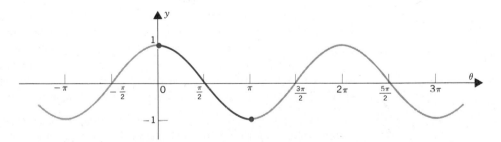

Figure 1

Thus, from (6) there is a unique angle θ such that

$$\cos \theta = \frac{\langle \mathbf{u}, \mathbf{v} \rangle}{\|\mathbf{u}\| \|\mathbf{v}\|} \qquad \text{and} \qquad 0 \le \theta \le \pi \tag{7}$$

We define θ to be the **angle between u *and* v**. Observe that in R^2 or R^3 with the Euclidean inner product, (7) agrees with the usual formula for the cosine of the angle between two nonzero vectors [Formula (2)].

Example 2   Let R^4 have the Euclidean inner product. Find the cosine of the angle θ between the vectors $\mathbf{u} = (4, 3, 1, -2)$ and $\mathbf{v} = (-2, 1, 2, 3)$.

Solution. We leave it for the reader to verify that

$$\|\mathbf{u}\| = \sqrt{30}, \qquad \|\mathbf{v}\| = \sqrt{18}, \qquad \text{and} \qquad \langle \mathbf{u}, \mathbf{v} \rangle = -9$$

so that

$$\cos \theta = \frac{\langle \mathbf{u}, \mathbf{v} \rangle}{\|\mathbf{u}\| \, \|\mathbf{v}\|} = -\frac{9}{\sqrt{30}\sqrt{18}} = -\frac{3}{2\sqrt{15}} \quad \blacktriangle$$

ORTHOGONALITY   Example 2 is primarily a mathematical exercise, for there is relatively little need to find angles between vectors, except in R^2 and R^3 with the Euclidean inner product. However, a problem of major importance in all inner product spaces is to determine whether two vectors are *orthogonal*—that is, whether the angle between them is $\theta = \pi/2$.

Example 3   Show that if M_{22} has the inner product of Example 7 of the previous section, then the angle between the matrices

$$U = \begin{bmatrix} 1 & 0 \\ 1 & 1 \end{bmatrix} \qquad \text{and} \qquad V = \begin{bmatrix} 0 & 2 \\ 0 & 0 \end{bmatrix}$$

is $\pi/2$.

Solution.

$$\cos \theta = \frac{\langle U, V \rangle}{\|U\| \, \|V\|} = \frac{1(0) + 0(2) + 1(0) + 1(0)}{\|U\| \, \|V\|} = 0$$

Thus, $\theta = \pi/2$. $\blacktriangle$

It follows from (7) that if $\mathbf{u}$ and $\mathbf{v}$ are *nonzero* vectors in an inner product space and θ is the angle between them, then $\cos \theta = 0$ if and only if $\langle \mathbf{u}, \mathbf{v} \rangle = 0$. Equivalently, for nonzero vectors we have $\theta = \pi/2$ if and only if $\langle \mathbf{u}, \mathbf{v} \rangle = 0$. If we agree to consider the angle between $\mathbf{u}$ and $\mathbf{v}$ to be $\pi/2$ when either or both of these vectors is $\mathbf{0}$, then we can state without exception that the angle between $\mathbf{u}$ and $\mathbf{v}$ is $\pi/2$ if and only if $\langle \mathbf{u}, \mathbf{v} \rangle = 0$. This suggests the following definition.

Definition. In an inner product space, two vectors $\mathbf{u}$ and $\mathbf{v}$ are called **orthogonal** if $\langle \mathbf{u}, \mathbf{v} \rangle = 0$. If $\mathbf{u}$ is orthogonal to each vector in a set W, we say that $\mathbf{u}$ is **orthogonal to W**.

We emphasize that orthogonality depends on the selection of the inner product; two vectors can be orthogonal with respect to one inner product but not another.

Example 4   (For readers who have studied calculus.)   Let P_2 have the inner product

$$\langle \mathbf{p}, \mathbf{q} \rangle = \int_{-1}^{1} p(x)q(x)\,dx$$

and let

$$\mathbf{p} = x, \qquad \mathbf{q} = x^2$$

Then

$$\|\mathbf{p}\| = \langle \mathbf{p}, \mathbf{p} \rangle^{1/2} = \left[\int_{-1}^{1} xx\,dx \right]^{1/2} = \left[\int_{-1}^{1} x^2\,dx \right]^{1/2} = \sqrt{\frac{2}{3}}$$

$$\|\mathbf{q}\| = \langle \mathbf{q}, \mathbf{q} \rangle^{1/2} = \left[\int_{-1}^{1} x^2 x^2\,dx \right]^{1/2} = \left[\int_{-1}^{1} x^4\,dx \right]^{1/2} = \sqrt{\frac{2}{5}}$$

$$\langle \mathbf{p}, \mathbf{q} \rangle = \int_{-1}^{1} xx^2\,dx = \int_{-1}^{1} x^3\,dx = 0$$

Because $\langle \mathbf{p}, \mathbf{q} \rangle = 0$, the vectors $\mathbf{p} = x$ and $\mathbf{q} = x^2$ are orthogonal relative to the given inner product.   ▲

We conclude this section with an interesting and useful generalization of a familiar result.

Theorem 5.2.3   (*Generalized Theorem of Pythagoras*).   *If* $\mathbf{u}$ *and* $\mathbf{v}$ *are orthogonal vectors in an inner product space, then*

$$\|\mathbf{u} + \mathbf{v}\|^2 = \|\mathbf{u}\|^2 + \|\mathbf{v}\|^2$$

Proof.

$$\|\mathbf{u} + \mathbf{v}\|^2 = \langle (\mathbf{u} + \mathbf{v}), (\mathbf{u} + \mathbf{v}) \rangle = \|\mathbf{u}\|^2 + 2\langle \mathbf{u}, \mathbf{v} \rangle + \|\mathbf{v}\|^2$$
$$= \|\mathbf{u}\|^2 + \|\mathbf{v}\|^2 \quad \blacksquare$$

Note that in R^2 or R^3 with the Euclidean inner product this theorem reduces to the ordinary Pythagorean theorem (Figure 2).

Figure 2 $\|\mathbf{u}\|^2 + \|\mathbf{v}\|^2 = \|\mathbf{u} + \mathbf{v}\|^2.$

Example 5 (**For readers who have studied calculus.**) In Example 4 we showed that $\mathbf{p} = x$ and $\mathbf{q} = x^2$ are orthogonal relative to the inner product

$$\langle \mathbf{p}, \mathbf{q} \rangle = \int_{-1}^{1} p(x)q(x)\,dx$$

on P_2. It follows from the Theorem of Pythagoras that

$$\|\mathbf{p} + \mathbf{q}\|^2 = \|\mathbf{p}\|^2 + \|\mathbf{q}\|^2$$

Thus, from the computations in Example 4 we have

$$\|\mathbf{p} + \mathbf{q}\|^2 = \left(\sqrt{\frac{2}{3}}\right)^2 + \left(\sqrt{\frac{2}{5}}\right)^2 = \frac{2}{3} + \frac{2}{5} = \frac{16}{15}$$

We can check this result by direct integration:

$$\|\mathbf{p} + \mathbf{q}\|^2 = \langle \mathbf{p} + \mathbf{q}, \mathbf{p} + \mathbf{q} \rangle = \int_{-1}^{1} (x + x^2)(x + x^2)\,dx$$

$$= \int_{-1}^{1} x^2\,dx + 2\int_{-1}^{1} x^3\,dx + \int_{-1}^{1} x^4\,dx = \frac{2}{3} + 0 + \frac{2}{5} = \frac{16}{15} \quad \blacktriangle$$

EXERCISE SET 5.2

1. In each part determine whether the given vectors are orthogonal with respect to the Euclidean inner product.
   (a) $\mathbf{u} = (-1, 3, 2)$, $\mathbf{v} = (4, 2, -1)$
   (b) $\mathbf{u} = (-2, -2, -2)$, $\mathbf{v} = (1, 1, 1)$
   (c) $\mathbf{u} = (u_1, u_2, u_3)$, $\mathbf{v} = (0, 0, 0)$
   (d) $\mathbf{u} = (-4, 6, -10, 1)$, $\mathbf{v} = (2, 1, -2, 9)$
   (e) $\mathbf{u} = (0, 3, -2, 1)$, $\mathbf{v} = (5, 2, -1, 0)$
   (f) $\mathbf{u} = (a, b)$, $\mathbf{v} = (-b, a)$

2. Let R^4 have the Euclidean inner product, and let $\mathbf{u} = (-1, 1, 0, 2)$. Determine whether the vector $\mathbf{u}$ is orthogonal to the set of vectors $W = \{\mathbf{w}_1, \mathbf{w}_2, \mathbf{w}_3\}$, where $\mathbf{w}_1 = (0, 0, 0, 0)$, $\mathbf{w}_2 = (1, -1, 3, 0)$, and $\mathbf{w}_3 = (4, 0, 9, 2)$.

3. Let R^2, R^3, and R^4 have the Euclidean inner product. In each part find the cosine of the angle between $\mathbf{u}$ and $\mathbf{v}$.
   (a) $\mathbf{u} = (1, -3)$, $\mathbf{v} = (2, 4)$
   (b) $\mathbf{u} = (-1, 0)$, $\mathbf{v} = (3, 8)$
   (c) $\mathbf{u} = (-1, 5, 2)$, $\mathbf{v} = (2, 4, -9)$
   (d) $\mathbf{u} = (4, 1, 8)$, $\mathbf{v} = (1, 0, -3)$
   (e) $\mathbf{u} = (1, 0, 1, 0)$, $\mathbf{v} = (-3, -3, -3, -3)$
   (f) $\mathbf{u} = (2, 1, 7, -1)$, $\mathbf{v} = (4, 0, 0, 0)$

4. Let P_2 have the inner product in Example 8 of Section 5.1. Find the cosine of the angle between $\mathbf{p}$ and $\mathbf{q}$.
   (a) $\mathbf{p} = -1 + 5x + 2x^2$, $\mathbf{q} = 2 + 4x - 9x^2$
   (b) $\mathbf{p} = x - x^2$, $\mathbf{q} = 7 + 3x + 3x^2$

5. Show that $\mathbf{p} = 1 - x + 2x^2$ and $\mathbf{q} = 2x + x^2$ are orthogonal with respect to the inner product in Exercise 4.

6. Let M_{22} have the inner product in Example 7 of Section 5.1. Find the cosine of the angle between A and B.

   (a) $A = \begin{bmatrix} 2 & 6 \\ 1 & -3 \end{bmatrix}$, $B = \begin{bmatrix} 3 & 2 \\ 1 & 0 \end{bmatrix}$
   (b) $A = \begin{bmatrix} 2 & 4 \\ -1 & 3 \end{bmatrix}$, $B = \begin{bmatrix} -3 & 1 \\ 4 & 2 \end{bmatrix}$

7. Let

$$A = \begin{bmatrix} 2 & 1 \\ -1 & 3 \end{bmatrix}$$

Which of the following matrices are orthogonal to A with respect to the inner product in Exercise 6?

(a) $\begin{bmatrix} -3 & 0 \\ 0 & 2 \end{bmatrix}$ (b) $\begin{bmatrix} 1 & 1 \\ 0 & -1 \end{bmatrix}$ (c) $\begin{bmatrix} 0 & 0 \\ 0 & 0 \end{bmatrix}$ (d) $\begin{bmatrix} 2 & 1 \\ 5 & 2 \end{bmatrix}$

8. Let R^3 have the Euclidean inner product. For which values of k are **u** and **v** orthogonal?
(a) $\mathbf{u} = (2, 1, 3)$, $\mathbf{v} = (1, 7, k)$ (b) $\mathbf{u} = (k, k, 1)$, $\mathbf{v} = (k, 5, 6)$

9. Let R^4 have the Euclidean inner product. Find two vectors of norm 1 that are orthogonal to the three vectors $\mathbf{u} = (2, 1, \ \ 4, 0)$, $\mathbf{v} = (-1, -1, 2, 2)$, and $\mathbf{w} = (3, 2, 5, 4)$.

10. In each part verify that the Cauchy-Schwarz inequality holds for the given vectors using the Euclidean inner product.
(a) $\mathbf{u} = (3, 2)$, $\mathbf{v} = (4, -1)$ (b) $\mathbf{u} = (-3, 1, 0)$, $\mathbf{v} = (2, -1, 3)$
(c) $\mathbf{u} = (-4, 2, 1)$, $\mathbf{v} = (8, -4, -2)$ (d) $\mathbf{u} = (0, -2, 2, 1)$, $\mathbf{v} = (-1, -1, 1, 1)$

11. In each part verify that the Cauchy-Schwarz inequality holds for the given vectors.
(a) $\mathbf{u} = (-2, 1)$ and $\mathbf{v} = (1, 0)$ using the inner product of Example 2 of Section 5.1.

(b) $U = \begin{bmatrix} -1 & 2 \\ 6 & 1 \end{bmatrix}$ and $V = \begin{bmatrix} 1 & 0 \\ 3 & 3 \end{bmatrix}$

using the inner product in Example 7 of Section 5.1
(c) $\mathbf{p} = -1 + 2x + x^2$ and $\mathbf{q} = 2 - 4x^2$ using the inner product given in Example 8 of Section 5.1

12. Let V be an inner product space. Show that if **u** and **v** are orthogonal vectors in V such that $\|\mathbf{u}\| = \|\mathbf{v}\| = 1$, then $\|\mathbf{u} - \mathbf{v}\| = \sqrt{2}$.

13. Let $\{\mathbf{v}_1, \mathbf{v}_2, \dots, \mathbf{v}_r\}$ be a basis for an inner product space V. Show that the zero vector is the only vector in V that is orthogonal to all of the basis vectors.

14. Let **v** be a vector in an inner product space V.
(a) Show that the set of vectors in V that are orthogonal to **v** forms a subspace of V.
(b) Describe the subspace geometrically in R^2 and R^3 with the Euclidean inner product.

15. Prove the following generalization of Theorem 5.2.3. If $\mathbf{v}_1, \mathbf{v}_2, \dots, \mathbf{v}_r$ are pairwise orthogonal vectors in an inner product space V, then

$$\|\mathbf{v}_1 + \mathbf{v}_2 + \cdots + \mathbf{v}_r\|^2 = \|\mathbf{v}_1\|^2 + \|\mathbf{v}_2\|^2 + \cdots + \|\mathbf{v}_r\|^2$$

16. Prove the following parts of Theorem 5.2.2.
(a) part $L1$ (b) part $L2$ (c) part $L3$ (d) part $D1$
(e) part $D2$ (f) part $D3$ (g) part $D4$

17. Use vector methods to prove that a triangle that is inscribed in a circle so it has a diameter for a side must be a right triangle. [**Hint.** Express the vectors $\overrightarrow{AB}$ and $\overrightarrow{BC}$ in Figure 3 in terms of **u** and **v**.]

Figure 3

18. Let V be an inner product space. Show that if $\mathbf{w}$ is orthogonal to both $\mathbf{u}_1$ and $\mathbf{u}_2$, it is orthogonal to $k_1\mathbf{u}_1 + k_2\mathbf{u}_2$ for all scalars k_1 and k_2. Interpret this result geometrically in the case where V is R^3 with the Euclidean inner product.

19. Let V be an inner product space. Show that if $\mathbf{w}$ is orthogonal to each of the vectors $\mathbf{u}_1, \mathbf{u}_2, \ldots, \mathbf{u}_r$, then it is orthogonal to every vector in $\lim\{\mathbf{u}_1, \mathbf{u}_2, \ldots, \mathbf{u}_r\}$.

20. Prove: If $\mathbf{u}$ and $\mathbf{v}$ are $n \times 1$ matrices and A is an invertible $n \times n$ matrix, then

$$[\mathbf{v}^t A^t A \mathbf{u}]^2 \leq (\mathbf{u}^t A^t A \mathbf{u})(\mathbf{v}^t A^t A \mathbf{v})$$

21. Use the Cauchy-Schwarz inequality to prove that for all real values of a, b, and θ,

$$[a \cos \theta + b \sin \theta]^2 \leq a^2 + b^2$$

22. Prove: If $w_1, w_2, \ldots, w_n$ are positive real numbers and if $\mathbf{u} = (u_1, u_2, \ldots, u_n)$ and $\mathbf{v} = (v_1, v_2, \ldots, v_n)$ are any two vectors R^n, then

$$|w_1 u_1 v_1 + w_2 u_2 v_2 + \cdots + w_n u_n v_n| \leq (w_1 u_1^2 + w_2 u_2^2 + \cdots + w_n u_n^2)^{1/2} (w_1 v_1^2 + w_2 v_2^2 + \cdots + w_n v_n^2)^{1/2}$$

23. Show that equality holds in the Cauchy-Schwarz inequality if and only if $\mathbf{u}$ and $\mathbf{v}$ are linearly dependent.

24. **(For readers who have studied calculus.)** Let $C[0, \pi]$ have the inner product

$$\langle \mathbf{f}, \mathbf{g} \rangle = \int_0^\pi f(x)g(x) \, dx$$

and let $\mathbf{f}_n = \cos nx$ $(n = 0, 1, 2, \ldots)$. Show that if $k \neq l$, then $\mathbf{f}_k$ and $\mathbf{f}_l$ are orthogonal with respect to the given inner product.

25. **(For readers who have studied calculus.)** Let $f(x)$ and $g(x)$ be continuous functions on $[0, 1]$. Prove:

(a) $\left[\int_0^1 f(x)g(x) \, dx \right]^2 \leq \left[\int_0^1 f^2(x) \, dx \right] \left[\int_0^1 g^2(x) \, dx \right]$

(b) $\left[\int_0^1 [f(x) + g(x)]^2 \, dx \right]^{1/2} \leq \left[\int_0^1 f^2(x) \, dx \right]^{1/2} + \left[\int_0^1 g^2(x) \, dx \right]^{1/2}$

[**Hint.** Use the Cauchy-Schwarz inequality.]

26. With respect to the Euclidean inner product, the vectors $\mathbf{u} = (1, \sqrt{3})$ and $\mathbf{v} = (-1, \sqrt{3})$ have norm 2, and the angle between them is $60°$ (Figure 4). Find a weighted Euclidean inner product with respect to which $\mathbf{u}$ and $\mathbf{v}$ are orthogonal unit vectors.

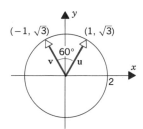

Figure 4

5.3 ORTHONORMAL BASES; GRAM-SCHMIDT PROCESS

In many problems involving vector spaces, the problem solver is free to choose any basis for the vector space that seems appropriate. In inner product spaces the solution of a problem is often greatly simplified by choosing a basis in which the vectors are orthogonal to one another. In this section we shall show how such bases can be obtained.

ORTHOGONAL AND ORTHONORMAL BASES

> **Definition.** A set of vectors in an inner product space is called an ***orthogonal set*** if all pairs of distinct vectors in the set are orthogonal. An orthogonal set in which each vector has norm 1 is called ***orthonormal.***

Example 1 Let
$$\mathbf{u}_1 = (0, 1, 0), \qquad \mathbf{u}_2 = (1, 0, 1), \qquad \mathbf{u}_3 = (1, 0, -1)$$
and assume that R^3 has the Euclidean inner product. The set of vectors $S = \{\mathbf{u}_1, \mathbf{u}_2, \mathbf{u}_3\}$ is orthogonal since $\langle \mathbf{u}_1, \mathbf{u}_2 \rangle = \langle \mathbf{u}_1, \mathbf{u}_3 \rangle = \langle \mathbf{u}_2, \mathbf{u}_3 \rangle = 0$. ▲

If $\mathbf{v}$ is a nonzero vector in an inner product space, then by property *L3* of Theorem 5.2.2 the vector
$$\frac{1}{\|\mathbf{v}\|} \mathbf{v}$$
has norm 1, since
$$\left\| \frac{1}{\|\mathbf{v}\|} \mathbf{v} \right\| = \frac{1}{\|\mathbf{v}\|} \|\mathbf{v}\| = 1$$

The process of multiplying a nonzero vector $\mathbf{v}$ by the reciprocal of its length to obtain a vector of norm 1 is called ***normalizing*** $\mathbf{v}$. An orthogonal set of *nonzero* vectors can always be converted to an orthonormal set by normalizing each of its vectors.

Example 2 The Euclidean norms of the vectors in the previous example are
$$\|\mathbf{u}_1\| = 1, \qquad \|\mathbf{u}_2\| = \sqrt{2}, \qquad \|\mathbf{u}_3\| = \sqrt{2}$$
Consequently, normalizing $\mathbf{u}_1$, $\mathbf{u}_2$, and $\mathbf{u}_3$ yields
$$\mathbf{v}_1 = \frac{\mathbf{u}_1}{\|\mathbf{u}_1\|} = (0, 1, 0) \qquad \mathbf{v}_2 = \frac{\mathbf{u}_2}{\|\mathbf{u}_2\|} = \left(\frac{1}{\sqrt{2}}, 0, \frac{1}{\sqrt{2}} \right)$$
$$\mathbf{v}_3 = \frac{\mathbf{u}_3}{\|\mathbf{u}_3\|} = \left(\frac{1}{\sqrt{2}}, 0, -\frac{1}{\sqrt{2}} \right)$$

We leave it for you to verify that the set $S = \{\mathbf{v}_1, \mathbf{v}_2, \mathbf{v}_3\}$ is orthonormal by showing that
$$\langle \mathbf{v}_1, \mathbf{v}_2 \rangle = \langle \mathbf{v}_1, \mathbf{v}_3 \rangle = \langle \mathbf{v}_2, \mathbf{v}_3 \rangle = 0$$
$$\|\mathbf{v}_1\| = \|\mathbf{v}_2\| = \|\mathbf{v}_3\| = 1 \quad ▲$$

In an inner product space, a basis consisting of orthonormal vectors is called an ***orthonormal basis***, and a basis consisting of orthogonal vectors is called an ***orthogonal basis***. A familiar example of an orthonormal basis is the standard basis for R^3 with the Euclidean inner product:

$$\mathbf{i} = (1, 0, 0), \qquad \mathbf{j} = (0, 1, 0), \qquad \mathbf{k} = (0, 0, 1)$$

This is the basis that is associated with rectangular coordinate systems (Figure 4 of Section 4.5). More generally, in R^n with the Euclidean inner product, the standard basis

$$\mathbf{e}_1 = (1, 0, 0, \dots, 0), \quad \mathbf{e}_2 = (0, 1, 0, \dots, 0), \quad \dots, \quad \mathbf{e}_n = (0, 0, 0, \dots, 1)$$

is orthonormal.

COORDINATES RELATIVE TO ORTHONORMAL BASES

The interest in finding orthonormal bases for inner product spaces is in part motivated by the following theorem, which shows that it is exceptionally simple to express a vector in terms of an orthonormal basis.

Theorem 5.3.1. *If $S = \{\mathbf{v}_1, \mathbf{v}_2, \dots, \mathbf{v}_n\}$ is an orthonormal basis for an inner product space V, and $\mathbf{u}$ is any vector in V, then*

$$\mathbf{u} = \langle \mathbf{u}, \mathbf{v}_1 \rangle \mathbf{v}_1 + \langle \mathbf{u}, \mathbf{v}_2 \rangle \mathbf{v}_2 + \cdots + \langle \mathbf{u}, \mathbf{v}_n \rangle \mathbf{v}_n$$

Proof. Since $S = \{\mathbf{v}_1, \mathbf{v}_2, \dots, \mathbf{v}_n\}$ is a basis, a vector $\mathbf{u}$ can be expressed in the form

$$\mathbf{u} = k_1\mathbf{v}_1 + k_2\mathbf{v}_2 + \cdots + k_n\mathbf{v}_n$$

We shall complete the proof by showing that $k_i = \langle \mathbf{u}, \mathbf{v}_i \rangle$ for $i = 1, 2, \dots, n$. For each vector $\mathbf{v}_i$ in S we have

$$\langle \mathbf{u}, \mathbf{v}_i \rangle = \langle k_1\mathbf{v}_1 + k_2\mathbf{v}_2 + \cdots + k_n\mathbf{v}_n, \mathbf{v}_i \rangle$$
$$= k_1\langle \mathbf{v}_1, \mathbf{v}_i \rangle + k_2\langle \mathbf{v}_2, \mathbf{v}_i \rangle + \cdots + k_n\langle \mathbf{v}_n, \mathbf{v}_i \rangle$$

Since $S = \{\mathbf{v}_1, \mathbf{v}_2, \dots, \mathbf{v}_n\}$ is an orthonormal set, we have

$$\langle \mathbf{v}_i, \mathbf{v}_i \rangle = \|\mathbf{v}_i\|^2 = 1 \qquad \text{and} \qquad \langle \mathbf{v}_j, \mathbf{v}_i \rangle = 0 \qquad \text{if } j \neq i$$

Therefore, the above expression for $\langle \mathbf{u}, \mathbf{v}_i \rangle$ simplifies to

$$\langle \mathbf{u}, \mathbf{v}_i \rangle = k_i \quad \blacksquare$$

Using the terminology and notation introduced in Section 4.5, the scalars

$$\langle \mathbf{u}, \mathbf{v}_1 \rangle, \langle \mathbf{u}, \mathbf{v}_2 \rangle, \dots, \langle \mathbf{u}, \mathbf{v}_n \rangle$$

in Theorem 5.3.1 are the coordinates of $\mathbf{u}$ relative to the orthonormal basis $S = \{\mathbf{v}_1, \mathbf{v}_2, \dots, \mathbf{v}_n\}$ and

$$(\mathbf{u})_S = (\langle \mathbf{u}, \mathbf{v}_1 \rangle, \langle \mathbf{u}, \mathbf{v}_2 \rangle, \dots, \langle \mathbf{u}, \mathbf{v}_n \rangle)$$

is the coordinate vector of $\mathbf{u}$ relative to this basis.

Example 3 Let

$$\mathbf{v}_1 = (0, 1, 0), \qquad \mathbf{v}_2 = (-\tfrac{4}{5}, 0, \tfrac{3}{5}), \qquad \mathbf{v}_3 = (\tfrac{3}{5}, 0, \tfrac{4}{5})$$

It is easy to check that $S = \{\mathbf{v}_1, \mathbf{v}_2, \mathbf{v}_3\}$ is an orthonormal basis for R^3 with the Euclidean inner product. Express the vector $\mathbf{u} = (1, 1, 1)$ as a linear combination of the vectors in S, and find the coordinate vector $(\mathbf{u})_S$.

Solution.

$$\langle \mathbf{u}, \mathbf{v}_1 \rangle = 1, \quad \langle \mathbf{u}, \mathbf{v}_2 \rangle = -\tfrac{1}{5}, \quad \text{and} \quad \langle \mathbf{u}, \mathbf{v}_3 \rangle = \tfrac{7}{5}$$

Therefore, by Theorem 5.3.1 we have

$$\mathbf{u} = \mathbf{v}_1 - \tfrac{1}{5}\mathbf{v}_2 + \tfrac{7}{5}\mathbf{v}_3$$

that is,

$$(1, 1, 1) = (0, 1, 0) - \tfrac{1}{5}(-\tfrac{4}{5}, 0, \tfrac{3}{5}) + \tfrac{7}{5}(\tfrac{3}{5}, 0, \tfrac{4}{5})$$

The coordinate vector of $\mathbf{u}$ relative to S is

$$(\mathbf{u})_S = (\langle \mathbf{u}, \mathbf{v}_1 \rangle, \langle \mathbf{u}, \mathbf{v}_2 \rangle, \langle \mathbf{u}, \mathbf{v}_3 \rangle) = (1, -\tfrac{1}{5}, \tfrac{7}{5}) \quad \blacktriangle$$

REMARK. The usefulness of Theorem 5.3.1 should be evident from this example if it is kept in mind that for nonorthonormal bases, it is usually necessary to solve a system of equations in order to express a vector in terms of the basis.

Orthonormal bases for inner product spaces are convenient because, as the following theorem shows, many familiar formulas hold for such bases.

Theorem 5.3.2. *If S is an orthonormal basis for an n-dimensional inner product space, and if*

$$(\mathbf{u})_S = (u_1, u_2, \ldots, u_n) \qquad \text{and} \qquad (\mathbf{v})_S = (v_1, v_2, \ldots, v_n)$$

then

(a) $\|\mathbf{u}\| = \sqrt{u_1^2 + u_2^2 + \cdots + u_n^2}$

(b) $d(\mathbf{u}, \mathbf{v}) = \sqrt{(u_1 - v_1)^2 + (u_2 - v_2)^2 + \cdots + (u_n - v_n)^2}$

(c) $\langle \mathbf{u}, \mathbf{v} \rangle = u_1 v_1 + u_2 v_2 + \cdots + u_n v_n$

The proof is left for the exercises.

Example 4 If R^3 has the Euclidean inner product, then the norm of the vector $\mathbf{u} = (1, 1, 1)$ is

$$\|\mathbf{u}\| = (\mathbf{u} \cdot \mathbf{u})^{1/2} = \sqrt{1^2 + 1^2 + 1^2} = \sqrt{3}$$

However, if we let R^3 have the orthonormal basis S in the last example, then we know from that example that the coordinate vector of $\mathbf{u}$ relative to S is

$$(\mathbf{u})_S = (1, -\tfrac{1}{5}, \tfrac{7}{5})$$

The norm of $\mathbf{u}$ can also be calculated from this vector using part (*a*) of Theorem 5.3.2. This yields

$$\|\mathbf{u}\| = \sqrt{1^2 + (-\tfrac{1}{5})^2 + (\tfrac{7}{5})^2} = \sqrt{\tfrac{75}{25}} = \sqrt{3} \quad \blacktriangle$$

COORDINATES RELATIVE TO ORTHOGONAL BASES

If $S = \{\mathbf{v}_1, \mathbf{v}_2, \ldots, \mathbf{v}_n\}$ is an *orthogonal* basis for a vector space V, then normalizing each of these vectors yields the orthnormal basis

$$S' = \left\{ \frac{\mathbf{v}_1}{\|\mathbf{v}_1\|}, \frac{\mathbf{v}_2}{\|\mathbf{v}_2\|}, \ldots, \frac{\mathbf{v}_n}{\|\mathbf{v}_n\|} \right\}$$

Thus, if $\mathbf{u}$ is any vector in V it follows from Theorem 5.3.1 that

$$\mathbf{u} = \left\langle \mathbf{u}, \frac{\mathbf{v}_1}{\|\mathbf{v}_1\|} \right\rangle \frac{\mathbf{v}_1}{\|\mathbf{v}_1\|} + \left\langle \mathbf{u}, \frac{\mathbf{v}_2}{\|\mathbf{v}_2\|} \right\rangle \frac{\mathbf{v}_2}{\|\mathbf{v}_2\|} + \cdots + \left\langle \mathbf{u}, \frac{\mathbf{v}_n}{\|\mathbf{v}_n\|} \right\rangle \frac{\mathbf{v}_n}{\|\mathbf{v}_n\|}$$

which by part (*a*) of Theorem 5.1.1 can be rewritten as

$$\mathbf{u} = \frac{\langle \mathbf{u}, \mathbf{v}_1 \rangle}{\|\mathbf{v}_1\|^2} \mathbf{v}_1 + \frac{\langle \mathbf{u}, \mathbf{v}_2 \rangle}{\|\mathbf{v}_2\|^2} \mathbf{v}_2 + \cdots + \frac{\langle \mathbf{u}, \mathbf{v}_n \rangle}{\|\mathbf{v}_n\|^2} \mathbf{v}_n \tag{1}$$

This formula expresses $\mathbf{u}$ as a linear combination of the vectors in the orthogonal basis S. Some problems requiring the use of this formula are given in the exercises.

Theorem 5.3.3. *If $S = \{\mathbf{v}_1, \mathbf{v}_2, \ldots, \mathbf{v}_n\}$ is an orthogonal set of nonzero vectors in an inner product space, then S is linearly independent.*

Proof. Assume that

$$k_1 \mathbf{v}_1 + k_2 \mathbf{v}_2 + \cdots + k_n \mathbf{v}_n = \mathbf{0} \tag{2}$$

To demonstrate that $S = \{\mathbf{v}_1, \mathbf{v}_2, \ldots, \mathbf{v}_n\}$ is linearly independent, we must prove that $k_1 = k_2 = \cdots = k_n = 0$.

For each $\mathbf{v}_i$ in S, it follows from (2) that

$$\langle k_1 \mathbf{v}_1 + k_2 \mathbf{v}_2 + \cdots + k_n \mathbf{v}_n, \mathbf{v}_i \rangle = \langle \mathbf{0}, \mathbf{v}_i \rangle = 0$$

or equivalently

$$k_1 \langle \mathbf{v}_1, \mathbf{v}_i \rangle + k_2 \langle \mathbf{v}_2, \mathbf{v}_i \rangle + \cdots + k_n \langle \mathbf{v}_n, \mathbf{v}_i \rangle = 0$$

From the orthogonality of S it follows that $\langle \mathbf{v}_j, \mathbf{v}_i \rangle = 0$ when $j \neq i$, so that this equation reduces to

$$k_i \langle \mathbf{v}_i, \mathbf{v}_i \rangle = 0$$

Since the vectors in S are assumed to be nonzero, $\langle \mathbf{v}_i, \mathbf{v}_i \rangle \neq 0$ by the positivity axiom for inner products. Therefore, $k_i = 0$. Since the subscript i is arbitrary, we have $k_1 = k_2 = \cdots = k_n = 0$; thus, S is linearly independent. ∎

Example 5   In Example 2 we showed that the vectors

$$\mathbf{v}_1 = (0, 1, 0), \qquad \mathbf{v}_2 = \left(\frac{1}{\sqrt{2}}, 0, \frac{1}{\sqrt{2}} \right), \qquad \text{and} \qquad \mathbf{v}_3 = \left(\frac{1}{\sqrt{2}}, 0, -\frac{1}{\sqrt{2}} \right)$$

form an orthonormal set with respect to the Euclidean inner product on R^3. By Theorem 5.3.3, these vectors form a linearly independent set, and since R^3 is three-dimensional, $S = \{\mathbf{v}_1, \mathbf{v}_2, \mathbf{v}_3\}$ is an orthonormal basis for R^3 by Theorem 4.5.4a. ▲

ORTHOGONAL PROJECTIONS

We shall now develop some results that will help us to construct orthogonal and orthonormal bases for inner product spaces.

In R^2 or R^3 with the Euclidean inner product, it is evident geometrically that if W is a line or a plane through the origin, then each vector $\mathbf{u}$ in the space can be expressed as a sum

$$\mathbf{u} = \mathbf{w}_1 + \mathbf{w}_2$$

where $\mathbf{w}_1$ is in W and $\mathbf{w}_2$ is perpendicular to W (Figure 1). This result is a special case of the following general theorem whose proof is given at the end of this section.

Figure 1

Theorem 5.3.4 (*Projection Theorem*). *If W is a finite-dimensional subspace of an inner product space V, then every vector $\mathbf{u}$ in V can be expressed in exactly one way as*

$$\mathbf{u} = \mathbf{w}_1 + \mathbf{w}_2$$

where $\mathbf{w}_1$ is in W and $\mathbf{w}_2$ is orthogonal to W.

The vector $\mathbf{w}_1$ in the foregoing theorem is called the ***orthogonal projection of*** $\mathbf{u}$ ***on*** W and is denoted by $\text{proj}_W \mathbf{u}$. The vector $\mathbf{w}_2 = \mathbf{u} - \text{proj}_W \mathbf{u}$ is called the ***component of*** $\mathbf{u}$ ***orthogonal to*** W. The following theorem, whose proof is discussed in the exercises, provides formulas for calculating orthogonal projections.

Theorem 5.3.5. *Let W be a finite-dimensional subspace of an inner product space V.*

(a) *If $\{\mathbf{v}_1, \mathbf{v}_2, \ldots, \mathbf{v}_r\}$ is an orthonormal basis for W, and $\mathbf{u}$ is any vector in V, then*

$$\text{proj}_W \mathbf{u} = \langle \mathbf{u}, \mathbf{v}_1 \rangle \mathbf{v}_1 + \langle \mathbf{u}, \mathbf{v}_2 \rangle \mathbf{v}_2 + \cdots + \langle \mathbf{u}, \mathbf{v}_r \rangle \mathbf{v}_r \qquad (3)$$

(b) *If $\{\mathbf{v}_1, \mathbf{v}_2, \ldots, \mathbf{v}_r\}$ is an orthogonal basis for W, and $\mathbf{u}$ is any vector in V, then*

$$\text{proj}_W \mathbf{u} = \frac{\langle \mathbf{u}, \mathbf{v}_1 \rangle}{\|\mathbf{v}_1\|^2} \mathbf{v}_1 + \frac{\langle \mathbf{u}, \mathbf{v}_2 \rangle}{\|\mathbf{v}_2\|^2} \mathbf{v}_2 + \cdots + \frac{\langle \mathbf{u}, \mathbf{v}_r \rangle}{\|\mathbf{v}_r\|^2} \mathbf{v}_r \qquad (4)$$

Example 6 Let R^3 have the Euclidean inner product, and let W be the subspace spanned by the orthonormal vectors $\mathbf{v}_1 = (0, 1, 0)$ and $\mathbf{v}_2 = (-\frac{4}{5}, 0, \frac{3}{5})$. From (3) the orthogonal projection of $\mathbf{u} = (1, 1, 1)$ on W is

$$\begin{aligned}
\text{proj}_W \mathbf{u} &= \langle \mathbf{u}, \mathbf{v}_1 \rangle \mathbf{v}_1 + \langle \mathbf{u}, \mathbf{v}_2 \rangle \mathbf{v}_2 \\
&= (1)(0, 1, 0) + (-\tfrac{1}{5})(-\tfrac{4}{5}, 0, \tfrac{3}{5}) \\
&= (\tfrac{4}{25}, 1, -\tfrac{3}{25})
\end{aligned}$$

The component of $\mathbf{u}$ orthogonal to W is

$$\mathbf{u} - \text{proj}_W \mathbf{u} = (1, 1, 1) - (\tfrac{4}{25}, 1, -\tfrac{3}{25}) = (\tfrac{21}{25}, 0, \tfrac{28}{25})$$

Observe that $\mathbf{u} - \text{proj}_W \mathbf{u}$ is orthogonal to both $\mathbf{v}_1$ and $\mathbf{v}_2$ so that this vector is orthogonal to each vector in the space W spanned by $\mathbf{v}_1$ and $\mathbf{v}_2$ as it should be. ▲

FINDING ORTHOGONAL AND ORTHONORMAL BASES

We are now in a position to prove the main result of this section.

Theorem 5.3.6. *Every nonzero finite-dimensional inner product space has an orthonormal basis.*

Proof. Let V be any nonzero finite-dimensional inner product space, and let $\{\mathbf{u}_1, \mathbf{u}_2, \ldots, \mathbf{u}_n\}$ be any basis for V. It suffices to show that V has an orthogonal basis, since the vectors in the orthogonal basis can be normalized to produce an orthonormal basis for V. The following sequence of steps will produce an orthogonal basis $\{\mathbf{v}_1, \mathbf{v}_2, \ldots, \mathbf{v}_n\}$ for V.

Step 1. Let $\mathbf{v}_1 = \mathbf{u}_1$.

Step 2. As illustrated in Figure 2, we can obtain a vector $\mathbf{v}_2$ that is orthogonal to $\mathbf{v}_1$ by computing the component of $\mathbf{u}_2$ that is orthogonal to the space W_1 spanned by $\mathbf{v}_1$. We use Formula (4):

$$\mathbf{v}_2 = \mathbf{u}_2 - \operatorname{proj}_{W_1} \mathbf{u}_2 = \mathbf{u}_2 - \frac{\langle \mathbf{u}_2, \mathbf{v}_1 \rangle}{\|\mathbf{v}_1\|^2} \mathbf{v}_1$$

Of course, if $\mathbf{v}_2 = \mathbf{0}$, then $\mathbf{v}_2$ is not a basis vector. But this cannot happen, since it would then follow from the foregoing formula for $\mathbf{v}_2$ that

$$\mathbf{u}_2 = \frac{\langle \mathbf{u}_2, \mathbf{v}_1 \rangle}{\|\mathbf{v}_1\|^2} \mathbf{v}_1 = \frac{\langle \mathbf{u}_2, \mathbf{v}_1 \rangle}{\|\mathbf{u}_1\|^2} \mathbf{u}_1$$

which says that $\mathbf{u}_2$ is a multiple of $\mathbf{u}_1$, contradicting the linear independence of the basis $S = \{\mathbf{u}_1, \mathbf{u}_2, \ldots, \mathbf{u}_n\}$.

Step 3. To construct a vector $\mathbf{v}_3$ that is orthogonal to both $\mathbf{v}_1$ and $\mathbf{v}_2$, we compute the component of $\mathbf{u}_3$ orthogonal to the space W_2 spanned by $\mathbf{v}_1$ and $\mathbf{v}_2$ (Figure 3). From (4)

$$\mathbf{v}_3 = \mathbf{u}_3 - \operatorname{proj}_{W_2} \mathbf{u}_3 = \mathbf{u}_3 - \frac{\langle \mathbf{u}_3, \mathbf{v}_1 \rangle}{\|\mathbf{v}_1\|^2} \mathbf{v}_1 - \frac{\langle \mathbf{u}_3, \mathbf{v}_2 \rangle}{\|\mathbf{v}_2\|^2} \mathbf{v}_2$$

As in Step 2, the linear independence of $\{\mathbf{u}_1, \mathbf{u}_2, \ldots, \mathbf{u}_n\}$ ensures that $\mathbf{v}_3 \neq \mathbf{0}$. We leave the details as an exercise.

Step 4. To determine a vector $\mathbf{v}_4$ that is orthogonal to $\mathbf{v}_1$, $\mathbf{v}_2$, and $\mathbf{v}_3$, we compute the component of $\mathbf{u}_4$ orthogonal to the space W_3 spanned by $\mathbf{v}_1$, $\mathbf{v}_2$, and $\mathbf{v}_3$. From (4)

$$\mathbf{v}_4 = \mathbf{u}_4 - \operatorname{proj}_{W_3} \mathbf{u}_4 = \mathbf{u}_4 - \frac{\langle \mathbf{u}_4, \mathbf{v}_1 \rangle}{\|\mathbf{v}_1\|^2} \mathbf{v}_1 - \frac{\langle \mathbf{u}_4, \mathbf{v}_2 \rangle}{\|\mathbf{v}_2\|^2} \mathbf{v}_2 - \frac{\langle \mathbf{u}_4, \mathbf{v}_3 \rangle}{\|\mathbf{v}_3\|^2} \mathbf{v}_3$$

Figure 2

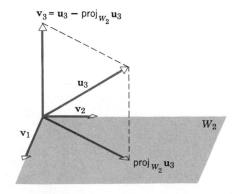

Figure 3

Continuing in this way, we will obtain after n steps an orthogonal set of vectors, $\{v_1, v_2, \ldots, v_n\}$. Since V is n-dimensional and every orthogonal set is linearly independent, the set $\{v_1, v_2, \ldots, v_n\}$ will be an orthogonal basis for V. ∎

The above step-by-step construction for converting an arbitrary basis into an orthogonal basis is called the ***Gram-Schmidt* process***. It can be shown that at each stage in this process, the vectors $v_1, v_2, \ldots, v_k$ form an orthogonal basis for the subspace spanned by $u_1, u_2, \ldots, u_k$.

Example 7 Consider the vector space R^3 with the Euclidean inner product. Apply the Gram-Schmidt process to transform the basis vectors $u_1 = (1, 1, 1)$, $u_2 = (0, 1, 1)$, $u_3 = (0, 0, 1)$ into an orthogonal basis; then normalize the orthogonal basis vectors to obtain an orthonormal basis.

Solution.

Step 1. $v_1 = u_1 = (1, 1, 1)$

Step 2. $v_2 = u_2 - \text{proj}_{W_1} u_2 = u_2 - \dfrac{\langle u_2, v_1 \rangle}{\|v_1\|^2} v_1$

$$= (0, 1, 1) - \frac{2}{3}(1, 1, 1) = \left(-\frac{2}{3}, \frac{1}{3}, \frac{1}{3}\right)$$

Step 3. $v_3 = u_3 - \text{proj}_{W_2} u_3 = u_3 - \dfrac{\langle u_3, v_1 \rangle}{\|v_1\|^2} v_1 - \dfrac{\langle u_3, v_2 \rangle}{\|v_2\|^2} v_2$

$$= (0, 0, 1) - \frac{1}{3}(1, 1, 1) - \frac{1/3}{2/3}\left(-\frac{2}{3}, \frac{1}{3}, \frac{1}{3}\right)$$

$$= \left(0, -\frac{1}{2}, \frac{1}{2}\right)$$

Thus,

$$v_1 = (1, 1, 1), \qquad v_2 = \left(-\frac{2}{3}, \frac{1}{3}, \frac{1}{3}\right), \qquad v_3 = \left(0, -\frac{1}{2}, \frac{1}{2}\right)$$

form an orthogonal basis for R^3. The norms of these vectors are

$$\|v_1\| = \sqrt{3}, \qquad \|v_2\| = \frac{\sqrt{6}}{3}, \qquad \|v_3\| = \frac{1}{\sqrt{2}}$$

---

* *Jörgen Pederson Gram (1850–1916)* was a Danish actuary.

Erhardt Schmidt (1876–1959) was a German mathematician. Schmidt received his doctoral degree from Göttingen University in 1905, where he studied under one of the giants of mathematics, David Hilbert. He eventually went to teach at Berlin University in 1917, where he stayed for the rest of his life. Schmidt made important contributions to a variety of mathematical fields, but is most noteworthy for fashioning many of Hilbert's diverse ideas into a general concept (called a Hilbert space), which is fundamental in the study of infinite-dimensional vector spaces. Schmidt first described the process which bears his name in a paper on integral equations published in 1907.

so an orthonormal basis for R^3 is

$$\frac{\mathbf{v}_1}{\|\mathbf{v}_1\|} = \left(\frac{1}{\sqrt{3}}, \frac{1}{\sqrt{3}}, \frac{1}{\sqrt{3}}\right), \qquad \frac{\mathbf{v}_2}{\|\mathbf{v}_2\|} = \left(-\frac{2}{\sqrt{6}}, \frac{1}{\sqrt{6}}, \frac{1}{\sqrt{6}}\right),$$

$$\frac{\mathbf{v}_3}{\|\mathbf{v}_3\|} = \left(0, -\frac{1}{\sqrt{2}}, \frac{1}{\sqrt{2}}\right) \quad \blacktriangle$$

ORTHOGONAL PROJECTIONS VIEWED AS APPROXIMATIONS

We conclude this section with an alternative viewpoint about orthogonal projections. If P is a point in ordinary 3-space and W is a plane through the origin, then the point Q in W closest to P is obtained by dropping a perpendicular from P to W (Figure 4a). Therefore, if we let $\mathbf{u} = \overrightarrow{OP}$, the distance between P and W is given by

$$\|\mathbf{u} - \text{proj}_W \mathbf{u}\|$$

In other words, among all vectors $\mathbf{w}$ in W the vector $\mathbf{w} = \text{proj}_W \mathbf{u}$ minimizes the distance $\|\mathbf{u} - \mathbf{w}\|$ (Figure 4b).

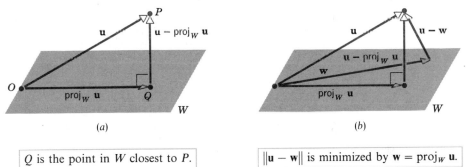

(a)   (b)

Figure 4 Q is the point in W closest to P. $\|\mathbf{u} - \mathbf{w}\|$ is minimized by $\mathbf{w} = \text{proj}_W \mathbf{u}$.

There is another way of thinking about this idea. View $\mathbf{u}$ as a fixed vector that we would like to approximate by a vector in W. Any such approximation $\mathbf{w}$ will result in an "error vector,"

$$\mathbf{u} - \mathbf{w}$$

which, unless $\mathbf{u}$ is in W, cannot be made equal to $\mathbf{0}$. However, by choosing

$$\mathbf{w} = \text{proj}_W \mathbf{u}$$

we can make the length of the error vector

$$\|\mathbf{u} - \mathbf{w}\| = \|\mathbf{u} - \text{proj}_W \mathbf{u}\|$$

as small as possible. Thus, we can describe $\text{proj}_W \mathbf{u}$ as the "best approximation" to $\mathbf{u}$ by vectors in W. The following theorem will make these intuitive ideas precise.

Theorem 5.3.7 (*Best Approximation Theorem*). *If W is a finite-dimensional subspace of an inner product space V, and if $\mathbf{u}$ is a vector in V, then $\text{proj}_W \mathbf{u}$ is the **best approximation** to $\mathbf{u}$ from W in the sense that*

$$\|\mathbf{u} - \text{proj}_W \mathbf{u}\| < \|\mathbf{u} - \mathbf{w}\|$$

for every vector $\mathbf{w}$ in W that is different from $\text{proj}_W \mathbf{u}$.

Proof. For every vector $\mathbf{w}$ in W we can write

$$\mathbf{u} - \mathbf{w} = (\mathbf{u} - \text{proj}_W \mathbf{u}) + (\text{proj}_W \mathbf{u} - \mathbf{w}) \tag{5}$$

But $\text{proj}_W \mathbf{u} - \mathbf{w}$, being a difference of vectors in W, is in W; and $\mathbf{u} - \text{proj}_W \mathbf{u}$ is orthogonal to W, so that the two terms on the right side of (5) are orthogonal. Thus, by the Theorem of Pythagoras (5.2.3),

$$\|\mathbf{u} - \mathbf{w}\|^2 = \|\mathbf{u} - \text{proj}_W \mathbf{u}\|^2 + \|\text{proj}_W \mathbf{u} - \mathbf{w}\|^2$$

If $\mathbf{w} \neq \text{proj}_W \mathbf{u}$, then the second term in this sum will be positive, so that

$$\|\mathbf{u} - \mathbf{w}\|^2 > \|\mathbf{u} - \text{proj}_W \mathbf{u}\|^2$$

or equivalently

$$\|\mathbf{u} - \mathbf{w}\| > \|\mathbf{u} - \text{proj}_W \mathbf{u}\| \quad \blacksquare$$

Applications of this theorem will be given later in the text.

OPTIONAL

Proof of Theorem 5.3.4. There are two parts to the proof. First we must find vectors $\mathbf{w}_1$ and $\mathbf{w}_2$ with the stated properties, and then we must show that these are the only such vectors.

By the Gram-Schmidt process there is an orthonormal basis $\{\mathbf{v}_1, \mathbf{v}_2, \ldots, \mathbf{v}_n\}$ for W. Let

$$\mathbf{w}_1 = \langle \mathbf{u}, \mathbf{v}_1 \rangle \mathbf{v}_1 + \langle \mathbf{u}, \mathbf{v}_2 \rangle \mathbf{v}_2 + \cdots + \langle \mathbf{u}, \mathbf{v}_n \rangle \mathbf{v}_n \tag{6}$$

and

$$\mathbf{w}_2 = \mathbf{u} - \mathbf{w}_1 \tag{7}$$

It follows that $\mathbf{w}_1 + \mathbf{w}_2 = \mathbf{w}_1 + (\mathbf{u} - \mathbf{w}_1) = \mathbf{u}$, so it remains to show that $\mathbf{w}_1$ is in W and $\mathbf{w}_2$ is orthogonal to W. But $\mathbf{w}_1$ lies in W because it is a linear combination of the basis vectors for W. To show that $\mathbf{w}_2$ is orthogonal to W, we must show that $\langle \mathbf{w}_2, \mathbf{w} \rangle = 0$ for every vector $\mathbf{w}$ in W. But if $\mathbf{w}$ is any vector in W, it can be expressed as a linear combination

$$\mathbf{w} = k_1 \mathbf{v}_1 + k_2 \mathbf{v}_2 + \cdots + k_n \mathbf{v}_n$$

of the basis vectors $\mathbf{v}_1, \mathbf{v}_2, \ldots, \mathbf{v}_n$. Thus,

$$\langle \mathbf{w}_2, \mathbf{w} \rangle = \langle \mathbf{u} - \mathbf{w}_1, \mathbf{w} \rangle = \langle \mathbf{u}, \mathbf{w} \rangle - \langle \mathbf{w}_1, \mathbf{w} \rangle \tag{8}$$

But

$$\langle \mathbf{u}, \mathbf{w} \rangle = \langle \mathbf{u}, k_1 \mathbf{v}_1 + k_2 \mathbf{v}_2 + \cdots + k_n \mathbf{v}_n \rangle = k_1 \langle \mathbf{u}, \mathbf{v}_1 \rangle + k_2 \langle \mathbf{u}, \mathbf{v}_2 \rangle + \cdots + k_n \langle \mathbf{u}, \mathbf{v}_n \rangle$$

and by part (c) of Theorem 5.3.2

$$\langle \mathbf{w}_1, \mathbf{w} \rangle = \langle \mathbf{u}, \mathbf{v}_1 \rangle k_1 + \langle \mathbf{u}, \mathbf{v}_2 \rangle k_2 + \cdots + \langle \mathbf{u}, \mathbf{v}_n \rangle k_n$$

Thus, $\langle \mathbf{u}, \mathbf{w} \rangle$ and $\langle \mathbf{w}_1, \mathbf{w} \rangle$ are equal, so (8) yields $\langle \mathbf{w}_2, \mathbf{w} \rangle = 0$, which is what we want to show.

To see that (6) and (7) are the only vectors with the properties stated in the theorem, suppose that we can also write

$$\mathbf{u} = \mathbf{w}_1' + \mathbf{w}_2' \tag{9}$$

where $\mathbf{w}_1'$ is in W and $\mathbf{w}_2'$ is orthogonal to W. If we subtract from (9) the equation

$$\mathbf{u} = \mathbf{w}_1 + \mathbf{w}_2$$

we obtain

$$\mathbf{0} = (\mathbf{w}_1' - \mathbf{w}_1) + (\mathbf{w}_2' - \mathbf{w}_2)$$

or

$$\mathbf{w}_1 - \mathbf{w}_1' = \mathbf{w}_2' - \mathbf{w}_2 \tag{10}$$

Since $\mathbf{w}_2$ and $\mathbf{w}_2'$ are orthogonal to W, their difference is also orthogonal to W, since for any vector $\mathbf{w}$ in W we can write

$$\langle \mathbf{w}, \mathbf{w}_2' - \mathbf{w}_2 \rangle = \langle \mathbf{w}, \mathbf{w}_2' \rangle - \langle \mathbf{w}, \mathbf{w}_2 \rangle = 0 - 0 = 0$$

But $\mathbf{w}_2' - \mathbf{w}_2$ is itself a vector in W, since from (10) it is the difference of the two vectors $\mathbf{w}_1$ and $\mathbf{w}_1'$ which lie in the subspace W. Thus, $\mathbf{w}_2' - \mathbf{w}_2$ must be orthogonal to itself; that is,

$$\langle \mathbf{w}_2' - \mathbf{w}_2, \mathbf{w}_2' - \mathbf{w}_2 \rangle = 0$$

But this implies that $\mathbf{w}_2' - \mathbf{w}_2 = 0$ by Axiom 4 for inner products. Thus, $\mathbf{w}_2' = \mathbf{w}_2$, and by (10), $\mathbf{w}_1' = \mathbf{w}_1$. ∎

EXERCISE SET 5.3

1. Which of the following sets of vectors are orthogonal with respect to the Euclidean inner product on R^2?
   (a) $(0, 1), (2, 0)$ (b) $(-1/\sqrt{2}, 1/\sqrt{2}), (1/\sqrt{2}, 1/\sqrt{2})$ (c) $(-1/\sqrt{2}, -1/\sqrt{2}), (1/\sqrt{2}, 1/\sqrt{2})$ (d) $(0, 0), (0, 1)$

2. Which of the sets in Exercise 1 are orthonormal with respect to the Euclidean inner product on R^2?

3. Which of the following sets of vectors are orthogonal with respect to the Euclidean inner product on R^3?

   (a) $\left(\frac{1}{\sqrt{2}}, 0, \frac{1}{\sqrt{2}}\right), \left(\frac{1}{\sqrt{3}}, \frac{1}{\sqrt{3}}, -\frac{1}{\sqrt{3}}\right), \left(-\frac{1}{\sqrt{2}}, 0, \frac{1}{\sqrt{2}}\right)$ (b) $\left(\frac{2}{3}, -\frac{2}{3}, \frac{1}{3}\right), \left(\frac{2}{3}, \frac{1}{3}, -\frac{2}{3}\right), \left(\frac{1}{3}, \frac{2}{3}, \frac{2}{3}\right)$

   (c) $(1, 0, 0), \left(0, \frac{1}{\sqrt{2}}, \frac{1}{\sqrt{2}}\right), (0, 0, 1)$ (d) $\left(\frac{1}{\sqrt{6}}, \frac{1}{\sqrt{6}}, -\frac{2}{\sqrt{6}}\right), \left(\frac{1}{\sqrt{2}}, -\frac{1}{\sqrt{2}}, 0\right)$

4. Which of the sets in Exercise 3 are orthonormal with respect to the Euclidean inner product on R^3?

5. Which of the following sets of polynomials are orthonormal with respect to the inner product on P_2 discussed in Example 8 of Section 5.1?

(a) $\frac{2}{3} - \frac{2}{3}x + \frac{1}{3}x^2$, $\frac{2}{3} + \frac{1}{3}x - \frac{2}{3}x^2$, $\frac{1}{3} + \frac{2}{3}x + \frac{2}{3}x^2$ (b) 1, $\frac{1}{\sqrt{2}}x + \frac{1}{\sqrt{2}}x^2$, x^2

6. Which of the following sets of matrices are orthonormal with respect to the inner product on M_{22} discussed in Example 7 of Section 5.1?

(a) $\begin{bmatrix} 1 & 0 \\ 0 & 0 \end{bmatrix}$ $\begin{bmatrix} 0 & \frac{2}{3} \\ \frac{1}{3} & -\frac{2}{3} \end{bmatrix}$ $\begin{bmatrix} 0 & \frac{2}{3} \\ -\frac{2}{3} & \frac{1}{3} \end{bmatrix}$ $\begin{bmatrix} 0 & \frac{1}{3} \\ \frac{2}{3} & \frac{2}{3} \end{bmatrix}$

(b) $\begin{bmatrix} 1 & 0 \\ 0 & 0 \end{bmatrix}$ $\begin{bmatrix} 0 & 1 \\ 0 & 0 \end{bmatrix}$ $\begin{bmatrix} 0 & 0 \\ 1 & 1 \end{bmatrix}$ $\begin{bmatrix} 0 & 0 \\ 1 & -1 \end{bmatrix}$

7. Verify that the given set of vectors is orthogonal with respect to the Euclidean inner product; then convert it to an orthonormal set by normalizing the vectors.

(a) $(-1, 2)$, $(6, 3)$ (b) $(1, 0, -1)$, $(2, 0, 2)$, $(0, 5, 0)$ (c) $(\frac{1}{5}, \frac{1}{5}, \frac{1}{5})$, $(-\frac{1}{2}, \frac{1}{2}, 0)$, $(\frac{1}{3}, \frac{1}{3}, -\frac{2}{3})$

8. Let $\mathbf{x} = \left(\dfrac{1}{\sqrt{5}}, -\dfrac{1}{\sqrt{5}}\right)$ and $\mathbf{y} = \left(\dfrac{2}{\sqrt{30}}, \dfrac{3}{\sqrt{30}}\right)$.

Show that $\{\mathbf{x}, \mathbf{y}\}$ is orthonormal if R^2 has the inner product $\langle \mathbf{u}, \mathbf{v} \rangle = 3u_1v_1 + 2u_2v_2$, but is not orthonormal if R^2 has the Euclidean inner product.

9. Verify that the vectors $\mathbf{v}_1 = (-\frac{3}{5}, \frac{4}{5}, 0)$, $\mathbf{v}_2 = (\frac{4}{5}, \frac{3}{5}, 0)$, $\mathbf{v}_3 = (0, 0, 1)$ form an orthonormal basis for R^3 with the Euclidean inner product; then use Theorem 5.3.1 to express each of the following as linear combinations of $\mathbf{v}_1$, $\mathbf{v}_2$, and $\mathbf{v}_3$.

(a) $(1, -1, 2)$ (b) $(3, -7, 4)$ (c) $(\frac{1}{7}, -\frac{3}{7}, \frac{5}{7})$

10. Verify that the vectors

$$\mathbf{v}_1 = (1, -1, 2, -1), \quad \mathbf{v}_2 = (-2, 2, 3, 2), \quad \mathbf{v}_3 = (1, 2, 0, -1), \quad \mathbf{v}_4 = (1, 0, 0, 1)$$

form an orthogonal basis for R^4 with the Euclidean inner product; then use Formula (1) to express each of the following as linear combinations of $\mathbf{v}_1$, $\mathbf{v}_2$, $\mathbf{v}_3$, and $\mathbf{v}_4$.

(a) $(1, 1, 1, 1)$ (b) $(\sqrt{2}, -3\sqrt{2}, 5\sqrt{2}, -\sqrt{2})$ (c) $(-\frac{1}{3}, \frac{2}{3}, -\frac{1}{3}, \frac{4}{3})$

11. In each part an orthonormal basis relative to the Euclidean inner product is given. Use Theorem 5.3.1 to find the coordinate vector of $\mathbf{w}$ with respect to that basis.

(a) $\mathbf{w} = (3, 7)$; $\mathbf{u}_1 = \left(\dfrac{1}{\sqrt{2}}, -\dfrac{1}{\sqrt{2}}\right)$, $\mathbf{u}_2 = \left(\dfrac{1}{\sqrt{2}}, \dfrac{1}{\sqrt{2}}\right)$

(b) $\mathbf{w} = (-1, 0, 2)$; $\mathbf{u}_1 = (\frac{2}{3}, -\frac{2}{3}, \frac{1}{3})$, $\mathbf{u}_2 = (\frac{2}{3}, \frac{1}{3}, -\frac{2}{3})$, $\mathbf{u}_3 = (\frac{1}{3}, \frac{2}{3}, \frac{2}{3})$

12. Let R^2 have the Euclidean inner product, and let $S = \{\mathbf{w}_1, \mathbf{w}_2\}$ be the orthonormal basis with $\mathbf{w}_1 = (\frac{3}{5}, -\frac{4}{5})$, $\mathbf{w}_2 = (\frac{4}{5}, \frac{3}{5})$.

(a) Find the vectors $\mathbf{u}$ and $\mathbf{v}$ that have coordinate vectors $(\mathbf{u})_S = (1, 1)$ and $(\mathbf{v})_S = (-1, 4)$.

(b) Compute $\|\mathbf{u}\|$, $d(\mathbf{u}, \mathbf{v})$, and $\langle \mathbf{u}, \mathbf{v} \rangle$ by applying Theorem 5.3.2 to the coordinate vectors $(\mathbf{u})_S$ and $(\mathbf{v})_S$; then check the results by performing the computations directly on $\mathbf{u}$ and $\mathbf{v}$.

13. Let R^3 have the Euclidean inner product, and let $S = \{w_1, w_2, w_3\}$ be the orthonormal basis with $w_1 = (0, -\frac{3}{5}, \frac{4}{5})$, $w_2 = (1, 0, 0)$, and $w_3 = (0, \frac{4}{5}, \frac{3}{5})$.
    (a) Find the vectors u, v, and w that have the coordinate vectors $(u)_S = (-2, 1, 2)$, $(v)_S = (3, 0, -2)$, and $(w)_S = (5, -4, 1)$.
    (b) Compute $\|v\|$, $d(u, w)$, and $\langle w, v \rangle$ by applying Theorem 5.3.2 to the coordinate vectors $(u)_S$, $(v)_S$, and $(w)_S$; then check the results by performing the computations directly on u and v.

14. In each part, S represents some orthonormal basis for a 4-dimensional inner product space. Use the given information to find $\|u\|$, $\|v - w\|$, $\|v + w\|$, and $\langle v, w \rangle$.
    (a) $(u)_S = (-1, 2, 1, 3)$, $(v)_S = (0, -3, 1, 5)$, $(w)_S = (-2, -4, 3, 1)$
    (b) $(u)_S = (0, 0, -1, -1)$, $(v)_S = (5, 5, -2, -2)$, $(w)_S = (3, 0, -3, 0)$

15. (a) Show that the vectors $v_1 = (1, -2, 3, -4)$, $v_2 = (2, 1, -4, -3)$, $v_3 = (-3, 4, 1, -2)$, and $v_4 = (4, 3, 2, 1)$ form an orthogonal basis for R^4 with the Euclidean inner product.
    (b) Use (1) to express $u = (-1, 2, 3, 7)$ as a linear combination of the vectors in (a).

16. Let R^2 have the Euclidean inner product. Use the Gram-Schmidt process to transform the basis $\{u_1, u_2\}$ into an orthonormal basis.
    (a) $u_1 = (1, -3)$, $u_2 = (2, 2)$ (b) $u_1 = (1, 0)$, $u_2 = (3, -5)$

17. Let R^3 have the Euclidean inner product. Use the Gram-Schmidt process to transform the basis $\{u_1, u_2, u_3\}$ into an orthonormal basis.
    (a) $u_1 = (1, 1, 1)$, $u_2 = (-1, 1, 0)$, $u_3 = (1, 2, 1)$
    (b) $u_1 = (1, 0, 0)$, $u_2 = (3, 7, -2)$, $u_3 = (0, 4, 1)$

18. Let R^4 have the Euclidean inner product. Use the Gram-Schmidt process to transform the basis $\{u_1, u_2, u_3, u_4\}$ into an orthonormal basis.

    $$u_1 = (0, 2, 1, 0), \quad u_2 = (1, -1, 0, 0), \quad u_3 = (1, 2, 0, 1), \quad u_4 = (1, 0, 0, 1)$$

19. Let R^3 have the Euclidean inner product. Find an orthonormal basis for the subspace spanned by $(0, 1, 2)$, $(-1, 0, 1)$.

20. Let R^3 have the inner product $\langle u, v \rangle = u_1 v_1 + 2u_2 v_2 + 3u_3 v_3$. Use the Gram-Schmidt process to transform $u_1 = (1, 1, 1)$, $u_2 = (1, 1, 0)$, $u_3 = (1, 0, 0)$ into an orthonormal basis.

21. The subspace of R^3 spanned by the vectors $u_1 = (\frac{4}{5}, 0, -\frac{3}{5})$ and $u_2 = (0, 1, 0)$ is a plane passing through the origin. Express $w = (1, 2, 3)$ in the form $w = w_1 + w_2$, where w_1 lies in the plane, and w_2 is perpendicular to the plane.

22. Repeat Exercise 21 with $u_1 = (1, 1, 1)$ and $u_2 = (2, 0, -1)$.

23. Let R^4 have the Euclidean inner product. Express $w = (-1, 2, 6, 0)$ in the form $w = w_1 + w_2$, where w_1 is in the space W spanned by $u_1 = (-1, 0, 1, 2)$ and $u_2 = (0, 1, 0, 1)$, and w_2 is orthogonal to W.

24. Let $\{v_1, v_2, v_3\}$ be an orthonormal basis for an inner product space V. Show that if w is a vector in V, then $\|w\|^2 = \langle w, v_1 \rangle^2 + \langle w, v_2 \rangle^2 + \langle w, v_3 \rangle^2$.

25. Let $\{v_1, v_2, \ldots, v_n\}$ be an orthonormal basis for an inner product space V. Show that if w is a vector in V, then $\|w\|^2 = \langle w, v_1 \rangle^2 + \langle w, v_2 \rangle^2 + \cdots + \langle w, v_n \rangle^2$.

26. In Step 3 of the proof of Theorem 5.3.6, it was stated that "the linear independence of $\{u_1, u_2, \ldots, u_n\}$ ensures that $v_3 \neq 0$." Prove this statement.

27. **(For readers who have studied calculus.)** Let the vector space P_2 have the inner product

$$\langle \mathbf{p}, \mathbf{q} \rangle = \int_{-1}^{1} p(x)q(x)\,dx$$

Apply the Gram-Schmidt process to transform the standard basis $S = \{1, x, x^2\}$ into an orthonormal basis. (The polynomials in the resulting basis are called the first three *normalized Legendre polynomials*.)

28. **(For readers who have studied calculus.)** Use Theorem 5.3.1 to express the following as linear combinations of the first three normalized Legendre polynomials (Exercise 27).
    (a) $1 + x + 4x^2$ (b) $2 - 7x^2$ (c) $4 + 3x$

29. **(For readers who have studied calculus.)** Let P_2 have the inner product

$$\langle \mathbf{p}, \mathbf{q} \rangle = \int_{0}^{1} p(x)q(x)\,dx$$

Apply the Gram-Schmidt process to transform the standard basis $S = \{1, x, x^2\}$ into an orthonormal basis.

30. **(For readers who have studied the optional material in this section.)** Find the point Q in the plane $5x - 3y + z = 0$ closest to $P(1, -2, 4)$, and determine the distance between the point P and the plane. [**Hint.** View the plane as a subspace W of R^3 with the Euclidean inner product and apply Theorem 5.3.7.]

31. **(For readers who have studied the optional material in this section.)** Find the point Q on the line

$$x = 2t$$
$$y = -t \qquad -\infty < t < +\infty$$
$$z = 4t$$

closest to $P(-4, 8, 1)$. [**Hint.** See the hint in the previous exercise.]

32. Prove Theorem 5.3.5.

33. Prove Theorem 5.3.2a.

34. Prove Theorem 5.3.2b.

35. Prove Theorem 5.3.2c.

5.4 CHANGE OF BASIS; ORTHOGONAL MATRICES

A basis that is suitable for one problem may not be suitable for another, so it is a common process in the study of vector spaces to change from one basis to another. Because a basis is the vector-space generalization of a coordinate system, changing bases is akin to changing coordinate axes in R^2 and R^3. In this section we shall study various problems relating to changes of basis.

COORDINATE
MATRICES

Recall from Theorem 4.5.1 that if $S = \{v_1, v_2, \ldots, v_n\}$ is a basis for a vector space V, then each vector v in V can be expressed uniquely as a linear combination of the basis vectors, say

$$v = k_1 v_1 + k_2 v_2 + \cdots + k_n v_n$$

The scalars $k_1, k_2, \ldots, k_n$ are the coordinates of v relative to S, and the vector

$$(v)_S = (k_1, k_2, \ldots, k_n)$$

is the coordinate vector of v relative to S. In this section it will be convenient to list the coordinates as entries of an $n \times 1$ matrix. Thus, we define

$$[v]_S = \begin{bmatrix} k_1 \\ k_2 \\ \vdots \\ k_n \end{bmatrix}$$

to be the ***coordinate matrix*** of v relative to S.

Example 1   Let

$$v_1 = (0, 1, 0), \quad v_2 = (-\tfrac{4}{5}, 0, \tfrac{3}{5}), \quad v_3 = (\tfrac{3}{5}, 0, \tfrac{4}{5})$$

We observed in Example 3 of Section 5.3 that $S = \{v_1, v_2, v_3\}$ is a basis for R^3 and that the coordinate vector of $u = (1, 1, 1)$ relative to this basis is

$$(u)_S = (1, -\tfrac{1}{5}, \tfrac{7}{5})$$

Thus, the coordinate matrix of u relative to S is

$$[u]_S = \begin{bmatrix} 1 \\ -\tfrac{1}{5} \\ \tfrac{7}{5} \end{bmatrix} \; \blacktriangle$$

CHANGE OF
BASIS

We now turn to the main problem in this section.

Change of Basis Problem.   If we change the basis for a vector space from some old basis B to some new basis B', how is the old coordinate matrix $[v]_B$ of a vector v related to the new coordinate matrix $[v]_{B'}$?

For simplicity, we will solve this problem for two-dimensional spaces. The solution for n-dimensional spaces is similar and will be left as an exercise. Let

$$B = \{u_1, u_2\} \quad \text{and} \quad B' = \{u_1', u_2'\}$$

be the old and new bases, respectively. We will need the coordinate matrices for the new basis vectors relative to the old basis. Suppose they are

$$[\mathbf{u}'_1]_B = \begin{bmatrix} a \\ b \end{bmatrix} \quad \text{and} \quad [\mathbf{u}'_2]_B = \begin{bmatrix} c \\ d \end{bmatrix} \tag{1}$$

That is,

$$\begin{aligned} \mathbf{u}'_1 &= a\mathbf{u}_1 + b\mathbf{u}_2 \\ \mathbf{u}'_2 &= c\mathbf{u}_1 + d\mathbf{u}_2 \end{aligned} \tag{2}$$

Now let $\mathbf{v}$ be any vector in V and let

$$[\mathbf{v}]_{B'} = \begin{bmatrix} k_1 \\ k_2 \end{bmatrix} \tag{3}$$

be the new coordinate matrix, so that

$$\mathbf{v} = k_1\mathbf{u}'_1 + k_2\mathbf{u}'_2 \tag{4}$$

In order to find the old coordinates of $\mathbf{v}$ we must express $\mathbf{v}$ in terms of the old basis B. To do this, we substitute (2) into (4). This yields

$$\mathbf{v} = k_1(a\mathbf{u}_1 + b\mathbf{u}_2) + k_2(c\mathbf{u}_1 + d\mathbf{u}_2)$$

or

$$\mathbf{v} = (k_1 a + k_2 c)\mathbf{u}_1 + (k_1 b + k_2 d)\mathbf{u}_2$$

Thus, the old coordinate matrix for $\mathbf{v}$ is

$$[\mathbf{v}]_B = \begin{bmatrix} k_1 a + k_2 c \\ k_1 b + k_2 d \end{bmatrix}$$

which can be written as

$$[\mathbf{v}]_B = \begin{bmatrix} a & c \\ b & d \end{bmatrix}\begin{bmatrix} k_1 \\ k_2 \end{bmatrix}$$

or, from (3),

$$[\mathbf{v}]_B = \begin{bmatrix} a & c \\ b & d \end{bmatrix}[\mathbf{v}]_{B'}$$

This equation states that the old coordinate matrix $[\mathbf{v}]_B$ results when we multiply the new coordinate matrix $[\mathbf{v}]_{B'}$ on the left by the matrix

$$P = \begin{bmatrix} a & c \\ b & d \end{bmatrix}$$

The columns of this matrix are the coordinates of the new basis vectors relative to the old basis [see (1)]. Thus, we have the following solution of the change of basis problem.

Solution of the Change of Basis Problem. If we change the basis for a vector space V from some old basis $B = \{\mathbf{u}_1, \mathbf{u}_2, \ldots, \mathbf{u}_n\}$ to some new basis $B' = \{\mathbf{u}'_1, \mathbf{u}'_2, \ldots, \mathbf{u}'_n\}$, then the old coordinate matrix $[\mathbf{v}]_B$ of a vector $\mathbf{v}$ is related to the new coordinate matrix $[\mathbf{v}]_{B'}$ of the same vector $\mathbf{v}$ by the equation

$$[\mathbf{v}]_B = P[\mathbf{v}]_{B'} \tag{5}$$

where the columns of P are the coordinate matrices of the new basis vectors relative to the old basis; that is, the column vectors of P are

$$[\mathbf{u}'_1]_B, [\mathbf{u}'_2]_B, \ldots, [\mathbf{u}'_n]_B$$

TRANSITION MATRICES

The matrix P is called the ***transition matrix*** from B' to B; it can be expressed in terms of its column vectors as

$$P = \left[[\mathbf{u}'_1]_B \mid [\mathbf{u}'_2]_B \mid \cdots \mid [\mathbf{u}'_n]_B \right] \tag{6}$$

Example 2   Consider the bases $B = \{\mathbf{u}_1, \mathbf{u}_2\}$ and $B' = \{\mathbf{u}'_1, \mathbf{u}'_2\}$ for R^2, where

$$\mathbf{u}_1 = (1, 0); \quad \mathbf{u}_2 = (0, 1); \quad \mathbf{u}'_1 = (1, 1); \quad \mathbf{u}'_2 = (2, 1)$$

(a)   Find the transition matrix from B' to B.
(b)   Use (5) to find $[\mathbf{v}]_B$ if

$$[\mathbf{v}]_{B'} = \begin{bmatrix} -3 \\ 5 \end{bmatrix}$$

Solution (a). First we must find the coordinate matrices for the new basis vectors $\mathbf{u}'_1$ and $\mathbf{u}'_2$ relative to the old basis B. By inspection

$$\mathbf{u}'_1 = \mathbf{u}_1 + \mathbf{u}_2$$
$$\mathbf{u}'_2 = 2\mathbf{u}_1 + \mathbf{u}_2$$

so that

$$[\mathbf{u}'_1]_B = \begin{bmatrix} 1 \\ 1 \end{bmatrix} \quad \text{and} \quad [\mathbf{u}'_2]_B = \begin{bmatrix} 2 \\ 1 \end{bmatrix}$$

Thus, the transition matrix from B' to B is

$$P = \begin{bmatrix} 1 & 2 \\ 1 & 1 \end{bmatrix}$$

Solution (b). Using (5) and the transition matrix in part (a),

$$[\mathbf{v}]_B = \begin{bmatrix} 1 & 2 \\ 1 & 1 \end{bmatrix} \begin{bmatrix} -3 \\ 5 \end{bmatrix} = \begin{bmatrix} 7 \\ 2 \end{bmatrix}$$

As a check, we should be able to recover the vector **v** either from $[\mathbf{v}]_B$ or $[\mathbf{v}]_{B'}$. We leave it for the reader to show that $-3\mathbf{u}_1' + 5\mathbf{u}_2' = 7\mathbf{u}_1 + 2\mathbf{u}_2 = \mathbf{v} = (7, 2)$. ▲

Example 3   Consider the vectors $\mathbf{u}_1 = (1, 0)$, $\mathbf{u}_2 = (0, 1)$, $\mathbf{u}_1' = (1, 1)$, $\mathbf{u}_2' = (2, 1)$. In Example 2 we found the transition matrix from the basis $B' = \{\mathbf{u}_1', \mathbf{u}_2'\}$ for R^2 to the basis $B = \{\mathbf{u}_1, \mathbf{u}_2\}$. However, we can just as well ask for the transition matrix from B to B'. To obtain this matrix, we simply change our point of view and regard B' as the old basis and B as the new basis. As usual, the columns of the transition matrix will be the coordinates of the new basis vectors relative to the old basis.

By equating corresponding components and solving the resulting linear system, the reader should be able to show that

$$\mathbf{u}_1 = -\mathbf{u}_1' + \mathbf{u}_2'$$
$$\mathbf{u}_2 = 2\mathbf{u}_1' - \mathbf{u}_2'$$

so that

$$[\mathbf{u}_1]_{B'} = \begin{bmatrix} -1 \\ 1 \end{bmatrix} \quad \text{and} \quad [\mathbf{u}_2]_{B'} = \begin{bmatrix} 2 \\ -1 \end{bmatrix}$$

Thus, the transition matrix from B to B' is

$$Q = \begin{bmatrix} -1 & 2 \\ 1 & -1 \end{bmatrix} \quad \text{▲}$$

If we multiply the transition matrix from B' to B obtained in Example 2 and the transition matrix from B to B' obtained in Example 3, we find

$$PQ = \begin{bmatrix} 1 & 2 \\ 1 & 1 \end{bmatrix} \begin{bmatrix} -1 & 2 \\ 1 & -1 \end{bmatrix} = \begin{bmatrix} 1 & 0 \\ 0 & 1 \end{bmatrix} = I$$

which shows that $Q = P^{-1}$. The following theorem shows that this is not accidental.

Theorem 5.4.1. *If P is the transition matrix from a basis B' to a basis B, then*

(a) P is invertible;
(b) P^{-1} is the transition matrix from B to B'.

(The proof is deferred to the end of the section.) To summarize, if P is the transition matrix from a basis B' to a basis B, then for every vector **v** the following relationships hold:

$$[\mathbf{v}]_B = P[\mathbf{v}]_{B'} \tag{7}$$

$$[\mathbf{v}]_{B'} = P^{-1}[\mathbf{v}]_B \tag{8}$$

**ROTATION
OF COORDINATE
AXES**

Example 4   (Application to Rotation of Coordinate Axes.)   In many problems a rectangular xy-coordinate system is given and a new $x'y'$-coordinate system is obtained by rotating the xy-system counterclockwise about the origin through an angle θ. When this is done, each point Q in the plane has two sets of coordinates: coordinates (x, y) relative to the xy-system and coordinates (x', y') relative to the $x'y'$-system (Figure 1a).

By introducing unit vectors $\mathbf{u}_1$ and $\mathbf{u}_2$ along the positive x and y axes and unit vectors $\mathbf{u}_1'$ and $\mathbf{u}_2'$ along the positive x' and y' axes, we can regard this rotation as a change from an old basis $B = \{\mathbf{u}_1, \mathbf{u}_2\}$ to a new basis $B' = \{\mathbf{u}_1', \mathbf{u}_2'\}$ (Figure 1b). Thus, the new coordinates (x', y') and the old coordinates (x, y) of a point Q will be related by

$$\begin{bmatrix} x' \\ y' \end{bmatrix} = P^{-1} \begin{bmatrix} x \\ y \end{bmatrix} \tag{9}$$

where P is the transition from B' to B. To find P we must determine the coordinate matrices of the new basis vectors $\mathbf{u}_1'$ and $\mathbf{u}_2'$ relative to the old basis. As indicated in Figure 1c, the components of $\mathbf{u}_1'$ in the old basis are $\cos\theta$ and $\sin\theta$ so that

$$[\mathbf{u}_1']_B = \begin{bmatrix} \cos\theta \\ \sin\theta \end{bmatrix}$$

Similarly, from Figure 1d, we see that the components of $\mathbf{u}_2'$ in the old basis are $\cos(\theta + \pi/2) = -\sin\theta$ and $\sin(\theta + \pi/2) = \cos\theta$, so that

$$[\mathbf{u}_2']_B = \begin{bmatrix} -\sin\theta \\ \cos\theta \end{bmatrix}$$

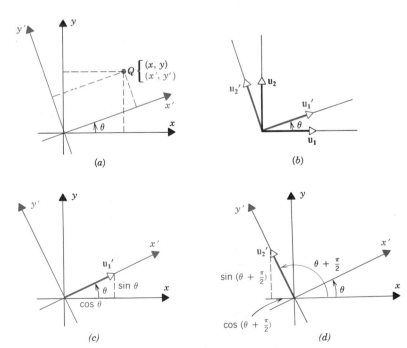

Figure 1

(a)   (b)

(c)   (d)

Thus, the transition matrix from B' to B is

$$P = \begin{bmatrix} \cos \theta & -\sin \theta \\ \sin \theta & \cos \theta \end{bmatrix}$$

and it is easy to verify that

$$P^{-1} = \begin{bmatrix} \cos \theta & \sin \theta \\ -\sin \theta & \cos \theta \end{bmatrix}$$

(Use the method of Example 7 in Section 1.5.) Thus, (8) yields

$$\begin{bmatrix} x' \\ y' \end{bmatrix} = \begin{bmatrix} \cos \theta & \sin \theta \\ -\sin \theta & \cos \theta \end{bmatrix} \begin{bmatrix} x \\ y \end{bmatrix} \tag{10}$$

or equivalently

$$x' = x \cos \theta + y \sin \theta$$
$$y' = -x \sin \theta + y \cos \theta$$

For example, if the axes are rotated $\theta = 45°$, then since

$$\sin 45° = \cos 45° = \frac{1}{\sqrt{2}}$$

equation (10) becomes

$$\begin{bmatrix} x' \\ y' \end{bmatrix} = \begin{bmatrix} \dfrac{1}{\sqrt{2}} & \dfrac{1}{\sqrt{2}} \\ -\dfrac{1}{\sqrt{2}} & \dfrac{1}{\sqrt{2}} \end{bmatrix} \begin{bmatrix} x \\ y \end{bmatrix}$$

Thus, if the old coordinates of a point Q are $(x, y) = (2, -1)$, then

$$\begin{bmatrix} x' \\ y' \end{bmatrix} = \begin{bmatrix} \dfrac{1}{\sqrt{2}} & \dfrac{1}{\sqrt{2}} \\ -\dfrac{1}{\sqrt{2}} & \dfrac{1}{\sqrt{2}} \end{bmatrix} \begin{bmatrix} 2 \\ -1 \end{bmatrix} = \begin{bmatrix} \dfrac{1}{\sqrt{2}} \\ -\dfrac{3}{\sqrt{2}} \end{bmatrix}$$

so the new coordinates of Q are $(x', y') = (1/\sqrt{2}, -3/\sqrt{2})$. ▲

Example 5 (Application to Rotation of Axes in 3-space.) Suppose a rectangular xyz-coordinate system is rotated around its z-axis counterclockwise (looking down the positive z-axis) through an angle θ (Figure 2). If we introduce unit vectors $\mathbf{u}_1$, $\mathbf{u}_2$, and $\mathbf{u}_3$ along the positive x, y, and z axes and unit vectors $\mathbf{u}'_1$, $\mathbf{u}'_2$, and $\mathbf{u}'_3$ along the positive x', y', and z' axes, we can regard the rotation as a change from the old basis $B = \{\mathbf{u}_1, \mathbf{u}_2, \mathbf{u}_3\}$ to the new basis $B' = \{\mathbf{u}'_1, \mathbf{u}'_2, \mathbf{u}'_3\}$. In light of Example 4 it should be evident that

$$[\mathbf{u}'_1]_B = \begin{bmatrix} \cos \theta \\ \sin \theta \\ 0 \end{bmatrix} \quad \text{and} \quad [\mathbf{u}'_2]_B = \begin{bmatrix} -\sin \theta \\ \cos \theta \\ 0 \end{bmatrix}$$

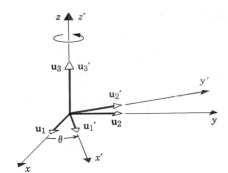

Figure 2

Moreover, since $\mathbf{u}'_3$ extends 1 unit up the positive z'-axis

$$[\mathbf{u}'_3]_B = \begin{bmatrix} 0 \\ 0 \\ 1 \end{bmatrix}$$

Thus, the transition matrix from B' to B is

$$P = \begin{bmatrix} \cos\theta & -\sin\theta & 0 \\ \sin\theta & \cos\theta & 0 \\ 0 & 0 & 1 \end{bmatrix}$$

and the transition matrix from B to B' is

$$P^{-1} = \begin{bmatrix} \cos\theta & \sin\theta & 0 \\ -\sin\theta & \cos\theta & 0 \\ 0 & 0 & 1 \end{bmatrix}$$

(verify). Thus, the new coordinates (x', y', z') of a point Q can be computed from its old coordinates (x, y, z) by

$$\begin{bmatrix} x' \\ y' \\ z' \end{bmatrix} = \begin{bmatrix} \cos\theta & \sin\theta & 0 \\ -\sin\theta & \cos\theta & 0 \\ 0 & 0 & 1 \end{bmatrix}\begin{bmatrix} x \\ y \\ z \end{bmatrix} \quad \triangle \tag{11}$$

CHANGE OF ORTHONORMAL BASIS

The next theorem shows that the transition matrix from one *orthonormal* basis to another has a special property that makes its inverse easy to find.

> **Theorem 5.4.2.** *If P is the transition matrix from one orthonormal basis to another orthonormal basis for an inner product space, then*
> $$P^{-1} = P^t$$

(We omit the proof.)

Example 6 For the transition matrix

$$P = \begin{bmatrix} \cos\theta & -\sin\theta \\ \sin\theta & \cos\theta \end{bmatrix}$$

in Example 4, we found

$$P^{-1} = \begin{bmatrix} \cos\theta & \sin\theta \\ -\sin\theta & \cos\theta \end{bmatrix}$$

by direct computation. However, P is the transition matrix from one orthonormal basis to another, so we can obtain P^{-1} more simply from Theorem 5.4.2.

$$P^{-1} = P^t = \begin{bmatrix} \cos\theta & \sin\theta \\ -\sin\theta & \cos\theta \end{bmatrix}$$

Similarly, the inverse of the matrix P in Example 5 can be found by transposing P (verify). ▲

ORTHOGONAL MATRICES

Matrices whose inverses can be obtained by transposition are sufficiently important that there is some terminology associated with them.

Definition. A square matrix A with the property

$$A^{-1} = A^t$$

is said to be an ***orthogonal matrix***.

To paraphrase this definition, a square matrix A is orthogonal if and only if

$$AA^t = A^tA = I$$

With the terminology in the foregoing definition, Theorem 5.4.2 states that a *transition matrix from one orthonormal basis to another is orthogonal.*

The following result, whose proof is discussed in the exercises, makes it easy to determine when an $n \times n$ matrix A is orthogonal.

Theorem 5.4.3. *The following are equivalent for an $n \times n$ matrix A:*

(a) *A is orthogonal.*
(b) *The row vectors of A form an orthonormal set in R^n with the Euclidean inner product.*
(c) *The column vectors of A form an orthonormal set in R^n with the Euclidean inner product.*

Example 7 Consider the matrix

$$
A = \begin{bmatrix} \dfrac{1}{\sqrt{2}} & \dfrac{1}{\sqrt{2}} & 0 \\[2mm] 0 & 0 & 1 \\[2mm] \dfrac{1}{\sqrt{2}} & -\dfrac{1}{\sqrt{2}} & 0 \end{bmatrix}
$$

The row vectors of A are

$$
\mathbf{r}_1 = \left(\frac{1}{\sqrt{2}}, \frac{1}{\sqrt{2}}, 0 \right), \qquad \mathbf{r}_2 = (0, 0, 1), \qquad \mathbf{r}_3 = \left(\frac{1}{\sqrt{2}}, -\frac{1}{\sqrt{2}}, 0 \right)
$$

Relative to the Euclidean inner product we have

$$
\|\mathbf{r}_1\| = \|\mathbf{r}_2\| = \|\mathbf{r}_3\| = 1
$$

and

$$
\mathbf{r}_1 \cdot \mathbf{r}_2 = \mathbf{r}_2 \cdot \mathbf{r}_3 = \mathbf{r}_1 \cdot \mathbf{r}_3 = 0
$$

so that the row vectors of A form an orthonormal set in R^3. Thus, A is orthogonal and

$$
A^{-1} = A^t = \begin{bmatrix} \dfrac{1}{\sqrt{2}} & 0 & \dfrac{1}{\sqrt{2}} \\[2mm] \dfrac{1}{\sqrt{2}} & 0 & -\dfrac{1}{\sqrt{2}} \\[2mm] 0 & 1 & 0 \end{bmatrix}
$$

(The reader should check that the column vectors of A also form an orthonormal set and that $A^t A = A A^t = I$.) ▲

It is shown in the exercises that the determinant of an orthogonal matrix is always $+1$ or -1. For example, if A is the orthogonal matrix in the foregoing example, then expanding $\det(A)$ by cofactors along the second row yields

$$
\det(A) = \begin{vmatrix} \dfrac{1}{\sqrt{2}} & \dfrac{1}{\sqrt{2}} & 0 \\[2mm] 0 & 0 & 1 \\[2mm] \dfrac{1}{\sqrt{2}} & -\dfrac{1}{\sqrt{2}} & 0 \end{vmatrix} = - \begin{vmatrix} \dfrac{1}{\sqrt{2}} & \dfrac{1}{\sqrt{2}} \\[2mm] \dfrac{1}{\sqrt{2}} & -\dfrac{1}{\sqrt{2}} \end{vmatrix} = 1
$$

Interchanging two rows or columns of A would produce an orthogonal matrix with determinant -1.

ORTHOGONAL COORDINATE TRANSFORMATIONS

In Examples 4 and 5, a rotation of coordinate axes was specified, and we found an algebraic relationship between coordinates in the two coordinate systems. In both R^2 and R^3 that relationship had the form

$$\mathbf{x} = P\mathbf{x}' \tag{12}$$

or equivalently

$$\mathbf{x}' = P^{-1}\mathbf{x}$$

where P was an orthogonal matrix [see (10) and (11)]. Sometimes the reverse problem occurs: An equation of form (12) is given, where P is a 2×2 or 3×3 orthogonal matrix, and it is of interest to determine how the coordinate axes in R^2 or R^3 are related geometrically.

An equation of form (12), where P is an orthogonal matrix, is called an *orthogonal coordinate transformation*. We shall now consider the effect of an orthogonal coordinate transformation

$$\begin{bmatrix} x \\ y \end{bmatrix} = \begin{bmatrix} a_1 & a_2 \\ b_1 & b_2 \end{bmatrix} \begin{bmatrix} x' \\ y' \end{bmatrix} \tag{13}$$

on a rectangular coordinate system in R^2. Suppose that

$$\mathbf{u}_1 = \begin{bmatrix} 1 \\ 0 \end{bmatrix}, \quad \mathbf{u}_2 = \begin{bmatrix} 0 \\ 1 \end{bmatrix}, \quad \mathbf{u}_1' = \begin{bmatrix} a_1 \\ b_1 \end{bmatrix}, \quad \mathbf{u}_2' = \begin{bmatrix} a_2 \\ b_2 \end{bmatrix}$$

are viewed as vectors in a rectangular xy-coordinate system, and let us introduce an $x'y'$-coordinate system with positive x'-axis along $\mathbf{u}_1'$ and positive y'-axis along $\mathbf{u}_2'$ (Figure 3a). Because the 2×2 matrix in (13) is orthogonal, its column vectors $\mathbf{u}_1'$ and $\mathbf{u}_2'$ are orthogonal, which ensures that the x' and y' axes are perpendicular. Since

$$\mathbf{u}_1' = a_1\mathbf{u}_1 + b_1\mathbf{u}_2$$

and

$$\mathbf{u}_2' = a_2\mathbf{u}_1 + b_2\mathbf{u}_2$$

the coefficient matrix

$$P = \begin{bmatrix} a_1 & a_2 \\ b_1 & b_2 \end{bmatrix}$$

in (13) is the transition matrix from the basis $\{\mathbf{u}_1', \mathbf{u}_2'\}$ to the basis $\{\mathbf{u}_1, \mathbf{u}_2\}$. Stated another way, P is the transition matrix from the basis formed by its column vectors to the standard basis.

Geometrically, there are two possible relationships between the xy-coordinate system and the $x'y'$-coordinate system: The $x'y'$-coordinate system can be obtained by rotating the xy-coordinate system about the origin (Figure 3b), or the $x'y'$-coordinate system can be obtained by first reflecting the xy-coordinate system about the x-axis and then rotating the reflected coordinate system about the origin (Figure 3c). The two possibilities can be distinguished by the value of

$$\begin{vmatrix} a_1 & a_2 \\ b_1 & b_2 \end{vmatrix}$$

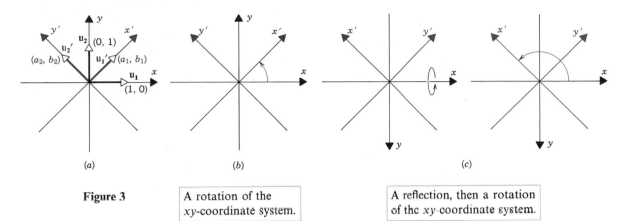

Figure 3

A rotation of the
xy-coordinate system.

A reflection, then a rotation
of the xy-coordinate system.

which is the determinant of the coefficient matrix in (13). The orthogonal coordinate transformation will be a rotation if $\det(P) = 1$ and a reflection followed by a rotation if $\det(P) = -1$. We omit the proof.

The situation in R^3 is similar. An orthogonal coordinate transformation.

$$\begin{bmatrix} x \\ y \\ z \end{bmatrix} = \begin{bmatrix} a_1 & a_2 & a_3 \\ b_1 & b_2 & b_3 \\ c_1 & c_2 & c_3 \end{bmatrix} \begin{bmatrix} x' \\ y' \\ z' \end{bmatrix}$$

is either a rotation about a line through the origin or a reflection in one of the coordinate planes followed by a rotation about a line through the origin, depending on whether the determinant

$$\begin{vmatrix} a_1 & a_2 & a_3 \\ b_1 & b_2 & b_3 \\ c_1 & c_2 & c_3 \end{vmatrix}$$

has a value of 1 or a value of -1, respectively.

Example 8 The matrix

$$P = \begin{bmatrix} 1/\sqrt{2} & -1/\sqrt{2} \\ 1/\sqrt{2} & 1/\sqrt{2} \end{bmatrix}$$

is orthogonal and $\det(P) = 1$ (verify). Thus, the orthogonal coordinate transformation

$$\begin{bmatrix} x \\ y \end{bmatrix} = \begin{bmatrix} 1/\sqrt{2} & -1/\sqrt{2} \\ 1/\sqrt{2} & 1/\sqrt{2} \end{bmatrix} \begin{bmatrix} x' \\ y' \end{bmatrix}$$

is a rotation. As shown in Figure 4, the positive x' and y' axes are along the column vectors

$$\mathbf{u}_1' = \begin{bmatrix} 1/\sqrt{2} \\ 1/\sqrt{2} \end{bmatrix} \quad \text{and} \quad \mathbf{u}_2' = \begin{bmatrix} -1/\sqrt{2} \\ 1/\sqrt{2} \end{bmatrix} \quad \blacktriangle$$

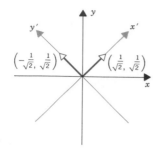

Figure 4

Proof of Theorem 5.4.1. Let Q be the transition matrix from B to B'. We shall show that $PQ = I$ and thus conclude that $Q = P^{-1}$ to complete the proof.

Assume that $B = \{\mathbf{u}_1, \mathbf{u}_2, \ldots, \mathbf{u}_n\}$ and suppose that

$$PQ = \begin{bmatrix} c_{11} & c_{12} & \cdots & c_{1n} \\ c_{21} & c_{22} & \cdots & c_{2n} \\ \vdots & \vdots & & \vdots \\ c_{n1} & c_{n2} & \cdots & c_{nn} \end{bmatrix}$$

From (5)

$$[\mathbf{x}]_B = P[\mathbf{x}]_{B'}$$

and

$$[\mathbf{x}]_{B'} = Q[\mathbf{x}]_B$$

for all $\mathbf{x}$ in V. Multiplying the bottom equation through on the left by P and substituting the top equation gives

$$[\mathbf{x}]_B = PQ[\mathbf{x}]_B \qquad (14)$$

for all $\mathbf{x}$ in V. Letting $\mathbf{x} = \mathbf{u}_1$ in (14) gives

$$\begin{bmatrix} 1 \\ 0 \\ \vdots \\ 0 \end{bmatrix} = \begin{bmatrix} c_{11} & c_{12} & \cdots & c_{1n} \\ c_{21} & c_{22} & \cdots & c_{2n} \\ \vdots & \vdots & & \vdots \\ c_{n1} & c_{n2} & \cdots & c_{nn} \end{bmatrix} \begin{bmatrix} 1 \\ 0 \\ \vdots \\ 0 \end{bmatrix}$$

or

$$\begin{bmatrix} 1 \\ 0 \\ \vdots \\ 0 \end{bmatrix} = \begin{bmatrix} c_{11} \\ c_{21} \\ \vdots \\ c_{n1} \end{bmatrix}$$

Similarly, successively substituting $x = u_2, \ldots, u_n$ in (14) yields

$$\begin{bmatrix} c_{12} \\ c_{22} \\ \vdots \\ c_{n2} \end{bmatrix} = \begin{bmatrix} 0 \\ 1 \\ \vdots \\ 0 \end{bmatrix}, \ldots, \begin{bmatrix} c_{1n} \\ c_{2n} \\ \vdots \\ c_{nn} \end{bmatrix} = \begin{bmatrix} 0 \\ 0 \\ \vdots \\ 1 \end{bmatrix}$$

Therefore, $PQ = I$. ▌

EXERCISE SET 5.4

1. Find the coordinate matrix for $\mathbf{w}$ relative to the basis $S = \{\mathbf{u}_1, \mathbf{u}_2\}$ for R^2.
   (a) $\mathbf{u}_1 = (1, 0)$, $\mathbf{u}_2 = (0, 1)$; $\mathbf{w} = (3, -7)$ (b) $\mathbf{u}_1 = (2, -4)$, $\mathbf{u}_2 = (3, 8)$; $\mathbf{w} = (1, 1)$
   (c) $\mathbf{u}_1 = (1, 1)$, $\mathbf{u}_2 = (0, 2)$; $\mathbf{w} = (a, b)$

2. Find the coordinate matrix for $\mathbf{v}$ relative to $S = \{\mathbf{v}_1, \mathbf{v}_2, \mathbf{v}_3\}$.
   (a) $\mathbf{v} = (2, -1, 3)$; $\mathbf{v}_1 = (1, 0, 0)$, $\mathbf{v}_2 = (2, 2, 0)$, $\mathbf{v}_3 = (3, 3, 3)$
   (b) $\mathbf{v} = (5, -12, 3)$; $\mathbf{v}_1 = (1, 2, 3)$, $\mathbf{v}_2 = (-4, 5, 6)$, $\mathbf{v}_3 = (7, -8, 9)$

3. Find the coordinate matrix for $\mathbf{p}$ relative to $S = \{\mathbf{p}_1, \mathbf{p}_2, \mathbf{p}_3\}$.
   (a) $\mathbf{p} = 4 - 3x + x^2$; $\mathbf{p}_1 = 1$, $\mathbf{p}_2 = x$, $\mathbf{p}_3 = x^2$
   (b) $\mathbf{p} = 2 - x + x^2$; $\mathbf{p}_1 = 1 + x$, $\mathbf{p}_2 = 1 + x^2$, $\mathbf{p}_3 = x + x^2$

4. Find the coordinate matrix for A relative to $S = \{A_1, A_2, A_3, A_4\}$

$$A = \begin{bmatrix} 2 & 0 \\ -1 & 3 \end{bmatrix}, \quad A_1 = \begin{bmatrix} -1 & 1 \\ 0 & 0 \end{bmatrix}, \quad A_2 = \begin{bmatrix} 1 & 1 \\ 0 & 0 \end{bmatrix}, \quad A_3 = \begin{bmatrix} 0 & 0 \\ 1 & 0 \end{bmatrix}, \quad A_4 = \begin{bmatrix} 0 & 0 \\ 0 & 1 \end{bmatrix}$$

5. Consider the coordinate matrices

$$[\mathbf{w}]_S = \begin{bmatrix} 6 \\ -1 \\ 4 \end{bmatrix}, \quad [\mathbf{q}]_S = \begin{bmatrix} 3 \\ 0 \\ 4 \end{bmatrix}, \quad [B]_S = \begin{bmatrix} -8 \\ 7 \\ 6 \\ 3 \end{bmatrix}$$

   (a) Find $\mathbf{w}$ if S is the basis in Exercise 2(a).
   (b) Find $\mathbf{q}$ if S is the basis in Exercise 3(a).
   (c) Find B if S is the basis in Exercise 4.

6. Consider the bases $B = \{\mathbf{u}_1, \mathbf{u}_2\}$ and $B' = \{\mathbf{v}_1, \mathbf{v}_2\}$ for R^2, where

$$\mathbf{u}_1 = \begin{bmatrix} 1 \\ 0 \end{bmatrix}, \quad \mathbf{u}_2 = \begin{bmatrix} 0 \\ 1 \end{bmatrix}, \quad \mathbf{v}_1 = \begin{bmatrix} 2 \\ 1 \end{bmatrix}, \quad \text{and} \quad \mathbf{v}_2 = \begin{bmatrix} -3 \\ 4 \end{bmatrix}$$

   (a) Find the transition matrix from B' to B.
   (b) Find the transition matrix from B to B'.
   (c) Compute the coordinate matrix $[\mathbf{w}]_B$, where

$$\mathbf{w} = \begin{bmatrix} 3 \\ -5 \end{bmatrix}$$

   and use (8) to compute $[\mathbf{w}]_{B'}$.
   (d) Check your work by computing $[\mathbf{w}]_{B'}$ directly.

7. Repeat the directions of Exercise 6 with

$$\mathbf{u}_1 = \begin{bmatrix} 2 \\ 2 \end{bmatrix}, \quad \mathbf{u}_2 = \begin{bmatrix} 4 \\ -1 \end{bmatrix}, \quad \mathbf{v}_1 = \begin{bmatrix} 1 \\ 3 \end{bmatrix}, \quad \mathbf{v}_2 = \begin{bmatrix} -1 \\ -1 \end{bmatrix}$$

8. Consider the bases $B = \{\mathbf{u}_1, \mathbf{u}_2, \mathbf{u}_3\}$ and $B' = \{\mathbf{v}_1, \mathbf{v}_2, \mathbf{v}_3\}$ for R^3, where

$$\mathbf{u}_1 = \begin{bmatrix} -3 \\ 0 \\ -3 \end{bmatrix}, \quad \mathbf{u}_2 = \begin{bmatrix} -3 \\ 2 \\ -1 \end{bmatrix}, \quad \mathbf{u}_3 = \begin{bmatrix} 1 \\ 6 \\ -1 \end{bmatrix}, \quad \mathbf{v}_1 = \begin{bmatrix} -6 \\ -6 \\ 0 \end{bmatrix}, \quad \mathbf{v}_2 = \begin{bmatrix} -2 \\ -6 \\ 4 \end{bmatrix}, \quad \mathbf{v}_3 = \begin{bmatrix} -2 \\ -3 \\ 7 \end{bmatrix}$$

(a) Find the transition matrix from B to B'.

(b) Compute the coordinate matrix $[\mathbf{w}]_B$, where

$$\mathbf{w} = \begin{bmatrix} -5 \\ 8 \\ -5 \end{bmatrix}$$

and use (8) to compute $[\mathbf{w}]_{B'}$.

(c) Check your work by computing $[\mathbf{w}]_{B'}$ directly.

9. Repeat the directions of Exercise 8 with the same vector $\mathbf{w}$, but with

$$\mathbf{u}_1 = \begin{bmatrix} 2 \\ 1 \\ 1 \end{bmatrix}, \quad \mathbf{u}_2 = \begin{bmatrix} 2 \\ -1 \\ 1 \end{bmatrix}, \quad \mathbf{u}_3 = \begin{bmatrix} 1 \\ 2 \\ 1 \end{bmatrix}, \quad \mathbf{v}_1 = \begin{bmatrix} 3 \\ 1 \\ -5 \end{bmatrix}, \quad \mathbf{v}_2 = \begin{bmatrix} 1 \\ 1 \\ -3 \end{bmatrix}, \quad \mathbf{v}_3 = \begin{bmatrix} -1 \\ 0 \\ 2 \end{bmatrix}$$

10. Consider the bases $B = \{\mathbf{p}_1, \mathbf{p}_2\}$ and $B' = \{\mathbf{q}_1, \mathbf{q}_2\}$ for P_1, where

$$\mathbf{p}_1 = 6 + 3x, \quad \mathbf{p}_2 = 10 + 2x, \quad \mathbf{q}_1 = 2, \quad \mathbf{q}_2 = 3 + 2x$$

(a) Find the transition matrix from B' to B.

(b) Find the transition matrix from B to B'.

(c) Compute the coordinate matrix $[\mathbf{p}]_B$, where $\mathbf{p} = -4 + x$, and use (8) to compute $[\mathbf{p}]_{B'}$.

(d) Check your work by computing $[\mathbf{p}]_{B'}$ directly.

11. Let V be the space spanned by $\mathbf{f}_1 = \sin x$ and $\mathbf{f}_2 = \cos x$.

(a) Show that $\mathbf{g}_1 = 2 \sin x + \cos x$ and $\mathbf{g}_2 = 3 \cos x$ form a basis for V.

(b) Find the transition matrix from $B' = \{\mathbf{g}_1, \mathbf{g}_2\}$ to $B = \{\mathbf{f}_1, \mathbf{f}_2\}$.

(c) Find the transition matrix from B to B'.

(d) Compute the coordinate matrix $[\mathbf{h}]_B$, where $\mathbf{h} = 2 \sin x - 5 \cos x$, and use (8) to obtain $[\mathbf{h}]_{B'}$.

(e) Check your work by computing $[\mathbf{h}]_{B'}$ directly.

12. Let a rectangular $x'y'$-coordinate system be obtained by rotating a rectangular xy-coordinate system counterclockwise through the angle $\theta = 3\pi/4$.

(a) Find the $x'y'$-coordinates of the point whose xy-coordinates are $(-2, 6)$.

(b) Find the xy-coordinates of the point whose $x'y'$-coordinates are $(5, 2)$.

13. Repeat Exercise 12 with $\theta = \pi/3$.

14. Let a rectangular $x'y'z'$-coordinate system be obtained by rotating a rectangular xyz-coordinate system counterclockwise about the z-axis (looking down the z-axis) through the angle $\theta = \pi/4$.

(a) Find the $x'y'z'$-coordinates of the point whose xyz-coordinates are $(-1, 2, 5)$.

(b) Find the xyz-coordinates of the point whose $x'y'z'$-coordinates are $(1, 6, -3)$.

15. Repeat Exercise 14 for a rotation of $\theta = \pi/3$ counterclockwise about the *y*-axis (looking along the positive *y*-axis toward the origin).

16. Repeat Exercise 14 for a rotation of $\theta = 3\pi/4$ counterclockwise about the *x*-axis (looking along the positive *x*-axis toward the origin).

17. Use Theorem 5.4.3 to determine which of the following are orthogonal.

(a) $\begin{bmatrix} 1 & 0 \\ 0 & 1 \end{bmatrix}$

(b) $\begin{bmatrix} 1/\sqrt{2} & -1/\sqrt{2} \\ 1/\sqrt{2} & 1/\sqrt{2} \end{bmatrix}$

(c) $\begin{bmatrix} 0 & 1 & 1/\sqrt{2} \\ 1 & 0 & 0 \\ 0 & 0 & 1/\sqrt{2} \end{bmatrix}$

(d) $\begin{bmatrix} -1/\sqrt{2} & 1/\sqrt{6} & 1/\sqrt{3} \\ 0 & -2/\sqrt{6} & 1/\sqrt{3} \\ 1/\sqrt{2} & 1/\sqrt{6} & 1/\sqrt{3} \end{bmatrix}$

(e) $\begin{bmatrix} \frac{1}{2} & \frac{1}{2} & \frac{1}{2} & \frac{1}{2} \\ \frac{1}{2} & -\frac{5}{6} & \frac{1}{6} & \frac{1}{6} \\ \frac{1}{2} & \frac{1}{6} & \frac{1}{6} & -\frac{5}{6} \\ \frac{1}{2} & \frac{1}{6} & -\frac{5}{6} & \frac{1}{6} \end{bmatrix}$

(f) $\begin{bmatrix} 1 & 0 & 0 & 0 \\ 0 & 1/\sqrt{3} & -1/2 & 0 \\ 0 & 1/\sqrt{3} & 0 & 1 \\ 0 & 1/\sqrt{3} & 1/2 & 0 \end{bmatrix}$

18. Find the inverse of those matrices in Exercise 17 that are orthogonal.

19. Show that the following are orthogonal matrices for every value of θ.

(a) $\begin{bmatrix} \cos\theta & -\sin\theta \\ \sin\theta & \cos\theta \end{bmatrix}$

(b) $\begin{bmatrix} \cos\theta & -\sin\theta & 0 \\ \sin\theta & \cos\theta & 0 \\ 0 & 0 & 1 \end{bmatrix}$

20. Find the inverses of the matrices in Exercise 19.

21. Consider the orthogonal coordinate transformation

$$\begin{bmatrix} x \\ y \end{bmatrix} = \begin{bmatrix} -\frac{3}{5} & -\frac{4}{5} \\ \frac{4}{5} & -\frac{3}{5} \end{bmatrix} \begin{bmatrix} x' \\ y' \end{bmatrix}$$

Find (x', y') for the points with the following (x, y) coordinates.
(a) $(2, -1)$ (b) $(4, 2)$ (c) $(-7, -8)$ (d) $(0, 0)$

22. Sketch the *xy*-axes and the *x'y'*-axes for the coordinate transformation in Exercise 21.

23. For which of the following is $\mathbf{x} = P\mathbf{x}'$ a rotation?

(a) $P = \begin{bmatrix} 1/\sqrt{2} & 1/\sqrt{2} \\ -1/\sqrt{2} & 1/\sqrt{2} \end{bmatrix}$

(b) $P = \begin{bmatrix} -1/\sqrt{2} & 1/\sqrt{2} \\ 1/\sqrt{2} & 1/\sqrt{2} \end{bmatrix}$

(c) $P = \begin{bmatrix} \frac{3}{5} & \frac{4}{5} \\ -\frac{4}{5} & \frac{3}{5} \end{bmatrix}$

(d) $P = \begin{bmatrix} -\frac{3}{5} & -\frac{4}{5} \\ -\frac{4}{5} & \frac{3}{5} \end{bmatrix}$

24. Sketch the *xy*-axes and the *x'y'*-axes for the coordinate transformations in Exercise 23.

25. Consider the orthogonal coordinate transformation

$$\begin{bmatrix} x \\ y \\ z \end{bmatrix} = \begin{bmatrix} \frac{4}{5} & -\frac{3}{5} & 0 \\ \frac{3}{5} & \frac{4}{5} & 0 \\ 0 & 0 & 1 \end{bmatrix} \begin{bmatrix} x' \\ y' \\ z' \end{bmatrix}$$

Find (x', y', z') for the points with the following (x, y, z) coordinates.
(a) $(3, 0, -7)$ (b) $(1, 2, 6)$ (c) $(-9, -2, -3)$ (d) $(0, 0, 0)$

26. Sketch the *xyz*-axes and the *x'y'z'*-axes for the coordinate transformation in Exercise 25.

27. For which of the following is $\mathbf{x} = P\mathbf{x}'$ a rotation?

(a) $P = \begin{bmatrix} \frac{4}{5} & 0 & -\frac{3}{5} \\ \frac{3}{5} & 0 & \frac{4}{5} \\ 0 & 1 & 0 \end{bmatrix}$ (b) $P = \begin{bmatrix} \frac{6}{7} & \frac{2}{7} & \frac{3}{7} \\ \frac{2}{7} & \frac{3}{7} & -\frac{6}{7} \\ -\frac{3}{7} & \frac{6}{7} & \frac{2}{7} \end{bmatrix}$

28. Sketch the xyz-axes and the $x'y'z'$-axes for the coordinate transformations in Exercise 27.

29. (a) A rectangular $x'y'z'$-coordinate system is obtained by rotating an xyz-coordinate system counterclockwise about the y-axis through an angle θ (looking along the positive y-axis toward the origin). Find a matrix A such that

$$\begin{bmatrix} x' \\ y' \\ z' \end{bmatrix} = A \begin{bmatrix} x \\ y \\ z \end{bmatrix}$$

where (x, y, z) and (x', y', z') are the coordinates of the same point in the xyz and $x'y'z'$-systems, respectively.
(b) Repeat part (a) for a rotation about the x-axis.

30. A rectangular $x''y''z''$-coordinate system is obtained by first rotating a rectangular xyz-coordinate system $60°$ counterclockwise about the z-axis (looking down the positive z-axis) to obtain an $x'y'z'$-coordinate system, and then rotating the $x'y'z'$-coordinate system $45°$ counterclockwise about the y'-axis (looking along the positive y'-axis toward the origin). Find a matrix A such that

$$\begin{bmatrix} x'' \\ y'' \\ z'' \end{bmatrix} = A \begin{bmatrix} x \\ y \\ z \end{bmatrix}$$

where (x, y, z) and (x'', y'', z'') are the xyz- and $x''y''z''$-coordinates of the same point.

31. Prove that if A is an orthogonal matrix, then A^t is also orthogonal.

32. Prove that an $n \times n$ matrix is orthogonal if and only if its rows form an orthonormal set in R^n with the Euclidean inner product.

33. Use Exercises 31 and 32 to prove that an $n \times n$ matrix is orthogonal if and only if its column vectors form an orthonormal set in R^n with the Euclidean inner product.

34. Prove that if P is an orthogonal matrix, then $\det(P) = 1$ or -1.

SUPPLEMENTARY EXERCISES

1. Let R^4 have the Euclidean inner product.
(a) Find a vector in R^4 that is orthogonal to $\mathbf{u}_1 = (1, 0, 0, 0)$ and $\mathbf{u}_4 = (0, 0, 0, 1)$ and makes equal angles with $\mathbf{u}_2 = (0, 1, 0, 0)$ and $\mathbf{u}_3 = (0, 0, 1, 0)$.
(b) Find a vector $\mathbf{x} = (x_1, x_2, x_3, x_4)$ of length 1 that is orthogonal to $\mathbf{u}_1$ and $\mathbf{u}_4$ above and such that the cosine of the angle between $\mathbf{x}$ and $\mathbf{u}_2$ is twice the cosine of the angle between $\mathbf{x}$ and $\mathbf{u}_3$.

2. (a) Let A be an $m \times n$ matrix with row vectors $\mathbf{r}_1, \mathbf{r}_2, \ldots, \mathbf{r}_m$ and B an $n \times p$ matrix with column vectors $\mathbf{c}_1, \mathbf{c}_2, \ldots, \mathbf{c}_p$. Show that the entry in row i and column j of AB is the Euclidean inner product of $\mathbf{r}_i$ and $\mathbf{c}_j$.

(b) Use the result in (a) to show that if A is an $n \times n$ matrix such that $AA^t = I$, then the row vectors of A form an orthonormal set in R^n with the Euclidean inner product.

(c) Show that if A is an $n \times n$ matrix and $AA^t = I$, then the column vectors of A form an orthonormal set in R^n with the Euclidean inner product. [*Hint.* Show that $A^t A = I$ and apply part (b) to A^t.]

3. Let $A\mathbf{x} = \mathbf{0}$ be a system of m equations in n unknowns. Show that

$$\mathbf{x} = \begin{bmatrix} x_1 \\ x_2 \\ \vdots \\ x_n \end{bmatrix}$$

is a solution of the system if and only if the vector $\mathbf{x} = (x_1, x_2, \ldots, x_n)$ is orthogonal to every row vector of A in the Euclidean inner product on R^n.

4. Use the Cauchy-Schwarz inequality to show that if $a_1, a_2, \ldots, a_n$ are positive real numbers, then

$$(a_1 + a_2 + \cdots + a_n)\left(\frac{1}{a_1} + \frac{1}{a_2} + \cdots + \frac{1}{a_n}\right) \geq n^2$$

5. Show that if $\mathbf{x}$ and $\mathbf{y}$ are vectors in an inner product space and c is any scalar, then

$$\|c\mathbf{x} + \mathbf{y}\|^2 = c^2\|\mathbf{x}\|^2 + 2c\langle\mathbf{x}, \mathbf{y}\rangle + \|\mathbf{y}\|^2$$

6. Let R^3 have the Euclidean inner product. Find two vectors of length 1 that are orthogonal to all three of the vectors $\mathbf{u}_1 = (1, 1, -1)$, $\mathbf{u}_2 = (-2, -1, 2)$, and $\mathbf{u}_3 = (-1, 0, 1)$.

7. Find a weighted Euclidean inner product on R^n such that the vectors

$$\mathbf{v}_1 = (1, 0, 0, \ldots, 0)$$
$$\mathbf{v}_2 = (0, \sqrt{2}, 0, \ldots, 0)$$
$$\mathbf{v}_3 = (0, 0, \sqrt{3}, \ldots, 0)$$
$$\vdots$$
$$\mathbf{v}_n = (0, 0, 0, \ldots, \sqrt{n})$$

form an orthonormal set.

8. Is there a weighted Euclidean inner product on R^2 for which the vectors $(1, 2)$ and $(3, -1)$ form an orthonormal set? Justify your answer.

9. (a) Prove: If $\mathbf{v}$ is orthogonal to $\mathbf{u}_1$ and $\mathbf{u}_2$, then $\mathbf{v}$ is orthogonal to every vector in the subspace $\text{lin}(\mathbf{u}_1, \mathbf{u}_2)$.

(b) Give a geometric interpretation of this result in R^3.

10. If $\mathbf{u}$ and $\mathbf{v}$ are vectors in an inner product space V, then $\mathbf{u}$, $\mathbf{v}$, and $\mathbf{u} - \mathbf{v}$ can be regarded as sides of a "triangle" in V (Figure 1). Prove that the law of cosines holds for any such triangle; that is, $\|\mathbf{u} - \mathbf{v}\|^2 = \|\mathbf{u}\|^2 + \|\mathbf{v}\|^2 - 2\|\mathbf{u}\|\,\|\mathbf{v}\| \cos \theta$, where θ is the angle between $\mathbf{u}$ and $\mathbf{v}$.

Figure 1

11. (a) In R^3 the vectors $(k, 0, 0)$, $(0, k, 0)$, and $(0, 0, k)$ form the edges of a cube with diagonal (k, k, k) (Figure 4 of Section 3.3). Similarly, in R^n the vectors

$$(k, 0, 0, \ldots, 0), \quad (0, k, 0, \ldots, 0) \quad, \ldots, (0, 0, 0, \ldots, k)$$

can be regarded as edges of a "cube" with diagonal $(k, k, k, \ldots, k)$. Show that each of the above edges makes an angle of θ with the diagonal, where

$$\cos \theta = \frac{1}{\sqrt{n}}$$

(b) **(For readers who have studied calculus.)** What happens to the angle θ in part (a) as the dimension of R^n approaches $+\infty$?

12. Let $\mathbf{u}$ and $\mathbf{v}$ be vectors in an inner product space.
(a) Prove that $\|\mathbf{u}\| = \|\mathbf{v}\|$ if and only if $\mathbf{u} + \mathbf{v}$ and $\mathbf{u} - \mathbf{v}$ are orthogonal.
(b) Give a geometric interpretation of this result in R^2 with the Euclidean inner product.

13. Let $\mathbf{u}$ be a vector in an inner product space V, and let $\{\mathbf{v}_1, \mathbf{v}_2, \ldots, \mathbf{v}_n\}$ be an orthonormal basis for V. Show that if α_i is the angle between $\mathbf{u}$ and $\mathbf{v}_i$, then

$$\cos^2 \alpha_1 + \cos^2 \alpha_2 + \cdots + \cos^2 \alpha_n = 1$$

14. Prove: If $\langle \mathbf{u}, \mathbf{v} \rangle_1$ and $\langle \mathbf{u}, \mathbf{v} \rangle_2$ are two inner products on a vector space V, then the quantity $\langle \mathbf{u}, \mathbf{v} \rangle = \langle \mathbf{u}, \mathbf{v} \rangle_1 + \langle \mathbf{u}, \mathbf{v} \rangle_2$ is also an inner product.

15. Show that the inner product on R^n generated by any orthogonal matrix is the Euclidean inner product.

CHAPTER 6

EIGENVALUES, EIGENVECTORS

6.1 EIGENVALUES AND EIGENVECTORS

If A is an $n \times n$ matrix and $\mathbf{x}$ is a vector in R^n, then there is usually no general geometric relationship between the vector $\mathbf{x}$ and the vector $A\mathbf{x}$ (Figure 1a). However, there are often certain nonzero vectors $\mathbf{x}$ such that $\mathbf{x}$ and $A\mathbf{x}$ are scalar multiples of one another (Figure 1b). Such vectors arise naturally in the study of vibrations, electrical systems, genetics, chemical reactions, quantum mechanics, mechanical stress, economics, and geometry. In this section we shall show how to find these vectors, and in later sections we shall touch on some of their applications.

Figure 1　　(a)　　　　　　(b)

**EIGENVALUES
AND
EIGENVECTORS**

> **Definition.** If A is an $n \times n$ matrix, then a nonzero vector $\mathbf{x}$ in R^n is called an *eigenvector* of A if $A\mathbf{x}$ is a scalar multiple of $\mathbf{x}$; that is,
>
> $$A\mathbf{x} = \lambda\mathbf{x}$$
>
> for some scalar λ. The scalar λ is called an *eigenvalue* of A, and $\mathbf{x}$ is said to be an eigenvector of A *corresponding* to λ.

REMARK. The word "eigenvector" is a mixture of German and English. The German prefix "eigen" can be translated as "proper" or "characteristic"; hence eigenvalues are also called *proper values* or *characteristic values*. In the older literature they are sometimes called *latent roots*.

Example 1   The vector $\mathbf{x} = \begin{bmatrix} 1 \\ 2 \end{bmatrix}$ is an eigenvector of

$$A = \begin{bmatrix} 3 & 0 \\ 8 & -1 \end{bmatrix}$$

corresponding to the eigenvalue $\lambda = 3$ since

$$A\mathbf{x} = \begin{bmatrix} 3 & 0 \\ 8 & -1 \end{bmatrix} \begin{bmatrix} 1 \\ 2 \end{bmatrix} = \begin{bmatrix} 3 \\ 6 \end{bmatrix} = 3\mathbf{x} \quad \blacktriangle$$

Eigenvalues and eigenvectors have a useful geometric interpretation in R^2 and R^3. If λ is an eigenvalue of A corresponding to $\mathbf{x}$, then $A\mathbf{x} = \lambda\mathbf{x}$, so that multiplication by A dilates $\mathbf{x}$, contracts $\mathbf{x}$, or reverses the direction of $\mathbf{x}$, depending on the value of λ (Figure 2).

Figure 2   (a) Dilation ($\lambda > 1$).   (b) Contraction ($0 < \lambda < 1$).   (c) Reversal of direction ($\lambda < 0$).

To find the eigenvalues of an $n \times n$ matrix A we rewrite $A\mathbf{x} = \lambda\mathbf{x}$ as

$$A\mathbf{x} = \lambda I\mathbf{x}$$

or equivalently

$$(\lambda I - A)\mathbf{x} = \mathbf{0} \tag{1}$$

For λ to be an eigenvalue, there must be a nonzero solution of this equation. However, by Theorem 4.7.3, Equation (1) will have a nonzero solution if and only if

$$\det(\lambda I - A) = 0$$

CHARACTERISTIC EQUATIONS AND POLYNOMIALS

This is called the *characteristic equation* of A; the scalars satisfying this equation are the eigenvalues of A. When expanded, the determinant $\det(\lambda I - A)$ is a polynomial in λ called the *characteristic polynomial* of A.

It can be shown (Exercise 16) that if A is an $n \times n$ matrix, then the characteristic polynomial of A has degree n and the coefficient of λ^n is 1. Thus, the characteristic polynomial of an $n \times n$ matrix has the form

$$\det(\lambda I - A) = \lambda^n + c_1 \lambda^{n-1} + \cdots + c_n$$

Example 2   Find the eigenvalues of the matrix

$$A = \begin{bmatrix} 3 & 2 \\ -1 & 0 \end{bmatrix}$$

Solution.   Since

$$\lambda I - A = \lambda \begin{bmatrix} 1 & 0 \\ 0 & 1 \end{bmatrix} - \begin{bmatrix} 3 & 2 \\ -1 & 0 \end{bmatrix} = \begin{bmatrix} \lambda - 3 & -2 \\ 1 & \lambda \end{bmatrix}$$

the characteristic polynomial of A is

$$\det(\lambda I - A) = \det \begin{bmatrix} \lambda - 3 & -2 \\ 1 & \lambda \end{bmatrix} = \lambda^2 - 3\lambda + 2$$

and the characteristic equation of A is

$$\lambda^2 - 3\lambda + 2 = 0$$

The solutions of this equation are $\lambda = 1$ and $\lambda = 2$; these are the eigenvalues of A.   ▲

Example 3   Find the eigenvalues of the matrix

$$A = \begin{bmatrix} -2 & -1 \\ 5 & 2 \end{bmatrix}$$

Solution.   Proceeding as in Example 2,

$$\det(\lambda I - A) = \det \begin{bmatrix} \lambda + 2 & 1 \\ -5 & \lambda - 2 \end{bmatrix} = \lambda^2 + 1$$

The eigenvalues of A must therefore satisfy the quadratic equation $\lambda^2 + 1 = 0$. Since the only solutions to this equation are the imaginary numbers $\lambda = i$ and $\lambda = -i$, and since we are assuming in this section that all of our scalars are real numbers, A has no eigenvalues.*   ▲

---

* As we pointed out in Section 4.2, there are some applications that require complex scalars and complex vector spaces. In such cases, matrices are allowed to have complex entries and complex eigenvalues. This will be discussed in Chapter 10; however, until then we will allow real eigenvalues only, and all matrices will have real entries.

Example 4 Find the eigenvalues of

$$A = \begin{bmatrix} 0 & 1 & 0 \\ 0 & 0 & 1 \\ 4 & -17 & 8 \end{bmatrix}$$

Solution. As in the preceding examples

$$\det(\lambda I - A) = \det \begin{bmatrix} \lambda & -1 & 0 \\ 0 & \lambda & -1 \\ -4 & 17 & \lambda - 8 \end{bmatrix} = \lambda^3 - 8\lambda^2 + 17\lambda - 4$$

The eigenvalues of A must therefore satisfy the cubic equation

$$\lambda^3 - 8\lambda^2 + 17\lambda - 4 = 0 \qquad (2)$$

To solve this equation, we shall begin by searching for integer solutions. This task can be greatly simplified by exploiting the fact that all integer solutions (if there are any) to a polynomial equation with integer coefficients

$$\lambda^n + c_1\lambda^{n-1} + \cdots + c_n = 0$$

must be divisors of the constant term, c_n. Thus, the only possible integer solutions of (2) are the divisors of -4, that is, ± 1, ± 2, ± 4. Successively substituting these values in (2) shows that $\lambda = 4$ is an integer solution. As a consequence, $\lambda - 4$ must be a factor of the left side of (2). Dividing $\lambda - 4$ into $\lambda^3 - 8\lambda^2 + 17\lambda - 4$ shows that (2) can be rewritten as

$$(\lambda - 4)(\lambda^2 - 4\lambda + 1) = 0$$

Thus, the remaining solutions of (2) satisfy the quadratic equation

$$\lambda^2 - 4\lambda + 1 = 0$$

which can be solved by the quadratic formula. Thus, the eigenvalues of A are

$$\lambda = 4, \qquad \lambda = 2 + \sqrt{3}, \qquad \text{and} \qquad \lambda = 2 - \sqrt{3} \quad \blacktriangle$$

REMARK. In practical problems, the matrix A is often so large that computing the characteristic equation is not practical. As a result, various approximation methods are used to obtain eigenvalues; some of these are discussed in Chapter 9.

The following theorem summarizes our results so far.

Theorem 6.1.1. *If A is an $n \times n$ matrix and λ is a real number, then the following are equivalent:*
(a) λ is an eigenvalue of A.
(b) The system of equations $(\lambda I - A)\mathbf{x} = \mathbf{0}$ has nontrivial solutions.
(c) There is a nonzero vector $\mathbf{x}$ in R^n such that $A\mathbf{x} = \lambda\mathbf{x}$.
(d) λ is a solution of the characteristic equation $\det(\lambda I - A) = 0$.

FINDING BASES FOR EIGENSPACES

Now that we know how to find eigenvalues, we turn to the problem of finding eigenvectors. The eigenvectors of A corresponding to an eigenvalue λ are the nonzero vectors $\mathbf{x}$ that satisfy $A\mathbf{x} = \lambda\mathbf{x}$. Equivalently, the eigenvectors corresponding to λ are the nonzero vectors in the solution space of $(\lambda I - A)\mathbf{x} = \mathbf{0}$. We call this solution space the *eigenspace* of A corresponding to λ.

Example 5 Find bases for the eigenspaces of

$$A = \begin{bmatrix} 0 & 0 & -2 \\ 1 & 2 & 1 \\ 1 & 0 & 3 \end{bmatrix}$$

Solution. The characteristic equation of A is $\lambda^3 - 5\lambda^2 + 8\lambda - 4 = 0$, or in factored form, $(\lambda - 1)(\lambda - 2)^2 = 0$ (verify); thus, the eigenvalues of A are $\lambda = 1$ and $\lambda = 2$, so there are two eigenspaces of A.

By definition,

$$\mathbf{x} = \begin{bmatrix} x_1 \\ x_2 \\ x_3 \end{bmatrix}$$

is an eigenvector of A corresponding to λ if and only if $\mathbf{x}$ is a nontrivial solution of $(\lambda I - A)\mathbf{x} = \mathbf{0}$, that is, of

$$\begin{bmatrix} \lambda & 0 & 2 \\ -1 & \lambda - 2 & -1 \\ -1 & 0 & \lambda - 3 \end{bmatrix} \begin{bmatrix} x_1 \\ x_2 \\ x_3 \end{bmatrix} = \begin{bmatrix} 0 \\ 0 \\ 0 \end{bmatrix} \tag{3}$$

If $\lambda = 2$, then (3) becomes

$$\begin{bmatrix} 2 & 0 & 2 \\ -1 & 0 & -1 \\ -1 & 0 & -1 \end{bmatrix} \begin{bmatrix} x_1 \\ x_2 \\ x_3 \end{bmatrix} = \begin{bmatrix} 0 \\ 0 \\ 0 \end{bmatrix}$$

Solving this system yields (verify)

$$x_1 = -s \qquad x_2 = t \qquad x_3 = s$$

Thus, the eigenvectors of A corresponding to $\lambda = 2$ are the nonzero vectors of the form

$$\mathbf{x} = \begin{bmatrix} -s \\ t \\ s \end{bmatrix} = \begin{bmatrix} -s \\ 0 \\ s \end{bmatrix} + \begin{bmatrix} 0 \\ t \\ 0 \end{bmatrix} = s\begin{bmatrix} -1 \\ 0 \\ 1 \end{bmatrix} + t\begin{bmatrix} 0 \\ 1 \\ 0 \end{bmatrix}$$

Since

$$\begin{bmatrix} -1 \\ 0 \\ 1 \end{bmatrix} \quad \text{and} \quad \begin{bmatrix} 0 \\ 1 \\ 0 \end{bmatrix}$$

are linearly independent, these vectors form a basis for the eigenspace corresponding to $\lambda = 2$.

If $\lambda = 1$, then (3) becomes

$$\begin{bmatrix} 1 & 0 & 2 \\ -1 & -1 & -1 \\ -1 & 0 & -2 \end{bmatrix} \begin{bmatrix} x_1 \\ x_2 \\ x_3 \end{bmatrix} = \begin{bmatrix} 0 \\ 0 \\ 0 \end{bmatrix}$$

Solving this system yields (verify)

$$x_1 = -2s \qquad x_2 = s \qquad x_3 = s$$

Thus, the eigenvectors corresponding to $\lambda = 1$ are the nonzero vectors of the form

$$\begin{bmatrix} -2s \\ s \\ s \end{bmatrix} = s \begin{bmatrix} -2 \\ 1 \\ 1 \end{bmatrix}$$

so that

$$\begin{bmatrix} -2 \\ 1 \\ 1 \end{bmatrix}$$

is a basis for the eigenspace corresponding to $\lambda = 1$. ▲

EIGENVALUES OF TRIANGULAR MATRICES

We conclude this section with some results that will simplify the computation of eigenvalues in special situations.

Example 6 Find the eigenvalues of the upper triangular matrix

$$A = \begin{bmatrix} a_{11} & a_{12} & a_{13} & a_{14} \\ 0 & a_{22} & a_{23} & a_{24} \\ 0 & 0 & a_{33} & a_{34} \\ 0 & 0 & 0 & a_{44} \end{bmatrix}$$

Solution. Recalling that the determinant of a triangular matrix is the product of the entries on the main diagonal (Theorem 2.2.2), we obtain

$$\det(\lambda I - A) = \det \begin{bmatrix} \lambda - a_{11} & -a_{12} & -a_{13} & -a_{14} \\ 0 & \lambda - a_{22} & -a_{23} & -a_{24} \\ 0 & 0 & \lambda - a_{33} & -a_{34} \\ 0 & 0 & 0 & \lambda - a_{44} \end{bmatrix}$$

$$= (\lambda - a_{11})(\lambda - a_{22})(\lambda - a_{33})(\lambda - a_{44})$$

Thus, the characteristic equation is

$$(\lambda - a_{11})(\lambda - a_{22})(\lambda - a_{33})(\lambda - a_{44}) = 0$$

and the eigenvalues are

$$\lambda = a_{11}, \qquad \lambda = a_{22}, \qquad \lambda = a_{33}, \qquad \lambda = a_{44}$$

which are precisely the diagonal entries of A. ▲

The following general theorem should be evident from the computations in the last example.

Theorem 6.1.2. *If A is an $n \times n$ triangular matrix (upper triangular, lower triangular, or diagonal), then the eigenvalues of A are the entries on the main diagonal of A.*

Example 7 By inspection, the eigenvalues of the lower triangular matrix

$$A = \begin{bmatrix} \dfrac{1}{2} & 0 & 0 \\ -1 & \dfrac{2}{3} & 0 \\ 5 & -8 & -\dfrac{1}{4} \end{bmatrix}$$

are $\lambda = \dfrac{1}{2}$, $\lambda = \dfrac{2}{3}$, and $\lambda = -\dfrac{1}{4}$. ▲

EIGENVALUES OF THE POWERS OF A MATRIX

Once the eigenvalues and eigenvectors of a matrix A are found, it is a simple matter to find the eigenvalues and eigenvectors of any positive integer power of A; for example, if λ is an eigenvalue of A and $\mathbf{x}$ is a corresponding eigenvector, then

$$A^2\mathbf{x} = A(A\mathbf{x}) = A(\lambda\mathbf{x}) = \lambda(A\mathbf{x}) = \lambda(\lambda\mathbf{x}) = \lambda^2\mathbf{x}$$

which shows that λ^2 is an eigenvalue of A^2 and $\mathbf{x}$ is a corresponding eigenvector. In general, we have the following result.

Theorem 6.1.3. *If k is a positive integer, λ is an eigenvalue of a matrix A, and $\mathbf{x}$ is a corresponding eigenvector, then λ^k is an eigenvalue of A^k and $\mathbf{x}$ is a corresponding eigenvector.*

Example 8 In Example 5 we showed that the eigenvalues of

$$A = \begin{bmatrix} 0 & 0 & -2 \\ 1 & 2 & 1 \\ 1 & 0 & 3 \end{bmatrix}$$

and $\lambda = 2$ and $\lambda = 1$, so from Theorem 6.1.3 both $\lambda = 2^7 = 128$ and $\lambda = 1^7 = 1$

are eigenvalues of A^7. We also showed that

$$\begin{bmatrix} -1 \\ 0 \\ 1 \end{bmatrix} \quad \text{and} \quad \begin{bmatrix} 0 \\ 1 \\ 0 \end{bmatrix}$$

are eigenvectors of A corresponding to the eigenvalue $\lambda = 2$, so from Theorem 6.1.3 they are also eigenvectors of A^7 corresponding to $\lambda = 2^7 = 128$. Similarly, the eigenvector

$$\begin{bmatrix} -2 \\ 1 \\ 1 \end{bmatrix}$$

of A corresponding to the eigenvalue $\lambda = 1$ is also an eigenvector of A^7 corresponding to $\lambda = 1^7 = 1$. ▲

EXERCISE SET 6.1

1. Find the characteristic equations of the following matrices:

(a) $\begin{bmatrix} 3 & 0 \\ 8 & -1 \end{bmatrix}$
(b) $\begin{bmatrix} 10 & -9 \\ 4 & -2 \end{bmatrix}$
(c) $\begin{bmatrix} 0 & 3 \\ 4 & 0 \end{bmatrix}$
(d) $\begin{bmatrix} -2 & -7 \\ 1 & 2 \end{bmatrix}$
(e) $\begin{bmatrix} 0 & 0 \\ 0 & 0 \end{bmatrix}$
(f) $\begin{bmatrix} 1 & 0 \\ 0 & 1 \end{bmatrix}$

2. Find the eigenvalues of the matrices in Exercise 1.

3. Find bases for the eigenspaces of the matrices in Exercise 1.

4. Find the characteristic equations of the following matrices:

(a) $\begin{bmatrix} 4 & 0 & 1 \\ -2 & 1 & 0 \\ -2 & 0 & 1 \end{bmatrix}$
(b) $\begin{bmatrix} 3 & 0 & -5 \\ \frac{1}{5} & -1 & 0 \\ 1 & 1 & -2 \end{bmatrix}$
(c) $\begin{bmatrix} -2 & 0 & 1 \\ -6 & -2 & 0 \\ 19 & 5 & -4 \end{bmatrix}$

(d) $\begin{bmatrix} -1 & 0 & 1 \\ -1 & 3 & 0 \\ -4 & 13 & -1 \end{bmatrix}$
(e) $\begin{bmatrix} 5 & 0 & 1 \\ 1 & 1 & 0 \\ -7 & 1 & 0 \end{bmatrix}$
(f) $\begin{bmatrix} 5 & 6 & 2 \\ 0 & -1 & -8 \\ 1 & 0 & -2 \end{bmatrix}$

5. Find the eigenvalues of the matrices in Exercise 4.

6. Find bases for the eigenspaces of the matrices in Exercise 4.

7. Find the characteristic equations of the following matrices:

(a) $\begin{bmatrix} 0 & 0 & 2 & 0 \\ 1 & 0 & 1 & 0 \\ 0 & 1 & -2 & 0 \\ 0 & 0 & 0 & 1 \end{bmatrix}$
(b) $\begin{bmatrix} 10 & -9 & 0 & 0 \\ 4 & -2 & 0 & 0 \\ 0 & 0 & -2 & -7 \\ 0 & 0 & 1 & 2 \end{bmatrix}$

8. Find the eigenvalues of the matrices in Exercise 7.

9. Find bases for the eigenspaces of the matrices in Exercise 7.

10. By inspection, find the eigenvalues of the following matrices:

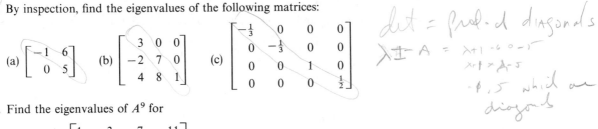

(a) $\begin{bmatrix} -1 & 6 \\ 0 & 5 \end{bmatrix}$ (b) $\begin{bmatrix} 3 & 0 & 0 \\ -2 & 7 & 0 \\ 4 & 8 & 1 \end{bmatrix}$ (c) $\begin{bmatrix} -\frac{1}{3} & 0 & 0 & 0 \\ 0 & -\frac{1}{3} & 0 & 0 \\ 0 & 0 & 1 & 0 \\ 0 & 0 & 0 & \frac{1}{2} \end{bmatrix}$

11. Find the eigenvalues of A^9 for

$$A = \begin{bmatrix} 1 & 3 & 7 & 11 \\ 0 & \frac{1}{2} & 3 & 8 \\ 0 & 0 & 0 & 4 \\ 0 & 0 & 0 & 2 \end{bmatrix}$$

12. Find the eigenvalues and bases for the eigenspaces of A^{25} for

$$A = \begin{bmatrix} -1 & -2 & -2 \\ 1 & 2 & 1 \\ -1 & -1 & 0 \end{bmatrix}$$

13. Let A be a 2×2 matrix, and call a line through the origin of R^2 *invariant* under A if $A\mathbf{x}$ lies on the line when $\mathbf{x}$ does. Find equations for all lines in R^2, if any, that are invariant under the given matrix.

(a) $A = \begin{bmatrix} 4 & -1 \\ 2 & 1 \end{bmatrix}$ (b) $A = \begin{bmatrix} 0 & 1 \\ -1 & 0 \end{bmatrix}$ (c) $A = \begin{bmatrix} 2 & 3 \\ 0 & 2 \end{bmatrix}$

14. Prove that $\lambda = 0$ is an eigenvalue of a matrix A if and only if A is not invertible.

15. Prove that the constant term in the characteristic polynomial of an $n \times n$ matrix A is $(-1)^n \det(A)$. [**Hint.** The constant term is the value of the characteristic polynomial when $\lambda = 0$.]

16. Let A be an $n \times n$ matrix.
    (a) Prove that the characteristic polynomial of A has degree n.
    (b) Prove that the coefficient of λ^n in the characteristic polynomial is 1.

17. Recall that the *trace* of a square matrix A is the sum of the elements on the main diagonal. Show that the characteristic equation of a 2×2 matrix A can be expressed as $\lambda^2 - \text{tr}(A)\lambda + \det(A) = 0$, where $\text{tr}(A)$ is the trace of A.

18. Use the result in Exercise 17 to show that if

$$A = \begin{bmatrix} a & b \\ c & d \end{bmatrix}$$

then the solutions of the characteristic equation of A are

$$\lambda = \tfrac{1}{2}\left[(a + d) \pm \sqrt{(a - d)^2 + 4bc}\right]$$

Use this result to show that A has

(a) two distinct real eigenvalues if $(a - d)^2 + 4bc > 0$ (b) one real eigenvalue if $(a - d)^2 + 4bc = 0$
(c) no real eigenvalues if $(a - d)^2 + 4bc < 0$

19. Let A be the matrix in Exercise 18. Show that if $(a - d)^2 + 4bc > 0$ and $b \neq 0$, then the eigenvectors of A corresponding to the eigenvalues

$$\lambda_1 = \tfrac{1}{2}\Big[(a + d) + \sqrt{(a - d)^2 + 4bc}\,\Big] \qquad \text{and} \qquad \lambda_2 = \tfrac{1}{2}\Big[(a + d) - \sqrt{(a - d)^2 + 4bc}\,\Big]$$

are

$$\begin{bmatrix} -b \\ a - \lambda_1 \end{bmatrix} \qquad \text{and} \qquad \begin{bmatrix} -b \\ a - \lambda_2 \end{bmatrix}$$

respectively.

20. Prove: If a, b, c, and d are integers such that $a + b = c + d$, then

$$A = \begin{bmatrix} a & b \\ c & d \end{bmatrix}$$

has integer eigenvalues, namely, $\lambda_1 = a + b$ and $\lambda_2 = a - c$. [**Hint.** See Exercise 18.]

6.2 DIAGONALIZATION

In this section we shall be concerned with the problem of finding a basis for R^n that consists of eigenvectors of a given $n \times n$ matrix A. Such bases can be used to study geometric properties of A and to simplify various numerical computations involving A. These bases are also of physical significance in a wide variety of applications, some of which will be considered later in this text.

THE MATRIX DIAGONALIZATION PROBLEM

Our first objective in this section is to show that the following two problems, which on the surface seem quite different, are actually equivalent.

> *The Eigenvector Problem.* Given an $n \times n$ matrix A, does there exist a basis for R^n consisting of eigenvectors of A?

> *The Diagonalization Problem (Matrix Form).* Given an $n \times n$ matrix A, does there exist an invertible matrix P such that $P^{-1}AP$ is a diagonal matrix?

The latter problem suggests the following terminology.

> **Definition.** A square matrix A is called **diagonalizable** if there is an invertible matrix P such that $P^{-1}AP$ is a diagonal matrix; the matrix P is said to **diagonalize** A.

The following theorem shows that the eigenvector problem and the diagonalization problem are equivalent.

Theorem 6.2.1. *If A is an $n \times n$ matrix, then the following are equivalent.*
(a) A is diagonalizable.
(b) A has n linearly independent eigenvectors.

Proof (a) $\Rightarrow$ (b). Since A is assumed diagonalizable, there is an invertible matrix

$$P = \begin{bmatrix} p_{11} & p_{12} & \cdots & p_{1n} \\ p_{21} & p_{22} & \cdots & p_{2n} \\ \vdots & \vdots & & \vdots \\ p_{n1} & p_{n2} & \cdots & p_{nn} \end{bmatrix}$$

such that $P^{-1}AP$ is diagonal, say $P^{-1}AP = D$, where

$$D = \begin{bmatrix} \lambda_1 & 0 & \cdots & 0 \\ 0 & \lambda_2 & \cdots & 0 \\ \vdots & \vdots & & \vdots \\ 0 & 0 & \cdots & \lambda_n \end{bmatrix}$$

It follows from the formula $P^{-1}AP = D$ that $AP = PD$; that is,

$$AP = \begin{bmatrix} p_{11} & p_{12} & \cdots & p_{1n} \\ p_{21} & p_{22} & \cdots & p_{2n} \\ \vdots & \vdots & & \vdots \\ p_{n1} & p_{n2} & \cdots & p_{nn} \end{bmatrix} \begin{bmatrix} \lambda_1 & 0 & \cdots & 0 \\ 0 & \lambda_2 & \cdots & 0 \\ \vdots & \vdots & & \vdots \\ 0 & 0 & \cdots & \lambda_n \end{bmatrix} = \begin{bmatrix} \lambda_1 p_{11} & \lambda_2 p_{12} & & \lambda_n p_{1n} \\ \lambda_1 p_{21} & \lambda_2 p_{22} & \cdots & \lambda_n p_{2n} \\ \vdots & \vdots & & \vdots \\ \lambda_1 p_{n1} & \lambda_2 p_{n2} & \cdots & \lambda_n p_{nn} \end{bmatrix}$$

(1)

If we now let $\mathbf{p}_1, \mathbf{p}_2, \ldots, \mathbf{p}_n$ denote the column vectors of P, then from (1) the successive columns of AP are $\lambda_1 \mathbf{p}_1, \lambda_2 \mathbf{p}_2, \ldots, \lambda_n \mathbf{p}_n$. However, from Example 9 of Section 1.4 the sucessive columns of AP are $A\mathbf{p}_1, A\mathbf{p}_2, \ldots, A\mathbf{p}_n$. Thus, we must have

$$A\mathbf{p}_1 = \lambda_1 \mathbf{p}_1, \quad A\mathbf{p}_2 = \lambda_2 \mathbf{p}_2, \quad \ldots, \quad A\mathbf{p}_n = \lambda_n \mathbf{p}_n \tag{2}$$

Since P is invertible, its column vectors are all nonzero; thus, it follows from (2) that $\lambda_1, \lambda_2, \ldots, \lambda_n$ are eigenvaluesof A, and $\mathbf{p}_1, \mathbf{p}_2, \ldots, \mathbf{p}_n$ are corresponding eigenvectors. Since P is invertible, it follows from Theorem 4.7.3 that $\mathbf{p}_1, \mathbf{p}_2, \ldots, \mathbf{p}_n$ are linearly independent. Thus, A has n linearly independent eigenvectors.

(b) $\Rightarrow$ (a). Assume that A has n linearly independent eigenvectors, $\mathbf{p}_1, \mathbf{p}_2, \ldots, \mathbf{p}_n$, with corresponding eigenvalues $\lambda_1, \lambda_2, \ldots, \lambda_n$, and let

$$P = \begin{bmatrix} p_{11} & p_{12} & \cdots & p_{1n} \\ p_{21} & p_{22} & \cdots & p_{2n} \\ \vdots & \vdots & & \vdots \\ p_{n1} & p_{n2} & \cdots & p_{nn} \end{bmatrix}$$

be the matrix whose column vectors are $\mathbf{p}_1, \mathbf{p}_2, \ldots, \mathbf{p}_n$. By Example 9 in Section 1.4, the columns of the product AP are

$$A\mathbf{p}_1, A\mathbf{p}_2, \ldots, A\mathbf{p}_n$$

But

$$A\mathbf{p}_1 = \lambda_1\mathbf{p}_1, \quad A\mathbf{p}_2 = \lambda_2\mathbf{p}_2, \quad \ldots, \quad A\mathbf{p}_n = \lambda_n\mathbf{p}_n$$

so that

$$AP = \begin{bmatrix} \lambda_1 p_{11} & \lambda_2 p_{12} & \cdots & \lambda_n p_{1n} \\ \lambda_1 p_{21} & \lambda_2 p_{22} & \cdots & \lambda_n p_{2n} \\ \vdots & \vdots & & \vdots \\ \lambda_1 p_{n1} & \lambda_2 p_{n2} & \cdots & \lambda_n p_{nn} \end{bmatrix}$$

$$= \begin{bmatrix} p_{11} & p_{12} & \cdots & p_{1n} \\ p_{21} & p_{22} & \cdots & p_{2n} \\ \vdots & \vdots & & \vdots \\ p_{n1} & p_{n2} & \cdots & p_{nn} \end{bmatrix} \begin{bmatrix} \lambda_1 & 0 & \cdots & 0 \\ 0 & \lambda_2 & \cdots & 0 \\ \vdots & \vdots & & \vdots \\ 0 & 0 & \cdots & \lambda_n \end{bmatrix} = PD \qquad (3)$$

where D is the diagonal matrix having the eigenvalues $\lambda_1, \lambda_2, \ldots, \lambda_n$ on the main diagonal. Since the column vectors of P are linearly independent, P is invertible; thus, (3) can be rewritten as $P^{-1}AP = D$; that is, A is diagonalizable. ∎

PROCEDURE FOR DIAGONALIZING A MATRIX

From the foregoing proof we obtain the following procedure for diagonalizing a diagonalizable $n \times n$ matrix A.

Step 1. Find n linearly independent eigenvectors of A, say, $\mathbf{p}_1, \mathbf{p}_2, \ldots, \mathbf{p}_n$.

Step 2. Form the matrix P having $\mathbf{p}_1, \mathbf{p}_2, \ldots, \mathbf{p}_n$ as its column vectors.

Step 3. The matrix $P^{-1}AP$ will then be diagonal with $\lambda_1, \lambda_2, \ldots, \lambda_n$ as its successive diagonal entries, where λ_i is the eigenvalue corresponding to $\mathbf{p}_i$, $i = 1, 2, \ldots, n$.

Example 1 Find a matrix P that diagonalizes

$$A = \begin{bmatrix} 0 & 0 & -2 \\ 1 & 2 & 1 \\ 1 & 0 & 3 \end{bmatrix}$$

Solution. From Example 5 of the previous section the eigenvalues of A are $\lambda = 1$ and $\lambda = 2$. Also from that example, the vectors

$$\mathbf{p}_1 = \begin{bmatrix} -1 \\ 0 \\ 1 \end{bmatrix} \quad \text{and} \quad \mathbf{p}_2 = \begin{bmatrix} 0 \\ 1 \\ 0 \end{bmatrix}$$

form a basis for the eigenspace corresponding to $\lambda = 2$ and

$$\mathbf{p}_3 = \begin{bmatrix} -2 \\ 1 \\ 1 \end{bmatrix}$$

is a basis for the eigenspace corresponding to $\lambda = 1$. It is easy to verify that $\{\mathbf{p}_1, \mathbf{p}_2, \mathbf{p}_3\}$ is linearly independent, so that

$$P = \begin{bmatrix} -1 & 0 & -2 \\ 0 & 1 & 1 \\ 1 & 0 & 1 \end{bmatrix}$$

diagonalizes A. As a check, the reader should verify that

$$P^{-1}AP = \begin{bmatrix} 1 & 0 & 2 \\ 1 & 1 & 1 \\ -1 & 0 & -1 \end{bmatrix}\begin{bmatrix} 0 & 0 & -2 \\ 1 & 2 & 1 \\ 1 & 0 & 3 \end{bmatrix}\begin{bmatrix} -1 & 0 & -2 \\ 0 & 1 & 1 \\ 1 & 0 & 1 \end{bmatrix} = \begin{bmatrix} 2 & 0 & 0 \\ 0 & 2 & 0 \\ 0 & 0 & 1 \end{bmatrix}$$

There is no preferred order for the columns of P. Since the ith diagonal entry of $P^{-1}AP$ is an eigenvalue for the ith column vector of P, changing the order of the columns of P just changes the order of the eigenvalues on the diagonal of $P^{-1}AP$. Thus, had we written

$$P = \begin{bmatrix} -1 & -2 & 0 \\ 0 & 1 & 1 \\ 1 & 1 & 0 \end{bmatrix}$$

in the last example, we would have obtained

$$P^{-1}AP = \begin{bmatrix} 2 & 0 & 0 \\ 0 & 1 & 0 \\ 0 & 0 & 2 \end{bmatrix}$$

Example 2   The characteristic equation of

$$A = \begin{bmatrix} -3 & 2 \\ -2 & 1 \end{bmatrix}$$

is

$$\det(\lambda I - A) = \det\begin{bmatrix} \lambda + 3 & -2 \\ 2 & \lambda - 1 \end{bmatrix} = (\lambda + 1)^2 = 0$$

Thus, $\lambda = -1$ is the only eigenvalue of A; the eigenvectors corresponding to $\lambda = -1$ are the solutions of $(-I - A)\mathbf{x} = \mathbf{0}$, that is, of

$$2x_1 - 2x_2 = 0$$
$$2x_1 - 2x_2 = 0$$

The solutions of this system are $x_1 = t$, $x_2 = t$ (verify); hence the eigenspace consists of all vectors of the form

$$\begin{bmatrix} t \\ t \end{bmatrix} = t \begin{bmatrix} 1 \\ 1 \end{bmatrix}$$

Since this space is 1-dimensional, A does not have two linearly independent eigenvectors and is therefore not diagonalizable. ▲

CONDITIONS FOR DIAGONALIZ-ABILITY

In many applications it is not important to actually compute a matrix P that diagonalizes a matrix A. Rather, one is concerned only with knowing whether A is diagonalizable, and if so, what the diagonal matrix is. Often, this information can be ascertained directly from the eigenvalues without going through the work of computing eigenvectors. To see why this is so we will need the following theorem, whose proof is deferred to the end of the section.

Theorem 6.2.2. *If $\mathbf{v}_1, \mathbf{v}_2, \ldots, \mathbf{v}_k$ are eigenvectors of A corresponding to distinct eigenvalues $\lambda_1, \lambda_2, \ldots, \lambda_k$, then $\{\mathbf{v}_1, \mathbf{v}_2, \ldots, \mathbf{v}_k\}$ is a linearly independent set.*

As a consequence of this theorem, we obtain the following useful result.

Theorem 6.2.3. *If an $n \times n$ matrix A has n distinct eigenvalues, then A is diagonalizable.*

Proof. If $\mathbf{v}_1, \mathbf{v}_2, \ldots, \mathbf{v}_n$ are eigenvectors corresponding to the distinct eigenvalues $\lambda_1, \lambda_2, \ldots, \lambda_n$, then by Theorem 6.2.2, $\mathbf{v}_1, \mathbf{v}_2, \ldots, \mathbf{v}_n$ are linearly independent. Thus, A is diagonalizable by Theorem 6.2.1. ▮

Example 3 We saw in Example 4 of the previous section that

$$A = \begin{bmatrix} 0 & 1 & 0 \\ 0 & 0 & 1 \\ 4 & -17 & 8 \end{bmatrix}$$

has three distinct eigenvalues, $\lambda = 4$, $\lambda = 2 + \sqrt{3}$, $\lambda = 2 - \sqrt{3}$. Therefore, A is diagonalizable. Further,

$$P^{-1}AP = \begin{bmatrix} 4 & 0 & 0 \\ 0 & 2 + \sqrt{3} & 0 \\ 0 & 0 & 2 - \sqrt{3} \end{bmatrix}$$

for some invertible matrix P. If desired, the matrix P can be found using the method shown in Example 1 of this section. ▲

REMARK. Theorem 6.2.2 is a special case of a more general result: Suppose that $\lambda_1, \lambda_2, \ldots, \lambda_k$ are distinct eigenvalues, and we choose a linearly independent set in each of the corresponding eigenspaces. If we then merge all these vectors into a single set, the result will still be a linearly independent set. For example, if we choose three linearly independent vectors from one eigenspace and two linearly independent vectors from another eigenspace, then the five vectors together form a linearly independent set. We omit the proof.

Example 4 The converse of Theorem 6.2.3 is false; that is, an $n \times n$ matrix A may be diagonalizable even if it does not have n distinct eigenvalues. For example, the only eigenvalue of the diagonal matrix

$$A = \begin{bmatrix} 3 & 0 \\ 0 & 3 \end{bmatrix}$$

is $\lambda = 3$ (see Theorem 6.1.2). Yet A is obviously diagonalizable; it is diagonalized by $P = I$, since

$$P^{-1}AP = I^{-1}AI = A = \begin{bmatrix} 3 & 0 \\ 0 & 3 \end{bmatrix} \quad \blacktriangle$$

REMARK. There are theorems studied in more advanced courses that can be used to determine whether an $n \times n$ matrix with fewer than n distinct eigenvalues is diagonalizable. However, even those theorems do not provide a simple *computational* procedure for making that determination.

COMPUTING POWERS OF A MATRIX

There are numerous problems in applied mathematics that require the computation of high powers of a square matrix. We shall conclude this section by showing how diagonalization can be used to simplify such computations for diagonalizable matrices.

If A is an $n \times n$ matrix and P is an invertible matrix, then

$$(P^{-1}AP)^2 = P^{-1}APP^{-1}AP = P^{-1}AIAP = P^{-1}A^2P$$

More generally, for any positive integer k

$$(P^{-1}AP)^k = P^{-1}A^kP \tag{4}$$

It follows from this equation that if A is diagonalizable, and $P^{-1}AP = D$ is a diagonal matrix, then

$$P^{-1}A^kP = (P^{-1}AP)^k = D^k \tag{5}$$

Solving this equation for A^k yields

$$A^k = PD^kP^{-1} \tag{6}$$

This last equation expresses the kth power of A in terms of the kth power of the diagonal matrix D. But D^k is easy to compute; for example, if

$$D = \begin{bmatrix} d_1 & 0 & \cdots & 0 \\ 0 & d_2 & \cdots & 0 \\ \vdots & \vdots & & \vdots \\ 0 & 0 & \cdots & d_n \end{bmatrix}$$

then

$$D^k = \begin{bmatrix} d_1^k & 0 & \cdots & 0 \\ 0 & d_2^k & \cdots & 0 \\ \vdots & \vdots & & \vdots \\ 0 & 0 & \cdots & d_n^k \end{bmatrix}$$

Example 5   Use (6) to find A^{13}, where

$$A = \begin{bmatrix} 0 & 0 & -2 \\ 1 & 2 & 1 \\ 1 & 0 & 3 \end{bmatrix}$$

Solution. We showed in Example 1 that the matrix A is diagonalized by

$$P = \begin{bmatrix} -1 & 0 & -2 \\ 0 & 1 & 1 \\ 1 & 0 & 1 \end{bmatrix}$$

and that

$$D = P^{-1}AP = \begin{bmatrix} 2 & 0 & 0 \\ 0 & 2 & 0 \\ 0 & 0 & 1 \end{bmatrix}$$

Thus, from (6)

$$A^{13} = PD^{13}P^{-1} = \begin{bmatrix} -1 & 0 & -2 \\ 0 & 1 & 1 \\ 1 & 0 & 1 \end{bmatrix} \begin{bmatrix} 2^{13} & 0 & 0 \\ 0 & 2^{13} & 0 \\ 0 & 0 & 1^{13} \end{bmatrix} \begin{bmatrix} 1 & 0 & 2 \\ 1 & 1 & 1 \\ -1 & 0 & -1 \end{bmatrix} \quad (7)$$

$$= \begin{bmatrix} -8190 & 0 & -16382 \\ 8191 & 8192 & 8191 \\ 8191 & 0 & 16383 \end{bmatrix} \blacktriangle$$

REMARK.   With the method in the last example, most of the work is in diagonalizing A. Once that work is done, it can be used to compute any power of A. Thus, to compute A^{1000} we need only change the exponents from 13 to 1000 in (7). The higher the power of A, the more efficient the method becomes.

OPTIONAL

We conclude this section with a proof of Theorem 6.2.2.

Proof. Let $v_1, v_2, \ldots, v_k$ be eigenvectors of A corresponding to distinct eigenvalues $\lambda_1, \lambda_2, \ldots, \lambda_k$. We shall assume that $v_1, v_2, \ldots, v_k$ are linearly dependent and obtain a contradiction. We can then conclude that $v_1, v_2, \ldots, v_k$ are linearly independent.

Since an eigenvector is nonzero by definition, $\{v_1\}$ is linearly independent. Let r be the largest integer such that $\{v_1, v_2, \ldots, v_r\}$ is linearly independent. Since we are assuming that $\{v_1, v_2, \ldots, v_k\}$ is linearly dependent, r satisfies $1 \le r < k$. Moreover, by definition of r, $\{v_1, v_2, \ldots, v_{r+1}\}$ is linearly dependent. Thus, there are scalars $c_1, c_2, \ldots, c_{r+1}$, not all zero, such that

$$c_1 v_1 + c_2 v_2 + \cdots + c_{r+1} v_{r+1} = 0 \tag{8}$$

Multiplying both sides of (8) by A and using

$$A v_1 = \lambda_1 v_1, \quad A v_2 = \lambda_2 v_2, \quad \ldots, \quad A v_{r+1} = \lambda_{r+1} v_{r+1}$$

we obtain

$$c_1 \lambda_1 v_1 + c_2 \lambda_2 v_2 + \cdots \mid c_{r+1} \lambda_{r+1} v_{r+1} = 0 \tag{9}$$

Multiplying both sides of (8) by λ_{r+1} and subtracting the resulting equation from (9) yields

$$c_1 (\lambda_1 - \lambda_{r+1}) v_1 + c_2 (\lambda_2 - \lambda_{r+1}) v_2 + \cdots + c_r (\lambda_r - \lambda_{r+1}) v_r = 0$$

Since $\{v_1, v_2, \ldots, v_r\}$ is a linearly independent sct, this equation implies that

$$c_1 (\lambda_1 - \lambda_{r+1}) = c_2 (\lambda_2 - \lambda_{r+1}) = \cdots = c_r (\lambda_r - \lambda_{r+1}) = 0$$

and since $\lambda_1, \lambda_2, \ldots, \lambda_{r+1}$ are distinct, it follows that

$$c_1 = c_2 = \cdots = c_r = 0 \tag{10}$$

Substituting these values in (8) yields

$$c_{r+1} v_{r+1} = 0$$

Since the eigenvector v_{r+1} is nonzero, it follows that

$$c_{r+1} = 0 \tag{11}$$

Equations (10) and (11) contradict the fact that $c_1, c_2, \ldots, c_{r+1}$ are not all zero; this completes the proof. ▮

EXERCISE SET 6.2

Show that the matrices in Exercises 1–4 are not diagonalizable.

1. $\begin{bmatrix} 2 & 0 \\ 1 & 2 \end{bmatrix}$ **2.** $\begin{bmatrix} 2 & -3 \\ 1 & -1 \end{bmatrix}$ **3.** $\begin{bmatrix} 3 & 0 & 0 \\ 0 & 2 & 0 \\ 0 & 1 & 2 \end{bmatrix}$ **4.** $\begin{bmatrix} -1 & 0 & 1 \\ -1 & 3 & 0 \\ -4 & 13 & -1 \end{bmatrix}$

In Exercises 5–8 find a matrix P that diagonalizes A, and determine $P^{-1}AP$.

5. $A = \begin{bmatrix} -14 & 12 \\ -20 & 17 \end{bmatrix}$ **6.** $A = \begin{bmatrix} 1 & 0 \\ 6 & -1 \end{bmatrix}$ **7.** $A = \begin{bmatrix} 1 & 0 & 0 \\ 0 & 1 & 1 \\ 0 & 1 & 1 \end{bmatrix}$ **8.** $A = \begin{bmatrix} 2 & 0 & -2 \\ 0 & 3 & 0 \\ 0 & 0 & 3 \end{bmatrix}$

In Exercises 9–14 determine whether A is diagonalizable. If so, find a matrix P that diagonalizes A, and determine $P^{-1}AP$.

9. $A = \begin{bmatrix} 19 & -9 & -6 \\ 25 & -11 & -9 \\ 17 & -9 & -4 \end{bmatrix}$ **10.** $A = \begin{bmatrix} -1 & 4 & -2 \\ -3 & 4 & 0 \\ -3 & 1 & 3 \end{bmatrix}$ **11.** $A = \begin{bmatrix} 5 & 0 & 0 \\ 1 & 5 & 0 \\ 0 & 1 & 5 \end{bmatrix}$

12. $A = \begin{bmatrix} 0 & 0 & 0 \\ 0 & 0 & 0 \\ 3 & 0 & 1 \end{bmatrix}$ **13.** $A = \begin{bmatrix} -2 & 0 & 0 & 0 \\ 0 & -2 & 0 & 0 \\ 0 & 0 & 3 & 0 \\ 0 & 0 & 1 & 3 \end{bmatrix}$ **14.** $A = \begin{bmatrix} -2 & 0 & 0 & 0 \\ 0 & -2 & 5 & -5 \\ 0 & 0 & 3 & 0 \\ 0 & 0 & 0 & 3 \end{bmatrix}$

15. Use the method of Example 5 to compute A^{10}, where

$$A = \begin{bmatrix} 1 & 0 \\ -1 & 2 \end{bmatrix}$$

16. Use the method of Example 5 to compute A^{11}, where

$$A = \begin{bmatrix} -1 & 7 & -1 \\ 0 & 1 & 0 \\ 0 & 15 & -2 \end{bmatrix}$$

17. In each part compute the stated power of

$$A = \begin{bmatrix} 1 & -2 & 8 \\ 0 & -1 & 0 \\ 0 & 0 & -1 \end{bmatrix}$$

(a) A^{1000} (b) A^{-1000} (c) A^{2301} (d) A^{-2301}

18. Find A^n if n is a positive integer and

$$A = \begin{bmatrix} 3 & -1 & 0 \\ -1 & 2 & -1 \\ 0 & -1 & 3 \end{bmatrix}$$

19. Let

$$A = \begin{bmatrix} a & b \\ c & d \end{bmatrix}$$

Show:
(a) A is diagonalizable if $(a - d)^2 + 4bc > 0$.
(b) A is not diagonalizable if $(a - d)^2 + 4bc < 0$.
  [*Hint.* See Exercises 18 and 19 of Section 6.1]

20. In the case where the matrix A in Exercise 19 is diagonalizable, find a matrix P that diagonalizes A.

21. Show that if A is a diagonalizable matrix, then the rank of A is the number of nonzero eigenvalues of A.

6.3 ORTHOGONAL DIAGONALIZATION; SYMMETRIC MATRICES

In this section we shall be concerned with the problem of finding an orthonormal basis for R^n with the Euclidean inner product that consists of eigenvectors of a given $n \times n$ matrix A. This will lead us to investigate an important new class of matrices, called symmetric matrices. Such matrices play an important role in numerous applications.

THE MATRIX ORTHOGONAL DIAGONAL-IZATION PROBLEM

Our first objective in this section is to show that the following two problems are equivalent.

The Orthonormal Eigenvector Problem. Given an $n \times n$ matrix A, does there exist an orthonormal basis for R^n with the Euclidean inner product consisting of eigenvectors of A?

The Orthogonal Diagonalization Problem (Matrix Form). Given an $n \times n$ matrix A, does there exist an orthogonal matrix P such that the matrix $P^{-1}AP(=P^tAP)$ is diagonal?

The latter problem suggests the following terminology.

Definition. A square matrix A is called ***orthogonally diagonalizable*** if there is an orthogonal matrix P such that $P^{-1}AP(=P^tAP)$ is a diagonal matrix; the matrix P is said to ***orthogonally diagonalize*** A.

Before proceeding further, we note that there is no hope of solving the orthogonal diagonalization problem unless A satisfies a special condition, namely,

$$A = A^t$$

To see why this is so, suppose that

$$P^tAP = D \tag{1}$$

where P is an orthogonal matrix and D is a diagonal matrix. Since P is orthogonal, $PP^t = P^tP = I$, so it follows that (1) can be rewritten as

$$A = PDP^t \tag{2}$$

Since D is a diagonal matrix, we have $D = D^t$, so transposing both sides of (2) yields

$$A^t = (PDP^t)^t = (P^t)^tD^tP^t = PDP^t = A$$

We make the following definition.

Definition. A square matrix A is called ***symmetric*** if $A = A^t$.

Example 1   If

$$A = \begin{bmatrix} 1 & 4 & 5 \\ 4 & -3 & 0 \\ 5 & 0 & 7 \end{bmatrix}$$

then

$$A^t = \begin{bmatrix} 1 & 4 & 5 \\ 4 & -3 & 0 \\ 5 & 0 & 7 \end{bmatrix} = A$$

so A is symmetric. ▲

It is easy to recognize symmetric matrices by inspection: The entries on the main diagonal are arbitrary, but "mirror images" of entries across the main diagonal are equal (Figure 1).

Figure 1

We have two questions to consider:

· Which matrices are orthogonally diagonalizable?
· How do we find an orthogonal matrix to carry out the diagonalization?

CONDITIONS FOR ORTHOGONAL DIAGONALIZ-ABILITY

The following theorem addresses the first of these questions. In this theorem and for the remainder of this section, *orthogonal* will mean orthogonal with respect to the Euclidean inner product on R^n.

Theorem 6.3.1. *If A is an n × n matrix, then the following are equivalent.*
(a) A is orthogonally diagonalizable.
(b) A has an orthonormal set of n eigenvectors.
(c) A is symmetric.

Proof (a) ⇒ (b). Since A is orthogonally diagonalizable, there is an orthogonal matrix P such that $P^{-1}AP$ is diagonal. As shown in the proof of Theorem 6.2.1, the n column vectors of P are eigenvectors of A. Since P is orthogonal, these column vectors are orthonormal (see Theorem 5.4.3) so that A has n orthonormal eigenvectors.

(b) ⇒ (a). Assume that A has an orthonormal set of n eigenvectors $\{\mathbf{p}_1, \mathbf{p}_2, \ldots, \mathbf{p}_n\}$. As shown in the proof of Theorem 6.2.1, the matrix P with these eigenvectors

as columns diagonalizes A. Since these eigenvectors are orthonormal, P is orthogonal and thus orthogonally diagonalizes A.

$(a) \Rightarrow (c)$. In the proof that $(a) \Rightarrow (b)$ we showed that an orthogonally diagonalizable $n \times n$ matrix A is orthogonally diagonalized by an $n \times n$ matrix P whose columns form an orthonormal set of eigenvectors of A. Let D be the diagonal matrix

$$D = P^{-1}AP$$

Thus,

$$A = PDP^{-1}$$

or, since P is orthogonal,

$$A = PDP^t$$

Therefore

$$A^t = (PDP^t)^t = PD^tP^t = PDP^t = A$$

which shows that A is symmetric.

$(c) \Rightarrow (a)$. The proof of this part is beyond the scope of this text and will be omitted. ▌

DIAGONALIZ-ATION OF SYMMETRIC MATRICES

We now turn to the problem of finding an orthogonal matrix P to diagonalize a symmetric matrix. The key is the following theorem, whose proof is given at the end of the section.

Theorem 6.3.2. *If A is a symmetric matrix, then eigenvectors from different eigenspaces are orthogonal.*

As a consequence of this theorem we obtain the following procedure for orthogonally diagonalizing a symmetric matrix.

Step 1. Find a basis for each eigenspace of A.

Step 2. Apply the Gram-Schmidt process to each of these bases to obtain an orthonormal basis for each eigenspace.

Step 3. Form the matrix P whose columns are the basis vectors constructed in Step 2; this matrix orthogonally diagonalizes A.

The justification of this procedure should be clear: Theorem 6.3.2 ensures that eigenvectors from *distinct* eigenspaces are orthogonal, while the application of the Gram-Schmidt process ensures that the eigenvectors obtained within the *same* eigenspace are orthonormal. Thus, the *entire* set of eigenvectors obtained by this procedure is orthonormal.

Example 2 Find an orthogonal matrix P that diagonalizes

$$A = \begin{bmatrix} 4 & 2 & 2 \\ 2 & 4 & 2 \\ 2 & 2 & 4 \end{bmatrix}.$$

Solution. The characteristic equation of A is

$$\det(\lambda I - A) = \det \begin{bmatrix} \lambda - 4 & -2 & -2 \\ -2 & \lambda - 4 & -2 \\ -2 & -2 & \lambda - 4 \end{bmatrix} = (\lambda - 2)^2(\lambda - 8) = 0$$

Thus, the eigenvalues of A are $\lambda = 2$ and $\lambda = 8$. By the method used in Example 5 of Section 6.1, it can be shown that

$$\mathbf{u}_1 = \begin{bmatrix} -1 \\ 1 \\ 0 \end{bmatrix} \quad \text{and} \quad \mathbf{u}_2 = \begin{bmatrix} -1 \\ 0 \\ 1 \end{bmatrix}$$

form a basis for the eigenspace corresponding to $\lambda = 2$. Applying the Gram-Schmidt process to $\{\mathbf{u}_1, \mathbf{u}_2\}$ yields the following orthornormal eigenvectors (verify):

$$\mathbf{v}_1 = \begin{bmatrix} -\dfrac{1}{\sqrt{2}} \\[2mm] \dfrac{1}{\sqrt{2}} \\[2mm] 0 \end{bmatrix} \quad \text{and} \quad \mathbf{v}_2 = \begin{bmatrix} -\dfrac{1}{\sqrt{6}} \\[2mm] -\dfrac{1}{\sqrt{6}} \\[2mm] \dfrac{2}{\sqrt{6}} \end{bmatrix}$$

The eigenspace corresponding to $\lambda = 8$ has

$$\mathbf{u}_3 = \begin{bmatrix} 1 \\ 1 \\ 1 \end{bmatrix}$$

as a basis. Applying the Gram-Schmidt process to $\{\mathbf{u}_3\}$ yields

$$\mathbf{v}_3 = \begin{bmatrix} \dfrac{1}{\sqrt{3}} \\[2mm] \dfrac{1}{\sqrt{3}} \\[2mm] \dfrac{1}{\sqrt{3}} \end{bmatrix}$$

Finally, using $\mathbf{v}_1$, $\mathbf{v}_2$, and $\mathbf{v}_3$ as column vectors we obtain

$$P = \begin{bmatrix} -\dfrac{1}{\sqrt{2}} & -\dfrac{1}{\sqrt{6}} & \dfrac{1}{\sqrt{3}} \\[2ex] \dfrac{1}{\sqrt{2}} & -\dfrac{1}{\sqrt{6}} & \dfrac{1}{\sqrt{3}} \\[2ex] 0 & \dfrac{2}{\sqrt{6}} & \dfrac{1}{\sqrt{3}} \end{bmatrix}$$

which orthogonally diagonalizes A. (As a check, the reader may wish to verify that $P^t A P$ is a diagonal matrix.) ▲

SOME PROPERTIES OF SYMMETRIC MATRICES

We conclude this section with two important properties of symmetric matrices.

Theorem 6.3.3

(a) *The characteristic equation of a symmetric matrix A has only real roots.*

(b) *If an eigenvalue λ of a symmetric matrix A is repeated k times as a root of the characteristic equation, then the eigenspace corresponding to the eigenvalue λ is k-dimensional.*

The proof of part (a), which requires results about complex vector spaces, is discussed in Section 10.6; the proof of part (b) is studied in more advanced courses.

Example 3 The characteristic equation of the symmetric matrix

$$A = \begin{bmatrix} 3 & 1 & 0 & 0 & 0 \\ 1 & 3 & 0 & 0 & 0 \\ 0 & 0 & 2 & 1 & 1 \\ 0 & 0 & 1 & 2 & 1 \\ 0 & 0 & 1 & 1 & 2 \end{bmatrix}$$

is

$$(\lambda - 4)^2(\lambda - 1)^2(\lambda - 2) = 0$$

so that the eigenvalues are $\lambda = 4$, $\lambda = 1$, and $\lambda = 2$, where $\lambda = 4$ and $\lambda = 1$ are repeated twice and $\lambda = 2$ occurs once. Thus, the eigenspaces corresponding to $\lambda = 4$ and $\lambda = 1$ are 2-dimensional, and the eigenspace corresponding to $\lambda = 1$ is 1-dimensional. ▲

REMARK. We remind the reader that we have assumed to this point that all of our matrices have real entries. Indeed, we shall see in Chapter 10 that part (*a*) of Theorem 6.3.3 is false for matrices with complex entries.

OPTIONAL

Proof of Theorem 6.3.2. Let λ_1 and λ_2 be two different eigenvalues of the $n \times n$ symmetric matrix A, and let

$$\mathbf{v}_1 = \begin{bmatrix} v_1 \\ v_2 \\ \vdots \\ v_n \end{bmatrix} \quad \text{and} \quad \mathbf{v}_2 = \begin{bmatrix} v'_1 \\ v'_2 \\ \vdots \\ v'_n \end{bmatrix}$$

be the corresponding eigenvectors. We want to show that

$$\mathbf{v}_1 \cdot \mathbf{v}_2 = v_1 v'_1 + v_2 v'_2 + \cdots + v_n v'_n = 0$$

Since $\mathbf{v}_1^t \mathbf{v}_2$ is a 1×1 matrix with $\mathbf{v}_1 \cdot \mathbf{v}_2$ as its only entry, we can complete the proof by showing that $\mathbf{v}_1^t \mathbf{v}_2 = 0$.

Since $\mathbf{v}_1$ and $\mathbf{v}_2$ are eigenvectors corresponding to the eigenvalues λ_1 and λ_2, we have

$$A\mathbf{v}_1 = \lambda_1 \mathbf{v}_1 \tag{3}$$

$$A\mathbf{v}_2 = \lambda_2 \mathbf{v}_2 \tag{4}$$

From (3)

$$(A\mathbf{v}_1)^t = (\lambda_1 \mathbf{v}_1)^t$$

or

$$\mathbf{v}_1^t A^t = \lambda_1 \mathbf{v}_1^t$$

or, since A is symmetric,

$$\mathbf{v}_1^t A = \lambda_1 \mathbf{v}_1^t$$

Multiplying both sides of this equation on the right by $\mathbf{v}_2$ yields

$$\mathbf{v}_1^t A \mathbf{v}_2 = \lambda_1 \mathbf{v}_1^t \mathbf{v}_2 \tag{5}$$

and multiplying both sides of (4) on the left by $\mathbf{v}_1^t$ yields

$$\mathbf{v}_1^t A \mathbf{v}_2 = \lambda_2 \mathbf{v}_1^t \mathbf{v}_2 \tag{6}$$

Thus, from (5) and (6)

$$\lambda_1 \mathbf{v}_1^t \mathbf{v}_2 = \lambda_2 \mathbf{v}_1^t \mathbf{v}_2$$

or

$$(\lambda_1 - \lambda_2)\mathbf{v}_1^t \mathbf{v}_2 = 0$$

But $\lambda_1 \neq \lambda_2$, so $\mathbf{v}_1^t \mathbf{v}_2 = 0$, which is what we wanted to prove. ▮

EXERCISE SET 6.3

1. Use part (b) of Theorem 6.3.3 to find the dimensions of the eigenspaces of the following symmetric matrices.

(a) $\begin{bmatrix} 1 & 2 \\ 2 & 4 \end{bmatrix}$
(b) $\begin{bmatrix} 1 & -4 & 2 \\ -4 & 1 & -2 \\ 2 & -2 & -2 \end{bmatrix}$
(c) $\begin{bmatrix} 1 & 1 & 1 \\ 1 & 1 & 1 \\ 1 & 1 & 1 \end{bmatrix}$

(d) $\begin{bmatrix} 4 & 2 & 2 \\ 2 & 4 & 2 \\ 2 & 2 & 4 \end{bmatrix}$
(e) $\begin{bmatrix} 4 & 4 & 0 & 0 \\ 4 & 4 & 0 & 0 \\ 0 & 0 & 0 & 0 \\ 0 & 0 & 0 & 0 \end{bmatrix}$
(f) $\begin{bmatrix} 2 & -1 & 0 & 0 \\ -1 & 2 & 0 & 0 \\ 0 & 0 & 2 & -1 \\ 0 & 0 & -1 & 2 \end{bmatrix}$

In Exercises 2–9 find a matrix P that orthogonally diagonalizes A, and determine $P^{-1}AP$.

2. $A = \begin{bmatrix} 3 & 1 \\ 1 & 3 \end{bmatrix}$
3. $A = \begin{bmatrix} 6 & 2\sqrt{3} \\ 2\sqrt{3} & 7 \end{bmatrix}$
4. $A = \begin{bmatrix} 6 & -2 \\ -2 & 3 \end{bmatrix}$
5. $A = \begin{bmatrix} -2 & 0 & -36 \\ 0 & -3 & 0 \\ -36 & 0 & -23 \end{bmatrix}$

6. $A = \begin{bmatrix} 1 & 1 & 0 \\ 1 & 1 & 0 \\ 0 & 0 & 0 \end{bmatrix}$
7. $A = \begin{bmatrix} 2 & -1 & -1 \\ -1 & 2 & -1 \\ -1 & -1 & 2 \end{bmatrix}$
8. $A = \begin{bmatrix} 3 & 1 & 0 & 0 \\ 1 & 3 & 0 & 0 \\ 0 & 0 & 0 & 0 \\ 0 & 0 & 0 & 0 \end{bmatrix}$
9. $A = \begin{bmatrix} -7 & 24 & 0 & 0 \\ 24 & 7 & 0 & 0 \\ 0 & 0 & -7 & 24 \\ 0 & 0 & 24 & 7 \end{bmatrix}$

10. Assuming that $b \neq 0$, find a matrix that orthogonally diagonalizes

$$\begin{bmatrix} a & b \\ b & a \end{bmatrix}$$

11. Show that if A is any $m \times n$ matrix, then $A^t A$ has an orthonormal set of n eigenvectors.

12. (a) Show that if $\mathbf{v}$ is any $n \times 1$ matrix and I is the $n \times n$ identity matrix, then $I - \mathbf{v}\mathbf{v}^t$ is orthogonally diagonalizable.

(b) Find a matrix P that orthogonally diagonalizes $I - \mathbf{v}\mathbf{v}^t$ if

$$\mathbf{v} = \begin{bmatrix} 1 \\ 0 \\ 1 \end{bmatrix}$$

13. Two $n \times n$ matrices, A and B, are called **orthogonally similar** if there is an orthogonal matrix P such that $B = P^t A P$. Show that if A is symmetric and the matrices A and B are orthogonally similar, then B is symmetric.

14. Prove Theorem 6.3.2 for 2×2 symmetric matrices.

15. Prove Theorem 6.3.3a for 2×2 symmetric matrices.

SUPPLEMENTARY EXERCISES

1. (a) Show that if $0 < \theta < \pi$, then

$$A = \begin{bmatrix} \cos \theta & -\sin \theta \\ \sin \theta & \cos \theta \end{bmatrix}$$

has no eigenvalues and consequently no eigenvectors.
(b) Give a geometric explanation of the result in (a).

2. Find the eigenvalues of

$$A = \begin{bmatrix} 0 & 1 & 0 \\ 0 & 0 & 1 \\ k^3 & -3k^2 & 3k \end{bmatrix}$$

3. (a) Show that if D is a diagonal matrix with nonnegative entries on the main diagonal, then there is a matrix S such that $S^2 = D$.
(b) Show that if A is a diagonalizable matrix with nonnegative eigenvalues, then there is a matrix S such that $S^2 = A$.
(c) Find a matrix S such that $S^2 = A$, if

$$A = \begin{bmatrix} 1 & 3 & 1 \\ 0 & 4 & 5 \\ 0 & 0 & 9 \end{bmatrix}$$

4. (a) Prove: If A is a square matrix, then A and A^t have the same eigenvalues.
(b) Show that A and A^t need not have the same eigenvectors. [**Hint.** Use Exercise 19 of Section 6.1 to find a 2×2 matrix A in which A and A^t have different eigenvectors.]

5. Prove: If A is a square matrix and $p(\lambda) = \det(\lambda I - A)$ is the characteristic polynomial of A, then the coefficient of λ^{n-1} in $p(\lambda)$ is the negative of the trace of A.

6. Prove: If $b \neq 0$, then

$$A = \begin{bmatrix} a & b \\ 0 & a \end{bmatrix}$$

is not diagonalizable.

7. In advanced linear algebra, one proves the **Cayley-Hamilton Theorem**, which states that a square matrix A satisfies its characteristic equation; that is, if

$$c_0 + c_1\lambda + c_2\lambda^2 + \cdots + c_{n-1}\lambda^{n-1} + \lambda^n = 0$$

is the characteristic equation of A, then

$$c_0 I + c_1 A + c_2 A^2 + \cdots + c_{n-1}A^{n-1} + A^n = 0.$$

Verify this result for

(a) $A = \begin{bmatrix} 3 & 6 \\ 1 & 2 \end{bmatrix}$ (b) $A = \begin{bmatrix} 0 & 1 & 0 \\ 0 & 0 & 1 \\ 1 & -3 & 3 \end{bmatrix}$

Exercises 8–10 use the Cayley-Hamilton Theorem, stated in Exercise 7.

8. Use Exercise 17 of Section 6.1 to prove the Cayley-Hamilton Theorem for 2×2 matrices.

9. The Cayley-Hamilton Theorem provides an efficient method for calculating powers of a matrix. For example, if A is a 2×2 matrix with characteristic equation

$$c_0 + c_1\lambda + \lambda^2 = 0$$

then $c_0 I + c_1 A + A^2 = 0$, so

$$A^2 = -c_1 A - c_0 I$$

Multiplying through by A yields $A^3 = -c_1 A^2 - c_0 A$, which expresses A^3 in terms of A^2 and A, and multiplying through by A^2 yields $A^4 = -c_1 A^3 - c_0 A^2$, which expresses A^4 in terms of A^3 and A^2. Continuing in this way, we can calculate successive powers of A simply by expressing them in terms of lower powers. Use this procedure to calculate

$$A^2, \quad A^3, \quad A^4, \quad \text{and} \quad A^5$$

for

$$A = \begin{bmatrix} 3 & 6 \\ 1 & 2 \end{bmatrix}$$

10. Use the method of the preceding exercise to calculate A^3 and A^4 for

$$A = \begin{bmatrix} 0 & 1 & 0 \\ 0 & 0 & 1 \\ 1 & -3 & 3 \end{bmatrix}$$

11. Find the eigenvalues of the matrix

$$\begin{bmatrix} c_1 & c_2 & \cdots & c_n \\ c_1 & c_2 & \cdots & c_n \\ \vdots & \vdots & & \vdots \\ c_1 & c_2 & \cdots & c_n \end{bmatrix}$$

12. (a) It was shown in Exercise 16 of Section 6.1 that if A is an $n \times n$ matrix, then the coefficient of λ^n in the characteristic polynomial of A is 1. (A polynomial with this property is called *monic*.) Show that the matrix

$$\begin{bmatrix} 0 & 0 & 0 & \cdots & 0 & -c_0 \\ 1 & 0 & 0 & \cdots & 0 & -c_1 \\ 0 & 1 & 0 & \cdots & 0 & -c_2 \\ \vdots & \vdots & \vdots & & \vdots & \vdots \\ 0 & 0 & 0 & \cdots & 1 & -c_{n-1} \end{bmatrix}$$

has characteristic polynomial $p(\lambda) = c_0 + c_1\lambda + \cdots + c_{n-1}\lambda^{n-1} + \lambda^n$. This shows that every monic polynomial is the characteristic polynomial of some matrix. The matrix in this example is called the *companion matrix* of $p(\lambda)$. [*Hint.* Evaluate all determinants in the problem by adding a multiple of the second row to the first to introduce a zero at the top of the first column, and then expanding by cofactors along the first column.]

(b) Find a matrix with characteristic polynomial $p(\lambda) = 1 - 2\lambda + \lambda^2 + 3\lambda^3 + \lambda^4$.

13. A square matrix A is called *nilpotent* if $A^n = 0$ for some positive integer n. What can you say about the eigenvalues of a nilpotent matrix?

14. Prove: If A is an $n \times n$ matrix and n is odd, then A has at least one real eigenvalue.

15. Prove: If λ is an eigenvalue of an invertible matrix A and $\mathbf{x}$ is a corresponding eigenvector, then $1/\lambda$ is an eigenvalue of A^{-1}, and $\mathbf{x}$ is a corresponding eigenvector.

16. Prove: If λ is an eigenvalue of A, $\mathbf{x}$ is a corresponding eigenvector, and s is a scalar, then $\lambda - s$ is an eigenvalue of $A - sI$, and $\mathbf{x}$ is a corresponding eigenvector.

17. Find the eigenvalues and bases for the eigenspaces of

$$A = \begin{bmatrix} -2 & 2 & 3 \\ -2 & 3 & 2 \\ -4 & 2 & 5 \end{bmatrix}$$

Then use Exercises 15 and 16 to find the eigenvalues and bases for the eigenspaces of
(a) A^{-1} (b) $A - 3I$ (c) $A + 2I$

18. Let A be a square matrix such that $A^3 = A$. What can you say about the eigenvalues of A?

CHAPTER 7

LINEAR TRANSFORMATIONS

7.1 INTRODUCTION TO LINEAR TRANSFORMATIONS

In this section we shall begin the study of vector-valued functions of a vector variable—that is, functions of the form **w** = *F*(**v**), *where the independent variable* **v** *and the dependent variable* **w** *are both vectors. We shall concentrate on a special class of vector functions called "linear transformations." These have many important applications in physics, engineering, social sciences, and various branches of mathematics.*

LINEAR TRANS-
FORMATIONS

If V and W are vector spaces and F is a function that associates a unique vector in W with each vector in V, we say F *maps* V into W, and write $F : V \to W$. Further, if F associates the vector **w** with the vector **v**, we write **w** = *F*(**v**) and say that **w** is the *image* of **v** under F. The vector space V is called the *domain* of F, and the vector space W is called the *image space* of F. For example, the function defined by the formula

$$F(x, y) = (x - y, x + y, 5x) \tag{1}$$

maps R^2 into R^3. For this function the image of a vector **v** = (x, y) in R^2 is the vector **w** = $(x - y, x + y, 5x)$ in R^3. In particular, if **v** = $(1, 3)$, then $x = 1$ and $y = 3$, so from (1) we have

$$\mathbf{w} = F(\mathbf{v}) = F(1, 3) = (-2, 4, 5)$$

(Figure 1). The domain of F is R^2 and the image space is R^3.

We will be interested in a special class of functions defined as follows:

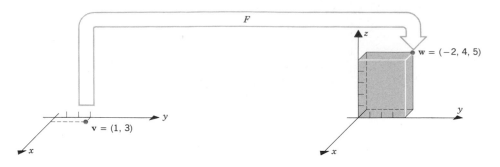

Figure 1

$w = (-2, 4, 5)$ is the image of $v = (1, 3)$ under F.

Definition. If $F:V \rightarrow W$ is a function from the vector space V into the vector space W, then F is called a **linear transformation** if

(a) $F(\mathbf{u} + \mathbf{v}) = F(\mathbf{u}) + F(\mathbf{v})$ for all vectors $\mathbf{u}$ and $\mathbf{v}$ in V;
(b) $F(k\mathbf{u}) = kF(\mathbf{u})$ for all vectors $\mathbf{u}$ in V and all scalars k.

Example 1 Let $F:R^2 \rightarrow R^3$ be the function defined by Formula (1). If $\mathbf{u} = (x_1, y_1)$ and $\mathbf{v} = (x_2, y_2)$, then $\mathbf{u} + \mathbf{v} = (x_1 + x_2, y_1 + y_2)$, so that

$$F(\mathbf{u}+\mathbf{v}) = F(x_1 + x_2, y_1 + y_2)$$

We let $x = x_1 + x_2$ and $y = y_1 + y_2$ in (1).

$$= ((x_1 + x_2) - (y_1 + y_2), (x_1 + x_2) + (y_1 + y_2), 5(x_1 + x_2))$$
$$= ((x_1 - y_1) + (x_2 - y_2), (x_1 + y_1) + (x_2 + y_2), 5x_1 + 5x_2)$$

See (1).

$$= (x_1 - y_1, x_1 + y_1, 5x_1) + (x_2 - y_2, x_2 + y_2, 5x_2)$$
$$= F(x_1, y_1) + F(x_2, y_2)$$
$$= F(\mathbf{u}) + F(\mathbf{v})$$

Moreover, if k is any scalar, then $k\mathbf{u} = (kx_1, ky_1)$, so that

$$F(k\mathbf{u}) = F(kx_1, ky_1)$$
$$= (kx_1 - ky_1, kx_1 + ky_1, 5(kx_1))$$

We let $x = kx_1$ and $y = ky_1$ in (1).

$$= (k(x_1 - y_1), k(x_1 + y_1), k(5x_1))$$
$$= k(x_1 - y_1, x_1 + y_1, 5x_1)$$
$$= kF(x_1, y_1)$$
$$= kF(\mathbf{u})$$

Thus, F is a linear transformation. ▲

If $F:V \rightarrow W$ is a linear transformation, then for any vectors $\mathbf{v}_1$ and $\mathbf{v}_2$ in V and any scalars k_1 and k_2, we have

$$F(k_1\mathbf{v}_1 + k_2\mathbf{v}_2) = F(k_1\mathbf{v}_1) + F(k_2\mathbf{v}_2) = k_1F(\mathbf{v}_1) + k_2F(\mathbf{v}_2)$$

Similarly, if $v_1, v_2, \ldots, v_n$ are vectors in V and $k_1, k_2, \ldots, k_n$ are scalars, then

$$F(k_1 v_1 + k_2 v_2 + \cdots + k_n v_n) = k_1 F(v_1) + k_2 F(v_2) + \cdots + k_n F(v_n) \tag{2}$$

Two linear transformations, F_1 and F_2, will be regarded as *equal*, written $F_1 = F_2$, if they have the same domain and

$$F_1(v) = F_2(v)$$

for every vector v in the domain.

MATRIX TRANS FORMATIONS

Example 1 is a special case of a more general class of linear transformations which can be motivated by rewriting the vectors in Formula (1) in matrix form; this yields

$$F\left(\begin{bmatrix} x \\ y \end{bmatrix}\right) = \begin{bmatrix} x - y \\ x + y \\ 5x \end{bmatrix} \tag{3}$$

which can be expressed as

$$F\left(\begin{bmatrix} x \\ y \end{bmatrix}\right) = \begin{bmatrix} 1 & -1 \\ 1 & 1 \\ 5 & 0 \end{bmatrix} \begin{bmatrix} x \\ y \end{bmatrix}$$

Thus, for each vector

$$x = \begin{bmatrix} x \\ y \end{bmatrix}$$

in R^2 we have $F(x) = Ax$, where

$$A = \begin{bmatrix} 1 & -1 \\ 1 & 1 \\ 5 & 0 \end{bmatrix}$$

More generally, if A is a fixed $m \times n$ matrix, say

$$A = \begin{bmatrix} a_{11} & a_{12} & \cdots & a_{1n} \\ a_{21} & a_{22} & \cdots & a_{2n} \\ \vdots & \vdots & & \vdots \\ a_{m1} & a_{m2} & \cdots & a_{mn} \end{bmatrix}$$

and

$$x = \begin{bmatrix} x_1 \\ x_2 \\ \vdots \\ x_n \end{bmatrix}$$

is a vector in R^n expressed in matrix notation, then $A\mathbf{x}$ is a vector in R^m, and the function $T: R^n \to R^m$ defined by the formula

$$T(\mathbf{x}) = A\mathbf{x} = \begin{bmatrix} a_{11} & a_{12} & \cdots & a_{1n} \\ a_{21} & a_{22} & \cdots & a_{2n} \\ \vdots & \vdots & & \vdots \\ a_{m1} & a_{m2} & \cdots & a_{mn} \end{bmatrix} \begin{bmatrix} x_1 \\ x_2 \\ \vdots \\ x_n \end{bmatrix} \tag{4}$$

is a linear transformation called *multiplication by A*. Such linear transformations are called *matrix transformations*. To see that (4) is a linear transformation, let **u** and **v** be $n \times 1$ matrices and let k be a scalar. Using properties of matrix multiplication, we obtain

$$A(\mathbf{u} + \mathbf{v}) = A\mathbf{u} + A\mathbf{v} \qquad \text{and} \qquad A(k\mathbf{u}) = k(A\mathbf{u})$$

or equivalently

$$T(\mathbf{u} + \mathbf{v}) = T(\mathbf{u}) + T(\mathbf{v}) \qquad \text{and} \qquad T(k\mathbf{u}) = kT(\mathbf{u})$$

which proves that (4) is a linear transformation.

REMARK. For the remainder of this text, the notations

$$\mathbf{x} = (x_1, x_2, \ldots, x_n) \qquad \text{and} \qquad \mathbf{x} = \begin{bmatrix} x_1 \\ x_2 \\ \vdots \\ x_n \end{bmatrix}$$

will both be used for vectors in R^n. However, in any given problem we will use one notation or the other; they will not be mixed.

Example 2 As an illustration of a matrix transformation, let θ be a fixed angle, and let $T: R^2 \to R^2$ be multiplication by the matrix

$$A = \begin{bmatrix} \cos \theta & -\sin \theta \\ \sin \theta & \cos \theta \end{bmatrix}$$

If **v** is the vector

$$\mathbf{v} = \begin{bmatrix} x \\ y \end{bmatrix}$$

then

$$T(\mathbf{v}) = A\mathbf{v} = \begin{bmatrix} \cos \theta & -\sin \theta \\ \sin \theta & \cos \theta \end{bmatrix} \begin{bmatrix} x \\ y \end{bmatrix} = \begin{bmatrix} x \cos \theta - y \sin \theta \\ x \sin \theta + y \cos \theta \end{bmatrix}$$

Geometrically, $T(\mathbf{v})$ is the vector that results if **v** is rotated counterclockwise through an angle θ. To see this, let ϕ be the angle between **v** and the positive x axis, and let

$$\mathbf{v}' = \begin{bmatrix} x' \\ y' \end{bmatrix}$$

be the vector that results when **v** is rotated through an angle θ (Figure 2). We will show that $\mathbf{v}' = T(\mathbf{v})$. If r denotes the length of **v**, then

$$x = r \cos \phi, \quad y = r \sin \phi$$

Similarly, since $\mathbf{v}'$ has the same length as **v**, we have

$$x' = r \cos(\theta + \phi), \quad y' = r \sin(\theta + \phi)$$

Therefore,

$$\mathbf{v}' = \begin{bmatrix} x' \\ y' \end{bmatrix} = \begin{bmatrix} r \cos(\theta + \phi) \\ r \sin(\theta + \phi) \end{bmatrix}$$

$$= \begin{bmatrix} r \cos \theta \cos \phi - r \sin \theta \sin \phi \\ r \sin \theta \cos \phi + r \cos \theta \sin \phi \end{bmatrix}$$

$$= \begin{bmatrix} x \cos \theta - y \sin \theta \\ x \sin \theta + y \cos \theta \end{bmatrix}$$

$$= \begin{bmatrix} \cos \theta & -\sin \theta \\ \sin \theta & \cos \theta \end{bmatrix} \begin{bmatrix} x \\ y \end{bmatrix}$$

$$= A\mathbf{v} = T(\mathbf{v})$$

The linear transformation in this example is called the ***rotation of R^2 through the angle θ.*** ▲

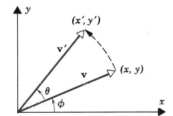

Figure 2

<table>
<tr><td>**EXAMPLES OF LINEAR TRANS- FORMATIONS**</td></tr>
</table>

Example 3 Let V and W be any two vector spaces. The mapping $T: V \to W$ such that $T(\mathbf{v}) = \mathbf{0}$ for every **v** in V is a linear transformation called the ***zero transformation.*** To see that T is linear, observe that

$$T(\mathbf{u} + \mathbf{v}) = \mathbf{0}, \quad T(\mathbf{u}) = \mathbf{0}, \quad T(\mathbf{v}) = \mathbf{0}, \quad \text{and} \quad T(k\mathbf{u}) = \mathbf{0}$$

Therefore,

$$T(\mathbf{u} + \mathbf{v}) = T(\mathbf{u}) + T(\mathbf{v}) \quad \text{and} \quad T(k\mathbf{u}) = kT(\mathbf{u})$$ ▲

Example 4 Let V be any vector space. The mapping $I: V \to V$ defined by $I(\mathbf{v}) = \mathbf{v}$ is called the ***identity transformation*** on V. The verification that I is linear is left as an exercise. ▲

A linear transformation that maps a vector space V into itself is called a *linear operator* on V. Thus, the linear transformation in Example 2 is a linear operator on R^2, and the identity transformation in Example 4 is a linear operator on V.

Example 5   Let V be any vector space and k any fixed scalar. We leave it as an exercise to check that the function $T: V \rightarrow V$ defined by

$$T(\mathbf{v}) = k\mathbf{v}$$

is a linear operator on V. This linear operator is called a *dilation* of V with factor k if $k > 1$ and is called a *contraction* of V with factor k if $0 < k < 1$. Geometrically, the dilation "stretches" each vector in V by a factor of k, and the contraction of V "compresses" each vector by a factor of k (Figure 3). ▲

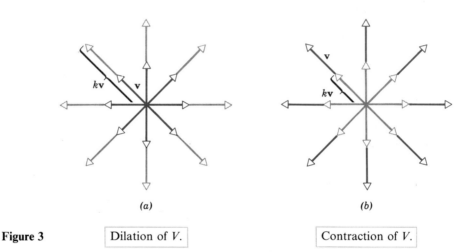

(a)

(b)

Figure 3

Dilation of V.

Contraction of V.

Example 6   Let V be an inner product space, and suppose that W is a finite dimensional subspace of V having

$$S = \{\mathbf{w}_1, \mathbf{w}_2, \ldots, \mathbf{w}_r\}$$

as an orthonormal basis. Let $T: V \rightarrow W$ be the function that maps a vector $\mathbf{v}$ in V into its orthogonal projection on W (Theorem 5.3.5); that is,

$$T(\mathbf{v}) = \text{proj}_W \mathbf{v} = \langle \mathbf{v}, \mathbf{w}_1 \rangle \mathbf{w}_1 + \langle \mathbf{v}, \mathbf{w}_2 \rangle \mathbf{w}_2 + \cdots + \langle \mathbf{v}, \mathbf{w}_r \rangle \mathbf{w}_r$$

(See Figure 4.) The mapping T is called the *orthogonal projection of V onto W*; its linearity follows from the basic properties of the inner product. For example,

$$T(\mathbf{u} + \mathbf{v}) = \langle \mathbf{u} + \mathbf{v}, \mathbf{w}_1 \rangle \mathbf{w}_1 + \langle \mathbf{u} + \mathbf{v}, \mathbf{w}_2 \rangle \mathbf{w}_2 + \cdots + \langle \mathbf{u} + \mathbf{v}, \mathbf{w}_r \rangle \mathbf{w}_r$$

$$= \langle \mathbf{u}, \mathbf{w}_1 \rangle \mathbf{w}_1 + \langle \mathbf{u}, \mathbf{w}_2 \rangle \mathbf{w}_2 + \cdots + \langle \mathbf{u}, \mathbf{w}_r \rangle \mathbf{w}_r$$

$$+ \langle \mathbf{v}, \mathbf{w}_1 \rangle \mathbf{w}_1 + \langle \mathbf{v}, \mathbf{w}_2 \rangle \mathbf{w}_2 + \cdots + \langle \mathbf{v}, \mathbf{w}_r \rangle \mathbf{w}_r$$

$$= T(\mathbf{u}) + T(\mathbf{v})$$

Similarly, $T(k\mathbf{u}) = kT(\mathbf{u})$. ▲

Figure 4 The orthogonal projection of V onto W.

Example 7   As a special case of the previous example, let $V = R^3$ have the Euclidean inner product. The vectors $\mathbf{w}_1 = (1, 0, 0)$ and $\mathbf{w}_2 = (0, 1, 0)$ form an orthonormal basis for the xy-plane. Thus, if $\mathbf{v} = (x, y, z)$ is any vector in R^3, the orthogonal projection of R^3 onto the xy-plane is given by

$$T(\mathbf{v}) = \langle \mathbf{v}, \mathbf{w}_1 \rangle \mathbf{w}_1 + \langle \mathbf{v}, \mathbf{w}_2 \rangle \mathbf{w}_2$$
$$= x(1, 0, 0) + y(0, 1, 0)$$
$$= (x, y, 0)$$

(See Figure 5.)

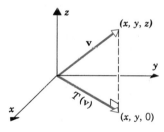

Figure 5 The orthogonal projection of R^3 onto the xy-plane.

Example 8   Let $S = \{\mathbf{w}_1, \mathbf{w}_2, \ldots, \mathbf{w}_n\}$ be a basis for an n-dimensional vector space V, and let

$$(\mathbf{v})_S = (k_1, k_2, \ldots, k_n)$$

be the coordinate vector relative to S of a vector $\mathbf{v}$ in V; thus

$$\mathbf{v} = k_1 \mathbf{w}_1 + k_2 \mathbf{w}_2 + \cdots + k_n \mathbf{w}_n$$

Define $T: V \to R^n$ to be the function that maps $\mathbf{v}$ into its coordinate vector relative to S; that is,

$$T(\mathbf{v}) = (\mathbf{v})_S = (k_1, k_2, \ldots, k_n)$$

The function T is a linear transformation. To see that this is so, suppose that $\mathbf{u}$ and $\mathbf{v}$ are vectors in V and that

$$\mathbf{u} = c_1 \mathbf{w}_1 + c_2 \mathbf{w}_2 + \cdots + c_n \mathbf{w}_n \qquad \text{and} \qquad \mathbf{v} = d_1 \mathbf{w}_1 + d_2 \mathbf{w}_2 + \cdots + d_n \mathbf{w}_n$$

Thus,

$$(\mathbf{u})_S = (c_1, c_2, \ldots, c_n)$$
$$(\mathbf{v})_S = (d_1, d_2, \ldots, d_n)$$

But

$$\mathbf{u} + \mathbf{v} = (c_1 + d_1)\mathbf{w}_1 + (c_2 + d_2)\mathbf{w}_2 + \cdots + (c_n + d_n)\mathbf{w}_n$$
$$k\mathbf{u} = (kc_1)\mathbf{w}_1 + (kc_2)\mathbf{w}_2 + \cdots + (kc_n)\mathbf{w}_n$$

so that

$$(\mathbf{u} + \mathbf{v})_S = (c_1 + d_1, c_2 + d_2, \ldots, c_n + d_n)$$
$$(k\mathbf{u})_S = (kc_1, kc_2, \ldots, kc_n)$$

Therefore,

$$(\mathbf{u} + \mathbf{v})_S = (\mathbf{u})_S + (\mathbf{v})_S \quad \text{and} \quad (k\mathbf{u})_S = k(\mathbf{u})_S$$

Expressing these equations in terms of T, we obtain

$$T(\mathbf{u} + \mathbf{v}) = T(\mathbf{u}) + T(\mathbf{v}) \quad \text{and} \quad T(k\mathbf{u}) = kT(\mathbf{u})$$

which shows that T is a linear transformation. ▲

REMARK. The computations in the foregoing example could just as well have been performed using coordinate matrices rather than coordinate vectors, that is,

$$[\mathbf{u} + \mathbf{v}]_S = [\mathbf{u}]_S + [\mathbf{v}]_S$$

and

$$[k\mathbf{u}]_S = k[\mathbf{u}]_S$$

Example 9   Let $\mathbf{p} = p(x) = c_0 + c_1 x + \cdots + c_n x^n$ be a polynomial in P_n, and define the function $T: P_n \to P_{n+1}$ by

$$T(\mathbf{p}) = T(p(x)) = xp(x) = c_0 x + c_1 x^2 + \cdots + c_n x^{n+1}$$

The function T is a linear transformation, since for any scalar k and any polynomials $\mathbf{p}_1$ and $\mathbf{p}_2$ in P_n we have

$$T(\mathbf{p}_1 + \mathbf{p}_2) = T(p_1(x) + p_2(x)) = x(p_1(x) + p_2(x))$$
$$= xp_1(x) + xp_2(x) = T(\mathbf{p}_1) + T(\mathbf{p}_2)$$

and

$$T(k\mathbf{p}) = T(kp(x)) = x(kp(x)) = k(xp(x)) = kT(\mathbf{p}) \quad ▲$$

Example 10   Let $\mathbf{p} = p(x) = c_0 + c_1 x + \cdots + c_n x^n$ be a polynomial in P_n, and let a and b be any scalars. We leave it as an exercise to show that the function T defined by

$$T(\mathbf{p}) = T(p(x)) = p(ax + b) = c_0 + c_1(ax + b) + \cdots + c_n(ax + b)^n$$

is a linear operator. For example, if $ax + b = 3x - 5$, then $T: P_2 \to P_2$ would be the linear operator given by the formula

$$T(c_0 + c_1 x + c_2 x^2) = c_0 + c_1(3x - 5) + c_2(3x - 5)^2 \quad ▲$$

Example 11 Let V be an inner product space and let $\mathbf{v}_0$ be any fixed vector in V. Let $T: V \rightarrow R$ be the transformation that maps a vector $\mathbf{v}$ into its inner product with $\mathbf{v}_0$; that is,

$$T(\mathbf{v}) = \langle \mathbf{v}, \mathbf{v}_0 \rangle$$

From the properties of an inner product

$$T(\mathbf{u} + \mathbf{v}) = \langle \mathbf{u} + \mathbf{v}, \mathbf{v}_0 \rangle = \langle \mathbf{u}, \mathbf{v}_0 \rangle + \langle \mathbf{v}, \mathbf{v}_0 \rangle = T(\mathbf{u}) + T(\mathbf{v})$$

and

$$T(k\mathbf{u}) = \langle k\mathbf{u}, \mathbf{v}_0 \rangle = k\langle \mathbf{u}, \mathbf{v}_0 \rangle = kT(\mathbf{u})$$

so that T is a linear transformation. ▲

Example 12 **(For readers who have studied calculus.)** Let $V = C[0, 1]$ be the vector space of all real-valued functions that are continuous on the interval $0 \leq x \leq 1$, and let W be the subspace of $C[0, 1]$ consisting of all functions with continuous first derivatives on the interval $0 \leq x \leq 1$.

Let $D: W \rightarrow V$ be the transformation that maps $\mathbf{f} = f(x)$ into its derivative; that is,

$$D(\mathbf{f}) = f'(x)$$

From the properties of differentiation, we have

$$D(\mathbf{f} + \mathbf{g}) = D(\mathbf{f}) + D(\mathbf{g})$$

and

$$D(k\mathbf{f}) = kD(\mathbf{f})$$

Thus, D is a linear transformation. ▲

Example 13 **(For readers who have studied calculus.)** Let $V = C[0, 1]$ be as in the previous example, and let $J: V \rightarrow R$ be defined by

$$J(\mathbf{f}) = \int_0^1 f(x)\, dx$$

For example, if $f(x) = x^2$, then

$$J(\mathbf{f}) = \int_0^1 x^2\, dx \doteq \frac{1}{3}$$

Since

$$\int_0^1 (f(x) + g(x))\, dx = \int_0^1 f(x)\, dx + \int_0^1 g(x)\, dx$$

and

$$\int_0^1 kf(x)\, dx = k \int_0^1 f(x)\, dx$$

for any constant k, it follows that

$$J(\mathbf{f} + \mathbf{g}) = J(\mathbf{f}) + J(\mathbf{g})$$
$$J(k\mathbf{f}) = kJ(\mathbf{f})$$

Thus, J is a linear transformation. ▲

COMPOSITION

The following example illustrates a common process in linear algebra, namely, performing two linear transformations in succession.

Example 14   Let $F: R^2 \to R^3$ and $T: R^3 \to R^3$ be the linear transformations given by the formulas

$$F(x, y) = (x - y, \ x + y, \ 5x)$$

and

$$T(x, y, z) = (2x, \ -y, \ 3z) \tag{5}$$

Because the image of (x, y) under F is a point in R^3 and the domain of T is R^3, we can compute

$$T(F(x, y)) = T(x - y, \ x + y, \ 5x)$$

The expression on the right side of this equation is calculated by making the following substitutions in Formula (5):

$$
\begin{array}{ccc}
x - y & \text{for} & x \\
x + y & \text{for} & y \\
5x & \text{for} & z
\end{array}
$$

This yields

$$T(F(x, y)) = (2(x - y), \ -(x + y), \ 3(5x)) = (2x - 2y, \ -x - y, \ 15x)$$

Thus, by successively performing the linear transformations F and T, we have created a new function $T(F(x, y))$ that maps R^2 into R^3. This function is called the *composition of T with F* and is denoted by $T \circ F$. We leave it as an exercise to show that $T \circ F$ is a linear transformation. Observe that we could not have reversed the order in which the transformations were performed: The expression $F(T(x, y, z))$ makes no sense, since $T(x, y, z)$ is a point in R^3 and the domain of F is R^2. ▲

The following definition generalizes the concept illustrated in the foregoing example.

Definition. If $T_1: U \to V$ and $T_2: V \to W$ are linear transformations, the *composition of T_2 with T_1*, denoted by $T_2 \circ T_1$ (read "T_2 circle T_1"), is the function defined by the formula

$$(T_2 \circ T_1)(\mathbf{u}) = T_2(T_1(\mathbf{u})) \tag{6}$$

where $\mathbf{u}$ is a vector in U.

REMARK.   Observe that this definition requires that the domain of T_2 (which is V) contain the range of T_1; this is essential for the formula $T_2(T_1(\mathbf{u}))$ to make sense (Figure 6).

Figure 6 | The composition of T_2 with T_1. |

The next result shows that the composition of two linear transformations is itself a linear transformation.

Theorem 7.1.1. *If $T_1: U \to V$ and $T_2: V \to W$ are linear transformations, then $(T_2 \circ T_1): U \to W$ is also a linear transformation.*

Proof. If **u** and **v** are vectors in U and k is a scalar, then if follows from (6) and the linearity of T_1 and T_2 that

$$(T_2 \circ T_1)(\mathbf{u} + \mathbf{v}) = T_2(T_1(\mathbf{u} + \mathbf{v})) = T_2(T_1(\mathbf{u}) + T_1(\mathbf{v}))$$
$$= T_2(T_1(\mathbf{u})) + T_2(T_1(\mathbf{v}))$$
$$= (T_2 \circ T_1)(\mathbf{u}) + (T_2 \circ T_1)(\mathbf{v})$$

and

$$(T_2 \circ T_1)(k\mathbf{u}) = T_2(T_1(k\mathbf{u})) = T_2(kT_1(\mathbf{u}))$$
$$= kT_2(T_1(\mathbf{u})) = k(T_2 \circ T_1)(\mathbf{u})$$

Thus, $T_2 \circ T_1$ satisfies the two requirements of a linear transformation. ∎

Example 15 Let $T_1: R^2 \to R^2$ and $T_2: R^2 \to R^2$ be the linear operators that rotate vectors through the angles θ_1 and θ_2, respectively. Thus, the operation

$$(T_2 \circ T_1)(\mathbf{x}) = T_2(T_1(\mathbf{x}))$$

first rotates **x** through the angle θ_1, then rotates $T_1(\mathbf{x})$ through the angle θ_2. It follows that the net effect of $T_2 \circ T_1$ is to rotate each vector in R^2 through the angle $\theta_1 + \theta_2$ (Figure 7). ▲

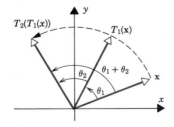

Figure 7

Example 16 Let $T_1 : P_1 \to P_2$ and $T_2 : P_2 \to P_2$ be the linear transformations given by the formulas

$$T_1(p(x)) = xp(x) \qquad \text{and} \qquad T_2(p(x)) = p(2x + 4)$$

Then the composition $(T_2 \circ T_1) : P_1 \to P_2$ is given by the formula

$$(T_2 \circ T_1)(p(x)) = T_2(T_1(p(x))) = T_2(xp(x)) = (2x + 4)p(2x + 4)$$

In particular, if $p(x) = c_0 + c_1 x$, then

$$(T_2 \circ T_1)(p(x)) = (T_2 \circ T_1)(c_0 + c_1 x) = (2x + 4)(c_0 + c_1(2x + 4))$$
$$= c_0(2x + 4) + c_1(2x + 4)^2 \quad \blacktriangle$$

Example 17 If $T : V \to V$ is any linear operator, and if $I : V \to V$ is the identity operator (Example 4), then for all vectors $\mathbf{v}$ in V we have

$$(T \circ I)(\mathbf{v}) = T(I(\mathbf{v})) = T(\mathbf{v})$$
$$(I \circ T)(\mathbf{v}) = I(T(\mathbf{v})) = T(\mathbf{v})$$

It follows that $T \circ I$ and $I \circ T$ are the same as T, that is,

$$T \circ I = T$$
$$I \circ T = T \quad \blacktriangle \tag{7}$$

We conclude this section by noting that compositions can be defined for more than two linear transformations. For example, if

$$T_1 : U \to V, \qquad T_2 : V \to W, \qquad \text{and} \qquad T_3 : W \to Y$$

are linear transformations, then the composition $T_3 \circ T_2 \circ T_1$ is defined by

$$(T_3 \circ T_2 \circ T_1)(\mathbf{u}) = T_3(T_2(T_1(\mathbf{u}))) \tag{8}$$

(Figure 8).

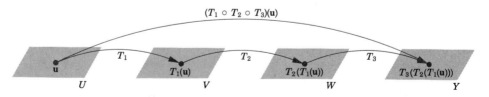

Figure 8   The composition of three linear transformations.

EXERCISE SET 7.1

In Exercises 1 and 2 determine whether $F:R^2 \to R^2$ is a linear operator.

1. (a) $F(x, y) = (2x, y)$ (b) $F(x, y) = (x^2, y)$ (c) $F(x, y) = (-y, x)$ (d) $F(x, y) = (x, 0)$

2. (a) $F(x, y) = (2x + y, x - y)$ (b) $F(x, y) = (x + 1, y)$ (c) $F(x, y) = (y, y)$ (d) $F(x, y) = (\sqrt[3]{x}, \sqrt[3]{y})$

In Exercises 3 and 4 determine whether $F:R^3 \to R^2$ is a linear transformation.

3. (a) $F(x, y, z) = (x, x + y + z)$ (b) $F(x, y, z) = (1, 1)$

4. (a) $F(x, y, z) = (0, 0)$ (b) $F(x, y, z) = (3x - 4y, 2x - 5z)$

In Exercises 5 and 6 determine whether $F:M_{22} \to R$ is a linear transformation.

5. (a) $F\left(\begin{bmatrix} a & b \\ c & d \end{bmatrix}\right) = b + c$ (b) $F\left(\begin{bmatrix} a & b \\ c & d \end{bmatrix}\right) = \det\begin{bmatrix} a & b \\ c & d \end{bmatrix}$

6. (a) $F\left(\begin{bmatrix} a & b \\ c & d \end{bmatrix}\right) = 3a - 4b + c - d$ (b) $F\left(\begin{bmatrix} a & b \\ c & d \end{bmatrix}\right) = a^2 + b^2$

In Exercises 7 and 8, determine whether $F:P_2 \to P_2$ is a linear operator.

7. (a) $F(a_0 + a_1x + a_2x^2) = a_0 + (a_1 - a_2)x + (2a_0 + 3a_1)x^2$
   (b) $F(a_0 + a_1x + a_2x^2) = a_0 + a_1(x + 1) + a_2(x + 1)^2$

8. (a) $F(a_0 + a_1x + a_2x^2) = 0$ (b) $F(a_0 + a_1x + a_2x^2) = (a_0 + 1) + a_1x + a_2x^2$

9. Let $F:R^2 \to R^2$ be the function that maps each point in the plane into its reflection about the y-axis. Find a formula for F, and show that F is a linear operator on R^2.

10. Let B be a fixed 2×3 matrix. Show that the function $T:M_{22} \to M_{23}$ defined by $T(A) = AB$ is a linear transformation.

11. Let A be a 2×3 matrix and $T:R^3 \to R^2$ the matrix transformation $T(\mathbf{x}) = A\mathbf{x}$. Given that

$$T\left(\begin{bmatrix} 1 \\ 0 \\ 0 \end{bmatrix}\right) = \begin{bmatrix} 1 \\ 1 \end{bmatrix}, \quad T\left(\begin{bmatrix} 0 \\ 1 \\ 0 \end{bmatrix}\right) = \begin{bmatrix} 3 \\ 0 \end{bmatrix}, \quad T\left(\begin{bmatrix} 0 \\ 0 \\ 1 \end{bmatrix}\right) = \begin{bmatrix} 4 \\ -7 \end{bmatrix}$$

find the following:

(a) $T\left(\begin{bmatrix} 1 \\ 3 \\ 8 \end{bmatrix}\right)$ (b) $T\left(\begin{bmatrix} x \\ y \\ z \end{bmatrix}\right)$ (c) A

12. Let $T:R^3 \to W$ be the orthogonal projection of R^3 onto the xz-plane W.
   (a) Find a formula for T.   (b) Find $T(2, 7, -1)$.

13. Let $T:R^3 \to W$ be the orthogonal projection of R^3 onto the plane W having the equation $x + y + z = 0$.
   (a) Find a formula for T.   (b) Find $T(3, 8, 4)$.

14. In each part let $T: R^2 \rightarrow R^2$ be the linear operator that rotates each vector in the plane through the angle θ. Find $T(-1, 2)$ and $T(x, y)$ when

   (a) $\theta = \dfrac{\pi}{4}$ (b) $\theta = \pi$ (c) $\theta = \dfrac{\pi}{6}$ (d) $\theta = -\dfrac{\pi}{3}$

15. Let $T: R^3 \rightarrow R^3$ be the orthogonal projection on the subspace of R^3 spanned by the vectors

$$\mathbf{v}_1 = (\tfrac{1}{3}, \tfrac{2}{3}, \tfrac{2}{3}) \qquad \text{and} \qquad \mathbf{v}_2 = (0, -\tfrac{1}{\sqrt{2}}, \tfrac{1}{\sqrt{2}})$$

Find

   (a) $T(1, -2, 4)$ (b) $T(1, 1, 1)$ (c) $T(x, y, z)$

16. Show that the function T in Example 10 is a linear operator.

17. In each part, formulas are given for linear operators $T_1: R^2 \rightarrow R^2$ and $T_2: R^2 \rightarrow R^2$. State the domain and image space of $T_2 \circ T_1$, and find $(T_2 \circ T_1)(x, y)$.
   (a) $T_1(x, y) = (2x, 3y)$, $T_2(x, y) = (x - y, x + y)$
   (b) $T_1(x, y) = (x - 3y, 0)$, $T_2(x, y) = (4x - 5y, 3x - 6y)$

18. In each part, formulas are given for linear transformations $T_1: R^2 \rightarrow R^3$ and $T_2: R^3 \rightarrow R^2$. State the domain and image space of $T_2 \circ T_1$, and find $(T_2 \circ T_1)(x, y)$.
   (a) $T_1(x, y) = (2x, -3y, x + y)$, $T_2(x, y, z) = (x - y, y + z)$
   (b) $T_1(x, y) = (x - y, y + z, x - z)$, $T_2(x, y, z) = (0, x + y + z)$

19. In each part, formulas are given for linear transformations $T_1: R^2 \rightarrow R^3$, $T_2: R^3 \rightarrow R^3$, and $T_3: R^3 \rightarrow R^2$. State the domain and image space of $T_3 \circ T_2 \circ T_1$, and find $(T_3 \circ T_2 \circ T_1)(x, y)$.
   (a) $T_1(x, y) = (-2y, 3x, x - 2y)$, $T_2(x, y, z) = (y, z, x)$, $T_3(x, y, z) = (x + z, y - z)$
   (b) $T_1(x, y) = (x + y, y, -x)$, $T_2(x, y, z) = (0, x + y + z, 3y)$, $T_3(x, y, z) = (3x + 2y, 4z - x - 3y)$

20. Just as it is *not* true, in general, for matrices that $AB = BA$, so it is *not* true, in general, for linear transformations that $T_1 \circ T_2 = T_2 \circ T_1$. Show that $T_1 \circ T_2 \neq T_2 \circ T_1$ for the linear operators $T_1: R^2 \rightarrow R^2$ and $T_2: R^2 \rightarrow R^2$ given by the formulas

$$T_1(x, y) = (y, x) \qquad \text{and} \qquad T_2(x, y) = (0, x)$$

21. Suppose that the linear transformations $T_1: P_2 \rightarrow P_2$ and $T_2: P_2 \rightarrow P_3$ are given by the formulas

$$T_1(p(x)) = p(x + 1) \qquad \text{and} \qquad T_2(p(x)) = xp(x)$$

   Find $(T_2 \circ T_1)(a_0 + a_1 x + a_2 x^2)$.

22. Let $T_1: R^2 \rightarrow R^2$ be the rotation of R^2 through $60°$ and $T_2: R^2 \rightarrow R^2$ the rotation of R^2 through $90°$. Find $T_1(\mathbf{v})$, $T_2(\mathbf{v})$, and $(T_2 \circ T_1)(\mathbf{v})$ for

   (a) $\mathbf{v} = \begin{bmatrix} 1 \\ -1 \end{bmatrix}$ (b) $\mathbf{v} = \begin{bmatrix} -2 \\ 3 \end{bmatrix}$ (c) $\mathbf{v} = \begin{bmatrix} x \\ y \end{bmatrix}$

23. Let $\mathbf{v}_1, \mathbf{v}_2$, and $\mathbf{v}_3$ be vectors in a vector space V and $T: V \rightarrow R^3$ a linear transformation for which

$$T(\mathbf{v}_1) = (1, -1, 2), \qquad T(\mathbf{v}_2) = (0, 3, 2), \qquad T(\mathbf{v}_3) = (-3, 1, 2)$$

   Find $T(2\mathbf{v}_1 - 3\mathbf{v}_2 + 4\mathbf{v}_3)$.

24. Let $q_0(x)$ be a fixed polynomial of degree m, and define a function T with domain P_n by the formula $T(p(x)) = p(q_0(x))$.
   (a) Show that T is a linear transformation.   (b) What is the image space of T?

25. Use the definition of $T_3 \circ T_2 \circ T_1$ given by Formula (8) to prove that
   (a) $T_3 \circ T_2 \circ T_1$ is a linear transformation
   (b) $T_3 \circ T_2 \circ T_1 = (T_3 \circ T_2) \circ T_1$
   (c) $T_3 \circ T_2 \circ T_1 = T_3 \circ (T_2 \circ T_1)$

26. Let $T:R^3 \to R^3$ be the orthogonal projection of R^3 onto the xy-plane. Show that $T \circ T = T$.

27. (a) Let $T:V \to W$ be a linear transformation and k a scalar. Define the function $(kT):V \to W$ by $(kT)(\mathbf{v}) = k(T(\mathbf{v}))$. Show that kT is a linear transformation.
   (b) Find $(3T)(x_1, x_2)$ if $T:R^2 \to R^2$ is given by the formula $T(x_1, x_2) = (2x_1 - x_2, x_2 + x_1)$.

28. (a) Let $T_1:V \to W$ and $T_2:V \to W$ be linear transformations. Define the functions $(T_1 + T_2):V \to W$ and $(T_1 - T_2):V \to W$ by

$$(T_1 + T_2)(\mathbf{v}) = T_1(\mathbf{v}) + T_2(\mathbf{v})$$
$$(T_1 - T_2)(\mathbf{v}) = T_1(\mathbf{v}) - T_2(\mathbf{v})$$

Show that $T_1 + T_2$ and $T_1 - T_2$ are linear transformations.

   (b) Find $(T_1 + T_2)(x, y)$ and $(T_1 - T_2)(x, y)$ if $T_1:R^2 \to R^2$ and $T_2:R^2 \to R^2$ are given by the formulas $T_1(x, y) = (2y, 3x)$ and $T_2(x, y) = (y, x)$.

29. Let (a_0, b_0, c_0) be a fixed vector in R^3, and define a function $T:R^3 \to R^3$ by

$$T(x, y, z) = \begin{vmatrix} \mathbf{i} & \mathbf{j} & \mathbf{k} \\ a_0 & b_0 & c_0 \\ x & y & z \end{vmatrix}$$

Show that T is a linear operator. [**Hint.** The determinant is the cross product of the vectors $\mathbf{v}_0 = (a_0, b_0, c_0)$ and $\mathbf{u} = (x, y, z)$.]

30. (a) Prove that if a_1, a_2, b_1, and b_2 are any scalars, then the formula

$$F(x, y) = (a_1 x + b_1 y, a_2 x + b_2 y)$$

defines a linear operator on R^2.
   (b) Does the formula $F(x, y) = (a_1 x^2 + b_1 y^2, a_2 x^2 + b_2 y^2)$ define a linear operator on R^2? Explain.

31. **(For readers who have studied calculus.)** Let

$$D(\mathbf{f}) = \mathbf{f}' \qquad \text{and} \qquad J(\mathbf{f}) = \int_0^1 \mathbf{f}(x)\,dx$$

be the linear transformations in Examples 12 and 13. Find $(J \circ D)(\mathbf{f})$ for
   (a) $\mathbf{f}(x) = x^2 + 3x + 2$ (b) $\mathbf{f}(x) = \sin x$ (c) $\mathbf{f}(x) = x$

32. Prove that if $T:V \to W$ is a linear transformation, then $T(\mathbf{u} - \mathbf{v}) = T(\mathbf{u}) - T(\mathbf{v})$ for all vectors $\mathbf{u}$ and $\mathbf{v}$ in V.

33. Let $\{\mathbf{v}_1, \mathbf{v}_2, \ldots, \mathbf{v}_n\}$ be a basis for a vector space V, and let $T:V \to W$ be a linear transformation. Show that if $T(\mathbf{v}_1) = T(\mathbf{v}_2) = \cdots = T(\mathbf{v}_n) = \mathbf{0}$, then T is the zero transformation.

34. Let $\{\mathbf{v}_1, \mathbf{v}_2, \ldots, \mathbf{v}_n\}$ be a basis for a vector space V, and let $T:V \to V$ be a linear operator. Show that if $T(\mathbf{v}_1) = \mathbf{v}_1, T(\mathbf{v}_2) = \mathbf{v}_2, \ldots, T(\mathbf{v}_n) = \mathbf{v}_n$, then T is the identity transformation on V.

7.2 KERNEL AND RANGE

In this section we shall discuss some basic properties of linear transformations, and explore various relationships between linear transformations and matrices.

**PROPERTIES OF
LINEAR TRANS-
FORMATIONS**

Theorem 7.2.1. *If $T:V \to W$ is a linear transformation, then*

(a) $T(\mathbf{0}) = \mathbf{0}$
(b) $T(-\mathbf{v}) = -T(\mathbf{v})$ *for all* $\mathbf{v}$ *in* V
(c) $T(\mathbf{v} - \mathbf{w}) = T(\mathbf{v}) - T(\mathbf{w})$ *for all* $\mathbf{v}$ *and* $\mathbf{w}$ *in* V

Proof. Let $\mathbf{v}$ be any vector in V. Since $0\mathbf{v} = \mathbf{0}$ we have

$$T(\mathbf{0}) = T(0\mathbf{v}) = 0T(\mathbf{v}) = \mathbf{0}$$

which proves (a). Also,

$$T(-\mathbf{v}) = T((-1)\mathbf{v}) = (-1)T(\mathbf{v}) = -T(\mathbf{v})$$

which proves (b). Finally, $\mathbf{v} - \mathbf{w} = \mathbf{v} + (-1)\mathbf{w}$; thus,

$$\begin{aligned} T(\mathbf{v} - \mathbf{w}) &= T(\mathbf{v} + (-1)\mathbf{w}) \\ &= T(\mathbf{v}) + (-1)T(\mathbf{w}) \\ &= T(\mathbf{v}) - T(\mathbf{w}) \end{aligned}$$

which proves (c). ▌

**KERNEL
AND RANGE**

Definition. If $T:V \to W$ is a linear transformation, then the set of vectors in V that T maps into $\mathbf{0}$ is called the **kernel** (or **nullspace**) of T; it is denoted by $\ker(T)$. The set of all vectors in W that are images under T of at least one vector in V is called the **range** of T; it is denoted by $R(T)$.

Example 1 Let $T:V \to W$ be the zero transformation (Example 3 of Section 7.1). Since T maps *every* vector in V into $\mathbf{0}$, it follows that $\ker(T) = V$. Moreover, since $\mathbf{0}$ is the *only* image under T of vectors in V, we have $R(T) = \{\mathbf{0}\}$. ▲

Example 2 Let $I:V \to V$ be the identity transformation (Example 4 of Section 7.1). Since $I(\mathbf{v}) = \mathbf{v}$ for all vectors in V, *every* vector in V is the image of some vector (namely, itself); thus, $R(I) = V$. Since the *only* vector that I maps into $\mathbf{0}$ is $\mathbf{0}$, it follows that $\ker(I) = \{\mathbf{0}\}$. ▲

Example 3 Let $T:R^3 \to R^3$ be the orthogonal projection on the xy-plane (Example 7 of Section 7.1). The kernel of T is the set of points that T maps into $\mathbf{0} = (0, 0, 0)$; these are the points on the z-axis (Figure 1a). Since T maps every point in R^3 into the xy-plane, the range of T must be some subset of this plane. But every point $(x_0, y_0, 0)$ in the xy-plane is the image under T of some point; in fact it is the image of all points on the vertical line that passes through $(x_0, y_0, 0)$ (Figure 1b). Thus $R(T)$ is the entire xy-plane. ▲

(a)

(b)

Figure 1   ker(T) is the z-axis. $R(T)$ is the entire xy-plane.

Example 4   Let $T:R^2 \rightarrow R^2$ be the linear operator that rotates each vector in the xy-plane through the angle θ (Figure 2). Since *every* vector in the xy-plane can be obtained by rotating some vector through the angle θ (why?), we have $R(T) = R^2$. Moreover, the only vector that rotates into $\mathbf{0}$ is $\mathbf{0}$, so $\ker(T) = \{\mathbf{0}\}$. ▲

Figure 2

PROPERTIES OF KERNEL AND RANGE

In all of the foregoing examples, $\ker(T)$ and $R(T)$ turned out to be *subspaces*. In Examples 1, 2, and 4 they were either the zero subspace or the entire vector space. In Example 3 the kernel was a line through the origin and the range was a plane through the origin, both of which are subspaces of R^3. All of this is not accidental; it is a consequence of the following general result.

> **Theorem 7.2.2.** *If $T:V \rightarrow W$ is a linear transformation, then*
>
> (a) *the kernel of T is a subspace of V;*
> (b) *the range of T is a subspace of W.*

Proof (a). To show that $\ker(T)$ is a subspace, we must show that it contains at least one vector and is closed under addition and scalar multiplication. By part (a) of Theorem 7.2.1, the vector $\mathbf{0}$ is in $\ker(T)$, so this set contains at least one vector. Let v_1 and v_2 be vectors in ker(T), and let k be any scalar. Then

$$T(v_1 + v_2) = T(v_1) + T(v_2) = 0 + 0 = 0$$

so that $v_1 + v_2$ is in ker(T). Also,

$$T(kv_1) = kT(v_1) = k0 = 0$$

so that kv_1 is in ker(T).

Proof (b). Let w_1 and w_2 be vectors in the range of T, and let k be any scalar. To prove this part we must show that $w_1 + w_2$ and kw_1 are in the range of T; that is, we must find vectors a and b in V such that $T(a) = w_1 + w_2$ and $T(b) = kw_1$.

Since w_1 and w_2 are in the range of T, there are vectors a_1 and a_2 in V such that $T(a_1) = w_1$ and $T(a_2) = w_2$. Let $a = a_1 + a_2$ and $b = ka_1$. Then

$$T(a) = T(a_1 + a_2) = T(a_1) + T(a_2) = w_1 + w_2$$

and

$$T(b) = T(ka_1) = kT(a_1) = kw_1$$

which completes the proof. ∎

Example 5 Let $T:R^n \to R^m$ be multiplication by an $m \times n$ matrix A. Since $T(x) = Ax$, the kernel of T consists of all x in R^n such that $Ax = 0$, which is the nullspace of A.

The range of T consists of all vectors b in R^m such that $T(x) = b$ for at least one vector x in R^n. Since $T(x) = Ax$, this is just the set of all vectors b in R^m such that $Ax = b$ is a consistent system of linear equations. By Theorem 4.7.4, this is the column space of A. ▲

Example 6 Let $T:R^3 \to R^3$ be the linear operator defined by the formula

$$T(x_1, x_2, x_3) = (x_1 + x_2 + 2x_3,\ x_1 + x_3,\ 2x_1 + x_2 + 3x_3)$$

Find bases for the kernel and range of T.

Solution. The formula for T can be expressed in matrix notation as

$$T\left(\begin{bmatrix} x_1 \\ x_2 \\ x_3 \end{bmatrix}\right) = \begin{bmatrix} x_1 + x_2 + 2x_3 \\ x_1 + x_3 \\ 2x_1 + x_2 + 3x_3 \end{bmatrix} = \begin{bmatrix} 1 & 1 & 2 \\ 1 & 0 & 1 \\ 2 & 1 & 3 \end{bmatrix}\begin{bmatrix} x_1 \\ x_2 \\ x_3 \end{bmatrix}$$

so T is the same as multiplication by

$$A = \begin{bmatrix} 1 & 1 & 2 \\ 1 & 0 & 1 \\ 2 & 1 & 3 \end{bmatrix}$$

It follows from the previous example that the kernel of T is the nullspace of A, and the range of T is the column space of A. Using the methods studied in Section 4.6, the reader should be able to show that a basis for the nullspace is

$$\mathbf{v} = \begin{bmatrix} -1 \\ -1 \\ 1 \end{bmatrix}$$

or in horizontal notation $\mathbf{v} = (-1, -1, 1)$; and a basis for the column space is

$$\mathbf{w}_1 = \begin{bmatrix} 1 \\ 0 \\ 1 \end{bmatrix} \quad \text{and} \quad \mathbf{w}_2 = \begin{bmatrix} 0 \\ 1 \\ 1 \end{bmatrix}$$

or in horizontal notation $\mathbf{w}_1 = (1, 0, 1)$ and $\mathbf{w}_2 = (0, 1, 1)$. ▲

The results in the two foregoing examples are so important that we shall summarize them in a theorem.

Theorem 7.2.3. *If $T:R^n \rightarrow R^m$ is multiplication by an $m \times n$ matrix A, then*

(a) the kernel of T is the nullspace of A;
(b) the range of T is the column space of A.

In Section 4.7 we defined the rank of a matrix to be the dimension of its column (or row) space and the nullity to be the dimension of its nullspace. The following definition extends these definitions to general linear transformations.

RANK AND NULLITY OF LINEAR TRANS-FORMATIONS

Definition. If $T:V \rightarrow W$ is a linear transformation, then the dimension of the range of T is called the **rank of T** and is denoted by rank(T); the dimension of the kernel is called the **nullity of T** and is denoted by nullity(T).

If $T:R^n \rightarrow R^m$ is multiplication by an $m \times n$ matrix A, then we have two definitions of rank and nullity, one for A and one for T. The following theorem shows that those definitions do not conflict.

Theorem 7.2.4. *If A is an $m \times n$ matrix and $T:R^n \rightarrow R^m$ is multiplication by A, then*
(a) nullity(T) = nullity(A)
(b) rank(T) = rank(A)

Proof. From Theorem 7.2.3 and the definitions of rank and nullity for T and A we have

$$\text{nullity}(A) = \dim(\text{nullspace of } A) = \dim(\ker(T)) = \text{nullity}(T)$$
$$\text{rank}(A) = \dim(\text{column space of } A) = \dim(R(T)) = \text{rank}(T) \quad \blacksquare$$

Example 7   Let $T:R^6 \to R^4$ be multiplication by

$$A = \begin{bmatrix} -1 & 2 & 0 & 4 & 5 & -3 \\ 3 & -7 & 2 & 0 & 1 & 4 \\ 2 & -5 & 2 & 4 & 6 & 1 \\ 4 & -9 & 2 & -4 & -4 & 7 \end{bmatrix}$$

Find the rank and nullity of T.

Solution. In Example 1 of Section 4.7 we showed that $\text{rank}(A) = 2$ and $\text{nullity}(A) = 4$. Thus, from Theorem 7.2.3 we have $\text{rank}(T) = 2$ and $\text{nullity}(T) = 4$.

Example 8   Let $T:R^3 \to R^3$ be the orthogonal projection on the xy-plane. From Example 3, the kernel of T is the z-axis, which is 1-dimensional; and the range of T is the xy-plane, which is 2-dimensional. Thus,

$$\text{nullity}(T) = 1 \quad \text{and} \quad \text{rank}(T) = 2$$

DIMENSION THEOREM FOR LINEAR TRANSFORMATIONS

In Section 4.7 we proved the dimension theorem for matrices (Theorem 4.7.2), which states that for a matrix A with n columns

$$\text{rank}(A) + \text{nullity}(A) = n$$

The following theorem, whose proof is deferred to the end of this section, extends this result to more general linear transformations.

Theorem 7.2.5. (*Dimension Theorem for Linear Transformations*). *If $T:V \to W$ is a linear transformation from an n-dimensional vector space V to a vector space W, then*

$$rank(T) + nullity(T) = n \tag{1}$$

In words, this theorem states that *for linear transformations the rank plus the nullity is equal to the dimension of the domain.*

Example 9   Let $T:R^2 \to R^2$ be the linear operator that rotates each vector in the xy-plane through an angle θ. We showed in Example 4 that $\ker(T) = \{\mathbf{0}\}$ and $R(T) = R^2$. Thus,

$$\text{rank}(T) + \text{nullity}(T) = 0 + 2 = 2$$

which is consistent with the fact that the domain of T is 2-dimensional.

We leave it as an exercise to show that Theorem 7.2.4 reduces to Theorem 4.7.2 when T is multiplication by A.

OPTIONAL

Proof of Theorem 7.2.5. We must show that

$$\dim(R(T)) + \dim(\ker(T)) = n$$

We shall give the proof for the case where $1 \le \dim(\ker(T)) < n$. The cases $\dim(\ker(T)) = 0$ and $\dim(\ker(T)) = n$ are left as exercises. Assume $\dim(\ker(T)) = r$, and let $v_1, \ldots, v_r$ be a basis for the kernel. Since $\{v_1, \ldots, v_r\}$ is linearly independent, Theorem 4.5.5a states that there are $n - r$ vectors, $v_{r+1}, \ldots, v_n$, such that $\{v_1, \ldots, v_r, v_{r+1}, \ldots, v_n\}$ is a basis for V. To complete the proof, we shall show that the $n - r$ vectors in the set $S = \{T(v_{r+1}), \ldots, T(v_n)\}$ form a basis for the range of T. It will then follow that

$$\dim(R(T)) + \dim(\ker(T)) = (n - r) + r = n$$

First we show that S spans the range of T. If b is any vector in the range of T, then $b = T(v)$ for some vector v in V. Since $\{v_1, \ldots, v_r, v_{r+1}, \ldots, v_n\}$ is a basis for V, the vector v can be written in the form

$$v = c_1 v_1 + \cdots + c_r v_r + c_{r+1} v_{r+1} + \cdots + c_n v_n$$

Since $v_1, \ldots, v_r$ lie in the kernel of T, we have $T(v_1) = \cdots = T(v_r) = 0$, so that

$$b - T(v) = c_{r+1} T(v_{r+1}) + \cdots + c_n T(v_n)$$

Thus, S spans the range of T.

Finally, we show that S is a linearly independent set and consequently forms a basis for the range of T. Suppose some linear combination of the vectors in S is zero; that is,

$$k_{r+1} T(v_{r+1}) + \cdots + k_n T(v_n) = 0 \qquad (2)$$

We must show $k_{r+1} = \cdots = k_n = 0$. Since T is linear, (2) can be rewritten as

$$T(k_{r+1} v_{r+1} + \cdots + k_n v_n) = 0$$

which says that $k_{r+1} v_{r+1} + \cdots + k_n v_n$ is in the kernel of T. This vector can therefore be written as a linear combination of the basis vectors $\{v_1, \ldots, v_r\}$, say

$$k_{r+1} v_{r+1} + \cdots + k_n v_n = k_1 v_1 + \cdots + k_r v_r$$

Thus,

$$k_1 v_1 + \cdots + k_r v_r - k_{r+1} v_{r+1} - \cdots - k_n v_n = 0$$

Since $\{v_1, \ldots, v_n\}$ is linearly independent, all of the k's are zero; in particular, $k_{r+1} = \cdots = k_n = 0$, which completes the proof. ∎

EXERCISE SET 7.2

1. Let $T: R^2 \to R^2$ be the linear operator given by the formula

   $$T(x, y) = (2x - y, -8x + 4y)$$

   Which of the following vectors are in $R(T)$?

   (a) $(1, -4)$ (b) $(5, 0)$ (c) $(-3, 12)$

2. Let $T:R^2 \to R^2$ be the linear operator in Exercise 1. Which of the following vectors are in ker(T)? $\quad$ Set $= 0$
(a) (5, 10) $\quad$ (b) (3, 2) $\quad$ (c) (1, 1)

3. Let $T:R^4 \to R^3$ be the linear transformation given by the formula

$$T(x_1, x_2, x_3, x_4) = (4x_1 + x_2 - 2x_3 - 3x_4, 2x_1 + x_2 + x_3 - 4x_4, 6x_1 - 9x_3 + 9x_4)$$

Which of the following are in $R(T)$?
(a) (0, 0, 6) $\quad$ (b) (1, 3, 0) $\quad$ (c) (2, 4, 1)

4. Let $T:R^4 \to R^3$ be the linear transformation in Exercise 3. Which of the following are in ker(T)?
(a) (3, -8, 2, 0) $\quad$ (b) (0, 0, 0, 1) $\quad$ (c) (0, -4, 1, 0)

5. Let $T:P_2 \to P_3$ be the linear transformation defined by $T(p(x)) = xp(x)$. Which of the following are in ker(T)?
(a) x^2 $\quad$ (b) 0 $\quad$ (c) $1 + x$

6. Let $T:P_2 \to P_3$ be the linear transformation in Exercise 5. Which of the following are in $R(T)$?
(a) $x + x^2$ $\quad$ (b) $1 + x$ $\quad$ (c) $3 - x^2$

7. Find a basis for the kernel of
(a) the linear operator in Exercise 1 $\quad$ (b) the linear transformation in Exercise 3
(c) the linear transformation in Exercise 5.

8. Find a basis for the range of
(a) the linear operator in Exercise 1 $\quad$ (b) the linear transformation in Exercise 3
(c) the linear transformation in Exercise 5.

9. Verify Formula (1) of the dimension theorem for
(a) the linear operator in Exercise 1 $\quad$ (b) the linear transformation in Exercise 3
(c) the linear transformation in Exercise 5.

In Exercises 10–13 let T be multiplication by the matrix A. Find
(a) a basis for the range of T $\quad$ (b) a basis for the kernel of T
(c) the rank and nullity of T $\quad$ (d) the rank and nullity of A.

10. $A = \begin{bmatrix} 1 & -1 & 3 \\ 5 & 6 & -4 \\ 7 & 4 & 2 \end{bmatrix}$ **11.** $A = \begin{bmatrix} 2 & 0 & -1 \\ 4 & 0 & -2 \\ 0 & 0 & 0 \end{bmatrix}$ **12.** $A = \begin{bmatrix} 4 & 1 & 5 & 2 \\ 1 & 2 & 3 & 0 \end{bmatrix}$ **13.** $A = \begin{bmatrix} 1 & 4 & 5 & 0 & 9 \\ 3 & -2 & 1 & 0 & -1 \\ -1 & 0 & -1 & 0 & -1 \\ 2 & 3 & 5 & 1 & 8 \end{bmatrix}$

14. Describe the range and nullspace of
(a) the orthogonal projection on the xz-plane
(b) the orthogonal projection on the yz-plane
(c) the orthogonal projection on the plane with the equation $y = x$

15. Let V be any vector space, and let $T:V \to V$ be defined by $T(\mathbf{v}) = 3\mathbf{v}$.
(a) What is the kernel of T? $\quad$ (b) What is the range of T?

16. In each part use the given information to find the nullity of T.
(a) $T:R^5 \to R^7$ has rank 3. $\quad$ (b) $T:P_4 \to P_3$ has rank 1.
(c) The range of $T:R^6 \to R^3$ is R^3. $\quad$ (d) $T:M_{22} \to M_{22}$ has rank 3.

17. Let A be a 7×6 matrix such that $A\mathbf{x} = \mathbf{0}$ has only the trivial solution, and let $T:R^6 \to R^7$ be multiplication by A. Find the rank and nullity of T.

18. Let A be a 5 × 7 matrix with rank 4.
    (a) What is the dimension of the solution space of $A\mathbf{x} = \mathbf{0}$?
    (b) Is $A\mathbf{x} = \mathbf{b}$ consistent for all vectors $\mathbf{b}$ in R^5? Explain.

19. Let $T:R^3 \to V$ be a linear transformation from R^3 to any vector space. Show that the kernel of T is a line through the origin, a plane through the origin, the origin only, or all of R^3.

20. Let $T:V \to R^3$ be a linear transformation from any vector space to R^3. Show that the range of T is a line through the origin, a plane through the origin, the origin only, or all of R^3.

21. Let $T:R^3 \to R^3$ be multiplication by

$$\begin{bmatrix} 1 & 3 & 4 \\ 3 & 4 & 7 \\ -2 & 2 & 0 \end{bmatrix}$$

    (a) Show that the kernel of T is a line through the origin, and find parametric equations for it.
    (b) Show that the range of T is a plane through the origin, and find an equation for it.

22. Prove: If $\{\mathbf{v}_1, \mathbf{v}_2, \ldots, \mathbf{v}_n\}$ is a basis for V and $\mathbf{w}_1, \mathbf{w}_2, \ldots, \mathbf{w}_n$ are vectors in W, not necessarily distinct, then there exists a linear transformation $T:V \to W$ such that $T(\mathbf{v}_1) = \mathbf{w}_1$, $T(\mathbf{v}_2) = \mathbf{w}_2, \ldots, T(\mathbf{v}_n) = \mathbf{w}_n$.

23. Prove the dimension theorem in the cases
    (a) $\dim(\ker(T)) = 0$ (b) $\dim(\ker(T)) = n$

24. Let $T:V \to V$ be a linear operator on a finite dimensional vector space V. Prove that $R(T) = V$ if and only if $\ker(T) = \{\mathbf{0}\}$.

25. **(For readers who have studied calculus.)** Let $D:P_3 \to P_2$ be the differentiation transformation $D(\mathbf{p}) = \mathbf{p}'$. Describe the kernel of D.

26. **(For readers who have studied calculus.)** Let $J:P_1 \to R$ be the integration transformation $J(\mathbf{p}) = \int_{-1}^{1} \mathbf{p}(x)\,dx$. Describe the kernel of J.

7.3 INVERSE LINEAR TRANSFORMATIONS

In this section we shall introduce the concept of the inverse of a linear transformation, and we shall explore the relationship between the inverse of a linear operator and the inverse of a matrix.

ONE-TO-ONE LINEAR TRANS-FORMATIONS

Linear transformations that map distinct vectors into distinct vectors are of special importance. One example of such a transformation is the linear operator $T:R^2 \to R^2$ that rotates each vector through an angle θ. It is obvious geometrically that if $\mathbf{v}$ and $\mathbf{w}$ are distinct vectors in R^2, then so are the rotated vectors $T(\mathbf{v})$ and $T(\mathbf{w})$ (Figure 1).

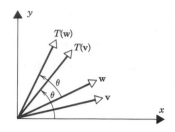

Distinct vectors **v** and **w** are rotated into distinct vectors $T(\mathbf{v})$ and $T(\mathbf{w})$.

In contrast, if $T: R^3 \to R^3$ is the orthogonal projection of R^3 on the xy-plane, then distinct points on the same vertical line get mapped into the same point in the xy-plane (Figure 2).

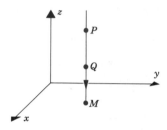

Figure 2   The distinct points P and Q get mapped into the same point M.

> **Definition.** A linear transformation $T: V \to W$ is said to be ***one-to-one*** if T maps distinct vectors in V into distinct vectors in W.

Example 1   In the terminology of the foregoing definition, the rotation operator of Figure 1 is one-to-one, but the orthogonal projection operator of Figure 2 is not. ▲

The following theorem establishes a relationship between a one-to-one linear transformation and its kernel.

> **Theorem 7.3.1.** *If $T: V \to W$ is a linear transformation, then the following are equivalent:*
>
> *(a) T is one-to-one;*
> *(b) the kernel of T contains only the zero vector, that is, $\ker(T) = \{\mathbf{0}\}$;*
> *(c) $\text{nullity}(T) = 0$.*

Proof. We leave it as a simple exercise to show the equivalence of (b) and (c); we shall complete the proof by proving the equivalence of (a) and (b).

(a) ⇒ (b). Assume that T is one-to-one, and let $\mathbf{v}$ be any vector in ker(T). Since $\mathbf{v}$ and $\mathbf{0}$ both lie in ker(T) we have $T(\mathbf{v}) = \mathbf{0}$ and $T(\mathbf{0}) = \mathbf{0}$, so $T(\mathbf{v}) = T(\mathbf{0})$. But this implies that $\mathbf{v} = \mathbf{0}$, since T is one-to-one; thus, ker(T) contains only the zero vector.

(b) ⇒ (a). Assume that ker(T) = $\{\mathbf{0}\}$ and that $\mathbf{v}$ and $\mathbf{w}$ are distinct vectors in V, that is,

$$\mathbf{v} - \mathbf{w} \neq \mathbf{0} \tag{1}$$

To prove that T is one-to-one we must show that $T(\mathbf{v})$ and $T(\mathbf{w})$ are distinct vectors. But if this were not so, then we would have

$$T(\mathbf{v}) = T(\mathbf{w})$$
$$T(\mathbf{v}) - T(\mathbf{w}) = \mathbf{0}$$
$$T(\mathbf{v} - \mathbf{w}) = \mathbf{0}$$

which implies that $\mathbf{v} - \mathbf{w}$ is in the kernel of T. Since ker(T) = $\{\mathbf{0}\}$, this implies that

$$\mathbf{v} - \mathbf{w} = \mathbf{0}$$

which contradicts (1). Thus, $T(\mathbf{v})$ and $T(\mathbf{w})$ must be distinct. ▌

Example 2 In each part determine whether the linear transformation is one-to-one by finding the kernel or the nullity and applying Theorem 7.3.1.

(a) $T:R^2 \rightarrow R^2$ rotates each vector through the angle θ.
(b) $T:R^3 \rightarrow R^3$ is the orthogonal projection on the xy-plane.
(c) $T:R^6 \rightarrow R^4$ is multiplication by the matrix

$$A = \begin{bmatrix} -1 & 2 & 0 & 4 & 5 & -3 \\ 3 & -7 & 2 & 0 & 1 & 4 \\ 2 & -5 & 2 & 4 & 6 & 1 \\ 4 & -9 & 2 & -4 & -4 & 7 \end{bmatrix}$$

Solution (a). From Example 4 of Section 7.2, ker(T) = $\{\mathbf{0}\}$, so T is one-to-one.

Solution (b). From Example 3 of Section 7.2, ker(T) contains nonzero vectors, so T is not one-to-one.

Solution (c). From Example 7 of Section 7.2, nullity(T) = 4, so T is not one-to-one.

In the special case where T is a *linear operator* on a *finite-dimensional* vector space, a fourth equivalent statement can be added to those in Theorem 7.3.1.

Theorem 7.3.2. *If V is a finite-dimensional vector space, and $T: V \rightarrow V$ is a linear operator, then the following are equivalent:*

(a) *T is one-to-one*;
(b) *$\ker(T) = \{\mathbf{0}\}$*;
(c) *nullity$(T) = 0$*;
(d) *the range of T is V, that is, $R(T) = V$.*

Proof. We already know that (a), (b), and (c) are equivalent, so we can complete the proof by proving the equivalence of (c) and (d).

$(c) \Rightarrow (d)$. Suppose that $\dim(V) = n$ and that nullity$(T) = 0$. It follows from Theorem 7.2.5 that

$$\text{rank}(T) = n - \text{nullity}(T) = n$$

By definition, rank(T) is the dimension of the range of T, so the range of T has dimension n. It now follows from Theorem 4.5.5 that the range of T is V, since the two spaces have the same dimension.

$(d) \Rightarrow (c)$. Suppose that $\dim(V) = n$ and that $R(T) = V$. It follows from these relationships that $\dim(R(T)) = n$, or equivalently, rank$(T) = n$. Thus, it follows from Theorem 7.2.5 that

$$\text{nullity}(T) = n - \text{rank}(T) = n - n = 0 \quad \blacksquare$$

The foregoing theorem provides us with an important link between invertible matrices and one-to-one linear operators.

Theorem 7.3.3. *If $T: R^n \rightarrow R^n$ is multiplication by an $n \times n$ matrix A, then the following are equivalent:*

(a) *T is one-to-one*;
(b) *A is invertible.*

Proof (a) $\Rightarrow$ (b). Assume that T is one-to-one. Thus, nullity$(T) = 0$ by Theorem 7.3.2, and consequently nullity$(A) = 0$ by Theorem 7.2.4. Thus, A is invertible by Theorem 4.7.3.

$(b) \Rightarrow (a)$. Assume that A is invertible. Thus, nullity$(A) = 0$ by Theorem 4.7.3, and consequently nullity$(T) = 0$ by Theorem 7.2.4. Thus, T is one-to-one by Theorem 7.3.2. $\blacksquare$

Example 3 Let $T:R^4 \to R^4$ be multiplication by

$$A = \begin{bmatrix} 1 & 3 & -2 & 4 \\ 2 & 6 & -4 & 8 \\ 3 & 9 & 1 & 5 \\ 1 & 1 & 4 & 8 \end{bmatrix}$$

Determine whether T is one-to-one.

Solution. By Theorem 7.3.3, the given problem is equivalent to determining whether A is invertible. But $\det(A) = 0$, since the first two rows of A are proportional, and consequently A is not invertible. Thus, T is not one-to-one. ▲

INVERSE LINEAR TRANS-FORMATIONS

Recall that if $T:V \to W$ is a linear transformation, then the range of T, denoted by $R(T)$, is the subspace of W consisting of all images under T of vectors in V. If T is one-to-one, then each vector $\mathbf{v}$ in V has a unique image $\mathbf{w} = T(\mathbf{v})$ in $R(T)$. This uniqueness of the image vector allows us to define a new function, called the *inverse of T*, and denoted by T^{-1}, which maps $\mathbf{w}$ back into $\mathbf{v}$ (Figure 3).

Figure 3   The inverse of T maps $T(\mathbf{v})$ back into $\mathbf{v}$.

It can be proved (Exercise 16) that $T^{-1}:R(T) \to V$ is a linear transformation. Moreover, it follows from the definition of T^{-1} that

$$T^{-1}(T(\mathbf{v})) = T^{-1}(\mathbf{w}) = \mathbf{v} \tag{2a}$$

$$T(T^{-1}(\mathbf{w})) = T(\mathbf{v}) = \mathbf{w} \tag{2b}$$

so that T and T^{-1}, when applied in succession in either order, cancel the effect of one another.

Example 4 Find T^{-1} if $T:R^2 \to R^2$ is the linear operator that rotates each vector in R^2 through an angle θ.

Solution. From Example 4 of the previous section, the range of T is R^2, so the domain of T^{-1} is also R^2. Moreover, it is evident geometrically that if a vector $\mathbf{w}$ is obtained by rotating a vector $\mathbf{v}$ through an angle θ, then $\mathbf{v}$ can be obtained by rotating $\mathbf{w}$ through the angle $-\theta$. Thus, $T^{-1}:R^2 \to R^2$ is the linear operator that rotates each vector in R^2 through the angle $-\theta$ (Figure 4). ▲

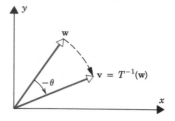

Figure 4

We conclude this section with an important theorem that relates inverses of matrices and linear operators.

Theorem 7.3.4. *If $T:R^n \to R^n$ is multiplication by an invertible $n \times n$ matrix A, then T has an inverse and $T^{-1}:R^n \to R^n$ is multiplication by A^{-1}.*

Proof. Since A is invertible, it follows from Theorem 7.3.3 that T is one-to-one, so T has an inverse. To complete the proof we must show that the domain of T^{-1} is R^n and that for each vector $\mathbf{w}$ in R^n, $T^{-1}(\mathbf{w}) = A^{-1}\mathbf{w}$.

By definition, the domain of T^{-1} is the range of T, so we must show that the range of T is R^n. Since T is one-to-one, this follows from parts (*a*) and (*d*) of Theorem 7.3.2.

Now let $\mathbf{w}$ be any vector in R^n. Since the range of T is R^n, there is some vector $\mathbf{v}$ such that

$$T(\mathbf{v}) = \mathbf{w} \tag{3}$$

or equivalently

$$T^{-1}(\mathbf{w}) = \mathbf{v} \tag{4}$$

Since T is multiplication by A, it follows from (3) that

$$A\mathbf{v} = \mathbf{w}$$
$$A^{-1}A\mathbf{v} = A^{-1}\mathbf{w}$$
$$\mathbf{v} = A^{-1}\mathbf{w} \tag{5}$$

Thus, from (4) and (5) we obtain

$$T^{-1}(\mathbf{w}) = A^{-1}\mathbf{w}$$

which shows that T^{-1} is multiplication by A^{-1}. ∎

Example 5 Let $T:R^2 \to R^2$ be the linear operator that rotates each vector in R^2 counterclockwise through an angle θ. We observed in Example 2 of Section 7.1 that T is multiplication by the matrix

$$A = \begin{bmatrix} \cos \theta & -\sin \theta \\ \sin \theta & \cos \theta \end{bmatrix} \tag{6}$$

This matrix is invertible since

$$\det(A) = \cos^2 \theta - (-\sin^2 \theta) = \cos^2 \theta + \sin^2 \theta = 1$$

so $\det(A) \neq 0$. Thus, it follows from Theorem 7.3.4 that $T^{-1}:R^2 \to R^2$ is multiplication by A^{-1}. It will be instructive to verify that this is so by direct computation. Using Formula (1) of Section 1.5 to invert (6) yields

$$A^{-1} = \begin{bmatrix} \cos \theta & \sin \theta \\ -\sin \theta & \cos \theta \end{bmatrix} \tag{7}$$

(verify). It follows from Example 4 that $T^{-1}:R^2 \to R^2$ rotates each vector in R^2 through the angle $-\theta$. To see that this is the same as multiplication by A^{-1}, we shall use the trigonometric identities

$$\sin(-\theta) = -\sin \theta \quad \text{and} \quad \cos(-\theta) = \cos \theta$$

to rewrite (7) as

$$A^{-1} = \begin{bmatrix} \cos(-\theta) & -\sin(-\theta) \\ \sin(-\theta) & \cos(-\theta) \end{bmatrix}$$

But this matrix has the same form as (6), except that $-\theta$ replaces θ. Thus, multiplication by A^{-1} rotates each vector in R^2 through the angle $-\theta$. ▲

EXERCISE SET 7.3

1. In each part find $\ker(T)$, and determine whether the linear transformation T is one-to-one.
   (a) $T:R^2 \to R^2$, where $T(x, y) = (y, x)$
   (b) $T:R^2 \to R^2$, where $T(x, y) = (0, 2x + 3y)$
   (c) $T:R^2 \to R^2$, where $T(x, y) = (x + y, x - y)$
   (d) $T:R^2 \to R^3$, where $T(x, y) = (x, y, x + y)$
   (e) $T:R^2 \to R^3$, where $T(x, y) = (x - y, y - x, 2x - 2y)$
   (f) $T:R^3 \to R^2$, where $T(x, y, z) = (x + y + z, x - y - z)$

2. In each part let $T:R^2 \to R^2$ be multiplication by A. Determine whether T has an inverse; if so, find

$$T^{-1}\left(\begin{bmatrix} x_1 \\ x_2 \end{bmatrix}\right)$$

(a) $A = \begin{bmatrix} 5 & 2 \\ 2 & 1 \end{bmatrix}$ (b) $A = \begin{bmatrix} 6 & -3 \\ 4 & -2 \end{bmatrix}$ (c) $A = \begin{bmatrix} 4 & 7 \\ -1 & 3 \end{bmatrix}$

3. In each part let $T:R^3 \to R^3$ be multiplication by A. Determine whether T has an inverse; if so, find

$$T^{-1}\left(\begin{bmatrix} x_1 \\ x_2 \\ x_3 \end{bmatrix}\right)$$

(a) $A = \begin{bmatrix} 1 & 5 & 2 \\ 1 & 2 & 1 \\ -1 & 1 & 0 \end{bmatrix}$ (b) $A = \begin{bmatrix} 1 & 4 & -1 \\ 1 & 2 & 1 \\ -1 & 1 & 0 \end{bmatrix}$ (c) $A = \begin{bmatrix} 1 & 0 & 1 \\ 0 & 1 & 1 \\ 1 & 1 & 0 \end{bmatrix}$ (d) $A = \begin{bmatrix} 1 & -1 & 1 \\ 0 & 2 & -1 \\ 2 & 3 & 0 \end{bmatrix}$

4. In each part determine whether multiplication by A is a one-to-one linear transformation.

(a) $A = \begin{bmatrix} 1 & -2 \\ 2 & -4 \\ -3 & 6 \end{bmatrix}$
(b) $A = \begin{bmatrix} 1 & 3 & 5 & 7 \\ 2 & -1 & 2 & 4 \\ -1 & 3 & 0 & 0 \end{bmatrix}$
(c) $A = \begin{bmatrix} 4 & -2 \\ 1 & 5 \\ 5 & 3 \end{bmatrix}$

5. Let $T:R^2 \to R^2$ be the orthogonal projection on the line $y = x$ (Figure 5).
(a) Find the kernel of T.
(b) Is T one-to-one? Justify your conclusion.

Figure 5

6. Let $T:R^2 \to R^2$ be the linear operator $T(x, y) = (-x, y)$ that reflects each point about the y-axis (Figure 6).
(a) Find the nullspace of T.
(b) Is T one-to-one? Justify your conclusion.

Figure 6

7. In each part use the given information to determine whether T is one-to-one.
(a) $T:R^m \to R^m$; nullity$(T) = 0$
(b) $T:R^n \to R^n$; rank$(T) = n - 1$
(c) $T:R^m \to R^n$; $n < m$
(d) $T:R^n \to R^n$; $R(T) = R^n$

8. In each part determine whether the linear transformation T is one-to-one.
(a) $T:P_2 \to P_3$, where $T(a_0 + a_1x + a_2x^2) = x(a_0 + a_1x + a_2x^2)$
(b) $T:P_2 \to P_2$, where $T(p(x)) = p(x + 1)$

9. Let A be a square matrix such that $\det(A) = 0$. Is multiplication by A a one-to-one linear transformation? Justify your conclusion.

10. In each part determine whether the linear operator $T:R^n \to R^n$ is one-to-one; if so, find $T^{-1}(x_1, x_2, \ldots, x_n)$.
(a) $T(x_1, x_2, \ldots, x_n) = (0, x_1, x_2, \ldots, x_{n-1})$
(b) $T(x_1, x_2, \ldots, x_n) = (x_n, x_{n-1}, \ldots, x_2, x_1)$
(c) $T(x_1, x_2, \ldots, x_n) = (x_2, x_3, \ldots, x_n, x_1)$

11. Let $T:R^n \to R^n$ be the linear operator defined by the formula

$$T(x_1, x_2, \ldots, x_n) = (a_1x_1, a_2x_2, \ldots, a_nx_n)$$

(a) Under what conditions will T have an inverse?
(b) Assuming that the conditions determined in part (a) are satisfied, find a formula for $T^{-1}(x_1, x_2, \ldots, x_n)$.

12. Let $T_1: V \rightarrow V$ and $T_2: V \rightarrow V$ be one-to-one linear operators.
    (a) Show that $(T_2 \circ T_1): V \rightarrow V$ is one-to-one.
    (b) Show that $(T_2 \circ T_1)^{-1} = T_1^{-1} \circ T_2^{-1}$.

13. Prove: If V and W are finite-dimensional vector spaces such that $\dim W < \dim V$, then there is no one-to-one linear transformation $T: V \rightarrow W$.

14. In each part determine whether the linear operator $T: M_{22} \rightarrow M_{22}$ is one-to-one. If so, find

$$T^{-1}\left(\begin{bmatrix} a & b \\ c & d \end{bmatrix}\right)$$

(a) $T\left(\begin{bmatrix} a & b \\ c & d \end{bmatrix}\right) = \begin{bmatrix} a & 0 \\ 0 & d \end{bmatrix}$ (b) $T\left(\begin{bmatrix} a & b \\ c & d \end{bmatrix}\right) = \begin{bmatrix} a & c \\ b & d \end{bmatrix}$ (c) $T\left(\begin{bmatrix} a & b \\ c & d \end{bmatrix}\right) = \begin{bmatrix} d & -b \\ -c & a \end{bmatrix}$

15. Let $T: R^2 \rightarrow R^2$ be the linear operator given by the formula $T(x, y) = (x + ky, -y)$. Show that T is one-to-one for every real value of k and that $T^{-1} = T$.

16. Prove that if $T: V \rightarrow W$ is a one-to-one linear transformation, then $T^{-1}: R(T) \rightarrow V$ is a linear transformation.

17. **(For readers who have studied calculus.)** Let $D: P_3 \rightarrow P_2$ be the differentiation transformation $D(\mathbf{p}) = \mathbf{p}'$. Determine whether D is one to one. Justify your conclusion.

18. **(For readers who have studied calculus.)** Let $J: P_1 \rightarrow R$ be the integration transformation $J(\mathbf{p}) = \int_{-1}^{1} \mathbf{p}(x)\,dx$. Determine whether J is one-to-one. Justify your conclusion.

7.4 LINEAR TRANSFORMATIONS FROM R^n TO R^m

In this section we shall obtain some additional properties of linear transformations, and we shall show that if $T: R^n \rightarrow R^m$ is any linear transformation, then there is some $m \times n$ matrix A such that T is multiplication by A.

FINDING LINEAR TRANSFORMATIONS FROM IMAGES OF BASIS VECTORS

Our first objective in this section is to show that a linear transformation is completely determined by its "values" at a basis. More precisely, we shall show that if $\{v_1, v_2, \ldots, v_n\}$ is a basis for a vector space V, and if $T: V \rightarrow W$ is a linear transformation, then the image $T(\mathbf{v})$ of any vector $\mathbf{v}$ in V can be calculated from the images

$$T(v_1), T(v_2), \ldots, T(v_n)$$

of the basis vectors. This can be done by first expressing $\mathbf{v}$ as a linear combination of the basis vectors, say

$$\mathbf{v} = k_1 v_1 + k_2 v_2 + \cdots + k_n v_n$$

and then using Formula (2) of Section 7.1 to write

$$T(\mathbf{v}) = k_1 T(v_1) + k_2 T(v_2) + \cdots + k_n T(v_n)$$

Example 1 Consider the basis $S = \{v_1, v_2, v_3\}$ for R^3, where $v_1 = (1, 1, 1)$, $v_2 = (1, 1, 0)$, $v_3 = (1, 0, 0)$; and let $T: R^3 \to R^2$ be the linear transformation such that

$$T(v_1) = (1, 0) \qquad T(v_2) = (2, -1) \qquad T(v_3) = (4, 3)$$

Find a formula for $T(x_1, x_2, x_3)$; then use this formula to compute $T(2, -3, 5)$.

Solution. We first express $x = (x_1, x_2, x_3)$ as a linear combination of $v_1 = (1, 1, 1)$, $v_2 = (1, 1, 0)$, and $v_3 = (1, 0, 0)$. If we write

$$(x_1, x_2, x_3) = k_1(1, 1, 1) + k_2(1, 1, 0) + k_3(1, 0, 0)$$

then on equating corresponding components we obtain

$$\begin{aligned} k_1 + k_2 + k_3 &= x_1 \\ k_1 + k_2 \quad\;\; &= x_2 \\ k_1 \qquad\quad &= x_3 \end{aligned}$$

which yields $k_1 = x_3$, $k_2 = x_2 - x_3$, $k_3 = x_1 - x_2$ so that

$$\begin{aligned} (x_1, x_2, x_3) &= x_3(1, 1, 1) + (x_2 - x_3)(1, 1, 0) + (x_1 - x_2)(1, 0, 0) \\ &= x_3 v_1 + (x_2 - x_3)v_2 + (x_1 - x_2)v_3 \end{aligned}$$

Thus,

$$\begin{aligned} T(x_1, x_2, x_3) &= x_3 T(v_1) + (x_2 - x_3)T(v_2) + (x_1 - x_2)T(v_3) \\ &= x_3(1, 0) + (x_2 - x_3)(2, -1) + (x_1 - x_2)(4, 3) \\ &= (4x_1 - 2x_2 - x_3, \; 3x_1 - 4x_2 + x_3) \end{aligned}$$

From this formula we obtain

$$T(2, -3, 5) = (9, 23) \quad \blacktriangle$$

ALL LINEAR TRANSFORMA-TIONS FROM R^n TO R^m ARE MATRIX TRANS-FORMATIONS

Our next objective is to show that *every* linear transformation from R^n to R^m can be carried out by performing an appropriate matrix multiplication. More precisely, we shall show that if

$$T: R^n \to R^m$$

is *any* linear transformation, then there is an $m \times n$ matrix A such that T is multiplication by A. For this discussion, we shall assume that all vectors are expressed in matrix notation. Thus, the formula

$$T(x_1, x_2, x_3) = (4x_1 - 2x_2 - x_3, \; 3x_1 - 4x_2 + x_3)$$

obtained in Example 1 would be written as

$$T\left(\begin{bmatrix} x_1 \\ x_2 \\ x_3 \end{bmatrix}\right) = \begin{bmatrix} 4x_1 - 2x_2 - x_3 \\ 3x_1 - 4x_2 + x_3 \end{bmatrix} \tag{1}$$

Suppose now that $T: R^n \to R^m$ is any linear transformation and that

$$\mathbf{e}_1 = \begin{bmatrix} 1 \\ 0 \\ 0 \\ \vdots \\ 0 \end{bmatrix}, \ \mathbf{e}_2 = \begin{bmatrix} 0 \\ 1 \\ 0 \\ \vdots \\ 0 \end{bmatrix}, \dots, \ \mathbf{e}_n = \begin{bmatrix} 0 \\ 0 \\ 0 \\ \vdots \\ 1 \end{bmatrix}$$

is the standard basis for R^n. Let us form the matrix A whose column vectors are the images under T of these standard basis vectors. More precisely, if

$$T(\mathbf{e}_1) = \begin{bmatrix} a_{11} \\ a_{21} \\ \vdots \\ a_{m1} \end{bmatrix}, \ T(\mathbf{e}_2) = \begin{bmatrix} a_{12} \\ a_{22} \\ \vdots \\ a_{m2} \end{bmatrix}, \dots, \ T(\mathbf{e}_n) = \begin{bmatrix} a_{1n} \\ a_{2n} \\ \vdots \\ a_{mn} \end{bmatrix}$$

then let

$$A = \begin{bmatrix} a_{11} & a_{12} & \cdots & a_{1n} \\ a_{21} & a_{22} & \cdots & a_{2n} \\ \vdots & \vdots & & \vdots \\ a_{m1} & a_{m2} & \cdots & a_{mn} \end{bmatrix}$$
$$\begin{matrix} \uparrow & \uparrow & & \uparrow \\ T(\mathbf{e}_1) & T(\mathbf{e}_2) & \cdots & T(\mathbf{e}_n) \end{matrix}$$

This matrix can be written more compactly as

$$A = [\,T(\mathbf{e}_1) \mid T(\mathbf{e}_2) \mid \cdots \mid T(\mathbf{e}_n)\,]$$

We shall show that the linear transformation $T: R^n \to R^m$ is multiplication by A. To see this, observe first that

$$\mathbf{x} = \begin{bmatrix} x_1 \\ x_2 \\ \vdots \\ x_n \end{bmatrix} = x_1\mathbf{e}_1 + x_2\mathbf{e}_2 + \cdots + x_n\mathbf{e}_n$$

Therefore, by the linearity of T,

$$T(\mathbf{x}) = x_1 T(\mathbf{e}_1) + x_2 T(\mathbf{e}_2) + \cdots + x_n T(\mathbf{e}_n) \tag{2}$$

Moreover,

$$A\mathbf{x} = \begin{bmatrix} a_{11} & a_{12} & \cdots & a_{1n} \\ a_{21} & a_{22} & \cdots & a_{2n} \\ \vdots & \vdots & & \vdots \\ a_{m1} & a_{m2} & \cdots & a_{mn} \end{bmatrix} \begin{bmatrix} x_1 \\ x_2 \\ \vdots \\ x_n \end{bmatrix} = \begin{bmatrix} a_{11}x_1 + a_{12}x_2 + \cdots + a_{1n}x_n \\ a_{21}x_1 + a_{22}x_2 + \cdots + a_{2n}x_n \\ \vdots & \vdots & \vdots \\ a_{m1}x_1 + a_{m2}x_2 + \cdots + a_{mn}x_n \end{bmatrix}$$

$$= x_1 \begin{bmatrix} a_{11} \\ a_{21} \\ \vdots \\ a_{m1} \end{bmatrix} + x_2 \begin{bmatrix} a_{12} \\ a_{22} \\ \vdots \\ a_{m2} \end{bmatrix} + \cdots + x_n \begin{bmatrix} a_{1n} \\ a_{2n} \\ \vdots \\ a_{mn} \end{bmatrix}$$

$$= x_1 T(\mathbf{e}_1) + x_2 T(\mathbf{e}_2) + \cdots + x_n T(\mathbf{e}_n) \tag{3}$$

Comparing (2) and (3) shows that $T(\mathbf{x}) = A\mathbf{x}$, so that T is multiplication by A. This is summarized in the following theorem.

Theorem 7.4.1. *If* $T: R^n \rightarrow R^m$ *is a linear transformation, and if* $S = \{\mathbf{e}_1, \mathbf{e}_2, \ldots, \mathbf{e}_n\}$ *is the standard basis for* R^n, *then* T *is multiplication by the matrix*

$$A = [T(\mathbf{e}_1) \mid T(\mathbf{e}_2) \mid \cdots \mid T(\mathbf{e}_n)] \qquad (4)$$

STANDARD MATRICES OF LINEAR TRANSFORMATIONS

The matrix A in this theorem is called the **standard matrix for** T; it is denoted by

$$[T]$$

Thus, (4) can be written as

$$[T] = [T(\mathbf{e}_1) \mid T(\mathbf{e}_2) \mid \cdots \mid T(\mathbf{e}_n)] \qquad (5)$$

Moreover, the fact that T is multiplication by $[T]$ is expressed by the mathematical equation

$$T(\mathbf{x}) = [T]\mathbf{x} \qquad (6)$$

REMARK. It is assumed in Formulas (4), (5), and (6) that the vectors are all written in matrix notation.

Example 2 Find the standard matrix for the linear transformation $T: R^3 \rightarrow R^2$ defined by the formula

$$T(x_1, x_2, x_3) = (4x_1 - 2x_2 - x_3, \; 3x_1 - 4x_2 + x_3)$$

Solution. In order to apply Formula (5), we must write the linear transformation in matrix notation. This yields

$$T\left(\begin{bmatrix} x_1 \\ x_2 \\ x_3 \end{bmatrix}\right) = \begin{bmatrix} 4x_1 - 2x_2 - x_3 \\ 3x_1 - 4x_2 + x_3 \end{bmatrix} \qquad (7)$$

From this formula we obtain

$$T(\mathbf{e}_1) = T\left(\begin{bmatrix} 1 \\ 0 \\ 0 \end{bmatrix}\right) = \begin{bmatrix} 4 \\ 3 \end{bmatrix}, \quad T(\mathbf{e}_2) = T\left(\begin{bmatrix} 0 \\ 1 \\ 0 \end{bmatrix}\right) = \begin{bmatrix} -2 \\ -4 \end{bmatrix}, \quad T(\mathbf{e}_3) = T\left(\begin{bmatrix} 0 \\ 0 \\ 1 \end{bmatrix}\right) = \begin{bmatrix} -1 \\ 1 \end{bmatrix}$$

Thus, the standard matrix for T is

$$[T] = [T(\mathbf{e}_1) \mid T(\mathbf{e}_2) \mid T(\mathbf{e}_3)] = \begin{bmatrix} 4 & -2 & -1 \\ 3 & -4 & 1 \end{bmatrix}$$

As a check, observe that

$$T(\mathbf{x}) = [T]\mathbf{x} = \begin{bmatrix} 4 & -2 & -1 \\ 3 & -4 & 1 \end{bmatrix} \begin{bmatrix} x_1 \\ x_2 \\ x_3 \end{bmatrix} = \begin{bmatrix} 4x_1 - 2x_2 - x_3 \\ 3x_1 - 4x_2 + x_3 \end{bmatrix}$$

which agrees with (7). ▲

STANDARD MATRICES FOR INVERSES AND COMPOSITIONS

We shall now consider two problems that are concerned with inverses and compositions of standard matrices.

Problem 1. If $T_1: R^n \to R^k$ and $T_2: R^k \to R^m$ are linear transformations, how is the standard matrix for the composition $(T_2 \circ T_1): R^n \to R^m$ related to the standard matrices for T_1 and T_2?

Problem 2. If $T: R^n \to R^n$ is a one-to-one linear operator, how is the standard matrix for $T^{-1}: R^n \to R^n$ related to the standard matrix for T?

The following basic fact will be helpful in solving these problems.

Theorem 7.4.2. *If A and B are $m \times n$ matrices such that $A\mathbf{x} = B\mathbf{x}$ for all $\mathbf{x}$ in R^n, then $A = B$.*

Proof. Since $A\mathbf{x} = B\mathbf{x}$ for all $\mathbf{x}$ in R^n, this equality holds for the standard basis vectors $\mathbf{e}_1, \mathbf{e}_2, \ldots, \mathbf{e}_n$; that is,

$$A\mathbf{e}_1 = B\mathbf{e}_1, \quad A\mathbf{e}_2 = B\mathbf{e}_2, \quad \ldots, \quad A\mathbf{e}_n = B\mathbf{e}_n$$

But $A\mathbf{e}_1, A\mathbf{e}_2, \ldots, A\mathbf{e}_n$ and $B\mathbf{e}_1, B\mathbf{e}_2, \ldots, B\mathbf{e}_n$ are the column vectors of A and B, respectively. Thus, A and B have the same column vectors and consequently $A = B$. ∎

REMARK. In the exercises, we shall ask you to prove a stronger version of the foregoing theorem, namely, if $A\mathbf{x} = B\mathbf{x}$ for all $\mathbf{x}$ in any *basis* for R^n, then $A = B$.

The following theorem solves Problem 1 posed above; it shows that the standard matrix for a composition is the product of the standard matrices for the individual linear transformations.

Theorem 7.4.3. *If $T_1: R^n \to R^k$ and $T_2: R^k \to R^m$ are linear transformations, then*

$$[T_2 \circ T_1] = [T_2][T_1] \tag{8}$$

Proof. **By Theorem 7.4.2, we can prove (8) by showing that**

$$[T_2 \circ T_1]\mathbf{x} = [T_2][T_1]\mathbf{x}$$

for all $\mathbf{x}$ in R^n. With the help of (6) this can be proved as follows:

$$[T_2 \circ T_1]\mathbf{x} = (T_2 \circ T_1)(\mathbf{x}) = T_2(T_1(\mathbf{x}))$$
$$= [T_2](T_1(\mathbf{x})) = [T_2][T_1]\mathbf{x} \quad \blacksquare$$

REMARK. Formula (8) can be extended to more than two linear transformations. For example,

$$[T_3 \circ T_2 \circ T_1] = [T_3][T_2][T_1]$$

provided that the domains and image spaces are such that the composition can be formed.

Example 3 Let $T_1 : R^2 \to R^2$ and $T_2 : R^2 \to R^2$ be the linear operators that rotate vectors in R^2 through the angles θ_1 and θ_2, respectively. Thus, $T_2 \circ T_1$ is the linear operator that rotates each vector through the angle $\theta_1 + \theta_2$ (Example 15 of Section 7.1). It follows from Example 2 of Section 7.1 that the standard matrices for these linear operators are

$$[T_1] = \begin{bmatrix} \cos \theta_1 & -\sin \theta_1 \\ \sin \theta_1 & \cos \theta_1 \end{bmatrix}, \qquad [T_2] = \begin{bmatrix} \cos \theta_2 & -\sin \theta_2 \\ \sin \theta_2 & \cos \theta_2 \end{bmatrix},$$

$$[T_2 \circ T_1] = \begin{bmatrix} \cos(\theta_1 + \theta_2) & -\sin(\theta_1 + \theta_2) \\ \sin(\theta_1 + \theta_2) & \cos(\theta_1 + \theta_2) \end{bmatrix}$$

These matrices should satisfy (8). With the help of some basic trigonometric identities, we can show that this is so as follows:

$$[T_2][T_1] = \begin{bmatrix} \cos \theta_2 & -\sin \theta_2 \\ \sin \theta_2 & \cos \theta_2 \end{bmatrix} \begin{bmatrix} \cos \theta_1 & -\sin \theta_1 \\ \sin \theta_1 & \cos \theta_1 \end{bmatrix}$$

$$= \begin{bmatrix} \cos \theta_2 \cos \theta_1 - \sin \theta_2 \sin \theta_1 & -(\cos \theta_2 \sin \theta_1 + \sin \theta_2 \cos \theta_1) \\ \sin \theta_2 \cos \theta_1 + \cos \theta_2 \sin \theta_1 & -\sin \theta_2 \sin \theta_1 + \cos \theta_2 \cos \theta_1 \end{bmatrix}$$

$$= \begin{bmatrix} \cos(\theta_1 + \theta_2) & -\sin(\theta_1 + \theta_2) \\ \sin(\theta_1 + \theta_2) & \cos(\theta_1 + \theta_2) \end{bmatrix}$$

$$= [T_2 \circ T_1] \quad \blacktriangle$$

The following theorem solves Problem 2 posed earlier in this section; it shows that the inverse of the standard matrix for a one-to-one linear operator is the standard matrix for the inverse of that operator.

Theorem 7.4.4. *If* $T: R^n \to R^n$ *is a linear operator, then the following are equivalent:*

(a) T *is one-to-one;*
(b) $[T]$ *is invertible.*

Moreover, when these equivalent conditions hold we have the following relationship:

$$[T^{-1}] = [T]^{-1} \qquad (9)$$

Proof. The equivalence of (*a*) and (*b*) is left as an exercise. By Theorem 7.4.2, we can prove (9) by showing that for all $\mathbf{x}$ in R^n

$$[T^{-1}]\mathbf{x} = [T]^{-1}\mathbf{x}$$

or equivalently (on multiplying by $[T]$)

$$[T][T^{-1}]\mathbf{x} = \mathbf{x}$$

But, from Formula (2b) of Section 7.3

$$[T][T^{-1}]\mathbf{x} = [T](T^{-1}(\mathbf{x})) = T(T^{-1}(\mathbf{x})) = \mathbf{x} \quad \blacksquare$$

Example 4   Let $T: R^3 \to R^3$ be the linear operator defined by the formula

$$T(x_1, x_2, x_3) = (3x_1 + x_2, \ -2x_1 - 4x_2 + 3x_3, \ 5x_1 + 4x_2 - 2x_3)$$

Determine whether T is one-to-one; if so, find $T^{-1}(x_1, x_2, x_3)$.

Solution.   Using the method of Example 2, the standard matrix for T is

$$[T] = \begin{bmatrix} 3 & 1 & 0 \\ -2 & -4 & 3 \\ 5 & 4 & -2 \end{bmatrix}$$

This matrix is invertible and

$$[T]^{-1} = \begin{bmatrix} 4 & -2 & -3 \\ -11 & 6 & 9 \\ -12 & 7 & 10 \end{bmatrix}$$

(verify). Thus, T is one-to-one by Theorem 7.4.4, and by Formula (9)

$$[T^{-1}] = \begin{bmatrix} 4 & -2 & -3 \\ -11 & 6 & 9 \\ -12 & 7 & 10 \end{bmatrix}$$

Applying Formula (6) with T^{-1} in place of T yields

$$T^{-1}\left(\begin{bmatrix} x_1 \\ x_2 \\ x_3 \end{bmatrix}\right) = [T^{-1}]\begin{bmatrix} x_1 \\ x_2 \\ x_3 \end{bmatrix} = \begin{bmatrix} 4 & -2 & -3 \\ -11 & 6 & 9 \\ -12 & 7 & 10 \end{bmatrix}\begin{bmatrix} x_1 \\ x_2 \\ x_3 \end{bmatrix} = \begin{bmatrix} 4x_1 - 2x_2 - 3x_3 \\ -11x_1 + 6x_2 + 9x_3 \\ -12x_1 + 7x_2 + 10x_3 \end{bmatrix}$$

Expressing this result in horizontal notation yields

$$T^{-1}(x_1, x_2, x_3) = (4x_1 - 2x_2 - 3x_3, \ -11x_1 + 6x_2 + 9x_3, \ -12x_1 + 7x_2 + 10x_3)$$

**STANDARD
MATRIX OF A
MATRIX
TRANSFOR-
MATION**

We conclude this section with an observation that is both interesting and important: Suppose we *start* with an $m \times n$ matrix B and *define* $T:R^n \rightarrow R^m$ to be multiplication by B. According to Theorem 7.4.1, T is also multiplication by

$$[T] = [T(\mathbf{e}_1) \mid T(\mathbf{e}_2) \mid \cdots \mid T(\mathbf{e}_n)]$$

Thus, T is multiplication by both B and $[T]$, that is,

$$B\mathbf{x} = [T]\mathbf{x}$$

for all $\mathbf{x}$ in R^n. But, by Theorem 7.4.2 this implies that $B = [T]$. In summary, *the standard matrix for a matrix transformation is the matrix itself*.

The foregoing observation suggests a new way of thinking about matrices: An arbitrary $m \times n$ matrix A can be viewed as the standard matrix for the linear transformation that maps the standard basis for R^n into the column vectors of A. Thus,

$$A = \begin{bmatrix} 1 & -2 & 1 \\ 3 & 4 & 6 \end{bmatrix}$$

is the standard matrix for the linear transformation of R^3 to R^2 that maps

$$\mathbf{e}_1 = \begin{bmatrix} 1 \\ 0 \\ 0 \end{bmatrix}, \qquad \mathbf{e}_2 = \begin{bmatrix} 0 \\ 1 \\ 0 \end{bmatrix}, \qquad \mathbf{e}_3 = \begin{bmatrix} 0 \\ 0 \\ 1 \end{bmatrix}$$

into

$$\begin{bmatrix} 1 \\ 3 \end{bmatrix}, \qquad \begin{bmatrix} -2 \\ 4 \end{bmatrix}, \qquad \begin{bmatrix} 1 \\ 6 \end{bmatrix}$$

respectively.

EXERCISE SET 7.4

1. Consider the basis $S = \{\mathbf{v}_1, \mathbf{v}_2\}$ for R^2, where $\mathbf{v}_1 = (1, 1)$ and $\mathbf{v}_2 = (2, 0)$; let $T:R^2 \rightarrow R^3$ be the linear transformation for which

$$T(\mathbf{v}_1) = (-1, 3, 4), \qquad T(\mathbf{v}_2) = (0, 2, -5)$$

Find a formula for $T(x_1, x_2)$ and use it to compute $T(3, -6)$.

2. Consider the basis $S = \{v_1, v_2, v_3\}$ for R^3, where $v_1 = (1, 2, 3)$, $v_2 = (2, 5, 3)$, and $v_3 = (1, 0, 10)$; let $T:R^3 \to R^2$ be the linear transformation for which

$$T(v_1) = (1, 0), \qquad T(v_2) = (1, 0), \qquad T(v_3) = (0, 1)$$

Find a formula for $T(x_1, x_2, x_3)$ and use it to compute $T(1, 1, 1)$.

3. Consider the linear transformation $T:P_2 \to P_2$ for which $T(1) = 1 + x$, $T(x) = 3 - x^2$, and $T(x^2) = 4 + 2x - 3x^2$. Find a formula for $T(a_0 + a_1 x + a_2 x^2)$ and use it to compute $T(2 - 2x + 3x^2)$.

4. Consider the basis $S = \{v_1, v_2, v_3\}$ for R^3, where $v_1 = (-1, 0, 1)$, $v_2 = (0, 1, -1)$, and $v_3 = (1, -1, 1)$; let $T:R^3 \to R^3$ be the linear transformation for which

$$T(v_1) = e_1, \qquad T(v_2) = e_2, \qquad T(v_3) = e_3$$

where $\{e_1, e_2, e_3\}$ is the standard basis for R^3. Find a formula for $T(x_1, x_2, x_3)$ and use it to compute $T(2, 1, -3)$.

5. Find the standard matrix for the linear operator T.
   (a) $T(x_1, x_2) = (2x_1 - x_2, x_1 + x_2)$
   (b) $T(x_1, x_2) = (x_1, x_2)$
   (c) $T(x_1, x_2, x_3) = (x_1 + 2x_2 + x_3, x_1 + 5x_2, x_3)$
   (d) $T(x_1, x_2, x_3) = (4x_1, 7x_2, -8x_3)$

6. Find the standard matrix for the linear transformation T.
   (a) $T(x_1, x_2) = (x_2, -x_1, x_1 + 3x_2, x_1 - x_2)$
   (b) $T(x_1, x_2, x_3, x_4) = (7x_1 + 2x_2 - x_3 + x_4, x_2 + x_3, -x_1)$
   (c) $T(x_1, x_2, x_3) = (0, 0, 0, 0, 0)$
   (d) $T(x_1, x_2, x_3, x_4) = (x_4, x_1, x_3, x_2, x_1 - x_3)$

7. In each part the standard matrix $[T]$ of a linear transformation T is given. Use it to find $T(x)$. [Express the answers in matrix form.]

(a) $[T] = \begin{bmatrix} 1 & 2 \\ 3 & 4 \end{bmatrix}$; $x = \begin{bmatrix} 3 \\ -2 \end{bmatrix}$

(b) $[T] = \begin{bmatrix} -1 & 2 & 0 \\ 3 & 1 & 5 \end{bmatrix}$; $x = \begin{bmatrix} -1 \\ 1 \\ 3 \end{bmatrix}$

(c) $[T] = \begin{bmatrix} -2 & 1 & 4 \\ 3 & 5 & 7 \\ 6 & 0 & -1 \end{bmatrix}$; $x = \begin{bmatrix} x_1 \\ x_2 \\ x_3 \end{bmatrix}$

(d) $[T] = \begin{bmatrix} -1 & 1 \\ 2 & 4 \\ 7 & 8 \end{bmatrix}$; $x = \begin{bmatrix} x_1 \\ x_2 \end{bmatrix}$

8. In each part use the standard matrix for T to find $T(x)$; then check the result by calculating $T(x)$ directly.
   (a) $T(x_1, x_2) = (-x_1 + x_2, x_2)$; $x = (-1, 4)$
   (b) $T(x_1, x_2, x_3) = (2x_1 - x_2 + x_3, x_2 + x_3, 0)$; $x = (2, 1, -3)$

9. Use Theorem 7.4.4 to determine whether $T:R^2 \to R^2$ is a one-to-one linear operator; if so, find $T^{-1}(x_1, x_2)$.
   (a) $T(x_1, x_2) = (x_1 + 2x_2, -x_1 + x_2)$
   (b) $T(x_1, x_2) = (4x_1 - 6x_2, -2x_1 + 3x_2)$
   (c) $T(x_1, x_2) = (-x_2, -x_1)$
   (d) $T(x_1, x_2) = (3x_1, -5x_1)$

10. Use Theorem 7.4.4 to determine whether $T:R^3 \to R^3$ is a one-to-one linear operator; if so, find $T^{-1}(x_1, x_2, x_3)$.
    (a) $T(x_1, x_2, x_3) = (x_1 - 2x_2 + 2x_3, 2x_1 + x_2 + x_3, x_1 + x_3)$
    (b) $T(x_1, x_2, x_3) = (x_1 - 3x_2 + 4x_3, -x_1 + x_2 + x_3, -2x_2 + 5x_3)$
    (c) $T(x_1, x_2, x_3) = (x_1 + 4x_2 - x_3, 2x_1 + 7x_2 + x_3, x_1 + 3x_2)$
    (d) $T(x_1, x_2, x_3) = (x_1 + 2x_2 + x_3, -2x_1 + x_2 + 4x_3, 7x_1 + 4x_2 - 5x_3)$

11. Find the standard matrix for T and use it to find the rank and nullity of T.

(a) $T(x_1, x_2, x_3, x_4) = (x_1 - 2x_2 + x_3 + 3x_4, 2x_1 + x_2 + 4x_4, -x_1 - 3x_2 + x_3 - x_4, 3x_1 + 3x_2 + x_3 + 7x_4)$

(b) $T(x_1, x_2, x_3, x_4, x_5) =$
$(x_1 + 4x_2 + 5x_3 + 9x_5, 3x_1 - 2x_2 + x_3 - x_5, -x_1 - x_3 - x_5, 2x_1 + 3x_2 + 5x_3 + x_4 + 8x_5)$

12. Verify Theorem 7.4.3 for the given linear transformations.

(a) $T_1(x_1, x_2) = (x_1 + x_2, x_1 - x_2)$ and $T_2(x_1, x_2) = (3x_1, 2x_1 + 4x_2)$

(b) $T_1(x_1, x_2) = (4x_1, -2x_1 + x_2, -x_1 - 3x_2)$ and $T_2(x_1, x_2, x_3) = (x_1 + 2x_2 - x_3, 4x_1 - x_3)$

(c) $T_1(x_1, x_2, x_3) = (-x_1 + x_2, -x_2 + x_3, -x_3 + x_1)$ and $T_2(x_1, x_2, x_3) = (-2x_1, 3x_3, -4x_2)$

13. Prove: If A and B are $m \times n$ matrices such that $A\mathbf{x} = B\mathbf{x}$ for all $\mathbf{x}$ in some basis for R^n, then $A = B$.

14. Let $T_1: R^2 \to R^2$, $T_2: R^2 \to R^2$, and $T_3: R^2 \to R^2$ be the linear operators that rotate vectors in R^2 through the angles θ_1, θ_2, and θ_3, respectively. Verify that

$$[T_3 \circ T_2 \circ T_1] = [T_3] \circ [T_2] \circ [T_1]$$

15. Let $T: R^n \to R^n$ be a linear operator.

(a) Prove: If T is one-to-one, then $[T]$ is an invertible matrix.

(b) Prove: If $[T]$ is an invertible matrix, then T is one-to-one.

16. Let $T: R^3 \to R^4$ be a linear transformation such that $T(1, 0, 0) = (1, 1, -2, -2)$ and $T(-1, 0, 1) = (3, 0, -4, 0)$.

(a) Find $T(-2, 0, 3)$.   (b) Can you find $T(1, 2, -1)$? Explain.

17. (a) Let $\mathbf{v}_0 = (a_0, b_0, c_0)$ be a fixed vector in R^3. Find the standard matrix for the "cross-product" operator $T: R^3 \to R^3$ defined by the formula

$$T(x, y, z) = \begin{vmatrix} \mathbf{i} & \mathbf{j} & \mathbf{k} \\ a_0 & b_0 & c_0 \\ x & y & z \end{vmatrix}$$

[*Note.* See Exercise 29 of Section 7.1.]

(b) Use the matrix $[T]$ obtained in part (a) to compute $T(x, y, z)$; then check your result by expanding the determinant in part (a).

7.5 MATRICES OF GENERAL LINEAR TRANSFORMATIONS

In this section we shall show that if V and W are finite-dimensional vector spaces (not necessarily R^n and R^m), then with a little ingenuity any linear transformation $T: V \to W$ can be regarded as a matrix transformation. The basic idea is to work with coordinate matrices of the vectors rather than with the vectors themselves.

MATRICES OF LINEAR TRANSFOR-MATIONS

Suppose that V is an n-dimensional vector space and W an m-dimensional vector space. If we choose bases B and B' for V and W, respectively, then for each $\mathbf{x}$ in V, the coordinate matrix $[\mathbf{x}]_B$ will be a vector in R^n, and the coordinate matrix $[T(\mathbf{x})]_{B'}$ will be a vector in R^m (Figure 1).

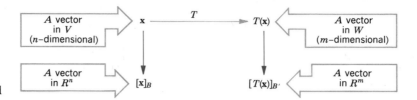

Figure 1

If, as illustrated in Figure 2, if we complete the rectangle suggested by Figure 1, we obtain a mapping from R^n to R^m, which can be shown to be a linear transformation. If we let A be the standard matrix for this transformation, then

$$A[\mathbf{x}]_B = [T(\mathbf{x})]_{B'} \tag{1}$$

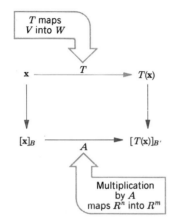

Figure 2

The matrix A in (1) is called the **matrix for T with respect to the bases B and B'**.

Later in this section, we shall give some of the uses of the matrix A in (1), but first, let us show how it can be computed. For this purpose, let us suppose that $B = \{\mathbf{u}_1, \mathbf{u}_2, \ldots, \mathbf{u}_n\}$ is a basis for the n-dimensional space V and $B' = \{\mathbf{v}_1, \mathbf{v}_2, \ldots, \mathbf{v}_m\}$ is a basis for the m-dimensional space W. We are looking for an $m \times n$ matrix

$$A = \begin{bmatrix} a_{11} & a_{12} & \cdots & a_{1n} \\ a_{21} & a_{22} & \cdots & a_{2n} \\ \vdots & \vdots & & \vdots \\ a_{m1} & a_{m2} & \cdots & a_{mn} \end{bmatrix}$$

such that (1) holds for all vectors $\mathbf{x}$ in V. In particular, we want this equation to hold for the basis vectors $\mathbf{u}_1, \mathbf{u}_2, \ldots, \mathbf{u}_n$, that is,

$$A[\mathbf{u}_1]_B = [T(\mathbf{u}_1)]_{B'}, \quad A[\mathbf{u}_2]_B = [T(\mathbf{u}_2)]_{B'}, \quad \ldots, \quad A[\mathbf{u}_n]_B = [T(\mathbf{u}_n)]_{B'} \tag{2}$$

But

$$[\mathbf{u}_1]_B = \begin{bmatrix} 1 \\ 0 \\ 0 \\ \vdots \\ 0 \end{bmatrix}, \quad [\mathbf{u}_2]_B = \begin{bmatrix} 0 \\ 1 \\ 0 \\ \vdots \\ 0 \end{bmatrix}, \quad \cdots, \quad [\mathbf{u}_n]_B = \begin{bmatrix} 0 \\ 0 \\ 0 \\ \vdots \\ 1 \end{bmatrix}$$

so

$$A[\mathbf{u}_1]_B = \begin{bmatrix} a_{11} & a_{12} & \cdots & a_{1n} \\ a_{21} & a_{22} & \cdots & a_{2n} \\ \vdots & \vdots & & \vdots \\ a_{m1} & a_{m2} & \cdots & a_{mn} \end{bmatrix} \begin{bmatrix} 1 \\ 0 \\ 0 \\ \vdots \\ 0 \end{bmatrix} = \begin{bmatrix} a_{11} \\ a_{21} \\ \vdots \\ a_{m1} \end{bmatrix}$$

$$A[\mathbf{u}_2]_B = \begin{bmatrix} a_{11} & a_{12} & \cdots & a_{1n} \\ a_{21} & a_{22} & \cdots & a_{2n} \\ \vdots & \vdots & & \vdots \\ a_{m1} & a_{m2} & \cdots & a_{mn} \end{bmatrix} \begin{bmatrix} 0 \\ 1 \\ 0 \\ \vdots \\ 0 \end{bmatrix} = \begin{bmatrix} a_{12} \\ a_{22} \\ \vdots \\ a_{m2} \end{bmatrix}$$

$$\vdots$$

$$A[\mathbf{u}_n]_B = \begin{bmatrix} a_{11} & a_{12} & \cdots & a_{1n} \\ a_{21} & a_{22} & \cdots & a_{2n} \\ \vdots & \vdots & & \vdots \\ a_{m1} & a_{m2} & \cdots & a_{mn} \end{bmatrix} \begin{bmatrix} 0 \\ 0 \\ 0 \\ \vdots \\ 1 \end{bmatrix} = \begin{bmatrix} a_{1n} \\ a_{2n} \\ \vdots \\ a_{mn} \end{bmatrix}$$

Substituting these results into (2) yields

$$\begin{bmatrix} a_{11} \\ a_{21} \\ \vdots \\ a_{m1} \end{bmatrix} = [T(\mathbf{u}_1)]_{B'}, \quad \begin{bmatrix} a_{12} \\ a_{22} \\ \vdots \\ a_{m2} \end{bmatrix} = [T(\mathbf{u}_2)]_{B'}, \quad \cdots, \quad \begin{bmatrix} a_{1n} \\ a_{2n} \\ \vdots \\ a_{mn} \end{bmatrix} = [T(\mathbf{u}_n)]_{B'}$$

which shows that the successive columns of A are the coordinate matrices of

$$T(\mathbf{u}_1), \ T(\mathbf{u}_2), \ldots, \ T(\mathbf{u}_n)$$

with respect to the basis B'. Thus, the matrix for T with respect to the bases B and B' is

$$A = \left[[T(\mathbf{u}_1)]_{B'} \ \vdots \ [T(\mathbf{u}_2)]_{B'} \ \vdots \ \cdots \ \vdots \ [T(\mathbf{u}_n)]_{B'} \right] \tag{3}$$

This matrix is commonly denoted by the symbol

$$[T]_{B',B}$$

so that the foregoing formula can also be written as

$$[T]_{B',B} = \left[[T(\mathbf{u}_1)]_{B'} \mid [T(\mathbf{u}_2)]_{B'} \mid \cdots \mid [T(\mathbf{u}_n)]_{B'} \right] \tag{4}$$

and from (1) this matrix has the property

$$[T]_{B',B}[\mathbf{x}]_B = [T(\mathbf{x})]_{B'} \tag{4a}$$

REMARK. Observe that in the notation $[T]_{B',B}$ the right subscript is a basis for the domain of T, and the left subscript is a basis for the image space of T (Figure 3).

Figure 3

Moreover, observe how the subscript B seems to "cancel out" in Formula (4a) (Figure 4).

Figure 4

MATRICES OF LINEAR OPERATORS

In the special case where $V = W$ (so that $T:V \rightarrow V$ is a linear operator) it is usual to take $B = B'$ when constructing a matrix for T. In this case the resulting matrix is called the **matrix for T with respect to the basis B** and is usually denoted by $[T]_B$ rather than $[T]_{B,B}$. If $B = \{\mathbf{u}_1, \mathbf{u}_2, \ldots, \mathbf{u}_n\}$, then in this case Formulas (4) and (4a) become

$$[T]_B = \left[[T(\mathbf{u}_1)]_B \mid [T(\mathbf{u}_2)]_B \mid \cdots \mid [T(\mathbf{u}_n)]_B \right] \tag{5}$$

and

$$[T]_B[\mathbf{x}]_B = [T(\mathbf{x})]_B \tag{5a}$$

Phrased informally, (4a) and (5a) state that *the matrix for T times the coordinate matrix for* $\mathbf{x}$ *is the coordinate matrix for* $T(\mathbf{x})$.

Example 1 Let $T:P_1 \rightarrow P_2$ be the linear transformation defined by

$$T(p(x)) = xp(x)$$

Find the matrix for T with respect to the standard bases

$$B = \{\mathbf{u}_1, \mathbf{u}_2\} \qquad \text{and} \qquad B' = \{\mathbf{v}_1, \mathbf{v}_2, \mathbf{v}_3\}$$

where

$$\mathbf{u}_1 = 1, \quad \mathbf{u}_2 = x; \quad \mathbf{v}_1 = 1, \quad \mathbf{v}_2 = x, \quad \mathbf{v}_3 = x^2$$

Solution. From the given formula for T we obtain

$$T(\mathbf{u}_1) = T(1) = (x)(1) = x$$
$$T(\mathbf{u}_2) = T(x) = (x)(x) = x^2$$

By inspection, we can determine the coordinate matrices for $T(\mathbf{u}_1)$ and $T(\mathbf{u}_2)$ relative to B'; they are

$$[T(\mathbf{u}_1)]_{B'} = \begin{bmatrix} 0 \\ 1 \\ 0 \end{bmatrix}, \qquad [T(\mathbf{u}_2)]_{B'} = \begin{bmatrix} 0 \\ 0 \\ 1 \end{bmatrix}$$

Thus, the matrix for T with respect to B and B' is

$$[T]_{B',B} = \left[[T(\mathbf{u}_1)]_{B'} \mid [T(\mathbf{u}_2)]_{B'} \right] = \begin{bmatrix} 0 & 0 \\ 1 & 0 \\ 0 & 1 \end{bmatrix} \quad \triangle$$

Example 2 Let $T : P_1 \to P_2$ be the linear transformation in Example 1. Show that the matrix

$$[T]_{B',B} = \begin{bmatrix} 0 & 0 \\ 1 & 0 \\ 0 & 1 \end{bmatrix}$$

(obtained in Example 1) satisfies (4a) for every vector $\mathbf{x} = a + bx$ in P_1.

Solution. Since $\mathbf{x} = p(x) = a + bx$, we have

$$T(\mathbf{x}) = xp(x) = ax + bx^2$$

For the bases B and B' in Example 1, if follows by inspection that

$$[\mathbf{x}]_B = [ax + b]_B = \begin{bmatrix} a \\ b \end{bmatrix}$$

$$[T(\mathbf{x})]_{B'} = [ax + bx^2] = \begin{bmatrix} 0 \\ a \\ b \end{bmatrix}$$

Thus,

$$[T]_{B',B}[\mathbf{x}]_B = \begin{bmatrix} 0 & 0 \\ 1 & 0 \\ 0 & 1 \end{bmatrix} \begin{bmatrix} a \\ b \end{bmatrix} = \begin{bmatrix} 0 \\ a \\ b \end{bmatrix} = [T(\mathbf{x})]_{B'}$$

so (4a) holds. $\triangle$

Example 3 Let $T: R^2 \to R^3$ be the linear transformation defined by

$$T\left(\begin{bmatrix} x_1 \\ x_2 \end{bmatrix}\right) = \begin{bmatrix} x_2 \\ -5x_1 + 13x_2 \\ -7x_1 + 16x_2 \end{bmatrix}$$

Find the matrix for T with respect to the bases $B = \{u_1, u_2\}$ for R^2 and $B' = \{v_1, v_2, v_3\}$ for R^3, where

$$u_1 = \begin{bmatrix} 3 \\ 1 \end{bmatrix}, \quad u_2 = \begin{bmatrix} 5 \\ 2 \end{bmatrix}; \quad v_1 = \begin{bmatrix} 1 \\ 0 \\ -1 \end{bmatrix}, \quad v_2 = \begin{bmatrix} -1 \\ 2 \\ 2 \end{bmatrix}, \quad v_3 = \begin{bmatrix} 0 \\ 1 \\ 2 \end{bmatrix}$$

Solution. From the formula for T,

$$T(u_1) = \begin{bmatrix} 1 \\ -2 \\ -5 \end{bmatrix}, \quad T(u_2) = \begin{bmatrix} 2 \\ 1 \\ -3 \end{bmatrix}$$

Expressing these vectors as linear combinations of v_1, v_2, and v_3 we obtain (verify)

$$T(u_1) = v_1 - 2v_3, \quad T(u_2) = 3v_1 + v_2 - v_3$$

Thus,

$$[T(u_1)]_{B'} = \begin{bmatrix} 1 \\ 0 \\ -2 \end{bmatrix}, \quad [T(u_2)]_{B'} = \begin{bmatrix} 3 \\ 1 \\ -1 \end{bmatrix}$$

so

$$[T]_{B',B} = \left[[T(u_1)]_{B'} \mid [T(u_2)]_{B'} \right] = \begin{bmatrix} 1 & 3 \\ 0 & 1 \\ -2 & -1 \end{bmatrix} \quad \blacktriangle$$

Example 4 Let $T: R^2 \to R^2$ be the linear operator defined by

$$T\left(\begin{bmatrix} x_1 \\ x_2 \end{bmatrix}\right) = \begin{bmatrix} x_1 + x_2 \\ -2x_1 + 4x_2 \end{bmatrix}$$

and let $B = \{u_1, u_2\}$ be the basis where

$$u_1 = \begin{bmatrix} 1 \\ 1 \end{bmatrix}, \quad u_2 = \begin{bmatrix} 1 \\ 2 \end{bmatrix}$$

(a) Find $[T]_B$
(b) Verify that (5a) holds for every vector x in R^2.

Solution (a). From the given formula for T,

$$T(\mathbf{u}_1) = \begin{bmatrix} 2 \\ 2 \end{bmatrix} = 2\mathbf{u}_1, \qquad T(\mathbf{u}_2) = \begin{bmatrix} 3 \\ 6 \end{bmatrix} = 3\mathbf{u}_2$$

Therefore,

$$[T(\mathbf{u}_1)]_B = \begin{bmatrix} 2 \\ 0 \end{bmatrix} \quad \text{and} \quad [T(\mathbf{u}_2)]_B = \begin{bmatrix} 0 \\ 3 \end{bmatrix}$$

Consequently,

$$[T]_B = \left[[T(\mathbf{u}_1)]_B \mid [T(\mathbf{u}_2)]_B \right] = \begin{bmatrix} 2 & 0 \\ 0 & 3 \end{bmatrix}$$

Solution (b). If

$$\mathbf{x} = \begin{bmatrix} x_1 \\ x_2 \end{bmatrix} \tag{6}$$

is any vector in R^2, then from the given formula for T

$$T(\mathbf{x}) = \begin{bmatrix} x_1 + x_2 \\ -2x_1 + 4x_2 \end{bmatrix} \tag{7}$$

To find $[\mathbf{x}]_B$ and $[T(\mathbf{x})]_B$, we must express (6) and (7) as linear combinations of $\mathbf{u}_1$ and $\mathbf{u}_2$. This yields the vector equations

$$\begin{bmatrix} x_1 \\ x_2 \end{bmatrix} = k_1 \begin{bmatrix} 1 \\ 1 \end{bmatrix} + k_2 \begin{bmatrix} 1 \\ 2 \end{bmatrix} \tag{8}$$

$$\begin{bmatrix} x_1 + x_2 \\ -2x_1 + 4x_2 \end{bmatrix} = k_1 \begin{bmatrix} 1 \\ 1 \end{bmatrix} + k_2 \begin{bmatrix} 1 \\ 2 \end{bmatrix} \tag{9}$$

Equating corresponding entries yields the linear systems

$$\begin{aligned} k_1 + \ k_2 &= x_1 \\ k_1 + 2k_2 &= x_2 \end{aligned} \tag{10}$$

and

$$\begin{aligned} k_1 + \ k_2 &= \ x_1 + \ x_2 \\ k_1 + 2k_2 &= -2x_1 + 4x_2 \end{aligned} \tag{11}$$

Solving (10) for k_1 and k_2 yields

$$k_1 = 2x_1 - x_2, \qquad k_2 = -x_1 + x_2$$

so that

$$[\mathbf{x}]_B = \begin{bmatrix} 2x_1 - x_2 \\ -x_1 + x_2 \end{bmatrix}$$

and solving (11) yields

$$k_1 = 4x_1 - 2x_2, \qquad k_2 = -3x_1 + 3x_2$$

so that

$$[T(\mathbf{x})]_B = \begin{bmatrix} 4x_1 - 2x_2 \\ -3x_1 + 3x_2 \end{bmatrix}$$

Thus,

$$[T]_B[\mathbf{x}]_B = \begin{bmatrix} 2 & 0 \\ 0 & 3 \end{bmatrix} \begin{bmatrix} 2x_1 - x_2 \\ -x_1 + x_2 \end{bmatrix} = \begin{bmatrix} 4x_1 - 2x_2 \\ -3x_1 + 3x_2 \end{bmatrix} = [T(\mathbf{x})]_B$$

so (5a) holds. ▲

**MATRICES OF
IDENTITY
OPERATORS**

Example 5 If $B = \{\mathbf{u}_1, \mathbf{u}_2, \ldots, \mathbf{u}_n\}$ is any basis for a finite-dimensional vector space V and $I : V \to V$ is the identity operator on V, then $I(\mathbf{u}_1) = \mathbf{u}_1$, $I(\mathbf{u}_2) = \mathbf{u}_2, \ldots,$ $I(\mathbf{u}_n) = \mathbf{u}_n$. Therefore,

$$[I(\mathbf{u}_1)]_B = \begin{bmatrix} 1 \\ 0 \\ 0 \\ \vdots \\ 0 \end{bmatrix}, \quad [I(\mathbf{u}_2)]_B = \begin{bmatrix} 0 \\ 1 \\ 0 \\ \vdots \\ 0 \end{bmatrix}, \quad \ldots, \quad [I(\mathbf{u}_n)]_B = \begin{bmatrix} 0 \\ 0 \\ 0 \\ \vdots \\ 1 \end{bmatrix}$$

Thus,

$$[I]_B = \begin{bmatrix} 1 & 0 & \cdots & 0 \\ 0 & 1 & \cdots & 0 \\ 0 & 0 & \cdots & 0 \\ \vdots & \vdots & & \vdots \\ 0 & 0 & \cdots & 1 \end{bmatrix} = I$$

Consequently, the matrix of the identity operator with respect to any basis is the $n \times n$ identity matrix. This result could have been anticipated from. Formula (5a), since that formula yields

$$[I]_B[\mathbf{x}]_B = [I(\mathbf{x})]_B = [\mathbf{x}]_B$$

which is consistent with the fact that $[I]_B = I$. ▲

We leave it as an exercise to prove the following result.

Theorem 7.5.1. *If $T : R^n \to R^m$ is a linear transformation and if B and B' are the standard bases for R^n and R^m, respectively, then*

$$[T]_{B',B} = [T] \tag{12}$$

This theorem tells us that in the special case where T maps R^n into R^m, the matrix for T with respect to the standard bases is the standard matrix for T. In this special case Formula (4a) of this section reduces to Formula (6) of Section 7.4, that is,

$$[T]_{B',B}[\mathbf{x}]_B = [T(\mathbf{x})]_{B'}$$

reduces to

$$[T]\mathbf{x} = T(\mathbf{x})$$

WHY MATRICES OF LINEAR TRANSFORMA-TIONS ARE IMPORTANT

There are two primary reasons for studying matrices of general linear transformations, one theoretical and the other quite practical:

· Answers to theoretical questions about the structure of general linear transformations on finite-dimensional vector spaces can often be obtained by studying just the matrix transformations. Such matters are considered in detail in more advanced linear algebra courses, but will be touched on in later sections.

· These matrices make it possible to compute images of vectors using matrix multiplication. Such computations can be performed rapidly on computers.

To focus on the latter idea, let $T:V \to W$ be a linear transformation. As shown in Figure 5, the matrix $[T]_{B',B}$ can be used to calculate $T(\mathbf{x})$ in three steps by the following *indirect* procedure:

(1) Compute the coordinate matrix $[\mathbf{x}]_B$.
(2) Multiply $[\mathbf{x}]_B$ on the left by $[T]_{B',B}$ to produce $[T(\mathbf{x})]_{B'}$.
(3) Reconstruct $T(\mathbf{x})$ from its coordinate matrix $[T(\mathbf{x})]_{B'}$.

Figure 5

Example 6 Let $T:P_2 \to P_2$ be the linear operator defined by

$$T(p(x)) = p(3x - 5)$$

that is, $T(c_0 + c_1x + c_2x^2) = c_0 + c_1(3x - 5) + c_2(3x - 5)^2$.

(a) Find $[T]_B$ with respect to the basis $B = \{1, x, x^2\}$.
(b) Use the indirect procedure to compute $T(1 + 2x + 3x^2)$.
(c) Check the result in (b) by computing $T(1 + 2x + 3x^2)$ directly.

Solution (a). From the formula for T

$$T(1) = 1, \qquad T(x) = 3x - 5, \qquad T(x^2) = (3x - 5)^2 = 9x^2 - 30x + 25$$

so that

$$[T(1)]_B = \begin{bmatrix} 1 \\ 0 \\ 0 \end{bmatrix}, \qquad [T(x)]_B = \begin{bmatrix} -5 \\ 3 \\ 0 \end{bmatrix}, \qquad [T(x^2)]_B = \begin{bmatrix} 25 \\ -30 \\ 9 \end{bmatrix}$$

Thus,

$$[T]_B = \begin{bmatrix} 1 & -5 & 25 \\ 0 & 3 & -30 \\ 0 & 0 & 9 \end{bmatrix}$$

Solution (b). The coordinate matrix relative to B for the vector $\mathbf{p} = 1 + 2x + 3x^2$ is

$$[\mathbf{p}]_B = \begin{bmatrix} 1 \\ 2 \\ 3 \end{bmatrix}$$

Thus, from (5a)

$$[T(1 + 2x + 3x^2)]_B = [T(\mathbf{p})]_B = [T]_B[\mathbf{p}]_B$$

$$= \begin{bmatrix} 1 & -5 & 25 \\ 0 & 3 & -30 \\ 0 & 0 & 9 \end{bmatrix} \begin{bmatrix} 1 \\ 2 \\ 3 \end{bmatrix} = \begin{bmatrix} 66 \\ -84 \\ 27 \end{bmatrix}$$

from which it follows that

$$T(1 + 2x + 3x^2) = 66 - 84x + 27x^2$$

Solution (c). By direct computation

$$\begin{aligned} T(1 + 2x + 3x^2) &= 1 + 2(3x - 5) + 3(3x - 5)^2 \\ &= 1 + 6x - 10 + 27x^2 - 90x + 75 \\ &= 66 - 84x + 27x^2 \end{aligned}$$

which agrees with the result in (b). ▲

MATRICES OF COMPOSITIONS AND INVERSE TRANSFORMATIONS

We shall now mention two theorems that are generalizations of Theorems 7.4.3 and 7.4.4. The proofs are left as exercises.

Theorem 7.5.2. *If $T_1: U \to V$ and $T_2: V \to W$ are linear transformations, and if B, B'', and B' are bases for U, V, and W, respectively, then*

$$[T_2 \circ T_1]_{B',B} = [T_2]_{B',B''}[T_1]_{B'',B} \qquad\qquad (13)$$

> **Theorem 7.5.3.** *If $T:V \to V$ is a linear operator, and if B is a basis for V, then the following are equivalent:*
>
> *(a)* T *is one-to-one;*
> *(b)* $[T]_B$ *is invertible.*
>
> *Moreover, when these equivalent conditions hold*
>
> $$[T^{-1}]_B = [T]_B^{-1} \tag{14}$$

REMARK. In (13), observe how the interior subscript B'' (the basis for the intermediate space V) seems to "cancel out," leaving only the bases for the domain and image space of the composition as subscripts (Figure 6).

$$[T_2 \circ T_1]_{B',B} = [T_2]_{B',B''} \quad [T_1]_{B'',B}$$

Cancellation

Figure 6

This cancellation of interior subscripts suggests the following extension of Formula (13) to compositions of three linear transformations (Figure 7).

$$[T_3 \circ T_2 \circ T_1]_{B',B} = [T_3]_{B',B'''}[T_2]_{B''',B''}[T_1]_{B'',B} \tag{15}$$

Figure 7   Basis B T_1 Basis B'' T_2 Basis B''' T_3 Basis B'

The following example illustrates Theorem 7.5.2.

Example 7   Let $T_1:P_1 \to P_2$ be the linear transformation defined by

$$T_1(p(x)) = xp(x)$$

and let $T_2:P_2 \to P_2$ be the linear operator defined by

$$T_2(p(x)) = p(3x - 5)$$

Then the composition $(T_2 \circ T_1):P_1 \to P_2$ is given by

$$(T_2 \circ T_1)(p(x)) = T_2(T_1(p(x))) = T_2(xp(x)) = (3x - 5)p(3x - 5)$$

Thus, if $p(x) = c_0 + c_1 x$, then

$$(T_2 \circ T_1)(c_0 + c_1 x) = (3x - 5)(c_0 + c_1(3x - 5))$$
$$= c_0(3x - 5) + c_1(3x - 5)^2 \tag{16}$$

In this example, P_1 plays the role of U in Theorem 7.5.2 and P_2 plays the roles of both V and W; thus we can take $B' = B''$ in (13) so that the formula simplifies to

$$[T_2 \circ T_1]_{B',B} = [T_2]_{B'}[T_1]_{B',B} \tag{17}$$

Let us choose $B = \{1, x\}$ to be the basis for P_1 and $B' = \{1, x, x^2\}$ to be the basis for P_2. We showed in Examples 1 and 6 that

$$[T_1]_{B',B} = \begin{bmatrix} 0 & 0 \\ 1 & 0 \\ 0 & 1 \end{bmatrix} \quad \text{and} \quad [T_2]_{B'} = \begin{bmatrix} 1 & -5 & 25 \\ 0 & 3 & -30 \\ 0 & 0 & 9 \end{bmatrix}$$

Thus, it follows from (17) that

$$[T_2 \circ T_1]_{B',B} = \begin{bmatrix} 1 & -5 & 25 \\ 0 & 3 & -30 \\ 0 & 0 & 9 \end{bmatrix} \begin{bmatrix} 0 & 0 \\ 1 & 0 \\ 0 & 1 \end{bmatrix} = \begin{bmatrix} -5 & 25 \\ 3 & -30 \\ 0 & 9 \end{bmatrix} \tag{18}$$

As a check, we will calculate $[T_2 \circ T_1]_{B',B}$ directly from Formula (4). Since $B = \{1, x\}$, it follows from Formula (4) with $\mathbf{u}_1 = 1$ and $\mathbf{u}_2 = x$ that

$$[T_2 \circ T_1]_{B',B} = \left[[(T_2 \circ T_1)(1)]_{B'} \mid [(T_2 \circ T_1)(x)]_{B'} \right] \tag{19}$$

Using (16) yields

$$(T_2 \circ T_1)(1) = 3x - 5 \quad \text{and} \quad (T_2 \circ T_1)(x) = (3x - 5)^2 = 9x^2 - 30x + 25$$

Since $B' = \{1, x, x^2\}$, it follows from this that

$$[(T_2 \circ T_1)(1)]_{B'} = \begin{bmatrix} -5 \\ 3 \\ 0 \end{bmatrix} \quad \text{and} \quad [(T_2 \circ T_1)(x)]_{B'} = \begin{bmatrix} 25 \\ -30 \\ 9 \end{bmatrix}$$

Substituting in (19) yields

$$[T_2 \circ T_1]_{B',B} = \begin{bmatrix} -5 & 25 \\ 3 & -30 \\ 0 & 9 \end{bmatrix}$$

which agrees with (18). ▲

EXERCISE SET 7.5

1. Let $T: P_2 \to P_3$ be the linear transformation defined by $T(p(x)) = xp(x)$.
   (a) Find the matrix for T with respect to the standard bases

$$B = \{\mathbf{u}_1, \mathbf{u}_2, \mathbf{u}_3\} \quad \text{and} \quad B' = \{\mathbf{v}_1, \mathbf{v}_2, \mathbf{v}_3, \mathbf{v}_4\}$$

where

$$\mathbf{u}_1 = 1, \quad \mathbf{u}_2 = x, \quad \mathbf{u}_3 = x^2$$
$$\mathbf{v}_1 = 1, \quad \mathbf{v}_2 = x, \quad \mathbf{v}_3 = x^2, \quad \mathbf{v}_4 = x^3$$

   (b) Verify that the matrix $[T]_{B',B}$ obtained in part (a) satisfies Formula (4a) for every vector $\mathbf{x} = c_0 + c_1 x + c_2 x^2$ in P_2.

2. Let $T:P_2 \to P_1$ be the linear transformation defined by

$$T(a_0 + a_1x + a_2x^2) = (a_0 + a_1) - (2a_1 + 3a_2)x$$

(a) Find the matrix for T with respect to the standard bases $B = \{1, x, x^2\}$ and $B' = \{1, x\}$ for P_2 and P_1.

(b) Verify that the matrix $[T]_{B',B}$ obtained in part (a) satisfies Formula (4a) for every vector $x = c_0 + c_1x + c_2x^2$ in P_2.

3. Let $T:P_2 \to P_2$ be the linear operator defined by

$$T(a_0 + a_1x + a_2x^2) = a_0 + a_1(x - 1) + a_2(x - 1)^2$$

(a) Find the matrix for T with respect to the standard basis $B = \{1, x, x^2\}$ for P_2.

(b) Verify that the matrix $[T]_B$ obtained in part (a) satisfies Formula (5a) for every vector $x = a_0 + a_1x + a_2x^2$ in P_2.

4. Let $T:R^2 \to R^2$ be the linear operator defined by

$$T\left(\begin{bmatrix} x_1 \\ x_2 \end{bmatrix}\right) = \begin{bmatrix} x_1 - x_2 \\ x_1 + x_2 \end{bmatrix}$$

and let $B = \{\mathbf{u}_1, \mathbf{u}_2\}$ be the basis for which

$$\mathbf{u}_1 = \begin{bmatrix} 1 \\ 1 \end{bmatrix} \quad \text{and} \quad \mathbf{u}_2 = \begin{bmatrix} -1 \\ 0 \end{bmatrix}$$

(a) Find $[T]_B$.

(b) Verify that Formula (5a) holds for every vector x in R^2.

5. Let $T:R^2 \to R^3$ be defined by

$$T\left(\begin{bmatrix} x_1 \\ x_2 \end{bmatrix}\right) = \begin{bmatrix} x_1 + 2x_2 \\ -x_1 \\ 0 \end{bmatrix}$$

(a) Find the matrix $[T]_{B',B}$ with respect to the bases $B = \{\mathbf{u}_1, \mathbf{u}_2\}$ and $B' = \{\mathbf{v}_1, \mathbf{v}_2, \mathbf{v}_3\}$, where

$$\mathbf{u}_1 = \begin{bmatrix} 1 \\ 3 \end{bmatrix} \quad \mathbf{u}_2 = \begin{bmatrix} -2 \\ 4 \end{bmatrix} \quad \mathbf{v}_1 = \begin{bmatrix} 1 \\ 1 \\ 1 \end{bmatrix} \quad \mathbf{v}_2 = \begin{bmatrix} 2 \\ 2 \\ 0 \end{bmatrix} \quad \mathbf{v}_3 = \begin{bmatrix} 3 \\ 0 \\ 0 \end{bmatrix}$$

(b) Verify that Formula (4a) holds for every vector

$$\mathbf{x} = \begin{bmatrix} x_1 \\ x_2 \end{bmatrix}$$

in R^2.

6. Let $T:R^3 \to R^3$ be defined by $T(x_1, x_2, x_3) = (x_1 - x_2, x_2 - x_1, x_1 - x_3)$.

(a) Find the matrix for T with respect to the basis $B = \{\mathbf{v}_1, \mathbf{v}_2, \mathbf{v}_3\}$, where

$$\mathbf{v}_1 = (1, 0, 1), \qquad \mathbf{v}_2 = (0, 1, 1), \qquad \mathbf{v}_3 = (1, 1, 0)$$

(b) Verify that Formula (5a) holds for every vector $\mathbf{x} = (x_1, x_2, x_3)$ in R^3.

7. Let $T:P_2 \to P_2$ be the linear operator defined by $T(p(x)) = p(2x + 1)$, that is,

$$T(c_0 + c_1x + c_2x^2) = c_0 + c_1(2x + 1) + c_2(2x + 1)^2$$

(a) Find $[T]_B$ with respect to the basis $B = \{1, x, x^2\}$.
(b) Use the indirect procedure illustrated in Figure 5 to compute $T(2 - 3x + 4x^2)$.
(c) Check the result obtained in part (b) by computing $T(2 - 3x + 4x^2)$ directly.

8. Let $T:P_2 \to P_3$ be the linear transformation defined by $T(p(x)) = xp(x - 3)$, that is,

$$T(c_0 + c_1x + c_2x^2) = x(c_0 + c_1(x - 3) + c_2(x - 3)^2)$$

(a) Find $[T]_{B',B}$ with respect to the bases $B = \{1, x, x^2\}$ and $B' = \{1, x, x^2, x^3\}$.
(b) Use the indirect procedure illustrated in Figure 5 to compute $T(1 + x - x^2)$.
(c) Check the result obtained in part (b) by computing $T(1 + x - x^2)$ directly.

9. Let $v_1 = \begin{bmatrix} 1 \\ 3 \end{bmatrix}$ and $v_2 = \begin{bmatrix} -1 \\ 4 \end{bmatrix}$, and let

$$A = \begin{bmatrix} 1 & 3 \\ -2 & 5 \end{bmatrix}$$

be the matrix for $T:R^2 \to R^2$ with respect to the basis $B = \{v_1, v_2\}$.
(a) Find $[T(v_1)]_B$ and $[T(v_2)]_B$.   (b) Find $T(v_1)$ and $T(v_2)$.

(c) Find a formula for $T\left(\begin{bmatrix} x_1 \\ x_2 \end{bmatrix}\right)$.

(d) Use the formula obtained in (c) to compute $T\left(\begin{bmatrix} 1 \\ 1 \end{bmatrix}\right)$.

10. Let $A = \begin{bmatrix} 3 & -2 & 1 & 0 \\ 1 & 6 & 2 & 1 \\ -3 & 0 & 7 & 1 \end{bmatrix}$ be the matrix of $T:R^4 \to R^3$ with respect to the

bases $B = \{v_1, v_2, v_3, v_4\}$ and $B' = \{w_1, w_2, w_3\}$, where

$$v_1 = \begin{bmatrix} 0 \\ 1 \\ 1 \\ 1 \end{bmatrix} \quad v_2 = \begin{bmatrix} 2 \\ 1 \\ -1 \\ -1 \end{bmatrix} \quad v_3 = \begin{bmatrix} 1 \\ 4 \\ -1 \\ 2 \end{bmatrix} \quad v_4 = \begin{bmatrix} 6 \\ 9 \\ 4 \\ 2 \end{bmatrix}$$

$$w_1 = \begin{bmatrix} 0 \\ 8 \\ 8 \end{bmatrix} \quad w_2 = \begin{bmatrix} -7 \\ 8 \\ 1 \end{bmatrix} \quad w_3 = \begin{bmatrix} -6 \\ 9 \\ 1 \end{bmatrix}$$

(a) Find $[T(v_1)]_{B'}$, $[T(v_2)]_{B'}$, $[T(v_3)]_{B'}$, and $[T(v_4)]_{B'}$.   (b) Find $T(v_1)$, $T(v_2)$, $T(v_3)$, and $T(v_4)$.

(c) Find a formula for $T\left(\begin{bmatrix} x_1 \\ x_2 \\ x_3 \\ x_4 \end{bmatrix}\right)$.   (d) Use the formula obtained in (c) to compute $T\left(\begin{bmatrix} 2 \\ 2 \\ 0 \\ 0 \end{bmatrix}\right)$.

11. Let $A = \begin{bmatrix} 1 & 3 & -1 \\ 2 & 0 & 5 \\ 6 & -2 & 4 \end{bmatrix}$ be the matrix of $T:P_2 \to P_2$ with respect to the basis

$B = \{v_1, v_2, v_3\}$, where $v_1 = 3x + 3x^2$, $v_2 = -1 + 3x + 2x^2$, $v_3 = 3 + 7x + 2x^2$.
(a) Find $[T(v_1)]_B$, $[T(v_2)]_B$, and $[T(v_3)]_B$.   (b) Find $T(v_1)$, $T(v_2)$, and $T(v_3)$.
(c) Find a formula for $T(a_0 + a_1x + a_2x^2)$.
(d) Use the formula obtained in (c) to compute $T(1 + x^2)$.

12. Let $T_1:P_1 \to P_2$ be the linear transformation defined by

$$T_1(p(x)) = xp(x)$$

and let $T_2:P_2 \to P_2$ be the linear operator defined by

$$T_2(p(x)) = p(2x + 1)$$

Let $B = \{1, x\}$ and $B' = \{1, x, x^2\}$ be the standard bases for P_1 and P_2.
(a) Find $[T_2 \circ T_1]_{B',B}$, $[T_2]_{B'}$, and $[T_1]_{B',B}$.
(b) State a formula relating the matrices in part (a).
(c) Verify that the matrices in part (a) satisfy the formula you stated in part (b).

13. Let $T_1:P_1 \to P_2$ be the linear transformation defined by

$$T_1(c_0 + c_1x) = 2c_0 - 3c_1x$$

and let $T_2:P_2 \to P_3$ be the linear transformation defined by

$$T_1(c_0 + c_1x + c_2x^2) = 3c_0x + 3c_1x^2 + 3c_2x^3$$

Let $B = \{1, x\}$, $B'' = \{1, x, x^2\}$, $B' = \{1, x, x^2, x^3\}$
(a) Find $[T_2 \circ T_1]_{B',B}$, $[T_2]_{B',B''}$, and $[T_1]_{B'',B}$.
(b) State a formula relating the matrices in part (a).
(c) Verify that the matrices in part (a) satisfy the formula you stated in part (b).

14. Show that if $T:V \to W$ is the zero transformation, then the matrix of T with respect to any bases for V and W is a zero matrix.

15. Show that if $T:V \to V$ is a contraction or a dilation of V (Example 5 of Section 7.1), then the matrix of T with respect to any basis for V is a diagonal matrix.

16. Let $B = \{v_1, v_2, v_3, v_4\}$ be a basis for a vector space V. Find the matrix with respect to B of the linear operator $T:V \to V$ defined by $T(v_1) = v_2$, $T(v_2) = v_3$, $T(v_3) = v_4$, $T(v_4) = v_1$.

17. **(For readers who have studied calculus.)** Let $D:P_2 \to P_2$ be the differentiation operator $D(\mathbf{p}) = \mathbf{p}'$. In parts (a) and (b) find the matrix of D with respect to the basis $B = \{\mathbf{p}_1, \mathbf{p}_2, \mathbf{p}_3\}$.
(a) $\mathbf{p}_1 = 1$, $\mathbf{p}_2 = x$, $\mathbf{p}_3 = x^2$ (b) $\mathbf{p}_1 = 2$, $\mathbf{p}_2 = 2 - 3x$, $\mathbf{p}_3 = 2 - 3x + 8x^2$
(c) Use the matrix in part (a) to compute $D(6 - 6x + 24x^2)$.
(d) Repeat the directions of part (c) for the matrix in part (b).

18. **(For readers who have studied calculus.)** In each part, $B = \{\mathbf{f}_1, \mathbf{f}_2, \mathbf{f}_3\}$ is a basis for a subspace V of the vector space of real-valued functions defined on the real line. Find the matrix with respect to B of the differentiation operator $D:V \to V$.
(a) $\mathbf{f}_1 = 1$, $\mathbf{f}_2 = \sin x$, $\mathbf{f}_3 = \cos x$ (b) $\mathbf{f}_1 = 1$, $\mathbf{f}_2 = e^x$, $\mathbf{f}_3 = e^{2x}$
(c) $\mathbf{f}_1 = e^{2x}$, $\mathbf{f}_2 = xe^{2x}$, $\mathbf{f}_3 = x^2e^{2x}$

19. Prove: If B and B' are the standard bases for R^n and R^m, respectively, then the matrix for a linear transformation $T:R^n \to R^m$ with respect to the bases B and B' is the standard matrix for T.

7.6 SIMILARITY

The matrix of a linear operator $T: V \to V$ depends on the basis selected for V. One of the fundamental problems of linear algebra is to choose a basis for V that makes the matrix for T as simple as possible—diagonal or triangular, for example. In this section we shall study this problem.

CHOOSING BASES TO PRODUCE SIMPLE MATRICES FOR LINEAR OPERATORS

Standard bases do not necessarily produce the simplest matrices for linear operators. For example, consider the linear operator $T: R^2 \to R^2$ defined by

$$T\left(\begin{bmatrix} x_1 \\ x_2 \end{bmatrix}\right) = \begin{bmatrix} x_1 + x_2 \\ -2x_1 + 4x_2 \end{bmatrix} \tag{1}$$

and the standard basis $B = \{\mathbf{e}_1, \mathbf{e}_2\}$ for R^2, where

$$\mathbf{e}_1 = \begin{bmatrix} 1 \\ 0 \end{bmatrix} \qquad \mathbf{e}_2 = \begin{bmatrix} 0 \\ 1 \end{bmatrix}$$

By Theorem 7.5.1, the matrix for T with respect to this basis is the standard matrix for T, that is,

$$[T]_B = [T] = [T(\mathbf{e}_1) \mid T(\mathbf{e}_2)]$$

From (1),

$$T(\mathbf{e}_1) = \begin{bmatrix} 1 \\ -2 \end{bmatrix}, \qquad T(\mathbf{e}_2) = \begin{bmatrix} 1 \\ 4 \end{bmatrix}$$

so

$$[T]_B = \begin{bmatrix} 1 & 1 \\ -2 & 4 \end{bmatrix} \tag{2}$$

In comparison, we showed in Example 4 of Section 7.5 that if

$$\mathbf{u}_1 = \begin{bmatrix} 1 \\ 1 \end{bmatrix}, \qquad \mathbf{u}_2 = \begin{bmatrix} 1 \\ 2 \end{bmatrix} \tag{3}$$

then the matrix for T with respect to the basis $B' = \{\mathbf{u}_1, \mathbf{u}_2\}$ is the diagonal matrix

$$[T]_{B'} = \begin{bmatrix} 2 & 0 \\ 0 & 3 \end{bmatrix} \tag{4}$$

This matrix is "simpler" than (2) in the sense that diagonal matrices enjoy special properties that more general matrices do not [see (2) in Section 1.5, for example].

One of the major themes in more advanced linear algebra courses is to determine the "simplest possible form" that can be obtained for the matrix of a linear operator by choosing the basis appropriately. Sometimes it is possible to obtain a diagonal matrix (as above, for example); other times one must settle for a triangular matrix or some other form. We will only be able to touch on this important topic in this text.

The problem of finding a basis that produces the simplest possible matrix for a linear operator $T: V \to V$ can be attacked by first finding a matrix for T relative to *any* basis, say a standard basis, where applicable, then changing the basis in a manner that simplifies the matrix. Before pursuing this idea, it will be helpful to review some concepts about changing bases.

Recall from Formula (6) in Section 5.4 that if $B = \{\mathbf{u}_1, \mathbf{u}_2, \ldots, \mathbf{u}_n\}$ and $B' = \{\mathbf{u}'_1, \mathbf{u}'_2, \ldots, \mathbf{u}'_n\}$ are bases for a vector space V, then the *transition matrix* from B' to B is given by the formula

$$P = \left[[\mathbf{u}'_1]_B \mid [\mathbf{u}'_2]_B \mid \cdots \mid [\mathbf{u}'_n]_B \right] \tag{5}$$

This matrix has the property that for every vector $\mathbf{v}$ in V

$$P[\mathbf{v}]_{B'} = [\mathbf{v}]_B \tag{6}$$

that is, multiplication by P maps the coordinate matrix for $\mathbf{v}$ relative to B' into the coordinate matrix for $\mathbf{v}$ relative to B [see Formula (5) in Section 5.4]. We showed in Theorem 5.4.1 that P is invertible and P^{-1} is the transition matrix from B to B'.

RELATIONSHIP
BETWEEN
TRANSITION
MATRICES AND
IDENTITY
OPERATORS

The following theorem gives a useful alternative viewpoint about transition matrices; it shows that the transition matrix from a basis B' to a basis B can be regarded as the matrix of an identity operator.

Theorem 7.6.1. *If B and B' are bases for a finite-dimensional vector space V, and if $I: V \to V$ is the identity operator, then $[I]_{B,B'}$ is the transition matrix from B' to B.*

Proof. Suppose that $B = \{\mathbf{u}_1, \mathbf{u}_2, \ldots, \mathbf{u}_n\}$ and $B' = \{\mathbf{u}'_1, \mathbf{u}'_2, \ldots, \mathbf{u}'_n\}$ are bases for V. Using the fact that $I(\mathbf{v}) = \mathbf{v}$ for all $\mathbf{v}$ in V, it follows from Formula (4) of Section 7.5 with B and B' *reversed* that

$$[I]_{B,B'} = \left[[I(\mathbf{u}'_1)]_B \mid [I(\mathbf{u}'_2)]_B \mid \cdots \mid [I(\mathbf{u}'_n)]_B \right]$$

$$= \left[[\mathbf{u}'_1]_B \mid [\mathbf{u}'_2]_B \mid \cdots \mid [\mathbf{u}'_n]_B \right]$$

Thus, from (5), we have $[I]_{B,B'} = P$, which shows that $[I]_{B,B'}$ is the transition from B' to B. ∎

The result in this theorem is illustrated in Figure 1.

Figure 1 $[I]_{B,B'}$ is the transition matrix from B' to B.

EFFECT OF CHANGING BASES ON MATRICES OF LINEAR OPERATORS

We are now ready to consider the main problem in this section.

> **Problem.** If B and B' are two bases for a finite-dimensional vector space V, and if $T:V \to V$ is a linear operator, what relationship, if any, exists between the matrices $[T]_B$ and $[T]_{B'}$?

The answer to this question can be obtained by considering the composition of the three linear operators on V pictured in Figure 2.

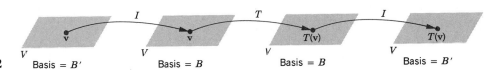

Figure 2

In this figure **v** is first mapped into itself by the identity operator, then **v** is mapped into $T(\mathbf{v})$ by T, then $T(\mathbf{v})$ is mapped into itself by the identity operator. All four vector spaces involved in the composition are the same (namely, V); however, the bases for the spaces vary. Since the starting vector is **v** and the final vector is $T(\mathbf{v})$, the composition is the same as T, that is,

$$T = I \circ T \circ I \tag{7}$$

If, as illustrated in Figure 2, the first and last vector spaces are assigned the basis B' and the middle two spaces are assigned the basis B, then it follows from (7) and Formula (15) of Section 7.5 (with an appropriate adjustment in the names of the bases) that

$$[T]_{B',B'} = [I \circ T \circ I]_{B',B'} = [I]_{B',B}[T]_{B,B}[I]_{B,B'} \tag{8}$$

or in simpler notation

$$[T]_{B'} = [I]_{B',B}[T]_B[I]_{B,B'} \tag{9}$$

But it follows from Theorem 7.6.1 that $[I]_{B,B'}$ is the transition matrix from B' to B and consequently $[I]_{B',B}$ is the transition matrix from B to B'. Thus, if we let $P = [I]_{B,B'}$, then $P^{-1} = [I]_{B',B}$, so (9) can be written as

$$[T]_{B'} = P^{-1}[T]_B P$$

In summary, we have the following theorem.

> **Theorem 7.6.2.** *Let $T:V \to V$ be a linear operator on a finite-dimensional vector space V, and let B and B' be bases for V. Then*
>
> $$[T]_{B'} = P^{-1}[T]_B P \tag{10}$$
>
> *where P is the transition matrix from B' to B.*

Warning. *When applying Theorem 7.6.2, it is easy to forget whether P is the transition matrix from B to B' (incorrect) or from B' to B (correct). It may help to write (10) in form (9), keeping in mind that the three "interior" subscripts are the same, and the two exterior subscripts are the same:*

$$[T]_{B'} = [I]_{B',B}\ [T]_B\ [I]_{B,B'}$$

```
      ↑↑    ↑    ↑↑
   ┌─────────────────┐
   │      The        │
   │    interior     │
   │   subscripts    │
   │      are        │
   │   the same      │
   ├─────────────────┤
   │  The exterior   │
   │ subscripts are  │
   │    the same     │
   └─────────────────┘
```

Once you master this pattern, you need only remember that $P = [I]_{B,B'}$ *is the transition matrix from B' to B and* $P^{-1} = [I]_{B',B}$ *is its inverse.*

Example 1 Let $T: R^2 \to R^2$ be defined by

$$T\left(\begin{bmatrix} x_1 \\ x_2 \end{bmatrix}\right) = \begin{bmatrix} x_1 + x_2 \\ -2x_1 + 4x_2 \end{bmatrix}$$

Find the matrix of T with respect to the standard basis $B = \{e_1, e_2\}$ for R^2, then use Theorem 7.6.2 to find the matrix of T with respect to the basis $B' = \{u'_1, u'_2\}$, where

$$u'_1 = \begin{bmatrix} 1 \\ 1 \end{bmatrix} \qquad \text{and} \qquad u'_2 = \begin{bmatrix} 1 \\ 2 \end{bmatrix}$$

Solution. We showed earlier in this section [see (2)] that

$$[T]_B = \begin{bmatrix} 1 & 1 \\ -2 & 4 \end{bmatrix}$$

To find $[T]_{B'}$ from (10), we will need to find the transition matrix

$$P = [I]_{B,B'} = \left[[u'_1]_B \mid [u'_2]_B \right]$$

[see (5)]. By inspection

$$u'_1 = e_1 + e_2$$
$$u'_2 = e_1 + 2e_2$$

so that

$$[u'_1]_B = \begin{bmatrix} 1 \\ 1 \end{bmatrix} \qquad \text{and} \qquad [u'_2]_B = \begin{bmatrix} 1 \\ 2 \end{bmatrix}$$

Thus, the transition matrix from B' to B is

$$P = \begin{bmatrix} 1 & 1 \\ 1 & 2 \end{bmatrix}$$

The reader can check that

$$P^{-1} = \begin{bmatrix} 2 & -1 \\ -1 & 1 \end{bmatrix}$$

so that by Theorem 7.6.2 the matrix of T relative to the basis B' is

$$[T]_{B'} = P^{-1}[T]_B P = \begin{bmatrix} 2 & -1 \\ -1 & 1 \end{bmatrix}\begin{bmatrix} 1 & 1 \\ -2 & 4 \end{bmatrix}\begin{bmatrix} 1 & 1 \\ 1 & 2 \end{bmatrix} = \begin{bmatrix} 2 & 0 \\ 0 & 3 \end{bmatrix}$$

which agrees with (4). ▲

SIMILARITY

The relationship in Formula (10) is of such importance that there is some terminology associated with it.

> **Definition.** If A and B are square matrices, we say that B *is similar to* A if there is an invertible matrix P such that $B = P^{-1}AP$.

REMARK. Note that the equation $B = P^{-1}AP$ can be rewritten as

$$A = PBP^{-1} = (P^{-1})^{-1}BP^{-1}$$

Letting $Q = P^{-1}$ yields

$$A = Q^{-1}BQ$$

which says that A is similar to B. Therefore, B is similar to A if and only if A is similar to B; consequently, we shall usually simply say that A *and* B *are similar*.

SIMILARITY INVARIANTS

Similar matrices often have properties in common; for example, if A and B are similar matrices, then A and B have the same determinant. To see that this is so, suppose that

$$B = P^{-1}AP$$

Thus,

$$\det(B) = \det(P^{-1}AP) = \det(P^{-1})\det(A)\det(P)$$

$$= \frac{1}{\det(P)}\det(A)\det(P) = \det(A)$$

We make the following definition.

> **Definition.** A property of square matrices is said to be a *similarity invariant* or *invariant under similarity* if that property is shared by any two similar matrices.

In the terminology of this definition, the determinant of a square matrix is a similarity invariant. Table 1 lists some other important similarity invariants. The proofs of some of the results in Table 1 are given in the exercises.

It follows from Theorem 7.6.2 that *two matrices representing the same linear operator $T:V \to V$ with respect to different bases are similar*. Thus, if B is a basis for V, and the matrix $[T]_B$ has some property that is invariant under similarity, then for every basis B', the matrix $[T]_{B'}$ has that same property. For example, for any two bases B and B' we must have

$$\det([T]_B) = \det([T]_{B'})$$

It follows from this equation that the value of the determinant depends on T, but not on the particular basis that is used to obtain the matrix for T. Thus, the determinant can be regarded as a property of the linear operator T; indeed, if V is a finite-dimensional vector space, then we can *define* the **determinant of the linear operator** T to be

$$\det(T) = \det([T]_B) \tag{11}$$

where B is any basis for V.

TABLE 1 *Similarity Invariants*

Property	Description
Determinant	A and $P^{-1}AP$ have the same determinant.
Invertibility	A is invertible if and only if $P^{-1}AP$ is invertible.
Rank	A and $P^{-1}AP$ have the same rank.
Nullity	A and $P^{-1}AP$ have the same nullity.
Trace	A and $P^{-1}AP$ have the same trace.
Characteristic polynomial	A and $P^{-1}AP$ have the same characteristic polynomial.
Eigenvalues	A and $P^{-1}AP$ have the same eigenvalues.
Eigenspace dimension	If λ is an eigenvalue of A and $P^{-1}AP$, then the eigenspace of A corresponding to λ and the eigenspace of $P^{-1}AP$ corresponding to λ have the same dimension.

Example 2 Let $T:R^2 \rightarrow R^2$ be defined by

$$T\left(\begin{bmatrix} x_1 \\ x_2 \end{bmatrix}\right) = \begin{bmatrix} x_1 + x_2 \\ -2x_1 + 4x_2 \end{bmatrix}$$

Find $\det(T)$.

Solution. We can choose any basis B and calculate $\det([T]_B)$. If we take the standard basis, then from Example 1

$$[T]_B = \begin{bmatrix} 1 & 1 \\ -2 & 4 \end{bmatrix}$$

so

$$\det(T) = \begin{vmatrix} 1 & 1 \\ -2 & 4 \end{vmatrix} = 6$$

Had we chosen the basis $B' = \{\mathbf{u}_1, \mathbf{u}_2\}$ of Example 1, then we would have obtained

$$[T]_{B'} = \begin{bmatrix} 2 & 0 \\ 0 & 3 \end{bmatrix}$$

so

$$\det(T) = \begin{vmatrix} 2 & 0 \\ 0 & 3 \end{vmatrix} = 6$$

which agrees with the previous computation. ▲

A GEOMETRIC APPLICATION

Example 3 In R^2 let l be the line in the xy-plane that passes through the origin and makes an angle θ with the positive x-axis, and let $T:R^2 \rightarrow R^2$ be the linear operator that maps each point (x, y) into its reflection about the line l (Figure 3).

(a) Find a formula for T in terms of θ.
(b) Use the formula obtained in part (a) to find the reflection of the point $(1, 2)$ about the line l that makes an angle of $30°$ with the positive x-axis.

Solution (a). It follows from Formula (6) of Section 7.4 that

$$T\left(\begin{bmatrix} x \\ y \end{bmatrix}\right) = [T]\begin{bmatrix} x \\ y \end{bmatrix} \tag{12}$$

so our objective is to find the standard matrix $[T]$ for the given linear transformation.

Figure 3

If we denote the standard basis by $B = \{\mathbf{u}_1, \mathbf{u}_2\}$ (Figure 4), then

$$[T] = [T]_B \tag{13}$$

We shall not attempt to find $[T]_B$ directly; rather, we shall first find the matrix $[T]_{B'}$, where

$$B' = \{\mathbf{u}'_1, \mathbf{u}'_2\}$$

is the basis consisting of a unit vector $\mathbf{u}'_1$ along l and a unit vector $\mathbf{u}'_2$ perpendicular to l (Figure 4).

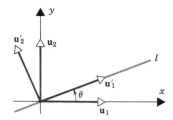

Figure 4

Once we have found $[T]_{B'}$, we shall perform a change of basis to find $[T]_B$. The computations are as follows:

$$T(\mathbf{u}'_1) = \mathbf{u}'_1 \quad \text{and} \quad T(\mathbf{u}'_2) = -\mathbf{u}'_2$$

so

$$[T(\mathbf{u}'_1)]_{B'} = \begin{bmatrix} 1 \\ 0 \end{bmatrix} \quad \text{and} \quad T[(\mathbf{u}'_2)]_{B'} = \begin{bmatrix} 0 \\ -1 \end{bmatrix}$$

Thus,

$$[T]_{B'} = \begin{bmatrix} 1 & 0 \\ 0 & -1 \end{bmatrix}$$

From the computations in Example 4 of Section 5.4, the transition matrix from B' to B is

$$P = \left[[\mathbf{u}'_1]_B \mid [\mathbf{u}'_2]_B \right] = \begin{bmatrix} \cos\theta & -\sin\theta \\ \sin\theta & \cos\theta \end{bmatrix} \tag{14}$$

It follows from Formula (10) that

$$[T]_B = P[T]_{B'}P^{-1}$$

Thus, from (13) and (14)

$$[T] = P[T]_{B'}P^{-1} = \begin{bmatrix} \cos\theta & -\sin\theta \\ \sin\theta & \cos\theta \end{bmatrix} \begin{bmatrix} 1 & 0 \\ 0 & -1 \end{bmatrix} \begin{bmatrix} \cos\theta & \sin\theta \\ -\sin\theta & \cos\theta \end{bmatrix}$$

$$= \begin{bmatrix} \cos^2\theta - \sin^2\theta & 2\sin\theta\cos\theta \\ 2\sin\theta\cos\theta & \sin^2\theta - \cos^2\theta \end{bmatrix}$$

$$= \begin{bmatrix} \cos 2\theta & \sin 2\theta \\ \sin 2\theta & -\cos 2\theta \end{bmatrix}$$

It now follows from (12) that the formula for T in matrix notation is

$$T\left(\begin{bmatrix} x \\ y \end{bmatrix}\right) = \begin{bmatrix} \cos 2\theta & \sin 2\theta \\ \sin 2\theta & -\cos 2\theta \end{bmatrix}\begin{bmatrix} x \\ y \end{bmatrix} \tag{15}$$

Solution (b). Letting $\theta = 30°$ in (15) yields

$$T\left(\begin{bmatrix} x \\ y \end{bmatrix}\right) = \begin{bmatrix} \frac{1}{2} & \frac{\sqrt{3}}{2} \\ \frac{\sqrt{3}}{2} & -\frac{1}{2} \end{bmatrix}\begin{bmatrix} x \\ y \end{bmatrix}$$

so

$$T\left(\begin{bmatrix} 1 \\ 2 \end{bmatrix}\right) = \begin{bmatrix} \frac{1}{2} & \frac{\sqrt{3}}{2} \\ \frac{\sqrt{3}}{2} & -\frac{1}{2} \end{bmatrix}\begin{bmatrix} 1 \\ 2 \end{bmatrix} = \begin{bmatrix} \frac{1}{2} + \sqrt{3} \\ \frac{\sqrt{3}}{2} - 1 \end{bmatrix}$$

Thus, $T(1, 2) = (\frac{1}{2} + \sqrt{3}, \frac{\sqrt{3}}{2} - 1)$. ▲

EIGENVALUES OF A LINEAR OPERATOR

Eigenvectors and eigenvalues can be defined for linear operators as well as matrices. A scalar λ is called an *eigenvalue* of a linear operator $T: V \rightarrow V$ if there is a nonzero vector $\mathbf{x}$ in V such that $T\mathbf{x} = \lambda\mathbf{x}$. The vector $\mathbf{x}$ is called an *eigenvector* of T corresponding to λ. Equivalently, the eigenvectors of T corresponding to λ are the nonzero vectors in the kernel of $\lambda I - T$ (Exercise 15). This kernel is called the *eigenspace* of T corresponding to λ.

It can be shown that if V is a finite-dimensional vector space, and B is *any* basis for V, then

1. The eigenvalues of T are the same as the eigenvalues of $[T]_B$.
2. A vector $\mathbf{x}$ is an eigenvector of T corresponding to λ if and only if its coordinate matrix $[\mathbf{x}]_B$ is an eigenvector of $[T]_B$ corresponding to λ.

We omit the proofs.

Example 4 Find the eigenvalues and bases for the eigenspaces of the linear operator $T: P_2 \rightarrow P_2$ defined by

$$T(a + bx + cx^2) = -2c + (a + 2b + c)x + (a + 3c)x^2$$

Solution. The matrix of T with respect to the standard basis $B = \{1, x, x^2\}$ is

$$[T]_B = \begin{bmatrix} 0 & 0 & -2 \\ 1 & 2 & 1 \\ 1 & 0 & 3 \end{bmatrix}$$

(verify). The eigenvalues of T are $\lambda = 1$ and $\lambda = 2$ (Example 5 of Section 6.1). Also from that example, the eigenspace of $[T]_B$ corresponding to $\lambda = 2$ has the basis $\{\mathbf{u}_1, \mathbf{u}_2\}$, where

$$\mathbf{u}_1 = \begin{bmatrix} -1 \\ 0 \\ 1 \end{bmatrix}, \qquad \mathbf{u}_2 = \begin{bmatrix} 0 \\ 1 \\ 0 \end{bmatrix}$$

and the eigenspace of $[T]_B$ corresponding to $\lambda = 1$ has the basis $\{\mathbf{u}_3\}$, where

$$\mathbf{u}_3 = \begin{bmatrix} -2 \\ 1 \\ 1 \end{bmatrix}$$

The matrices $\mathbf{u}_1$, $\mathbf{u}_2$, and $\mathbf{u}_3$ are the coordinate matrices relative to B of

$$\mathbf{p}_1 = -1 + x^2, \qquad \mathbf{p}_2 = x, \qquad \mathbf{p}_3 = -2 + x + x^2$$

Thus, the eigenspace of T corresponding to $\lambda = 2$ has the basis

$$\{\mathbf{p}_1, \mathbf{p}_2\} = \{-1 + x^2, x\}$$

and that corresponding to $\lambda = 1$ has the basis

$$\{\mathbf{p}_3\} = \{-2 + x + x^2\}$$

As a check, the reader should use the given formula for T to verify that $T(\mathbf{p}_1) = 2\mathbf{p}_1$, $T(\mathbf{p}_2) = 2\mathbf{p}_2$, and $T(\mathbf{p}_3) = \mathbf{p}_3$. ▲

Example 5 Let $T: R^3 \to R^3$ be the linear operator given by

$$T\left(\begin{bmatrix} x_1 \\ x_2 \\ x_3 \end{bmatrix} \right) = \begin{bmatrix} -2x_3 \\ x_1 + 2x_2 + x_3 \\ x_1 + 3x_3 \end{bmatrix}$$

Find a basis for R^3 relative to which the matrix for T is diagonal.

Solution. First we will find the standard matrix for T; then we will look for a change of basis that diagonalizes the standard matrix.

If $B = \{\mathbf{e}_1, \mathbf{e}_2, \mathbf{e}_3\}$ denotes the standard basis for R^3, then

$$T(\mathbf{e}_1) = T\left(\begin{bmatrix} 1 \\ 0 \\ 0 \end{bmatrix} \right) = \begin{bmatrix} 0 \\ 1 \\ 1 \end{bmatrix} \qquad T(\mathbf{e}_2) = T\left(\begin{bmatrix} 0 \\ 1 \\ 0 \end{bmatrix} \right) = \begin{bmatrix} 0 \\ 2 \\ 0 \end{bmatrix} \qquad T(\mathbf{e}_3) = T\left(\begin{bmatrix} 0 \\ 0 \\ 1 \end{bmatrix} \right) = \begin{bmatrix} -2 \\ 1 \\ 3 \end{bmatrix}$$

so that the standard matrix for T is

$$[T] = \begin{bmatrix} 0 & 0 & -2 \\ 1 & 2 & 1 \\ 1 & 0 & 3 \end{bmatrix} \tag{16}$$

We now want to change from the standard basis B to a new basis $B' = \{\mathbf{u}_1', \mathbf{u}_2', \mathbf{u}_3'\}$ in order to obtain a diagonal matrix for T. If we let P be the transition matrix from the unknown basis B' to the standard basis B, then by Theorem 7.6.2 the matrices $[T]$ and $[T]_{B'}$ will be related by

$$[T]_{B'} = P^{-1}[T]P \tag{17}$$

In Example 1 of Section 6.2 we found that the matrix in (16) is diagonalized by

$$P = \begin{bmatrix} -1 & 0 & -2 \\ 0 & 1 & 1 \\ 1 & 0 & 1 \end{bmatrix}$$

Since P represents the transition matrix from the basis $B' = \{\mathbf{u}_1', \mathbf{u}_2', \mathbf{u}_3'\}$ to the standard basis $B = \{\mathbf{e}_1, \mathbf{e}_2, \mathbf{e}_3\}$, the columns of P are $[\mathbf{u}_1']_B$, $[\mathbf{u}_2']_B$, and $[\mathbf{u}_3']_B$, so that

$$[\mathbf{u}_1']_B = \begin{bmatrix} -1 \\ 0 \\ 1 \end{bmatrix}, \qquad [\mathbf{u}_2']_B = \begin{bmatrix} 0 \\ 1 \\ 0 \end{bmatrix}, \qquad [\mathbf{u}_3']_B = \begin{bmatrix} -2 \\ 1 \\ 1 \end{bmatrix}$$

Thus,

$$\mathbf{u}_1' = (-1)\mathbf{e}_1 + (0)\mathbf{e}_2 + (1)\mathbf{e}_3 = \begin{bmatrix} -1 \\ 0 \\ 1 \end{bmatrix}$$

$$\mathbf{u}_2' = (0)\mathbf{e}_1 + (1)\mathbf{e}_2 + (0)\mathbf{e}_3 = \begin{bmatrix} 0 \\ 1 \\ 0 \end{bmatrix}$$

$$\mathbf{u}_3' = (-2)\mathbf{e}_1 + (1)\mathbf{e}_2 + (1)\mathbf{e}_3 = \begin{bmatrix} -2 \\ 1 \\ 1 \end{bmatrix}$$

are basis vectors that produce a diagonal matrix for $[T]_{B'}$. As a check, let us compute $[T]_{B'}$ directly. From the given formula for T we have

$$T(\mathbf{u}_1') = \begin{bmatrix} -2 \\ 0 \\ 2 \end{bmatrix} = 2\mathbf{u}_1', \qquad T(\mathbf{u}_2') = \begin{bmatrix} 0 \\ 2 \\ 0 \end{bmatrix} = 2\mathbf{u}_2', \qquad T(\mathbf{u}_3') = \begin{bmatrix} -2 \\ 1 \\ 1 \end{bmatrix} = \mathbf{u}_3'$$

so that

$$[T(\mathbf{u}_1')]_{B'} = \begin{bmatrix} 2 \\ 0 \\ 0 \end{bmatrix}, \qquad [T(\mathbf{u}_2')]_{B'} = \begin{bmatrix} 0 \\ 2 \\ 0 \end{bmatrix}, \qquad [T(\mathbf{u}_3')]_{B'} = \begin{bmatrix} 0 \\ 0 \\ 1 \end{bmatrix}$$

Thus,

$$[T]_{B'} = \left[[T(\mathbf{u}_1')]_{B'} \mid [T(\mathbf{u}_2')]_{B'} \mid [T(\mathbf{u}_3')]_{B'} \right] = \begin{bmatrix} 2 & 0 & 0 \\ 0 & 2 & 0 \\ 0 & 0 & 1 \end{bmatrix}$$

This is consistent with (17) since

$$P^{-1}[T]P = \begin{bmatrix} 1 & 0 & 2 \\ 1 & 1 & 1 \\ -1 & 0 & -1 \end{bmatrix} \begin{bmatrix} 0 & 0 & -2 \\ 1 & 2 & 1 \\ 1 & 0 & 3 \end{bmatrix} \begin{bmatrix} -1 & 0 & -2 \\ 0 & 1 & 1 \\ 1 & 0 & 1 \end{bmatrix} = \begin{bmatrix} 2 & 0 & 0 \\ 0 & 2 & 0 \\ 0 & 0 & 1 \end{bmatrix}$$

EXERCISE SET 7.6

In Exercises 1–7 find the matrix of T with respect to B, and use Theorem 7.6.2 to compute the matrix of T with respect to B'.

1. $T: R^2 \to R^2$ is defined by

$$T\left(\begin{bmatrix} x_1 \\ x_2 \end{bmatrix} \right) = \begin{bmatrix} x_1 - 2x_2 \\ -x_2 \end{bmatrix}$$

$B = \{\mathbf{u}_1, \mathbf{u}_2\}$ and $B' = \{\mathbf{v}_1, \mathbf{v}_2\}$, where

$$\mathbf{u}_1 = \begin{bmatrix} 1 \\ 0 \end{bmatrix} \quad \mathbf{u}_2 = \begin{bmatrix} 0 \\ 1 \end{bmatrix} \quad \mathbf{v}_1 = \begin{bmatrix} 2 \\ 1 \end{bmatrix} \quad \text{and} \quad \mathbf{v}_2 = \begin{bmatrix} -3 \\ 4 \end{bmatrix}$$

2. $T: R^2 \to R^2$ is defined by

$$T\left(\begin{bmatrix} x_1 \\ x_2 \end{bmatrix} \right) = \begin{bmatrix} x_1 + 7x_2 \\ 3x_1 - 4x_2 \end{bmatrix}$$

$B = \{\mathbf{u}_1, \mathbf{u}_2\}$ and $B' = \{\mathbf{v}_1, \mathbf{v}_2\}$, where

$$\mathbf{u}_1 = \begin{bmatrix} 2 \\ 2 \end{bmatrix} \quad \mathbf{u}_2 = \begin{bmatrix} 4 \\ -1 \end{bmatrix} \quad \mathbf{v}_1 = \begin{bmatrix} 1 \\ 3 \end{bmatrix} \quad \mathbf{v}_2 = \begin{bmatrix} -1 \\ -1 \end{bmatrix}$$

3. $T: R^2 \to R^2$ is the rotation about the origin through $45°$; B and B' are the bases in Exercise 1.

4. $T: R^3 \to R^3$ is defined by

$$T\left(\begin{bmatrix} x_1 \\ x_2 \\ x_3 \end{bmatrix} \right) = \begin{bmatrix} x_1 + 2x_2 - x_3 \\ -x_2 \\ x_1 + 7x_3 \end{bmatrix}$$

B is the standard basis for R^3 and $B' = \{\mathbf{v}_1, \mathbf{v}_2, \mathbf{v}_3\}$, where

$$\mathbf{v}_1 = \begin{bmatrix} 1 \\ 0 \\ 0 \end{bmatrix}, \quad \mathbf{v}_2 = \begin{bmatrix} 1 \\ 1 \\ 0 \end{bmatrix}, \quad \text{and} \quad \mathbf{v}_3 = \begin{bmatrix} 1 \\ 1 \\ 1 \end{bmatrix}$$

5. $T:R^3 \rightarrow R^3$ is the orthogonal projection on the xy-plane; B and B' are as in Exercise 4.

6. $T:R^2 \rightarrow R^2$ is defined by $T(\mathbf{x}) = 5\mathbf{x}$; B and B' are the bases in Exercise 2.

7. $T:P_1 \rightarrow P_1$ is defined by $T(a_0 + a_1 x) = a_0 + a_1(x + 1)$; $B = \{\mathbf{p}_1, \mathbf{p}_2\}$ and $B' = \{\mathbf{q}_1, \mathbf{q}_2\}$, where $\mathbf{p}_1 = 6 + 3x$, $\mathbf{p}_2 = 10 + 2x$, $\mathbf{q}_1 = 2$, $\mathbf{q}_2 = 3 + 2x$.

8. Find $\det(T)$.
   (a) $T:R^2 \rightarrow R^2$, where $T(x_1, x_2) = (3x_1 - 4x_2, -x_1 + 7x_2)$
   (b) $T:R^3 \rightarrow R^3$, where $T(x_1, x_2, x_3) = (x_1 - x_2, x_2 - x_3, x_3 - x_1)$
   (c) $T:P_2 \rightarrow P_2$, where $T(p(x)) = p(x - 1)$

9. Prove that the following are similarity invariants:
   (a) rank   (b) nullity   (c) invertibility

10. Let $T:P_4 \rightarrow P_4$ be the linear operator given by the formula $T(p(x)) = p(2x + 1)$.
    (a) Find a matrix for T with respect to some convenient basis; then use the result in Exercise 9 to find the rank and nullity of T.
    (b) Use the result in part (a) to determine whether T is one-to-one.

11. In each part find a basis for R^2 relative to which the matrix for T is diagonal.

   (a) $T\left(\begin{bmatrix} x_1 \\ x_2 \end{bmatrix}\right) = \begin{bmatrix} x_1 - x_2 \\ 2x_1 + 4x_2 \end{bmatrix}$ (b) $T\left(\begin{bmatrix} x_1 \\ x_2 \end{bmatrix}\right) = \begin{bmatrix} 4x_1 - x_2 \\ -3x_1 + x_2 \end{bmatrix}$

12. In each part find a basis for R^3 relative to which the matrix for T is diagonal.

   (a) $T\left(\begin{bmatrix} x_1 \\ x_2 \\ x_3 \end{bmatrix}\right) = \begin{bmatrix} -2x_1 + x_2 - x_3 \\ x_1 - 2x_2 - x_3 \\ -x_1 - x_2 - 2x_3 \end{bmatrix}$ (b) $T\left(\begin{bmatrix} x_1 \\ x_2 \\ x_3 \end{bmatrix}\right) = \begin{bmatrix} -x_2 + x_3 \\ -x_1 + x_3 \\ x_1 + x_2 \end{bmatrix}$

   (c) $T\left(\begin{bmatrix} x_1 \\ x_2 \\ x_3 \end{bmatrix}\right) = \begin{bmatrix} 4x_1 + x_3 \\ 2x_1 + 3x_2 + 2x_3 \\ x_1 + 4x_3 \end{bmatrix}$

13. Let $T:P_2 \rightarrow P_2$ be defined by

$$T(a_0 + a_1 x + a_2 x^2) = (5a_0 + 6a_1 + 2a_2) - (a_1 + 8a_2)x + (a_0 - 2a_2)x^2$$

   (a) Find the eigenvalues of T.   (b) Find bases for the eigenspaces of T.

14. Let $T:M_{22} \rightarrow M_{22}$ be defined by

$$T\left(\begin{bmatrix} a & b \\ c & d \end{bmatrix}\right) = \begin{bmatrix} 2c & a + c \\ b - 2c & d \end{bmatrix}$$

   (a) Find the eigenvalues of T.   (b) Find bases for the eigenspaces of T.

15. Let λ be an eigenvalue of a linear operator $T:V \rightarrow V$. Prove that the eigenvectors of T corresponding to λ are the nonzero vectors in the kernel of $\lambda I - T$.

16. Prove that if A and B are similar matrices, then A^2 and B^2 are also similar. More generally, prove that A^k and B^k are similar, where k is any positive integer.

17. Let C and D be $m \times n$ matrices, and let $B = \{\mathbf{v}_1, \mathbf{v}_2, \ldots, \mathbf{v}_n\}$ be a basis for a vector space V. Show that if $C[\mathbf{x}]_B = D[\mathbf{x}]_B$ for all $\mathbf{x}$ in V, then $C = D$.

18. Let l be a line in the xy-plane that passes through the origin and makes an angle θ with the positive x-axis. As illustrated in Figure 5, let $T:R^2 \to R^2$ be the orthogonal projection of R^2 onto l.

(a) Use the method of Example 3 to show that

$$T\left(\begin{bmatrix} x \\ y \end{bmatrix}\right) = \begin{bmatrix} \cos^2 \theta & \sin \theta \cos \theta \\ \sin \theta \cos \theta & \sin^2 \theta \end{bmatrix} \begin{bmatrix} x \\ y \end{bmatrix}$$

(b) Use the result in (a) to find the orthogonal projection of the point $(1, 5)$ onto the line through the origin that makes an angle $\theta = 30°$ with the positive x-axis.

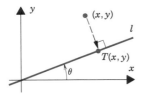

Figure 5

SUPPLEMENTARY EXERCISES

1. Let A be an $n \times n$ matrix, B a nonzero $n \times 1$ matrix, and $\mathbf{x}$ a vector in R^n expressed in matrix notation. Is $T(\mathbf{x}) = A\mathbf{x} + B$ a linear operator on R^n? Justify your answer.

2. Let

$$A = \begin{bmatrix} \cos \theta & -\sin \theta \\ \sin \theta & \cos \theta \end{bmatrix}$$

(a) Show that

$$A^2 = \begin{bmatrix} \cos 2\theta & -\sin 2\theta \\ \sin 2\theta & \cos 2\theta \end{bmatrix} \quad \text{and} \quad A^3 = \begin{bmatrix} \cos 3\theta & -\sin 3\theta \\ \sin 3\theta & \cos 3\theta \end{bmatrix}$$

(b) Guess the form of the matrix A^n for any positive integer n.

(c) By considering the geometric effect of $T:R^2 \to R^2$, where T is multiplication by A, obtain the result in (b) geometrically.

3. Let $\mathbf{v}_0$ be a fixed vector in an inner product space V, and let $T:V \to V$ be defined by $T(\mathbf{v}) = \langle \mathbf{v}, \mathbf{v}_0 \rangle \mathbf{v}_0$. Show that T is a linear operator on V.

4. Let $\mathbf{v}_1, \mathbf{v}_2, \ldots, \mathbf{v}_m$ be fixed vectors in R^n, and let $T:R^n \to R^m$ be the function defined by $T(\mathbf{x}) = (\mathbf{x} \cdot \mathbf{v}_1, \mathbf{x} \cdot \mathbf{v}_2, \ldots, \mathbf{x} \cdot \mathbf{v}_m)$, where $\mathbf{x} \cdot \mathbf{v}_i$ is the Euclidean inner product on R^n.

(a) Show that T is a linear transformation.

(b) Show that the matrix with row vectors $\mathbf{v}_1, \mathbf{v}_2, \ldots, \mathbf{v}_m$ is the standard matrix for T.

5. Let $\{\mathbf{e}_1, \mathbf{e}_2, \mathbf{e}_3, \mathbf{e}_4\}$ be the standard basis for R^4 and $T:R^4 \to R^3$ the linear transformation for which

$$T(\mathbf{e}_1) = (1, 2, 1) \qquad T(\mathbf{e}_2) = (0, 1, 0)$$

$$T(\mathbf{e}_3) = (1, 3, 0) \qquad T(\mathbf{e}_4) = (1, 1, 1)$$

(a) Find bases for the range and kernel of T.   (b) Find the rank and nullity of T.

6. Let $A\mathbf{x} = \mathbf{0}$ be a homogeneous linear system with n unknowns, and let r be the rank of A. Find the dimension of the solution space if
(a) $n = 5$, $r = 3$ (b) $n = 4$, $r = 4$

7. Let $B = \{\mathbf{v}_1, \mathbf{v}_2, \mathbf{v}_3, \mathbf{v}_4\}$ be a basis for a vector space V and $T: V \to V$ the linear operator for which

$$T(\mathbf{v}_1) = \mathbf{v}_1 + \mathbf{v}_2 + \mathbf{v}_3 + 3\mathbf{v}_4$$
$$T(\mathbf{v}_2) = \mathbf{v}_1 - \mathbf{v}_2 + 2\mathbf{v}_3 + 2\mathbf{v}_4$$
$$T(\mathbf{v}_3) = 2\mathbf{v}_1 - 4\mathbf{v}_2 + 5\mathbf{v}_3 + 3\mathbf{v}_4$$
$$T(\mathbf{v}_4) = -2\mathbf{v}_1 + 6\mathbf{v}_2 - 6\mathbf{v}_3 - 2\mathbf{v}_4$$

(a) Find the rank and nullity of T.
(b) Determine whether T is one-to-one.

8. Let V and W be vector spaces, T, T_1, and T_2 linear transformations from V to W, and k a scalar. Define new transformations, $T_1 + T_2$ and kT, by the formulas

$$(T_1 + T_2)(\mathbf{x}) = T_1(\mathbf{x}) + T_2(\mathbf{x})$$
$$(kT)(\mathbf{x}) = k(T(\mathbf{x}))$$

(a) Show that $(T_1 + T_2): V \to W$ and $kT: V \to W$ are linear transformations.
(b) Show that the set of all linear transformations from V to W with the operations in (a) forms a vector space.

9. Let A and B be similar matrices. Prove:
(a) A^t and B^t are similar.
(b) If A and B are invertible, then A^{-1} and B^{-1} are similar.

10. (**Fredholm Alternative Theorem.**) Let $T: V \to V$ be a linear operator on an n-dimensional vector space. Prove that exactly one of the following statements holds:
(i) The equation $T(\mathbf{x}) = \mathbf{b}$ has a solution for all vectors $\mathbf{b}$ in V.
(ii) Nullity of $T > 0$.

11. Let $T: M_{22} \to M_{22}$ be the linear operator defined by

$$T(X) = \begin{bmatrix} 1 & 1 \\ 0 & 0 \end{bmatrix} X + X \begin{bmatrix} 0 & 0 \\ 1 & 1 \end{bmatrix}$$

Find the rank and nullity of T.

12. Prove: If A and B are similar matrices, and if B and C are similar matrices, then A and C are similar matrices.

13. Let $T: M_{22} \to M_{22}$ be the linear operator defined by $T(M) = M^t$. Find the matrix for T with respect to the standard basis for M_{22}.

14. Let $B = \{\mathbf{u}_1, \mathbf{u}_2, \mathbf{u}_3\}$ and $B' = \{\mathbf{v}_1, \mathbf{v}_2, \mathbf{v}_3\}$ be bases for a vector space V, and let

$$P = \begin{bmatrix} 2 & -1 & 3 \\ 1 & 1 & 4 \\ 0 & 1 & 2 \end{bmatrix}$$

be the transition matrix from B' to B.
(a) Express $\mathbf{v}_1$, $\mathbf{v}_2$, $\mathbf{v}_3$ as linear combinations of $\mathbf{u}_1$, $\mathbf{u}_2$, $\mathbf{u}_3$.
(b) Express $\mathbf{u}_1$, $\mathbf{u}_2$, $\mathbf{u}_3$ as linear combinations of $\mathbf{v}_1$, $\mathbf{v}_2$, $\mathbf{v}_3$.

15. Let $B = \{u_1, u_2, u_3\}$ be a basis for a vector space V and $T: V \to V$ a linear operator such that

$$[T]_B = \begin{bmatrix} -3 & 4 & 7 \\ 1 & 0 & -2 \\ 0 & 1 & 0 \end{bmatrix}$$

Find $[T]_{B'}$, where $B' = \{v_1, v_2, v_3\}$ is the basis for V defined by

$$v_1 = u_1, \qquad v_2 = u_1 + u_2, \qquad v_3 = u_1 + u_2 + u_3$$

16. Show that the matrices

$$\begin{bmatrix} 1 & 1 \\ -1 & 4 \end{bmatrix} \quad \text{and} \quad \begin{bmatrix} 2 & 1 \\ 1 & 3 \end{bmatrix}$$

are similar, but that

$$\begin{bmatrix} 3 & 1 \\ -6 & -2 \end{bmatrix} \quad \text{and} \quad \begin{bmatrix} -1 & 2 \\ 1 & 0 \end{bmatrix}$$

are not.

17. Suppose that $T: V \to V$ is a linear operator and B is a basis for V such that for any vector x in V

$$[T(x)]_B = \begin{bmatrix} x_1 - x_2 + x_3 \\ x_2 \\ x_1 - x_3 \end{bmatrix} \quad \text{if} \quad [x]_B = \begin{bmatrix} x_1 \\ x_2 \\ x_3 \end{bmatrix}$$

Find $[T]_B$.

18. Let $T: V \to V$ be a linear operator. Prove that T is one-to-one if and only if $\det(T) \neq 0$.

19. (For readers who have studied calculus.)
    (a) Show that the function $J: P_{n-1} \to P_n$ defined by the formula

$$J(p(x)) = \int_0^x p(t)\, dt$$

    is a linear transformation.
    (b) Find $(D \circ J)(p(x))$, where $D: P_n \to P_{n-1}$ is the differentiation transformation

$$D(f(x)) = f'(x)$$

20. Let $T: P_2 \to R^3$ be the function defined by the formula

$$T(p(x)) = \begin{bmatrix} p(-1) \\ p(0) \\ p(1) \end{bmatrix}$$

    (a) Find $T(x^2 + 5x + 6)$.
    (b) Show that T is a linear transformation.
    (c) Show that T is one-to-one.
    (d) Find

$$T^{-1}\left(\begin{bmatrix} 0 \\ 3 \\ 0 \end{bmatrix} \right)$$

    (e) Sketch the graph of the polynomial in part (d).

21. Let x_1, x_2, and x_3 be distinct real numbers such that $x_1 < x_2 < x_3$, and let $T:P_2 \to R^3$ be the function defined by the formula

$$T(p(x)) = \begin{bmatrix} p(x_1) \\ p(x_2) \\ p(x_3) \end{bmatrix}$$

(a) Show that T is a linear transformation.

(b) Show that T is one-to-one.

(c) Verify that if a_1, a_2, and a_3 are any real numbers, then

$$T^{-1}\left(\begin{bmatrix} a_1 \\ a_2 \\ a_3 \end{bmatrix}\right) = a_1 P_1(x) + a_2 P_2(x) + a_3 P_3(x)$$

where

$$P_1(x) = \frac{(x - x_2)(x - x_3)}{(x_1 - x_2)(x_1 - x_3)}, \qquad P_2(x) = \frac{(x - x_1)(x - x_3)}{(x_2 - x_1)(x_2 - x_3)},$$

$$P_3(x) = \frac{(x - x_1)(x \quad x_2)}{(x_3 - x_1)(x_3 - x_2)}$$

(d) What relationship exists between the graph of the function

$$a_1 P_1(x) + a_2 P_2(x) + a_3 P_3(x)$$

and the points (x_1, a_1), (x_2, a_2), and (x_3, a_3)?

22. (For readers who have studied calculus.) Let $p(x)$ and $q(x)$ be continuous functions, and let V be the subspace of $C(-\infty, +\infty)$ consisting of all twice differentiable functions. Define $L:V \to V$ by

$$L(y(x)) = y''(x) + p(x)y'(x) + q(x)y(x)$$

(a) Show that L is a linear operator.

(b) Consider the special case where $p(x) = 0$ and $q(x) = 1$. Show that the function $\phi(x) = c_1 \sin x + c_2 \cos x$ is in the nullspace of L for all real values of c_1 and c_2.

23. (For readers who have studied calculus.) Let $D:P_n \to P_n$ be the differentiation operator $D(\mathbf{p}) = \mathbf{p}'$. Show that the matrix for D with respect to the basis $B = \{1, x, x^2, \ldots, x^n\}$ is

$$\begin{bmatrix} 0 & 1 & 0 & 0 & \cdots & 0 \\ 0 & 0 & 2 & 0 & \cdots & 0 \\ 0 & 0 & 0 & 3 & \cdots & 0 \\ \vdots & \vdots & \vdots & \vdots & & \vdots \\ 0 & 0 & 0 & 0 & \cdots & n \\ 0 & 0 & 0 & 0 & \cdots & 0 \end{bmatrix}$$

24. (For readers who have studied calculus.) It can be shown that for any real number c, the vectors

$$1, \, x - c, \, \frac{(x - c)^2}{2!}, \ldots, \frac{(x - c)^n}{n!}$$

form a basis for P_n. Find the matrix for the differentiation operator of Exercise 23 with respect to this basis.

25. (For readers who have studied calculus.) Let $J : P_n \to P_{n+1}$ be the integration transformation defined by

$$J(\mathbf{p}) = \int (a_0 + a_1 x + \cdots + a_n x^n)\, dx = a_0 x + \frac{a_1}{2} x^2 + \cdots + \frac{a_n}{n+1} x^{n+1}$$

where $\mathbf{p} = a_0 + a_1 x + \cdots + a_n x^n$. Find the matrix for T with respect to the standard bases for P_n and P_{n+1}.

CHAPTER 8

APPLICATIONS

8.1 APPLICATION TO DIFFERENTIAL EQUATIONS

Many laws of physics, chemistry, biology, and economics are described in terms of differential equations, that is, equations involving functions and their derivatives. The purpose of this section is to illustrate one way in which linear algebra can be applied to solve certain systems of differential equations. The scope of this section is narrow, but it illustrates an important area of application of linear algebra.

TERMINOLOGY

One of the simplest differential equations is

$$y' = ay \qquad (1)$$

where $y = f(x)$ is an unknown function to be determined, $y' = dy/dx$ is its derivative, and a is a constant. Like most differential equations, (1) has infinitely many solutions; they are the functions of the form

$$y = ce^{ax} \qquad (2)$$

where c is an arbitrary constant. Each function of this form is a solution of $y' = ay$ since

$$y' = cae^{ax} = ay$$

Conversely, every solution of $y' = ay$ must be a function of the form ce^{ax} (Exercise 7), so that (2) describes all solutions of $y' = ay$. We call (2) the ***general solution*** of $y' = ay$.

Sometimes the physical problem that generates a differential equation imposes some added condition that enables us to isolate one ***particular solution*** from the general solution. For example, if we require that the solution of $y' = ay$ satisfy the added condition

$$y(0) = 3 \qquad (3)$$

that is, $y = 3$ when $x = 0$, then on substituting these values in the general solution $y = ce^{ax}$ we obtain a value for c, namely

$$3 = ce^0 = c$$

Thus,

$$y = 3e^{ax}$$

is the only solution of $y' = ay$ that satisfies the added condition. A condition such as (3), which specifies the value of the solution at a point, is called an ***initial condition***, and the problem of solving a differential equation subject to an initial condition is called an ***initial-value problem***.

LINEAR SYSTEMS OF FIRST ORDER EQUATIONS

In this section we will be concerned with solving systems of differential equations having the form

$$
\begin{aligned}
y'_1 &= a_{11}y_1 + a_{12}y_2 + \cdots + a_{1n}y_n \\
y'_2 &= a_{21}y_1 + a_{22}y_2 + \cdots + a_{2n}y_n \\
&\vdots \qquad \vdots \qquad \vdots \qquad \qquad \vdots \\
y'_n &= a_{n1}y_1 + a_{n2}y_2 + \cdots + a_{nn}y_n
\end{aligned}
\tag{4}
$$

where $y_1 = f_1(x)$, $y_2 = f_2(x)$, ..., $y_n = f_n(x)$ are functions to be determined, and the a_{ij}'s are constants. In matrix notation (4) can be written

$$
\begin{bmatrix} y'_1 \\ y'_2 \\ \vdots \\ y'_n \end{bmatrix}
=
\begin{bmatrix}
a_{11} & a_{12} & \cdots & a_{1n} \\
a_{21} & a_{22} & \cdots & a_{2n} \\
\vdots & \vdots & & \vdots \\
a_{n1} & a_{n2} & \cdots & a_{nn}
\end{bmatrix}
\begin{bmatrix} y_1 \\ y_2 \\ \vdots \\ y_n \end{bmatrix}
$$

or more briefly

$$Y' = AY$$

Example 1

(a) Write the following system in matrix form:

$$
\begin{aligned}
y'_1 &= 3y_1 \\
y'_2 &= -2y_2 \\
y'_3 &= 5y_3
\end{aligned}
$$

(b) Solve the system.
(c) Find a solution of the system that satisfies the initial conditions $y_1(0) = 1$, $y_2(0) = 4$, and $y_3(0) = -2$.

Solution (a).

$$
\begin{bmatrix} y'_1 \\ y'_2 \\ y'_3 \end{bmatrix}
=
\begin{bmatrix}
3 & 0 & 0 \\
0 & -2 & 0 \\
0 & 0 & 5
\end{bmatrix}
\begin{bmatrix} y_1 \\ y_2 \\ y_3 \end{bmatrix}
\tag{5}
$$

or

$$Y' = \begin{bmatrix} 3 & 0 & 0 \\ 0 & -2 & 0 \\ 0 & 0 & 5 \end{bmatrix} Y$$

Solution (b). Because each equation involves only one unknown function, we can solve the equations individually. From (2), we obtain

$$y_1 = c_1 e^{3x}$$
$$y_2 = c_2 e^{-2x}$$
$$y_3 = c_3 e^{5x}$$

or in matrix notation

$$Y = \begin{vmatrix} y_1 \\ y_2 \\ y_3 \end{vmatrix} = \begin{vmatrix} c_1 e^{3x} \\ c_2 e^{-2x} \\ c_3 e^{5x} \end{vmatrix}$$

Solution (c). From the given initial conditions, we obtain

$$1 = y_1(0) = c_1 e^0 = c_1$$
$$4 = y_2(0) = c_2 e^0 = c_2$$
$$-2 = y_3(0) = c_3 e^0 = c_3$$

so that the solution satisfying the initial conditions is

$$y_1 = e^{3x}, \qquad y_2 = 4e^{-2x}, \qquad y_3 = -2e^{5x}$$

or, in matrix notation,

$$Y = \begin{bmatrix} y_1 \\ y_2 \\ y_3 \end{bmatrix} = \begin{bmatrix} e^{3x} \\ 4e^{-2x} \\ -2e^{5x} \end{bmatrix}$$

The system in the foregoing example is easy to solve because each equation involves only one unknown function, and this is the case because the matrix of coefficients (5) for the system is diagonal. But how do we handle a system

$$Y' = AY$$

in which the matrix A is not diagonal? The idea is simple: try to make a substitution for Y that will yield a new system with a diagonal coefficient matrix; solve this new simpler system, and then use this solution to determine the solution of the original system.

The kind of substitution we have in mind is

$$y_1 = p_{11}u_1 + p_{12}u_2 + \cdots + p_{1n}u_n$$
$$y_2 = p_{21}u_1 + p_{22}u_2 + \cdots + p_{2n}u_n$$
$$\vdots \qquad \vdots \qquad \vdots \qquad \qquad \vdots$$
$$y_n = p_{n1}u_1 + p_{n2}u_2 + \cdots + p_{nn}u_n$$

(6)

or, in matrix notation,

$$
\begin{bmatrix} y_1 \\ y_2 \\ \vdots \\ y_n \end{bmatrix} = \begin{bmatrix} p_{11} & p_{12} & \cdots & p_{1n} \\ p_{21} & p_{22} & \cdots & p_{2n} \\ \vdots & \vdots & & \vdots \\ p_{n1} & p_{n2} & \cdots & p_{nn} \end{bmatrix} \begin{bmatrix} u_1 \\ u_2 \\ \vdots \\ u_n \end{bmatrix}
$$

or more briefly

$$Y = PU$$

In this substitution the p_{ij}'s are constants to be determined in such a way that the new system involving the unknown functions $u_1, u_2, \ldots, u_n$ has a diagonal coefficient matrix. We leave it as an exercise for the reader to differentiate each equation in (6) and deduce

$$Y' = PU'$$

If we make the substitutions $Y = PU$ and $Y' = PU'$ in the original system

$$Y' = AY$$

and if we assume P to be invertible, we obtain

$$PU' = A(PU)$$

or

$$U' = (P^{-1}AP)U$$

or

$$U' = DU$$

where $D = P^{-1}AP$. The choice for P is now clear; if we want the new coefficient matrix D to be diagonal, we must choose P to be a matrix that diagonalizes A.

PROCEDURE FOR SOLVING A SYSTEM OF FIRST ORDER LINEAR DIFFERENTIAL EQUATIONS

This suggests the following procedure for solving a system

$$Y' = AY$$

with a diagonalizable coefficient matrix A.

Step 1. Find a matrix P that diagonalizes A.

Step 2. Make the substitutions $Y = PU$ and $Y' = PU'$ to obtain a new "diagonal system" $U' = DU$, where $D = P^{-1}AP$.

Step 3. Solve $U' = DU$.

Step 4. Determine Y from the equation $Y = PU$.

Example 2

(a) Solve the system

$$
\begin{aligned}
y_1' &= y_1 + y_2 \\
y_2' &= 4y_1 - 2y_2
\end{aligned}
$$

(b)   Find the solution that satisfies the initial conditions $y_1(0) = 1$, $y_2(0) = 6$.

Solution (a). The coefficient matrix for the system is

$$A = \begin{bmatrix} 1 & 1 \\ 4 & -2 \end{bmatrix}$$

As discussed in Section 6.2, A will be diagonalized by any matrix P whose columns are linearly independent eigenvectors of A. Since

$$\det(\lambda I - A) = \begin{vmatrix} \lambda - 1 & -1 \\ -4 & \lambda + 2 \end{vmatrix} = \lambda^2 + \lambda - 6 = (\lambda + 3)(\lambda - 2)$$

the eigenvalues of A are $\lambda = 2$, $\lambda = -3$. By definition,

$$\mathbf{x} = \begin{bmatrix} x_1 \\ x_2 \end{bmatrix}$$

is an eigenvector of A corresponding to λ if and only if $\mathbf{x}$ is a nontrivial solution of $(\lambda I - A)\mathbf{x} = \mathbf{0}$, that is, of

$$\begin{bmatrix} \lambda - 1 & -1 \\ -4 & \lambda + 2 \end{bmatrix} \begin{bmatrix} x_1 \\ x_2 \end{bmatrix} = \begin{bmatrix} 0 \\ 0 \end{bmatrix}$$

If $\lambda = 2$, this system becomes

$$\begin{bmatrix} 1 & -1 \\ -4 & 4 \end{bmatrix} \begin{bmatrix} x_1 \\ x_2 \end{bmatrix} = \begin{bmatrix} 0 \\ 0 \end{bmatrix}$$

Solving this system yields

$$x_1 = t, \qquad x_2 = t$$

so that

$$\begin{bmatrix} x_1 \\ x_2 \end{bmatrix} = \begin{bmatrix} t \\ t \end{bmatrix} = t \begin{bmatrix} 1 \\ 1 \end{bmatrix}$$

Thus,

$$\mathbf{p}_1 = \begin{bmatrix} 1 \\ 1 \end{bmatrix}$$

is a basis for the eigenspace corresponding to $\lambda = 2$. Similarly, the reader can show that

$$\mathbf{p}_2 = \begin{bmatrix} -\frac{1}{4} \\ 1 \end{bmatrix}$$

is a basis for the eigenspace corresponding to $\lambda = -3$. Thus,

$$P = \begin{bmatrix} 1 & -\frac{1}{4} \\ 1 & 1 \end{bmatrix}$$

diagonalizes A, and

$$D = P^{-1}AP = \begin{bmatrix} 2 & 0 \\ 0 & -3 \end{bmatrix}$$

Therefore, the substitution

$$Y = PU \quad \text{and} \quad Y' = PU'$$

yields the new "diagonal system"

$$U' = DU = \begin{bmatrix} 2 & 0 \\ 0 & -3 \end{bmatrix} U \quad \text{or} \quad \begin{array}{l} u'_1 = 2u_1 \\ u'_2 = -3u_2 \end{array}$$

From (2) the solution of this system is

$$\begin{array}{l} u_1 = c_1 e^{2x} \\ u_2 = c_2 e^{-3x} \end{array} \quad \text{or} \quad U = \begin{bmatrix} c_1 e^{2x} \\ c_2 e^{-3x} \end{bmatrix}$$

so that the equation $Y = PU$ yields as the solution for Y

$$Y = \begin{bmatrix} y_1 \\ y_2 \end{bmatrix} = \begin{bmatrix} 1 & -\frac{1}{4} \\ 1 & 1 \end{bmatrix} \begin{bmatrix} c_1 e^{2x} \\ c_2 e^{-3x} \end{bmatrix} = \begin{bmatrix} c_1 e^{2x} - \frac{1}{4} c_2 e^{-3x} \\ c_1 e^{2x} + c_2 e^{-3x} \end{bmatrix}$$

or

$$\begin{array}{l} y_1 = c_1 e^{2x} - \frac{1}{4} c_2 e^{-3x} \\ y_2 = c_1 e^{2x} + c_2 e^{-3x} \end{array} \tag{7}$$

Solution (b). **If we substitute the given initial conditions in (7), we obtain**

$$c_1 - \tfrac{1}{4} c_2 = 1$$
$$c_1 + c_2 = 6$$

Solving this system we obtain

$$c_1 = 2, \quad c_2 = 4$$

so that from (7) the solution satisfying the initial conditions is

$$y_1 = 2e^{2x} - e^{-3x}$$
$$y_2 = 2e^{2x} + 4e^{-3x} \quad \blacktriangle$$

We have assumed in this section that the coefficient matrix of $Y' = AY$ is diagonalizable. If this is not the case, other methods must be used to solve the system. Such methods are discussed in more advanced texts.

EXERCISE SET 8.1

1. (a) Solve the system

$$y'_1 = y_1 + 4y_2$$
$$y'_2 = 2y_1 + 3y_2$$

   (b) Find the solution that satisfies the initial conditions $y_1(0) = 0$, $y_2(0) = 0$.

2. (a) Solve the system

$$y'_1 = y_1 + 3y_2$$
$$y'_2 = 4y_1 + 5y_2$$

   (b) Find the solution that satisfies the conditions $y_1(0) = 2$, $y'_2(0) = 1$.

3. (a) Solve the system

$$y'_1 = 4y_1 + y_3$$
$$y'_2 = -2y_1 + y_2$$
$$y'_3 = -2y_1 + y_3$$

(b) Find the solution that satisfies the initial conditions $y_1(0) = -1$, $y_2(0) = 1$, $y_3(0) = 0$.

4. Solve the system

$$y'_1 = 4y_1 + 2y_2 + 2y_3$$
$$y'_2 = 2y_1 + 4y_2 + 2y_3$$
$$y'_3 = 2y_1 + 2y_2 + 4y_3$$

5. Solve the differential equation $y'' - y' - 6y = 0$. [***Hint.*** Let $y_1 = y$, $y_2 = y'$ and then show that

$$y'_1 = y_2$$
$$y'_2 = y'' = y' + 6y = 6y_1 + y_2]$$

6. Solve the differential equation $y''' - 6y'' + 11y' - 6y = 0$. [***Hint.*** Let $y_1 = y$, $y_2 = y'$, $y_3 = y''$ and then show that

$$y'_1 = y_2$$
$$y'_2 = y_3$$
$$y'_3 = 6y_1 - 11y_2 + 6y_3]$$

7. Prove: Every solution of $y' = ay$ has the form $y = ce^{ax}$. [***Hint.*** Let $y = f(x)$ be a solution and show that $f(x)e^{-ax}$ is constant.]

8. Prove: If A is diagonalizable and

$$Y = \begin{bmatrix} y_1 \\ y_2 \\ \vdots \\ y_n \end{bmatrix}$$

satisfies $Y' = AY$, then each y_i is a linear combination of $e^{\lambda_1 x}, e^{\lambda_2 x}, \ldots, e^{\lambda_n x}$, where λ_1, $\lambda_2, \ldots, \lambda_n$ are eigenvalues of A.

8.2   GEOMETRY OF LINEAR TRANSFORMATIONS FROM R^2 TO R^2

In this section we shall study the geometric properties of linear operators on R^2. Such operators have important applications in the field of computer graphics.

A linear operator $T:R^2 \rightarrow R^2$ is commonly called a **plane linear transformation**. If T is such a transformation and

$$A = \begin{bmatrix} a & b \\ c & d \end{bmatrix}$$

is the standard matrix for T, then

$$T\left(\begin{bmatrix} x \\ y \end{bmatrix}\right) = \begin{bmatrix} a & b \\ c & d \end{bmatrix}\begin{bmatrix} x \\ y \end{bmatrix} = \begin{bmatrix} ax + by \\ cx + dy \end{bmatrix}. \tag{1}$$

There are two equally good geometric interpretations of this formula. We may view the entries in the matrices

$$\begin{bmatrix} x \\ y \end{bmatrix} \quad \text{and} \quad \begin{bmatrix} ax + by \\ cx + dy \end{bmatrix}$$

either as components of vectors or coordinates of points. With the first interpretation, T maps arrows to arrows, and with the second, points to points (Figure 1). The choice is a matter of taste. In the discussion to follow, we shall view plane linear transformations as mapping points to points.

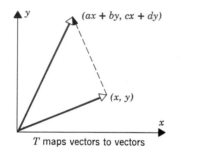

Figure 1 T maps vectors to vectors T maps points to points

Example 1   Let $T: R^2 \to R^2$ be the linear operator that maps each point into its symmetric image about the y-axis (Figure 2). Find the standard matrix for T.

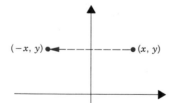

Figure 2

Solution.

$$T(\mathbf{e}_1) = T\left(\begin{bmatrix} 1 \\ 0 \end{bmatrix}\right) = \begin{bmatrix} -1 \\ 0 \end{bmatrix} \qquad T(\mathbf{e}_2) = T\left(\begin{bmatrix} 0 \\ 1 \end{bmatrix}\right) = \begin{bmatrix} 0 \\ 1 \end{bmatrix}$$

Using $T(\mathbf{e}_1)$ and $T(\mathbf{e}_2)$ as column vectors we obtain the standard matrix

$$A = \begin{bmatrix} -1 & 0 \\ 0 & 1 \end{bmatrix}$$

As a check,

$$\begin{bmatrix} -1 & 0 \\ 0 & 1 \end{bmatrix}\begin{bmatrix} x \\ y \end{bmatrix} = \begin{bmatrix} -x \\ y \end{bmatrix}$$

so that multiplication by A maps the point (x, y) into its symmetric image $(-x, y)$ about the y-axis. ▲

We will now focus our attention on five types of plane linear transformations that have special importance: rotations, reflections, expansions, compressions, and shears.

ROTATIONS   If $T:R^2 \rightarrow R^2$ *rotates* each point in the plane about the origin counterclockwise through an angle θ, then it follows from Example 2 of Section 7.1 that the standard matrix for T is

$$\begin{bmatrix} \cos \theta & -\sin \theta \\ \sin \theta & \cos \theta \end{bmatrix}$$

REFLECTIONS   A *reflection about a line* l through the origin is a transformation that maps each point in the plane into its mirror image about l. It can be shown that reflections are linear transformations. The simplest cases are the reflections about the coordinate axes and about the line $y = x$. Following the method of Example 1, the reader should be able to show that the standard matrices for these transformations are as shown in Figure 3.

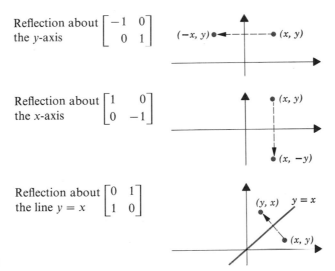

Reflection about the y-axis $\begin{bmatrix} -1 & 0 \\ 0 & 1 \end{bmatrix}$

Reflection about the x-axis $\begin{bmatrix} 1 & 0 \\ 0 & -1 \end{bmatrix}$

Reflection about the line $y = x$ $\begin{bmatrix} 0 & 1 \\ 1 & 0 \end{bmatrix}$

Figure 3

EXPANSIONS AND COMPRESSIONS   If the x-coordinate of each point in the plane is multiplied by a positive constant k, then the effect is to expand or compress each plane figure in the x-direction. If $0 < k < 1$, the result is a compression, and if $k > 1$, an expansion (Figure 4). We call such a transformation an *expansion* (or *compression*) *in the x-direction with*

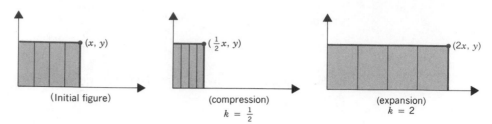

Figure 4

(Initial figure)   (compression) $k = \frac{1}{2}$ (expansion) $k = 2$

factor k. Similarly, if the y-coordinate of each point is multiplied by a positive constant k, we obtain an **expansion** (or **compression**) **in the y-direction with factor** k. It can be shown that expansions and compressions along the coordinate axes are linear transformations.

If $T:R^2 \to R^2$ is an expansion or compression in the x-direction with factor k, then

$$T(\mathbf{e}_1) = T\left(\begin{bmatrix} 1 \\ 0 \end{bmatrix}\right) = \begin{bmatrix} k \\ 0 \end{bmatrix} \qquad T(\mathbf{e}_2) = T\left(\begin{bmatrix} 0 \\ 1 \end{bmatrix}\right) = \begin{bmatrix} 0 \\ 1 \end{bmatrix}$$

so the standard matrix for T is

$$\begin{bmatrix} k & 0 \\ 0 & 1 \end{bmatrix}$$

Similarly, the standard matrix for an expansion or compression in the y-direction is

$$\begin{bmatrix} 1 & 0 \\ 0 & k \end{bmatrix}$$

SHEARS

A **shear in the x-direction with factor k** is a transformation that moves each point (x, y) parallel to the x-axis by an amount ky to the new position $(x + ky, y)$. Under such a transformation, points on the x-axis are unmoved since $y = 0$. However, as we progress away from the x-axis, the magnitude of y increases, so that points farther from the x-axis move a greater distance than those closer (Figure 5).

A **shear in the y-direction with factor k** is a transformation that moves each point (x, y) parallel to the y-axis by an amount kx to the new position $(x, y + kx)$. Under such a transformation, points on the y-axis remain fixed, and points farther

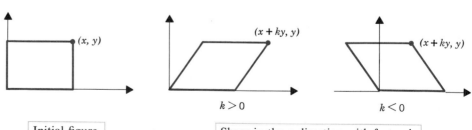

$k > 0$ $k < 0$

Figure 5   Initial figure.   Shear in the x-direction with factor k.

from the y-axis move a greater distance than do those closer.

It can be shown that shears are linear transformations. If $T:R^2 \to R^2$ is a shear of factor k in the x-direction, then

$$T(e_1) = T\left(\begin{bmatrix} 1 \\ 0 \end{bmatrix}\right) = \begin{bmatrix} 1 \\ 0 \end{bmatrix} \qquad T(e_2) = T\left(\begin{bmatrix} 0 \\ 1 \end{bmatrix}\right) = \begin{bmatrix} k \\ 1 \end{bmatrix}$$

so the standard matrix for T is

$$\begin{bmatrix} 1 & k \\ 0 & 1 \end{bmatrix}$$

Similarly, the standard matrix for a shear in the y-direction of factor k is

$$\begin{bmatrix} 1 & 0 \\ k & 1 \end{bmatrix}$$

REMARK. Multiplication by the 2×2 identity matrix maps each point into itself. This is called the *identity transformation*. If desired, this transformation can be viewed as a rotation through $0°$, or as a shear along either axis with $k = 0$, or as a compression or expansion along either axis with factor $k = 1$.

If finitely many matrix transformations from R^n to R^n are performed in succession, then the same result can be achieved by a single matrix transformation. The following example illustrates this point.

Example 2 Suppose that the plane is rotated through an angle θ and then subjected to a shear of factor k in the x-direction. Find a single matrix transformation that produces the same effect as the two successive transformations.

Solution. Under the rotation, the point (x, y) is transformed into the point (x', y') with coordinates given by

$$\begin{bmatrix} x' \\ y' \end{bmatrix} = \begin{bmatrix} \cos \theta & -\sin \theta \\ \sin \theta & \cos \theta \end{bmatrix} \begin{bmatrix} x \\ y \end{bmatrix} \tag{2}$$

Under the shear, the point (x', y') is then transformed into the point (x'', y'') with coordinates given by

$$\begin{bmatrix} x'' \\ y'' \end{bmatrix} = \begin{bmatrix} 1 & k \\ 0 & 1 \end{bmatrix} \begin{bmatrix} x' \\ y' \end{bmatrix} \tag{3}$$

Substituting (2) in (3) yields

$$\begin{bmatrix} x'' \\ y'' \end{bmatrix} = \begin{bmatrix} 1 & k \\ 0 & 1 \end{bmatrix} \begin{bmatrix} \cos \theta & -\sin \theta \\ \sin \theta & \cos \theta \end{bmatrix} \begin{bmatrix} x \\ y \end{bmatrix}$$

or

$$\begin{bmatrix} x'' \\ y'' \end{bmatrix} = \begin{bmatrix} \cos \theta + k \sin \theta & -\sin \theta + k \cos \theta \\ \sin \theta & \cos \theta \end{bmatrix} \begin{bmatrix} x \\ y \end{bmatrix}$$

Thus, the rotation followed by the shear can be performed by the matrix transformation with matrix

$$\begin{bmatrix} \cos\theta + k\sin\theta & -\sin\theta + k\cos\theta \\ \sin\theta & \cos\theta \end{bmatrix} \quad \blacktriangle$$

In general, if the matrix transformations

$$T_1(\mathbf{x}) = A_1\mathbf{x}, \quad T_2(\mathbf{x}) = A_2\mathbf{x}, \ldots, T_k(\mathbf{x}) = A_k\mathbf{x}$$

from R^n to R^n are performed in succession (first T_1, then T_2, etc.), then the same result is achieved by the single matrix transformation $T(x) = Ax$, where

$$A = A_k \cdots A_2 A_1 \tag{4}$$

Note that the order in which the transformations are performed is obtained by reading from right to left in (4).

Example 3

(a) Find a matrix transformation from R^2 to R^2 that first shears by a factor of 2 in the x-direction and then reflects about $y = x$.
(b) Find a matrix transformation from R^2 to R^2 that first reflects about $y = x$ and then shears by a factor of 2 in the x-direction.

Solution (a). The standard matrix for the shear is

$$A_1 = \begin{bmatrix} 1 & 2 \\ 0 & 1 \end{bmatrix}$$

and for the reflection is

$$A_2 = \begin{bmatrix} 0 & 1 \\ 1 & 0 \end{bmatrix}$$

Thus, the standard matrix for the shear followed by the reflection is

$$A_2 A_1 = \begin{bmatrix} 0 & 1 \\ 1 & 0 \end{bmatrix}\begin{bmatrix} 1 & 2 \\ 0 & 1 \end{bmatrix} = \begin{bmatrix} 0 & 1 \\ 1 & 2 \end{bmatrix}$$

Solution (b). The reflection followed by the shear is represented by

$$A_1 A_2 = \begin{bmatrix} 1 & 2 \\ 0 & 1 \end{bmatrix}\begin{bmatrix} 0 & 1 \\ 1 & 0 \end{bmatrix} = \begin{bmatrix} 2 & 1 \\ 1 & 0 \end{bmatrix} \quad \blacktriangle$$

In the last example, note that $A_1 A_2 \neq A_2 A_1$, so that the effect of shearing and then reflecting is different from reflecting and then shearing. This is illustrated geometrically in Figure 6, where we show the effect of the transformations on a rectangle.

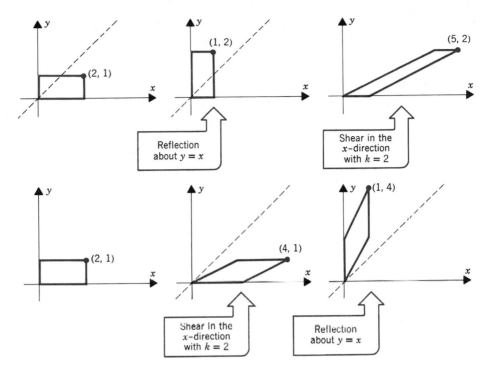

Figure 6

Example 4 Show that if $T:R^2 \to R^2$ is multiplication by an *elementary matrix*, then the transformation is one of the following:

(a) a shear along a coordinate axis
(b) a reflection about $y = x$
(c) a compression along a coordinate axis
(d) an expansion along a coordinate axis
(e) a reflection about a coordinate axis
(f) a compression or expansion along a coordinate axis followed by a reflection about a coordinate axis

Solution. Because a 2×2 elementary matrix results from performing a single elementary row operation on the 2×2 identity matrix, it must have one of the following forms (verify):

$$\begin{bmatrix} 1 & 0 \\ k & 1 \end{bmatrix} \quad \begin{bmatrix} 1 & k \\ 0 & 1 \end{bmatrix} \quad \begin{bmatrix} 0 & 1 \\ 1 & 0 \end{bmatrix} \quad \begin{bmatrix} k & 0 \\ 0 & 1 \end{bmatrix} \quad \begin{bmatrix} 1 & 0 \\ 0 & k \end{bmatrix}$$

The first two matrices represent shears along coordinate axes, and the third, a reflection about $y = x$. If $k > 0$, the last two matrices represent compressions or expansions along coordinate axes depending on whether $0 \le k \le 1$ or $k \ge 1$. If

$k < 0$ and if we express k in the form $k = -k_1$ where $k_1 > 0$, then the last two matrices can be written as

$$\begin{bmatrix} k & 0 \\ 0 & 1 \end{bmatrix} = \begin{bmatrix} -k_1 & 0 \\ 0 & 1 \end{bmatrix} = \begin{bmatrix} -1 & 0 \\ 0 & 1 \end{bmatrix}\begin{bmatrix} k_1 & 0 \\ 0 & 1 \end{bmatrix} \tag{5}$$

$$\begin{bmatrix} 1 & 0 \\ 0 & k \end{bmatrix} = \begin{bmatrix} 1 & 0 \\ 0 & -k_1 \end{bmatrix} = \begin{bmatrix} 1 & 0 \\ 0 & -1 \end{bmatrix}\begin{bmatrix} 1 & 0 \\ 0 & k_1 \end{bmatrix} \tag{6}$$

Since $k_1 > 0$, the product in (5) represents a compression or expansion along the x-axis followed by a reflection about the y-axis, and (6) represents a compression or expansion along the y-axis followed by a reflection about the x-axis. In the case where $k = -1$, transformations (5) and (6) are simply reflections about the y- and x-axis, respectively. ▲

Let $T:R^2 \to R^2$ be multiplication by an *invertible* matrix A, and suppose that T maps the point (x, y) to the point (x', y'). Then

$$\begin{bmatrix} x' \\ y' \end{bmatrix} = A\begin{bmatrix} x \\ y \end{bmatrix}$$

and

$$\begin{bmatrix} x \\ y \end{bmatrix} = A^{-1}\begin{bmatrix} x' \\ y' \end{bmatrix}$$

so multiplication by A^{-1} maps (x', y') back to its original position (x, y). We will use this fact in the following examples.

Example 5 If $T:R^2 \to R^2$ compresses the plane by a factor of $\frac{1}{2}$ in the y-direction, then it is intuitively clear that we must expand the plane by a factor of 2 in the y-direction to move each point back to its original position. This is indeed the case, for

$$A = \begin{bmatrix} 1 & 0 \\ 0 & \frac{1}{2} \end{bmatrix}$$

represents a compression of factor $\frac{1}{2}$ in the y-direction, and

$$A^{-1} = \begin{bmatrix} 1 & 0 \\ 0 & 2 \end{bmatrix}$$

is an expansion of factor 2 in the y-direction. ▲

Example 6 Multiplication by

$$A = \begin{bmatrix} \cos\theta & -\sin\theta \\ \sin\theta & \cos\theta \end{bmatrix}$$

rotates the points in the plane by an angle θ. To bring a point back to its original position, it must be rotated through an angle $-\theta$. This can be achieved by multiplying by the rotation matrix

$$\begin{bmatrix} \cos(-\theta) & -\sin(-\theta) \\ \sin(-\theta) & \cos(-\theta) \end{bmatrix}$$

Using the identities, $\cos(-\theta) = \cos\theta$ and $\sin(-\theta) = -\sin\theta$, this can be rewritten as

$$\begin{bmatrix} \cos\theta & \sin\theta \\ -\sin\theta & \cos\theta \end{bmatrix}$$

The reader should verify that this matrix is the inverse of A. ▲

GEOMETRIC PROPERTIES OF LINEAR OPERATORS ON R²

We conclude this section with two theorems that provide some insight into the geometric properties of plane linear operators.

Theorem 8.2.1. *If $T:R^2 \to R^2$ is multiplication by an invertible matrix A, then the geometric effect of T is the same as an appropriate succession of shears, compressions, expansions, and reflections.*

Proof. Since A is invertible, it can be reduced to the identity by a finite sequence of elementary row operations. An elementary row operation can be performed by multiplying on the left by an elementary matrix, and so there exist elementary matrices $E_1, E_2, \ldots, E_k$ such that

$$E_k \cdots E_2 E_1 A = I$$

Solving for A yields

$$A = E_1^{-1} E_2^{-1} \cdots E_k^{-1} I$$

or equivalently

$$A = E_1^{-1} E_2^{-1} \cdots E_k^{-1} \tag{7}$$

This equation expresses A as a product of elementary matrices (since the inverse of an elementary matrix is also elementary by Theorem 1.6.2). The result now follows from Example 4. ▮

Example 7 Let $T:R^2 \to R^2$ be the linear operator given by the formula

$$T\left(\begin{bmatrix} x \\ y \end{bmatrix}\right) = k\begin{bmatrix} x \\ y \end{bmatrix} = \begin{bmatrix} kx \\ ky \end{bmatrix}$$

As discussed in Example 5 of Section 7.1, T is a dilation of R^2 if $k > 1$ and a contraction of R^2 if $0 < k < 1$. The standard matrix for this linear operator is

$$A = \begin{bmatrix} k & 0 \\ 0 & k \end{bmatrix}$$

(verify), which can be written as the product

$$\begin{bmatrix} k & 0 \\ 0 & k \end{bmatrix} = \begin{bmatrix} k & 0 \\ 0 & 1 \end{bmatrix} \begin{bmatrix} 1 & 0 \\ 0 & k \end{bmatrix}$$

This shows that a dilation (contraction) of R^2 with factor k can be viewed as a composition of an expansion (compression) in the x-direction with factor k and an expansion (compression) in the y direction with factor k.

Example 8 Express

$$A = \begin{bmatrix} 1 & 2 \\ 3 & 4 \end{bmatrix}$$

as a product of elementary matrices, and then describe the geometric effect of multiplication by A in terms of shears, compressions, expansions, and reflections.

Solution. A can be reduced to I as follows:

$$\begin{bmatrix} 1 & 2 \\ 3 & 4 \end{bmatrix} \longrightarrow \begin{bmatrix} 1 & 2 \\ 0 & -2 \end{bmatrix} \longrightarrow \begin{bmatrix} 1 & 2 \\ 0 & 1 \end{bmatrix} \longrightarrow \begin{bmatrix} 1 & 0 \\ 0 & 1 \end{bmatrix}$$

Add -3 times the first row to the second.	Multiply the second row by $-\frac{1}{2}$.	Add -2 times the second row to the first.

The three successive row operations can be performed by multiplying on the left successively by

$$E_1 = \begin{bmatrix} 1 & 0 \\ -3 & 1 \end{bmatrix} \qquad E_2 = \begin{bmatrix} 1 & 0 \\ 0 & -\frac{1}{2} \end{bmatrix} \qquad E_3 = \begin{bmatrix} 1 & -2 \\ 0 & 1 \end{bmatrix}$$

Inverting these matrices and using (7) yields

$$A = E_1^{-1} E_2^{-1} E_3^{-1} = \begin{bmatrix} 1 & 0 \\ 3 & 1 \end{bmatrix} \begin{bmatrix} 1 & 0 \\ 0 & -2 \end{bmatrix} \begin{bmatrix} 1 & 2 \\ 0 & 1 \end{bmatrix}$$

Reading from right to left and noting that

$$\begin{bmatrix} 1 & 0 \\ 0 & -2 \end{bmatrix} = \begin{bmatrix} 1 & 0 \\ 0 & -1 \end{bmatrix} \begin{bmatrix} 1 & 0 \\ 0 & 2 \end{bmatrix}$$

it follows that the effect of multiplying by A is equivalent to

(1) shearing by a factor of 2 in the x-direction,
(2) then expanding by a factor of 2 in the y-direction,
(3) then reflecting about the x-axis,
(4) then shearing by a factor of 3 in the y-direction. ▲

The proofs for parts of the following theorem are discussed in the exercises.

> **Theorem 8.2.2.** *If* $T:R^2 \to R^2$ *is multiplication by an invertible matrix, then:*
>
> (a) *The image of a straight line is a straight line.*
> (b) *The image of a straight line through the origin is a straight line through the origin.*
> (c) *The images of parallel straight lines are parallel straight lines.*
> (d) *The image of the line segment joining points P and Q is the line segment joining the images of P and Q.*
> (e) *The images of three points lie on a line if and only if the points themselves do.*

REMARK. It follows from parts (c), (d), and (e) that multiplication by an invertible 2×2 matrix A maps triangles into triangles and parallelograms into parallelograms.

Example 9   Sketch the image of the square with vertices $P_1(0, 0)$, $P_2(1, 0)$, $P_3(0, 1)$, and $P_4(1, 1)$ under multiplication by

$$A = \begin{bmatrix} -1 & 2 \\ 2 & -1 \end{bmatrix}$$

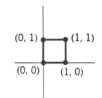

Solution. Since

$$\begin{bmatrix} -1 & 2 \\ 2 & -1 \end{bmatrix}\begin{bmatrix} 0 \\ 0 \end{bmatrix} = \begin{bmatrix} 0 \\ 0 \end{bmatrix} \qquad \begin{bmatrix} -1 & 2 \\ 2 & -1 \end{bmatrix}\begin{bmatrix} 1 \\ 0 \end{bmatrix} = \begin{bmatrix} -1 \\ 2 \end{bmatrix}$$

$$\begin{bmatrix} -1 & 2 \\ 2 & -1 \end{bmatrix}\begin{bmatrix} 0 \\ 1 \end{bmatrix} = \begin{bmatrix} 2 \\ -1 \end{bmatrix} \qquad \begin{bmatrix} -1 & 2 \\ 2 & -1 \end{bmatrix}\begin{bmatrix} 1 \\ 1 \end{bmatrix} = \begin{bmatrix} 1 \\ 1 \end{bmatrix}$$

the image is a parallelogram with vertices $(0, 0)$, $(-1, 2)$, $(2, -1)$, and $(1, 1)$ (Figure 7). ▲

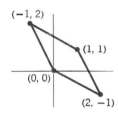

Figure 7

Example 10   According to Theorem 8.2.2, the invertible matrix

$$A = \begin{bmatrix} 3 & 1 \\ 2 & 1 \end{bmatrix}$$

maps the line $y = 2x + 1$ into another line. Find its equation.

Solution. Let (x, y) be a point on the line $y = 2x + 1$ and let (x', y') be its image under multiplication by A. Then

$$\begin{bmatrix} x' \\ y' \end{bmatrix} = \begin{bmatrix} 3 & 1 \\ 2 & 1 \end{bmatrix}\begin{bmatrix} x \\ y \end{bmatrix} \qquad \text{and} \qquad \begin{bmatrix} x \\ y \end{bmatrix} = \begin{bmatrix} 3 & 1 \\ 2 & 1 \end{bmatrix}^{-1}\begin{bmatrix} x' \\ y' \end{bmatrix} = \begin{bmatrix} 1 & -1 \\ -2 & 3 \end{bmatrix}\begin{bmatrix} x' \\ y' \end{bmatrix}$$

so

$$\begin{aligned} x &= x' - y' \\ y &= -2x' + 3y' \end{aligned}$$

Substituting in $y = 2x + 1$ yields

$$-2x' + 3y' = 2(x' - y') + 1$$

or equivalently

$$y' = \tfrac{4}{5}x' + \tfrac{1}{5}$$

Thus, (x', y') satisfies

$$y = \tfrac{4}{5}x + \tfrac{1}{5}$$

which is the equation we want. ▲

EXERCISE SET 8.2

1. Find the standard matrix for the plane linear transformation $T:R^2 \to R^2$ that maps a point (x, y) into (see Figure 8)
   (a) its reflection about the line $y = -x$
   (b) its reflection through the origin
   (c) its orthogonal projection on the x-axis
   (d) its orthogonal projection on the y-axis

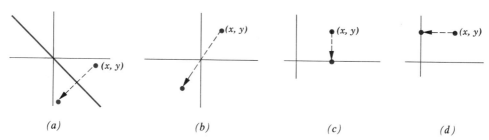

(a)   (b)   (c)   (d)   **Figure 8**

2. For each part of Exercise 1, use the matrix you have obtained to compute $T(2, 1)$. Check your answers geometrically by plotting the points $(2, 1)$ and $T(2, 1)$.

3. Find the standard matrix for the linear operator $T:R^3 \to R^3$ that maps a point (x, y, z) into
   (a) its reflection through the xy-plane
   (b) its reflection through the xz-plane
   (c) its reflection through the yz-plane

4. For each part of Exercise 3, use the matrix you have obtained to compute $T(1, 1, 1)$. Check your answers geometrically by sketching the vectors $(1, 1, 1)$ and $T(1, 1, 1)$.

5. Find the standard matrix for the linear operator $T:R^3 \to R^3$ that
   (a) rotates each vector $90°$ counterclockwise about the z-axis (looking along the positive z-axis toward the origin)
   (b) rotates each vector $90°$ counterclockwise about the x-axis (looking along the positive x-axis toward the origin)
   (c) rotates each vector $90°$ counterclockwise about the y-axis (looking along the positive y-axis toward the origin)

6. Sketch the image of the rectangle with vertices $(0, 0), (1, 0), (1, 2)$, and $(0, 2)$ under
   (a) a reflection about the x-axis
   (b) a reflection about the y-axis
   (c) a compression of factor $k = \frac{1}{4}$ in the y-direction
   (d) an expansion of factor $k = 2$ in the x-direction
   (e) a shear of factor $k = 3$ in the x-direction
   (f) a shear of factor $k = 2$ in the y-direction

7. Sketch the image of the square with vertices $(0, 0), (1, 0), (0, 1)$, and $(1, 1)$ under multiplication by

   $$A = \begin{bmatrix} -3 & 0 \\ 0 & 1 \end{bmatrix}$$

8. Find the matrix that rotates a point (x, y) about the origin through
   (a) $45°$ (b) $90°$ (c) $180°$ (d) $270°$ (e) $-30°$

9. Find the matrix that shears by
   (a) a factor of $k = 4$ in the y-direction   (b) a factor of $k = -2$ in the x-direction

10. Find the matrix that compresses or expands by
    (a) a factor of $\frac{1}{3}$ in the y-direction   (b) a factor of 6 in the x-direction

11. In each part, describe the geometric effect of multiplication by the given matrix.

    (a) $\begin{bmatrix} 3 & 0 \\ 0 & 1 \end{bmatrix}$ (b) $\begin{bmatrix} 1 & 0 \\ 0 & -5 \end{bmatrix}$ (c) $\begin{bmatrix} 1 & 4 \\ 0 & 1 \end{bmatrix}$

12. Express the matrix as a product of elementary matrices, and then describe the effect of multiplication by the given matrix in terms of compressions, expansions, reflections, and shears.

    (a) $\begin{bmatrix} 2 & 0 \\ 0 & 3 \end{bmatrix}$ (b) $\begin{bmatrix} 1 & 4 \\ 2 & 9 \end{bmatrix}$ (c) $\begin{bmatrix} 0 & -2 \\ 4 & 0 \end{bmatrix}$ (d) $\begin{bmatrix} 1 & -3 \\ 4 & 6 \end{bmatrix}$

13. In each part, find a single matrix that performs the indicated succession of operations:
    (a) compresses by a factor of $\frac{1}{2}$ in the x-direction, then expands by a factor of 5 in the y-direction
    (b) expands by a factor of 5 in the y-direction, then shears by a factor of 2 in the y-direction
    (c) reflects about $y = x$, then rotates through an angle of $180°$

14. In each part, find a single matrix that performs the indicated succession of operations:
    (a) reflects about the y-axis, then expands by a factor of 5 in the x-direction, and then reflects about $y = x$
    (b) rotates through $30°$, then shears by a factor of -2 in the y-direction, and then expands by a factor of 3 in the y-direction

15. By matrix inversion, show the following:
    (a) The inverse transformation for a reflection about $y = x$ is a reflection about $y = x$.
    (b) The inverse transformation for a compression along an axis is an expansion along that axis.
    (c) The inverse transformation for a reflection about a coordinate axis is a reflection about that axis.
    (d) The inverse transformation for a shear along a coordinate axis is a shear along that axis.

16. Find the equation of the image of the line $y = -4x + 3$ under multiplication by

$$A = \begin{bmatrix} 4 & -3 \\ 3 & -2 \end{bmatrix}$$

17. In parts (a) through (e) find the equation of the image of the line $y = 2x$ under
(a) a shear of factor 3 in the x-direction (b) a compression of factor $\frac{1}{2}$ in the y-direction
(c) a reflection about $y = x$ (d) a reflection about the y-axis
(e) a rotation of $60°$

18. Find the matrix for a shear in the x-direction that transforms the triangle with vertices $(0, 0)$, $(2, 1)$, and $(3, 0)$ into a right triangle with the right angle at the origin.

19. (a) Show that multiplication by

$$A = \begin{bmatrix} 3 & 1 \\ 6 & 2 \end{bmatrix}$$

maps every point in the plane onto the line $y = 2x$.
(b) It follows from (a) that the noncollinear points $(1, 0)$, $(0, 1)$, $(-1, 0)$ are mapped on a line. Does this violate part (e) of Theorem 8.2.2?

20. Prove part (a) of Theorem 8.2.2. [**Hint.** A line in the plane has an equation $Ax + By + C = 0$, where A and B are not both zero. Use the method of Example 10 to show that the image of this line under multiplication by the invertible matrix

$$\begin{bmatrix} a & b \\ c & d \end{bmatrix}$$

has the equation $A'x + B'y + C = 0$, where $A' = (dA - cB)/(ad - bc)$ and $B' = (-bA + aB)/(ad - bc)$. Then show that A' and B' are not both zero to conclude that the image is a line.]

21. Use the hint in Exercise 20 to prove parts (b) and (c) of Theorem 8.2.2.

22. In each part find the standard matrix for the linear operator $T:R^3 \to R^3$ described by Figure 9.

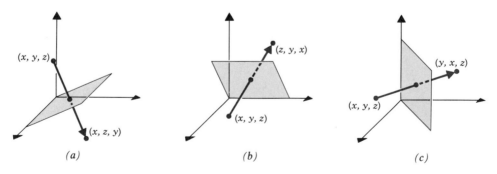

(a) (b) (c) **Figure 9**

23. In R^3 the **shear in the xy-direction with factor k** is the linear transformation that moves each point (x, y, z) parallel to the xy-plane to the new position $(x + kz, y + kz, z)$. (See Figure 10.)
(a) Find the standard matrix for the shear in the xy-direction with factor k.
(b) How would you define the shear in the xz-direction with factor k and the shear in the yz-direction with factor k? Find the standard matrix for each of these linear transformations.

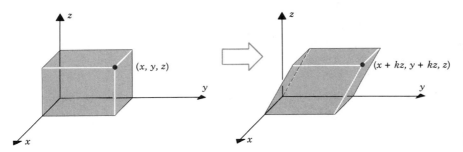

Figure 10

24. In each part find as many linearly independent eigenvectors as you can by inspection (by visualizing the geometric effect of the transformation on R^2). For each of your eigenvectors find the corresponding eigenvalue by inspection, then check your results by computing the eigenvalues and bases for the eigenspaces from the standard matrix for the transformation.

(a) reflection about the x-axis

(b) reflection about the y-axis

(c) reflection about $y = x$

(d) shear in the x-direction with factor k

(e) shear in the y-direction with factor k

(f) rotation through the angle θ

8.3 LEAST SQUARES FITTING TO DATA

In this section we shall use results about orthogonal projections in inner product spaces to obtain a technique for fitting a line or other polynomial curve to a set of experimentally determined points in the plane.

FITTING A CURVE TO EXPERIMENTAL DATA

A common problem in experimental work is to obtain a mathematical relationship $y = f(x)$ between two variables x and y by "fitting" a curve to points in the plane corresponding to various experimentally determined values of x and y, say,

$$(x_1, y_1), (x_2, y_2), \ldots, (x_n, y_n)$$

On the basis of theoretical considerations or simply by the pattern of the points, one decides on the general form of the curve $y = f(x)$ to be fitted. Some possibilities are (Figure 1)

(a) A straight line: $y = a + bx$

(b) A quadratic polynomial: $y = a + bx + cx^2$

(c) A cubic polynomial: $y = a + bx + cx^2 + dx^3$

Because the points are obtained experimentally, there is usually some measurement "error" in the data making it impossible to find a curve of the desired form that passes through all the points. Thus, the idea is to choose the curve (by determining its coefficients) that "best" fits the data. We begin with the simplest case: fitting a straight line to the data points.

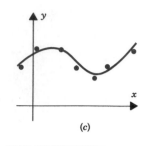

(a)   (b)   (c)

Figure 1

    $y = a + bx$

$y = a + bx + cx^2$

    $y = a + bx + cx^2 + dx^3$

**LEAST SQUARES
FIT OF A
STRAIGHT LINE**

Suppose we want to fit a straight line

$$y = a + bx$$

to the experimentally determined points

$$(x_1, y_1), (x_2, y_2), \ldots, (x_n, y_n)$$

If the data points are collinear, the line would pass through all n points, and so the unknown coefficients a and b would satisfy

$$y_1 = a + bx_1$$
$$y_2 = a + bx_2$$
$$\vdots$$
$$y_n = a + bx_n$$

We can write this system in matrix form as

$$\begin{bmatrix} y_1 \\ y_2 \\ \vdots \\ y_n \end{bmatrix} = \begin{bmatrix} 1 & x_1 \\ 1 & x_2 \\ \vdots & \vdots \\ 1 & x_n \end{bmatrix} \begin{bmatrix} a \\ b \end{bmatrix}$$

or, more compactly, as

$$\mathbf{y} = M\mathbf{v} \tag{1}$$

where

$$\mathbf{y} = \begin{bmatrix} y_1 \\ y_2 \\ \vdots \\ y_n \end{bmatrix}, \qquad M = \begin{bmatrix} 1 & x_1 \\ 1 & x_2 \\ \vdots & \vdots \\ 1 & x_n \end{bmatrix}, \qquad \mathbf{v} = \begin{bmatrix} a \\ b \end{bmatrix} \tag{2}$$

If the data points are not collinear, it is impossible to find coefficients a and b that satisfy (1) exactly; thus, no matter how we choose $\mathbf{v}$, the difference

$$\mathbf{y} - M\mathbf{v}$$

between the two sides of (1) will not be zero. In this case our objective will be to find a vector $\mathbf{v}$ that minimizes the Euclidean length of this difference

$$\|\mathbf{y} - M\mathbf{v}\| \tag{3}$$

If

$$\mathbf{v} = \mathbf{v}^* = \begin{bmatrix} a^* \\ b^* \end{bmatrix}$$

is such a minimizing vector, we call the line $y = a^* + b^*x$ a **least squares straight line fit** to the data. To explain the motivation for this terminology, let us express the square of (3) in terms of components. This yields

$$\|\mathbf{y} - M\mathbf{v}\|^2 = (y_1 - a - bx_1)^2 + (y_2 - a - bx_2)^2 + \cdots + (y_n - a - bx_n)^2 \tag{4}$$

If we now let

$$d_1 = |y_1 - a - bx_1|, d_2 = |y_2 - a - bx_2|, \ldots, d_n = |y_n - a - bx_n|$$

then (4) can be written as

$$\|\mathbf{y} - M\mathbf{v}\|^2 = d_1^2 + d_2^2 + \cdots + d_n^2 \tag{5}$$

As illustrated in Figure 2, d_i can be interpreted as the vertical distance between the line $y = a + bx$ and the data point (x_i, y_i). This distance is a measure of the "error" at the point (x_i, y_i) resulting from the inexact fit of $y = a + bx$ to the data points. Since (3) and (5) are minimized by the same vector $\mathbf{v}^*$, the least squares straight line fit minimizes the sum of the squares of these errors, hence the name *least squares straight line fit.*

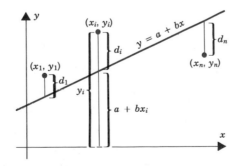

Figure 2 d_i measures the vertical error in the least squares straight line fit.

NORMAL EQUATIONS

A method for finding v^* can be motivated by considering Figure 3. As v varies over R^2, the vector Mv varies over the column space of M, which is a subspace of R^n [Formula (1) of Section 4.6]. By Theorem 5.3.7, (3) will be minimized by v^* when Mv^* is the orthogonal projection of y on the column space of M (relative to the Euclidean inner product). This implies that $y - Mv^*$ must be orthogonal to the column space of M (Figure 3) or equivalently

$$(y - Mv^*) \cdot Mv = 0 \tag{6}$$

for every vector v in R^2.

Figure 3

By Formula (3) of Section 4.1, Formula (6) can be rewritten as

$$(Mv)^t(y - Mv^*) = 0$$
$$v^t M^t(y - Mv^*) = 0$$
$$v^t(M^t y - M^t Mv^*) = 0$$
$$(M^t y - M^t Mv^*) \cdot v = 0$$

which states that $M^t y - M^t Mv^*$ is orthogonal to every vector v in R^2. But this implies that $M^t y - M^t Mv^*$ must be the zero vector (why?). Thus

$$M^t Mv^* = M^t y$$

which implies that v^* satisfies the linear system

$$M^t Mv = M^t y \tag{7}$$

of two equations in two unknowns (M is an $n \times 2$ matrix); these are called the *normal equations*. If $M^t M$ is invertible, then (5) has a unique solution $v = v^*$, which is given by

$$v^* = (M^t M)^{-1} M^t y \tag{8}$$

It can be shown that (7) is always consistent, and each solution produces coefficients of a least squares straight line fit to the data points. In the exercises it will be shown that the solution is unique if and only if the n data points do not lie on a vertical line in the xy-plane. In summary, we have the following theorem.

Theorem 8.3.1. *Let* $(x_1, y_1), (x_2, y_2), \ldots, (x_n, y_n)$ *be a set of two or more data points, not all lying on a vertical line, and let*

$$M = \begin{bmatrix} 1 & x_1 \\ 1 & x_2 \\ \vdots & \vdots \\ 1 & x_n \end{bmatrix}, \qquad and \qquad \mathbf{y} = \begin{bmatrix} y_1 \\ y_2 \\ \vdots \\ y_n \end{bmatrix}$$

Then there is a unique least squares straight line fit

$$y = a* + b*x$$

to the data points. Moreover,

$$\mathbf{v}* = \begin{bmatrix} a* \\ b* \end{bmatrix}$$

is given by the formula

$$\mathbf{v}* = (M^tM)^{-1}M^t\mathbf{y} \tag{9}$$

which expresses the fact that $\mathbf{v} = \mathbf{v}*$ is the unique solution of the normal equations

$$M^tM\mathbf{v} = M^t\mathbf{y} \tag{10}$$

Example 1 Find the least squares straight line fit to the four points $(0, 1)$, $(1, 3)$, $(2, 4)$, and $(3, 4)$. (See Figure 4.)

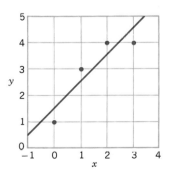

Figure 4

Solution. We have

$$M = \begin{bmatrix} 1 & 0 \\ 1 & 1 \\ 1 & 2 \\ 1 & 3 \end{bmatrix}$$

$$M^t M = \begin{bmatrix} 4 & 6 \\ 6 & 14 \end{bmatrix}$$

$$(M^t M)^{-1} = \frac{1}{10} \begin{bmatrix} 7 & -3 \\ -3 & 2 \end{bmatrix}$$

$$\mathbf{v}^* = (M^t M)^{-1} M^t \mathbf{y} = \frac{1}{10} \begin{bmatrix} 7 & -3 \\ -3 & 2 \end{bmatrix} \begin{bmatrix} 1 & 1 & 1 & 1 \\ 0 & 1 & 2 & 3 \end{bmatrix} \begin{bmatrix} 1 \\ 3 \\ 4 \\ 4 \end{bmatrix} = \begin{bmatrix} 1.5 \\ 1 \end{bmatrix}$$

And so the desired line is $y = 1.5 + x$.

Example 2 Hooke's law in physics states that the length x of a uniform spring is a linear function of the force y applied to it. If we write $y = a + bx$, then the coefficient b is called the spring constant. Suppose a particular unstretched spring has a measured length of 6.1 inches (i.e., $x = 6.1$ when $y = 0$). Forces of 2 pounds, 4 pounds, and 6 pounds are then applied to the spring, and the corresponding lengths are found to be 7.6 inches, 8.7 inches, and 10.4 inches (see Figure 5). Find the spring constant of this spring.

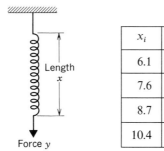

x_i	y_i
6.1	0
7.6	2
8.7	4
10.4	6

Figure 5 Force y

Solution. We have

$$M = \begin{bmatrix} 1 & 6.1 \\ 1 & 7.6 \\ 1 & 8.7 \\ 1 & 10.4 \end{bmatrix}, \qquad \mathbf{y} = \begin{bmatrix} 0 \\ 2 \\ 4 \\ 6 \end{bmatrix}$$

and

$$\mathbf{v}^* = \begin{bmatrix} a^* \\ b^* \end{bmatrix} = (M^t M)^{-1} M^t \mathbf{y} = \begin{bmatrix} -8.6 \\ 1.4 \end{bmatrix}$$

Thus, the estimated value of the spring constant is $b^* = 1.4$ pounds/inch.

LEAST SQUARES FIT OF A POLYNOMIAL

The technique described for fitting a straight line to data points generalizes easily to fitting a polynomial of any specified degree to data points. Let us attempt to fit a polynomial of fixed degree m

$$y = a_0 + a_1 x + \cdots + a_m x^m \tag{11}$$

to n points

$$(x_1, y_1), (x_2, y_2), \ldots, (x_n, y_n)$$

Substituting these n values of x and y into (11) yields the n equations

$$y_1 = a_0 + a_1 x_1 + \cdots + a_m x_1^m$$
$$y_2 = a_0 + a_1 x_2 + \cdots + a_m x_2^m$$
$$\vdots \qquad \vdots \qquad \vdots \qquad \vdots$$
$$y_n = a_0 + a_1 x_n + \cdots + a_m x_n^m$$

or, more simply,

$$\mathbf{y} = M\mathbf{v} \tag{12}$$

where

$$\mathbf{y} = \begin{bmatrix} y_1 \\ y_2 \\ \vdots \\ y_n \end{bmatrix}, \qquad M = \begin{bmatrix} 1 & x_1 & x_1^2 & \cdots & x_1^m \\ 1 & x_2 & x_2^2 & \cdots & x_2^m \\ \vdots & \vdots & \vdots & & \vdots \\ 1 & x_n & x_n^2 & & x_n^m \end{bmatrix}, \qquad \mathbf{v} = \begin{bmatrix} a_0 \\ a_1 \\ \vdots \\ a_m \end{bmatrix}$$

As in the derivation of (7), the solutions of the **normal equations**

$$M^t M \mathbf{v} = M^t \mathbf{y}$$

determine coefficients of polynomials that minimize

$$\|\mathbf{y} - M\mathbf{v}\|$$

Conditions that guarantee a unique solution of the normal equations, or equivalently the invertibility of $M^t M$ are discussed in the exercises. If $M^t M$ is invertible, then the normal equations have a unique solution $\mathbf{v} = \mathbf{v}^*$ which is given by

$$\mathbf{v}^* = (M^t M)^{-1} M^t \mathbf{y}$$

Example 3 According to Newton's second law of motion, a body near the earth's surface falls vertically downward according to the equation

$$s = s_0 + v_0 t + \tfrac{1}{2} g t^2 \tag{13}$$

where

s = vertical displacement downward relative to some fixed point

s_0 = initial displacement at time $t = 0$

v_0 = initial velocity at time $t = 0$

g = acceleration of gravity at earth's surface

Suppose a laboratory experiment is performed to evaluate g using this equation. A weight is released with unknown initial displacement and velocity, and at certain times the distances fallen from some fixed reference point are measured. In particular, suppose that at times $t = .1, .2, .3, .4,$ and $.5$ seconds it is found that the weight has fallen $s = -0.18, 0.31, 1.03, 2.48,$ and 3.73 feet, respectively, from the reference point. Find an approximate value of g using these data.

Solution. The mathematical problem is to fit a quadratic curve

$$s = a_0 + a_1 t + a_2 t^2 \tag{14}$$

to the five data points:

$$(.1, -0.18), \quad (.2, 0.31), \quad (.3, 1.03), \quad (.4, 2.48), \quad (.5, 3.73)$$

The necessary calculations are

$$M = \begin{bmatrix} 1 & t_1 & t_1^2 \\ 1 & t_2 & t_2^2 \\ 1 & t_3 & t_3^2 \\ 1 & t_4 & t_4^2 \\ 1 & t_5 & t_5^2 \end{bmatrix} = \begin{bmatrix} 1 & .1 & .01 \\ 1 & .2 & .04 \\ 1 & .3 & .09 \\ 1 & .4 & .16 \\ 1 & .5 & .25 \end{bmatrix}$$

$$\mathbf{y} = \begin{bmatrix} s_1 \\ s_2 \\ s_3 \\ s_4 \\ s_5 \end{bmatrix} = \begin{bmatrix} -0.18 \\ 0.31 \\ 1.03 \\ 2.48 \\ 3.73 \end{bmatrix}$$

and

$$\mathbf{v}^* = \begin{bmatrix} a_0^* \\ a_1^* \\ a_2^* \end{bmatrix} = (M^t M)^{-1} M^t \mathbf{y} = \begin{bmatrix} -0.40 \\ 0.35 \\ 16.1 \end{bmatrix}$$

From (13) and (14) we have $a_2 = \frac{1}{2}g$, and so the estimated value of g is

$$g = 2a_2^* = 2(16.1) = 32.2 \text{ feet/second}^2$$

If desired, we can also estimate the initial displacement and initial velocity of the weight:

$$s_0 = a_0^* = -0.40 \text{ feet}$$

$$v_0 = a_1^* = \quad 0.35 \text{ feet/second}$$

In Figure 6 we have plotted the five data points and the approximating polynomial. ▲

Figure 6

EXERCISE SET 8.3

1. Find the least squares straight line fit to the three points $(0, 0)$, $(1, 2)$, and $(2, 7)$.

2. Find the least squares straight line fit to the four points $(0, 1)$, $(2, 0)$, $(3, 1)$, and $(3, 2)$.

3. Find the quadratic polynomial that best fits the four points $(2, 0)$, $(3, -10)$, $(5, -48)$, and $(6, -76)$.

4. Find the cubic polynomial that best fits the five points $(-1, -14)$, $(0, -5)$, $(1, -4)$, $(2, 1)$, and $(3, 22)$.

5. Show that if M is an $m \times n$ matrix with linearly independent columns, then $M^t M$ is an $n \times n$ invertible matrix.

6. Show that the matrix M in Equation (2) has linearly independent columns if and only if at least two of the numbers $x_1, x_2, \ldots, x_n$ are distinct. Conclude from this and from Exercise 5 that the matrix $M^t M$ in Equation (8) is invertible if and only if the n data points do not lie on a vertical line in the xy-plane.

7. Show that the columns of the $n \times (m + 1)$ matrix M in Equation (12) are linearly independent if $n > m$ and at least $m + 1$ of the numbers $x_1, x_2, \ldots, x_n$ are distinct.

8. Let M be the matrix in Equation (12). Using Exercise 7, show that a sufficient condition that the matrix $M^t M$ is invertible is that $n > m$ and at least $m + 1$ of the numbers $x_1, x_2, \ldots, x_n$ are distinct.

9. The owner of a rapidly expanding business finds that for the first five months of the year the sales (in thousands) are $4.0, $4.4, $5.2, $6.4, and $8.0. The owner plots these figures on a graph and conjectures that for the rest of the year the sales curve can be approximated by a quadratic polynomial. Find the least squares quadratic polynomial fit to the sales curve, and use it to project the sales for the twelfth month of the year.

8.4 APPROXIMATION PROBLEMS; FOURIER SERIES

In this section we shall use results about orthogonal projections in inner product spaces to solve problems that involve approximating a given function by simpler functions. Such problems arise in a variety of engineering and scientific applications.

BEST APPROXI-MATIONS

All of the problems that we will study in this section will be special cases of the following general problem.

> **Approximation Problem.** Given a function f that is continuous on an interval $[a, b]$, find the "best possible approximation" to f using only functions from a specified subspace W of $C[a, b]$.

Here are some examples of such problems:

(a) Find the best possible approximation to e^x over $[0, 1]$ by a polynomial of the form $a_0 + a_1 x + a_2 x^2$.

(b) Find the best possible approximation to $\sin \pi x$ over $[-1, 1]$ by a function of the form $a_0 + a_1 e^x + a_2 e^{2x} + a_3 e^{3x}$.

(c) Find the best possible approximation to x over $[0, 2\pi]$ by a function of the form $a_0 + a_1 \sin x + a_2 \sin 2x + b_1 \cos x + b_2 \cos 2x$.

In the first example W is the subspace of $C[0, 1]$ spanned by 1, x, and x^2; in the second example W is the subspace of $C[-1, 1]$ spanned by 1, e^x, e^{2x}, and e^{3x}; and in the third example W is the subspace of $C[0, 2\pi]$ spanned by 1, $\sin x$, $\sin 2x$, $\cos x$, and $\cos 2x$.

MEASUREMENTS OF ERROR

To solve approximation problems of the foregoing types, we must make the phrase "best approximation over $[a, b]$" mathematically precise; to do this we need a precise way of measuring the error that results when one continuous function is approximated by another over $[a, b]$. If we were concerned only with approximating $f(x)$ at a single point x_0, then the error at x_0 by an approximation $g(x)$ would be simply

$$\text{error} = |f(x_0) - g(x_0)|$$

sometimes called the **deviation** between f and g at x_0 (Figure 1). However, we are

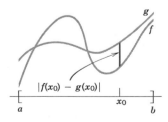

Figure 1 The deviation between f and g at x_0.

concerned with approximation over the entire interval $[a, b]$, not at a single point. Consequently, in one part of the interval an approximation g_1 to f may have smaller deviations from f than an approximation g_2 to f, and in another part of the interval it might be the other way around. How does one decide which is the better overall approximation? What we need is some way of measuring the overall error in an approximation $g(x)$. One possible measure of overall error is obtained by integrating the deviation $|f(x) - g(x)|$ over the entire interval $[a, b]$; that is,

$$\text{error} = \int_a^b |f(x) - g(x)|\, dx \tag{1}$$

Geometrically (1) is the area between the graphs of $f(x)$ and $g(x)$ over the interval $[a, b]$ (Figure 2); the greater the area, the greater the overall error.

The area between the graphs of f and g over $[a, b]$ measures the error in approximating f by g over $[a, b]$.

While (1) is natural and appealing geometrically, the occurrence of an absolute value sign is sufficiently bothersome in computations that most mathematicians and scientists generally favor the following alternative measure of error, called the *mean square error*.

$$\text{mean square error} = \int_a^b [f(x) - g(x)]^2\, dx$$

Mean square error has the added advantage that it allows us to bring to bear the theory of inner product spaces. To see how, suppose that $\mathbf{f}$ is a continuous function on $[a, b]$ that we want to approximate by a function $\mathbf{g}$ from a subspace W of $C[a, b]$, and suppose that $C[a, b]$ is given the inner product

$$\langle \mathbf{f}, \mathbf{g} \rangle = \int_a^b f(x)g(x)\, dx$$

It follows that

$$\|\mathbf{f} - \mathbf{g}\|^2 = \langle \mathbf{f} - \mathbf{g}, \mathbf{f} - \mathbf{g} \rangle = \int_a^b [f(x) - g(x)]^2\, dx = \text{mean square error}$$

so that minimizing the mean square error is the same as minimizing $\|\mathbf{f} - \mathbf{g}\|^2$. Thus, the approximation problem posed informally at the beginning of this section can be restated more precisely as follows:

LEAST SQUARES APPROXIMATION

Least Squares Approximation Problem. Let **f** be a function that is continuous on an interval $[a, b]$, let $C[a, b]$ have the inner product

$$\langle \mathbf{f}, \mathbf{g} \rangle = \int_a^b f(x)g(x)\,dx$$

and let W be a finite-dimensional subspace of $C[a, b]$. Find a function **g** in W that minimizes

$$\|\mathbf{f} - \mathbf{g}\|^2 = \int_a^b [f(x) - g(x)]^2\,dx$$

Since $\|\mathbf{f} - \mathbf{g}\|^2$ and $\|\mathbf{f} - \mathbf{g}\|$ are minimized by the same function **g**, the foregoing problem is equivalent to looking for a function **g** in W that is closest to **f**. But we know from Theorem 5.3.7 that $\mathbf{g} = \text{proj}_W \mathbf{f}$ is such a function (Figure 3).

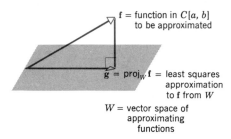

f = function in $C[a, b]$ to be approximated

$\mathbf{g} = \text{proj}_W \mathbf{f}$ = least squares approximation to f from W

W = vector space of approximating functions

Figure 3

Thus, we have the following result.

Solution of the Least Squares Approximation Problem. If **f** is a continuous function on $[a, b]$, and W is a finite dimensional subspace of $C[a, b]$, then the function **g** in W that minimizes the mean square error

$$\int_a^b [f(x) - g(x)]^2\,dx$$

is $\mathbf{g} = \text{proj}_W \mathbf{f}$, where the orthogonal projection is relative to the inner product

$$\langle \mathbf{f}, \mathbf{g} \rangle = \int_a^b f(x)g(x)\,dx$$

The function $\mathbf{g} = \text{proj}_W \mathbf{f}$ is called the *least squares approximation* to **f** from W.

FOURIER SERIES

A function of the form

$$t(x) = c_0 + c_1 \cos x + c_2 \cos 2x + \cdots + c_n \cos nx$$
$$+ d_1 \sin x + d_2 \sin 2x + \cdots + d_n \sin nx \tag{2}$$

is called a *trigonometric polynomial*; if c_n and d_n are not both zero, then $t(x)$ is said to have *order n*. For example,

$$t(x) = 2 + \cos x - 3 \cos 2x + 7 \sin 4x$$

is a trigonometric polynomial with

$$c_0 = 2, \quad c_1 = 1, \quad c_2 = -3, \quad d_1 = 0, \quad d_2 = 0, \quad d_3 = 0, \quad d_4 = 7$$

The order of $t(x)$ is 4.

It is evident from (2) that the trigonometric polynomials of order n or less are the various possible linear combinations of

$$1, \quad \cos x, \quad \cos 2x, \quad \ldots, \cos nx, \quad \sin x, \quad \sin 2x, \quad \ldots, \sin nx \qquad (3)$$

It can be shown that these $2n + 1$ functions are linearly independent and that consequently for any interval $[a, b]$ they form a basis for a $(2n + 1)$-dimensional subspace of $C[a, b]$.

Let us now consider the problem of finding the least squares approximation of a continuous function $f(x)$ over the interval $[0, 2\pi]$ by a trigonometric polynomial of order n or less. As noted above, the least squares approximation to $\mathbf{f}$ from W is the orthogonal projection of $\mathbf{f}$ on W. To find this orthogonal projection, we must find an orthonormal basis $\mathbf{g}_0, \mathbf{g}_1, \ldots, \mathbf{g}_{2n}$ for W, after which we can compute the orthogonal projection on W from the formula

$$\text{proj}_W \mathbf{f} = \langle \mathbf{f}, \mathbf{g}_0 \rangle \mathbf{g}_0 + \langle \mathbf{f}, \mathbf{g}_1 \rangle \mathbf{g}_1 + \cdots + \langle \mathbf{f}, \mathbf{g}_{2n} \rangle \mathbf{g}_{2n} \qquad (4)$$

[see Theorem 5.3.5]. An orthonormal basis for W can be obtained by applying the Gram-Schmidt process to the basis (3), using the inner product

$$\langle \mathbf{f}, \mathbf{g} \rangle = \int_0^{2\pi} f(x)g(x)\,dx$$

This yields (Exercise 6) the orthonormal basis

$$\mathbf{g}_0 = \frac{1}{\sqrt{2\pi}}, \quad \mathbf{g}_1 = \frac{1}{\sqrt{\pi}} \cos x, \quad \ldots, \mathbf{g}_n = \frac{1}{\sqrt{\pi}} \cos nx, \qquad (5)$$

$$\mathbf{g}_{n+1} = \frac{1}{\sqrt{\pi}} \sin x, \quad \ldots, \mathbf{g}_{2n} = \frac{1}{\sqrt{\pi}} \sin nx$$

If we introduce the notation

$$a_0 = \frac{2}{\sqrt{2\pi}} \langle \mathbf{f}, \mathbf{g}_0 \rangle, \quad a_1 = \frac{1}{\sqrt{\pi}} \langle \mathbf{f}, \mathbf{g}_1 \rangle, \quad \ldots, a_n = \frac{1}{\sqrt{\pi}} \langle \mathbf{f}, \mathbf{g}_n \rangle$$

$$b_1 = \frac{1}{\sqrt{\pi}} \langle \mathbf{f}, \mathbf{g}_{n+1} \rangle, \quad \ldots, b_n = \frac{1}{\sqrt{\pi}} \langle \mathbf{f}, \mathbf{g}_{2n} \rangle$$

then on substituting (5) in (4) we obtain

$$\text{proj}_W \mathbf{f} = \frac{a_0}{2} + [a_1 \cos x + \cdots + a_n \cos nx] + [b_1 \sin x + \cdots + b_n \sin nx]$$

where

$$a_0 = \frac{2}{\sqrt{2\pi}} \langle \mathbf{f}, \mathbf{g}_0 \rangle = \frac{2}{\sqrt{2\pi}} \int_0^{2\pi} f(x) \frac{1}{\sqrt{2\pi}}\,dx = \frac{1}{\pi} \int_0^{2\pi} f(x)\,dx$$

$$a_1 = \frac{1}{\sqrt{\pi}} \langle \mathbf{f}, \mathbf{g}_1 \rangle = \frac{1}{\sqrt{\pi}} \int_0^{2\pi} f(x) \frac{1}{\sqrt{\pi}} \cos x \, dx = \frac{1}{\pi} \int_0^{2\pi} f(x) \cos x \, dx$$

$$\vdots$$

$$a_n = \frac{1}{\sqrt{\pi}} \langle \mathbf{f}, \mathbf{g}_n \rangle = \frac{1}{\sqrt{\pi}} \int_0^{2\pi} f(x) \frac{1}{\sqrt{\pi}} \cos nx \, dx = \frac{1}{\sqrt{\pi}} \int_0^{2\pi} f(x) \cos nx \, dx$$

$$b_1 = \frac{1}{\sqrt{\pi}} \langle \mathbf{f}, \mathbf{g}_{n+1} \rangle = \frac{1}{\sqrt{\pi}} \int_0^{2\pi} f(x) \frac{1}{\sqrt{\pi}} \sin x \, dx = \frac{1}{\pi} \int_0^{2\pi} f(x) \sin x \, dx$$

$$\vdots$$

$$b_n = \frac{1}{\sqrt{\pi}} \langle \mathbf{f}, \mathbf{g}_{2n} \rangle = \frac{1}{\sqrt{\pi}} \int_0^{2\pi} f(x) \frac{1}{\sqrt{\pi}} \sin nx \, dx = \frac{1}{\pi} \int_0^{2\pi} f(x) \sin nx \, dx$$

In short

$$a_k = \frac{1}{\pi} \int_0^{2\pi} f(x) \cos kx \, dx, \qquad b_k = \frac{1}{\pi} \int_0^{2\pi} f(x) \sin kx \, dx \qquad (6)$$

The numbers $a_0, a_1, \ldots, a_n, b_1, \ldots, b_n$ are called the ***Fourier* coefficient*** of **f**.

Example 1 Find the least squares approximation of $f(x) = x$ on $[0, 2\pi]$ by

(a) a trigonometric polynomial of order 2 or less;
(b) a trigonometric polynomial of order n or less.

Solution (a).

$$a_0 = \frac{1}{\pi} \int_0^{2\pi} f(x) \, dx = \frac{1}{\pi} \int_0^{2\pi} x \, dx = 2\pi \qquad (7a)$$

For $k = 1, 2, \ldots$ integration by parts yields (verify)

$$a_k = \frac{1}{\pi} \int_0^{2\pi} f(x) \cos kx \, dx = \frac{1}{\pi} \int_0^{2\pi} x \cos kx \, dx = 0 \qquad (7b)$$

$$b_k = \frac{1}{\pi} \int_0^{2\pi} f(x) \sin kx \, dx = \frac{1}{\pi} \int_0^{2\pi} x \sin kx \, dx = -\frac{2}{k} \qquad (7c)$$

---

* *Jean Baptiste Joseph Fourier (1768–1830)* was a French mathematician and physicist who discovered the Fourier series and related ideas while working on problems of heat diffusion. This discovery is one of the most influential in the history of mathematics; it is the cornerstone of many fields of mathematical research and a basic tool in many branches of engineering. Fourier, a political activist during the French revolution, spent time in jail for his defense of many victims during the Terror. He later became a favorite of Napoleon and was named both a baron and a count.

Thus, the least squares approximation to x on $[0, 2\pi]$ by a trigonometric polynomial of order 2 or less is

$$x \simeq \frac{a_0}{2} + a_1 \cos x + a_2 \cos 2x + b_1 \sin x + b_2 \sin 2x$$

or from (7a), (7b), and (7c)

$$x \simeq \pi - 2 \sin x - \sin 2x$$

Solution (b). The least squares approximation to x on $[0, 2\pi]$ by a trigonometric polynomial of order n or less is

$$x \simeq \frac{a_0}{2} + [a_1 \cos x + \cdots + a_n \cos nx] + [b_1 \sin x + \cdots + b_n \sin nx]$$

or from (7a), (7b), and (7c)

$$x \simeq \pi - 2\left(\sin x + \frac{\sin 2x}{2} + \frac{\sin 3x}{3} + \cdots + \frac{\sin nx}{n} \right)$$

The graphs of $y = x$ and some of these approximations are shown in Figure 4.

It is natural to expect that the mean square error will diminish as the number of terms in the least squares approximation

$$f(x) \simeq \frac{a_0}{2} + \sum_{k=1}^{n} (a_k \cos kx + b_k \sin kx)$$

increases. It can be proved that for functions f in $C[0, \pi]$ the mean square error approaches zero as $n \to +\infty$; this is denoted by writing

$$f(x) = \frac{a_0}{2} + \sum_{k=1}^{\infty} (a_k \cos kx + b_k \sin kx)$$

The right side of this equation is called the **Fourier series** for f over the interval $[0, \pi]$. Such series are of major importance in engineering, science, and mathematics. ▲

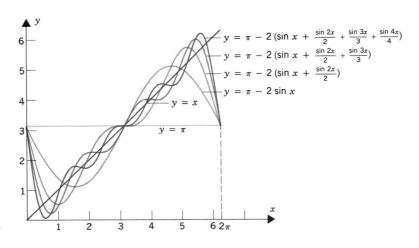

$y = \pi - 2 (\sin x + \frac{\sin 2x}{2} + \frac{\sin 3x}{3} + \frac{\sin 4x}{4})$

$y = \pi - 2 (\sin x + \frac{\sin 2x}{2} + \frac{\sin 3x}{3})$

$y = \pi - 2 (\sin x + \frac{\sin 2x}{2})$

$y = \pi - 2 \sin x$

$y = x$

$y = \pi$

Figure 4

EXERCISE SET 8.4

1. Find the least squares approximation of $f(x) = 1 + x$ over the interval $[0, 2\pi]$ by
   (a) a trigonometric polynomial of order 2 or less
   (b) a trigonometric polynomial of order n or less

2. Find the least squares approximation of $f(x) = x^2$ over the interval $[0, 2\pi]$ by
   (a) a trigonometric polynomial of order 3 or less
   (b) a trigonometric polynomial of order n or less

3. (a) Find the least squares approximation of x over the interval $[0, 1]$ by a function of the form $a + be^x$.
   (b) Find the mean square error of the approximation.

4. (a) Find the least squares approximation of e^x over the interval $[0, 1]$ by a polynomial of the form $a_0 + a_1 x$.
   (b) Find the mean square error of the approximation.

5. (a) Find the least squares approximation of $\sin \pi x$ over the interval $[-1, 1]$ by a polynomial of the form $a_0 + a_1 x + a_2 x^2$.
   (b) Find the mean square error of the approximation.

6. Use the Gram-Schmidt process to obtain the orthonormal basis (5) from the basis (3).

7. Carry out the integrations in (7a), (7b), and (7c).

8. Find the Fourier series of $f(x) = \pi - x$ over the interval $[0, \pi]$.

8.5 QUADRATIC FORMS

Up to now, our emphasis in this text has been on linear equations, that is, equations of the form

$$a_1 x_1 + a_2 x_2 + \cdots + a_n x_n = b$$

The left side of this equation

$$a_1 x_1 + a_2 x_2 + \cdots + a_n x_n$$

*is a function of n variables, called a **linear form**. In a linear form all variables occur to the first power, and there are no products of variables in the expression. In this section we shall study functions in which the terms are squares of variables or products of two variables. Such functions arise in a variety of applications, including geometry, vibrations of mechanical systems, statistics, and electrical engineering.*

QUADRATIC FORMS IN TWO VARIABLES

A **quadratic form in two variables**, x and y, is defined to be an expression that can be written as

$$ax^2 + 2bxy + cy^2 \tag{1}$$

Example 1   The following are quadratic forms in x and y:

$$2x^2 + 6xy - 7y^2 \qquad (a = 2, b = 3, c = -7)$$
$$4x^2 - 5y^2 \qquad (a = 4, b = 0, c = -5)$$
$$xy \qquad (a = 0, b = 1/2, c = 0) \quad \blacktriangle$$

If we agree to omit the brackets on 1×1 matrices, then (1) can be written in matrix form as

$$ax^2 + 2bxy + cy^2 = \begin{bmatrix} x & y \end{bmatrix} \begin{bmatrix} a & b \\ b & c \end{bmatrix} \begin{bmatrix} x \\ y \end{bmatrix} \tag{2}$$

(Verify by multiplying the matrices.) Note that the 2×2 matrix in (2) is symmetric, the diagonal entries are the coefficients of the squared terms, and the entries off the main diagonal are each half the coefficient of the product term xy.

Example 2

$$2x^2 + 6xy - 7y^2 = \begin{bmatrix} x & y \end{bmatrix} \begin{bmatrix} 2 & 3 \\ 3 & -7 \end{bmatrix} \begin{bmatrix} x \\ y \end{bmatrix}$$

$$4x^2 - 5y^2 = \begin{bmatrix} x & y \end{bmatrix} \begin{bmatrix} 4 & 0 \\ 0 & -5 \end{bmatrix} \begin{bmatrix} x \\ y \end{bmatrix}$$

$$xy = \begin{bmatrix} x & y \end{bmatrix} \begin{bmatrix} 0 & \frac{1}{2} \\ \frac{1}{2} & 0 \end{bmatrix} \begin{bmatrix} x \\ y \end{bmatrix} \quad \blacktriangle$$

QUADRATIC FORMS IN n VARIABLES

Quadratic forms are not limited to two variables. A general quadratic form is defined as follows.

Definition. A *quadratic form* in the n variables $x_1, x_2, \ldots, x_n$ is an expression that can be written as

$$\begin{bmatrix} x_1 & x_2 & \cdots & x_n \end{bmatrix} A \begin{bmatrix} x_1 \\ x_2 \\ \vdots \\ x_n \end{bmatrix} \tag{3}$$

where A is a symmetric $n \times n$ matrix.

If we let

$$\mathbf{x} = \begin{bmatrix} x_1 \\ x_2 \\ \vdots \\ x_n \end{bmatrix}$$

then (3) can be written more compactly as

$$\mathbf{x}^t A \mathbf{x} \tag{4}$$

Moreover, it can be shown that if the matrices in (4) are multiplied out, the resulting expression has the form

$$\mathbf{x}^t A \mathbf{x} = a_{11}x_1^2 + a_{22}x_2^2 + \cdots + a_{nn}x_n^2 + \sum_{i \neq j} a_{ij}x_i x_j$$

where

$$\sum_{i \neq j} a_{ij}x_i x_j$$

denotes the sum of all terms of the form $a_{ij}x_i x_j$, where x_i and x_j are different variables. The terms $a_{ij}x_i x_j$ are called the ***cross-product terms*** of the quadratic form.

Symmetric matrices are useful, but not essential, for representing quadratic forms in matrix notation. For example, for the quadratic form $2x^2 + 6xy - 7y^2$ in Example 2, we might split the coefficient of the cross-product term into $5 + 1$ or $4 + 2$ and write

$$2x^2 + 6xy - 7y^2 = \begin{bmatrix} x & y \end{bmatrix} \begin{bmatrix} 2 & 5 \\ 1 & -7 \end{bmatrix} \begin{bmatrix} x \\ y \end{bmatrix}$$

or

$$2x^2 + 6xy - 7y^2 = \begin{bmatrix} x & y \end{bmatrix} \begin{bmatrix} 2 & 4 \\ 2 & -7 \end{bmatrix} \begin{bmatrix} x \\ y \end{bmatrix}$$

However, symmetric matrices generally produce the simplest results, so we will always use them. Thus, when we denote a quadratic form by $\mathbf{x}^t A \mathbf{x}$ it will be understood that A is symmetric, even if this is not stated explicitly.

REMARK. If we use the fact that A is symmetric, that is, $A = A^t$, then (4) can be expressed in terms of the Euclidean inner product by writing

$$\mathbf{x}^t A \mathbf{x} = \mathbf{x}^t(A\mathbf{x}) = \langle A\mathbf{x}, \mathbf{x} \rangle = \langle \mathbf{x}, A\mathbf{x} \rangle \tag{5}$$

Example 3 The following is a quadratic form in x_1, x_2, and x_3:

$$x_1^2 + 7x_2^2 - 3x_3^2 + 4x_1x_2 - 2x_1x_3 + 6x_2x_3 = \begin{bmatrix} x_1 & x_2 & x_3 \end{bmatrix} \begin{bmatrix} 1 & 2 & -1 \\ 2 & 7 & 3 \\ -1 & 3 & -3 \end{bmatrix} \begin{bmatrix} x_1 \\ x_2 \\ x_3 \end{bmatrix}$$

Note that the coefficients of the squared terms appear on the main diagonal of the 3×3 matrix, and the coefficients of the cross-product terms are each split in half and appear in the off-diagonal positions as follows:

Coefficient of	Positions in Matrix A		
x_1x_2	a_{12}	and	a_{21}
x_1x_3	a_{13}	and	a_{31}
x_2x_3	a_{23}	and	a_{32}

**PROBLEMS
INVOLVING
QUADRATIC
FORMS**

The study of quadratic forms is an extensive topic that we can only touch on in this section. The following are some of the important mathematical problems relating to quadratic forms.

· Find the maximum and minimum values of the quadratic form $\mathbf{x}^t A\mathbf{x}$ if $\mathbf{x}$ is constrained so that

$$\|\mathbf{x}\| = (x_1^2 + x_2^2 + \cdots + x_n^2)^{1/2} = 1$$

· What conditions must A satisfy in order for a quadratic form to satisfy the inequality $\mathbf{x}^t A\mathbf{x} > 0$ for all $\mathbf{x} \neq \mathbf{0}$?

· If $\mathbf{x}^t A\mathbf{x}$ is a quadratic form in two or three variables and c is a constant, what does the graph of the equation $\mathbf{x}^t A\mathbf{x} = c$ look like?

· If P is an orthogonal matrix, the change of variables $\mathbf{x} = P\mathbf{y}$ converts the quadratic form $\mathbf{x}^t A\mathbf{x}$ to $(P\mathbf{y})^t A(P\mathbf{y}) = \mathbf{y}^t (P^t AP)\mathbf{y}$. But $P^t AP$ is a symmetric matrix if A is (verify), so $\mathbf{y}^t (P^t AP)\mathbf{y}$ is a new quadratic form in the variables of $\mathbf{y}$. It is important to know if P can be chosen so that this new quadratic form has no cross-product terms.

In this section we shall study the first two problems, and in the following sections we shall study the last two. The following theorem provides a solution to the first problem. The proof is deferred to the end of this section.

Theorem 8.5.1. *Let A be a symmetric n × n matrix whose eigenvalues in decreasing size order are $\lambda_1 \geq \lambda_2 \geq \cdots \geq \lambda_n$. If x is constrained so that $\|\mathbf{x}\| = 1$ relative to the Euclidean inner product on R^n, then*

(a) $\lambda_1 \geq \mathbf{x}^t A\mathbf{x} \geq \lambda_n$

(b) $\mathbf{x}^t A\mathbf{x} = \lambda_n$ if x is an eigenvector of A corresponding to λ_n and $\mathbf{x}^t A\mathbf{x} = \lambda_1$ if x is an eigenvector of A corresponding to λ_1.

It follows from this theorem that subject to the constraint

$$\|\mathbf{x}\| = (x_1^2 + x_2^2 + \cdots + x_n^2)^{1/2} = 1$$

the quadratic form $\mathbf{x}^t A\mathbf{x}$ has a maximum value of λ_1 (the largest eigenvalue) and a minimum value of λ_n (the smallest eigenvalue).

Example 4   Find the maximum and minimum values of the quadratic form

$$x_1^2 + x_2^2 + 4x_1x_2$$

subject to the constraint $x_1^2 + x_2^2 = 1$, and determine values of x_1 and x_2 at which the maximum and minimum occur.

Solution. The quadratic form can be written as

$$x_1^2 + x_2^2 + 4x_1x_2 = \mathbf{x}^t A\mathbf{x} = \begin{bmatrix} x_1 & x_2 \end{bmatrix} \begin{bmatrix} 1 & 2 \\ 2 & 1 \end{bmatrix} \begin{bmatrix} x_1 \\ x_2 \end{bmatrix}$$

The characteristic equation of A is

$$\det(\lambda I - A) = \det\begin{bmatrix} \lambda - 1 & -2 \\ -2 & \lambda - 1 \end{bmatrix} = \lambda^2 - 2\lambda - 3 = (\lambda - 3)(\lambda + 1) = 0$$

Thus, the eigenvalues of A are $\lambda = 3$ and $\lambda = -1$, which are the maximum and minimum values, respectively, of the quadratic form subject to the constraint. To find values of x_1 and x_2 at which these extreme values occur, we must find eigenvectors corresponding to these eigenvalues and then normalize these eigenvectors to satisfy the condition $x_1^2 + x_2^2 = 1$.

We leave it for the reader to show that bases for the eigenspaces are

$$\lambda = 3: \ \begin{bmatrix} 1 \\ 1 \end{bmatrix}, \qquad \lambda = -1: \ \begin{bmatrix} 1 \\ -1 \end{bmatrix}$$

Normalizing these eigenvectors yields

$$\begin{bmatrix} 1/\sqrt{2} \\ 1/\sqrt{2} \end{bmatrix}, \qquad \begin{bmatrix} 1/\sqrt{2} \\ -1\sqrt{2} \end{bmatrix}$$

Thus, subject to the constraint $x_1^2 + x_2^2 = 1$, the maximum value of the quadratic form is $\lambda = 3$, which occurs if $x_1 = 1/\sqrt{2}$, $x_2 = 1/\sqrt{2}$; and the minimum value is $\lambda = -1$, which occurs if $x_1 = 1/\sqrt{2}$, $x_2 = -1/\sqrt{2}$. Moreover, alternative bases for the eigenspaces can be obtained by multiplying the basis vectors above by -1. Thus, the maximum value, $\lambda = 3$, also occurs if $x_1 = -1/\sqrt{2}$, $x_2 = -1/\sqrt{2}$, and the minimum value, $\lambda = -1$, also occurs if $x_1 = -1/\sqrt{2}$, $x_2 = 1/\sqrt{2}$. ▲

POSITIVE DEFINITE MATRICES AND QUADRATIC FORMS

Definition. A quadratic form $x^t A x$ is called *positive definite* if $x^t A x > 0$ for all $x \neq 0$, and a symmetric matrix A is called a *positive definite matrix* if $x^t A x$ is a positive definite quadratic form.

The following theorem is the main result about positive definite matrices.

Theorem 8.5.2. *A symmetric matrix A is positive definite if and only if all the eigenvalues of A are positive.*

Proof. Assume that A is positive definite, and let λ be any eigenvalue of A. If x is an eigenvector of A corresponding to λ, then $x \neq 0$ and $Ax = \lambda x$, so

$$0 < x^t A x = x^t \lambda x = \lambda x^t x = \lambda \|x\|^2 \tag{6}$$

where $\|x\|$ is the Euclidean norm of x. Since $\|x\|^2 > 0$ it follows that $\lambda > 0$, which is what we wanted to show.

Conversely, assume that all eigenvalues of A are positive. We must show that $x^t A x > 0$ for all $x \neq 0$. But if $x \neq 0$, we can normalize x to obtain the vector $y = x/\|x\|$ with the property $\|y\| = 1$. It now follows from Theorem 8.5.1 that

$$y^t A y \geq \lambda_n > 0$$

where λ_n is the smallest eigenvalue of A. Thus,

$$\mathbf{y}^t A \mathbf{y} = \left(\frac{\mathbf{x}}{\|\mathbf{x}\|}\right)^t A\left(\frac{\mathbf{x}}{\|\mathbf{x}\|}\right) = \frac{1}{\|\mathbf{x}\|^2}\,\mathbf{x}^t A \mathbf{x} > 0$$

Multiplying through by $\|\mathbf{x}\|^2$ yields

$$\mathbf{x}^t A \mathbf{x} > 0$$

which is what we wanted to show. ▌

Example 5 In Example 2 of Section 6.3 we showed that the symmetric matrix

$$A = \begin{bmatrix} 4 & 2 & 2 \\ 2 & 4 & 2 \\ 2 & 2 & 4 \end{bmatrix}$$

has eigenvalues $\lambda = 2$ and $\lambda = 8$. Since these are positive, the matrix A is positive definite, and for all $\mathbf{x} \neq \mathbf{0}$

$$\mathbf{x}^t A \mathbf{x} = 4x_1^2 + 4x_2^2 + 4x_3^2 + 4x_1 x_2 + 4x_1 x_3 + 4x_2 x_3 > 0 \quad \blacktriangle$$

Our next objective is to give a criterion that can be used to determine whether a symmetric matrix is positive definite without finding its eigenvalues. To do this it will be helpful to introduce some terminology. If

$$A = \begin{bmatrix} a_{11} & a_{12} & \cdots & a_{1n} \\ a_{21} & a_{22} & \cdots & a_{2n} \\ \vdots & \vdots & & \vdots \\ a_{n1} & a_{n2} & \cdots & a_{nn} \end{bmatrix}$$

is a square matrix, then the ***principal submatrices*** of A are the submatrices formed from the first r rows and r columns of A for $r = 1, 2, \ldots, n$. These submatrices are

$$A_1 = [a_{11}], \quad A_2 = \begin{bmatrix} a_{11} & a_{12} \\ a_{21} & a_{22} \end{bmatrix}, \quad A_3 = \begin{bmatrix} a_{11} & a_{12} & a_{13} \\ a_{21} & a_{22} & a_{23} \\ a_{31} & a_{32} & a_{33} \end{bmatrix}, \ldots,$$

$$A_n = A = \begin{bmatrix} a_{11} & a_{12} & \cdots & a_{1n} \\ a_{21} & a_{22} & \cdots & a_{2n} \\ \vdots & \vdots & & \vdots \\ a_{n1} & a_{n2} & \cdots & a_{nn} \end{bmatrix}$$

Theorem 8.5.3. *A symmetric matrix A is positive definite if and only if the determinant of every principal submatrix is positive.*

We omit the proof.

Example 6 The matrix

$$A = \begin{bmatrix} 2 & -1 & -3 \\ -1 & 2 & 4 \\ -3 & 4 & 9 \end{bmatrix}$$

is positive definite since

$$|2| = 2, \quad \begin{vmatrix} 2 & -1 \\ -1 & 2 \end{vmatrix} = 3, \quad \begin{vmatrix} 2 & -1 & -3 \\ -1 & 2 & 4 \\ -3 & 4 & 9 \end{vmatrix} = 1$$

all of which are positive. Thus, we are guaranteed that all eigenvalues of A are positive and $\mathbf{x}^t A \mathbf{x} > 0$ for all $\mathbf{x} \neq \mathbf{0}$. ▲

REMARK. A symmetric matrix A and the quadratic form $\mathbf{x}^t A \mathbf{x}$ are called

positive semidefinite	if $\mathbf{x}^t A \mathbf{x} \geq 0$ for all $\mathbf{x}$
negative definite	if $\mathbf{x}^t A \mathbf{x} < 0$ for $\mathbf{x} \neq \mathbf{0}$
negative semidefinite	if $\mathbf{x}^t A \mathbf{x} \leq 0$ for all $\mathbf{x}$
indefinite	if $\mathbf{x}^t A \mathbf{x}$ has both positive and negative values

Theorems 8.5.2 and 8.5.3 can be modified in an obvious way to apply to matrices of the first three types. For example, a symmetric matrix A is positive semidefinite if and only if all of its eigenvalues are *nonnegative*. Also, A is positive semidefinite if and only if all its principal submatrices have *nonnegative* determinants.

OPTIONAL

Proof of Theorem 8.5.1a. Since A is symmetric, if follows from Theorem 6.3.1 that there is an orthonormal basis for R^n consisting of eigenvectors of A. Let $S = \{\mathbf{v}_1, \mathbf{v}_2, \ldots, \mathbf{v}_n\}$ be such a basis, where $\mathbf{v}_i$ is the eigenvector corresponding to the eigenvalue λ_i. If $\langle \, , \, \rangle$ denotes the Euclidean inner product, then it follows from Theorem 5.3.1 that for any $\mathbf{x}$ in R^n

$$\mathbf{x} = \langle \mathbf{x}, \mathbf{v}_1 \rangle \mathbf{v}_1 + \langle \mathbf{x}, \mathbf{v}_2 \rangle \mathbf{v}_2 + \cdots + \langle \mathbf{x}, \mathbf{v}_n \rangle \mathbf{v}_n$$

Thus,

$$\begin{aligned} A\mathbf{x} &= \langle \mathbf{x}, \mathbf{v}_1 \rangle A\mathbf{v}_1 + \langle \mathbf{x}, \mathbf{v}_2 \rangle A\mathbf{v}_2 + \cdots + \langle \mathbf{x}, \mathbf{v}_n \rangle A\mathbf{v}_n \\ &= \langle \mathbf{x}, \mathbf{v}_1 \rangle \lambda_1 \mathbf{v}_1 + \langle \mathbf{x}, \mathbf{v}_2 \rangle \lambda_2 \mathbf{v}_2 + \cdots + \langle \mathbf{x}, \mathbf{v}_n \rangle \lambda_n \mathbf{v}_n \\ &= \lambda_1 \langle \mathbf{x}, \mathbf{v}_1 \rangle \mathbf{v}_1 + \lambda_2 \langle \mathbf{x}, \mathbf{v}_2 \rangle \mathbf{v}_2 + \cdots + \lambda_n \langle \mathbf{x}, \mathbf{v}_n \rangle \mathbf{v}_n \end{aligned}$$

It follows that the coordinate vectors for $\mathbf{x}$ and $A\mathbf{x}$ relative to the basis S are

$$(\mathbf{x})_S = (\langle \mathbf{x}, \mathbf{v}_1 \rangle, \langle \mathbf{x}, \mathbf{v}_2 \rangle, \ldots, \langle \mathbf{x}, \mathbf{v}_n \rangle)$$
$$(A\mathbf{x})_S = (\lambda_1 \langle \mathbf{x}, \mathbf{v}_1 \rangle, \lambda_2 \langle \mathbf{x}, \mathbf{v}_2 \rangle, \ldots, \lambda_n \langle \mathbf{x}, \mathbf{v}_n \rangle)$$

Thus, from Theorem 5.3.2a and the fact that $\|\mathbf{x}\| = 1$, we obtain

$$\|\mathbf{x}\|^2 = \langle \mathbf{x}, \mathbf{v}_1 \rangle^2 + \langle \mathbf{x}, \mathbf{v}_2 \rangle^2 + \cdots + \langle \mathbf{x}, \mathbf{v}_n \rangle^2 = 1$$
$$\langle \mathbf{x}, A\mathbf{x} \rangle = \lambda_1 \langle \mathbf{x}, \mathbf{v}_1 \rangle^2 + \lambda_2 \langle \mathbf{x}, \mathbf{v}_2 \rangle^2 + \cdots + \lambda_n \langle \mathbf{x}, \mathbf{v}_n \rangle^2$$

Using these two equations and Formula (5), we can prove that $x^t Ax \le \lambda_1$ as follows:

$$\begin{aligned}
x^t Ax = \langle x, Ax \rangle &= \lambda_1 \langle x, v_1 \rangle^2 + \lambda_2 \langle x, v_2 \rangle^2 + \cdots + \lambda_n \langle x, v_n \rangle^2 \\
&\le \lambda_1 \langle x, v_1 \rangle^2 + \lambda_1 \langle x, v_2 \rangle^2 + \cdots + \lambda_1 \langle x, v_n \rangle^2 \\
&= \lambda_1 (\langle x, v_1 \rangle^2 + \langle x, v_2 \rangle^2 + \cdots + \langle x, v_n \rangle^2) \\
&= \lambda_1
\end{aligned}$$

The proof that $\lambda_n \le x^t Ax$ is similar and is left as an exercise.

Proof of Theorem 8.5.1b. If x is an eigenvector of A corresponding to λ_1 and $\|x\| = 1$, then

$$x^t Ax = \langle x, Ax \rangle = \langle x, \lambda_1 x \rangle = \lambda_1 \langle x, x \rangle = \lambda_1 \|x\|^2 = \lambda_1$$

Similarly, $x^t Ax = \lambda_n$ if $\|x\| = 1$ and x is an eigenvector of A corresponding to λ_n. $\blacksquare$

EXERCISE SET 8.5

1. Which of the following are quadratic forms?
   (a) $x^2 - \sqrt{2}xy$
   (b) $5x_1^2 - 2x_2^3 + 4x_1 x_2$
   (c) $4x_1^2 - 3x_2^2 + x_3^2 - 5x_1 x_3$
   (d) $x_1^2 - 7x_2^2 + x_3^2 + 4x_1 x_2 x_3$
   (e) $x_1 x_2 - 3x_1 x_3 + 2x_2 x_3$
   (f) $x_1^2 - 6x_2^2 + x_1 - 5x_2$
   (g) $(x_1 - 3x_2)^2$
   (h) $(x_1 - x_3)^2 + 2(x_1 + 4x_2)^2$

2. Express the following quadratic forms in the matrix notation $x^t Ax$, where A is a symmetric matrix.
   (a) $3x_1^2 + 7x_2^2$ (b) $4x_1^2 - 9x_2^2 - 6x_1 x_2$ (c) $5x_1^2 + 5x_1 x_2$ (d) $-7x_1 x_2$

3. Express the following quadratic forms in the matrix notation $x^t Ax$, where A is a symmetric matrix.
   (a) $9x_1^2 - x_2^2 + 4x_3^2 + 6x_1 x_2 - 8x_1 x_3 + x_2 x_3$
   (b) $x_1^2 + x_2^2 - 3x_3^2 - 5x_1 x_2 + 9x_1 x_3$
   (c) $x_1 x_2 + x_1 x_3 + x_2 x_3$
   (d) $\sqrt{2}x_1^2 - \sqrt{3}x_3^2 + 2\sqrt{2}x_1 x_2 - 8\sqrt{3}x_1 x_3$
   (e) $x_1^2 + x_2^2 - x_3^2 - x_4^2 + 2x_1 x_2 - 10x_1 x_4 + 4x_3 x_4$

4. In each part find a formula for the quadratic form that does not use matrices.

   (a) $[x \quad y] \begin{bmatrix} 2 & -3 \\ -3 & 5 \end{bmatrix} \begin{bmatrix} x \\ y \end{bmatrix}$ (b) $[x_1 \quad x_2] \begin{bmatrix} 7 & \frac{5}{2} \\ \frac{5}{2} & 0 \end{bmatrix} \begin{bmatrix} x_1 \\ x_2 \end{bmatrix}$ (c) $[x \quad y \quad z] \begin{bmatrix} 1 & 0 & 0 \\ 0 & -3 & 0 \\ 0 & 0 & 5 \end{bmatrix} \begin{bmatrix} x \\ y \\ z \end{bmatrix}$

   (d) $[x_1 \quad x_2 \quad x_3] \begin{bmatrix} -2 & \frac{7}{2} & \frac{1}{2} \\ \frac{7}{2} & 0 & 6 \\ \frac{1}{2} & 6 & 3 \end{bmatrix} \begin{bmatrix} x_1 \\ x_2 \\ x_3 \end{bmatrix}$ (e) $[x_1 \quad x_2 \quad x_3 \quad x_4] \begin{bmatrix} 0 & 1 & 1 & 1 \\ 1 & 0 & 1 & 1 \\ 1 & 1 & 0 & 1 \\ 1 & 1 & 1 & 0 \end{bmatrix} \begin{bmatrix} x_1 \\ x_2 \\ x_3 \\ x_4 \end{bmatrix}$

5. In each part find the maximum and minimum values of the quadratic form subject to the constraint $x_1^2 + x_2^2 = 1$, and determine the values of x_1 and x_2 at which the maximum and minimum occur.
   (a) $5x_1^2 - x_2^2$ (b) $7x_1^2 + 4x_2^2 + x_1x_2$ (c) $5x_1^2 + 2x_2^2 - x_1x_2$ (d) $2x_1^2 + x_2^2 + 3x_1x_2$

6. In each part find the maximum and minimum values of the quadratic form subject to the constraint $x_1^2 + x_2^2 + x_3^2 = 1$, and determine the values of x_1, x_2, and x_3 at which the maximum and minimum occur.
   (a) $x_1^2 + x_2^2 + 2x_3^2 - 2x_1x_2 + 4x_1x_3 + 4x_2x_3$ (b) $2x_1^2 + x_2^2 + x_3^2 + 2x_1x_3 + 2x_1x_2$
   (c) $3x_1^2 + 2x_2^2 + 3x_3^2 + 2x_1x_3$

7. Use Theorem 8.5.2 to determine which of the following matrices are positive definite.
   (a) $\begin{bmatrix} 2 & 3 \\ 3 & 2 \end{bmatrix}$ (b) $\begin{bmatrix} 5 & -1 \\ -1 & 5 \end{bmatrix}$ (c) $\begin{bmatrix} 2 & -2 \\ -2 & -1 \end{bmatrix}$

8. Use Theorem 8.5.3 to determine which of the matrices in Exercise 7 are positive definite.

9. Use Theorem 8.5.2 to determine which of the following matrices are positive definite.
   (a) $\begin{bmatrix} 3 & -1 & 0 \\ -1 & 2 & -1 \\ 0 & -1 & 3 \end{bmatrix}$ (b) $\begin{bmatrix} 0 & 1 & 1 \\ 1 & 0 & 1 \\ 1 & 1 & 0 \end{bmatrix}$ (c) $\begin{bmatrix} 1 & 2 & 1 \\ 2 & 1 & 1 \\ 1 & 1 & 3 \end{bmatrix}$

10. Use Theorem 8.5.3 to determine which of the matrices in Exercise 9 are positive definite.

11. In each part classify the quadratic form as positive definite, positive semidefinite, negative definite, negative semidefinite, or indefinite.
    (a) $x_1^2 + x_2^2$ (b) $-x_1^2 - 3x_2^2$ (c) $(x_1 - x_2)^2$
    (d) $-(x_1 - x_2)^2$ (e) $x_1^2 - x_2^2$ (f) x_1x_2

12. In each part classify the matrix as positive definite, positive semidefinite, negative definite, negative semidefinite, or indefinite.
    (a) $\begin{bmatrix} 3 & 0 & 0 \\ 0 & -2 & 0 \\ 0 & 0 & 1 \end{bmatrix}$ (b) $\begin{bmatrix} -5 & 0 & 0 \\ 0 & 0 & 0 \\ 0 & 0 & 1 \end{bmatrix}$ (c) $\begin{bmatrix} 6 & 7 & 1 \\ 0 & 9 & 2 \\ 0 & 0 & 1 \end{bmatrix}$

    (d) $\begin{bmatrix} -4 & 0 & 0 \\ 7 & -3 & 0 \\ 8 & 9 & -1 \end{bmatrix}$ (e) $\begin{bmatrix} 0 & 0 & 0 \\ 0 & 0 & 0 \\ 0 & 0 & 0 \end{bmatrix}$ (f) $\begin{bmatrix} 1 & 0 & 0 \\ 0 & 1 & 0 \\ 0 & 0 & 1 \end{bmatrix}$

13. Let $x^t A x$ be a quadratic form in $x_1, x_2, \ldots, x_n$ and define $T: R^n \to R$ by $T(x) = x^t A x$.
    (a) Show that $T(x + y) = T(x) + 2x^t A y + T(y)$.   (b) Show that $T(kx) = k^2 T(x)$.
    (c) Is T a linear transformation? Explain.

14. In each part find all values of k for which the quadratic form is positive definite.
    (a) $x_1^2 + kx_2^2 - 4x_1x_2$ (b) $5x_1^2 + x_2^2 + kx_3^2 + 4x_1x_2 - 2x_1x_3 - 2x_2x_3$
    (c) $3x_1^2 + x_2^2 + 2x_3^2 + 2x_1x_3 + 2kx_2x_3$

15. Express the quadratic form $(c_1x_1 + c_2x_2 + \cdots + c_nx_n)^2$ in the matrix notation $x^t A x$, where A is symmetric.

16. Let $\mathbf{x} = (x_1, x_2, \ldots, x_n)$. In statistics the quantity

$$\bar{\mathbf{x}} = \frac{1}{n}(x_1 + x_2 + \cdots + x_n)$$

is called the ***sample mean*** of $x_1, x_2, \ldots, x_n$, and

$$s_{\mathbf{x}}^2 = \frac{1}{n-1}[(x_1 - \bar{x})^2 + (x_2 - \bar{x})^2 + \cdots + (x_n - \bar{x})^2]$$

is called the ***sample variance***.
   (a) Express the quadratic form $s_{\mathbf{x}}^2$ in the matrix notation $\mathbf{x}^t A \mathbf{x}$, where A is symmetric.
   (b) Is $s_{\mathbf{x}}^2$ a positive definite quadratic form? Explain.

17. Complete the proof of Theorem 8.5.1 by showing that $\lambda_n \leq \mathbf{x}^t A \mathbf{x}$ if $\|\mathbf{x}\| = 1$ and $\lambda_n = \mathbf{x}^t A \mathbf{x}$ if $\mathbf{x}$ is an eigenvector of A corresponding to λ_n.

8.6   DIAGONALIZING QUADRATIC FORMS; CONIC SECTIONS

In this section we shall show how to remove the cross-product terms from a quadratic form by changing variables, and we shall use our results to study the graphs of the conic sections.

**DIAGONALIZA-
TION OF
QUADRATIC
FORMS**

Let

$$\mathbf{x}^t A \mathbf{x} = \begin{bmatrix} x_1 & x_2 & \cdots & x_n \end{bmatrix} \begin{bmatrix} a_{11} & a_{12} & \cdots & a_{1n} \\ a_{21} & a_{22} & \cdots & a_{2n} \\ \vdots & \vdots & & \vdots \\ a_{n1} & a_{n2} & \cdots & a_{nn} \end{bmatrix} \begin{bmatrix} x_1 \\ x_2 \\ \vdots \\ x_n \end{bmatrix} \tag{1}$$

be a quadratic form, where A is a symmetric matrix. We know from Theorem 6.3.1 that there is an orthogonal matrix P that diagonalizes A; that is,

$$P^t A P = D = \begin{bmatrix} \lambda_1 & 0 & \cdots & 0 \\ 0 & \lambda_2 & \cdots & 0 \\ \vdots & \vdots & & \vdots \\ 0 & 0 & \cdots & \lambda_n \end{bmatrix}$$

where $\lambda_1, \lambda_2, \ldots, \lambda_n$ are the eigenvalues of A. If we let

$$\mathbf{y} = \begin{bmatrix} y_1 \\ y_2 \\ \vdots \\ y_n \end{bmatrix}$$

where $y_1, y_2, \ldots, y_n$ are new variables, and if we make the substitution $\mathbf{x} = P\mathbf{y}$ in (1), then we obtain

$$\mathbf{x}^t A\mathbf{x} = (P\mathbf{y})^t AP\mathbf{y} = \mathbf{y}^t P^t AP\mathbf{y} = \mathbf{y}^t D\mathbf{y}$$

But

$$\mathbf{y}^t D\mathbf{y} = \begin{bmatrix} y_1 & y_2 & \cdots & y_n \end{bmatrix} \begin{bmatrix} \lambda_1 & 0 & \cdots & 0 \\ 0 & \lambda_2 & \cdots & 0 \\ \vdots & \vdots & & \vdots \\ 0 & 0 & \cdots & \lambda_n \end{bmatrix} \begin{bmatrix} y_1 \\ y_2 \\ \vdots \\ y_n \end{bmatrix}$$

$$= \lambda_1 y_1^2 + \lambda_2 y_2^2 + \cdots + \lambda_n y_n^2$$

which is a quadratic form with no cross-product terms.

In summary, we have the following result.

Theorem 8.6.1. *Let* $\mathbf{x}^t A\mathbf{x}$ *be a quadratic form in the variables* $x_1, x_2, \ldots, x_n$, *where A is symmetric. If P orthogonally diagonalizes A, and if the new variables* $y_1, y_2, \ldots, y_n$ *are defined by the equation* $\mathbf{x} = P\mathbf{y}$, *then substituting this equation in* $\mathbf{x}^t A\mathbf{x}$ *yields*

$$\mathbf{x}^t A\mathbf{x} = \mathbf{y}^t D\mathbf{y} = \lambda_1 y_1^2 + \lambda_2 y_2^2 + \cdots + \lambda_n y_n^2$$

where $\lambda_1, \lambda_2, \ldots, \lambda_n$ *are the eigenvalues of A and*

$$D = P^t AP = \begin{bmatrix} \lambda_1 & 0 & \cdots & 0 \\ 0 & \lambda_2 & \cdots & 0 \\ \vdots & \vdots & & \vdots \\ 0 & 0 & \cdots & \lambda_n \end{bmatrix}$$

The matrix P in this theorem is said to **orthogonally diagonalize** the quadratic form or **reduce the quadratic form to a sum of squares**.

Example 1 Find a change of variables that will reduce the quadratic form $x_1^2 - x_3^2 - 4x_1 x_2 + 4x_2 x_3$ to a sum of squares, and express the quadratic form in terms of the new variables.

Solution. The quadratic form can be written as

$$\begin{bmatrix} x_1 & x_2 & x_3 \end{bmatrix} \begin{bmatrix} 1 & -2 & 0 \\ -2 & 0 & 2 \\ 0 & 2 & -1 \end{bmatrix} \begin{bmatrix} x_1 \\ x_2 \\ x_3 \end{bmatrix}$$

The characteristic equation of the 3×3 matrix is

$$\begin{vmatrix} \lambda - 1 & 2 & 0 \\ 2 & \lambda & -2 \\ 0 & -2 & \lambda + 1 \end{vmatrix} = \lambda^3 - 9\lambda = \lambda(\lambda + 3)(\lambda - 3) = 0$$

so the eigenvalues are $\lambda = 0$, $\lambda = -3$, $\lambda = 3$. We leave it for the reader to show that orthonormal bases for the three eigenspaces are

$$\lambda = 0: \begin{bmatrix} \frac{2}{3} \\ \frac{1}{3} \\ \frac{2}{3} \end{bmatrix} \qquad \lambda = -3: \begin{bmatrix} -\frac{1}{3} \\ -\frac{2}{3} \\ \frac{2}{3} \end{bmatrix} \qquad \lambda = 3: \begin{bmatrix} -\frac{2}{3} \\ \frac{2}{3} \\ \frac{1}{3} \end{bmatrix}$$

Thus, the substitution $\mathbf{x} = P\mathbf{y}$ that eliminates cross-product terms is

$$\begin{bmatrix} x_1 \\ x_2 \\ x_3 \end{bmatrix} = \begin{bmatrix} \frac{2}{3} & -\frac{1}{3} & -\frac{2}{3} \\ \frac{1}{3} & -\frac{2}{3} & \frac{2}{3} \\ \frac{2}{3} & \frac{2}{3} & \frac{1}{3} \end{bmatrix} \begin{bmatrix} y_1 \\ y_2 \\ y_3 \end{bmatrix}$$

or equivalently

$$x_1 = \tfrac{2}{3}y_1 - \tfrac{1}{3}y_2 - \tfrac{2}{3}y_3$$
$$x_2 = \tfrac{1}{3}y_1 - \tfrac{2}{3}y_2 + \tfrac{2}{3}y_3$$
$$x_3 = \tfrac{2}{3}y_1 + \tfrac{2}{3}y_2 + \tfrac{1}{3}y_3$$

The new quadratic form is

$$\begin{bmatrix} y_1 & y_2 & y_3 \end{bmatrix} \begin{bmatrix} 0 & 0 & 0 \\ 0 & -3 & 0 \\ 0 & 0 & 3 \end{bmatrix} \begin{bmatrix} y_1 \\ y_2 \\ y_3 \end{bmatrix}$$

or equivalently

$$-3y_2^2 + 3y_3^2 \quad \blacktriangle$$

REMARK. There are other methods for eliminating the cross-product terms from a quadratic form, which we shall not discuss here. Two such methods, *Lagrange's reduction* and *Kronecker's reduction*, are discussed in more advanced books.

CONIC SECTIONS   We shall now apply our work on quadratic forms to the study of equations of the form

$$ax^2 + 2bxy + cy^2 + dx + ey + f = 0 \tag{2}$$

where $a, b, \ldots, f$ are real numbers, and at least one of the numbers a, b, c is not zero. An equation of this type is called a *quadratic equation* in x and y, and

$$ax^2 + 2bxy + cy^2$$

is called the *associated quadratic form*.

Example 2   In the quadratic equation

$$3x^2 + 5xy - 7y^2 + 2x + 7 = 0$$

the constants in (2) are

$$a = 3, \quad b = \tfrac{5}{2}, \quad c = -7, \quad d = 2, \quad e = 0, \quad f = 7 \quad \blacktriangle$$

Example 3

Quadratic Equation	Associated Quadratic Form
$3x^2 + 5xy - 7y^2 + 2x + 7 = 0$	$3x^2 + 5xy - 7y^2$
$4x^2 - 5y^2 + 8y + 9 = 0$	$4x^2 - 5y^2$
$xy + y = 0$	xy

▲

Graphs of quadratic equations in x and y are called **conics** or **conic sections**. The most important conics are ellipses, circles, hyperbolas, and parabolas; these are called the **nondegenerate** conics. The remaining conics are called **degenerate** and include single points and pairs of lines (see Exercise 15).

A nondegenerate conic is said to be in **standard position** relative to the co-ordinate axes if its equation can be expressed in one of the forms given in Figure 1.

Example 4 The equation

$$\frac{x^2}{4} + \frac{y^2}{9} = 1 \text{ is of the form } \frac{x^2}{k^2} + \frac{y^2}{l^2} = 1 \text{ with } k = 2, l = 3$$

Thus, its graph is an ellipse in standard position intersecting the x-axis at $(-2, 0)$ and $(2, 0)$ and intersecting the y-axis at $(0, -3)$ and $(0, 3)$.

The equation $x^2 - 8y^2 = -16$ can be rewritten as $y^2/2 - x^2/16 = 1$, which is of the form $y^2/k^2 - x^2/l^2 = 1$ with $k = \sqrt{2}, l = 4$. Its graph is thus a hyperbola in standard position intersecting the y-axis at $(0, -\sqrt{2})$ and $(0, \sqrt{2})$.

The equation $5x^2 + 2y = 0$ can be rewritten as $x^2 = -\frac{2}{5}y$, which is of the form $x^2 = ky$ with $k = -\frac{2}{5}$. Since $k < 0$, its graph is a parabola in standard position opening downward. ▲

SIGNIFICANCE OF THE CROSS-PRODUCT TERM

Observe that no conic in standard position has an xy-term (that is, a cross-product term) in its equation; the presence of an xy-term in the equation of a nondegenerate conic indicates that the conic is rotated out of standard position (Figure 2a). Also, no conic in standard position has both an x^2 and x term or both a y^2 and y term. If there is no cross-product term, the occurrence of either of these pairs in the equation of a nondegenerate conic indicates that the conic is translated out of standard position (Figure 2b).

One technique for identifying the graph of a nondegenerate conic that is not in standard position consists of rotating and translating the xy-coordinate axes to obtain an $x'y'$-coordinate system relative to which the conic is in standard position. Once this is done, the equation of the conic in the $x'y'$-system will have one of the forms given in Figure 1 and can then easily be identified.

Graph

$$\frac{x^2}{k^2} + \frac{y^2}{l^2} = 1; k, l > 0$$

Ellipse or Circle

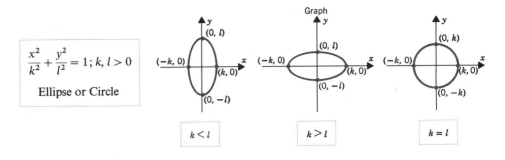

$k < l$ $k > l$ $k = l$

$$\frac{x^2}{k^2} - \frac{y^2}{l^2} = 1; k, l > 0$$

Hyperbola

$$\frac{y^2}{k^2} - \frac{x^2}{l^2} = 1; k, l > 0$$

Hyperbola

$y^2 = kx$
Parabola

$x^2 = ky$
Parabola

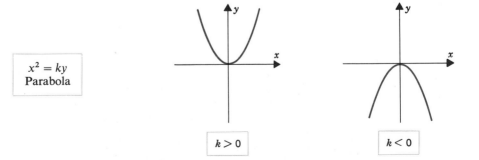

$k > 0$ $k < 0$

Figure 1

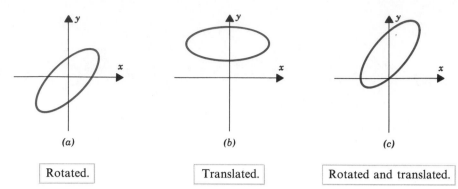

(a)   (b)   (c)

Figure 2   Rotated.   Translated.   Rotated and translated.

Example 5   Since the quadratic equation

$$2x^2 + y^2 - 12x - 4y + 18 = 0$$

contains x^2, x, y^2, and y terms but no cross-product term, its graph is a conic that is translated out of standard position but not rotated. This conic can be brought into standard position by suitably translating the coordinate axes. To do this, first collect x and y terms. This yields

$$(2x^2 - 12x) + (y^2 - 4y) + 18 = 0$$

or

$$2(x^2 - 6x) + (y^2 - 4y) = -18$$

By completing the squares* on the two expressions in parentheses, we obtain

$$2(x^2 - 6x + 9) + (y^2 - 4y + 4) = -18 + 18 + 4$$

or

$$2(x - 3)^2 + (y - 2)^2 = 4 \tag{3}$$

If we translate the coordinate axes by means of the translation equations

$$x' = x - 3 \qquad y' = y - 2$$

then (3) becomes

$$2x'^2 + y'^2 = 4$$

or

$$\frac{x'^2}{2} + \frac{y'^2}{4} = 1$$

which is the equation of an ellipse in standard position in the $x'y'$-system. This ellipse is sketched in Figure 3. ▲

---

* To complete the square on an expression of the form $x^2 + px$, add and subtract the constant $(p/2)^2$ to obtain

$$x^2 + px = x^2 + px + \left(\frac{p}{2}\right)^2 - \left(\frac{p}{2}\right)^2 = \left(x + \frac{p}{2}\right)^2 - \left(\frac{p}{2}\right)^2$$

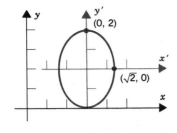

Figure 3

$$\frac{x'^2}{2} + \frac{y'^2}{4} = 1$$

**ELIMINATING
THE CROSS-
PRODUCT TERM**

We shall now show how to identify conics that are rotated out of standard position. If we omit the brackets on 1×1 matrices, then (2) can be written in the matrix form

$$[x \quad y]\begin{bmatrix} a & b \\ b & c \end{bmatrix}\begin{bmatrix} x \\ y \end{bmatrix} + [d \quad e]\begin{bmatrix} x \\ y \end{bmatrix} + f = 0$$

or

$$\mathbf{x}^t A \mathbf{x} + K \mathbf{x} + f = 0$$

where

$$\mathbf{x} = \begin{bmatrix} x \\ y \end{bmatrix} \qquad A = \begin{bmatrix} a & b \\ b & c \end{bmatrix} \qquad K = [d \quad e]$$

Now consider a conic C whose equation in xy-coordinates is

$$\mathbf{x}^t A \mathbf{x} + K \mathbf{x} + f = 0 \tag{4}$$

We would like to rotate the xy-coordinate axes so that the equation of the conic in the new $x'y'$-coordinate system has no cross-product term. This can be done as follows.

Step 1. Find a matrix

$$P = \begin{bmatrix} p_{11} & p_{12} \\ p_{21} & p_{22} \end{bmatrix}$$

that orthogonally diagonalizes the quadratic form $\mathbf{x}^t A \mathbf{x}$.

Step 2. Interchange the columns of P, if necessary, to make $\det(P) = 1$. This assures that the orthogonal coordinate transformation

$$\mathbf{x} = P\mathbf{x}', \text{ that is,} \quad \begin{bmatrix} x \\ y \end{bmatrix} = P\begin{bmatrix} x' \\ y' \end{bmatrix} \tag{5}$$

is a rotation (see the discussion preceding Example 8 in Section 5.4).

Step 3. To obtain the equation for C in the $x'y'$-system, substitute (5) into (4). This yields

$$(Px')^t A(Px') + K(Px') + f = 0$$

or

$$(x')^t(P^t AP)x' + (KP)x' + f = 0 \tag{6}$$

Since P orthogonally diagonalizes A

$$P^t AP = \begin{bmatrix} \lambda_1 & 0 \\ 0 & \lambda_2 \end{bmatrix}$$

where λ_1 and λ_2 are eigenvalues of A. Thus, (6) can be rewritten as

$$\begin{bmatrix} x' & y' \end{bmatrix} \begin{bmatrix} \lambda_1 & 0 \\ 0 & \lambda_2 \end{bmatrix} \begin{bmatrix} x' \\ y' \end{bmatrix} + \begin{bmatrix} d & e \end{bmatrix} \begin{bmatrix} p_{11} & p_{12} \\ p_{21} & p_{22} \end{bmatrix} \begin{bmatrix} x' \\ y' \end{bmatrix} + f = 0$$

or

$$\lambda_1 x'^2 + \lambda_2 y'^2 + d'x' + e'y' + f = 0$$

(where $d' = dp_{11} + ep_{21}$ and $e' = dp_{12} + ep_{22}$). This equation has no cross-product term.

The following theorem summarizes this discussion.

> **Theorem 8.6.2 (*Principal Axes Theorem for R^2*).** *Let*
> $$ax^2 + 2bxy + cy^2 + dx + ey + f = 0$$
> *be the equation of a conic C, and let*
> $$\mathbf{x}^t A\mathbf{x} = ax^2 + 2bxy + cy^2$$
> *be the associated quadratic form. Then the coordinate axes can be rotated so that the equation for C in the new $x'y'$-coordinate system has the form*
> $$\lambda_1 x'^2 + \lambda_2 y'^2 + d'x' + e'y' + f = 0$$
> *where λ_1 and λ_2 are the eigenvalues of A. The rotation can be accomplished by the substitution*
> $$\mathbf{x} = P\mathbf{x}'$$
> *where P orthogonally diagonalizes $\mathbf{x}^t A\mathbf{x}$ and $\det(P) = 1$.*

Example 6 Describe the conic C whose equation is $5x^2 - 4xy + 8y^2 - 36 = 0$.

Solution. The matrix form of this equation is

$$\mathbf{x}^t A\mathbf{x} - 36 = 0 \tag{7}$$

where

$$A = \begin{bmatrix} 5 & -2 \\ -2 & 8 \end{bmatrix}$$

The characteristic equation of A is

$$\det(\lambda I - A) = \det\begin{bmatrix} \lambda - 5 & 2 \\ 2 & \lambda - 8 \end{bmatrix} = (\lambda - 9)(\lambda - 4) = 0$$

so the eigenvalues of A are $\lambda = 4$ and $\lambda = 9$. We leave it for the reader to show that orthonormal bases for the eigenspaces are

$$\lambda = 4: \quad \mathbf{v}_1 = \begin{bmatrix} 2/\sqrt{5} \\ 1/\sqrt{5} \end{bmatrix}, \qquad \lambda = 9: \quad \mathbf{v}_2 = \begin{bmatrix} -1/\sqrt{5} \\ 2/\sqrt{5} \end{bmatrix}$$

Thus,

$$P = \begin{bmatrix} 2/\sqrt{5} & -1/\sqrt{5} \\ 1/\sqrt{5} & 2/\sqrt{5} \end{bmatrix}$$

orthogonally diagonalizes $\mathbf{x}^t A\mathbf{x}$. Moreover, $\det(P) = 1$ so that the orthogonal coordinate transformation

$$\mathbf{x} = P\mathbf{x}' \tag{8}$$

is a rotation. Substituting (8) into (7) yields

$$(P\mathbf{x}')^t A(P\mathbf{x}') - 36 = 0$$

or

$$(\mathbf{x}')^t (P^t A P)\mathbf{x}' - 36 = 0$$

Since

$$P^t A P = \begin{bmatrix} 4 & 0 \\ 0 & 9 \end{bmatrix}$$

this equation can be written as

$$[x' \quad y']\begin{bmatrix} 4 & 0 \\ 0 & 9 \end{bmatrix}\begin{bmatrix} x' \\ y' \end{bmatrix} - 36 = 0$$

or

$$4x'^2 + 9y'^2 - 36 = 0$$

or

$$\frac{x'^2}{9} + \frac{y'^2}{4} = 1$$

which is the equation of the ellipse sketched in Figure 4. In that figure, the vectors $\mathbf{v}_1$ and $\mathbf{v}_2$ are the column vectors of P.   ▲

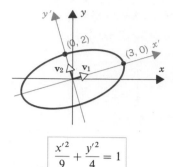

Figure 4

$$\frac{x'^2}{9} + \frac{y'^2}{4} = 1$$

Example 7 Describe the conic C whose equation is

$$5x^2 - 4xy + 8y^2 + \frac{20}{\sqrt{5}}x - \frac{80}{\sqrt{5}}y + 4 = 0$$

Solution. The matrix form of this equation is

$$\mathbf{x}^t A \mathbf{x} + K \mathbf{x} + 4 = 0 \tag{9}$$

where

$$A = \begin{bmatrix} 5 & -2 \\ -2 & 8 \end{bmatrix} \quad \text{and} \quad K = [20/\sqrt{5} \quad -80/\sqrt{5}]$$

As shown in Example 6

$$P = \begin{bmatrix} 2/\sqrt{5} & -1/\sqrt{5} \\ 1/\sqrt{5} & 2/\sqrt{5} \end{bmatrix}$$

orthogonally diagonalizes $\mathbf{x}^t A \mathbf{x}$. Substituting $\mathbf{x} = P\mathbf{x}'$ into (9) gives

$$(P\mathbf{x}')^t A(P\mathbf{x}') + K(P\mathbf{x}') + 4 = 0$$

or

$$(\mathbf{x}')^t (P^t A P)\mathbf{x}' + (KP)\mathbf{x}' + 4 = 0 \tag{10}$$

Since

$$P^t A P = \begin{bmatrix} 4 & 0 \\ 0 & 9 \end{bmatrix} \quad \text{and} \quad KP = [20/\sqrt{5} \quad -80/\sqrt{5}]\begin{bmatrix} 2/\sqrt{5} & -1/\sqrt{5} \\ 1/\sqrt{5} & 2/\sqrt{5} \end{bmatrix} = [-8 \quad -36]$$

(10) can be written as

$$4x'^2 + 9y'^2 - 8x' - 36y' + 4 = 0 \tag{11}$$

To bring the conic into standard position, the $x'y'$-axes must be translated. Proceeding as in Example 5, we rewrite (11) as

$$4(x'^2 - 2x') + 9(y'^2 - 4y') = -4$$

Completing the squares yields

$$4(x'^2 - 2x' + 1) + 9(y'^2 - 4y' + 4) = -4 + 4 + 36$$

or

$$4(x' - 1)^2 + 9(y' - 2)^2 = 36 \tag{12}$$

If we translate the coordinate axes by means of the translation equations

$$x'' = x' - 1 \qquad y'' = y' - 2$$

then (12) becomes

$$4x''^2 + 9y''^2 = 36$$

or

$$\frac{x''^2}{9} + \frac{y''^2}{4} = 1$$

which is the equation of the ellipse sketched in Figure 5. In that figure, the vectors v_1 and v_2 are the column vectors of P. ▲

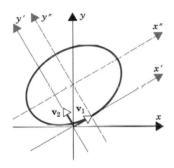

Figure 5

$$\boxed{\dfrac{x''^2}{9} + \dfrac{y''^2}{4} = 1}$$

EXERCISE SET 8.6

1. In each part find a change of variables that reduces the quadratic form to a sum of squares, and express the quadratic form in terms of the new variables.
   (a) $2x_1^2 + 2x_2^2 - 2x_1x_2$ (b) $5x_1^2 + 2x_2^2 + 4x_1x_2$ (c) $2x_1x_2$ (d) $-3x_1^2 + 5x_2^2 + 2x_1x_2$

2. In each part find a change of variables that reduces the quadratic form to a sum of squares, and express the quadratic form in terms of the new variables.
   (a) $3x_1^2 + 4x_2^2 + 3x_3^2 + 4x_1x_2 - 4x_2x_3$ (b) $2x_1^2 + 5x_2^2 + 5x_3^2 + 4x_1x_2 - 4x_1x_3 - 8x_2x_3$
   (c) $-5x_1^2 + x_2^2 - x_3^2 + 6x_1x_3 + 4x_1x_2$ (d) $2x_1x_3 + 6x_2x_3$

3. Find the quadratic forms associated with the following quadratic equations.
   (a) $2x^2 - 3xy + 4y^2 - 7x + 2y + 7 = 0$ (b) $x^2 - xy + 5x + 8y - 3 = 0$ (c) $5xy = 8$
   (d) $4x^2 - 2y^2 = 7$ (e) $y^2 + 7x - 8y - 5 = 0$

4. Find the matrices of the quadratic forms in Exercise 3.

5. Express each of the quadratic equations in Exercise 3 in the matrix form
$\mathbf{x}^t A \mathbf{x} + K \mathbf{x} + f = 0$.

6. Name the following conics.
   (a) $2x^2 + 5y^2 = 20$ (b) $4x^2 + 9y^2 = 1$ (c) $x^2 - y^2 - 8 = 0$ (d) $4y^2 - 5x^2 = 20$
   (e) $x^2 + y^2 - 25 = 0$ (f) $7y^2 - 2x = 0$ (g) $-x^2 = 2y$ (h) $3x - 11y^2 = 0$
   (i) $y - x^2 = 0$ (j) $x^2 - 3 = -y^2$

7. In each part a translation will put the conic in standard position. Name the conic and
   give its equation in the translated coordinate system.
   (a) $9x^2 + 4y^2 - 36x - 24y + 36 = 0$ (b) $x^2 - 16y^2 + 8x + 128y = 256$
   (c) $y^2 - 8x - 14y + 49 = 0$ (d) $x^2 + y^2 + 6x - 10y + 18 = 0$
   (e) $2x^2 - 3y^2 + 6x + 20y = -41$ (f) $x^2 + 10x + 7y = -32$

8. The following nondegenerate conics are rotated out of standard position. In each part
   rotate the coordinate axes to remove the xy-term. Name the conic and give its equation
   in the rotated coordinate system.
   (a) $2x^2 - 4xy - y^2 + 8 = 0$ (b) $x^2 + 2xy + y^2 + 8x + y = 0$
   (c) $5x^2 + 4xy + 5y^2 = 9$ (d) $11x^2 + 24xy + 4y^2 - 15 = 0$

In Exercises 9–14 translate and rotate the coordinate axes, if necessary, to put the conic
in standard position. Name the conic and give its equation in the final coordinate system.

9. $9x^2 - 4xy + 6y^2 - 10x - 20y = 5$ 10. $3x^2 - 8xy - 12y^2 - 30x - 64y = 0$

11. $2x^2 - 4xy - y^2 - 4x - 8y = -14$ 12. $21x^2 + 6xy + 13y^2 - 114x + 34y + 73 = 0$

13. $x^2 - 6xy - 7y^2 + 10x + 2y + 9 = 0$ 14. $4x^2 - 20xy + 25y^2 - 15x - 6y = 0$

15. The graph of a quadratic equation in x and y can, in certain cases, be a point, a line,
   or a pair of lines. These are called **degenerate** conics. It is also possible that the equation
   is not satisfied by any real values of x and y. In such cases the equation has no graph;
   it is said to represent an **imaginary conic**. Each of the following represents a degenerate
   or imaginary conic. Where possible, sketch the graph.
   (a) $x^2 - y^2 = 0$ (b) $x^2 + 3y^2 + 7 = 0$ (c) $8x^2 + 7y^2 = 0$
   (d) $x^2 - 2xy + y^2 = 0$ (e) $9x^2 + 12xy + 4y^2 - 52 = 0$ (f) $x^2 + y^2 - 2x - 4y = -5$

8.7 QUADRIC SURFACES

*In this section we shall apply the diagonalization techniques developed in the previous
section to quadratic equations in three variables, and we shall use our results to study
quadric surfaces.*

**QUADRIC
SURFACES**

An equation of the form

$$ax^2 + by^2 + cz^2 + 2dxy + 2exz + 2fyz + gx + hy + iz + j = 0 \qquad (1)$$

where $a, b, \ldots, f$ are not all zero, is called a **quadratic equation in x, y, and z**; the
expression

$$ax^2 + by^2 + cz^2 + 2dxy + 2exz + 2fyz$$

is called the **associated quadratic form**.

Equation (1) can be written in the matrix form

$$\begin{bmatrix} x & y & z \end{bmatrix} \begin{bmatrix} a & d & e \\ d & b & f \\ e & f & c \end{bmatrix} \begin{bmatrix} x \\ y \\ z \end{bmatrix} + \begin{bmatrix} g & h & i \end{bmatrix} \begin{bmatrix} x \\ y \\ z \end{bmatrix} + j = 0$$

or

$$\mathbf{x}^t A \mathbf{x} + K \mathbf{x} + j = 0$$

where

$$\mathbf{x} = \begin{bmatrix} x \\ y \\ z \end{bmatrix} \qquad A = \begin{bmatrix} a & d & e \\ d & b & f \\ e & f & c \end{bmatrix} \qquad K = \begin{bmatrix} g & h & i \end{bmatrix}$$

Example 1   The quadratic form associated with the qudratic equation

$$3x^2 + 2y^2 - z^2 + 4xy + 3xz - 8yz + 7x + 2y + 3z - 7 = 0$$

is

$$3x^2 + 2y^2 - z^2 + 4xy + 3xz - 8yz \quad \blacktriangle$$

Graphs of quadratic equations in x, y, and z are called **quadrics** or **quadric surfaces**. We now give some examples of quadrics and their equations.

A quadric whose equation can be expressed in one of the forms in Figure 1 is said to be in **standard position** relative to the coordinate axes. The presence of one or more of the cross-product terms xy, xz, and yz in the equation of a quadric indicates that the quadric is rotated out of standard position; the presence of both x^2 and x terms, y^2 and y terms, or z^2 and z terms in a quadric with no cross-product term indicates the quadric is translated out of standard position.

Example 2   Describe the quadric surface whose equation is

$$4x^2 + 36y^2 - 9z^2 - 16x - 216y + 304 = 0$$

Solution.   Rearranging terms gives

$$4(x^2 - 4x) + 36(y^2 - 6y) - 9z^2 = -304$$

Completing the squares yields

$$4(x^2 - 4x + 4) + 36(y^2 - 6y + 9) - 9z^2 = -304 + 16 + 324$$

or

$$4(x - 2)^2 + 36(y - 3)^2 - 9z^2 = 36$$

or

$$\frac{(x - 2)^2}{9} + (y - 3)^2 - \frac{z^2}{4} = 1$$

Translating the axes by means of the translation equations

$$x' = x - 2 \qquad y' = y - 3 \qquad z' = z$$

$$\frac{x^2}{l^2} + \frac{y^2}{m^2} + \frac{z^2}{n^2} = 1$$

Ellipsoid

$$\frac{x^2}{l^2} + \frac{y^2}{m^2} - \frac{z^2}{n^2} = 1$$

Hyperboloid of One Sheet

$$\frac{x^2}{l^2} - \frac{y^2}{m^2} - \frac{z^2}{n^2} = 1$$

Hyperboloid of Two Sheets

$$\frac{x^2}{l^2} + \frac{y^2}{m^2} - \frac{z^2}{n^2} = 0$$

Elliptic Cone

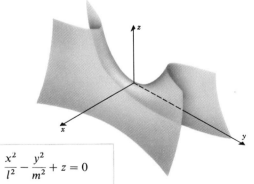

$$\frac{x^2}{l^2} + \frac{y^2}{m^2} - z = 0$$

Elliptic Paraboloid

$$\frac{x^2}{l^2} - \frac{y^2}{m^2} + z = 0$$

Hyperbolic Paraboloid

Figure 1

yields

$$\frac{x'^2}{9} + y'^2 - \frac{z'^2}{4} = 1$$

which is the equation of a hyperboloid of one sheet. ▲

ELIMINATING CROSS-PRODUCT TERMS

The procedure for identifying quadrics that are rotated out of standard position is similar to the procedure for conics. Let Q be a quadric surface whose equation in xyz-coordinates is

$$\mathbf{x}^t A\mathbf{x} + K\mathbf{x} + j = 0 \tag{2}$$

We want to rotate the xyz-coordinate axes so that the equation of the quadric in the new $x'y'z'$-coordinate system has no cross-product terms. This can be done as follows:

Step 1. Find a matrix P that orthogonally diagonalizes $\mathbf{x}^t A\mathbf{x}$.

Step 2. Interchange two columns of P, if necessary, to make $\det(P) = 1$. This assures that the orthogonal coordinate transformation

$$\mathbf{x} = P\mathbf{x}', \text{ that is,} \quad \begin{bmatrix} x \\ y \\ z \end{bmatrix} = P \begin{bmatrix} x' \\ y' \\ z' \end{bmatrix} \tag{3}$$

is a rotation.

Step 3. Substitute (3) into (2). This will produce an equation for the quadric in $x'y'z'$-coordinates with no cross-product terms. (The proof is similar to that for conics and is left as an exercise.)

The following theorem summarizes this discussion.

Theorem 8.7.1 (*Principal Axes Theorem for R^3*). *Let*

$$ax^2 + by^2 + cz^2 + 2dxy + 2exz + 2fyz + gx + hy + iz + j = 0$$

be the equation of a quadric Q, and let

$$\mathbf{x}^t A\mathbf{x} = ax^2 + by^2 + cz^2 + 2dxy + 2exz + 2fyz$$

be the associated quadratic form. The coordinate axes can be rotated so that the equation of Q in the $x'y'z'$-coordinate system has the form

$$\lambda_1 x'^2 + \lambda_2 y'^2 + \lambda_3 z'^2 + g'x' + h'y' + i'z' + j = 0$$

where λ_1, λ_2, and λ_3 are the eigenvalues of A. The rotation can be accomplished by the substitution

$$\mathbf{x} = P\mathbf{x}'$$

where P orthogonally diagonalizes $\mathbf{x}^t A\mathbf{x}$ and $\det(P) = 1$.

Example 3 Describe the quadric surface whose equation is

$$4x^2 + 4y^2 + 4z^2 + 4xy + 4xz + 4yz - 3 = 0$$

Solution. The matrix form of the above quadratic equation is

$$\mathbf{x}^t A \mathbf{x} - 3 = 0 \qquad (4)$$

where

$$A = \begin{bmatrix} 4 & 2 & 2 \\ 2 & 4 & 2 \\ 2 & 2 & 4 \end{bmatrix}$$

As shown in Example 2 of Section 6.3, the eigenvalues of A are $\lambda = 2$ and $\lambda = 8$, and A is orthogonally diagonalized by the matrix

$$P = \begin{bmatrix} -1/\sqrt{2} & -1/\sqrt{6} & 1/\sqrt{3} \\ 1/\sqrt{2} & -1/\sqrt{6} & 1/\sqrt{3} \\ 0 & 2/\sqrt{6} & 1/\sqrt{3} \end{bmatrix}$$

where the first two column vectors in P are eigenvectors corresponding to $\lambda = 2$, and the third column vector is an eigenvector corresponding to $\lambda = 8$.

Since $\det(P) = 1$ (verify), the orthogonal coordinate transformation $\mathbf{x} = P\mathbf{x}'$ is a rotation. Substituting this expression in (4) yields

$$(P\mathbf{x}')^t A (P\mathbf{x}') - 3 = 0$$

or equivalently

$$(\mathbf{x}')^t (P^t A P)\mathbf{x}' - 3 = 0 \qquad (5)$$

But

$$P^t A P = \begin{bmatrix} 2 & 0 & 0 \\ 0 & 2 & 0 \\ 0 & 0 & 8 \end{bmatrix}$$

so (5) becomes

$$\begin{bmatrix} x' & y' & z' \end{bmatrix} \begin{bmatrix} 2 & 0 & 0 \\ 0 & 2 & 0 \\ 0 & 0 & 8 \end{bmatrix} \begin{bmatrix} x' \\ y' \\ z' \end{bmatrix} - 3 = 0$$

or

$$2x'^2 + 2y'^2 + 8z'^2 = 3$$

This can be rewritten as

$$\frac{x'^2}{3/2} + \frac{y'^2}{3/2} + \frac{z'^2}{3/8} = 1$$

which is the equation of an ellipsoid. ▲

EXERCISE SET 8.7

1. Find the quadratic forms associated with the following quadratic equations.
   (a) $x^2 + 2y^2 - z^2 + 4xy - 5yz + 7x + 2z = 3$ (b) $3x^2 + 7z^2 + 2xy - 3xz + 4yz - 3x = 4$
   (c) $xy + xz + yz = 1$ (d) $x^2 + y^2 - z^2 = 7$
   (e) $3z^2 + 3xz - 14y + 9 = 0$ (f) $2z^2 + 2xz + y^2 + 2x - y + 3z = 0$

2. Find the matrices of the quadratic forms in Exercise 1.

3. Express each of the quadratic equations given in Exercise 1 in the matrix form
   $\mathbf{x}^t A \mathbf{x} + K\mathbf{x} + j = 0$.

4. Name the following quadrics.
   (a) $36x^2 + 9y^2 + 4z^2 - 36 = 0$ (b) $2x^2 + 6y^2 - 3z^2 = 18$ (c) $6x^2 - 3y^2 - 2z^2 - 6 = 0$
   (d) $9x^2 + 4y^2 - z^2 = 0$ (e) $16x^2 + y^2 = 16z$ (f) $7x^2 - 3y^2 + z = 0$
   (g) $x^2 + y^2 + z^2 = 25$

5. In each part determine the translation equations that will put the quadric in standard
   position. Name the quadric.
   (a) $9x^2 + 36y^2 + 4z^2 - 18x - 144y - 24z + 153 = 0$ (b) $6x^2 + 3y^2 - 2z^2 + 12x - 18y - 8z = -7$
   (c) $3x^2 - 3y^2 - z^2 + 42x + 144 = 0$ (d) $4x^2 + 9y^2 - z^2 - 54y - 50z = 544$
   (e) $x^2 + 16y^2 + 2x - 32y - 16z - 15 = 0$ (f) $7x^2 - 3y^2 + 126x + 72y + z + 135 = 0$
   (g) $x^2 + y^2 + z^2 - 2x + 4y - 6z - 11$

6. In each part find a rotation $\mathbf{x} = P\mathbf{x}'$ that removes the cross-product terms. Name the
   quadric and give its equation in the $x'y'z'$-system.
   (a) $2x^2 + 3y^2 + 23z^2 + 72xz + 150 = 0$ (b) $4x^2 + 4y^2 + 4z^2 + 4xy + 4xz + 4yz - 5 = 0$
   (c) $144x^2 + 100y^2 + 81z^2 - 216xz - 540x - 720z = 0$ (d) $2xy + z = 0$

In Exercises 7–10 translate and rotate the coordinate axes to put the quadric in standard
position. Name the quadric and give its equation in the final coordinate system.

7. $2xy + 2xz + 2yz - 6x - 6y - 4z = -9$

8. $7x^2 + 7y^2 + 10z^2 - 2xy - 4xz + 4yz - 12x + 12y + 60z = 24$

9. $2xy - 6x + 10y + z - 31 = 0$

10. $2x^2 + 2y^2 + 5z^2 - 4xy - 2xz + 2yz + 10x - 26y - 2z = 0$

11. Prove Theorem 8.7.1.

CHAPTER 9

INTRODUCTION TO NUMERICAL METHODS OF LINEAR ALGEBRA

9.1 COMPARISON OF PROCEDURES FOR SOLVING LINEAR SYSTEMS

In this chapter we shall discuss some practical aspects of solving systems of linear equations, inverting matrices, and finding eigenvalues. Although we have previously discussed methods for performing these computations, those methods are not directly applicable to the computer solution of the large-scale problems that arise in real-world applications.

COUNTING OPERATIONS

Since computers are limited in the number of decimal places they can carry, they round off or truncate most numerical quantities. For example, a computer designed to store eight decimal places might record $\frac{2}{3}$ either as .66666667 (rounded off) or .66666666 (truncated). In either case, an error is introduced that we shall call *roundoff error* or *rounding error*.

The main practical considerations in solving linear algebra problems on digital computers are minimizing the computer time (and thus cost) needed to obtain the solution, and minimizing inaccuracies due to roundoff errors. Thus, a good computer algorithm uses as few operations as possible and performs those operations in a way that minimizes the effect of roundoff errors.

In this text we have studied four methods for solving a linear system, $Ax = b$, of n equations in n unknowns:

1. Gaussian elimination with back-substitution
2. Gauss-Jordan elimination
3. Computing A^{-1}, then $\mathbf{x} = A^{-1}\mathbf{b}$
4. Cramer's Rule

To determine how these methods compare as computational tools, we need to know how many arithmetic operations each requires. On a large modern computer, typical execution times in microseconds (1 microsecond = 10^{-6} second) for the basic arithmetic operations are

$$\text{Multiplication} = 1.0 \text{ microsecond}$$
$$\text{Division} = 3.0 \text{ microseconds}$$
$$\text{Addition} = 0.5 \text{ microsecond}$$
$$\text{Subtraction} = 0.5 \text{ microsecond}$$

In our analysis we shall group divisions and multiplications together (average execution time = 2.0 microseconds), and we shall also group additions and subtractions together (average execution time = 0.5 microsecond). We shall refer to either multiplications or divisions as "multiplications" and to additions or subtractions as "additions."

In Table 1 we list the number of operations required to solve a linear system $A\mathbf{x} = \mathbf{b}$ of n equations in n unknowns by each of the four methods discussed in the text, as well as the number of operations required to invert A or to compute its determinant by row reduction.

Note that the text methods of Gauss-Jordan elimination and Gaussian elimination have the same operation counts. It is not hard to see why this is so. Both methods begin by reducing the augmented matrix to row-echelon form. This is called the *forward phase* or *forward pass*. Then the solution is completed by back-substitution in Gaussian elimination and by continued reduction to reduced row-echelon form in Gauss-Jordan elimination. This is called the *backward phase* or *backward pass*. It turns out that the number of operations required for the backward

TABLE 1 *Operation Counts for an Invertible $n \times n$ Matrix A*

Method	Number of Additions	Number of Multiplications
Solve $A\mathbf{x} = \mathbf{b}$ by Gauss-Jordan elimination	$\frac{1}{3}n^3 + \frac{1}{2}n^2 - \frac{5}{6}n$	$\frac{1}{3}n^3 + n^2 - \frac{1}{3}n$
Solve $A\mathbf{x} = \mathbf{b}$ by Gaussian elimination	$\frac{1}{3}n^3 + \frac{1}{2}n^2 - \frac{5}{6}n$	$\frac{1}{3}n^3 + n^2 - \frac{1}{3}n$
Find A^{-1} by reducing $[A \mid I]$ to $[I \mid A^{-1}]$	$n^3 - 2n^2 + n$	n^3
Solve $A\mathbf{x} = \mathbf{b}$ as $\mathbf{x} = A^{-1}\mathbf{b}$	$n^3 - n^2$	$n^3 + n^2$
Find $\det(A)$ by row reduction	$\frac{1}{3}n^3 - \frac{1}{2}n^2 + \frac{1}{6}n$	$\frac{1}{3}n^3 + \frac{2}{3}n - 1$
Solve $A\mathbf{x} = \mathbf{b}$ by Cramer's Rule	$\frac{1}{3}n^4 - \frac{1}{6}n^3 - \frac{1}{3}n^2 + \frac{1}{6}n$	$\frac{1}{3}n^4 + \frac{1}{3}n^3 + \frac{2}{3}n^2 + \frac{2}{3}n - 1$

phase is the same whether one uses back-substitution or continued reduction to reduce row-echelon form. Thus, the text methods of Gaussian elimination and Gauss-Jordan elimination have the same operation counts.

REMARK. There is a common variation of Gauss-Jordan elimination that is less efficient than the one presented in this text. In our method the augmented matrix is first reduced to reduced row-echelon form by introducing zeros *below* the leading ones; then the reduction is completed by introducing zeros above the leading ones. An alternative procedure is to introduce zeros *above* and *below* a leading one as soon as it is obtained. This method requires

$$\frac{n^3}{2} - \frac{n}{2} \quad \text{additions} \quad \text{and} \quad \frac{n^3}{2} + \frac{n^2}{2} \quad \text{multiplications}$$

both of which are larger than our values for all $n \geq 3$.

To illustrate how the results in Table 1 are computed, we shall derive the operation counts for Gauss-Jordan elimination. For this discussion we need the following formulas for the sum of the first n positive integers and the sum of the squares of the first n positive integers:

$$1 + 2 + 3 + \cdots + n = \frac{n(n + 1)}{2} \tag{1}$$

$$1^2 + 2^2 + 3^2 + \cdots + n^2 = \frac{n(n + 1)(2n + 1)}{6} \tag{2}$$

Derivations of these formulas are discussed in the exercises. We also need formulas for the sum of the first $n - 1$ positive integers and the sum of the squares of the first $n - 1$ positive integers. These can be obtained by substituting $n - 1$ for n in (1) and (2).

$$1 + 2 + 3 + \cdots + (n - 1) = \frac{(n - 1)n}{2} \tag{3}$$

$$1^2 + 2^2 + 3^2 + \cdots + (n - 1)^2 = \frac{(n - 1)n(2n - 1)}{6} \tag{4}$$

OPERATION COUNT FOR GAUSS-JORDAN ELIMINATION

Let $A\mathbf{x} = \mathbf{b}$ be a system of n linear equations in n unknowns, and assume that A is invertible, so the system has a unique solution. Also assume, for simplicity, that no row interchanges are required to put the augmented matrix $[A \mid \mathbf{b}]$ in reduced row-echelon form. This assumption is justified by the fact that row interchanges are performed as bookkeeping operations on a computer and require much less time than arithmetic operations.

elements in that row by the reciprocal of the leftmost entry in the row. We shall represent this step schematically as follows:

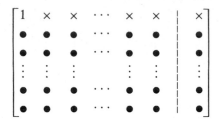

× denotes a quantity to be computed
● denotes a quantity that is not computed
The matrix size is $n \times (n + 1)$

Note that the leading 1 is simply recorded and requires no computation; only the remaining n entries in the first row must be computed.

The following is a schematic description of the steps and the number of operations required to reduce $[A \mid \mathbf{b}]$ to row-echelon form.

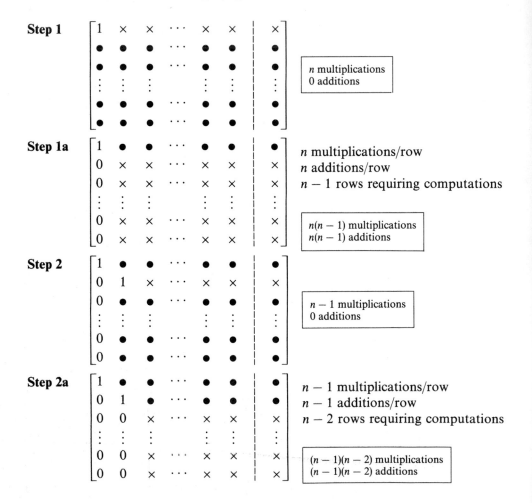

Step 1

n multiplications
0 additions

Step 1a

n multiplications/row
n additions/row
$n - 1$ rows requiring computations

$n(n - 1)$ multiplications
$n(n - 1)$ additions

Step 2

$n - 1$ multiplications
0 additions

Step 2a

$n - 1$ multiplications/row
$n - 1$ additions/row
$n - 2$ rows requiring computations

$(n - 1)(n - 2)$ multiplications
$(n - 1)(n - 2)$ additions

Step 3
$$\left[\begin{array}{ccccccc|c}
1 & \bullet & \bullet & \cdots & \bullet & \bullet & & \bullet \\
0 & 1 & \bullet & \cdots & \bullet & \bullet & & \bullet \\
0 & 0 & 1 & & \times & \times & & \times \\
\vdots & \vdots & \vdots & \cdots & \vdots & \vdots & & \vdots \\
0 & 0 & \bullet & \cdots & \bullet & \bullet & & \bullet \\
0 & 0 & \bullet & \cdots & \bullet & \bullet & & \bullet
\end{array}\right]$$

$n - 2$ multiplications
0 additions

Step 3a
$$\left[\begin{array}{ccccccc|c}
1 & \bullet & \bullet & \cdots & \bullet & \bullet & & \bullet \\
0 & 1 & \bullet & \cdots & \bullet & \bullet & & \bullet \\
0 & 0 & 1 & & \bullet & \bullet & & \bullet \\
\vdots & \vdots & \vdots & \cdots & \vdots & \vdots & & \vdots \\
0 & 0 & 0 & \cdots & \times & \times & & \times \\
0 & 0 & 0 & \cdots & \times & \times & & \times
\end{array}\right]$$

$n - 2$ multiplications/row
$n - 2$ additions/row
$n - 3$ rows requiring computations

$(n - 2)(n - 3)$ multiplications
$(n - 2)(n - 3)$ additions

Step $(n - 1)$
$$\left[\begin{array}{ccccccc|c}
1 & \bullet & \bullet & \cdots & \bullet & \bullet & & \bullet \\
0 & 1 & \bullet & \cdots & \bullet & \bullet & & \bullet \\
0 & 0 & 1 & & \bullet & \bullet & & \bullet \\
\vdots & \vdots & \vdots & \cdots & \vdots & \vdots & & \vdots \\
0 & 0 & 0 & \cdots & 1 & \times & & \times \\
0 & 0 & 0 & \cdots & \bullet & \bullet & & \bullet
\end{array}\right]$$

2 multiplications
0 additions

Step $(n - 1)$a
$$\left[\begin{array}{ccccccc|c}
1 & \bullet & \bullet & \cdots & \bullet & \bullet & & \bullet \\
0 & 1 & \bullet & \cdots & \bullet & \bullet & & \bullet \\
0 & 0 & 1 & & \bullet & \bullet & & \bullet \\
\vdots & \vdots & \vdots & \cdots & \vdots & \vdots & & \vdots \\
0 & 0 & 0 & \cdots & 1 & \bullet & & \bullet \\
0 & 0 & 0 & \cdots & 0 & \times & & \times
\end{array}\right]$$

2 multiplications/row
2 additions/row
1 row requiring computations

2 multiplications
2 additions

Step n
$$\left[\begin{array}{ccccccc|c}
1 & \bullet & \bullet & \cdots & \bullet & \bullet & & \bullet \\
0 & 1 & \bullet & \cdots & \bullet & \bullet & & \bullet \\
0 & 0 & 1 & & \bullet & \bullet & & \bullet \\
\vdots & \vdots & \vdots & \cdots & \vdots & \vdots & & \vdots \\
0 & 0 & 0 & \cdots & 1 & \bullet & & \bullet \\
0 & 0 & 0 & \cdots & 0 & 1 & & \times
\end{array}\right]$$

1 multiplication
0 additions

Thus, the number of operations required to complete successive steps is as follows:

Steps 1 and 1a
Multiplications: $n + n(n - 1) = n^2$
Additions: $n(n - 1) = n^2 - n$

Steps 2 and 2a
Multiplications: $(n - 1) + (n - 1)(n - 2) = (n - 1)^2$
Additions: $(n - 1)(n - 2) = (n - 1)^2 - (n - 1)$

Steps 3 and 3a

Multiplications: $(n - 2) + (n - 2)(n - 3) = (n - 2)^2$

Additions: $(n - 2)(n - 3) = (n - 2)^2 - (n - 2)$

$\vdots$

Steps $(n - 1)$ and $(n - 1)$a

Multiplications: $4(= 2^2)$

Additions: $2(= 2^2 - 2)$

Step n

Multiplications: $1(= 1^2)$

Additions: $0(= 1^2 - 1)$

Therefore, the total number of operations required to reduce $[A \mid \mathbf{b}]$ to row-echelon form is

Multiplications: $n^2 + (n - 1)^2 + (n - 2)^2 + \cdots + 1^2$

Additions: $[n^2 + (n - 1)^2 + (n - 2)^2 + \cdots + 1^2]$
$$- [n + (n - 1) + (n - 2) + \cdots + 1]$$

or, on applying Formulas (1) and (2),

Multiplications:
$$\frac{n(n + 1)(2n + 1)}{6} = \frac{n^3}{3} + \frac{n^2}{2} + \frac{n}{6} \tag{5}$$

Additions:
$$\frac{n(n + 1)(2n + 1)}{6} - \frac{n(n + 1)}{2} = \frac{n^3}{3} - \frac{n}{3} \tag{6}$$

This completes the operation count for the forward phase. For the backward phase we must put the row-echelon form of $[A \mid \mathbf{b}]$ into reduced row-echelon form by introducing zeros above the leading 1's. The operations are as follows:

Step 1

$$\begin{bmatrix} 1 & \bullet & \bullet & \cdots & \bullet & 0 & \vdots & \times \\ 0 & 1 & \bullet & \cdots & \bullet & 0 & \vdots & \times \\ 0 & 0 & 1 & \cdots & \bullet & 0 & \vdots & \times \\ \vdots & \vdots & \vdots & & \vdots & \vdots & \vdots & \vdots \\ 0 & 0 & 0 & \cdots & 1 & 0 & \vdots & \times \\ 0 & 0 & 0 & \cdots & 0 & 1 & \vdots & \bullet \end{bmatrix}$$

$n - 1$ multiplications
$n - 1$ additions

Step 2

$$\begin{bmatrix} 1 & \bullet & \bullet & \cdots & 0 & 0 & \vdots & \times \\ 0 & 1 & \bullet & \cdots & 0 & 0 & \vdots & \times \\ 0 & 0 & 1 & \cdots & 0 & 0 & \vdots & \times \\ \vdots & \vdots & \vdots & & \vdots & \vdots & \vdots & \vdots \\ 0 & 0 & 0 & \cdots & 1 & 0 & \vdots & \bullet \\ 0 & 0 & 0 & \cdots & 0 & 1 & \vdots & \bullet \end{bmatrix}$$

$n - 2$ multiplications
$n - 2$ additions

$\vdots$

Step
(n − 2)

$$\begin{bmatrix} 1 & \bullet & 0 & \cdots & 0 & 0 & | & \times \\ 0 & 1 & 0 & \cdots & 0 & 0 & | & \times \\ 0 & 0 & 1 & \cdots & 0 & 0 & | & \bullet \\ \vdots & \vdots & \vdots & & \vdots & \vdots & | & \vdots \\ 0 & 0 & 0 & \cdots & 1 & 0 & | & \bullet \\ 0 & 0 & 0 & \cdots & 0 & 1 & | & \bullet \end{bmatrix}$$

2 multiplications
2 additions

Step
(n − 1)

$$\begin{bmatrix} 1 & 0 & 0 & \cdots & 0 & 0 & | & \times \\ 0 & 1 & 0 & \cdots & 0 & 0 & | & \bullet \\ 0 & 0 & 1 & \cdots & 0 & 0 & | & \bullet \\ \vdots & \vdots & \vdots & & \vdots & \vdots & | & \vdots \\ 0 & 0 & 0 & \cdots & 1 & 0 & | & \bullet \\ 0 & 0 & 0 & \cdots & 0 & 1 & | & \bullet \end{bmatrix}$$

1 multiplication
1 addition

Thus, the number of operations required for the backward phase is

$$\text{Multiplications:}\quad (n-1) + (n-2) + \cdots + 2 + 1$$
$$\text{Additions:}\quad (n-1) + (n-2) + \cdots + 2 + 1$$

or, on applying formulas (3) and (4),

$$\text{Multiplications:}\quad \frac{(n-1)n}{2} = \frac{n^2}{2} - \frac{n}{2} \tag{7}$$

$$\text{Additions:}\quad \frac{(n-1)n}{2} = \frac{n^2}{2} - \frac{n}{2} \tag{8}$$

Thus, from (5), (6), (7), and (8) the total operation count for Gauss-Jordan elimination is

$$\text{Multiplications:}\quad \left(\frac{n^3}{3} + \frac{n^2}{2} + \frac{n}{6}\right) + \left(\frac{n^2}{2} - \frac{n}{2}\right) = \frac{n^3}{3} + n^2 - \frac{n}{3} \tag{9}$$

$$\text{Additions:}\quad \left(\frac{n^3}{3} - \frac{n}{3}\right) + \left(\frac{n^2}{2} - \frac{n}{2}\right) = \frac{n^3}{3} + \frac{n^2}{2} - \frac{5n}{6} \tag{10}$$

**COMPARISON OF
METHODS FOR
SOLVING LINEAR
SYSTEMS**

In practical applications it is not uncommon to encounter linear systems with thousands of equations in thousands of unknowns. Thus, we shall be interested in Table 1 for large values of n. It is a fact about polynomials that for large values of the variable, a polynomial can be approximated well by its term of highest degree; that is, if $a_k \neq 0$, then

$$a_0 + a_1 x + \cdots + a_k x^k \approx a_k x^k \qquad \text{for large } x$$

(Exercise 12).

Thus, for large values of n the operation counts in Table 1 can be approximated as shown in Table 2.

It follows from Table 2 that for large n the best methods for solving $A\mathbf{x} = \mathbf{b}$ are Gaussian elimination and Gauss-Jordan elimination. The method of multiplying by A^{-1} is much worse than these (it requires three times as many operations), and the poorest of the four methods is Cramer's Rule.

TABLE 2 *Approximate Operation Counts for an Invertible n × n Matrix with Large n*

Method	Number of Additions	Number of Multiplications
Solve $A\mathbf{x} = \mathbf{b}$ by Gauss-Jordan elimination	$\approx \dfrac{n^3}{3}$	$\approx \dfrac{n^3}{3}$
Solve $A\mathbf{x} = \mathbf{b}$ by Gaussian elimination	$\approx \dfrac{n^3}{3}$	$\approx \dfrac{n^3}{3}$
Find A^{-1} by reducing $[A \mid I]$ to $[I \mid A^{-1}]$	$\approx n^3$	$\approx n^3$
Solve $A\mathbf{x} = \mathbf{b}$ as $\mathbf{x} = A^{-1}\mathbf{b}$	$\approx n^3$	$\approx n^3$
Find $\det(A)$ by row reduction	$\approx \dfrac{n^3}{3}$	$\approx \dfrac{n^3}{3}$
Solve $A\mathbf{x} = \mathbf{b}$ by Cramer's Rule	$\approx \dfrac{n^4}{3}$	$\approx \dfrac{n^4}{3}$

REMARK. We observed in the remark following Table 1 that if Gauss-Jordan elimination is performed by introducing zeros above and below leading 1's as soon as they are obtained, then the operation count is

$$\frac{n^3}{2} - \frac{n}{2} \quad \text{additions} \quad \text{and} \quad \frac{n^3}{2} + \frac{n^2}{2} \quad \text{multiplications}$$

Thus, for large n this procedure requires $\approx n^3/2$ multiplications, which is 50% greater than the $n^3/3$ multiplications required by the text method. Similarly for additions.

It is reasonable to ask if it is possible to devise other methods for solving linear systems that might require significantly fewer than the $\approx n^3/3$ additions and multiplications needed in Gaussian elimination and Gauss-Jordan elimination. The answer is a qualified "yes." In recent years methods have been devised that require $\approx Cn^q$ multiplications, where q is slightly larger than 2.5. However, these methods have little practical value because the programming is complicated, the constant C is very large, and the number of additions required is excessive. In short, there is currently no practical method for solving general linear systems that significantly improves on the operation counts for Gaussian elimination and the text method of Gauss-Jordan elimination.

EXERCISE SET 9.1

1. Find the number of additions and multiplications required to compute AB if A is an $m \times n$ matrix and B is an $n \times p$ matrix.

2. Use the result in Exercise 1 to find the number of additions and multiplications required to compute A^k by direct multiplication if A is an $n \times n$ matrix.

3. Assuming A to be an $n \times n$ matrix, use the formulas in Table 1 to determine the number of operations required for the procedures in Table 3.

TABLE 3

	$n = 5$		$n = 10$		$n = 100$		$n = 1000$	
	+	×	+	×	+	×	+	×
Solve $A\mathbf{x} = \mathbf{b}$ by Gauss-Jordan elimination								
Solve $A\mathbf{x} = \mathbf{b}$ by Gaussian elimination								
Find A^{-1} by reducing $[A \mid I]$ to $[I \mid A^{-1}]$								
Solve $A\mathbf{x} = \mathbf{b}$ as $\mathbf{x} = A^{-1}\mathbf{b}$								
Find $\det(A)$ by row reduction								
Solve $A\mathbf{x} = \mathbf{b}$ by Cramer's Rule								

4. Assuming a computer execution time of 2.0 microseconds for multiplications and 0.5 microsecond for additions, use the results in Exercise 3 to fill in the execution times in seconds for the procedures in Table 4.

TABLE 4

	$n = 5$	$n = 10$	$n = 100$	$n = 1000$
	Execution Time (sec)	Execution Time (sec)	Execution Time (sec)	Execution Time (sec)
Solve $Ax = b$ by Gauss-Jordan elimination				
Solve $Ax = b$ by Gaussian elimination				
Find A^{-1} by reducing $[A \mid I]$ to $[I \mid A^{-1}]$				
Solve $Ax = b$ as $x = A^{-1}b$				
Find $\det(A)$ by row reduction				
Solve $Ax = b$ by Cramer's Rule				

5. Derive the formula

$$1 + 2 + 3 + \cdots + n = \frac{n(n + 1)}{2}$$

[*Hint.* Let $S_n = 1 + 2 + 3 + \cdots + n$. Write the terms of S_n in reverse order and add the two expressions for S_n.]

6. Use the result in Exercise 5 to show that

$$1 + 2 + 3 + \cdots + (n - 1) = \frac{(n - 1)n}{2}$$

7. Derive the formula

$$1^2 + 2^2 + 3^2 + \cdots + n^2 = \frac{n(n + 1)(2n + 1)}{6}$$

using the following steps.
(a) Show that $(k + 1)^3 - k^3 = 3k^2 + 3k + 1$.

(b) Show that

$$[2^3 - 1^3] + [3^3 - 2^3] + [4^3 - 3^3] + \cdots + [(n + 1)^3 - n^3] = (n + 1)^3 - 1$$

(c) Apply (a) to each term on the left side of (b) to show that

$$(n + 1)^3 - 1 = 3[1^2 + 2^2 + 3^2 + \cdots + n^2] + 3[1 + 2 + 3 + \cdots + n] + n$$

(d) Solve the equation in (c) for $1^2 + 2^2 + 3^2 + \cdots + n^2$, use the result of Exercise 5, and then simplify.

8. Use the result in Exercise 7 to show that

$$1^2 + 2^2 + 3^2 + \cdots + (n - 1)^2 = \frac{(n - 1)n(2n - 1)}{6}$$

9. Let R be a row-echelon form of an invertible $n \times n$ matrix. Show that solving the linear system $Rx = b$ by back-substitution requires

$$\frac{n^2}{2} - \frac{n}{2} \quad \text{multiplications}$$

$$\frac{n^2}{2} - \frac{n}{2} \quad \text{additions}$$

10. Show that to reduce an invertible $n \times n$ matrix to I_n by the text method requires

$$\frac{n^3}{3} - \frac{n}{3} \quad \text{multiplications}$$

$$\frac{n^3}{3} - \frac{n^2}{2} + \frac{n}{6} \quad \text{additions}$$

[*Note.* Assume that no row interchanges are required.]

11. Consider the variation of Gauss-Jordan elimination in which zeros are introduced above and below a leading 1 as soon as it is obtained, and let A be an invertible $n \times n$ matrix. Show that to solve a linear system $Ax = b$ using this version of Gauss-Jordan elimination requires

$$\frac{n^3}{2} + \frac{n^2}{2} \quad \text{multiplications}$$

$$\frac{n^3}{2} - \frac{n}{2} \quad \text{additions}$$

[*Note.* Assume that no row interchanges are required.]

12. (**For readers who have studied calculus.**) Show that if $p(x) = a_0 + a_1 x + \cdots + a_k x^k$, where $a_k \neq 0$, then

$$\lim_{x \to +\infty} \frac{p(x)}{a_k x^k} = 1$$

[*Note.* This result justifies the approximation $a_0 + a_1 x + \cdots + a_k x^k \approx a_k x^k$ for large x.]

9.2 *LU*-DECOMPOSITIONS

With Gaussian elimination and Gauss-Jordan elimination, a linear system is solved by operating systematically on the augmented matrix. In this section we shall discuss a different approach, one based on factoring the coefficient matrix into a product of lower and upper triangular matrices. This method is well suited for digital computers and is the basis for many practical computer programs.*

SOLVING LINEAR SYSTEMS BY FACTORIZATION

We shall proceed in two stages. First we shall show how a linear system $A\mathbf{x} = \mathbf{b}$ can be readily solved once A is factored into a product of lower and upper triangular matrices. Then we shall show how to construct such factorizations.

If an $n \times n$ matrix A can be factored into a product of $n \times n$ matrices as

$$A = LU$$

where L is lower triangular and U is upper triangular, then the linear system $A\mathbf{x} = \mathbf{b}$ can be solved as follows:

Step 1. Rewrite the system $A\mathbf{x} = \mathbf{b}$ as

$$LU\mathbf{x} = \mathbf{b} \tag{1}$$

Step 2. Define a new $n \times 1$ matrix $\mathbf{y}$ by

$$U\mathbf{x} = \mathbf{y} \tag{2}$$

Step 3. Use (2) to rewrite (1) as $L\mathbf{y} = \mathbf{b}$ and solve this system for $\mathbf{y}$.

Step 4. Substitute $\mathbf{y}$ in (2) and solve for $\mathbf{x}$.

Although this procedure replaces the problem of solving the single system $A\mathbf{x} = \mathbf{b}$ by the problem of solving the two systems, $L\mathbf{y} = \mathbf{b}$ and $U\mathbf{x} = \mathbf{y}$, the latter systems are easy to solve because the coefficient matrices are triangular. The following example illustrates this procedure.

Example 1 Later in this section we will derive the factorization

$$\begin{bmatrix} 2 & 6 & 2 \\ -3 & -8 & 0 \\ 4 & 9 & 2 \end{bmatrix} = \begin{bmatrix} 2 & 0 & 0 \\ -3 & 1 & 0 \\ 4 & -3 & 7 \end{bmatrix} \begin{bmatrix} 1 & 3 & 1 \\ 0 & 1 & 3 \\ 0 & 0 & 1 \end{bmatrix}$$

---

* In 1979 an important library of machine-independent linear algebra programs, called LINPAK, was developed at Argonne National Laboratories. Many of the programs in that library are based on the methods discussed in this section.

Use this result and the method described above to solve the system

$$\begin{bmatrix} 2 & 6 & 2 \\ -3 & -8 & 0 \\ 4 & 9 & 2 \end{bmatrix} \begin{bmatrix} x_1 \\ x_2 \\ x_3 \end{bmatrix} = \begin{bmatrix} 2 \\ 2 \\ 3 \end{bmatrix} \tag{3}$$

Solution. Rewrite (3) as

$$\begin{bmatrix} 2 & 0 & 0 \\ -3 & 1 & 0 \\ 4 & -3 & 7 \end{bmatrix} \begin{bmatrix} 1 & 3 & 1 \\ 0 & 1 & 3 \\ 0 & 0 & 1 \end{bmatrix} \begin{bmatrix} x_1 \\ x_2 \\ x_3 \end{bmatrix} = \begin{bmatrix} 2 \\ 2 \\ 3 \end{bmatrix} \tag{4}$$

As specified in Step 2 above, define y_1, y_2, and y_3 by the equation

$$\begin{bmatrix} 1 & 3 & 1 \\ 0 & 1 & 3 \\ 0 & 0 & 1 \end{bmatrix} \begin{bmatrix} x_1 \\ x_2 \\ x_3 \end{bmatrix} = \begin{bmatrix} y_1 \\ y_2 \\ y_3 \end{bmatrix} \tag{5}$$

so (3) can be rewritten as

$$\begin{bmatrix} 2 & 0 & 0 \\ -3 & 1 & 0 \\ 4 & -3 & 7 \end{bmatrix} \begin{bmatrix} y_1 \\ y_2 \\ y_3 \end{bmatrix} = \begin{bmatrix} 2 \\ 2 \\ 3 \end{bmatrix}$$

or equivalently

$$\begin{aligned} 2y_1 &= 2 \\ -3y_1 + y_2 &= 2 \\ 4y_1 - 3y_2 + 7y_3 &= 3 \end{aligned}$$

The procedure for solving this system is similar to back-substitution except that the equations are solved from the top down instead of from the bottom up. This procedure, called *forward-substitution*, yields

$$y_1 = 1, \qquad y_2 = 5, \qquad y_3 = 2$$

(Verify.) Substituting these values in (5) yields the linear system

$$\begin{bmatrix} 1 & 3 & 1 \\ 0 & 1 & 3 \\ 0 & 0 & 1 \end{bmatrix} \begin{bmatrix} x_1 \\ x_2 \\ x_3 \end{bmatrix} = \begin{bmatrix} 1 \\ 5 \\ 2 \end{bmatrix}$$

or equivalently

$$\begin{aligned} x_1 + 3x_2 + x_3 &= 1 \\ x_2 + 3x_3 &= 5 \\ x_3 &= 2 \end{aligned}$$

Solving this system by back-substitution yields the solution

$$x_1 = 2, \qquad x_2 = -1, \qquad x_3 = 2$$

(Verify.) ▲

LU-DECOMPOSI-TIONS

Now that we have seen how a linear system of n equations in n unknowns can be solved by factoring the coefficient matrix, we shall turn to the problem of constructing such factorizations. To motivate the method, suppose that an $n \times n$ matrix A has been reduced to a row-echelon form U by a sequence of elementary row operations. By Theorem 1.6.1 each of these operations can be accomplished by multiplying on the left by an appropriate elementary matrix. Thus, we can find elementary matrices $E_1, E_2, \ldots, E_k$ such that

$$E_k \cdots E_2 E_1 A = U \tag{6}$$

By Theorem 1.6.2, $E_1, E_2, \ldots, E_k$ are invertible, so we can multiply both sides of Equation (6) on the left successively by

$$E_k^{-1}, \ldots, E_2^{-1}, E_1^{-1}$$

to obtain

$$A = E_1^{-1} E_2^{-1} \cdots E_k^{-1} U \tag{7}$$

In Exercise 15 we will help the reader to show that the matrix L defined by

$$L = E_1^{-1} E_2^{-1} \cdots E_k^{-1} \tag{8}$$

is lower triangular provided that *no row interchanges are used in reducing A to U*. Assuming this to be so, substituting (8) into (7) yields

$$A = LU$$

which is a factorization of A into a product of a lower triangular matrix and an upper triangular matrix.

The following theorem summarizes the above result.

Theorem 9.2.1. *If A is a square matrix that can be reduced to a row-echelon form U without using row interchanges, then A can be factored as $A = LU$ where L is a lower triangular matrix.*

Definition. A factorization of a square matrix A as $A = LU$, where L is lower triangular and U is upper triangular, is called an ***LU-decomposition*** or ***triangular decomposition*** of A.

Example 2 Find an LU-decomposition of

$$A = \begin{bmatrix} 2 & 6 & 2 \\ -3 & -8 & 0 \\ 4 & 9 & 2 \end{bmatrix}$$

Solution. To obtain an LU-decomposition, $A = LU$, we shall reduce A to a row-echelon form U, then calculate L from (8). The steps are as follows:

	Reduction to Row-Echelon Form	Elementary Matrix Corresponding to the Row Operation	Inverse of the Elementary Matrix
	$\begin{bmatrix} 2 & 6 & 2 \\ -3 & -8 & 0 \\ 4 & 9 & 2 \end{bmatrix}$		
Step 1		$E_1 = \begin{bmatrix} \frac{1}{2} & 0 & 0 \\ 0 & 1 & 0 \\ 0 & 0 & 1 \end{bmatrix}$	$E_1^{-1} = \begin{bmatrix} 2 & 0 & 0 \\ 0 & 1 & 0 \\ 0 & 0 & 1 \end{bmatrix}$
	$\begin{bmatrix} 1 & 3 & 1 \\ -3 & -8 & 0 \\ 4 & 9 & 2 \end{bmatrix}$		
Step 2		$E_2 = \begin{bmatrix} 1 & 0 & 0 \\ 3 & 1 & 0 \\ 0 & 0 & 1 \end{bmatrix}$	$E_2^{-1} = \begin{bmatrix} 1 & 0 & 0 \\ -3 & 1 & 0 \\ 0 & 0 & 1 \end{bmatrix}$
	$\begin{bmatrix} 1 & 3 & 1 \\ 0 & 1 & 3 \\ 4 & 9 & 2 \end{bmatrix}$		
Step 3		$E_3 = \begin{bmatrix} 1 & 0 & 0 \\ 0 & 1 & 0 \\ -4 & 0 & 1 \end{bmatrix}$	$E_3^{-1} = \begin{bmatrix} 1 & 0 & 0 \\ 0 & 1 & 0 \\ 4 & 0 & 1 \end{bmatrix}$
	$\begin{bmatrix} 1 & 3 & 1 \\ 0 & 1 & 3 \\ 0 & -3 & -2 \end{bmatrix}$		
Step 4		$E_4 = \begin{bmatrix} 1 & 0 & 0 \\ 0 & 1 & 0 \\ 0 & 3 & 1 \end{bmatrix}$	$E_4^{-1} = \begin{bmatrix} 1 & 0 & 0 \\ 0 & 1 & 0 \\ 0 & -3 & 1 \end{bmatrix}$
	$\begin{bmatrix} 1 & 3 & 1 \\ 0 & 1 & 3 \\ 0 & 0 & 7 \end{bmatrix}$		
Step 5		$E_5 = \begin{bmatrix} 1 & 0 & 0 \\ 0 & 1 & 0 \\ 0 & 0 & \frac{1}{7} \end{bmatrix}$	$E_5^{-1} = \begin{bmatrix} 1 & 0 & 0 \\ 0 & 1 & 0 \\ 0 & 0 & 7 \end{bmatrix}$
	$\begin{bmatrix} 1 & 3 & 1 \\ 0 & 1 & 3 \\ 0 & 0 & 1 \end{bmatrix}$		

Thus,

$$U = \begin{bmatrix} 1 & 3 & 1 \\ 0 & 1 & 3 \\ 0 & 0 & 1 \end{bmatrix}$$

and, from (8),

$$L = \begin{bmatrix} 2 & 0 & 0 \\ 0 & 1 & 0 \\ 0 & 0 & 1 \end{bmatrix} \begin{bmatrix} 1 & 0 & 0 \\ -3 & 1 & 0 \\ 0 & 0 & 1 \end{bmatrix} \begin{bmatrix} 1 & 0 & 0 \\ 0 & 1 & 0 \\ 4 & 0 & 1 \end{bmatrix} \begin{bmatrix} 1 & 0 & 0 \\ 0 & 1 & 0 \\ 0 & -3 & 1 \end{bmatrix} \begin{bmatrix} 1 & 0 & 0 \\ 0 & 1 & 0 \\ 0 & 0 & 7 \end{bmatrix}$$

$$= \begin{bmatrix} 2 & 0 & 0 \\ -3 & 1 & 0 \\ 4 & -3 & 7 \end{bmatrix}$$

so

$$\begin{bmatrix} 2 & 6 & 2 \\ -3 & -8 & 0 \\ 4 & 9 & 2 \end{bmatrix} = \begin{bmatrix} 2 & 0 & 0 \\ -3 & 1 & 0 \\ 4 & -3 & 7 \end{bmatrix} \begin{bmatrix} 1 & 3 & 1 \\ 0 & 1 & 3 \\ 0 & 0 & 1 \end{bmatrix}$$

is an *LU*-decomposition of *A*. ▲

PROCEDURE FOR FINDING *LU*-DECOMPOSITIONS

As this example shows, most of the work in constructing an *LU*-decomposition is expended in the calculation of *L*. However, *all* this work can be eliminated by some careful bookkeeping of the operations used to reduce *A* to *U*. Because we are assuming that no row interchanges are required to reduce *A* to *U*, there are only two types of operations involved: multiplying a row by a nonzero constant, and adding a multiple of one row to another. The first operation is used to introduce the leading ones and the second to introduce zeros below the leading ones.

In Example 2, the multipliers needed to introduce the leading 1's in successive rows were as follows:

$\frac{1}{2}$ for the first row

1 for the second row

$\frac{1}{7}$ for the third row

Note that the successive diagonal entries in *L* were precisely the reciprocals of these multipliers (Figure 1).

$$L = \begin{bmatrix} ② & 0 & 0 \\ -3 & ① & 0 \\ 4 & -3 & ⑦ \end{bmatrix}$$

Figure 1

Next, observe that to introduce zeros below the leading 1 in the first row we used the following operations:

add 3 times the first row to the second

add -4 times the first row to the third

and to introduce the zero below the leading 1 in the second row, we used the operation

add 3 times the second row to the third

Now note that in each position below the main diagonal of L the entry is the *negative* of the multiplier in the operation that introduced the zero in that position in U (Figure 2).

$$L = \begin{bmatrix} 2 & 0 & 0 \\ -3 & 1 & 0 \\ 4 & -3 & 7 \end{bmatrix}$$

Figure 2

In summary, we have the following procedure for constructing an LU-decomposition of a square matrix A provided that A can be reduced to row-echelon form without row interchanges.

Step 1. Reduce A to a row-echelon form U without using row interchanges, keeping track of the multipliers used to introduce the leading 1's and the multipliers used to introduce the zeros below the leading 1's.

Step 2. In each position along the main diagonal of L, place the reciprocal of the multiplier that introduced the leading 1 in that position in U.

Step 3. In each position below the main diagonal of L, place the negative of the multiplier used to introduce the zero in that position in U.

Step 4. Form the decomposition $A = LU$.

Example 3 Find an LU-decomposition of

$$A = \begin{bmatrix} 6 & -2 & 0 \\ 9 & -1 & 1 \\ 3 & 7 & 5 \end{bmatrix}$$

Solution. We begin by reducing A to row-echelon form, keeping track of all multipliers.

$$\begin{bmatrix} 6 & -2 & 0 \\ 9 & -1 & 1 \\ 3 & 7 & 5 \end{bmatrix}$$

$$\begin{bmatrix} ① & -\frac{1}{3} & 0 \\ 9 & -1 & 1 \\ 3 & 7 & 5 \end{bmatrix} \longleftarrow \text{multiplier} = \frac{1}{6}$$

$$\begin{bmatrix} 1 & -\frac{1}{3} & 0 \\ ⓪ & 2 & 1 \\ ⓪ & 8 & 5 \end{bmatrix} \begin{matrix} \\ \longleftarrow \text{multiplier} = -9 \\ \longleftarrow \text{multiplier} = -3 \end{matrix}$$

$$\begin{bmatrix} 1 & -\frac{1}{3} & 0 \\ 0 & ① & \frac{1}{2} \\ 0 & 8 & 5 \end{bmatrix} \begin{matrix} \\ \longleftarrow \text{multiplier} = \frac{1}{2} \\ \\ \end{matrix}$$

$$\begin{bmatrix} 1 & -\frac{1}{3} & 0 \\ 0 & 1 & \frac{1}{2} \\ 0 & ⓪ & 1 \end{bmatrix} \begin{matrix} \\ \\ \longleftarrow \text{multiplier} = -8 \end{matrix}$$

$$\begin{bmatrix} 1 & -\frac{1}{3} & 0 \\ 0 & 1 & \frac{1}{2} \\ 0 & 0 & ① \end{bmatrix} \begin{matrix} \\ \\ \longleftarrow \text{multiplier} = 1 \end{matrix}$$

> No actual operation is performed here, since there is already a leading 1 in the third row.

Constructing L from the multipliers yields the LU-decomposition.

$$A = LU = \begin{bmatrix} 6 & 0 & 0 \\ 9 & 2 & 0 \\ 3 & 8 & 1 \end{bmatrix} \begin{bmatrix} 1 & -\frac{1}{3} & 0 \\ 0 & 1 & \frac{1}{2} \\ 0 & 0 & 1 \end{bmatrix} \quad \blacktriangle$$

We conclude this section by briefly discussing two fundamental questions about LU-decompositions:

> **1.** Does every square matrix have an LU-decomposition?
> **2.** Can a square matrix have more than one LU-decomposition?

We already know that if a square matrix A can be reduced to row-echelon form without using row interchanges, then A has an LU-decomposition. In general, if row interchanges are required to reduce A to row-echelon form, then there is no LU-decomposition of A. However, in such cases it is possible to factor A in the form

$$A = PLU$$

where L is lower triangular, U is upper triangular, and P is a matrix obtained by interchanging the rows of I_n appropriately (see Exercise 17).

In the absence of additional restrictions, LU-decompositions are not unique. For example, if

$$A = LU = \begin{bmatrix} a_{11} & 0 & 0 \\ a_{21} & a_{22} & 0 \\ a_{31} & a_{32} & a_{33} \end{bmatrix} \begin{bmatrix} 1 & u_{12} & u_{13} \\ 0 & 1 & u_{23} \\ 0 & 0 & 1 \end{bmatrix}$$

and L has nonzero diagonal entries, then we can shift the diagonal entries from the left factor to the right factor by writing

$$
A = \begin{bmatrix} 1 & 0 & 0 \\ \dfrac{a_{21}}{a_{11}} & 1 & 0 \\ \dfrac{a_{31}}{a_{11}} & \dfrac{a_{32}}{a_{22}} & 1 \end{bmatrix} \begin{bmatrix} a_{11} & 0 & 0 \\ 0 & a_{22} & 0 \\ 0 & 0 & a_{33} \end{bmatrix} \begin{bmatrix} 1 & u_{12} & u_{13} \\ 0 & 1 & u_{23} \\ 0 & 0 & 1 \end{bmatrix}
$$

$$
= \begin{bmatrix} 1 & 0 & 0 \\ \dfrac{a_{21}}{a_{11}} & 1 & 0 \\ \dfrac{a_{31}}{a_{11}} & \dfrac{a_{32}}{a_{22}} & 1 \end{bmatrix} \begin{bmatrix} a_{11} & a_{11}u_{12} & a_{11}u_{13} \\ 0 & a_{22} & a_{22}u_{23} \\ 0 & 0 & a_{33} \end{bmatrix}
$$

which is another triangular decomposition of A.

EXERCISE SET 9.2

1. Use the method of Example 1 and the LU-decomposition

$$
\begin{bmatrix} 3 & -6 \\ -2 & 5 \end{bmatrix} = \begin{bmatrix} 3 & 0 \\ -2 & 1 \end{bmatrix} \begin{bmatrix} 1 & -2 \\ 0 & 1 \end{bmatrix}
$$

to solve the system

$$
\begin{aligned}
3x_1 - 6x_2 &= 0 \\
-2x_1 + 5x_2 &= 1
\end{aligned}
$$

2. Use the method of Example 1 and the LU-decomposition

$$
\begin{bmatrix} 3 & -6 & -3 \\ 2 & 0 & 6 \\ -4 & 7 & 4 \end{bmatrix} = \begin{bmatrix} 3 & 0 & 0 \\ 2 & 4 & 0 \\ -4 & -1 & 2 \end{bmatrix} \begin{bmatrix} 1 & -2 & -1 \\ 0 & 1 & 2 \\ 0 & 0 & 1 \end{bmatrix}
$$

to solve the system

$$
\begin{aligned}
3x_1 - 6x_2 - 3x_3 &= -3 \\
2x_1 \qquad\quad + 6x_3 &= -22 \\
-4x_1 + 7x_2 + 4x_3 &= 3
\end{aligned}
$$

In Exercises 3–10 find an LU-decomposition of the coefficient matrix; then use the method of Example 1 to solve the system.

3. $\begin{bmatrix} 2 & 8 \\ -1 & -1 \end{bmatrix} \begin{bmatrix} x_1 \\ x_2 \end{bmatrix} = \begin{bmatrix} -2 \\ -2 \end{bmatrix}$

4. $\begin{bmatrix} -5 & -10 \\ 6 & 5 \end{bmatrix} \begin{bmatrix} x_1 \\ x_2 \end{bmatrix} = \begin{bmatrix} -10 \\ 19 \end{bmatrix}$

5. $\begin{bmatrix} 2 & -2 & -2 \\ 0 & -2 & 2 \\ -1 & 5 & 2 \end{bmatrix} \begin{bmatrix} x_1 \\ x_2 \\ x_3 \end{bmatrix} = \begin{bmatrix} -4 \\ -2 \\ 6 \end{bmatrix}$ **6.** $\begin{bmatrix} -3 & 12 & -6 \\ 1 & -2 & 2 \\ 0 & 1 & 1 \end{bmatrix} \begin{bmatrix} x_1 \\ x_2 \\ x_3 \end{bmatrix} = \begin{bmatrix} -33 \\ 7 \\ -1 \end{bmatrix}$

7. $\begin{bmatrix} 5 & 5 & 10 \\ -8 & -7 & -9 \\ 0 & 4 & 26 \end{bmatrix} \begin{bmatrix} x_1 \\ x_2 \\ x_3 \end{bmatrix} = \begin{bmatrix} 0 \\ 1 \\ 4 \end{bmatrix}$ **8.** $\begin{bmatrix} -1 & -3 & -4 \\ 3 & 10 & -10 \\ -2 & -4 & 11 \end{bmatrix} \begin{bmatrix} x_1 \\ x_2 \\ x_3 \end{bmatrix} = \begin{bmatrix} -6 \\ -3 \\ 9 \end{bmatrix}$

9. $\begin{bmatrix} -1 & 0 & 1 & 0 \\ 2 & 3 & -2 & 6 \\ 0 & -1 & 2 & 0 \\ 0 & 0 & 1 & 5 \end{bmatrix} \begin{bmatrix} x_1 \\ x_2 \\ x_3 \\ x_4 \end{bmatrix} = \begin{bmatrix} 5 \\ -1 \\ 3 \\ 7 \end{bmatrix}$ **10.** $\begin{bmatrix} 2 & -4 & 0 & 0 \\ 1 & 2 & 12 & 0 \\ 0 & -1 & -4 & -5 \\ 0 & 0 & 2 & 11 \end{bmatrix} \begin{bmatrix} x_1 \\ x_2 \\ x_3 \\ x_4 \end{bmatrix} = \begin{bmatrix} 8 \\ 0 \\ 1 \\ 0 \end{bmatrix}$

11. Let

$$A = \begin{bmatrix} 2 & 1 & -1 \\ -2 & -1 & 2 \\ 2 & 1 & 0 \end{bmatrix}$$

(a) Find an LU-decomposition of A.

(b) Express A in the form $A = L_1 D U_1$, where L_1 is lower triangular with 1's along the main diagonal, U_1 is upper triangular, and D is a diagonal matrix.

(c) Express A in the form $A = L_2 U_2$, where L_2 is lower triangular with 1's along the main diagonal and U_2 is upper triangular.

12. Show that the matrix

$$\begin{bmatrix} 0 & 1 \\ 1 & 0 \end{bmatrix}$$

has no LU-decomposition.

13. Let

$$A = \begin{bmatrix} a & b \\ c & d \end{bmatrix}$$

(a) Prove: If $A \neq 0$ then A has a unique LU-decomposition with 1's along the main diagonal of L.

(b) Find the LU-decomposition described in (a).

14. Let $A\mathbf{x} = \mathbf{b}$ be a linear system of n equations in n unknowns, and assume that A is an invertible matrix that can be reduced to row-echelon form without row interchanges. How many additions and multiplications are required to solve the system by the method of Example 1? [**Note.** Count subtractions as additions and divisions as multiplications.]

15. (a) Prove: If L_1 and L_2 are $n \times n$ lower triangular matrices, then so is $L_1 L_2$.

(b) The result in (a) is a special case of a general result that states that a product of finitely many lower triangular matrices is lower triangular. Use this fact to prove that the matrix L in (8) is lower triangular. [**Hint.** See Exercise 27 of Section 2.4.]

16. Use the result in Exercise 15(b) to prove that a product of finitely many upper triangular matrices is upper triangular. [**Hint.** Take transposes.]

17. Prove: If A is any $n \times n$ matrix, then A can be factored as $A = PLU$, where L is lower triangular, U is upper triangular, and P can be obtained by interchanging the rows of I_n appropriately. [**Hint.** Let U be a row-echelon form of A, and let all row interchanges required in the reduction of A to U be performed first.]

18. Factor

$$A = \begin{bmatrix} 3 & -1 & 0 \\ 3 & -1 & 1 \\ 0 & 2 & 1 \end{bmatrix}$$

as $A = PLU$, where P is obtained from I_3 by interchanging rows appropriately, L is lower triangular, and U is upper triangular.

9.3   THE GAUSS-SEIDEL AND JACOBI METHODS

*Although Gaussian elimination (or the text version of Gauss-Jordan elimination) is generally the method of choice for solving a linear system of n equations in n unknowns, there are other approaches to solving linear systems, called **iterative** or **indirect** methods, that are better in certain situations. These methods start with an initial approximation to a solution and then generate a succession of better and better approximations that tend toward an exact solution. In this section we shall study two iterative methods and discuss situations in which they should be used.*

JACOBI ITERATION

We shall begin by discussing the simplest iterative method, which is called *Jacobi iteration* or the *method of simultaneous displacements*. This method applies to linear systems of n equations in n unknowns. Suppose that the system

$$\begin{aligned} a_{11}x_1 + a_{12}x_2 + \cdots + a_{1n}x_n &= b_1 \\ a_{21}x_1 + a_{22}x_2 + \cdots + a_{2n}x_n &= b_2 \\ \vdots \qquad \vdots \qquad\qquad \vdots \qquad &\ \ \vdots \\ a_{n1}x_1 + a_{n2}x_2 + \cdots + a_{nn}x_n &= b_n \end{aligned} \tag{1}$$

has exactly one solution and that the diagonal entries $a_{11}, a_{22}, \ldots, a_{nn}$ are nonzero.

To start, rewrite system (1) by solving the first equation for x_1 in terms of the remaining unknowns, solving the second equation for x_2 in terms of the remaining unknowns, solving the third equation for x_3 in terms of the remaining unknowns, and so on. This yields

$$\begin{aligned} x_1 &= \frac{1}{a_{11}}(b_1 - a_{12}x_2 - a_{13}x_3 - \cdots - a_{1n}x_n) \\[2mm] x_2 &= \frac{1}{a_{22}}(b_2 - a_{21}x_1 - a_{23}x_3 - \cdots - a_{2n}x_n) \\[1mm] &\vdots \\ x_n &= \frac{1}{a_{nn}}(b_n - a_{n1}x_1 - a_{n2}x_2 - \cdots - a_{n(n-1)}x_{n-1}) \end{aligned} \tag{2}$$

For example, the system

$$20x_1 + x_2 - x_3 = 17$$
$$x_1 - 10x_2 + x_3 = 13 \qquad (3)$$
$$-x_1 + x_2 + 10x_3 = 18$$

would be rewritten as

$$x_1 = \tfrac{17}{20} - \tfrac{1}{20}x_2 + \tfrac{1}{20}x_3$$
$$x_2 = -\tfrac{13}{10} + \tfrac{1}{10}x_1 + \tfrac{1}{10}x_3$$
$$x_3 = \tfrac{18}{10} + \tfrac{1}{10}x_1 - \tfrac{1}{10}x_2$$

or

$$x_1 = .850 - .05x_2 + .05x_3$$
$$x_2 = -1.3 + .1x_1 + .1x_3 \qquad (4)$$
$$x_3 = 1.8 + .1x_1 - .1x_2$$

If an approximation to the solution of (1) is known, and these approximate values of the unknowns are substituted into the right side of (2), it is often the case that the values of $x_1, x_2, \ldots, x_n$ that result on the left side form an even better approximation to the solution. This observation is the key to the Jacobi method.

To solve system (1) by Jacobi iteration, make an initial approximation to the solution. When no better choice is available use $x_1 = 0, x_2 = 0, x_3 = 0, \ldots$.

Substitute this initial approximation into the right side of (2) and use the values of $x_1, x_2, \ldots$ that result on the left side as a new approximation to the solution.

For example, to solve (3) by the Jacobi method, we would substitute the initial approximation $x_1 = 0, x_2 = 0, x_3 = 0$ into the right side of (4), and calculate the new approximation

$$x_1 = .850 \qquad x_2 = -1.3 \qquad x_3 = 1.8 \qquad (5)$$

To improve the approximation, we can repeat the substitution process. For example, in solving (3), we would substitute the approximation (5) into the right side of (4) to obtain the next approximation

$$x_1 = .850 - .05(-1.3) + .05(1.8) = 1.005$$
$$x_2 = -1.3 + .1(.850) + .1(1.8) = -1.035$$
$$x_3 = 1.8 + .1(.850) - .1(-1.3) = 2.015$$

In this way a succession of approximations can be generated that, under certain conditions, get closer and closer to the exact solution of the system. In Table 1 we have summarized the results obtained in solving system (3) by Jacobi iteration. All computations were rounded off to five significant digits. At the end of six substitutions (called *iterations*), the exact solution $x_1 = 1, x_2 = -1, x_3 = 2$ is accurately known to five significant digits.

TABLE 1

	Initial Approx- imation	First Approx- imation	Second Approx- imation	Third Approx- imation	Fourth Approx- imation	Fifth Approx- imation	Sixth Approx- imation
x_1	0	.850	1.005	1.0025	1.0001	.99997	1.0000
x_2	0	−1.3	−1.035	−.9980	−.99935	.99999	−1.0000
x_3	0	1.8	2.015	2.004	2.0000	1.9999	2.0000

**GAUSS-SEIDEL
ITERATION**

We now discuss a minor modification of the Jacobi method that often reduces the number of iterations needed to obtain a given degree of accuracy. The technique is called *Gauss-Seidel iteration* or the method of *successive displacements*.

In each iteration of the Jacobi method, the new approximation is obtained by substituting the previous approximation into the right side of (2) and solving for new values of x_1, x_2, ... These new x-values are not all computed simultaneously; first x_1 is obtained from the top equation, then x_2 is obtained from the second equation, then x_3, and so on. Since the new x-values are generally closer to the exact solution, this suggests that better accuracy might be obtained by using the new x-values as soon as they are known. For example, consider system (3). In the first iteration of the Jacobi method, the initial approximation $x_1 = 0$, $x_2 = 0$, $x_3 = 0$ was substituted into each equation on the right side of (4) to obtain the new approximation

$$x_1 = .850 \qquad x_2 = -1.3 \qquad x_3 = 1.8 \tag{6}$$

In the first iteration of the Gauss-Seidel method, the new approximation would be computed as follows. Substitute the initial approximation $x_1 = 0, x_2 = 0, x_3 = 0$ into the right side of the first equation in (4). This yields the new estimate $x_1 = .850$.
Use this new value of x_1 immediately by substituting

$$x_1 = .850 \qquad x_2 = 0 \qquad x_3 = 0$$

into the right side of the second equation in (4). This yields the new estimate $x_2 = -1.215$.
Use this new value of x_2 immediately by substituting

$$x_1 = .850 \qquad x_2 = -1.215 \qquad x_3 = 0$$

into the right side of the third equation in (4). This yields the new estimate $x_3 = 2.0065$.
Thus, at the end of the first iteration of the Gauss-Seidel method the new approximation is

$$x_1 = .850 \qquad x_2 = -1.215 \qquad x_3 = 2.0065 \tag{7}$$

The computations for the second iteration would be carried out as follows.

Substituting (7) into the right side of the first equation in (4) and rounding off to five significant digits yields

$$x_1 = .850 - .05(-1.215) + .05(2.0065) = 1.0111$$

Substituting

$$x_1 = 1.0111 \qquad x_2 = -1.215 \qquad x_3 = 2.0065$$

into the right side of the second equation in (4) and rounding off to five significant digits yields

$$x_2 = -1.3 + .1(1.0111) + .1(2.0065) = -.99824$$

Substituting

$$x_1 = 1.0111 \qquad x_2 = -.99824 \qquad x_3 = 2.0065$$

into the right side of the third equation in (4) and rounding off to five significant digits yields

$$x_3 = 1.8 + .1(1.0111) - .1(-.99824) = 2.0009$$

Thus, at the end of the second iteration of the Gauss-Seidel method, the new approximation is

$$x_1 = 1.0111 \qquad x_2 = -.99824 \qquad x_3 = 2.0009$$

In Table 2 we have summarized the results obtained using four iterations of the Gauss-Seidel method to solve (3). All numbers were rounded off to five significant digits.

TABLE 2

	Initial Approximation	First Approximation	Second Approximation	Third Approximation	Fourth Approximation
x_1	0	.850	1.0111	.99995	1.0000
x_2	0	-1.215	$-.99824$	$-.99992$	-1.0000
x_3	0	2.0065	2.0009	2.0000	2.0000

Comparing Tables 1 and 2, we see that the Gauss-Seidel method produces the solution of (3) (accurate to five significant digits) in four iterations, while six iterations are needed to attain the same accuracy with the Jacobi method.

Although it may seem surprising, it is possible to construct examples in which Jacobi iteration is better than Gauss-Seidel iteration. However, for most practical problems Gauss-Seidel iteration is a better choice.

CONVERGENCE The Gauss-Seidel and Jacobi methods do not always work. In some cases, one or both of these methods can fail to produce a good approximation to the solution, regardless of the number of iterations performed. In such cases the approximations are said to **diverge**. However, if by performing sufficiently many iterations, the solution can be obtained to any desired degree of accuracy, the approximations are said to *converge*. We shall now discuss some conditions that ensure convergence.

A square matrix

$$A = \begin{bmatrix} a_{11} & a_{12} & \cdots & a_{1n} \\ a_{21} & a_{22} & \cdots & a_{2n} \\ \vdots & \vdots & & \vdots \\ a_{n1} & a_{n2} & \cdots & a_{nn} \end{bmatrix}$$

is called *strictly diagonally dominant* if the absolute value of each diagonal entry is greater than the sum of the absolute values of the remaining entries in the same row; that is,

$$|a_{11}| > |a_{12}| + |a_{13}| + \cdots + |a_{1n}|$$
$$|a_{22}| > |a_{21}| + |a_{23}| + \cdots + |a_{2n}|$$
$$\vdots \qquad \vdots \qquad \vdots \qquad \qquad \vdots$$
$$|a_{nn}| > |a_{n1}| + |a_{n2}| + \cdots + |a_{n(n-1)}|$$

Example 1

$$\begin{bmatrix} 7 & -2 & 3 \\ 4 & 1 & -6 \\ 5 & 12 & -4 \end{bmatrix}$$

is not strictly diagonally dominant since in the second row, $|1|$ is not greater than $|4| + |-6|$, and in the third row, $|-4|$ is not greater than $|5| + |12|$.

If the second and third rows are interchanged, however, the resulting matrix

$$\begin{bmatrix} 7 & -2 & 3 \\ 5 & 12 & -4 \\ 4 & 1 & -6 \end{bmatrix}$$

is strictly diagonally dominant since

$$|7| > |-2| + |3|$$
$$|12| > |5| + |-4|$$
$$|-6| > |4| + |1| \quad \blacktriangle$$

Two results on convergence of iterative methods are proved in some of the references at the end of this chapter.

- If A is strictly diagonally dominant, then the Gauss-Seidel and Jacobi approximations to the solution of $A\mathbf{x} = \mathbf{b}$ both converge to the exact solution of the system for all choices of the initial approximation.

- If A is a positive definite symmetric matrix, then the Gauss-Seidel approximations to the solution of $A\mathbf{x} = \mathbf{b}$ converge to the exact solution of the system for all choices of the initial approximation.

In practical problems we are concerned not only with the convergence of iterative methods but also with how fast they converge. For example, there are linear systems in which the Gauss-Seidel approximations converge, but the convergence is so slow that millions of iterations would be required to obtain any reasonable accuracy. Thus, numerical analysts have devised various methods to improve the convergence rate of Gauss-Seidel iteration. The most important of these methods is known as **extrapolated Gauss-Seidel iteration** or the **method of successive over-relaxation** (abbreviated as SOR in the literature). Readers interested in this topic are referred to the references at the end of this chapter.

Faced with a linear system in practice, one must decide whether to solve the system by a direct method, such as Gaussian elimination, or by an indirect method such as Gauss-Seidel iteration. Here are some considerations that enter into making that choice:

- When the coefficient matrix has a high proportion of zeros (such matrices are called *sparse*), iterative methods can be used to advantage because these zeros simplify the iteration equations, thereby reducing the amount of calculation. Gaussian elimination and Gauss-Jordan elimination generally destroy the sparseness by replacing zeros with nonzero entries, especially above the main diagonal.
- It may happen that a good estimate of the solution is known. If this is used as a starting value in an iterative method, then there is a good chance of obtaining a satisfactory approximate solution with fewer computations than a direct method would require.
- With clever programming, less computer memory is needed for iterative methods than direct methods. Thus, if memory space is a problem, iterative methods may be essential.
- If the iterative methods diverge or the rate of convergence is too slow, direct methods may be essential.

EXERCISE SET 9.3

In Exercises 1–4 solve the systems by Jacobi iteration. Start with $x_1 = 0$, $x_2 = 0$. Use four iterations, and round off the computations to three significant digits. Compare your results to the exact solutions.

1. $2x_1 + \ x_2 = 7$
$\quad x_1 - 2x_2 = 1$

2. $3x_1 - \ x_2 = \ \ 5$
$\quad 2x_1 + 3x_2 = -4$

3. $5x_1 - 2x_2 = -13$
$\quad x_1 + 7x_2 = -10$

4. $4x_1 + .1x_2 = \ .2$
$\quad .3x_1 + .7x_2 = 1.4$

In Exercises 5–8 solve the systems by Gauss-Seidel iteration. Start with $x_1 = 0$, $x_2 = 0$. Use three iterations, and round off the computations to three significant digits. Compare your results to the exact solutions.

5. Solve the system in Exercise 1.

6. Solve the system in Exercise 2.

7. Solve the system in Exercise 3.

8. Solve the system in Exercise 4.

In Exercises 9 and 10 solve the systems by Jacobi iteration. Start with $x_1 = 0$, $x_2 = 0$, $x_3 = 0$. Use three iterations, and round off the computations to three significant digits. Compare your results to the exact solutions.

9. $10x_1 + \ \ x_2 + 2x_3 = \ \ 3$
$\quad x_1 + 10x_2 - \ \ x_3 = \ \frac{3}{2}$
$\quad 2x_1 + \ \ x_2 + 10x_3 = -9$

10. $20x_1 - \ \ x_2 + \ \ x_3 = \ \ 20$
$\quad 2x_1 + 10x_2 - \ \ x_3 = \ \ 11$
$\quad x_1 + \ \ x_2 - 20x_3 = -18$

In Exercises 11 and 12 solve the systems by Gauss-Seidel iteration. Start with $x_1 = 0$, $x_2 = 0$, $x_3 = 0$. Use three iterations, and round off the computations to three significant digits. Compare your results to the exact solutions.

11. Solve the system in Exercise 9.   **12.** Solve the system in Exercise 10.

13. Which of the following are strictly diagonally dominant?

(a) $\begin{bmatrix} 2 & 1 \\ -1 & 4 \end{bmatrix}$ (b) $\begin{bmatrix} 3 & -5 \\ 1 & 2 \end{bmatrix}$ (c) $\begin{bmatrix} 6 & 0 & 1 \\ 3 & 5 & 3 \\ 0 & 0 & 1 \end{bmatrix}$ (d) $\begin{bmatrix} 4 & 1 & 2 \\ 0 & 3 & 2 \\ 4 & 1 & -7 \end{bmatrix}$ (e) $\begin{bmatrix} 5 & 1 & 2 & 0 \\ 3 & -7 & 2 & 1 \\ 0 & 2 & 5 & 1 \\ 1 & 1 & 2 & -5 \end{bmatrix}$

14. Consider the system

$$x_1 + 3x_2 = 4$$
$$x_1 - x_2 = 0$$

(a) Show that the approximations obtained by Jacobi iteration starting with $x_1 = 0$ and $x_2 = 0$ diverge.

(b) Is the coefficient matrix

$$A = \begin{bmatrix} 1 & 3 \\ 1 & -1 \end{bmatrix}$$

strictly diagonally dominant?

15. Show that if one or more of the diagonal entries $a_{11}, a_{22}, \ldots, a_{nn}$ in (1) is zero, then it is possible to interchange equations and relabel the unknowns so that the diagonal entries in the resulting system are all nonzero.

9.4 PARTIAL PIVOTING; REDUCTION OF ROUNDOFF ERROR

We noted earlier that minimizing roundoff error is a primary consideration when solving linear systems on a computer. In this section we shall discuss a procedure that is designed to minimize the cumulative effect of roundoff error in solving linear systems.

ROUNDING AND SIGNIFICANT DIGITS

Most computer arithmetic is performed using **normalized floating point numbers**. This means the numbers are expressed in the form*

$$\pm M \times 10^k \tag{1}$$

where k is an integer and M is a fraction that satisfies

$$.1 \leq M < 1$$

The fraction M is called the **mantissa**.

---

* Most computers convert decimal numbers (base 10) to binary numbers (base 2). For simplicity, however, we shall think in terms of decimals.

Example 1 The following numbers are expressed in normalized floating point form.

$$73 = .73 \times 10^2$$
$$-.000152 = -.152 \times 10^{-3}$$
$$1{,}579 = .1579 \times 10^4$$
$$-1/4 = -.25 \times 10^0 \quad \blacktriangle$$

The number of decimal places in the mantissa and the allowable size of the exponent k in (1) depend on the computer being used. For example, the INTEL 8087 coprocessor chip stores the equivalent of fifteen decimal digits in the mantissa and allows 10^k to range from 10^{-308} to 10^{308}. A computer that uses n decimal places in the mantissa is said to round off numbers to **n significant digits**.

Example 2 The following numbers are rounded off to three significant digits.

Number	Normalized Floating Point Form	Rounded Value
7/3	$.233 \times 10^1$	2.33
1,758	$.176 \times 10^4$	1,760
.0000092143	$.921 \times 10^{-5}$	.00000921
$-.12$	$-.120 \times 10^0$	$-.12$
13.850	$.138 \times 10^2$	13.8
$-.08495$	$-.850 \times 10^{-1}$	$-.085$

(If, as in the last two cases, the portion of decimal to be discarded in the rounding process is exactly half a unit, we shall adopt the convention of rounding so that the last retained digit is even. In practice, the treatment of this situation varies from computer to computer.) $\blacktriangle$

PARTIAL PIVOTING

We shall now introduce a technique, called **partial pivoting** (or **pivotal condensation**), that can be used to reduce roundoff error in Gaussian elimination, Gauss-Jordan elimination, or any other computational procedure that uses elementary row operations. To explain the method we shall describe its steps and illustrate them by solving the system

$$3x_1 + 2x_2 - x_3 = 1$$
$$6x_1 + 6x_2 + 2x_3 = 12$$
$$3x_1 - 2x_2 + x_3 = 11$$

using Gaussian elimination with partial pivoting applied to the augmented matrix.

Step 1. In the leftmost column find an entry that has the largest absolute value. This is called the *pivot entry*.

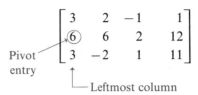

$$\begin{bmatrix} 3 & 2 & -1 & 1 \\ 6 & 6 & 2 & 12 \\ 3 & -2 & 1 & 11 \end{bmatrix}$$

Pivot entry

Leftmost column

Step 2. Perform a row interchange, if necessary, to bring the pivot entry to the top of the column.

$$\begin{bmatrix} 6 & 6 & 2 & 12 \\ 3 & 2 & -1 & 1 \\ 3 & -2 & 1 & 11 \end{bmatrix}$$

> The first and second rows of the previous matrix were interchanged.

Step 3. If the pivot entry is a, multiply the top row by $1/a$.

$$\begin{bmatrix} 1 & 1 & \frac{1}{3} & 2 \\ 3 & 2 & -1 & 1 \\ 3 & -2 & 1 & 11 \end{bmatrix}$$

> The first row of the previous matrix was multiplied by 1/6.

Step 4. Add suitable multiples of the top row to the rows below so that in the column located in Step 1, all the entries below the top become zeros.

$$\begin{bmatrix} 1 & 1 & \frac{1}{3} & 2 \\ 0 & -1 & -2 & -5 \\ 0 & -5 & 0 & 5 \end{bmatrix}$$

> -3 times the first row of the previous matrix was added to the second and third rows.

Step 5. Cover the top row in the matrix, and begin again with Step 1, applied to the submatrix that remains. Continue in this way until the *entire* matrix is in row-echelon form.

$$\begin{bmatrix} 1 & 1 & \frac{1}{3} & 2 \\ 0 & -1 & -2 & -5 \\ 0 & -5 & 0 & 5 \end{bmatrix}$$

Pivot entry

Leftmost nonzero column in the *submatrix*

$$\begin{bmatrix} 1 & 1 & \frac{1}{3} & 2 \\ 0 & -5 & 0 & 5 \\ 0 & -1 & -2 & -5 \end{bmatrix}$$

> The first and second rows of the *submatrix* were interchanged.

$$\begin{bmatrix} 1 & 1 & \frac{1}{3} & 2 \\ 0 & 1 & 0 & -1 \\ 0 & -1 & -2 & -5 \end{bmatrix}$$

> The first row of the *submatrix* was multiplied by $-1/5$.

$$\begin{bmatrix} 1 & 1 & \frac{1}{3} & 2 \\ 0 & 1 & 0 & -1 \\ 0 & 0 & -2 & -6 \end{bmatrix}$$

> The first row of the *submatrix* was added to the second row.

$$\begin{bmatrix} 1 & 1 & \frac{1}{3} & 2 \\ 0 & 1 & 0 & -1 \\ 0 & 0 & \boxed{-2} & -6 \end{bmatrix}$$

> The first row of the *submatrix* was covered and we returned again to Step 1.

Pivot entry   Leftmost nonzero column in the new submatrix

$$\begin{bmatrix} 1 & 1 & \frac{1}{3} & 2 \\ 0 & 1 & 0 & -1 \\ 0 & 0 & 1 & 3 \end{bmatrix}$$

> The first row of the *new* submatrix was multiplied by $-1/2$.

The entire matrix is now in row-echelon form.

Step 6. Solve the corresponding system of equations by back-substitution. The corresponding system of equations is

$$\begin{aligned} x_1 + x_2 + \tfrac{1}{3}x_3 &= 2 \\ x_2 &= -1 \\ x_3 &= 3 \end{aligned}$$

Solving by back-substitution yields

$$x_3 = 3 \qquad x_2 = -1 \qquad x_1 = 2$$

Since the above computations are exact, they do not illustrate the effectiveness of partial pivoting in reducing rounding error; the next example does.

Example 3 Solve the following system by Gaussian elimination with partial pivoting. After each calculation round off the result to three significant digits.

$$\begin{aligned} .00044x_1 + .0003x_2 - .0001x_3 &= .00046 \\ 4x_1 + x_2 + x_3 &= 1.5 \\ 3x_1 - 9.2x_2 - .5x_3 &= -8.2 \end{aligned} \tag{2}$$

Solution (with partial pivoting). The augmented matrix is

$$\begin{bmatrix} .00044 & .0003 & -.0001 & .00046 \\ 4 & 1 & 1 & 1.5 \\ 3 & -9.2 & -.5 & -8.2 \end{bmatrix}$$

To bring the pivot entry to the top of the first column, we interchange the first and second rows; this yields

$$\begin{bmatrix} 4 & 1 & 1 & 1.5 \\ .00044 & .0003 & -.0001 & .00046 \\ 3 & -9.2 & -.5 & -8.2 \end{bmatrix}$$

Dividing each entry in the first row by 4 yields

$$\begin{bmatrix} 1 & .25 & .25 & .375 \\ .00044 & .0003 & -.0001 & .00046 \\ 3 & -9.2 & -.5 & -8.2 \end{bmatrix}$$

Adding $-.00044$ times the first row to the second and -3 times the first row to the third yields (after rounding to three significant digits)

$$\begin{bmatrix} 1 & .25 & .25 & .375 \\ 0 & .000190 & -.00021 & .000295 \\ 0 & -9.95 & -1.25 & -9.32 \end{bmatrix}$$

Interchanging the second and third rows yields

$$\begin{bmatrix} 1 & .25 & .25 & .375 \\ 0 & -9.95 & -1.25 & -9.32 \\ 0 & .000190 & -.00021 & .000295 \end{bmatrix}$$

Dividing each entry in the second row by -9.95 yields

$$\begin{bmatrix} 1 & .25 & .25 & .375 \\ 0 & 1 & .126 & .937 \\ 0 & .000190 & -.00021 & .000295 \end{bmatrix}$$

Adding $-.000190$ times the second row to the third yields

$$\begin{bmatrix} 1 & .25 & .25 & .375 \\ 0 & 1 & .126 & .937 \\ 0 & 0 & -.000234 & .000117 \end{bmatrix}$$

Dividing each entry in the third row by $-.000234$ yields the row-echelon form

$$\begin{bmatrix} 1 & .25 & .25 & .375 \\ 0 & 1 & .126 & .937 \\ 0 & 0 & 1 & -.5 \end{bmatrix}$$

The corresponding system of equations is

$$x_1 + .25x_2 + .25x_3 = .375$$
$$x_2 + .126x_3 = .937$$
$$x_3 = -.5$$

Solving by back-substitution yields (to three significant digits)

$$x_1 = .250 \qquad x_2 = 1.00 \qquad x_3 = -.500 \tag{3}$$

If (2) is solved by Gaussian elimination *without* partial pivoting and each calculation is rounded to three significant digits, one obtains (details omitted)

$$x_1 = .245 \qquad x_2 = 1.01 \qquad x_3 = -.492 \tag{4}$$

Comparing (3) and (4) to the exact solution

$$x_1 = \tfrac{1}{4} \qquad x_2 = 1 \qquad x_3 = -\tfrac{1}{2}$$

shows that partial pivoting yields more accurate results. ▲

In spite of the fact that partial pivoting can reduce the cumulative effect of rounding error, there are certain systems of equations, called ***ill-conditioned*** systems, that are so extremely sensitive that even the slightest errors in the coefficients can result in major inaccuracies in the solution. For example, consider the system

$$x_1 + \quad x_2 = -3$$
$$x_1 + 1.016x_2 = \quad 5 \tag{5}$$

If we assume that this system is to be solved on a computer that rounds off to three significant digits, the computer will store this system as

$$x_1 + \quad x_2 = -3$$
$$x_1 + 1.02x_2 = \quad 5 \tag{6}$$

The exact solution to (5) is $x_1 = -503$, $x_2 = 500$, and the exact solution to (6) is $x_1 = -403$, $x_2 = 400$. Thus, a rounding error of only .004 in one coefficient of (5) results in a gross error in the solution.

There is little that can be done computationally to avoid large errors in the solutions of ill-conditioned linear systems. However, in physical problems, where ill-conditioned systems arise, it is sometimes possible to reformulate the problem giving rise to the system to avoid ill-conditioning. Some of the texts referenced at the end of this chapter explain how to recognize ill-conditioned systems.

In this section we have explained how to reduce roundoff error by partial pivoting, but have not explained why the method works. Suffice it to say that when a quantity has some error in it resulting from roundoff, division of that quantity by a number near zero amplifies the roundoff error—the closer the divisor is to zero, the greater the amplification. Partial pivoting is designed to keep all divisors in the reduction process as far from zero as possible.

Finally, we note that there is a procedure, called ***complete pivoting***, that uses both row and column interchanges to reduce roundoff error. Although this method is more effective than partial pivoting in reducing roundoff error, it is so time consuming that its advantage is often lost. For most purposes, partial pivoting is adequate.

EXERCISE SET 9.4

1. Express the following in normalized floating point form.
   (a) $\frac{14}{5}$ (b) 3,452   (c) .000003879   (d) $-.135$ (e) 17.921   (f) $-.0863$

2. Round off the numbers in Exercise 1 to three significant digits.

3. Round off the numbers in Exercise 1 to two significant digits.

In Exercises 4–7 use Gaussian elimination with partial pivoting to solve the system exactly. Check your work by using Gaussian elimination without partial pivoting to solve the system.

4. $\begin{aligned} 3x_1 + x_2 &= -2 \\ -5x_1 + x_2 &= 22 \end{aligned}$

5. $\begin{aligned} x_1 + x_2 + x_3 &= 6 \\ 2x_1 - x_2 + 4x_3 &= 12 \\ -3x_1 + 2x_2 - x_3 &= -4 \end{aligned}$

6. $\begin{aligned} 2x_1 + 3x_2 - x_3 &= 5 \\ 4x_1 + 4x_2 - 3x_3 &= 3 \\ 2x_1 - 3x_2 + x_3 &= -1 \end{aligned}$

7. $\begin{aligned} 5x_1 + 6x_2 - x_3 + 2x_4 &= -3 \\ 2x_1 - x_2 + x_3 + x_4 &= 0 \\ -8x_1 + x_2 + 2x_3 - x_4 &= 3 \\ 5x_1 + 2x_2 + 3x_3 - x_4 &= 4 \end{aligned}$

In Exercises 8 and 9 solve the system by Gaussian elimination with partial pivoting. Round off all calculations to three significant digits.

8. $\begin{aligned} .21x_1 + .33x_2 &= .54 \\ .70x_1 + .24x_2 &= .94 \end{aligned}$

9. $\begin{aligned} .11x_1 - .13x_2 + .20x_3 &- = -.02 \\ .10x_1 + .36x_2 + .45x_3 &= .25 \\ .50x_1 - .01x_2 + .30x_3 &= -.70 \end{aligned}$

10. Solve
$$\begin{aligned} .0001x_1 + x_2 &= 1 \\ x_1 + x_2 &= 2 \end{aligned}$$

by Gaussian elimination with and without partial pivoting. Round off all computations to three significant digits. Compare the results with the exact solution.

9.5 APPROXIMATING EIGENVALUES BY THE POWER METHOD

*We learned earlier that the eigenvalues of a square matrix can be found by solving its characteristic equation. In practical problems, especially those involving large matrices, that procedure for finding eigenvalues has computational difficulties, so other methods for finding eigenvalues are used. In this section we shall study a simple algorithm, called the **power method**, that produces an approximation to the eigenvalue with greatest absolute value and a corresponding eigenvector. In the next section we will discuss methods for approximating the remaining eigenvalues and eigenvectors.*

**DOMINANT
EIGENVALUES
AND
EIGENVECTORS**

Definition. An eigenvalue of an $n \times n$ matrix A is called the **dominant eigenvalue** of A if its absolute value is larger than the absolute values of the remaining $n - 1$ eigenvalues. An eigenvector corresponding to the dominant eigenvalue is called a **dominant eigenvector** of A.

Example 1 If a 4×4 matrix A has eigenvalues

$$\lambda_1 = -4 \qquad \lambda_2 = 3 \qquad \lambda_3 = -2 \qquad \lambda_4 = 2$$

then $\lambda_1 = -4$ is the dominant eigenvalue since

$$|-4| > |3| \qquad |-4| > |-2| \qquad \text{and} \qquad |-4| > |2| \quad \blacktriangle$$

Example 2 A 3×3 matrix A with eigenvalues

$$\lambda_1 = 7 \qquad \lambda_2 = -7 \qquad \lambda_3 = 2$$

has no dominant eigenvalue. $\blacktriangle$

Let A be a diagonalizable $n \times n$ matrix with a dominant eigenvalue. We shall show at the end of this section that if $\mathbf{x}_0$ is an arbitrary nonzero vector in R^n, then the vector

$$A^p \mathbf{x}_0 \tag{1}$$

is usually a good approximation to a dominant eigenvector of A when the exponent p is large. The following example illustrates this idea.

Example 3 As shown in Example 2 of Section 6.1, the matrix

$$A = \begin{bmatrix} 3 & 2 \\ -1 & 0 \end{bmatrix}$$

has eigenvalues $\lambda_1 = 2$ and $\lambda_2 = 1$.

The eigenspace corresponding to the dominant eigenvalue $\lambda_1 = 2$ is the solution space of the system

$$(2I - A)\mathbf{x} = \mathbf{0}$$

That is,

$$\begin{bmatrix} -1 & -2 \\ 1 & 2 \end{bmatrix} \begin{bmatrix} x_1 \\ x_2 \end{bmatrix} = \begin{bmatrix} 0 \\ 0 \end{bmatrix}$$

Solving this system yields $x_1 = -2t$, $x_2 = t$. Thus, the eigenvectors corresponding to $\lambda_1 = 2$ are the nonzero vectors of the form

$$\mathbf{x} = \begin{bmatrix} -2t \\ t \end{bmatrix} \tag{2}$$

We now illustrate a procedure for using (1) to estimate a dominant eigenvector of A. To start, we arbitrarily select

$$\mathbf{x}_0 = \begin{bmatrix} 1 \\ 1 \end{bmatrix}$$

as an initial approximation to a dominant eigenvector. Repeatedly multiplying $\mathbf{x}_0$ by A yields

$$A\mathbf{x}_0 = \begin{bmatrix} 3 & 2 \\ -1 & 0 \end{bmatrix}\begin{bmatrix} 1 \\ 1 \end{bmatrix} = \begin{bmatrix} 5 \\ -1 \end{bmatrix}$$

$$A^2\mathbf{x}_0 = A(A\mathbf{x}_0) = \begin{bmatrix} 3 & 2 \\ -1 & 0 \end{bmatrix}\begin{bmatrix} 5 \\ -1 \end{bmatrix} = \begin{bmatrix} 13 \\ -5 \end{bmatrix} = 5\begin{bmatrix} 2.6 \\ -1 \end{bmatrix}$$

$$A^3\mathbf{x}_0 = A(A^2\mathbf{x}_0) = \begin{bmatrix} 3 & 2 \\ -1 & 0 \end{bmatrix}\begin{bmatrix} 13 \\ -5 \end{bmatrix} = \begin{bmatrix} 29 \\ -13 \end{bmatrix} \approx 13\begin{bmatrix} 2.23 \\ -1 \end{bmatrix}$$

$$A^4\mathbf{x}_0 = A(A^3\mathbf{x}_0) = \begin{bmatrix} 3 & 2 \\ -1 & 0 \end{bmatrix}\begin{bmatrix} 29 \\ -13 \end{bmatrix} = \begin{bmatrix} 61 \\ -29 \end{bmatrix} \approx 29\begin{bmatrix} 2.10 \\ -1 \end{bmatrix}$$

$$A^5\mathbf{x}_0 = A(A^4\mathbf{x}_0) = \begin{bmatrix} 3 & 2 \\ -1 & 0 \end{bmatrix}\begin{bmatrix} 61 \\ -29 \end{bmatrix} = \begin{bmatrix} 125 \\ -61 \end{bmatrix} \approx 61\begin{bmatrix} 2.05 \\ -1 \end{bmatrix}$$

$$A^6\mathbf{x}_0 = A(A^5\mathbf{x}_0) = \begin{bmatrix} 3 & 2 \\ -1 & 0 \end{bmatrix}\begin{bmatrix} 125 \\ -61 \end{bmatrix} = \begin{bmatrix} 253 \\ -125 \end{bmatrix} \approx 125\begin{bmatrix} 2.02 \\ -1 \end{bmatrix}$$

$$A^7\mathbf{x}_0 = A(A^6\mathbf{x}_0) = \begin{bmatrix} 3 & 2 \\ -1 & 0 \end{bmatrix}\begin{bmatrix} 253 \\ -125 \end{bmatrix} = \begin{bmatrix} 509 \\ -253 \end{bmatrix} \approx 253\begin{bmatrix} 2.01 \\ -1 \end{bmatrix}$$

It is evident from these calculations that the products are getting closer and closer to scalar multiples of

$$\begin{bmatrix} 2 \\ -1 \end{bmatrix}$$

which is the dominant eigenvector of A obtained by letting $t = -1$ in (2). Since a scalar multiple of a dominant eigenvector is also a dominant eigenvector, the above calculations are producing better and better approximations to a dominant eigenvector of A. ▲

We now show how to approximate the dominant eigenvalue once an approximation to a dominant eigenvector is known. Let λ be an eigenvalue of A and $\mathbf{x}$ a corresponding eigenvector. If $\langle \ , \ \rangle$ denotes the Euclidean inner product, then

$$\frac{\langle \mathbf{x}, A\mathbf{x} \rangle}{\langle \mathbf{x}, \mathbf{x} \rangle} = \frac{\langle \mathbf{x}, \lambda\mathbf{x} \rangle}{\langle \mathbf{x}, \mathbf{x} \rangle} = \frac{\lambda\langle \mathbf{x}, \mathbf{x} \rangle}{\langle \mathbf{x}, \mathbf{x} \rangle} = \lambda$$

Thus, if $\tilde{\mathbf{x}}$ is an approximation to a dominant eigenvector, the dominant eigenvalue λ_1 can be approximated by

$$\lambda_1 \approx \frac{\langle \tilde{\mathbf{x}}, A\tilde{\mathbf{x}} \rangle}{\langle \tilde{\mathbf{x}}, \tilde{\mathbf{x}} \rangle} \tag{3}$$

The ratio in (3) is called the **_Rayleigh quotient_**.*

---

* *John William Strutt Rayleigh* (1842–1919) was a British physicist awarded the Nobel prize in physics in 1904 for his part in the discovery of argon in 1894. His research ranged over almost the entire field of physics, including sound, wave theory, optics, color vision, electrodynamics, electromagnetism, scattering of light, viscosity, and photography.

Example 4   In Example 3 we obtained

$$\tilde{\mathbf{x}} = \begin{bmatrix} 509 \\ -253 \end{bmatrix}$$

as an approximation to a dominant eigenvector corresponding to the dominant eigenvalue $\lambda_1 = 2$. Substituting,

$$A\tilde{\mathbf{x}} = \begin{bmatrix} 3 & 2 \\ -1 & 0 \end{bmatrix} \begin{bmatrix} 509 \\ -253 \end{bmatrix} = \begin{bmatrix} 1021 \\ -509 \end{bmatrix}$$

in (3) we obtain

$$\lambda_1 \approx \frac{\langle \tilde{\mathbf{x}}, A\tilde{\mathbf{x}} \rangle}{\langle \tilde{\mathbf{x}}, \tilde{\mathbf{x}} \rangle} = \frac{(509)(1021) + (-253)(-509)}{(509)(509) + (-253)(-253)} \approx 2.007$$

which is a relatively good approximation to the dominant eigenvalue $\lambda_1 = 2$. ▲

THE POWER METHOD

The technique illustrated in Examples 3 and 4 for approximating dominant eigenvalues and eigenvectors is often called the *power method* or *iteration method*. As illustrated in Example 3, the power method often generates vectors that have inconveniently large components. To remedy this problem it is usual to "scale down" the approximate eigenvector at each step so its components lie between $+1$ and -1. This can be achieved by multiplying the approximate eigenvector by the reciprocal of the component having largest absolute value. To illustrate, in the first step of Example 3, the approximation to a dominant eigenvector is

$$\begin{bmatrix} 5 \\ -1 \end{bmatrix}$$

The component with largest absolute value is 5; thus, the scaled-down eigenvector is

$$\frac{1}{5} \begin{bmatrix} 5 \\ -1 \end{bmatrix} = \begin{bmatrix} 1 \\ -.2 \end{bmatrix}$$

We now summarize the steps in the power method with scaling.

Step 0.   Pick an arbitrary nonzero vector $\mathbf{x}_0$.

Step 1.   Compute $A\mathbf{x}_0$ and scale down to obtain the first approximation to a dominant eigenvector. Call it $\mathbf{x}_1$.

Step 2.   Compute $A\mathbf{x}_1$ and scale down to obtain the second approximation, $\mathbf{x}_2$.

Step 3.   Compute $A\mathbf{x}_2$ and scale down to obtain the third approximation, $\mathbf{x}_3$.

Continuing in this way, a succession, $\mathbf{x}_0, \mathbf{x}_1, \mathbf{x}_2, \ldots$, of (generally) better and better approximations to a dominant eigenvector will be obtained.

Example 5   Use the power method with scaling to approximate a dominant eigenvector and the dominant eigenvalue of the matrix A in Example 3.

Solution.   We arbitrarily select

$$\mathbf{x}_0 = \begin{bmatrix} 1 \\ 1 \end{bmatrix}$$

as an initial approximation. Multiplying $\mathbf{x}_0$ by A and scaling down yields

$$A\mathbf{x}_0 = \begin{bmatrix} 3 & 2 \\ -1 & 0 \end{bmatrix}\begin{bmatrix} 1 \\ 1 \end{bmatrix} = \begin{bmatrix} 5 \\ -1 \end{bmatrix} \qquad \mathbf{x}_1 = \frac{1}{5}\begin{bmatrix} 5 \\ -1 \end{bmatrix} = \begin{bmatrix} 1 \\ -.2 \end{bmatrix}$$

Multiplying $\mathbf{x}_1$ by A and scaling down yields

$$A\mathbf{x}_1 = \begin{bmatrix} 3 & 2 \\ -1 & 0 \end{bmatrix}\begin{bmatrix} 1 \\ -.2 \end{bmatrix} = \begin{bmatrix} 2.6 \\ -1 \end{bmatrix} \qquad \mathbf{x}_2 = \frac{1}{2.6}\begin{bmatrix} 2.6 \\ -1 \end{bmatrix} = \begin{bmatrix} 1 \\ -.385 \end{bmatrix}$$

From the Rayleigh quotient the first estimate of the dominant eigenvalue is

$$\lambda_1 \approx \frac{\langle \mathbf{x}_1, A\mathbf{x}_1 \rangle}{\langle \mathbf{x}_1, \mathbf{x}_1 \rangle} = \frac{(1)(2.6) + (-.2)(-1)}{(1)(1) + (-.2)(-.2)} = 2.692$$

Multiplying $\mathbf{x}_2$ by A and scaling down yields

$$A\mathbf{x}_2 = \begin{bmatrix} 3 & 2 \\ -1 & 0 \end{bmatrix}\begin{bmatrix} 1 \\ -.385 \end{bmatrix} = \begin{bmatrix} 2.23 \\ -1 \end{bmatrix} \qquad \mathbf{x}_3 = \frac{1}{2.23}\begin{bmatrix} 2.23 \\ -1 \end{bmatrix} = \begin{bmatrix} 1 \\ -.448 \end{bmatrix}$$

From the Rayleigh quotient the second estimate of the dominant eigenvalue is

$$\lambda_1 \approx \frac{\langle \mathbf{x}_2, A\mathbf{x}_2 \rangle}{\langle \mathbf{x}_2, \mathbf{x}_2 \rangle} = \frac{(1)(2.23) + (-.385)(-1)}{(1)(1) + (-.385)(-.385)} = 2.278$$

Multiplying $\mathbf{x}_3$ by A and scaling down gives

$$A\mathbf{x}_3 = \begin{bmatrix} 3 & 2 \\ -1 & 0 \end{bmatrix}\begin{bmatrix} 1 \\ -.448 \end{bmatrix} = \begin{bmatrix} 2.104 \\ -1 \end{bmatrix} \qquad \mathbf{x}_4 = \frac{1}{2.104}\begin{bmatrix} 2.104 \\ -1 \end{bmatrix} = \begin{bmatrix} 1 \\ -.475 \end{bmatrix}$$

The third estimate of the dominant eigenvalue is

$$\lambda_1 \approx \frac{\langle \mathbf{x}_3, A\mathbf{x}_3 \rangle}{\langle \mathbf{x}_3, \mathbf{x}_3 \rangle} = \frac{(1)(2.104) + (-.448)(-1)}{(1)(1) + (-.448)(-.448)} = 2.125$$

Continuing in this way, we generate a succession of approximations to a dominant eigenvector and the dominant eigenvalue. These values, together with results of further estimates, are tabulated in Table 1.

TABLE 1

Step i	0	1	2	3	4	5	6	7
$x_i =$ the scaled down approximation to a dominant eigenvector	$\begin{bmatrix} 1 \\ 1 \end{bmatrix}$	$\begin{bmatrix} 1 \\ -.2 \end{bmatrix}$	$\begin{bmatrix} 1 \\ -.385 \end{bmatrix}$	$\begin{bmatrix} 1 \\ -.488 \end{bmatrix}$	$\begin{bmatrix} 1 \\ -.475 \end{bmatrix}$	$\begin{bmatrix} 1 \\ -.488 \end{bmatrix}$	$\begin{bmatrix} 1 \\ -.494 \end{bmatrix}$	$\begin{bmatrix} 1 \\ -.497 \end{bmatrix}$
Ax_i	$\begin{bmatrix} 5 \\ -1 \end{bmatrix}$	$\begin{bmatrix} 2.6 \\ -1 \end{bmatrix}$	$\begin{bmatrix} 2.23 \\ -1 \end{bmatrix}$	$\begin{bmatrix} 2.104 \\ -1 \end{bmatrix}$	$\begin{bmatrix} 2.050 \\ -1 \end{bmatrix}$	$\begin{bmatrix} 2.024 \\ -1 \end{bmatrix}$	$\begin{bmatrix} 2.012 \\ -1 \end{bmatrix}$	—
Approximation to λ_1	—	2.692	2.278	2.125	2.060	2.029	2.014	—

ERROR CONSIDERATIONS

There are no hard and fast rules for determining how many steps to use in the power method. We shall consider one possible procedure that is widely used. If $\tilde{q}$ denotes an approximation to a quantity q, then the **relative error** in the approximation is defined to be

$$\left| \frac{q - \tilde{q}}{q} \right| \tag{4}$$

and the **percentage error** in the approximation is defined to be

$$\left| \frac{q - \tilde{q}}{q} \right| \times 100\%$$

Example 6 If the exact value of a certain eigenvalue is $\lambda = 5$, and if $\tilde{\lambda} = 5.1$ is an approximation to λ, then the relative error is

$$\left| \frac{\lambda - \tilde{\lambda}}{\lambda} \right| = \left| \frac{5 - 5.1}{5} \right| = |-.02| = .02$$

and the percentage error is

$$(.02) \times 100\% = 2\% \quad \blacktriangle$$

In the power method one would ideally like to decide in advance the relative error E that can be tolerated in the eigenvalue, and then stop the computations once the relative error is less than E. Thus, if $\tilde{\lambda}(i)$ denotes the approximation to the dominant eigenvalue λ_1 at the ith step, the computations would be stopped once the condition

$$\left| \frac{\lambda_1 - \tilde{\lambda}(i)}{\lambda_1} \right| < E$$

is satisfied. Unfortunately, it is not possible to carry out this idea since the exact value λ_1 is unknown. To remedy this, it is usual to estimate λ_1 by $\tilde{\lambda}(i)$ and stop the computations at the ith step if

$$\left|\frac{\tilde{\lambda}(i) - \tilde{\lambda}(i-1)}{\tilde{\lambda}(i)}\right| < E \tag{5}$$

The quantity on the left side of (5) is called the **estimated relative error**. When multiplied by 100%, it is called the **estimated percentage error**.

Example 7 In Example 5, how many steps should be used to assure that the estimated percentage error in the dominant eigenvalue is less than 2%?

Solution. Let $\tilde{\lambda}(i)$ denote the approximation to the dominant eigenvalue at the ith step. Thus, from Table 1

$$\tilde{\lambda}(1) = 2.692, \qquad \tilde{\lambda}(2) = 2.278, \qquad \tilde{\lambda}(3) = 2.125, \quad \text{etc.}$$

From (5) the estimated relative error after two steps is

$$\left|\frac{\tilde{\lambda}(2) - \tilde{\lambda}(1)}{\tilde{\lambda}(2)}\right| = \left|\frac{2.278 - 2.692}{2.278}\right| \approx |-.182| = .182$$

so the estimated percentage error after two steps is 18.2%. the estimated relative error after three steps is

$$\left|\frac{\tilde{\lambda}(3) - \tilde{\lambda}(2)}{\tilde{\lambda}(3)}\right| = \left|\frac{2.125 - 2.278}{2.125}\right| \approx |-.072| = .072$$

and the estimated percentage error is 7.2%. The remaining percentage errors are listed in Table 2. From this table we see that the estimated percentage error is less than 2% at the end of the fifth step.

TABLE 2

i = step number	2	3	4	5	6
$\tilde{\lambda}(i)$	2.278	2.125	2.060	2.029	2.014
Estimated relative error after i steps	.182	.072	.032	.015	.007
Estimated percentage error after i steps	18.2%	7.2%	3.2%	1.5%	.7%

OPTIONAL

We conclude this section with a proof that the power method works when A is a diagonalizable matrix with a dominant eigenvalue.

Let A be a diagonalizable $n \times n$ matrix. By Theorem 6.2.1, A has n linearly independent eigenvectors $\mathbf{v}_1, \mathbf{v}_2, \ldots, \mathbf{v}_n$. Let $\lambda_1, \lambda_2, \ldots, \lambda_n$ be the corresponding eigenvalues, and assume that

$$|\lambda_1| > |\lambda_2| \geq \cdots \geq |\lambda_n| \tag{6}$$

By Theorem 4.5.4 the eigenvectors $\mathbf{v}_1, \mathbf{v}_2, \ldots, \mathbf{v}_n$ form a basis for R^n; thus, an arbitrary vector $\mathbf{x}_0$ in R^n can be expressed in the form

$$\mathbf{x}_0 = k_1\mathbf{v}_1 + k_2\mathbf{v}_2 + \cdots + k_n\mathbf{v}_n \tag{7}$$

Multiplying both sides on the left by A gives

$$\begin{aligned} A\mathbf{x}_0 &= A(k_1\mathbf{v}_1 + k_2\mathbf{v}_2 + \cdots + k_n\mathbf{v}_n) \\ &= k_1(A\mathbf{v}_1) + k_2(A\mathbf{v}_2) + \cdots + k_n(A\mathbf{v}_n) \\ &= k_1\lambda_1\mathbf{v}_1 + k_2\lambda_2\mathbf{v}_2 + \cdots + k_n\lambda_n\mathbf{v}_n \end{aligned}$$

Multiplying by A again gives

$$\begin{aligned} A^2\mathbf{x}_0 &= A(k_1\lambda_1\mathbf{v}_1 + k_2\lambda_2\mathbf{v}_2 + \cdots + k_n\lambda_n\mathbf{v}_n) \\ &= k_1\lambda_1(A\mathbf{v}_1) + k_2\lambda_2(A\mathbf{v}_2) + \cdots + k_n\lambda_n(A\mathbf{v}_n) \\ &= k_1\lambda_1^2\mathbf{v}_1 + k_2\lambda_2^2\mathbf{v}_2 + \cdots + k_n\lambda_n^2\mathbf{v}_n \end{aligned}$$

Continuing, we would obtain after p multiplications by A

$$A^p\mathbf{x}_0 = k_1\lambda_1^p\mathbf{v}_1 + k_2\lambda_2^p\mathbf{v}_2 + \cdots + k_n\lambda_n^p\mathbf{v}_n \tag{8}$$

Since $\lambda_1 \neq 0$ [see (6)], (8) can be rewritten as

$$A^p\mathbf{x}_0 = \lambda_1^p\left(k_1\mathbf{v}_1 + k_2\left(\frac{\lambda_2}{\lambda_1}\right)^p\mathbf{v}_2 + \cdots + k_n\left(\frac{\lambda_n}{\lambda_1}\right)^p\mathbf{v}_n\right) \tag{9}$$

It follows from (6) that

$$\frac{\lambda_2}{\lambda_1}, \ldots, \frac{\lambda_n}{\lambda_1}$$

are all less than one in absolute value; thus, $(\lambda_2/\lambda_1)^p, \ldots, (\lambda_n/\lambda_1)^p$ get steadily closer to zero as p increases, and from (9), the approximation

$$A^p\mathbf{x}_0 \approx \lambda_1^p k_1\mathbf{v}_1 \tag{10}$$

gets better and better.

If $k_1 \neq 0$,* then $\lambda_1^p k_1\mathbf{v}_1$ is a nonzero scalar multiple of the dominant eigenvector $\mathbf{v}_1$; thus, $\lambda_1^p k_1\mathbf{v}_1$ is also a dominant eigenvector. Therefore, by (10), $A^p\mathbf{x}_0$ becomes a better and better estimate of a dominant eigenvector as p is increased. ▌

---

* One cannot usually tell by inspection of the $\mathbf{x}_0$ selected whether $k_1 \neq 0$. If by accident $k_1 = 0$, the power method still works in practical problems, since computer roundoff errors generally build up to make k_1 small but nonzero. This is one instance where errors help to obtain correct results!

EXERCISE SET 9.5

1. Find the dominant eigenvalue (if it exists).

(a) $\begin{bmatrix} -1 & 4 \\ 1 & -1 \end{bmatrix}$
(b) $\begin{bmatrix} 0 & 1 \\ 4 & 0 \end{bmatrix}$
(c) $\begin{bmatrix} 4 & 2 & 1 \\ 0 & -5 & 3 \\ 0 & 0 & 6 \end{bmatrix}$
(d) $\begin{bmatrix} 1 & -12 & 0 \\ 1 & 0 & 0 \\ 0 & 0 & 3 \end{bmatrix}$

2. Let

$$A = \begin{bmatrix} 3 & 4 \\ 1 & 3 \end{bmatrix}$$

(a) Use the power method with scaling to approximate a dominant eigenvector of A. Start with

$$\mathbf{x}_0 = \begin{bmatrix} 1 \\ 1 \end{bmatrix}$$

Round off all computations to three significant digits, and stop after three iterations (that is, three multiplications by A).
(b) Use the result of part (a) and the Rayleigh quotient to approximate the dominant eigenvalue of A.
(c) Find the exact values of the dominant eigenvector and eigenvalue.
(d) Find the percentage error in the approximation of the dominant eigenvalue.

Exercises 3 and 4 follow the directions given in Exercise 2.

3. $A = \begin{bmatrix} 5 & 4 \\ 3 & 4 \end{bmatrix}$ 4. $A = \begin{bmatrix} 3 & 2 \\ 2 & 0 \end{bmatrix}$

5. Let

$$A = \begin{bmatrix} 18 & 17 \\ 2 & 3 \end{bmatrix}$$

(a) Use the power method with scaling to approximate the dominant eigenvalue and a dominant eigenvector of A. Start with

$$\mathbf{x}_0 = \begin{bmatrix} 1 \\ 1 \end{bmatrix}$$

Round off all computations to three significant digits, and stop when the estimated percentage error in the dominant eigenvalue is less than 2%.
(b) Find the exact values of the dominant eigenvalue and eigenvector.

6. Repeat the directions of Exercise 5 with

$$A = \begin{bmatrix} -5 & 5 \\ 6 & -4 \end{bmatrix}$$

7. Let

$$A = \begin{bmatrix} 2 & 1 & 0 \\ 1 & 2 & 0 \\ 0 & 0 & 10 \end{bmatrix}$$

(a) Use the power method with scaling to approximate a dominant eigenvector of A. Start with

$$\mathbf{x}_0 = \begin{bmatrix} 1 \\ 1 \\ 1 \end{bmatrix}$$

Round off all computations to three significant digits, and stop after three iterations.
(b) Use the result of part (a) and the Rayleigh quotient to approximate the dominant eigenvalue of A.
(c) Find the exact values for the dominant eigenvalue and eigenvector.
(d) Find the percentage error in the approximation of the dominant eigenvalue.

9.6 APPROXIMATING NONDOMINANT EIGENVALUES; DEFLATION AND INVERSE POWER METHODS

In many applications only the dominant eigenvalue and eigenvector of a matrix are needed, in which case the power method alone can be tried. However, if additional eigenvalues and eigenvectors are needed, then other methods are required. In this section we shall briefly discuss some ways in which the power method can be adapted to find nondominant eigenvalues and eigenvectors.

DEFLATION   We shall need the following theorem, which is proved in the exercises (Exercise 6).

Theorem 9.6.1. *Let A be a symmetric $n \times n$ matrix with eigenvalues $\lambda_1, \lambda_2, \ldots, \lambda_n$. If $\mathbf{v}_1$ is an eigenvector corresponding to λ_1, and $\|\mathbf{v}_1\| = 1$, then:*

(a) *The matrix $B = A - \lambda_1 \mathbf{v}_1 \mathbf{v}_1^t$ has eigenvalues $0, \lambda_2, \ldots, \lambda_n$.*

(b) *If $\mathbf{v}$ is an eigenvector of B corresponding to a nonzero eigenvalue in the set $\{\lambda_2, \ldots, \lambda_n\}$, then $\mathbf{v}$ is also an eigenvector of A corresponding to this eigenvalue.*

REMARK. In this theorem the eigenvalues $\lambda_1, \lambda_2, \ldots, \lambda_n$ need not be distinct. Also, we are assuming that $\mathbf{v}_1$ is expressed as an $n \times 1$ matrix, so $\mathbf{v}_1 \mathbf{v}_1^t$ is an $n \times n$ matrix.

Example 1   We leave it for you to show that

$$A = \begin{bmatrix} 3 & -2 & 0 \\ -2 & 3 & 0 \\ 0 & 0 & 5 \end{bmatrix}$$

has eigenvalues $\lambda_1 = 5$, $\lambda_2 = 5$, $\lambda_3 = 1$, and

$$\mathbf{v} = \begin{bmatrix} -1 \\ 1 \\ 0 \end{bmatrix}$$

is an eigenvector corresponding to $\lambda_1 = 5$. Normalizing $\mathbf{v}$ yields

$$\mathbf{v}_1 = \frac{1}{\sqrt{2}} \begin{bmatrix} -1 \\ 1 \\ 0 \end{bmatrix} = \begin{bmatrix} -\dfrac{1}{\sqrt{2}} \\ \dfrac{1}{\sqrt{2}} \\ 0 \end{bmatrix}$$

which is an eigenvector of norm 1 corresponding to $\lambda_1 = 5$

By Theorem 9.6.1 the matrix

$$B = A - \lambda_1 \mathbf{v}_1 \mathbf{v}_1^t = \begin{bmatrix} 3 & -2 & 0 \\ -2 & 3 & 0 \\ 0 & 0 & 5 \end{bmatrix} - 5 \begin{bmatrix} -\dfrac{1}{\sqrt{2}} \\ \dfrac{1}{\sqrt{2}} \\ 0 \end{bmatrix} \begin{bmatrix} -\dfrac{1}{\sqrt{2}} & \dfrac{1}{\sqrt{2}} & 0 \end{bmatrix}$$

$$= \begin{bmatrix} 3 & -2 & 0 \\ -2 & 3 & 0 \\ 0 & 0 & 5 \end{bmatrix} - 5 \begin{bmatrix} \frac{1}{2} & -\frac{1}{2} & 0 \\ -\frac{1}{2} & \frac{1}{2} & 0 \\ 0 & 0 & 0 \end{bmatrix} = \begin{bmatrix} \frac{1}{2} & \frac{1}{2} & 0 \\ \frac{1}{2} & \frac{1}{2} & 0 \\ 0 & 0 & 5 \end{bmatrix}$$

should have eigenvalues $\lambda = 0$, 5, and 1. As a check, the characteristic equation of B is

$$\det(\lambda I - B) = \det \begin{bmatrix} \lambda - \frac{1}{2} & -\frac{1}{2} & 0 \\ -\frac{1}{2} & \lambda - \frac{1}{2} & 0 \\ 0 & 0 & \lambda - 5 \end{bmatrix} = \lambda(\lambda - 5)(\lambda - 1) = 0$$

Hence, the eigenvalues of B are $\lambda = 0$, $\lambda = 5$, $\lambda = 1$ as predicted by Theorem 9.6.1. The eigenspace of B corresponding to $\lambda = 5$ is the solution space of the system

$$(5I - B)\mathbf{x} = \mathbf{0}$$

That is,

$$\begin{bmatrix} \frac{9}{2} & -\frac{1}{2} & 0 \\ -\frac{1}{2} & \frac{9}{2} & 0 \\ 0 & 0 & 0 \end{bmatrix} \begin{bmatrix} x_1 \\ x_2 \\ x_3 \end{bmatrix} = \begin{bmatrix} 0 \\ 0 \\ 0 \end{bmatrix}$$

Solving this system yields $x_1 = 0$, $x_2 = 0$, $x_3 = t$. Thus, the eigenvectors of B corresponding to $\lambda = 5$ are the nonzero vectors of the form

$$\mathbf{x} = \begin{bmatrix} 0 \\ 0 \\ t \end{bmatrix}$$

As predicted by part (*b*) of Theorem 9.6.1, these are also eigenvectors of A corresponding to $\lambda = 5$, since

$$A \begin{bmatrix} 0 \\ 0 \\ t \end{bmatrix} = \begin{bmatrix} 3 & -2 & 0 \\ -2 & 3 & 0 \\ 0 & 0 & 5 \end{bmatrix} \begin{bmatrix} 0 \\ 0 \\ t \end{bmatrix} = \begin{bmatrix} 0 \\ 0 \\ 5t \end{bmatrix}$$

That is,

$$A \begin{bmatrix} 0 \\ 0 \\ t \end{bmatrix} = 5 \begin{bmatrix} 0 \\ 0 \\ t \end{bmatrix}$$

Similarly, the eigenvectors of B corresponding to $\lambda = 1$ are also eigenvectors of A corresponding to $\lambda = 1$. ▲

Theorem 9.6.1, to a limited extent, makes it possible to determine the nondominant eigenvalues and eigenvectors of a *symmetric* $n \times n$ matrix A. To see how, assume the eigenvalues of A can be ordered according to the size of their absolute values as follows.

$$|\lambda_1| > |\lambda_2| > |\lambda_3| \geq \cdots \geq |\lambda_n| \tag{1}$$

Suppose the dominant eigenvalue and a dominant eigenvector of A have been obtained by the power method. By normalizing the dominant eigenvector, we can obtain a dominant eigenvector $\mathbf{v}_1$ having norm one. By Theorem 9.6.1 the eigenvalues of $B = A - \lambda_1 \mathbf{v}_1 \mathbf{v}_1'$ will be $0, \lambda_2, \lambda_3, \ldots, \lambda_n$. From (1) these eigenvalues will be ordered according to their absolute values as follows:

$$|\lambda_2| > |\lambda_3| \geq \cdots \geq |\lambda_n| \geq 0$$

Thus, λ_2 is the dominant eigenvalue of B. Now, by applying the power method to B, we can approximate the eigenvalue λ_2 and a corresponding eigenvector. This technique for approximating the eigenvalue with the second largest absolute value is called **deflation**.

Unfortunately, there are practical limitations to the deflation method. Since λ_1 and $\mathbf{v}_1$ are only approximated in the power method, an error is introduced into B when deflation is used. If the deflation process is applied again, the next matrix has additional errors introduced through the approximation of λ_2 and $\mathbf{v}_2$. As the process continues, this compounding of errors can affect the accuracy of the results so seriously that the approximations to the eigenvalues with smaller absolute values are worthless. In practice, one should avoid finding more than two or three eigenvalues by deflation.

**THE INVERSE
POWER METHOD**

There is a way of using the power method to approximate the eigenvalue of smallest absolute value when the matrix is invertible. It uses the fact that inverting a matrix takes reciprocals of the eigenvalues but leaves the corresponding eigenvectors unchanged. More precisely, if

$$\lambda_1, \lambda_2, \ldots, \lambda_n$$

are the eigenvalues of an invertible matrix A, then

$$\frac{1}{\lambda_1}, \frac{1}{\lambda_2}, \ldots, \frac{1}{\lambda_n}$$

are the eigenvalues of A^{-1}. Moreover, $\mathbf{x}$ is an eigenvector of A corresponding to λ if and only if $\mathbf{x}$ is an eigenvector of A^{-1} corresponding to $1/\lambda$ (see Exercise 5). It follows from this result that if the eigenvalues of A can be ordered according to the size of their absolute values as follows,

$$|\lambda_1| \geq |\lambda_2| \geq \cdots \geq |\lambda_{n-1}| > |\lambda_n|$$

then $1/\lambda_n$ will be the dominant eigenvalue of A^{-1}, which can be approximated by the power method. Once obtained, the reciprocal of the dominant eigenvalue of A^{-1} can be taken to find the eigenvalue of A with smallest absolute value. This procedure is called the ***inverse power method***.

Example 2   In Example 3 of Section 9.5 we noted that the eigenvalues of

$$A = \begin{bmatrix} 3 & 2 \\ -1 & 0 \end{bmatrix}$$

are $\lambda_1 = 2$ and $\lambda_2 = 1$, and we used the power method to approximate the dominant eigenvalue ($\lambda_1 = 2$) and a corresponding eigenvector. Now we shall use the inverse power method to approximate the eigenvalue with smallest absolute value ($\lambda_2 = 1$) and a corresponding eigenvector.

If we start with the initial vector

$$\mathbf{x}_0 = \begin{bmatrix} 1 \\ 1 \end{bmatrix}$$

and multiply repeatedly by

$$A^{-1} = \begin{bmatrix} 0 & -1 \\ \frac{1}{2} & \frac{3}{2} \end{bmatrix} = \begin{bmatrix} 0 & -1 \\ .5 & 1.5 \end{bmatrix}$$

the power method yields the results in Table 1. Thus, after six iterations with the inverse power method, the approximations to λ_2 and a corresponding eigenvector are

$$\lambda_2 \approx .999, \qquad \mathbf{x} \approx \begin{bmatrix} -.989 \\ 1 \end{bmatrix}$$

The exact values are

$$\lambda_2 = 1, \qquad \mathbf{x} = \begin{bmatrix} -1 \\ 1 \end{bmatrix} \ \blacktriangle$$

When the ratio $|\lambda_2/\lambda_1|$ is close to one, the power method has a slow rate of convergence; that is, many steps are needed to obtain a reasonable degree of accuracy. Readers interested in studying techniques for "speeding up" this rate of convergence and learning more about the numerical methods of linear algebra may want to consult the references that follow the exercise set.

TABLE 1

Step i	0	1	2	3	4	5	6
x_i = the scaled-down approximation to a dominant eigenvector of A^{-1}	$\begin{bmatrix} 1 \\ 1 \end{bmatrix}$	$\begin{bmatrix} -.5 \\ 1 \end{bmatrix}$	$\begin{bmatrix} -.8 \\ 1 \end{bmatrix}$	$\begin{bmatrix} -.909 \\ 1 \end{bmatrix}$	$\begin{bmatrix} -.956 \\ 1 \end{bmatrix}$	$\begin{bmatrix} -.978 \\ 1 \end{bmatrix}$	$\begin{bmatrix} -.989 \\ 1 \end{bmatrix}$
$A^{-1}x_i$		$\begin{bmatrix} -1 \\ 2 \end{bmatrix}$	$\begin{bmatrix} -1 \\ 1.25 \end{bmatrix}$	$\begin{bmatrix} -1 \\ 1.1 \end{bmatrix}$	$\begin{bmatrix} -1 \\ 1.046 \end{bmatrix}$	$\begin{bmatrix} -1 \\ 1.022 \end{bmatrix}$	$\begin{bmatrix} -1 \\ 1.011 \end{bmatrix}$	$\begin{bmatrix} -1 \\ 1.006 \end{bmatrix}$
Approximation to $1/\lambda_2$	—	1.4	1.159	1.070	1.033	1.016	1.001
Approximation to λ_2	—	.714	.863	.935	.968	.984	.999

EXERCISE SET 9.6

1. Let

$$A = \begin{bmatrix} 6 & 2 \\ 2 & 3 \end{bmatrix}$$

(a) Use the power method to approximate a dominant eigenvector. Start with

$$x_0 = \begin{bmatrix} 1 \\ 1 \end{bmatrix}$$

Round off all computations to three significant digits, and stop after three iterations (three multiplications by A).

(b) Use the result of part (a) and the Rayleigh quotient to approximate the dominant eigenvalue of A.

(c) Use deflation to approximate the remaining eigenvalue and a corresponding eigenvector; that is, apply the power method to

$$B = A - \tilde{\lambda}_1 \tilde{v}_1 \tilde{v}_1{}^t$$

where $\tilde{v}_1$ and $\tilde{\lambda}_1$ are the approximations obtained in parts (a) and (b). Start with

$$x_0 = \begin{bmatrix} 1 \\ 1 \end{bmatrix}$$

Round off all computations to three significant digits and stop after three iterations.

(d) Find the exact values of the eigenvalues and eigenvectors.

2. Follow the directions given in Exercise 1 with

$$A = \begin{bmatrix} 10 & 4 \\ 4 & 4 \end{bmatrix}$$

3. Use the power method and the inverse power method to approximate both eigenvalues and corresponding eigenvectors of

$$A = \begin{bmatrix} -9 & 11 \\ -1 & 3 \end{bmatrix}$$

Start with

$$\mathbf{x}_0 = \begin{bmatrix} 1 \\ 0 \end{bmatrix}$$

Round off all computations to three significant digits and stop after five iterations.

4. Follow the directions of Exercise 3 with

$$A = \begin{bmatrix} -3 & 2 \\ 2 & -3 \end{bmatrix}$$

5. Let A be an invertible matrix.
   (a) Prove that $1/\lambda$ is an eigenvalue of A^{-1} if and only if λ is an eigenvalue of A.
   (b) Prove that $\mathbf{x}$ is an eigenvector of A corresponding to the eigenvalue λ if and only if $\mathbf{x}$ is an eigenvector of A^{-1} corresponding to $1/\lambda$.

6. In this exercise we will help the reader to prove Theorem 9.6.1. Observe first that since A is a symmetric $n \times n$ matrix, there exists an orthonormal set of eigenvectors $\mathbf{v}_1, \mathbf{v}_2, \ldots, \mathbf{v}_n$ corresponding to the eigenvalues $\lambda_1, \lambda_2, \ldots, \lambda_n$, respectively (by Theorem 6.3.1).
   (a) Prove that $B\mathbf{v}_1 = \mathbf{0}$ and $B\mathbf{v}_i = \lambda_i \mathbf{v}_i$ for $i = 2, \ldots, n$. This shows that $0, \lambda_2, \ldots, \lambda_n$ are eigenvalues of B, which proves part (a) of Theorem 9.6.1.
   (b) Prove that B is a symmetric matrix.
   (c) Let $\mathbf{v}$ be an eigenvector of B corresponding to a nonzero eigenvalue λ in the set $\{\lambda_1, \lambda_2, \ldots, \lambda_n\}$. As shown in part (a) above, the eigenvector $\mathbf{v}_1$ corresponds to the zero eigenvalue of B, so $\mathbf{v}_1$ and $\mathbf{v}$ lie in distinct eigenspaces of B. Use this fact and part (b) of this exercise to prove that $\mathbf{v}$ is an eigenvector of A corresponding to λ. This proves part (b) of Theorem 9.6.1.

REFERENCES

Faddeev, D. K., and Faddeeva, V. N. (1960). *Computational Methods of Linear Algebra* (Russian ed.), Freeman, San Francisco.

Forsythe, G., and Moler, C. B. (1967). *Computer Solution of Linear Algebraic Systems*, Prentice-Hall, Englewood Cliffs, New Jersey.

Golub, G. H., and Van Loan, C. F. (1983). *Matrix Computations*, Johns Hopkins, Baltimore.

Householder, A. S. (1964). *The Theory of Matrices in Numerical Analysis*, Ginn (Blaisdell), Boston.

ISSACSON, E., and KELLER, H. B. (1966). *Analysis of Numerical Methods*, Wiley, New York.

MARON, M. (1982). *Numerical Analysis: A Practical Approach*, Macmillan, New York.

RICE, J. R. (1981). *Matrix Computations and Mathematical Software*, McGraw-Hill, New York.

STEWART, G. W. (1973). *Introduction to Matrix Computations*, Academic Press, New York.

VARGA, R. S. (1962). *Matrix Iterative Analysis*, Prentice-Hall, Englewood Cliffs, New Jersey.

WILKINSON, J. H. (1963). *Rounding Errors in Algebraic Processes*, Prentice-Hall, Englewood Cliffs, New Jersey.

WILKINSON, J. M., and REINSCH, C. (eds.) (1971). *Handbook for Automatic Computation, Vol. II, Linear Algebra*, Springer-Verlag, New York.

YOUNG, D. M. (1971). *Iterative Solution of Linear Systems*, Academic Press, New York.

CHAPTER 10

COMPLEX VECTOR SPACES

10.1 COMPLEX NUMBERS

*Up to now we have considered only vector spaces for which the scalars are real numbers. However, for many important applications of vectors it is desirable to allow the scalars to be complex numbers. A vector space that allows complex scalars is called a **complex vector space**, and one that allows real scalars only is called a **real vector space**. One advantage of allowing complex scalars is that all matrices with scalar entries have eigenvalues, which is not true if only real scalars are allowed. For example, the matrix*

$$A = \begin{bmatrix} -2 & -1 \\ 5 & 2 \end{bmatrix}$$

has characteristic polynomial

$$\det(\lambda I - A) = \det \begin{bmatrix} \lambda + 2 & 1 \\ -5 & \lambda - 2 \end{bmatrix} = \lambda^2 + 1$$

so the characteristic equation, $\lambda^2 + 1 = 0$, has no real solutions and therefore no real eigenvalues.

In the first three sections of this chapter we will review some of the basic properties of complex numbers, and in subsequent sections we will discuss complex vector spaces.

COMPLEX NUMBERS

Since $x^2 \geq 0$ for every real number x, the equation

$$x^2 = -1$$

has no real solutions. To deal with this problem, mathematicians of the eighteenth century introduced the "imaginary" number,

$$i = \sqrt{-1}$$

which they assumed had the property

$$i^2 = (\sqrt{-1})^2 = -1$$

but which otherwise could be treated as a real number. Expressions of the form

$$a + bi$$

where a and b are real numbers were called "complex numbers," and these were manipulated according to the standard rules of arithmetic with the added property that $i^2 = -1$.

By the beginning of the nineteenth century it was recognized that a complex number,

$$a + bi$$

could be regarded as an alternative symbol for the ordered pair

$$(a, b)$$

of real numbers, and that operations of addition, subtraction, multiplication, and division could be defined on these ordered pairs so that the familiar laws of arithmetic hold and $i^2 = -1$. This is the approach we will follow.

Definition. A ***complex number*** is an ordered pair of real numbers, denoted either by (a, b) or $a + bi$.

Example 1 Some examples of complex numbers in both notations are as follows:

Ordered Pair	Equivalent Notation
$(3, 4)$	$3 + 4i$
$(-1, 2)$	$-1 + 2i$
$(0, 1)$	$0 + i$
$(2, 0)$	$2 + 0i$
$(4, -2)$	$4 + (-2)i$

For simplicity, the last three complex numbers would usually be abbreviated as

$$0 + i = i \qquad 2 + 0i = 2 \qquad 4 + (-2)i = 4 - 2i \quad \blacktriangle$$

Geometrically, a complex number can be viewed either as a point or a vector in the xy-plane (Figure 1).

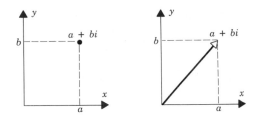

Figure 1 A complex number can be viewed as a point or a vector.

Example 2 Some complex numbers are shown as points in Figure 2*a* and as vectors in Figure 2*b*. ▲

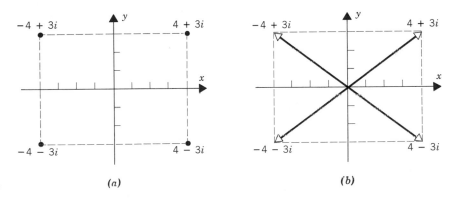

Figure 2   (*a*)   (*b*)

THE COMPLEX PLANE

Sometimes it is convenient to use a single letter, such as z, to denote a complex number. Thus, we might write

$$z = a + bi$$

The real number a is called the **real part of z** and the real number b the **imaginary part of z**. These numbers are denoted by $\text{Re}(z)$ and $\text{Im}(z)$, respectively. Thus,

$$\text{Re}(4 - 3i) = 4 \quad \text{and} \quad \text{Im}(4 - 3i) = -3$$

When complex numbers are represented geometrically in an xy-coordinate system, the x-axis is called the **real axis**, the y-axis the **imaginary axis**, and the plane is called the **complex plane** (Figure 3).

Figure 3

**OPERATIONS
ON COMPLEX
NUMBERS**

Just as two vectors in R^2 are defined to be equal if they have the same components, so we define two complex numbers to be equal if their real parts are equal and their imaginary parts are equal:

Definition. Two complex numbers, $a + bi$ and $c + di$, are defined to be *equal*, written

$$a + bi = c + di$$

if $a = c$ and $b = d$.

If $b = 0$, then the complex number $a + bi$ reduces to $a + 0i$, which we write simply as a. Thus, for any real number a,

$$a = a + 0i$$

so that the real numbers can be regarded as complex numbers with an imaginary part of zero. Geometrically, the real numbers correspond to points on the real axis. If $a = 0$, then $a + bi$ reduces to $0 + bi$, which we usually write as bi. These complex numbers, which correspond to points on the imaginary axis, are called *pure imaginary numbers*.

Just as vectors in R^2 are added by adding corresponding components, so complex numbers are added by adding their real parts and adding their imaginary parts:

$$(a + bi) + (c + di) = (a + c) + (b + d)i \tag{1}$$

The operations of subtraction and multiplication by a *real* number are also similar to the corresponding vector operations in R^2:

$$(a + bi) - (c + di) = (a - c) + (b - d)i \tag{2}$$

$$k(a + bi) = (ka) + (kb)i, \quad k \text{ real} \tag{3}$$

Because the operations of addition, subtraction, and multiplication of a complex number by a real number parallel the corresponding operations for vectors in R^2, the familiar geometric interpretations of these operations hold for complex numbers (Figure 4).

It follows from (3) that $(-1)z + z = 0$ (verify), so we denote $(-1)z$ as $-z$ and call it the *negative of z*.

The sum of two complex numbers.	The difference of two complex numbers.

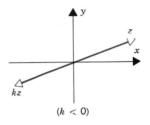

Figure 4 | The product of a complex number z and a real number k.

Example 3 If $z_1 = 4 - 5i$ and $z_2 = -1 + 6i$, find $z_1 + z_2$, $z_1 - z_2$, $3z_1$, and $-z_2$.

Solution.

$$z_1 + z_2 = (4 - 5i) + (-1 + 6i) = (4 - 1) + (-5 + 6)i = 3 + i$$
$$z_1 - z_2 = (4 - 5i) - (-1 + 6i) = (4 + 1) + (-5 - 6)i = 5 - 11i$$
$$3z_1 = 3(4 - 5i) = 12 - 15i$$
$$-z_2 = (-1)z_2 = (-1)(-1 + 6i) = 1 - 6i \quad \blacktriangle$$

So far, there has been a parallel between complex numbers and vectors in R^2. However, we now define multiplication of complex numbers, an operation with no vector analog in R^2. To motivate the definition, we expand the product

$$(a + bi)(c + di)$$

following the usual rules of algebra, but treating i^2 as -1. This yields

$$(a + bi)(c + di) = ac + bdi^2 + adi + bci$$
$$= (ac - bd) + (ad + bc)i$$

which suggests the following *definition*:

$$(a + bi)(c + di) = (ac - bd) + (ad + bc)i \tag{4}$$

Example 4

$$(3 + 2i)(4 + 5i) = (3 \cdot 4 - 2 \cdot 5) + (3 \cdot 5 + 2 \cdot 4)i$$
$$= 2 + 23i$$
$$(4 - i)(2 - 3i) = [4 \cdot 2 - (-1)(-3)] + [(4)(-3) + (-1)(2)]i$$
$$= 5 - 14i$$
$$i^2 = (0 + i)(0 + i) = (0 \cdot 0 - 1 \cdot 1) + (0 \cdot 1 + 1 \cdot 0)i = -1 \quad \blacktriangle$$

We leave it as an exercise to verify the following rules of complex arithmetic:

$$z_1 + z_2 = z_2 + z_1$$
$$z_1 z_2 = z_2 z_1$$
$$z_1 + (z_2 + z_3) = (z_1 + z_2) + z_3$$
$$z_1(z_2 z_3) = (z_1 z_2)z_3$$
$$z_1(z_2 + z_3) = z_1 z_2 + z_1 z_3$$
$$0 + z = z$$
$$z + (-z) = 0$$
$$1 \cdot z = z$$

These rules make it possible to multiply complex numbers without using Formula (4) directly. Following the procedure used to motivate this formula, we can simply multiply each term of $a + bi$ by each term of $c + di$, set $i^2 = -1$, and simplify.

Example 5

$$(3 + 2i)(4 + i) = 12 + 3i + 8i + 2i^2 = 12 + 11i - 2 = 10 + 11i$$
$$(5 - \tfrac{1}{2}i)(2 + 3i) = 10 + 15i - i - \tfrac{3}{2}i^2 = 10 + 14i + \tfrac{3}{2} = \tfrac{23}{2} + 14i$$
$$i(1 + i)(1 - 2i) = i(1 - 2i + i - 2i^2) = i(3 - i) = 3i - i^2 = 1 + 3i \quad \blacktriangle$$

REMARK. Unlike the real numbers, there is no size ordering for the complex numbers. Thus, the order symbols $<$, $\leq$, $>$, and $\geq$ are not used with complex numbers.

Now that we have defined addition, subtraction, and multiplication of complex numbers, it is possible to add, subtract, and multiply matrices with complex entries and multiply a matrix by a complex number. Without going into detail, we note that the matrix operations and terminology discussed in Chapter 1 carry over without change to matrices with complex entries.

Example 6 If

$$A = \begin{bmatrix} 1 & -i \\ 1 + i & 4 - i \end{bmatrix} \quad \text{and} \quad B = \begin{bmatrix} i & 1 - i \\ 2 - 3i & 4 \end{bmatrix}$$

then

$$A + B = \begin{bmatrix} 1 + i & 1 - 2i \\ 3 - 2i & 8 - i \end{bmatrix} \qquad A - B = \begin{bmatrix} 1 - i & -1 \\ -1 + 4i & -i \end{bmatrix}$$

$$iA = \begin{bmatrix} i & -i^2 \\ i + i^2 & 4i - i^2 \end{bmatrix} = \begin{bmatrix} i & 1 \\ -1 + i & 1 + 4i \end{bmatrix}$$

$$AB = \begin{bmatrix} 1 & -i \\ 1 + i & 4 - i \end{bmatrix} \begin{bmatrix} i & 1 - i \\ 2 - 3i & 4 \end{bmatrix}$$

$$= \begin{bmatrix} 1 \cdot i + (-i) \cdot (2 - 3i) & 1 \cdot (1 - i) + (-i) \cdot 4 \\ (1 + i) \cdot i + (4 - i) \cdot (2 - 3i) & (1 + i) \cdot (1 - i) + (4 - i) \cdot 4 \end{bmatrix}$$

$$= \begin{bmatrix} -3 - i & 1 - 5i \\ 4 - 13i & 18 - 4i \end{bmatrix}$$

EXERCISE SET 10.1

1. In each part plot the point and sketch the vector that corresponds to the given complex number.
   (a) $2 + 3i$ (b) -4 (c) $-3 - 2i$ (d) $-5i$

2. Express each complex number in Exercise 1 as an ordered pair of real numbers.

3. In each part use the given information to find the real numbers x and y.
   (a) $x - iy = -2 + 3i$ (b) $(x + y) + (x - y)i = 3 + i$

4. Given that $z_1 = 1 - 2i$ and $z_2 = 4 + 5i$, find
   (a) $z_1 + z_2$ (b) $z_1 - z_2$ (c) $4z_1$ (d) $-z_2$ (e) $3z_1 + 4z_2$ (f) $\frac{1}{2}z_1 - \frac{3}{2}z_2$

5. In each part solve for z.
   (a) $z + (1 - i) = 3 + 2i$ (b) $-5z = 5 + 10i$ (c) $(i - z) + (2z - 3i) = -2 + 7i$

6. In each part sketch the vectors z_1, z_2, $z_1 + z_2$, and $z_1 - z_2$.
   (a) $z_1 = 3 + i$, $z_2 = 1 + 4i$ (b) $z_1 = -2 + 2i$, $z_2 = 4 + 5i$

7. In each part sketch the vectors z and kz.
   (a) $z = 1 + i$, $k = 2$ (b) $z = -3 - 4i$, $k = -2$ (c) $z = 4 + 6i$, $k = \frac{1}{2}$

8. In each part find real numbers k_1 and k_2 that satisfy the equation.
   (a) $k_1 i + k_2(1 + i) = 3 - 2i$ (b) $k_1(2 + 3i) + k_2(1 - 4i) = 7 + 5i$

9. In each part find $z_1 z_2$, z_1^2, and z_2^2.
   (a) $z_1 = 3i$, $z_2 = 1 - i$ (b) $z_1 = 4 + 6i$, $z_2 = 2 - 3i$ (c) $z_1 = \frac{1}{3}(2 + 4i)$, $z_2 = \frac{1}{2}(1 - 5i)$

10. Given that $z_1 = 2 - 5i$ and $z_2 = -1 - i$, find
(a) $z_1 - z_1 z_2$ (b) $(z_1 + 3z_2)^2$ (c) $[z_1 + (1 + z_2)]^2$ (d) $iz_2 - z_1^2$

In Exercises 11–18 perform the calculations and express the result in the form $a + bi$.

11. $(1 + 2i)(4 - 6i)^2$

12. $(2 - i)(3 + i)(4 - 2i)$

13. $(1 - 3i)^3$

14. $i(1 + 7i) - 3i(4 + 2i)$

15. $[(2 + i)(\frac{1}{2} + \frac{3}{4}i)]^2$

16. $(\sqrt{2} + i) - i\sqrt{2}(1 + \sqrt{2}i)$

17. $(1 + i + i^2 + i^3)^{100}$

18. $(3 - 2i)^2 - (3 + 2i)^2$

19. Let

$$A = \begin{bmatrix} 1 & i \\ -i & 3 \end{bmatrix} \qquad B = \begin{bmatrix} 2 & 2+i \\ 3-i & 4 \end{bmatrix}$$

Find
(a) $A + 3iB$ (b) BA (c) AB (d) $B^2 - A^2$

20. Let

$$A = \begin{bmatrix} 3 + 2i & 0 \\ -i & 2 \\ 1 + i & 1 - i \end{bmatrix} \qquad B = \begin{bmatrix} -i & 2 \\ 0 & i \end{bmatrix} \qquad C = \begin{bmatrix} -1 - i & 0 & -i \\ 3 & 2i & -5 \end{bmatrix}$$

Find
(a) $A(BC)$ (b) $(BC)A$ (c) $(CA)B^2$ (d) $(1 + i)(AB) + (3 - 4i)A$

21. Show that
(a) $\text{Im}(iz) = \text{Re}(z)$ (b) $\text{Re}(iz) = -\text{Im}(z)$

22. In each part solve the equation by the quadratic formula and check your results by substituting the solutions into the given equation.
(a) $z^2 + 2z + 2 = 0$ (b) $z^2 - z + 1 = 0$

23. (a) Show that if n is a positive integer, then the only possible values for i^n are $1, -1, i$, and $-i$.
(b) Find i^{2509}. [**Hint.** The value of i^n can be determined from the remainder when n is divided by 4.]

24. Prove: If $z_1 z_2 = 0$, then $z_1 = 0$ or $z_2 = 0$.

25. Use the result of Exercise 24 to prove: If $zz_1 = zz_2$ and $z \neq 0$, then $z_1 = z_2$.

26. Prove that for all complex numbers z_1, z_2, and z_3
(a) $z_1 + z_2 = z_2 + z_1$ (b) $z_1 + (z_2 + z_3) = (z_1 + z_2) + z_3$

27. Prove that for all complex numbers z_1, z_2, and z_3
(a) $z_1 z_2 = z_2 z_1$ (b) $z_1(z_2 z_3) = (z_1 z_2)z_3$

28. Prove that $z_1(z_2 + z_3) = z_1 z_2 + z_1 z_3$ for all complex numbers z_1, z_2, and z_3.

29. In quantum mechanics the **Dirac* matrices** are

$$\beta = \begin{bmatrix} 1 & 0 & 0 & 0 \\ 0 & 1 & 0 & 0 \\ 0 & 0 & -1 & 0 \\ 0 & 0 & 0 & -1 \end{bmatrix} \qquad \alpha_x = \begin{bmatrix} 0 & 0 & 0 & 1 \\ 0 & 0 & 1 & 0 \\ 0 & 1 & 0 & 0 \\ 1 & 0 & 0 & 0 \end{bmatrix}$$

$$\alpha_y = \begin{bmatrix} 0 & 0 & 0 & -i \\ 0 & 0 & i & 0 \\ 0 & -i & 0 & 0 \\ i & 0 & 0 & 0 \end{bmatrix} \qquad \alpha_z = \begin{bmatrix} 0 & 0 & 1 & 0 \\ 0 & 0 & 0 & -1 \\ 1 & 0 & 0 & 0 \\ 0 & -1 & 0 & 0 \end{bmatrix}$$

(a) Prove that $\beta^2 = \alpha_x^2 = \alpha_y^2 = \alpha_z^2 = I$.

(b) Two matrices A and B are called **anticommutative** if $AB = -BA$. Prove that any two Dirac matrices are anticommutative.

10.2 MODULUS; COMPLEX CONJUGATE; DIVISION

Our main objective in this section is to define the division of complex numbers.

COMPLEX CONJUGATES

We will begin with some preliminary ideas.

If $z = a + bi$ is any complex number, then the **conjugate of z**, denoted by $\bar{z}$ (read "z bar"), is defined by

$$\bar{z} = a - bi$$

In words, $\bar{z}$ is obtained by reversing the sign of the imaginary part of z. Geometrically, $\bar{z}$ is the reflection of z about the real axis (Figure 1).

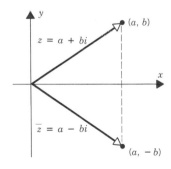

Figure 1 | The conjugate of a complex number.

---

* *Paul Adrien Maurice Dirac* (1902–1984) was a British theoretical physicist who devised a new form of quantum mechanics and a theory that predicted electron spin and the existence of a fundamental atomic particle called a positron. In 1933 he received the Nobel Prize for Physics and in 1939, the medal of the Royal Society.

Example 1

$$z = 3 + 2i \qquad \bar{z} = 3 - 2i$$
$$z = -4 - 2i \qquad \bar{z} = -4 + 2i$$
$$z = i \qquad \bar{z} = -i$$
$$z = 4 \qquad z = 4 \; \blacktriangle$$

REMARK. The last line in Example 1 illustrates the fact that a real number is the same as its conjugate. More precisely, it can be shown (Exercise 22) that $z = \bar{z}$ if and only if z is a real number.

If a complex number z is viewed as a vector in R^2, then the norm or length of the vector is called the modulus (or *absolute value*) of z. More precisely:

MODULUS

> **Definition.** The **modulus** of a complex number $z = a + bi$, denoted by $|z|$, is defined by
>
> $$|z| = \sqrt{a^2 + b^2} \qquad (1)$$

If $b = 0$, then $z = a$ is a real number, and

$$|z| = \sqrt{a^2 + 0^2} = \sqrt{a^2} = |a|$$

so the modulus of a real number is simply its absolute value. Thus, the modulus of z is also called the **absolute value** of z.

Example 2 Find $|z|$ if $z = 3 - 4i$.

Solution. From (1) with $a = 3$ and $b = -4$, $|z| = \sqrt{(3)^2 + (-4)^2} = \sqrt{25} = 5$. $\blacktriangle$

The following theorem establishes a basic relationship between $\bar{z}$ and $|z|$.

> **Theorem 10.2.1.** *For any complex number z,*
>
> $$z\bar{z} = |z|^2$$

Proof. If $z = a + bi$, then

$$z\bar{z} = (a + bi)(a - bi) = a^2 - abi + bai \quad b^2 i^2$$
$$= a^2 + b^2 = |z|^2 \quad \blacksquare$$

DIVISION OF COMPLEX NUMBERS

We now turn to the division of complex numbers. Our objective is to define division as the inverse of multiplication. Thus, if $z_2 \neq 0$, then our definition of $z = z_1/z_2$ should be such that

$$z_1 = z_2 z \tag{2}$$

Our procedure will be to prove that (2) has a unique solution for z if $z_2 \neq 0$, and then define z_1/z_2 to be this value of z. As with real numbers, division by zero is not allowed.

Theorem 10.2.2. *If $z_2 \neq 0$, then Equation (2) has a unique solution, which is*

$$z = \frac{1}{|z_2|^2} z_1 \bar{z}_2 \tag{3}$$

Proof. Let $z = x + iy$, $z_1 = x_1 + iy_1$, and $z_2 = x_2 + iy_2$. Then (2) can be written as

$$x_1 + iy_1 = (x_2 + iy_2)(x + iy)$$

or

$$x_1 + iy_1 = (x_2 x - y_2 y) + i(y_2 x + x_2 y)$$

or, on equating real and imaginary parts,

$$x_2 x - y_2 y = x_1$$
$$y_2 x + x_2 y = y_1$$

or

$$\begin{bmatrix} x_2 & -y_2 \\ y_2 & x_2 \end{bmatrix} \begin{bmatrix} x \\ y \end{bmatrix} = \begin{bmatrix} x_1 \\ y_1 \end{bmatrix} \tag{4}$$

Since $z_2 = x_2 + iy_2 \neq 0$, it follows that x_2 and y_2 are not both zero, so

$$\begin{vmatrix} x_2 & -y_2 \\ y_2 & x_2 \end{vmatrix} = x_2^2 + y_2^2 \neq 0$$

Thus, by Cramer's Rule (Theorem 2.4.3), system (4) has the unique solution

$$x = \frac{\begin{vmatrix} x_1 & -y_2 \\ y_1 & x_2 \end{vmatrix}}{\begin{vmatrix} x_2 & -y_2 \\ y_2 & x_2 \end{vmatrix}} = \frac{x_1 x_2 + y_1 y_2}{x_2^2 + y_2^2} = \frac{x_1 x_2 + y_1 y_2}{|z_2|^2}$$

$$y = \frac{\begin{vmatrix} x_2 & x_1 \\ y_2 & y_1 \end{vmatrix}}{\begin{vmatrix} x_2 & -y_2 \\ y_2 & x_2 \end{vmatrix}} = \frac{y_1 x_2 - x_1 y_2}{x_2^2 + y_2^2} = \frac{y_1 x_2 - x_1 y_2}{|z_2|^2}$$

Thus,

$$z = x + iy = \frac{1}{|z_2|^2}\left[(x_1x_2 + y_1y_2) + i(y_1x_2 - x_1y_2)\right]$$

$$= \frac{1}{|z_2|^2}(x_1 + iy_1)(x_2 - iy_2)$$

$$= \frac{1}{|z_2|^2}z_1\bar{z}_2 \quad \blacksquare$$

Thus, for $z_2 \neq 0$ we define

$$\frac{z_1}{z_2} = \frac{1}{|z_2|^2}z_1\bar{z}_2 \tag{5}$$

REMARK. To remember this formula, multiply numerator and denominator of z_1/z_2 by $\bar{z}_2$:

$$\frac{z_1}{z_2} = \frac{z_1\bar{z}_2}{z_2\bar{z}_2} = \frac{z_1\bar{z}_2}{|z_2|^2} = \frac{1}{|z_2|^2}z_1\bar{z}_2$$

Example 3 Express

$$\frac{3 + 4i}{1 - 2i}$$

in the form $a + bi$.

Solution. From (5) with $z_1 = 3 + 4i$ and $z_2 = 1 - 2i$,

$$\frac{3 + 4i}{1 - 2i} = \frac{1}{|1 - 2i|^2}(3 + 4i)(\overline{1 - 2i})$$

$$= \frac{1}{5}(3 + 4i)(1 + 2i)$$

$$= \frac{1}{5}(-5 + 10i)$$

$$= -1 + 2i$$

Alternative Solution. As in the remark above, multiply numerator and denominator by the conjugate of the denominator:

$$\frac{3 + 4i}{1 - 2i} = \frac{3 + 4i}{1 - 2i} \cdot \frac{1 + 2i}{1 + 2i} = \frac{-5 + 10i}{5} = -1 + 2i \quad \blacktriangle$$

Systems of linear equations with complex coefficients arise in various applications. Without going into detail we note that all the results about linear systems studied in Chapters 1 and 2 carry over without change to systems with complex coefficients.

Example 4   Use Cramer's Rule to solve

$$ix + 2y = 1 - 2i$$
$$4x - iy = -1 + 3i$$

Solution.

$$x = \frac{\begin{vmatrix} 1 - 2i & 2 \\ -1 + 3i & -i \end{vmatrix}}{\begin{vmatrix} i & 2 \\ 4 & -i \end{vmatrix}} = \frac{(-i)(1 - 2i) - 2(-1 + 3i)}{i(-i) - 2(4)} = \frac{-7i}{-7} = i$$

$$y = \frac{\begin{vmatrix} i & 1 - 2i \\ 4 & -1 + 3i \end{vmatrix}}{\begin{vmatrix} i & 2 \\ 4 & -i \end{vmatrix}} = \frac{(i)(-1 + 3i) - 4(1 - 2i)}{i(-i) - 2(4)} = \frac{-7 + 7i}{-7} = 1 - i$$

Thus, the solution is $x = i$, $y = 1 - i$. ▲

PROPERTIES OF THE COMPLEX CONJUGATE

We conclude this section by listing some properties of the complex conjugate that will be useful in later sections.

Theorem 10.2.3.   *For any complex numbers* z, z_1, *and* z_2

(a) $\overline{z_1 + z_2} = \bar{z}_1 + \bar{z}_2$

(b) $\overline{z_1 - z_2} = \bar{z}_1 - \bar{z}_2$

(c) $\overline{z_1 z_2} = \bar{z}_1 \bar{z}_2$

(d) $\overline{(z_1/z_2)} = \bar{z}_1/\bar{z}_2$

(e) $\bar{\bar{z}} = z$

We prove (a) and leave the rest as exercises.

Proof (a).   Let $z_1 = a_1 + b_1 i$ and $z_2 = a_2 + b_2 i$; then

$$\overline{z_1 + z_2} = \overline{(a_1 + a_2) + (b_1 + b_2)i}$$
$$= (a_1 + a_2) - (b_1 + b_2)i$$
$$= (a_1 - b_1 i) + (a_2 - b_2 i)$$
$$= \bar{z}_1 + \bar{z}_2 \quad \blacksquare$$

REMARK.   It is possible to extend part (a) of Theorem 10.2.3 to n terms and part (c) to n factors. More precisely,

$$\overline{z_1 + z_2 + \cdots + z_n} = \bar{z}_1 + \bar{z}_2 + \cdots + \bar{z}_n$$
$$\overline{z_1 z_2 \cdots z_n} = \bar{z}_1 \bar{z}_2 \cdots \bar{z}_n$$

EXERCISE SET 10.2

1. In each part find $\bar{z}$.
  (a) $z = 2 + 7i$ (b) $z = -3 - 5i$ (c) $z = 5i$ (d) $z = -i$ (e) $z = -9$ (f) $z = 0$

2. In each part find $|z|$.
  (a) $z = i$ (b) $z = -7i$ (c) $z = -3 - 4i$ (d) $z = 1 + i$ (e) $z = -8$ (f) $z = 0$

3. Verify that $z\bar{z} = |z|^2$ for
  (a) $z = 2 - 4i$ (b) $z = -3 + 5i$ (c) $z = \sqrt{2} - \sqrt{2}i$

4. Given that $z_1 = 1 - 5i$ and $z_2 = 3 + 4i$, find
  (a) z_1/z_2 (b) $\bar{z}_1/z_2$ (c) $z_1/\bar{z}_2$ (d) $\overline{(z_1/z_2)}$ (e) $z_1/|z_2|$ (f) $|z_1/z_2|$

5. In each part find $1/z$.

  (a) $z = i$ (b) $z = 1 - 5i$ (c) $z = \dfrac{-i}{7}$

6. Given that $z_1 = 1 + i$ and $z_2 = 1 - 2i$, find

  (a) $z_1 - \left(\dfrac{z_1}{z_2}\right)$ (b) $\dfrac{z_1 - 1}{z_2}$ (c) $z_1^2 - \left(\dfrac{iz_1}{z_2}\right)$ (d) $\dfrac{z_1}{iz_2}$

In Exercises 7–14 perform the calculations and express the result in the form $a + bi$.

7. $\dfrac{i}{1+i}$ **8.** $\dfrac{2}{(1-i)(3+i)}$ **9.** $\dfrac{1}{(3+4i)^2}$

10. $\dfrac{2+i}{i(-3+4i)}$ **11.** $\dfrac{\sqrt{3}+i}{(1-i)(\sqrt{3}-i)}$ **12.** $\dfrac{1}{i(3-2i)(1+i)}$

13. $\dfrac{i}{(1-i)(1-2i)(1+2i)}$ **14.** $\dfrac{1-2i}{3+4i} - \dfrac{2+i}{5i}$

15. In each part solve for z.
  (a) $iz = 2 - i$ (b) $(4 - 3i)\bar{z} = i$

16. Use Theorem 10.2.3 to prove the following identities:

  (a) $\overline{\bar{z} + 5i} = z - 5i$ (b) $\overline{iz} = -i\bar{z}$ (c) $\dfrac{\overline{i + \bar{z}}}{i - z} = -1$

17. In each part sketch the set of points in the complex plane that satisfy the equation.

18. In each part sketch the set of points in the complex plane that satisfy the given condition(s).
  (a) $|z + i| \leq 1$ (b) $1 < |z| < 2$ (c) $|2z - 4i| < 1$ (d) $|z| \leq |z + i|$

19. Given that $z = x + iy$, find
  (a) $\text{Re}(\overline{iz})$ (b) $\text{Im}(\overline{iz})$ (c) $\text{Re}(i\bar{z})$ (d) $\text{Im}(i\bar{z})$

20. (a) Show that if n is a positive integer, then the only possible values for $(1/i)^n$ are 1, -1, i, and $-i$.
  (b) Find $(1/i)^{2509}$. [**Hint.** See Exercise 23(b) of Section 10.1.]

21. Prove:

(a) $\frac{1}{2}(z + \bar{z}) = \text{Re}(z)$ (b) $\frac{1}{2i}(z - \bar{z}) = \text{Im}(z)$

22. Prove: $z = \bar{z}$ if and only if z is a real number.

23. Given that $z_1 = x_1 + iy_1$ and $z_2 = x_2 + iy_2$, find

(a) $\text{Re}\left(\dfrac{z_1}{z_2}\right)$ (b) $\text{Im}\left(\dfrac{z_1}{z_2}\right)$

24. Prove: If $(\bar{z})^2 = z^2$, then z is either real or pure imaginary.

25. Prove that $|z| = |\bar{z}|$.

26. Prove:
(a) $\overline{z_1 - z_2} = \bar{z}_1 - \bar{z}_2$ (b) $\overline{z_1 z_2} = \bar{z}_1 \bar{z}_2$ (c) $\overline{(z_1/z_2)} = \bar{z}_1/\bar{z}_2$ (d) $\bar{\bar{z}} = z$

27. (a) Prove that $\overline{z^2} = (\bar{z})^2$.
(b) Prove that if n is a positive integer, then $\overline{z^n} = (\bar{z})^n$.
(c) Is the result in (b) true if n is a negative integer? Explain.

In Exercises 28–31 solve the system of linear equations by Cramer's Rule.

28. $ix_1 - ix_2 = -2$
$2x_1 + x_2 = i$

29. $x_1 + x_2 = 2$
$x_1 - x_2 = 2i$

30. $x_1 + x_2 + x_3 = 3$
$x_1 + x_2 - x_3 = 2 + 2i$
$x_1 - x_2 + x_3 = -1$

31. $ix_1 + 3x_2 + (1 + i)x_3 = -i$
$x_1 + ix_2 + 3x_3 = -2i$
$x_1 + x_2 + x_3 = 0$

In Exercises 32 and 33 solve the system of linear equations by Gauss-Jordan elimination.

32. $\begin{bmatrix} -1 & -1 - i \\ -1 + i & -2 \end{bmatrix} \begin{bmatrix} x_1 \\ x_2 \end{bmatrix} - \begin{bmatrix} 0 \\ 0 \end{bmatrix}$

33. $\begin{bmatrix} 2 & -1 - i \\ -1 + i & 1 \end{bmatrix} \begin{bmatrix} x_1 \\ x_2 \end{bmatrix} = \begin{bmatrix} 0 \\ 0 \end{bmatrix}$

34. Solve the following system of linear equations by Gauss-Jordan elimination.

$$x_1 + ix_2 - ix_3 = 0$$
$$-x_1 + (1 - i)x_2 + 2ix_3 = 0$$
$$2x_1 + (-1 + 2i)x_2 - 3ix_3 = 0$$

35. In each part use the formula given in Example 7 of Section 1.5 to compute the inverse of the matrix, and check your result by showing that $AA^{-1} = A^{-1}A = I$.

(a) $A = \begin{bmatrix} i & -2 \\ 1 & i \end{bmatrix}$ (b) $A = \begin{bmatrix} 2 & i \\ 1 & 0 \end{bmatrix}$

36. Let $p(x) = a_0 + a_1 x + a_2 x^2 + \cdots + a_n x^n$ be a polynomial for which the coefficients $a_0, a_1, a_2, \ldots, a_n$ are real. Prove that if z is a solution of the equation $p(x) = 0$, then so is $\bar{z}$.

37. Prove: For any complex number z, $|\text{Re}(z)| \le |z|$ and $|\text{Im}(z)| \le |z|$.

38. Prove that

$$\frac{|\text{Re}(z)| + |\text{Im}(z)|}{\sqrt{2}} \leq |z|$$

[**Hint.** Let $z = x + iy$ and use the fact that $(|x| - |y|)^2 \geq 0$.]

39. In each part use the method of Example 4 in Section 1.6 to find A^{-1}, and check your result by showing that $AA^{-1} = A^{-1}A = I$.

(a) $A = \begin{bmatrix} 1 & 1+i & 0 \\ 0 & 1 & i \\ -i & 1-2i & 2 \end{bmatrix}$ (b) $A = \begin{bmatrix} i & 0 & -i \\ 0 & 1 & -1-4i \\ 2-i & i & 3 \end{bmatrix}$

10.3 POLAR FORM; DEMOIVRE'S THEOREM

In this section we shall discuss a way to represent complex numbers using trigonometric properties. Our work will lead to a fundamental formula for powers of complex numbers and a method for finding nth roots of complex numbers.

POLAR FORM OF A COMPLEX NUMBER

If $z = x + iy$ is a nonzero complex number, $r = |z|$, and θ measures the angle from the positive real axis to the vector z, then, as suggested by Figure 1,

$$\begin{aligned} x &= r \cos \theta \\ y &= r \sin \theta \end{aligned} \tag{1}$$

so that $z = x + iy$ can be written as

$$z = r \cos \theta + ir \sin \theta$$

or

$$z = r(\cos \theta + i \sin \theta) \tag{2}$$

This is called a *polar form of z*.

Figure 1

ARGUMENT OF A COMPLEX NUMBER

The angle θ is called an **argument of z** and is denoted by

$$\theta = \arg z$$

The argument of z is not uniquely determined because we can add or subtract any multiple of 2π from θ to produce another value of the argument. However, there is only one value of the argument in radians that satisfies

$$-\pi < \theta \le \pi$$

This is called the **principal argument of z** and is denoted by

$$\theta = \text{Arg } z$$

Example 1 Express the following complex numbers in polar form using their principal arguments:

(a) $z = 1 + \sqrt{3}i$ (b) $z = -1 - i$

Solution (a). The value of r is

$$r = |z| = \sqrt{1^2 + (\sqrt{3})^2} = \sqrt{4} = 2$$

and since $x = 1$ and $y = \sqrt{3}$, it follows from (1) that

$$1 = 2 \cos \theta$$
$$\sqrt{3} = 2 \sin \theta$$

so $\cos \theta = 1/2$ and $\sin \theta = \sqrt{3}/2$. The only value of θ that satisfies these relations and meets the requirement $-\pi < \theta \le \pi$ is $\theta = \pi/3 \, (= 60°)$ (see Figure 2a). Thus, a polar form of z is

$$z = 2\left(\cos \frac{\pi}{3} + i \sin \frac{\pi}{3}\right)$$

Solution (b). The value of r is

$$r = |z| = \sqrt{(-1)^2 + (-1)^2} = \sqrt{2}$$

and since $x = -1$, $y = -1$, it follows from (1) that

$$-1 = \sqrt{2} \cos \theta$$
$$-1 = \sqrt{2} \sin \theta$$

so $\cos \theta = -1/\sqrt{2}$ and $\sin \theta = -1/\sqrt{2}$. The only value of θ that satisfies these relations and meets the requirement $-\pi < \theta \le \pi$ is $\theta = -3\pi/4 \, (= -135°)$ (Figure 2b). Thus, a polar form of z is

$$z = \sqrt{2}\left(\cos \frac{-3\pi}{4} + i \sin \frac{-3\pi}{4}\right)$$

Figure 2 (a) (b)

GEOMETRIC INTERPRETATION OF MULTIPLICATION AND DIVISION

We now show how polar forms can be used to give geometric interpretations of multiplication and division of complex numbers. Let

$$z_1 = r_1(\cos \theta_1 + i \sin \theta_1) \qquad \text{and} \qquad z_2 = r_2(\cos \theta_2 + i \sin \theta_2)$$

Multiplying, we obtain

$$z_1 z_2 = r_1 r_2[(\cos \theta_1 \cos \theta_2 - \sin \theta_1 \sin \theta_2) + i(\sin \theta_1 \cos \theta_2 + \cos \theta_1 \sin \theta_2)]$$

Recalling the trigonometric identities

$$\cos(\theta_1 + \theta_2) = \cos \theta_1 \cos \theta_2 - \sin \theta_1 \sin \theta_2$$
$$\sin(\theta_1 + \theta_2) = \sin \theta_1 \cos \theta_2 + \cos \theta_1 \sin \theta_2$$

We obtain

$$z_1 z_2 = r_1 r_2[\cos(\theta_1 + \theta_2) + i \sin(\theta_1 + \theta_2)] \tag{3}$$

which is a polar form of the complex number with modulus $r_1 r_2$ and argument $\theta_1 + \theta_2$. Thus, we have shown that

$$|z_1 z_2| = |z_1||z_2| \tag{4}$$

and

$$\arg(z_1 z_2) = \arg z_1 + \arg z_2$$

(Why?)

In words, *the product of two complex numbers is obtained by multiplying their moduli and adding their arguments* (Figure 3).

We leave it as an exercise to show that if $z_2 \neq 0$, then

$$\frac{z_1}{z_2} = \frac{r_1}{r_2}[\cos(\theta_1 - \theta_2) + i \sin(\theta_1 - \theta_2)] \tag{5}$$

from which it follows that

$$\left|\frac{z_1}{z_2}\right| = \frac{|z_1|}{|z_2|} \qquad \text{if } z_2 \neq 0$$

and

$$\arg\left(\frac{z_1}{z_2}\right) = \arg z_1 - \arg z_2$$

In words, *the quotient of two complex numbers is obtained by dividing their moduli and subtracting their arguments* (*in the appropriate order*).

Figure 3   The product of two complex numbers.

Example 2   Let

$$z_1 = 1 + \sqrt{3}i \quad \text{and} \quad z_2 = \sqrt{3} + i$$

Polar forms of these complex numbers are

$$z_1 = 2\left(\cos\frac{\pi}{3} + i\sin\frac{\pi}{3}\right)$$

$$z_2 = 2\left(\cos\frac{\pi}{6} + i\sin\frac{\pi}{6}\right)$$

(verify) so that from (3)

$$z_1 z_2 = 4\left[\cos\left(\frac{\pi}{3} + \frac{\pi}{6}\right) + i\sin\left(\frac{\pi}{3} + \frac{\pi}{6}\right)\right]$$

$$= 4\left[\cos\frac{\pi}{2} + i\sin\frac{\pi}{2}\right]$$

$$= 4[0 + i] = 4i$$

and from (5)

$$\frac{z_1}{z_2} = 1 \cdot \left[\cos\left(\frac{\pi}{3} - \frac{\pi}{6}\right) + i\sin\left(\frac{\pi}{3} - \frac{\pi}{6}\right)\right]$$

$$= \cos\frac{\pi}{6} + i\sin\frac{\pi}{6}$$

$$= \frac{\sqrt{3}}{2} + \frac{1}{2}i$$

As a check, we calculate $z_1 z_2$ and z_1/z_2 directly without using polar forms for z_1 and z_2:

$$z_1 z_2 = (1 + \sqrt{3}i)(\sqrt{3} + i) = (\sqrt{3} - \sqrt{3}) + (3 + 1)i = 4i$$

$$\frac{z_1}{z_2} = \frac{1 + \sqrt{3}i}{\sqrt{3} + i} \cdot \frac{\sqrt{3} - i}{\sqrt{3} - i} = \frac{(\sqrt{3} + \sqrt{3}) + (-i + 3i)}{4} = \frac{\sqrt{3}}{2} + \frac{1}{2}i$$

which agrees with our previous results.   ▲

The complex number i has a modulus of 1 and an argument of $\pi/2$ ($= 90°$), so the product iz has the same modulus as z, but its argument is $90°$ greater than that of z. In short, *multiplying z by i rotates z counterclockwise by $90°$* (Figure 4).

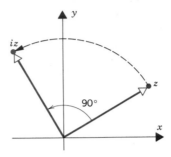

Figure 4 | Multiplying by i rotates z counterclockwise by $90°$.

DEMOIVRE'S FORMULA

If n is a positive integer and $z = r(\cos\theta + i\sin\theta)$, then from Formula (3),

$$z^n = \underbrace{z \cdot z \cdot z \cdots z}_{n\text{-factors}} = r^n[\cos(\underbrace{\theta + \theta + \cdots + \theta}_{n\text{-terms}}) + i\sin(\underbrace{\theta + \theta + \cdots + \theta}_{n\text{-terms}})]$$

or

$$z^n = r^n(\cos n\theta + i\sin n\theta) \qquad (6)$$

In the special case where $r = 1$, we have $z = \cos\theta + i\sin\theta$, so (6) becomes

$$(\cos\theta + i\sin\theta)^n = \cos n\theta + i\sin n\theta \qquad (7)$$

which is called *DeMoivre's* Formula*. Although we derived (7) assuming n to be a positive integer, it will be shown in the exercises that this formula is valid for all integers n.

FINDING nTH ROOTS

We now show how DeMoivre's Formula can be used to obtain roots of complex numbers. If n is a positive integer, and z is any complex number, then we define an *nth root of z* to be any complex number w that satisfies the equation

$$w^n = z \qquad (8)$$

We denote an nth root of z by $z^{1/n}$. If $z \neq 0$, then we can derive formulas for the nth roots of z as follows. Let

$$w = \rho(\cos\alpha + i\sin\alpha) \qquad \text{and} \qquad z = r(\cos\theta + i\sin\theta)$$

---

* *Abraham DeMoivre* (1667–1754) was a French mathematician who made important contributions to probability, statistics, and trigonometry. He developed the concept of statistically independent events, wrote a major influential treatise on probability, and helped transform trigonometry from a branch of geometry into a branch of analysis through his use of complex numbers. In spite of his important work, he barely managed to eke out a living as a tutor and a consultant on gambling and insurance.

If we assume that w satisfies (8), then it follows from (7) that

$$\rho^n(\cos n\alpha + i \sin n\alpha) = r(\cos \theta + i \sin \theta) \tag{9}$$

Comparing the moduli of the two sides, we see that $\rho^n = r$ or

$$\rho = \sqrt[n]{r}$$

where $\sqrt[n]{r}$ denotes the real positive nth root of r. Moreover, in order to have $\cos n\alpha = \cos \theta$ and $\sin n\alpha = \sin \theta$ in (9), the angles $n\alpha$ and θ must either be equal or differ by a multiple of 2π. That is,

$$n\alpha = \theta + 2k\pi, \qquad k = 0, \pm 1, \pm 2, \ldots$$

or

$$\alpha = \frac{\theta}{n} + \frac{2k\pi}{n}, \qquad k = 0, \pm 1, \pm 2, \ldots$$

Thus, the values of $w = \rho(\cos \alpha + i \sin \alpha)$ that satisfy (8) are given by

$$w = \sqrt[n]{r}\left[\cos\left(\frac{\theta}{n} + \frac{2k\pi}{n}\right) + i \sin\left(\frac{\theta}{n} + \frac{2k\pi}{n}\right)\right], \qquad k = 0, \pm 1, \pm 2, \ldots$$

Although there are infinitely many values of k, it can be shown (Exercise 16) that $k = 0, 1, 2, \ldots, n - 1$ produce distinct values of w satisfying (8), but all other choices of k yield duplicates of these. Therefore, there are exactly n different nth roots of $z = r(\cos \theta + i \sin \theta)$, and these are given by

$$z^{1/n} = \sqrt[n]{r}\left[\cos\left(\frac{\theta}{n} + \frac{2k\pi}{n}\right) + i \sin\left(\frac{\theta}{n} + \frac{2k\pi}{n}\right)\right], \qquad k = 0, 1, 2, \ldots, n - 1 \tag{10}$$

Example 3   Find all cube roots of -8.

Solution. Since -8 lies on the negative real axis, we can use $\theta = \pi$ as an argument. Moreover, $r = |z| = |-8| = 8$, so a polar form of -8 is

$$-8 = 8(\cos \pi + i \sin \pi)$$

From (10) with $n = 3$ it follows that

$$(-8)^{1/3} = \sqrt[3]{8}\left[\cos\left(\frac{\pi}{3} + \frac{2k\pi}{3}\right) + i \sin\left(\frac{\pi}{3} + \frac{2k\pi}{3}\right)\right], \qquad k = 0, 1, 2$$

Thus, the cube roots of -8 are

$$2\left(\cos \frac{\pi}{3} + i \sin \frac{\pi}{3}\right) = 2\left(\frac{1}{2} + \frac{\sqrt{3}}{2}i\right) = 1 + \sqrt{3}i$$

$$2(\cos \pi + i \sin \pi) = 2(-1) = -2$$

$$2\cos\left(\frac{5\pi}{3} + i \sin \frac{5\pi}{3}\right) = 2\left(\frac{1}{2} - \frac{\sqrt{3}}{2}i\right) = 1 - \sqrt{3}i \quad \blacktriangle$$

As shown in Figure 5, the three cube roots of -8 obtained in Example 3 are equally spaced $\pi/3$ radians $(= 120°)$ apart around the circle of radius 2 centered at the origin. This is not accidental. In general, it follows from Formula (10) that the nth roots of z lie on the circle of radius $\sqrt[n]{r}$ $(= \sqrt[n]{|z|})$, and are equally spaced $2\pi/n$ radians apart. (Can you see why?) Thus, once one nth root of z is found, the remaining $n - 1$ roots can be generated by rotating this root successively through increments of $2\pi/n$ radians.

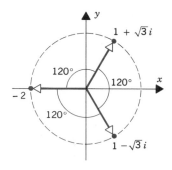

Figure 5   The cube roots of -8.

Example 4   Find all fourth roots of 1.

Solution. We could apply Formula (10). Instead, we observe that $w = 1$ is one fourth root of 1, so that the remaining three roots can be generated by rotating this root through increments of $2\pi/4 = \pi/2$ radians $(= 90°)$. From Figure 6 we see that the fourth roots of 1 are

$$1, \quad i, \quad -1, \quad -i \quad \blacktriangle$$

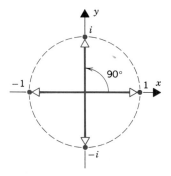

Figure 6   The fourth roots of 1.

**COMPLEX
EXPONENTS**

We conclude this section with some comments on notation.

In more detailed studies of complex numbers, complex exponents are defined, and it is proved that

$$\cos \theta + i \sin \theta = e^{i\theta} \qquad (11)$$

where e is an irrational real number given approximately by $e \approx 2.71828\ldots$. (For readers who have studied calculus, a proof of this result is given in Exercise 18.)

It follows from (11) that the polar form

$$z = r(\cos \theta + i \sin \theta)$$

can be written more briefly as

$$z = re^{i\theta} \qquad (12)$$

Example 5   In Example 1 it was shown that

$$1 + \sqrt{3}i = 2\left(\cos \frac{\pi}{3} + i \sin \frac{\pi}{3}\right)$$

From (12) this can also be written as

$$1 + \sqrt{3}i = 2e^{i\pi/3} \quad ▲$$

It can be proved that complex exponents follow the same laws as real exponents, so that if

$$z_1 = r_1 e^{i\theta_1} \qquad \text{and} \qquad z_2 = r_2 e^{i\theta_2}$$

then

$$z_1 z_2 = r_1 r_2 e^{i\theta_1 + i\theta_2} = r_1 r_2 e^{i(\theta_1 + \theta_2)}$$

$$\frac{z_1}{z_2} = \frac{r_1}{r_2} e^{i\theta_1 - i\theta_2} = \frac{r_1}{r_2} e^{i(\theta_1 - \theta_2)}$$

But these formulas are just Formulas (3) and (5) in a different notation.

We conclude this section with a useful formula for $\bar{z}$ in polar notation. If

$$z = re^{i\theta} = r(\cos \theta + i \sin \theta)$$

then

$$\bar{z} = r(\cos \theta - i \sin \theta) \qquad (13)$$

Recalling the trigonometric identities

$$\sin(-\theta) = -\sin \theta \qquad \text{and} \qquad \cos(-\theta) = \cos \theta$$

we can rewrite (13) as

$$\bar{z} = r[\cos(-\theta) + i \sin(-\theta)] = re^{i(-\theta)}$$

or equivalently

$$\bar{z} = re^{-i\theta} \tag{14}$$

In the special case where $r = 1$, the polar form of z is $z = e^{i\theta}$, and (14) yields the formula

$$\overline{e^{i\theta}} = e^{-i\theta} \tag{15}$$

EXERCISE SET 10.3

1. In each part find the principal argument of z.

(a) $z = 1$ (b) $z = i$ (c) $z = -i$ (d) $z = 1 + i$ (e) $z = -1 + \sqrt{3}i$ (f) $z = 1 - i$

2. In each part find the value of $\theta = \arg(1 - \sqrt{3}i)$ that satisfies the given condition.

(a) $0 \le \theta < 2\pi$ (b) $-\pi < \theta \le \pi$ (c) $-\dfrac{\pi}{6} \le \theta < \dfrac{11\pi}{6}$

3. In each part express the complex number in polar form using its principal argument.

(a) $2i$ (b) -4 (c) $5 + 5i$ (d) $-6 + 6\sqrt{3}i$ (e) $-3 - 3i$ (f) $2\sqrt{3} - 2i$

4. Given that $z_1 = 2(\cos \pi/4 + i \sin \pi/4)$ and $z_2 = 3(\cos \pi/6 + i \sin \pi/6)$, find a polar form of

(a) $z_1 z_2$ (b) $\dfrac{z_1}{z_2}$ (c) $\dfrac{z_2}{z_1}$ (d) $\dfrac{z_1^5}{z_2^2}$

5. Express $z_1 = i, z_2 = 1 - \sqrt{3}i$, and $z_3 = \sqrt{3} + i$ in polar form, and use your results to find $z_1 z_2 / z_3$. Check your results by performing the calculations without using polar forms.

6. Use Formula (6) to find

(a) $(1 + i)^{12}$ (b) $\left(\dfrac{1}{\sqrt{2}} - \dfrac{1}{\sqrt{2}}i\right)^{-6}$ (c) $(\sqrt{3} + i)^7$ (d) $(1 - i\sqrt{3})^{-10}$

7. In each part find all the roots and sketch them as vectors in the complex plane.

(a) $(-i)^{1/2}$ (b) $(1 + \sqrt{3}i)^{1/2}$ (c) $(-27)^{1/3}$ (d) $(i)^{1/3}$ (e) $(-1)^{1/4}$ (f) $(-8 + 8\sqrt{3}i)^{1/4}$

8. Use the method of Example 4 to find all cube roots of 1.

9. Use the method of Example 4 to find all sixth roots of 1.

10. Find all square roots of $1 + i$ and express your results in polar form.

11. In each part find all solutions of the equation.

(a) $z^4 - 16 = 0$ (b) $z^{4/3} = -4$

12. Find four solutions of the equation $z^4 + 8 = 0$ and use your results to factor $z^4 + 8$ into two quadratic factors with real coefficients.

13. It was shown in the text that multiplying z by i rotates z counterclockwise by $90°$. What is the geometric effect of dividing z by i?

14. In each part use (6) to calculate the given power.
    (a) $(1 + i)^8$ (b) $(-2\sqrt{3} + 2i)^{-9}$

15. In each part find Re(z) and Im(z).
    (a) $z = 3e^{i\pi}$ (b) $z = 3e^{-i\pi}$ (c) $\bar{z} = \sqrt{2}e^{\pi i/2}$ (d) $\bar{z} = -3e^{-2\pi i}$

16. (a) Show that the values of $z^{1/n}$ in Formula (10) are all different.
    (b) Show that integer values of k other than $k = 0, 1, 2, \ldots, n - 1$ produce values of $z^{1/n}$ that are duplicates of those in Formula (10).

17. Show that Formula (7) is valid if $n = 0$ or n is a negative integer.

18. **(For readers who have studied calculus.)** To prove Formula (11) recall the Maclaurin series for e^x:

$$e^x = 1 + x + \frac{x^2}{2!} + \cdots + \frac{x^n}{n!} + \cdots$$

(a) By substituting $x = i\theta$ in this series and simplifying, obtain the formula

$$e^{i\theta} = \left(1 - \frac{\theta^2}{2!} + \frac{\theta^4}{4!} - \frac{\theta^6}{6!} + \cdots\right) + i\left(\theta - \frac{\theta^3}{3!} + \frac{\theta^5}{5!} - \frac{\theta^7}{7!} + \cdots\right)$$

(b) Use the result in (a) to obtain (11).

19. Derive Formula (5).

10.4 COMPLEX VECTOR SPACES

In this section we shall develop the basic properties of complex vector spaces and discuss some of the ways in which they differ from real vector spaces. However, before going further the reader should review the definition of a complex vector space given at the beginning of Section 4.2.

BASIC PROPERTIES OF COMPLEX VECTOR SPACES

In complex vector spaces, linear combinations are defined exactly as in real vector spaces except the scalars are complex. More precisely, a vector $\mathbf{w}$ is called a **linear combination** of the vectors $\mathbf{v}_1, \mathbf{v}_2, \ldots, \mathbf{v}_r$ if $\mathbf{w}$ can be expressed in the form

$$\mathbf{w} = k_1\mathbf{v}_1 + k_2\mathbf{v}_2 + \cdots + k_r\mathbf{v}_r$$

where $k_1, k_2, \ldots, k_r$ are complex numbers.

The notions of **linear independence, spanning, basis, dimension,** and **subspace** carry over without change to complex vector spaces, and the theorems developed in Chapter 4 continue to hold.

Among the real vector spaces the most important one is R^n, the space of *n*-tuples of real numbers, with addition and scalar multiplication performed coordinatewise. Among the complex vector spaces the most important one is C^n, the space of *n*-tuples of complex numbers, with addition and scalar multiplication performed coordinatewise. A vector **u** in C^n can be written either in vector notation

$$\mathbf{u} = (u_1, u_2, \ldots, u_n)$$

or in matrix notation

$$\mathbf{u} = \begin{bmatrix} u_1 \\ u_2 \\ \vdots \\ u_n \end{bmatrix}$$

where

$$u_1 = a_1 + b_1 i, \quad u_2 = a_2 + b_2 i, \ldots, \quad u_n = a_n + b_n i$$

Example 1 If

$$\mathbf{u} = (i, 1 + i, -2) \qquad \text{and} \qquad \mathbf{v} = (2 + i, 1 - i, 3 + 2i)$$

then

$$\mathbf{u} + \mathbf{v} = (i, 1 + i, -2) + (2 + i, 1 - i, 3 + 2i) = (2 + 2i, 2, 1 + 2i)$$

and

$$i\mathbf{u} = i(i, 1 + i, -2) = (i^2, i + i^2, -2i) = (-1, -1 + i, -2i) \quad \blacktriangle$$

In C^n as in R^n, the vectors

$$\mathbf{e}_1 = (1, 0, 0, \ldots, 0), \quad \mathbf{e}_2 = (0, 1, 0, \ldots, 0), \ldots, \quad \mathbf{e}_n = (0, 0, 0, \ldots, 1)$$

form a basis. It is called the **standard basis** for C^n. Since there are *n* vectors in this basis, C^n is an *n*-dimensional vector space.

REMARK. Do not confuse the complex number $i = \sqrt{-1}$ with the vector $\mathbf{i} = (1, 0, 0)$ from the standard basis for R^3 (see Example 3, Section 3.4). The complex number i will always be set in lightface type and the vector **i** in boldface.

Example 2 In Example 3 of Section 4.2 we defined the vector space M_{mn} of $m \times n$ matrices with real entries. The complex analog of this space is the vector space of matrices with complex entries and the operations of matrix addition and scalar multiplication. We refer to this space as complex M_{mn}. ▲

Example 3 If $f_1(x)$ and $f_2(x)$ are real-valued functions of the real variable x, then the expression

$$f(x) = f_1(x) + if_2(x)$$

is called a *complex-valued function of the real variable x*. Some examples are

$$f(x) = 2x + ix^3 \quad \text{and} \quad g(x) = 2 \sin x + i \cos x \tag{1}$$

Let V be the set of all complex-valued functions that are defined on the entire line. If $\mathbf{f} = f_1(x) + if_2(x)$ and $\mathbf{g} = g_1(x) + ig_2(x)$ are two such functions and k is any complex number, then we define the *sum* function $\mathbf{f} + \mathbf{g}$ and the *scalar multiple* $k\mathbf{f}$ by

$$(\mathbf{f} + \mathbf{g})(x) = [f_1(x) + g_1(x)] + i[f_2(x) + g_2(x)]$$
$$(k\mathbf{f})(x) = kf_1(x) + ikf_2(x)$$

In words, to form $\mathbf{f} + \mathbf{g}$ add the real parts of $\mathbf{f}$ and $\mathbf{g}$ and add the imaginary parts. To form $k\mathbf{f}$ multiply the real and imaginary parts of $\mathbf{f}$ by k. For example, if $\mathbf{f} = f(x)$ and $\mathbf{g} = g(x)$ are the functions in (1), then

$$(\mathbf{f} + \mathbf{g})(x) = (2x + 2 \sin x) + i(x^3 + \cos x)$$
$$(i\mathbf{f})(x) = 2xi + i^2 x^3 = -x^3 + 2xi$$

It can be shown that V together with the stated operations is a complex vector space. It is the complex analog of the vector space of real-valued functions discussed in Example 4 of Section 4.2. ▲

Example 4   (For readers who have studied calculus.)   If $f(x) = f_1(x) + if_2(x)$ is a complex-valued function of the real variable x, then f is said to be *continuous* if $f_1(x)$ and $f_2(x)$ are continuous. We leave it as an exercise to show that the set of all continuous complex-valued functions of a real variable x is a subspace of the vector space of all complex-valued functions of x. This space is the complex analog of the vector space $C(-\infty, +\infty)$ discussed in Example 6 of Section 4.3 and is called complex $C(-\infty, +\infty)$. A closely related example is complex $C[a, b]$, the vector space of all complex-valued functions that are continuous on the closed interval $[a, b]$. ▲

Recall that in R^n the Euclidean inner product of two vectors

$$\mathbf{u} = (u_1, u_2, \ldots, u_n) \quad \text{and} \quad \mathbf{v} = (v_1, v_2, \ldots, v_n)$$

was defined as

$$\mathbf{u} \cdot \mathbf{v} = u_1 v_1 + u_2 v_2 + \cdots + u_n v_n \tag{2}$$

and the Euclidean norm (or length) of $\mathbf{u}$ as

$$\|\mathbf{u}\| = (\mathbf{u} \cdot \mathbf{u})^{1/2} = \sqrt{u_1^2 + u_2^2 + \cdots + u_n^2} \tag{3}$$

Unfortunately, these definitions are not appropriate for vectors in C^n. For example, if (3) were applied to the vector $\mathbf{u} = (i, 1)$ in C^2, we would obtain

$$\|\mathbf{u}\| = \sqrt{i^2 + 1} = \sqrt{0} = 0$$

so $\mathbf{u}$ would be a *nonzero* vector with zero length—a situation that is clearly unsatisfactory.

COMPLEX EUCLIDEAN INNER PRODUCTS

To extend the notions of norm, distance, and angle to C^n properly, we must modify the inner product slightly.

Definition. If $\mathbf{u} = (u_1, u_2, \ldots, u_n)$ and $\mathbf{v} = (v_1, v_2, \ldots, v_n)$ are vectors in C^n, then their **complex Euclidean inner product** $\mathbf{u} \cdot \mathbf{v}$ is defined by

$$\mathbf{u} \cdot \mathbf{v} = u_1\bar{v}_1 + u_2\bar{v}_2 + \cdots + u_n\bar{v}_n$$

---

where $\bar{v}_1, \bar{v}_2, \ldots, \bar{v}_n$ are the conjugates of $v_1, v_2, \ldots, v_n$.

Example 5 The complex Euclidean inner product of the vectors

$$\mathbf{u} = (-i, 2, 1 + 3i) \qquad \text{and} \qquad \mathbf{v} = (1 - i, 0, 1 + 3i)$$

is

$$
\begin{aligned}
\mathbf{u} \cdot \mathbf{v} &= (-i)\overline{(1 - i)} + (2)(\bar{0}) + (1 + 3i)\overline{(1 + 3i)} \\
&= (-i)(1 + i) + (2)(0) + (1 + 3i)(1 - 3i) \\
&= -i - i^2 + 1 - 9i^2 = 11 - i \quad \blacktriangle
\end{aligned}
$$

Theorem 4.1.2 listed the four main properties of the Euclidean inner product on R^n. The following theorem is the corresponding result for the complex Euclidean inner product on C^n.

Theorem 10.4.1. *If $\mathbf{u}$, $\mathbf{v}$, and $\mathbf{w}$ are vectors in C^n, and k is any complex number, then*

(a) $\mathbf{u} \cdot \mathbf{v} = \overline{\mathbf{v} \cdot \mathbf{u}}$

(b) $(\mathbf{u} + \mathbf{v}) \cdot \mathbf{w} = \mathbf{u} \cdot \mathbf{w} + \mathbf{v} \cdot \mathbf{w}$

(c) $(k\mathbf{u}) \cdot \mathbf{v} = k(\mathbf{u} \cdot \mathbf{v})$

(d) $\mathbf{v} \cdot \mathbf{v} \geq 0$. *Further*, $\mathbf{v} \cdot \mathbf{v} = 0$ *if and only if* $\mathbf{v} = \mathbf{0}$.

Note the difference between part (*a*) of this theorem and part (*a*) of Theorem 4.1.2. We will prove parts (*a*) and (*d*) and leave the rest as exercises.

Proof (a). Let $\mathbf{u} = (u_1, u_2, \ldots, u_n)$ and $\mathbf{v} = (v_1, v_2, \ldots, v_n)$. Then

$$\mathbf{u} \cdot \mathbf{v} = u_1\bar{v}_1 + u_2\bar{v}_2 + \cdots + u_n\bar{v}_n$$

and

$$\mathbf{v} \cdot \mathbf{u} = v_1\bar{u}_1 + v_2\bar{u}_2 + \cdots + v_n\bar{u}_n$$

so

$$
\begin{aligned}
\overline{\mathbf{v} \cdot \mathbf{u}} &= \overline{v_1\bar{u}_1 + v_2\bar{u}_2 + \cdots + v_n\bar{u}_n} \\
&= \bar{v}_1\bar{\bar{u}}_1 + \bar{v}_2\bar{\bar{u}}_2 + \cdots + \bar{v}_n\bar{\bar{u}}_n \qquad \text{[Theorem 10.2.3, parts (}a\text{) and (}c\text{)]}\\
&= \bar{v}_1 u_1 + \bar{v}_2 u_2 + \cdots + \bar{v}_n u_n \qquad \text{[Theorem 10.2.3, part (}e\text{)]}\\
&= u_1\bar{v}_1 + u_2\bar{v}_2 + \cdots + u_n\bar{v}_n \\
&= \mathbf{u} \cdot \mathbf{v}
\end{aligned}
$$

Proof (d).

$$\mathbf{v} \cdot \mathbf{v} = v_1\bar{v}_1 + v_2\bar{v}_2 + \cdots + v_n\bar{v}_n = |v_1|^2 + |v_2|^2 + \cdots + |v_n|^2 \geq 0$$

Moreover, equality holds if and only if $|v_1| = |v_2| = \cdots = |v_n| = 0$. But this is true if and only if $v_1 = v_2 = \cdots = v_n = 0$, that is, if and only if $\mathbf{v} = \mathbf{0}$. ∎

REMARK. We leave it as an exercise to prove that

$$\mathbf{u} \cdot (k\mathbf{v}) = \bar{k}(\mathbf{u} \cdot \mathbf{v})$$

for vectors in C^n. Compare this to the corresponding formula

$$\mathbf{u} \cdot (k\mathbf{v}) = k(\mathbf{u} \cdot \mathbf{v})$$

for vectors in R^n

NORM AND DISTANCE IN C^n

By analogy with (3), we define the **Euclidean norm** (or **Euclidean length**) of a vector $\mathbf{u} = (u_1, u_2, \ldots, u_n)$ in C^n by

$$\|\mathbf{u}\| = (\mathbf{u} \cdot \mathbf{u})^{1/2} = \sqrt{|u_1|^2 + |u_2|^2 + \cdots + |u_n|^2}$$

and we define the **Euclidean distance** between the points $\mathbf{u} = (u_1, u_2, \ldots, u_n)$ and $\mathbf{v} = (v_1, v_2, \ldots, v_n)$ by

$$d(\mathbf{u}, \mathbf{v}) = \|\mathbf{u} - \mathbf{v}\| = \sqrt{|u_1 - v_1|^2 + |u_2 - v_2|^2 + \cdots + |u_n - v_n|^2}$$

Example 6 If $\mathbf{u} = (i, 1 + i, 3)$ and $\mathbf{v} = (1 - i, 2, 4i)$, then

$$\|\mathbf{u}\| = \sqrt{|i|^2 + |1 + i|^2 + |3|^2} = \sqrt{1 + 2 + 9} = \sqrt{12} = 2\sqrt{3}$$

and

$$d(\mathbf{u}, \mathbf{v}) = \sqrt{|i - (1 - i)|^2 + |(1 + i) - 2|^2 + |3 - 4i|^2}$$
$$= \sqrt{|-1 + 2i|^2 + |-1 + i|^2 + |3 - 4i|^2} = \sqrt{5 + 2 + 25} = \sqrt{32} = 4\sqrt{2}$$ ▲

The complex vector space C^n with norm and inner product defined above is called **complex Euclidean n-space**.

EXERCISE SET 10.4

1. Let $\mathbf{u} = (2i, 0, -1, 3)$, $\mathbf{v} = (-i, i, 1 + i, -1)$, and $\mathbf{w} = (1 + i, -i, -1 + 2i, 0)$. Find
   (a) $\mathbf{u} - \mathbf{v}$ (b) $i\mathbf{v} + 2\mathbf{w}$ (c) $-\mathbf{w} + \mathbf{v}$ (d) $3(\mathbf{u} - (1 + i)\mathbf{v})$ (e) $-i\mathbf{v} + 2i\mathbf{w}$ (f) $2\mathbf{v} - (\mathbf{u} + \mathbf{w})$

2. Let $\mathbf{u}$, $\mathbf{v}$, and $\mathbf{w}$ be the vectors in Exercise 1. Find the vector $\mathbf{x}$ that satisfies
   $\mathbf{u} - \mathbf{v} + i\mathbf{x} = 2i\mathbf{x} + \mathbf{w}$.

3. Let $\mathbf{u}_1 = (1 - i, i, 0)$, $\mathbf{u}_2 = (2i, 1 + i, 1)$, and $\mathbf{u}_3 = (0, 2i, 2 - i)$. Find scalars c_1, c_2, and c_3 such that $c_1\mathbf{u}_1 + c_2\mathbf{u}_2 + c_3\mathbf{u}_3 = (-3 + i, 3 + 2i, 3 - 4i)$.

4. Show that there do not exist scalars c_1, c_2, and c_3 such that

   $$c_1(i, 2 - i, 2 + i) + c_2(1 + i, -2i, 2) + c_3(3, i, 6 + i) = (i, i, i)$$

5. Find the Euclidean norm of $\mathbf{v}$ if
(a) $\mathbf{v} = (1, i)$ (b) $\mathbf{v} = (1 + i, 3i, 1)$ (c) $\mathbf{v} = (2i, 0, 2i + 1, -1)$ (d) $\mathbf{v} = (-i, i, i, 3, 3 + 4i)$

6. Let $\mathbf{u} = (3i, 0, -i)$, $\mathbf{v} = (0, 3 + 4i, -2i)$, and $\mathbf{w} = (1 + i, 2i, 0)$. Find

(a) $\|\mathbf{u} + \mathbf{v}\|$ (b) $\|\mathbf{u}\| + \|\mathbf{v}\|$ (c) $\|-i\mathbf{u}\| + i\|\mathbf{u}\|$ (d) $\|3\mathbf{u} - 5\mathbf{v} + \mathbf{w}\|$ (e) $\dfrac{1}{\|\mathbf{w}\|}\mathbf{w}$ (f) $\left\|\dfrac{1}{\|\mathbf{w}\|}\mathbf{w}\right\|$

7. Show that if $\mathbf{v}$ is a nonzero vector in C^n, then $(1/\|\mathbf{v}\|)\mathbf{v}$ has Euclidean norm 1.

8. Find all scalars k such that $\|k\mathbf{v}\| = 1$, where $\mathbf{v} = (3i, 4i)$.

9. Find the Euclidean inner product $\mathbf{u} \cdot \mathbf{v}$ if
(a) $\mathbf{u} = (-i, 3i)$, $\mathbf{v} = (3i, 2i)$
(b) $\mathbf{u} = (3 - 4i, 2 + i, -6i)$, $\mathbf{v} = (1 + i, 2 - i, 4)$
(c) $\mathbf{u} = (1 - i, 1 + i, 2i, 3)$, $\mathbf{v} = (4 + 6i, -5i, -1 + i, i)$

In Exercises 10 and 11 a set of objects is given together with operations of addition and scalar multiplication. Determine which sets are complex vector spaces under the given operations. For those that are not, list all axioms that fail to hold.

10. The set of all triples of complex numbers (z_1, z_2, z_3) with the operations

$$(z_1, z_2, z_3) + (z_1', z_2', z_3') = (z_1 + z_1', z_2 + z_2', z_3 + z_3')$$

and

$$k(z_1, z_2, z_3) = (\bar{k}z_1, \bar{k}z_2, \bar{k}z_3)$$

11. The set of all complex 2×2 matrices of the form

$$\begin{bmatrix} z & 0 \\ 0 & \bar{z} \end{bmatrix}$$

with the standard matrix operations of addition and scalar multiplication.

12. Is R^n a subspace of C^n? Explain.

13. Use Theorem 4.3.1 to determine which of the following are subspaces of C^3:
(a) all vectors of the form $(z, 0, 0)$
(b) all vectors of the form (z, i, i)
(c) all vectors of the form (z_1, z_2, z_3), where $z_3 = \bar{z}_1 + \bar{z}_2$
(d) all vectors of the form (z_1, z_2, z_3), where $z_3 = z_1 + z_2 + i$

14. Use Theorem 4.3.1 to determine which of the following are subspaces of complex M_{22}:
(a) All complex matrices of the form

$$\begin{bmatrix} z_1 & z_2 \\ z_3 & z_4 \end{bmatrix}$$

where z_1 and z_2 are real.
(b) All complex matrices of the form

$$\begin{bmatrix} z_1 & z_2 \\ z_3 & z_4 \end{bmatrix}$$

where $z_1 + z_4 = 0$.
(c) All 2×2 complex matrices A such that $(\bar{A})^t = A$, where $\bar{A}$ is the matrix whose entries are the conjugates of the corresponding entries in A.

15. Use Theorem 4.3.1 to determine which of the following are subspaces of the vector space of complex-valued functions of the real variable x:
    (a) all f such that $f(1) = 0$
    (b) all f such that $f(0) = i$
    (c) all f such that $f(-x) = \overline{f(x)}$
    (d) all f of the form $k_1 + k_2 e^{ix}$, where k_1 and k_2 are complex numbers

16. Which of the following are linear combinations of $\mathbf{u} = (i, -i, 3i)$ and $\mathbf{v} = (2i, 4i, 0)$?
    (a) $(3i, 3i, 3i)$ (b) $(4i, 2i, 6i)$ (c) $(i, 5i, 6i)$ (d) $(0, 0, 0)$

17. Express the following as linear combinations of $\mathbf{u} = (1, 0, -i)$, $\mathbf{v} = (1 + i, 1, 1 - 2i)$, and $\mathbf{w} = (0, i, 2)$.
    (a) $(1, 1, 1)$ · (b) $(i, 0, -i)$ (c) $(0, 0, 0)$ (d) $(2 - i, 1, 1 + i)$

18. In each part determine whether the given vectors span C^3.
    (a) $\mathbf{v}_1 = (i, i, i)$, $\mathbf{v}_2 = (2i, 2i, 0)$, $\mathbf{v}_3 = (3i, 0, 0)$
    (b) $\mathbf{v}_1 = (1 + i, 2 - i, 3 + i)$, $\mathbf{v}_2 = (2 + 3i, 0, 1 - i)$
    (c) $\mathbf{v}_1 = (1, 0, -i)$, $\mathbf{v}_2 = (1 + i, 1, 1 - 2i)$, $\mathbf{v}_3 = (0, i, 2)$
    (d) $\mathbf{v}_1 = (1, i, 0)$, $\mathbf{v}_2 = (0, -i, 1)$, $\mathbf{v}_3 = (1, 0, 1)$

19. Determine which of the following lie in the space spanned by

    $$\mathbf{f} = e^{ix} \qquad \text{and} \qquad \mathbf{g} = e^{-ix}$$

    (a) $\cos x$ (b) $\sin x$ (c) $\cos x + 3i \sin x$

20. Explain why the following are linearly dependent sets of vectors. (Solve this problem by inspection.)
    (a) $\mathbf{u}_1 = (1 - i, i)$ and $\mathbf{u}_2 = (1 + i, -1)$ in C^2
    (b) $\mathbf{u}_1 = (1, -i)$, $\mathbf{u}_2 = (2 + i, -1)$, $\mathbf{u}_3 = (4, 0)$ in C^2

    (c) $A = \begin{bmatrix} i & 3i \\ 2i & 0 \end{bmatrix}$ and $B = \begin{bmatrix} 1 & 3 \\ 2 & 0 \end{bmatrix}$ in complex M_{22}

21. Which of the following sets of vectors in C^3 are linearly independent?
    (a) $\mathbf{u}_1 = (1 - i, 1, 0)$, $\mathbf{u}_2 = (2, 1 + i, 0)$, $\mathbf{u}_3 = (1 + i, i, 0)$
    (b) $\mathbf{u}_1 = (1, 0, -i)$, $\mathbf{u}_2 = (1 + i, 1, 1 - 2i)$, $\mathbf{u}_3 = (0, i, 2)$
    (c) $\mathbf{u}_1 = (i, 0, 2 - i)$, $\mathbf{u}_2 = (0, 1, i)$, $\mathbf{u}_3 = (-i, -1 - 4i, 3)$

22. Let V be the vector space of all complex-valued functions of the real variable x. Show that the following vectors are linearly dependent.

    $$\mathbf{f} = 3 + 3i \cos 2x, \qquad \mathbf{g} = \sin^2 x + i \cos^2 x, \qquad \mathbf{h} = \cos^2 x - i \sin^2 x$$

23. Explain why the following sets of vectors are not bases for the indicated vector spaces. (Solve this problem by inspection.)
    (a) $\mathbf{u}_1 = (i, 2i)$, $\mathbf{u}_2 = (0, 3i)$, $\mathbf{u}_3 = (1, 7i)$ for C^2
    (b) $\mathbf{u}_1 = (-1 + i, 0, 2 - i)$, $\mathbf{u}_2 = (1, -i, 1 + i)$ for C^3

24. Which of the following sets of vectors are bases for C^2?
    (a) $(2i, -i)$, $(4i, 0)$ (b) $(1 + i, 1)$, $(1 + i, i)$
    (c) $(0, 0)$, $(1 + i, 1 - i)$ (d) $(2 - 3i, i)$, $(3 + 2i, -1)$

25. Which of the following sets of vectors are bases for C^3?

(a) $(i, 0, 0)$, $(i, i, 0)$, (i, i, i)

(b) $(1, 0, -i)$, $(1 + i, 1, 1 - 2i)$, $(0, i, 2)$

(c) $(i, 0, 2 - i)$, $(0, 1, i)$, $(-i, -1 - 4i, 3)$

(d) $(1, 0, i)$, $(2 - i, 1, 2 + i)$, $(0, 3i, 3i)$

In Exercises 26–29 determine the dimension of and a basis for the solution space of the system.

26.
$$x_1 + (1 + i)x_2 = 0$$
$$(1 - i)x_1 + 2x_2 = 0$$

27.
$$2x_1 - (1 + i)x_2 = 0$$
$$(-1 + i)x_1 + x_2 = 0$$

28.
$$x_1 + (2 - i)x_2 = 0$$
$$x_2 + 3ix_3 = 0$$
$$ix_1 + (2 + 2i)x_2 + 3ix_3 = 0$$

29.
$$x_1 + ix_2 - 2ix_3 + x_4 = 0$$
$$ix_1 + 3x_2 + 4x_3 - 2ix_4 = 0$$

30. Prove: If **u** and **v** are vectors in complex Euclidean n-space, then

$$\mathbf{u} \cdot (k\mathbf{v}) = \bar{k}(\mathbf{u} \cdot \mathbf{v})$$

31. (a) Prove part (b) of Theorem 10.4.1.   (b) Prove part (c) of Theorem 10.4.1.

32. **(For readers who have studied calculus.)** Prove that complex $C(-\infty, +\infty)$ is a subspace of the vector space of complex-valued functions of a real variable.

33. Establish the identity

$$\mathbf{u} \cdot \mathbf{v} = \frac{1}{4}\|\mathbf{u} + \mathbf{v}\|^2 - \frac{1}{4}\|\mathbf{u} - \mathbf{v}\|^2 + \frac{i}{4}\|\mathbf{u} + i\mathbf{v}\|^2 - \frac{i}{4}\|\mathbf{u} - i\mathbf{v}\|^2$$

for vectors in complex Euclidean n-space.

10.5 COMPLEX INNER PRODUCT SPACES

In Section 5.1 we defined the notion of an inner product on a real vector space by using the basic properties of the Euclidean inner product on R^n as axioms. In this section we shall define inner products on complex vector spaces by using the properties of the Euclidean inner product on C^n as axioms.

UNITARY SPACES

Motivated by Theorem 10.4.1 we make the following definition.

Definition. An ***inner product on a complex vector space*** V is a function that associates a complex number $\langle \mathbf{u}, \mathbf{v} \rangle$ with each pair of vectors **u** and **v** in V in such a way that the following axioms are satisfied for all vectors **u**, **v**, and **w** in V and all scalars k.

(1) $\langle \mathbf{u}, \mathbf{v} \rangle = \overline{\langle \mathbf{v}, \mathbf{u} \rangle}$

(2) $\langle \mathbf{u} + \mathbf{v}, \mathbf{w} \rangle = \langle \mathbf{u}, \mathbf{w} \rangle + \langle \mathbf{v}, \mathbf{w} \rangle$

(3) $\langle k\mathbf{u}, \mathbf{v} \rangle = k\langle \mathbf{u}, \mathbf{v} \rangle$

(4) $\langle \mathbf{v}, \mathbf{v} \rangle \geq 0$ and $\langle \mathbf{v}, \mathbf{v} \rangle = 0$ if and only if $\mathbf{v} = \mathbf{0}$

A complex vector space with an inner product is called a **complex inner product space** or a **unitary space**.

The following additional properties follow immediately from the four inner product axioms:

(i) $\langle \mathbf{0}, \mathbf{v} \rangle = \langle \mathbf{v}, \mathbf{0} \rangle = 0$
(ii) $\langle \mathbf{u}, \mathbf{v} + \mathbf{w} \rangle = \langle \mathbf{u}, \mathbf{v} \rangle + \langle \mathbf{u}, \mathbf{w} \rangle$
(iii) $\langle \mathbf{u}, k\mathbf{v} \rangle = \bar{k}\langle \mathbf{u}, \mathbf{v} \rangle$

Since only (iii) differs from the corresponding results for real inner products, we will prove it and leave the other proofs as exercises.

$$
\begin{aligned}
\langle \mathbf{u}, k\mathbf{v} \rangle &= \overline{\langle k\mathbf{v}, \mathbf{u} \rangle} && \text{[Axiom 1]}\\
&= \overline{k\langle \mathbf{v}, \mathbf{u} \rangle} && \text{[Axiom 3]}\\
&= \bar{k}\overline{\langle \mathbf{v}, \mathbf{u} \rangle} && \text{[Property of conjugates]}\\
&= \bar{k}\langle \mathbf{u}, \mathbf{v} \rangle && \text{[Axiom 1]}
\end{aligned}
$$

Example 1   Let $\mathbf{u} = (u_1, u_2, \ldots, u_n)$ and $\mathbf{v} = (v_1, v_2, \ldots, v_n)$ be vectors in C^n. The Euclidean inner product $\langle \mathbf{u}, \mathbf{v} \rangle = \mathbf{u} \cdot \mathbf{v} = u_1\bar{v}_1 + u_2\bar{v}_2 + \cdots + u_n\bar{v}_n$ satisfies all the inner product axioms by Theorem 10.4.1.   ▲

Example 2   If

$$
U = \begin{bmatrix} u_1 & u_2 \\ u_3 & u_4 \end{bmatrix} \quad \text{and} \quad V = \begin{bmatrix} v_1 & v_2 \\ v_3 & v_4 \end{bmatrix}
$$

are any 2×2 matrices with complex entries, then the following formula defines a complex inner product on complex M_{22} (verify):

$$
\langle U, V \rangle = u_1\bar{v}_1 + u_2\bar{v}_2 + u_3\bar{v}_3 + u_4\bar{v}_4
$$

For example, if

$$
U = \begin{bmatrix} 0 & i \\ 1 & 1+i \end{bmatrix} \quad \text{and} \quad V = \begin{bmatrix} 1 & -i \\ 0 & 2i \end{bmatrix}
$$

then

$$
\begin{aligned}
\langle U, V \rangle &= (0)(\bar{1}) + i(\overline{-i}) + (1)(\bar{0}) + (1+i)(\overline{2i})\\
&= (0)(1) + i(i) + (1)(0) + (1+i)(-2i)\\
&= 0 + i^2 + 0 - 2i - 2i^2\\
&= 1 - 2i \quad ▲
\end{aligned}
$$

Example 3   (For readers who have studied calculus.)   If $f(x) = f_1(x) + if_2(x)$ is a complex-valued function of the real variable x, and if $f_1(x)$ and $f_2(x)$ are continuous on $[a, b]$, then we define

$$
\int_a^b f(x)\,dx = \int_a^b [f_1(x) + if_2(x)]\,dx = \int_a^b f_1(x)\,dx + i\int_a^b f_2(x)\,dx
$$

In words, *the integral of $f(x)$ is the integral of the real part of f plus i times the integral of the imaginary part of f.*

We leave it as an exercise to show that if the functions $\mathbf{f} = f_1(x) + if_2(x)$ and $\mathbf{g} = g_1(x) + ig_2(x)$ are vectors in complex $C[a, b]$, then the following formula defines an inner product on complex $C[a, b]$:

$$\langle \mathbf{f}, \mathbf{g} \rangle = \int_a^b [f_1(x) + if_2(x)]\overline{[g_1(x) + ig_2(x)]}\,dx$$

$$= \int_a^b [f_1(x) + if_2(x)][g_1(x) - ig_2(x)]\,dx$$

$$= \int_a^b [f_1(x)g_1(x) + f_2(x)g_2(x)]\,dx + i\int_a^b [f_2(x)g_1(x) - f_1(x)g_2(x)]\,dx \quad \blacktriangle$$

In complex inner product spaces, as in real inner product spaces, the **norm** (or **length**) of a vector $\mathbf{u}$ is defined by

$$\|\mathbf{u}\| = \langle \mathbf{u}, \mathbf{u} \rangle^{1/2}$$

and the **distance** between two vectors $\mathbf{u}$ and $\mathbf{v}$ is defined by

$$d(\mathbf{u}, \mathbf{v}) = \|\mathbf{u} - \mathbf{v}\|$$

It can be shown that with these definitions Theorem 5.2.2 remains true in complex inner product spaces (Exercise 35).

Example 4   If $\mathbf{u} = (u_1, u_2, \ldots, u_n)$ and $\mathbf{v} = (v_1, v_2, \ldots, v_n)$ are vectors in C^n with the Euclidean inner product, then

$$\|\mathbf{u}\| = \langle \mathbf{u}, \mathbf{u} \rangle^{1/2} = \sqrt{|u_1|^2 + |u_2|^2 + \cdots + |u_n|^2}$$

and

$$d(\mathbf{u}, \mathbf{v}) = \|\mathbf{u} - \mathbf{v}\| = \langle \mathbf{u} - \mathbf{v}, \mathbf{u} - \mathbf{v} \rangle^{1/2}$$
$$= \sqrt{|u_1 - v_1|^2 + |u_2 - v_2|^2 + \cdots + |u_n - v_n|^2}$$

Observe that these are just the formulas for the Euclidean norm and distance discussed in Section 10.4. $\blacktriangle$

Example 5   **(For readers who have studied calculus.)**   If complex $C[0, 2\pi]$ has the inner product of Example 3, and if $\mathbf{f} = e^{imx}$, where m is any integer, then with the help of Formula (15) of Section 10.3 we obtain

$$\|\mathbf{f}\| = \langle \mathbf{f}, \mathbf{f} \rangle^{1/2} = \left[\int_0^{2\pi} e^{imx}\overline{e^{imx}}\,dx\right]^{1/2}$$

$$= \left[\int_0^{2\pi} e^{imx}e^{-imx}\,dx\right]^{1/2} = \left[\int_0^{2\pi} dx\right]^{1/2} = \sqrt{2\pi} \quad \blacktriangle$$

ORTHOGONAL SETS

The definitions of such terms as **orthogonal vectors**, **orthogonal set**, **orthonormal set**, and **orthonormal basis** carry over to unitary spaces without change. Moreover, Theorem 5.2.3, and the theorems in Section 5.3, and Theorem 5.4.1

remain valid in complex inner product spaces, and the Gram-Schmidt process can be used to convert an arbitrary basis for a complex inner product space into an orthonormal basis.

Example 6   The vectors

$$\mathbf{u} = (i, 1) \qquad \text{and} \qquad \mathbf{v} = (1, i)$$

in C^2 are orthogonal with respect to the Euclidean inner product, since

$$\mathbf{u} \cdot \mathbf{v} = (i)(\bar{1}) + (1)(\bar{i}) = (i)(1) + (1)(-i) = 0 \quad \blacktriangle$$

Example 7   Consider the vector space C^3 with the Euclidean inner product. Apply the Gram-Schmidt process to transform the basis vectors $\mathbf{u}_1 = (i, i, i)$, $\mathbf{u}_2 = (0, i, i)$, $\mathbf{u}_3 = (0, 0, i)$ into an orthonormal basis.

Solution.

Step 1. $\mathbf{v}_1 = \mathbf{u}_1 = (i, i, i)$

Step 2. $\mathbf{v}_2 = \mathbf{u}_2 - \text{proj}_{W_1} \mathbf{u}_2 = \mathbf{u}_2 - \dfrac{\langle \mathbf{u}_2, \mathbf{v}_1 \rangle}{\|\mathbf{v}_1\|^2} \mathbf{v}_1$

$$= (0, i, i) - \frac{2}{3}(i, i, i) = \left(-\frac{2}{3}i, \frac{1}{3}i, \frac{1}{3}i \right)$$

Step 3. $\mathbf{v}_3 = \mathbf{u}_3 - \text{proj}_{W_2} \mathbf{u}_3 = \mathbf{u}_3 - \dfrac{\langle \mathbf{u}_3, \mathbf{v}_1 \rangle}{\|\mathbf{v}_1\|^2} \mathbf{v}_1 - \dfrac{\langle \mathbf{u}_3, \mathbf{v}_2 \rangle}{\|\mathbf{v}_2\|^2} \mathbf{v}_2$

$$= (0, 0, i) - \frac{1}{3}(i, i, i) - \frac{\frac{1}{3}}{\frac{2}{3}} \left(-\frac{2}{3}i, \frac{1}{3}i, \frac{1}{3}i \right)$$

$$= \left(0, -\frac{1}{2}i, \frac{1}{2}i \right)$$

Thus,

$$\mathbf{v}_1 = (i, i, i), \ \mathbf{v}_2 = \left(-\frac{2}{3}i, \frac{1}{3}i, \frac{1}{3}i \right), \ \mathbf{v}_3 = \left(0, -\frac{1}{2}i, \frac{1}{2}i \right)$$

form an orthogonal basis for C^3. The norms of these vectors are

$$\|\mathbf{v}_1\| = \sqrt{3}, \qquad \|\mathbf{v}_2\| = \frac{\sqrt{6}}{3}, \qquad \|\mathbf{v}_3\| = \frac{1}{\sqrt{2}}$$

so an orthonormal basis for C^3 is

$$\frac{\mathbf{v}_1}{\|\mathbf{v}_1\|} = \left(\frac{i}{\sqrt{3}}, \frac{i}{\sqrt{3}}, \frac{i}{\sqrt{3}} \right), \qquad \frac{\mathbf{v}_2}{\|\mathbf{v}_2\|} = \left(-\frac{2i}{\sqrt{6}}, \frac{i}{\sqrt{6}}, \frac{i}{\sqrt{6}} \right),$$

$$\frac{\mathbf{v}_3}{\|\mathbf{v}_3\|} = \left(0, -\frac{i}{\sqrt{2}}, \frac{i}{\sqrt{2}} \right) \quad \blacktriangle$$

Example 8 **(For readers who have studied calculus.)** Let complex $C[0, 2\pi]$ have the inner product of Example 3, and let W be the set of vectors in $C[0, 2\pi]$ of the form

$$e^{im\pi} = \cos m\pi + i \sin m\pi$$

where m is an integer. The set W is orthogonal because if

$$\mathbf{f} = e^{ikx} \quad \text{and} \quad \mathbf{g} = e^{ilx}$$

are distinct vectors in W, then

$$\langle \mathbf{f}, \mathbf{g} \rangle = \int_0^{2\pi} e^{ikx}\overline{e^{ilx}}\, dx$$

$$= \int_0^{2\pi} e^{ikx} e^{-ilx}\, dx$$

$$= \int_0^{2\pi} e^{i(k-l)x}\, dx$$

$$= \int_0^{2\pi} \cos(k-l)x\, dx + i \int_0^{2\pi} \sin(k-l)x\, dx$$

$$= \left[\frac{1}{k-l} \sin(k-l)x \right]_0^{2\pi} - i \left[\frac{1}{k-l} \cos(k-l)x \right]_0^{2\pi}$$

$$= (0) - i(0) = 0$$

If we normalize each vector in the orthogonal set W, we obtain an orthonormal set. But in Example 5 we showed that each vector in W has norm $\sqrt{2\pi}$, so the vectors

$$\frac{1}{\sqrt{2\pi}} e^{imx} \qquad m = 0, \pm 1, \pm 2, \ldots$$

form an orthonormal set in complex $C[0, 2\pi]$. ▲

EXERCISE SET 10.5

1. Let $\mathbf{u} = (u_1, u_2)$ and $\mathbf{v} = (v_1, v_2)$. Show that $\langle \mathbf{u}, \mathbf{v} \rangle = 3u_1\bar{v}_1 + 2u_2\bar{v}_2$ defines an inner product on C^2.

2. Compute $\langle \mathbf{u}, \mathbf{v} \rangle$ using the inner product in Exercise 1.
(a) $\mathbf{u} = (2i, -i)$, $\mathbf{v} = (-i, 3i)$ (b) $\mathbf{u} = (0, 0)$, $\mathbf{v} = (1 - i, 7 - 5i)$
(c) $\mathbf{u} = (1 + i, 1 - i)$, $\mathbf{v} = (1 - i, 1 + i)$ (d) $\mathbf{u} = (3i, -1 + 2i)$, $\mathbf{v} = (3i, -1 + 2i)$

3. Let $\mathbf{u} = (u_1, u_2)$ and $\mathbf{v} = (v_1, v_2)$. Show that

$$\langle \mathbf{u}, \mathbf{v} \rangle = u_1\bar{v}_1 + (1 + i)u_1\bar{v}_2 + (1 - i)u_2\bar{v}_1 + 3u_2\bar{v}_2$$

defines an inner product on C^2.

4. Compute $\langle \mathbf{u}, \mathbf{v} \rangle$ using the inner product in Exercise 3.
(a) $\mathbf{u} = (2i, -i)$, $\mathbf{v} = (-i, 3i)$ (b) $\mathbf{u} = (0, 0)$, $\mathbf{v} = (1 - i, 7 - 5i)$
(c) $\mathbf{u} = (1 + i, 1 - i)$, $\mathbf{v} = (1 - i, 1 + i)$ (d) $\mathbf{u} = (3i, -1 + 2i)$, $\mathbf{v} = (3i, -1 + 2i)$

5. Let $\mathbf{u} = (u_1, u_2)$ and $\mathbf{v} = (v_1, v_2)$. Determine which of the following are inner products on C^2. For those that are not, list the axioms that do not hold.
(a) $\langle \mathbf{u}, \mathbf{v} \rangle = u_1 \bar{v}_1$ (b) $\langle \mathbf{u}, \mathbf{v} \rangle = u_1 \bar{v}_1 - u_2 \bar{v}_2$ (c) $\langle \mathbf{u}, \mathbf{v} \rangle = |u_1|^2 |v_1|^2 + |u_2|^2 |v_2|^2$
(d) $\langle \mathbf{u}, \mathbf{v} \rangle = 2u_1 \bar{v}_1 + iu_1 \bar{v}_2 + iu_2 \bar{v}_1 + 2u_2 \bar{v}_2$ (e) $\langle \mathbf{u}, \mathbf{v} \rangle = 2u_1 \bar{v}_1 + iu_1 \bar{v}_2 - iu_2 \bar{v}_1 + 2u_2 \bar{v}_2$

6. Use the inner product of Example 2 to find $\langle U, V \rangle$ if

$$U = \begin{bmatrix} -i & 1+i \\ 1-i & i \end{bmatrix} \quad \text{and} \quad V = \begin{bmatrix} 3 & -2-3i \\ 4i & 1 \end{bmatrix}$$

7. Let $\mathbf{u} = (u_1, u_2, u_3)$ and $\mathbf{v} = (v_1, v_2, v_3)$. Does $\langle \mathbf{u}, \mathbf{v} \rangle = u_1 \bar{v}_1 + u_2 \bar{v}_2 + u_3 \bar{v}_3 - iu_3 \bar{v}_1$ define an inner product on C^3? If not, list all axioms that fail to hold.

8. Let V be the vector space of complex-valued functions of the real variable x, and let $\mathbf{f} = f_1(x) + if_2(x)$ and $\mathbf{g} = g_1(x) + ig_2(x)$ be vectors in V. Does

$$\langle \mathbf{f}, \mathbf{g} \rangle = (f_1(0) + if_2(0))\overline{(g_1(0) + ig_2(0))}$$

define an inner product on V? If not, list all axioms that fail to hold.

9. Let C^2 have the inner product of Exercise 1. Find $\|\mathbf{w}\|$ if
(a) $\mathbf{w} = (-i, 3i)$ (b) $\mathbf{w} = (1 - i, 1 + i)$ (c) $\mathbf{w} = (0, 2 - i)$ (d) $\mathbf{w} = (0, 0)$

10. For each vector in Exercise 9, use the Euclidean inner product to find $\|\mathbf{w}\|$.

11. Use the inner product of Exercise 3 to find $\|\mathbf{w}\|$ if
(a) $\mathbf{w} = (1, -i)$ (b) $\mathbf{w} = (1 - i, 1 + i)$ (c) $\mathbf{w} = (3 - 4i, 0)$ (d) $\mathbf{w} = (0, 0)$

12. Use the inner product of Example 2 to find $\|A\|$ if

(a) $A = \begin{bmatrix} -i & 7i \\ 6i & 2i \end{bmatrix}$ (b) $A = \begin{bmatrix} -1 & 1+i \\ 1-i & 3 \end{bmatrix}$

13. Let C^2 have the inner product of Exercise 1. Find $d(\mathbf{x}, \mathbf{y})$ if
(a) $\mathbf{x} = (1, 1)$, $\mathbf{y} = (i, -i)$ (b) $\mathbf{x} = (1 - i, 3 + 2i)$, $\mathbf{y} = (1 + i, 3)$

14. Repeat the directions of Exercise 13 using the Euclidean inner product on C^2.

15. Repeat the directions of Exercise 13 using the inner product of Exercise 3.

16. Let complex M_{22} have the inner product of Example 2. Find $d(A, B)$ if

(a) $A = \begin{bmatrix} i & 5i \\ 8i & 3i \end{bmatrix} \quad \text{and} \quad B = \begin{bmatrix} -5i & 0 \\ 7i & -3i \end{bmatrix}$

(b) $A = \begin{bmatrix} -1 & 1-i \\ 1+i & 2 \end{bmatrix} \quad \text{and} \quad B = \begin{bmatrix} 2i & 2-3i \\ i & 1 \end{bmatrix}$

17. Let C^3 have the Euclidean inner product. For which complex values of k are $\mathbf{u}$ and $\mathbf{v}$ orthogonal?
(a) $\mathbf{u} = (2i, i, 3i)$, $\mathbf{v} = (i, 6i, k)$ (b) $\mathbf{u} = (k, k, 1 + i)$, $\mathbf{v} = (1, -1, 1 - i)$

18. Let M_{22} have the inner product of Example 2. Determine which of the following are orthogonal to

$$A = \begin{bmatrix} 2i & i \\ -i & 3i \end{bmatrix}$$

(a) $\begin{bmatrix} -3 & 1-i \\ 1-i & 2 \end{bmatrix}$ (b) $\begin{bmatrix} 1 & 1 \\ 0 & -1 \end{bmatrix}$ (c) $\begin{bmatrix} 0 & 0 \\ 0 & 0 \end{bmatrix}$ (d) $\begin{bmatrix} 0 & 1 \\ 3-i & 0 \end{bmatrix}$

19. Let C^3 have the Euclidean inner product. Show that for all values of θ the vector
$$\mathbf{x} = e^{i\theta}\left(\frac{i}{\sqrt{3}}, \frac{1}{\sqrt{3}}, \frac{1}{\sqrt{3}}\right) \text{ has norm 1 and is orthogonal to both } (1, i, 0) \text{ and } (0, i, -i).$$

20. Let C^2 have the Euclidean inner product. Which of the following form orthonormal sets?

(a) $(i, 0)$, $(0, 1 - i)$ (b) $\left(\dfrac{i}{\sqrt{2}}, -\dfrac{i}{\sqrt{2}}\right)$, $\left(\dfrac{i}{\sqrt{2}}, \dfrac{i}{\sqrt{2}}\right)$ (c) $\left(\dfrac{i}{\sqrt{2}}, \dfrac{i}{\sqrt{2}}\right)$, $\left(-\dfrac{i}{\sqrt{2}}, -\dfrac{i}{\sqrt{2}}\right)$ (d) $(i, 0)$, $(0, 0)$

21. Let C^3 have the Euclidean inner product. Which of the following form orthonormal sets?

(a) $\left(\dfrac{i}{\sqrt{2}}, 0, \dfrac{i}{\sqrt{2}}\right)$, $\left(\dfrac{i}{\sqrt{3}}, \dfrac{i}{\sqrt{3}}, -\dfrac{i}{\sqrt{3}}\right)$, $\left(-\dfrac{i}{\sqrt{2}}, 0, \dfrac{i}{\sqrt{2}}\right)$ (b) $\left(\dfrac{2}{3}i, -\dfrac{2}{3}i, \dfrac{1}{3}i\right)$, $\left(\dfrac{2}{3}i, \dfrac{1}{3}i, -\dfrac{2}{3}i\right)$, $\left(\dfrac{1}{3}i, \dfrac{2}{3}i, \dfrac{2}{3}i\right)$

(c) $\left(\dfrac{i}{\sqrt{6}}, \dfrac{i}{\sqrt{6}}, -\dfrac{2i}{\sqrt{6}}\right)$, $\left(\dfrac{i}{\sqrt{2}}, -\dfrac{i}{\sqrt{2}}, 0\right)$

22. Let
$$\mathbf{x} = \left(\frac{i}{\sqrt{5}}, -\frac{i}{\sqrt{5}}\right) \quad \text{and} \quad \mathbf{y} = \left(\frac{2i}{\sqrt{30}}, \frac{3i}{\sqrt{30}}\right)$$

Show that $\{\mathbf{x}, \mathbf{y}\}$ is an orthonormal set if C^2 has the inner product

$$\langle \mathbf{u}, \mathbf{v}\rangle = 3u_1\bar{v}_1 + 2u_2\bar{v}_2$$

but is not orthonormal if C^2 has the Euclidean inner product.

23. Show that
$$\mathbf{u}_1 = (i, 0, 0, i), \quad \mathbf{u}_2 = (-i, 0, 2i, i), \quad \mathbf{u}_3 = (2i, 3i, 2i, -2i), \quad \mathbf{u}_4 = (-i, 2i, -i, i)$$

is an orthogonal set in C^4 with the Euclidean inner product. By normalizing each of these vectors, obtain an orthonormal set.

24. Let C^2 have the Euclidean inner product. Use the Gram-Schmidt process to transform the basis $\{\mathbf{u}_1, \mathbf{u}_2\}$ into an orthonormal basis.
(a) $\mathbf{u}_1 = (i, -3i)$, $\mathbf{u}_2 = (2i, 2i)$ (b) $\mathbf{u}_1 = (i, 0)$, $\mathbf{u}_2 = (3i, -5i)$

25. Let C^3 have the Euclidean inner product. Use the Gram-Schmidt process to transform the basis $\{\mathbf{u}_1, \mathbf{u}_2, \mathbf{u}_3\}$ into an orthonormal basis.
(a) $\mathbf{u}_1 = (i, i, i)$, $\mathbf{u}_2 = (-i, i, 0)$, $\mathbf{u}_3 = (i, 2i, i)$ (b) $\mathbf{u}_1 = (i, 0, 0)$, $\mathbf{u}_2 = (3i, 7i, -2i)$, $\mathbf{u}_3 = (0, 4i, i)$

26. Let C^4 have the Euclidean inner product. Use the Gram-Schmidt process to transform the basis $\{\mathbf{u}_1, \mathbf{u}_2, \mathbf{u}_3, \mathbf{u}_4\}$ into an orthonormal basis.

$$\mathbf{u}_1 = (0, 2i, i, 0), \quad \mathbf{u}_2 = (i, -i, 0, 0), \quad \mathbf{u}_3 = (i, 2i, 0, -i), \quad \mathbf{u}_4 = (i, 0, i, i)$$

27. Let C^3 have the Euclidean inner product. Find an orthonormal basis for the subspace spanned by $(0, i, 1 - i)$ and $(-i, 0, 1 + i)$.

28. Let C^4 have the Euclidean inner product. Express $\mathbf{w} = (-i, 2i, 6i, 0)$ in the form $\mathbf{w} = \mathbf{w}_1 + \mathbf{w}_2$, where the vector $\mathbf{w}_1$ is in the space W spanned by $\mathbf{u}_1 = (-i, 0, i, 2i)$ and $\mathbf{u}_2 = (0, i, 0, i)$, and $\mathbf{w}_2$ is orthogonal to W.

29. (a) Prove: If k is a complex number and $\langle \mathbf{u}, \mathbf{v}\rangle$ is an inner product on a complex vector space, then $\langle \mathbf{u} - k\mathbf{v}, \mathbf{u} - k\mathbf{v}\rangle = \langle \mathbf{u}, \mathbf{u}\rangle - \bar{k}\langle \mathbf{u}, \mathbf{v}\rangle - k\overline{\langle \mathbf{u}, \mathbf{v}\rangle} + k\bar{k}\langle \mathbf{v}, \mathbf{v}\rangle$.
(b) Use the result in (a) to prove that $0 \le \langle \mathbf{u}, \mathbf{u}\rangle - \bar{k}\langle \mathbf{u}, \mathbf{v}\rangle - k\overline{\langle \mathbf{u}, \mathbf{v}\rangle} + k\bar{k}\langle \mathbf{v}, \mathbf{v}\rangle$.

30. Prove that if **u** and **v** are vectors in a complex inner product space, then

$$|\langle \mathbf{u}, \mathbf{v} \rangle|^2 \leq \langle \mathbf{u}, \mathbf{u} \rangle \langle \mathbf{v}, \mathbf{v} \rangle$$

This result, called the ***Cauchy-Schwarz inequality for complex inner product spaces,*** differs from its real analog (Theorem 5.2.1) in that an absolute value sign must be included in the left side. [***Hint.*** Let $k = \langle \mathbf{u}, \mathbf{v} \rangle / \langle \mathbf{v}, \mathbf{v} \rangle$ in the inequality of Exercise 29(b).]

31. Prove: If $\mathbf{u} = (u_1, u_2, \ldots, u_n)$ and $\mathbf{v} = (v_1, v_2, \ldots, v_n)$ are vectors in C^n, then

$$|u_1 \bar{v}_1 + u_2 \bar{v}_2 + \cdots + u_n \bar{v}_n| \leq (|u_1|^2 + |u_2|^2 + \cdots + |u_n|^2)^{1/2} (|v_1|^2 + |v_2|^2 + \cdots + |v_n|^2)^{1/2}$$

This is the complex version of ***Cauchy's inequality*** discussed in Example 1 of Section 5.2. [***Hint.*** Use Exercise 30.]

32. Prove that equality holds in the Cauchy-Schwarz inequality for complex vector spaces if and only if **u** and **v** are linearly dependent.

33. Prove that if $\langle \mathbf{u}, \mathbf{v} \rangle$ is an inner product on a complex vector space, then

$$\langle \mathbf{0}, \mathbf{v} \rangle = \langle \mathbf{v}, \mathbf{0} \rangle = 0$$

34. Prove that if $\langle \mathbf{u}, \mathbf{v} \rangle$ is an inner product on a complex vector space, then

$$\langle \mathbf{u}, \mathbf{v} + \mathbf{w} \rangle = \langle \mathbf{u}, \mathbf{v} \rangle + \langle \mathbf{u}, \mathbf{w} \rangle$$

35. Theorem 5.2.2 remains true in complex inner product spaces. In each part prove that this is so.
(a) part $L1$ (b) part $L2$ (c) part $L3$ (d) part $L4$
(e) part $D1$ (f) part $D2$ (g) part $D3$ (h) part $D4$

36. In Example 7 it was shown that the vectors

$$\mathbf{v}_1 = \left(\frac{i}{\sqrt{3}}, \frac{i}{\sqrt{3}}, \frac{i}{\sqrt{3}} \right) \qquad \mathbf{v}_2 = \left(-\frac{2i}{\sqrt{6}}, \frac{i}{\sqrt{6}}, \frac{i}{\sqrt{6}} \right) \qquad \mathbf{v}_3 = \left(0, -\frac{i}{\sqrt{2}}, \frac{i}{\sqrt{2}} \right)$$

form an orthonormal basis for C^3. Use Theorem 5.3.1 to express $\mathbf{u} = (1 - i, 1 + i, 1)$ as a linear combination of these vectors.

37. Prove that if **u** and **v** are vectors in a complex inner product space, then

$$\langle \mathbf{u}, \mathbf{v} \rangle = \frac{1}{4} \|\mathbf{u} + \mathbf{v}\|^2 - \frac{1}{4} \|\mathbf{u} - \mathbf{v}\|^2 + \frac{i}{4} \|\mathbf{u} + i\mathbf{v}\|^2 - \frac{i}{4} \|\mathbf{u} - i\mathbf{v}\|^2$$

38. Prove: If $\mathbf{v}_1, \mathbf{v}_2, \ldots, \mathbf{v}_n$ is an orthonormal basis for a complex inner product space V, and if **u** and **w** are any vectors in V, then

$$\langle \mathbf{u}, \mathbf{w} \rangle = \langle \mathbf{u}, \mathbf{v}_1 \rangle \overline{\langle \mathbf{w}, \mathbf{v}_1 \rangle} + \langle \mathbf{u}, \mathbf{v}_2 \rangle \overline{\langle \mathbf{w}, \mathbf{v}_2 \rangle} + \cdots + \langle \mathbf{u}, \mathbf{v}_n \rangle \overline{\langle \mathbf{w}, \mathbf{v}_n \rangle}$$

[***Hint.*** Use Theorem 5.3.1 to express **u** and **w** as linear combinations of the basis vectors.]

39. (**For readers who have studied calculus.**) Prove that if $\mathbf{f} = f_1(x) + if_2(x)$ and $\mathbf{g} = g_1(x) + ig_2(x)$ are vectors in complex $C[a, b]$, then the formula

$$\langle \mathbf{f}, \mathbf{g} \rangle = \int_a^b [f_1(x) + if_2(x)] \overline{[g_1(x) + ig_2(x)]} \, dx$$

defines a complex inner product on $C[a, b]$.

40. (For readers who have studied calculus.)

(a) Let $\mathbf{f} = f_1(x) + if_2(x)$ and $\mathbf{g} = g_1(x) + ig_2(x)$ be vectors in $C[0, 1]$ and let this space have the inner product

$$\langle \mathbf{f}, \mathbf{g} \rangle = \int_0^1 [f_1(x) + if_2(x)]\overline{[g_1(x) + ig_2(x)]}\, dx$$

Show that the vectors

$$e^{2\pi imx}, \quad m = 0, \pm 1, \pm 2, \ldots$$

form an orthogonal set.

(b) Obtain an orthonormal set by normalizing the vectors in part (a).

10.6 UNITARY, NORMAL, AND HERMITIAN MATRICES

*For matrices with real entries, the orthogonal matrices $(A^{-1} = A^t)$ and the symmetric matrices $(A = A^t)$ played an important role in the orthogonal diagonalization problem (Section 6.3). For matrices with complex entries, the orthogonal and symmetric matrices are of relatively little importance; they are superseded by two new classes of matrices, the **unitary** and **Hermitian** matrices, which we shall discuss in this section.*

UNITARY MATRICES

If A is a matrix with complex entries, then the **conjugate transpose** of A, denoted by A^*, is defined by

$$A^* = \bar{A}^t$$

where $\bar{A}$ is the matrix whose entries are the complex conjugates of the corresponding entries in A and $\bar{A}^t$ is the transpose of $\bar{A}$.

Example 1 If

$$A = \begin{bmatrix} 1 + i & -i & 0 \\ 2 & 3 - 2i & i \end{bmatrix}$$

then

$$\bar{A} = \begin{bmatrix} 1 - i & i & 0 \\ 2 & 3 + 2i & -i \end{bmatrix}$$

So

$$A^* = \bar{A}^t = \begin{bmatrix} 1 - i & 2 \\ i & 3 + 2i \\ 0 & -i \end{bmatrix} \quad \blacktriangle$$

The basic properties of the conjugate transpose operation are similar to those of the transpose:

Theorem 10.6.1. *If A and B are matrices with complex entries and k is any complex number, then*

(a) $(A^*)^* = A$
(b) $(A + B)^* = A^* + B^*$
(c) $(kA)^* = \bar{k}A^*$
(d) $(AB)^* = B^*A^*$

The proofs are left as exercises.

Recall that a matrix with real entries is called *orthogonal* if $A^{-1} = A^t$. The complex analogs of the orthogonal matrices are called *unitary* matrices. They are defined as follows:

Definition. A square matrix A with complex entries is called **unitary** if

$$A^{-1} = A^*$$

The following theorem parallels Theorem 5.4.3.

Theorem 10.6.2. *If A is an $n \times n$ matrix with complex entries, then the following are equivalent:*

(a) *A is unitary.*
(b) *The row vectors of A form an orthonormal set in C^n with the Euclidean inner product.*
(c) *The column vectors of A form an orthonormal set in C^n with the Euclidean inner product.*

Example 2   The matrix

$$A = \begin{bmatrix} \dfrac{1+i}{2} & \dfrac{1+i}{2} \\ \dfrac{1-i}{2} & \dfrac{-1+i}{2} \end{bmatrix} \tag{1}$$

has row vectors

$$\mathbf{r}_1 = \left(\frac{1+i}{2}, \frac{1+i}{2} \right) \qquad \mathbf{r}_2 = \left(\frac{1-i}{2}, \frac{-1+i}{2} \right)$$

Relative to the Euclidean inner product on C^n, we have

$$\|\mathbf{r}_1\| = \sqrt{\left|\frac{1+i}{2}\right|^2 + \left|\frac{1+i}{2}\right|^2} = \sqrt{\frac{1}{2} + \frac{1}{2}} = 1$$

$$\|\mathbf{r}_2\| = \sqrt{\left|\frac{1-i}{2}\right|^2 + \left|\frac{-1+i}{2}\right|^2} = \sqrt{\frac{1}{2} + \frac{1}{2}} = 1$$

and

$$\mathbf{r}_1 \cdot \mathbf{r}_2 = \left(\frac{1+i}{2}\right)\overline{\left(\frac{1-i}{2}\right)} + \left(\frac{1+i}{2}\right)\overline{\left(\frac{-1+i}{2}\right)}$$

$$= \left(\frac{1+i}{2}\right)\left(\frac{1+i}{2}\right) + \left(\frac{1+i}{2}\right)\left(\frac{-1-i}{2}\right)$$

$$= i - i = 0$$

so the row vectors form an orthonormal set in C^2. Thus, A is unitary and

$$A^{-1} = A^* = \begin{bmatrix} \dfrac{1-i}{2} & \dfrac{1+i}{2} \\ \dfrac{1-i}{2} & \dfrac{-1-i}{2} \end{bmatrix} \tag{2}$$

The reader should verify that matrix (2) is the inverse of matrix (1) by showing that $AA^* = A^*A = I$. ▲

UNITARY DIAGONAL-IZATION

Recall that a square matrix A with real entries is called *orthogonally diagonalizable* if there is an orthogonal matrix P such that $P^{-1}AP\ (=P^tAP)$ is diagonal. For complex matrices we have an analogous concept.

> **Definition.** A square matrix A with complex entries is called **unitarily diagonalizable** if there is a unitary P such that $P^{-1}AP(=P^*AP)$ is diagonal; the matrix P is said to **unitarily diagonalize** A.

We have two questions to consider:
· Which matrices are unitarily diagonalizable?
· How do we find a unitary matrix P to carry out the diagonalization?

Before pursuing these questions, we note that our earlier definitions of the terms **eigenvector, eigenvalue, eigenspace, characteristic equation,** and **characteristic polynomial** carry over without change to complex vector spaces.

HERMITIAN MATRICES

In Section 6.3 we saw that the symmetric matrices played a fundamental role in the problem of orthogonally diagonalizing a matrix with real entries. The most natural complex analogs of the real symmetric matrices are the *Hermitian** matrices, which are defined as follows:

> **Definition.** A square matrix A with complex entries is called **Hermitian** if
>
> $$A = A^*$$

---

* *Charles Hermite* (1822–1901) was a French mathematician who made fundamental contributions to algebra, matrix theory, and various branches of analysis. He is noted for giving the first solution of a general fifth-degree polynomial equation and proving that the number e (the base for natural logarithms) is not the root of any polynomial equation with rational coefficients.

Example 3   If

$$A = \begin{bmatrix} 1 & i & 1+i \\ -i & -5 & 2-i \\ 1-i & 2+i & 3 \end{bmatrix}$$

then

$$\bar{A} = \begin{bmatrix} 1 & -i & 1-i \\ i & -5 & 2+i \\ 1+i & 2-i & 3 \end{bmatrix}$$

so

$$A^* = \bar{A}^t = \begin{bmatrix} 1 & i & 1+i \\ i & 5 & 2 & i \\ 1-i & 2+i & 3 \end{bmatrix} = A$$

which means that A is Hermitian. ▲

It is easy to recognize Hermitian matrices by inspection: The entries on the main diagonal are real numbers (Exercise 17), and the "mirror image" of each entry across the main diagonal is its complex conjugate (Figure 1).

Figure 1

NORMAL MATRICES

Hermitian matrices enjoy many but not all of the properties of real symmetric matrices. For example, just as the real symmetric matrices are orthogonally diagonalizable, so we shall see that the Hermitian matrices are unitarily diagonalizable. However, whereas the real symmetric matrices are the only matrices with real entries that are orthogonally diagonalizable (Theorem 6.3.1), the Hermitian matrices do not constitute the entire class of unitarily diagonalizable matrices; that is, there are unitarily diagonalizable matrices that are not Hermitian. To explain why this is so we shall need the following definition:

Definition.   A square matrix A with complex entries is called **normal** if

$$AA^* = A^*A$$

Example 4   Every Hermitian matrix A is normal since $AA^* = AA = A^*A$, and every unitary matrix A is normal since $AA^* = I = A^*A$. ▲

The following two theorems are the complex analogs of Theorems 6.3.1 and 6.3.2. The proofs will be omitted.

Theorem 10.6.3. *If A is a square matrix with complex entries, then the following are equivalent*:

(a) *A is unitarily diagonalizable.*
(b) *A has an orthonormal set of n eigenvectors.*
(c) *A is normal.*

Theorem 10.6.4. *If A is a normal matrix, then eigenvectors from different eigenspaces of A are orthogonal.*

Theorem 10.6.3 tells us that a square matrix A with complex entries is unitarily diagonalizable if and only if it is normal. Theorem 10.6.4 will be the key to constructing a matrix that unitarily diagonalizes a normal matrix.

DIAGONAL-IZATION PROCEDURE

We saw in Section 6.3 that a symmetric matrix A is orthogonally diagonalized by any orthogonal matrix whose column vectors are eigenvectors of A. Similarly, a normal matrix A is diagonalized by any unitary matrix whose column vectors are eigenvectors of A. The procedure for diagonalizing a normal matrix is as follows:

Step 1. Find a basis for each eigenspace of A.

Step 2. Apply the Gram-Schmidt process to each of these bases to obtain an orthonormal basis for each eigenspace.

Step 3. Form the matrix P whose columns are the basis vectors constructed in Step 2. This matrix unitarily diagonalizes A.

The justification of this procedure should be clear. Theorem 10.6.4 ensures that eigenvectors from *different* eigenspaces are orthogonal, and the application of the Gram-Schmidt process ensures that the eigenvectors within the *same* eigenspace are orthonormal. Thus, the *entire* set of eigenvectors obtained by this procedure is orthonormal. Theorem 10.6.3 ensures that this orthonormal set of eigenvectors is a basis.

Example 5 The matrix

$$A = \begin{bmatrix} 2 & 1+i \\ 1-i & 3 \end{bmatrix}$$

is unitarily diagonalizable because it is Hermitian and therefore normal. Find a matrix P that unitarily diagonalizes A.

Solution. The characteristic polynomial of A is

$$\det(\lambda I - A) = \det \begin{bmatrix} \lambda - 2 & -1 - i \\ -1 + i & \lambda - 3 \end{bmatrix} = (\lambda - 2)(\lambda - 3) - 2 = \lambda^2 - 5\lambda + 4$$

so the characteristic equation is

$$\lambda^2 - 5\lambda + 4 = (\lambda - 1)(\lambda - 4) = 0$$

and the eigenvalues are $\lambda = 1$ and $\lambda = 4$.

By definition,

$$\mathbf{x} = \begin{bmatrix} x_1 \\ x_2 \end{bmatrix}$$

will be an eigenvector of A corresponding to λ if and only if $\mathbf{x}$ is a nontrivial solution of

$$\begin{bmatrix} \lambda - 2 & -1 - i \\ -1 + i & \lambda - 3 \end{bmatrix} \begin{bmatrix} x_1 \\ x_2 \end{bmatrix} = \begin{bmatrix} 0 \\ 0 \end{bmatrix} \tag{3}$$

To find the eigenvectors corresponding to $\lambda = 1$, we substitute this value in (3):

$$\begin{bmatrix} -1 & -1 - i \\ -1 + i & -2 \end{bmatrix} \begin{bmatrix} x_1 \\ x_2 \end{bmatrix} = \begin{bmatrix} 0 \\ 0 \end{bmatrix}$$

Solving this system by Gauss-Jordan elimination yields (verify)

$$x_1 = (-1 - i)s, \qquad x_2 = s$$

Thus, the eigenvectors of A corresponding to $\lambda = 1$ are the nonzero vectors in C^2 of the form

$$\mathbf{x} = \begin{bmatrix} (-1 - i)s \\ s \end{bmatrix} = s \begin{bmatrix} -1 - i \\ 1 \end{bmatrix}$$

Thus, this eigenspace is one-dimensional with basis

$$\mathbf{u} = \begin{bmatrix} -1 - i \\ 1 \end{bmatrix} \tag{4}$$

In this case the Gram-Schmidt process involves only one step: normalizing this vector. Since

$$\|\mathbf{u}\| = \sqrt{|-1 - i|^2 + |1|^2} = \sqrt{2 + 1} = \sqrt{3}$$

the vector

$$\mathbf{p}_1 = \frac{\mathbf{u}}{\|\mathbf{u}\|} = \begin{bmatrix} \dfrac{-1 - i}{\sqrt{3}} \\ \dfrac{1}{\sqrt{3}} \end{bmatrix}$$

is an orthonormal basis for the eigenspace corresponding to $\lambda = 1$.

To find the eigenvectors corresponding to $\lambda = 4$, we substitute this value in (3):

$$\begin{bmatrix} 2 & -1 - i \\ -1 + i & 1 \end{bmatrix} \begin{bmatrix} x_1 \\ x_2 \end{bmatrix} = \begin{bmatrix} 0 \\ 0 \end{bmatrix}$$

Solving this system by Gauss-Jordan elimination yields (verify)

$$x_1 = \left(\frac{1 + i}{2} \right)s, \qquad x_2 = s$$

so the eigenvectors of A corresponding to $\lambda = 4$ are the nonzero vectors in C^2 of the form

$$\mathbf{x} = \begin{bmatrix} \left(\dfrac{1+i}{2}\right)s \\ s \end{bmatrix} = s\begin{bmatrix} \dfrac{1+i}{2} \\ 1 \end{bmatrix}$$

Thus, the eigenspace is one-dimensional with basis

$$\mathbf{u} = \begin{bmatrix} \dfrac{1+i}{2} \\ 1 \end{bmatrix}$$

Applying the Gram-Schmidt process (that is, normalizing this vector) yields

$$\mathbf{p}_2 = \frac{\mathbf{u}}{\|\mathbf{u}\|} = \begin{bmatrix} \dfrac{1+i}{\sqrt{6}} \\ \dfrac{2}{\sqrt{6}} \end{bmatrix}$$

Thus,

$$P = [\mathbf{p}_1 \mid \mathbf{p}_2] = \begin{bmatrix} \dfrac{-1-i}{\sqrt{3}} & \dfrac{1+i}{\sqrt{6}} \\ \dfrac{1}{\sqrt{3}} & \dfrac{2}{\sqrt{6}} \end{bmatrix}$$

diagonalizes A and

$$P^{-1}AP = \begin{bmatrix} 1 & 0 \\ 0 & 4 \end{bmatrix}$$

EIGENVALUES OF HERMITIAN AND SYMMETRIC MATRICES

In Theorem 6.3.3 it was stated that the characteristic equation of a symmetric matrix with real entries has only real roots, or equivalently, the eigenvalues of a symmetric matrix with real entries are real numbers. This important result is a corollary of the following more general theorem.

Theorem 10.6.5. *The eigenvalues of a Hermitian matrix are real numbers.*

Proof. If λ is an eigenvalue and $\mathbf{v}$ a corresponding eigenvector of an $n \times n$ Hermitian matrix A, then

$$A\mathbf{v} = \lambda\mathbf{v}$$

Multiplying each side of this equation on the left by the conjugate transpose of $\mathbf{v}$ yields

$$\mathbf{v}^*A\mathbf{v} = \mathbf{v}^*(\lambda\mathbf{v}) = \lambda\mathbf{v}^*\mathbf{v} \tag{5}$$

We will show that the 1×1 matrices $\mathbf{v}^*A\mathbf{v}$ and $\mathbf{v}^*\mathbf{v}$ both have real entries, so it will follow from (5) that λ must be a real number.

Both $\mathbf{v}^*A\mathbf{v}$ and $\mathbf{v}^*\mathbf{v}$ are Hermitian, since

$$(\mathbf{v}^*A\mathbf{v})^* = \mathbf{v}^*A^*(\mathbf{v}^*)^* = \mathbf{v}^*A\mathbf{v}$$

and

$$(\mathbf{v}^*\mathbf{v})^* = \mathbf{v}^*(\mathbf{v}^*)^* = \mathbf{v}^*\mathbf{v}$$

Since Hermitian matrices have real entries on the main diagonal, and since $\mathbf{v}^*A\mathbf{v}$ and $\mathbf{v}^*\mathbf{v}$ are 1×1, it follows that these matrices have real entries, which completes the proof.

Theorem 10.6.6. *The eigenvalues of a symmetric matrix with real entries are real numbers.*

Proof. Let A be a symmetric matrix with real entries. Because the entries in A are real, it follows that

$$\bar{A} = A$$

But this implies that A is Hermitian, since

$$A^* = (\bar{A})^t = A^t = A$$

Thus, A has real eigenvalues by Theorem 10.6.5.

EXERCISE SET 10.6

1. In each part find A^*.

(a) $A = \begin{bmatrix} 2i & 1-i \\ 4 & 3+i \\ 5+i & 0 \end{bmatrix}$ (b) $A = \begin{bmatrix} 2i & 1-i & -1+i \\ 4 & 5-7i & -i \\ i & 3 & 1 \end{bmatrix}$

(c) $A = \begin{bmatrix} 7i & 0 & -3i \end{bmatrix}$ (d) $A = \begin{bmatrix} a_{11} & a_{12} & a_{13} \\ a_{21} & a_{22} & a_{23} \end{bmatrix}$

2. Which of the following are Hermitian matrices?

(a) $\begin{bmatrix} 0 & i \\ i & 2 \end{bmatrix}$ (b) $\begin{bmatrix} 1 & 1+i \\ 1-i & -3 \end{bmatrix}$ (c) $\begin{bmatrix} i & i \\ -i & i \end{bmatrix}$ (d) $\begin{bmatrix} -2 & 1-i & -1+i \\ 1+i & 0 & 3 \\ -1-i & 3 & 5 \end{bmatrix}$ (e) $\begin{bmatrix} 1 & 0 & 0 \\ 0 & 1 & 0 \\ 0 & 0 & 1 \end{bmatrix}$

3. Find k, l, and m to make A a Hermitian matrix.

$$A = \begin{bmatrix} -1 & k & -i \\ 3-5i & 0 & m \\ l & 2+4i & 2 \end{bmatrix}$$

4. Use Theorem 10.6.2 to determine which of the following are unitary matrices.

(a) $\begin{bmatrix} i & 0 \\ 0 & i \end{bmatrix}$
(b) $\begin{bmatrix} \dfrac{i}{\sqrt{2}} & \dfrac{1}{\sqrt{2}} \\ -\dfrac{i}{\sqrt{2}} & \dfrac{1}{\sqrt{2}} \end{bmatrix}$
(c) $\begin{bmatrix} 1+i & 1+i \\ 1-i & -1+i \end{bmatrix}$
(d) $\begin{bmatrix} -\dfrac{i}{\sqrt{2}} & \dfrac{i}{\sqrt{6}} & \dfrac{i}{\sqrt{3}} \\ 0 & -\dfrac{i}{\sqrt{6}} & \dfrac{i}{\sqrt{3}} \\ \dfrac{i}{\sqrt{2}} & \dfrac{i}{\sqrt{6}} & \dfrac{i}{\sqrt{3}} \end{bmatrix}$

5. In each part verify that the matrix is unitary and find its inverse.

(a) $\begin{bmatrix} \dfrac{3}{5} & \dfrac{4}{5}i \\ -\dfrac{4}{5} & \dfrac{3}{5}i \end{bmatrix}$
(b) $\begin{bmatrix} \dfrac{1}{\sqrt{2}} & \dfrac{1}{\sqrt{2}} \\ -\dfrac{1+i}{2} & \dfrac{1+i}{2} \end{bmatrix}$
(c) $\begin{bmatrix} \dfrac{1}{4}(\sqrt{3}+i) & \dfrac{1}{4}(1-i\sqrt{3}) \\ \dfrac{1}{4}(1+i\sqrt{3}) & \dfrac{1}{4}(i-\sqrt{3}) \end{bmatrix}$
(d) $\begin{bmatrix} \dfrac{1+i}{2} & -\dfrac{1}{2} & \dfrac{1}{2} \\ \dfrac{i}{\sqrt{3}} & \dfrac{1}{\sqrt{3}} & -\dfrac{i}{\sqrt{3}} \\ \dfrac{3+i}{2\sqrt{15}} & \dfrac{4+3i}{2\sqrt{15}} & \dfrac{5i}{2\sqrt{15}} \end{bmatrix}$

6. Show that the matrix

$$\frac{1}{\sqrt{2}}\begin{bmatrix} e^{i\theta} & e^{-i\theta} \\ ie^{i\theta} & -ie^{-i\theta} \end{bmatrix}$$

is unitary for every real value of θ.

In Exercises 7–12 find a unitary matrix P that diagonalizes A, and determine $P^{-1}AP$.

7. $A = \begin{bmatrix} 4 & 1-i \\ 1+i & 5 \end{bmatrix}$
8. $A = \begin{bmatrix} 3 & -i \\ i & 3 \end{bmatrix}$
9. $A = \begin{bmatrix} 6 & 2+2i \\ 2-2i & 4 \end{bmatrix}$

10. $\begin{bmatrix} 0 & 3+i \\ 3-i & -3 \end{bmatrix}$
11. $A = \begin{bmatrix} 5 & 0 & 0 \\ 0 & -1 & -1+i \\ 0 & -1-i & 0 \end{bmatrix}$
12. $A = \begin{bmatrix} 2 & \dfrac{i}{\sqrt{2}} & -\dfrac{i}{\sqrt{2}} \\ -\dfrac{i}{\sqrt{2}} & 2 & 0 \\ \dfrac{i}{\sqrt{2}} & 0 & 2 \end{bmatrix}$

13. Show that the eigenvalues of the symmetric matrix

$$A = \begin{bmatrix} 1 & 4i \\ 4i & 3 \end{bmatrix}$$

are not real. Does this violate Theorem 10.6.6?

14. Find a 2×2 matrix that is both Hermitian and unitary and whose entries are not all real numbers.

15. Prove: If A is an $n \times n$ matrix with complex entries, then $\det(\bar{A}) = \overline{\det(A)}$. [***Hint.*** First show that the signed elementary products from $\bar{A}$ are the conjugates of the signed elementary products from A.]

16. (a) Use the result of Exercise 15 to prove that if A is an $n \times n$ matrix with complex entries, then $\det(A^*) = \overline{\det(A)}$.
    (b) Prove: If A is Hermitian, then $\det(A)$ is real.
    (c) Prove: If A is unitary, then $|\det(A)| = 1$.

17. Prove that the entries on the main diagonal of a Hermitian matrix are real numbers.

18. Let

$$A = \begin{bmatrix} a_{11} & a_{12} & a_{13} \\ a_{21} & a_{22} & a_{23} \\ a_{31} & a_{32} & a_{33} \end{bmatrix} \quad \text{and} \quad B = \begin{bmatrix} b_{11} & b_{12} & b_{13} \\ b_{21} & b_{22} & b_{23} \\ b_{31} & b_{32} & b_{33} \end{bmatrix}$$

be matrices with complex entries. Show that
(a) $(A^*)^* = A$ (b) $(A + B)^* = A^* + B^*$ (c) $(kA)^* = \bar{k}A^*$ (d) $(AB)^* = B^*A^*$

19. Prove: If A is invertible, then so is A^* in which case $(A^*)^{-1} = (A^{-1})^*$.

20. Show that if A is a unitary matrix, then A^* is also unitary.

21. Prove that an $n \times n$ matrix with complex entries is unitary if and only if its rows form an orthonormal set in C^n with the Euclidean inner product.

22. Use Exercises 20 and 21 to show that an $n \times n$ matrix is unitary if and only if its columns form an orthonormal set in C^n with the Euclidean inner product.

23. Prove: If $A = A^*$, then for every vector $\mathbf{x}$ in C^n, the entry in the 1×1 matrix $\mathbf{x}^*A\mathbf{x}$ is real.

24. Let λ and μ be distinct eigenvalues of a Hermitian matrix A.
    (a) Prove that if $\mathbf{x}$ is an eigenvector corresponding to λ and $\mathbf{y}$ an eigenvector corresponding to μ, then $\mathbf{x}^*A\mathbf{y} = \lambda\mathbf{x}^*\mathbf{y}$ and $\mathbf{x}^*A\mathbf{y} = \mu\mathbf{x}^*\mathbf{y}$.
    (b) Prove Theorem 10.6.4. [**Hint.** Subtract the equations in part (a).]

SUPPLEMENTARY EXERCISES

1. Let $\mathbf{u} = (u_1, u_2, \ldots, u_n)$ and $\mathbf{v} = (v_1, v_2, \ldots, v_n)$ be vectors in C^n, and let $\bar{\mathbf{u}} = (\bar{u}_1, \bar{u}_2, \ldots, \bar{u}_n)$ and $\bar{\mathbf{v}} = (\bar{v}_1, \bar{v}_2, \ldots, \bar{v}_n)$.
   (a) Prove: $\overline{\mathbf{u} \cdot \mathbf{v}} = \bar{\mathbf{u}} \cdot \bar{\mathbf{v}}$.
   (b) Prove: $\mathbf{u}$ and $\mathbf{v}$ are orthogonal if and only if $\bar{\mathbf{u}}$ and $\bar{\mathbf{v}}$ are orthogonal.

2. Show that if the matrix

$$\begin{bmatrix} a & b \\ -b & \bar{a} \end{bmatrix}$$

   is nonzero, then it is invertible.

3. Find a basis for the solution space of the system

$$\begin{bmatrix} -1 & -i & 1 \\ -i & 1 & i \\ 1 & i & -1 \end{bmatrix} \begin{bmatrix} x_1 \\ x_2 \\ x_3 \end{bmatrix} = \begin{bmatrix} 0 \\ 0 \\ 0 \end{bmatrix}$$

4. Prove: If a and b are complex numbers such that $|a|^2 + |b|^2 = 1$, and if θ is a real number, then

$$A = \begin{bmatrix} a & b \\ -e^{i\theta}\bar{b} & e^{i\theta}\bar{a} \end{bmatrix}$$

is a unitary matrix.

5. Find the eigenvalues of the matrix

$$\begin{bmatrix} 0 & 0 & 1 \\ 1 & 0 & \omega + 1 + \dfrac{1}{\omega} \\ 0 & 1 & -\omega - 1 - \dfrac{1}{\omega} \end{bmatrix}$$

where $\omega = e^{2\pi i/3}$.

6. (a) Prove that if z is a complex number other than 1, then

$$1 + z + z^2 + \cdots + z^n = \frac{1 - z^{n+1}}{1 - z}$$

[**Hint.** Let S be the sum on the left side of the equation and consider the quantity $S - zS$.]

(b) Use the result in (a) to prove that if $z^n = 1$ and $z \neq 1$, then $1 + z + z^2 + \cdots + z^{n-1} = 0$.

(c) Use the result in (a) to obtain Lagrange's trigonometric identity

$$1 + \cos\theta + \cos 2\theta + \cdots + \cos n\theta = \frac{1}{2} + \frac{\sin[(n + \frac{1}{2})\theta]}{2\sin(\theta/2)}$$

for $0 < \theta < 2\pi$. [**Hint.** Let $z = \cos\theta + i\sin\theta$.]

7. Let $\omega = e^{2\pi i/3}$. Show that the vectors $\mathbf{v}_1 = (1/\sqrt{3})(1, 1, 1)$, $\mathbf{v}_2 = (1/\sqrt{3})(1, \omega, \omega^2)$, and $\mathbf{v}_3 = (1/\sqrt{3})(1, \omega^2, \omega^4)$ form an orthonormal set in C^3. [**Hint.** Use part (b) of Exercise 6.]

8. Show that if U is an $n \times n$ unitary matrix and $|z_1| = |z_2| = \cdots = |z_n| = 1$, then the product

$$U \begin{bmatrix} z_1 & 0 & 0 & \cdots & 0 \\ 0 & z_2 & 0 & \cdots & 0 \\ \vdots & \vdots & \vdots & & \vdots \\ 0 & 0 & 0 & \cdots & z_n \end{bmatrix}$$

is also unitary.

9. Suppose that $A^* = -A$.
(a) Show that iA is Hermitian.
(b) Show that A is unitarily diagonalizable and has pure imaginary eigenvalues.

10. (a) Show that the set of complex numbers with the operations

$$(a + bi) + (c + di) = (a + c) + (b + d)i \text{ and } k(a + bi) = ka + kbi$$

where k is a real number, is a *real* vector space.

(b) What is the dimension of this space?

CHAPTER 11

APPLICATIONS OF LINEAR ALGEBRA

11.1 CONSTRUCTING CURVES AND SURFACES THROUGH SPECIFIED POINTS

This section describes a technique for using determinants to construct lines, circles, and general conic sections through specified points in the plane. The procedure is also used to pass planes and spheres in three-dimensional space through fixed points.

PREREQUISITES: Linear systems
Determinants
Analytic geometry

The following theorem follows from Theorem 1.7.3 and Theorem 2.3.4:

Theorem 11.1.1. *A homogeneous linear system with as many equations as unknowns has a nontrivial solution if and only if the determinant of the coefficient matrix is zero.*

We shall now show how this result can be used to determine the equations of various curves and surfaces through specified points.

A LINE THROUGH TWO POINTS

Suppose we are given two distinct points in the plane, (x_1, y_1) and (x_2, y_2). There is a unique line

$$c_1 x + c_2 y + c_3 = 0 \tag{1}$$

that passes through these two points (Figure 1). Notice that c_1, c_2, and c_3 are not all zero, and that these coefficients are unique only up to a multiplicative constant. Because (x_1, y_1) and (x_2, y_2) lie on the line, substituting them in (1) gives the two equations

$$c_1 x_1 + c_2 y_1 + c_3 = 0 \tag{2}$$
$$c_1 x_2 + c_2 y_2 + c_3 = 0 \tag{3}$$

The three equations, (1), (2), and (3), can be grouped together and rewritten as

$$x c_1 + y c_2 + c_3 = 0$$
$$x_1 c_1 + y_1 c_2 + c_3 = 0$$
$$x_2 c_1 + y_2 c_2 + c_3 = 0$$

which is a homogeneous linear system of three equations for c_1, c_2, and c_3. Because c_1, c_2, and c_3 are not all zero, this system has a nontrivial solution, and so the determinant of the system must be zero. That is,

$$\begin{vmatrix} x & y & 1 \\ x_1 & y_1 & 1 \\ x_2 & y_2 & 1 \end{vmatrix} = 0 \tag{4}$$

Consequently, every point (x, y) on the line satisfies (4); conversely, it can be shown that every point (x, y) that satisfies (4) lies on the line.

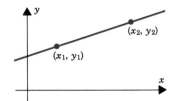

Figure 1

Example 1 Find the equation of the line that passes through the two points $(2, 1)$ and $(3, 7)$.

Solution. Substituting the coordinates of the two points into Equation (4) gives

$$\begin{vmatrix} x & y & 1 \\ 2 & 1 & 1 \\ 3 & 7 & 1 \end{vmatrix} = 0$$

The cofactor expansion of this determinant along the first row then gives

$$-6x + y + 11 = 0 \quad \blacktriangle$$

A CIRCLE THROUGH THREE POINTS

Suppose there are three distinct points in the plane, (x_1, y_1), (x_2, y_2), and (x_3, y_3), not all lying on a straight line. From analytic geometry, we know that there is a unique circle, say,

$$c_1(x^2 + y^2) + c_2x + c_3y + c_4 = 0 \tag{5}$$

which passes through them (Figure 2). Substituting the coordinates of the three points into this equation gives

$$c_1(x_1^2 + y_1^2) + c_2x_1 + c_3y_1 + c_4 = 0 \tag{6}$$
$$c_1(x_2^2 + y_2^2) + c_2x_2 + c_3y_2 + c_4 = 0 \tag{7}$$
$$c_1(x_3^2 + y_3^2) + c_2x_3 + c_3y_3 + c_4 = 0 \tag{8}$$

As before, Equations (5) through (8) form a homogeneous linear system with a nontrivial solution for c_1, c_2, c_3, and c_4. Thus, the determinant of the coefficient matrix is zero:

$$\begin{vmatrix} x^2 + y^2 & x & y & 1 \\ x_1^2 + y_1^2 & x_1 & y_1 & 1 \\ x_2^2 + y_2^2 & x_2 & y_2 & 1 \\ x_3^2 + y_3^2 & x_3 & y_3 & 1 \end{vmatrix} = 0 \tag{9}$$

This is a determinant form for the equation of the circle.

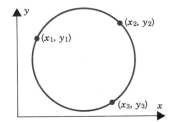

Figure 2

Example 2   Find the equation of the circle that passes through the three points $(1, 7)$, $(6, 2)$, and $(4, 6)$.

Solution. Substituting the coordinates of the three points into Equation (9) gives

$$\begin{vmatrix} x^2 + y^2 & x & y & 1 \\ 50 & 1 & 7 & 1 \\ 40 & 6 & 2 & 1 \\ 52 & 4 & 6 & 1 \end{vmatrix} = 0$$

which reduces to

$$10(x^2 + y^2) - 20x - 40y - 200 = 0$$

In standard form this is

$$(x - 1)^2 + (y - 2)^2 = 5^2$$

Thus, the circle has center $(1, 2)$ and radius 5. ▲

**A GENERAL
CONIC SECTION
THROUGH FIVE
POINTS**

The general equation of a conic section in the plane (a parabola, hyperbola, or ellipse, and degenerate forms of these three curves) is given by

$$c_1 x^2 + c_2 xy + c_3 y^2 + c_4 x + c_5 y + c_6 = 0$$

This equation contains six coefficients, but we can reduce the number to five if we divide through by any one of them which is not zero. Thus, only five coefficients must be determined, so that five distinct points in the plane are sufficient to determine the equation of the conic section (Figure 3). As before, the equation can be put in determinant form (see Exercise 6):

$$\begin{vmatrix} x^2 & xy & y^2 & x & y & 1 \\ x_1^2 & x_1 y_1 & y_1^2 & x_1 & y_1 & 1 \\ x_2^2 & x_2 y_2 & y_2^2 & x_2 & y_2 & 1 \\ x_3^2 & x_3 y_3 & y_3^2 & x_3 & y_3 & 1 \\ x_4^2 & x_4 y_4 & y_4^2 & x_4 & y_4 & 1 \\ x_5^2 & x_5 y_5 & y_5^2 & x_5 & y_5 & 1 \end{vmatrix} = 0 \qquad (10)$$

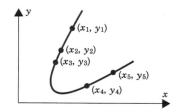

Figure 3

Example 3 An astronomer who wants to determine the orbit of an asteroid about the sun sets up a Cartesian coordinate system in the plane of the orbit with the sun at the origin. Astronomical units of measurement are used along the axes (1 astronomical unit = mean distance of earth to sun = 93 million miles). By Kepler's first law the orbit must be an ellipse, so the astronomer makes five observations of the asteroid at five different times and finds five points along the orbit to be

$$(8.025, 8.310), \quad (10.170, 6.355), \quad (11.202, 3.212), \quad (10.736, 0.375), \quad (9.092, -2.267)$$

Find the equation of the orbit.

Solution. Substituting the coordinates of the five given points into (10) gives

$$\begin{vmatrix} x^2 & xy & y^2 & x & y & 1 \\ 64.401 & 66.688 & 69.056 & 8.025 & 8.310 & 1 \\ 103.429 & 64.630 & 40.386 & 10.170 & 6.355 & 1 \\ 125.485 & 35.981 & 10.317 & 11.202 & 3.212 & 1 \\ 115.262 & 4.026 & 0.141 & 10.736 & 0.375 & 1 \\ 82.664 & -20.612 & 5.139 & 9.092 & -2.267 & 1 \end{vmatrix} = 0$$

The cofactor expansion of this determinant along the first row is

$$386.87x^2 - 103.46xy + 447.30y^2 - 2462.8x - 1429.5y - 17260 = 0$$

Figure 4 is an accurate diagram of the orbit, together with the five given points.

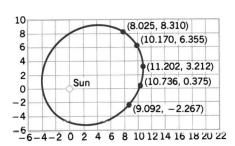

Figure 4

A PLANE THROUGH THREE POINTS

In Exercise 7 we ask the reader to show the following: The plane in 3-space with equation

$$c_1x + c_2y + c_3z + c_4 = 0$$

that passes through three noncollinear points (x_1, y_1, z_1), (x_2, y_2, z_2), and (x_3, y_3, z_3) is given by the determinant equation

$$\begin{vmatrix} x & y & z & 1 \\ x_1 & y_1 & z_1 & 1 \\ x_2 & y_2 & z_2 & 1 \\ x_3 & y_3 & z_3 & 1 \end{vmatrix} = 0 \tag{11}$$

Example 4   The equation of the plane that passes through the three noncollinear points $(1, 1, 0)$, $(2, 0, -1)$, and $(2, 9, 2)$ is

$$\begin{vmatrix} x & y & z & 1 \\ 1 & 1 & 0 & 1 \\ 2 & 0 & -1 & 1 \\ 2 & 9 & 2 & 1 \end{vmatrix} = 0$$

which reduces to

$$2x - y + 3z - 1 = 0 \quad \blacktriangle$$

A SPHERE THROUGH FOUR POINTS

In Exercise 8 we ask the reader to show the following: The sphere in 3-space with equation

$$c_1(x^2 + y^2 + z^2) + c_2x + c_3y + c_4z + c_5 = 0$$

that passes through four noncoplanar points (x_1, y_1, z_1), (x_2, y_2, z_2), (x_3, y_3, z_3), and (x_4, y_4, z_4) is given by the following determinant equation:

$$\begin{vmatrix} x^2 + y^2 + z^2 & x & y & z & 1 \\ x_1^2 + y_1^2 + z_1^2 & x_1 & y_1 & z_1 & 1 \\ x_2^2 + y_2^2 + z_2^2 & x_2 & y_2 & z_2 & 1 \\ x_3^2 + y_3^2 + z_3^2 & x_3 & y_3 & z_3 & 1 \\ x_4^2 + y_4^2 + z_4^2 & x_4 & y_4 & z_4 & 1 \end{vmatrix} = 0 \tag{12}$$

Example 5 The equation of the sphere that passes through the four points $(0, 3, 2)$, $(1, -1, 1)$, $(2, 1, 0)$, and $(5, 1, 3)$ is

$$\begin{vmatrix} x^2 + y^2 + z^2 & x & y & z & 1 \\ 13 & 0 & 3 & 2 & 1 \\ 3 & 1 & -1 & 1 & 1 \\ 5 & 2 & 1 & 0 & 1 \\ 35 & 5 & 1 & 3 & 1 \end{vmatrix} = 0$$

This reduces to

$$x^2 + y^2 + z^2 - 4x - 2y - 6z + 5 = 0$$

which in standard form is

$$(x - 2)^2 + (y - 1)^2 + (z - 3)^2 = 9 \quad \blacktriangle$$

EXERCISE SET 11.1

1. Find the equations of the lines that pass through the following points:
(a) $(1, -1), (2, 2)$ (b) $(0, 1), (1, -1)$

2. Find the equations of the circles that pass through the following points:
(a) $(2, 6), (2, 0), (5, 3)$ (b) $(2, -2), (3, 5), (-4, 6)$

3. Find the equation of the conic section that passes through the points $(0, 0), (0, -1), (2, 0)$, $(2, -5)$, and $(4, -1)$.

4. Find the equations of the planes in 3-space that pass through the following points:
(a) $(1, 1, -3), (1, -1, 1), (0, -1, 2)$ (b) $(2, 3, 1), (2, -1, -1), (1, 2, 1)$

5. Find the equations of the spheres in 3-space that pass through the following points:
(a) $(1, 2, 3), (-1, 2, 1), (1, 0, 1), (1, 2, -1)$ (b) $(0, 1, -2), (1, 3, 1), (2, -1, 0), (3, 1, -1)$

6. Show that Equation (10) is the equation of the conic section that passes through five given distinct points in the plane.

7. Show that Equation (11) is the equation of the plane in 3-space that passes through three given noncollinear points.

8. Show that Equation (12) is the equation of the sphere in 3-space that passes through four given noncoplanar points.

9. Find a determinant equation for the parabola of the form

$$c_1 y + c_2 x^2 + c_3 x + c_4 = 0$$

that passes through three given noncollinear points in the plane.

11.2   PLANE GEOMETRY

This section uses vector algebra to prove theorems in plane geometry.

PREREQUISITES: Plane geometry
Vector algebra in the plane
Dot product

The basic properties of Euclidean space were first systematically studied 23 centuries ago by Euclid, who began with accepted axioms and deduced theorems in plane and solid geometry from them. In this section we prove some elementary theorems in plane geometry using vector algebra. This method usually leads to simpler and more elegant proofs than the methods studied in a first course in geometry—the same methods that Euclid himself used. This is because the basic equations of vector algebra, such as

$$\mathbf{a} + \mathbf{b} = \mathbf{b} + \mathbf{a}$$
$$\mathbf{a} + (\mathbf{b} + \mathbf{c}) = (\mathbf{a} + \mathbf{b}) + \mathbf{c}$$
$$\mathbf{a} + (-\mathbf{a}) = \mathbf{0}$$
$$\vdots$$

are vector-space axioms and therefore apply to Euclidean space. Thus, any new equation we can generate from these basic equations constitutes a theorem. Because it is easier to manipulate equations than to write out in words how one is applying axioms, this method results in a certain economy of thought and a cleaner presentation of proofs.

Before giving some examples of such vector techniques applied to plane geometry, we introduce some notation and review some basic definitions.

NOTATION AND DEFINITIONS

For our applications to plane geometry, a vector may be viewed as a directed line segment or an arrow. We shall use the following notation (Figure 1): If A and B are two points, then AB denotes the (undirected) line segment connecting A and B and $\overrightarrow{AB}$ denotes the directed line segment or vector from point A to point B. We shall use $\mathbf{a}$ and $\mathbf{b}$ to denote vectors from an arbitrary fixed reference point O to the points A and B, respectively. Notice that we may write

$$\overrightarrow{AB} = \mathbf{b} - \mathbf{a}$$

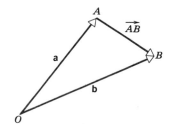

Figure 1

We assume that the reader is familiar with the rules of vector algebra discussed in Section 3.2. In addition, we shall need the following criterion for determining when three points are collinear:

Theorem 11.2.1. *Let A, B, and C be three points and let* **a, b,** *and* **c** *be the vectors from a fixed point O to A, B, and C, respectively (Figure 2). Then B lies on the line segment AC if and only if*

$$\mathbf{b} = \lambda\mathbf{a} + (1 - \lambda)\mathbf{c}$$

for some number λ satisfying

$$0 \le \lambda \le 1$$

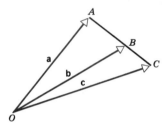

Figure 2 | B lies on the line segment AC |

When the criterion in Theorem 11.2.1 is satisfied, the point B cuts the line segment AC in the ratio $\lambda:(1 - \lambda)$. For example, if $\lambda = \frac{1}{2}$, that is, if $\mathbf{b} = \frac{1}{2}(\mathbf{a} + \mathbf{c})$, then B is the midpoint of AC; if $\lambda = \frac{2}{3}$, that is, if $\mathbf{b} = \frac{2}{3}\mathbf{a} + \frac{1}{3}\mathbf{c}$, then B cuts AC in the ratio 2:1.

Recall that the dot product $\mathbf{a} \cdot \mathbf{b}$ of two nonzero vectors $\mathbf{a}$ and $\mathbf{b}$ is given by

$$\mathbf{a} \cdot \mathbf{b} = \|\mathbf{a}\|\,\|\mathbf{b}\|\cos\theta$$

where $\|\mathbf{a}\|$ is the length of vector $\mathbf{a}$, $\|\mathbf{b}\|$ is the length of vector $\mathbf{b}$, and θ is the angle between $\mathbf{a}$ and $\mathbf{b}$ (Figure 3). It follows from this formula that $\mathbf{a}$ and $\mathbf{b}$ are perpendicular if and only if $\mathbf{a} \cdot \mathbf{b} = 0$.

Figure 3 | θ is the angle between $\mathbf{a}$ and $\mathbf{b}$ |

In parts (*a*), (*b*), and (*c*) of Figure 4, seven definitions from plane geometry are listed for reference. These are followed by six examples in which six basic theorems are proved using vector algebra.

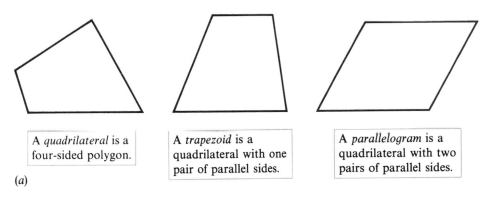

A *quadrilateral* is a four-sided polygon.

A *trapezoid* is a quadrilateral with one pair of parallel sides.

A *parallelogram* is a quadrilateral with two pairs of parallel sides.

(*a*)

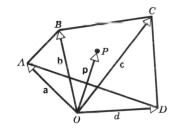

The *centroid* or *mean center* of a triangle *ABC* is the point *P* determined by $\overrightarrow{OP} = \frac{1}{3}(\mathbf{a} + \mathbf{b} + \mathbf{c})$.

The *centroid* or *mean center* of a quadrilateral *ABCD* is the point *P* determined by $\overrightarrow{OP} = \frac{1}{4}(\mathbf{a} + \mathbf{b} + \mathbf{c} + \mathbf{d})$.

(*b*)

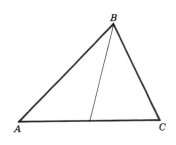

An *altitude* of a triangle *ABC* is a line segment from one of the vertices perpendicular to the opposite side.

A *median* of a triangle *ABC* is a line segment from one of the vertices to the midpoint of the opposite side.

Figure 4   (*c*)

Example 1 Prove that the three medians of a triangle intersect at the centroid of the triangle.

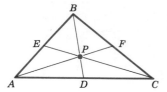

Figure 5

Proof. If D, E, and F are the midpoints of the three sides of the triangle ABC (Figure 5), we have

$$\mathbf{d} = \tfrac{1}{2}(\mathbf{a} + \mathbf{c})$$
$$\mathbf{e} = \tfrac{1}{2}(\mathbf{a} + \mathbf{b})$$
$$\mathbf{f} = \tfrac{1}{2}(\mathbf{b} + \mathbf{c})$$

The centroid P of the triangle is determined by

$$\mathbf{p} = \tfrac{1}{3}(\mathbf{a} + \mathbf{b} + \mathbf{c})$$

Substituting $\mathbf{a} + \mathbf{c} = 2\mathbf{d}$ into this last equation gives

$$\mathbf{p} = \tfrac{1}{3}\mathbf{b} + \tfrac{2}{3}\mathbf{d}$$

from which it follows that P lies on the median BD. Similarly, the equations

$$\mathbf{p} = \tfrac{1}{3}\mathbf{c} + \tfrac{2}{3}\mathbf{e}$$
$$\mathbf{p} = \tfrac{1}{3}\mathbf{a} + \tfrac{2}{3}\mathbf{f}$$

show that P also lies on the medians CE and AF. ▲

REMARK. From the last three equations for $\mathbf{p}$ in Example 1, it follows that the centroid divides each median in the ratio 2:1.

Example 2 Prove that the quadrilateral $EFGH$ whose vertices are the midpoints of the sides of an arbitrary quadrilateral $ABCD$ is a parallelogram (Figure 6). In addition, prove that the two quadrilaterals have the same centroid.

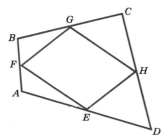

Figure 6

Proof. We have that

$$\mathbf{e} = \tfrac{1}{2}(\mathbf{a} + \mathbf{d})$$
$$\mathbf{f} = \tfrac{1}{2}(\mathbf{a} + \mathbf{b})$$
$$\mathbf{g} = \tfrac{1}{2}(\mathbf{b} + \mathbf{c})$$
$$\mathbf{h} = \tfrac{1}{2}(\mathbf{c} + \mathbf{d})$$

Thus,

$$\mathbf{g} - \mathbf{f} = \tfrac{1}{2}(\mathbf{c} - \mathbf{a}) = \mathbf{h} - \mathbf{e}$$

and so *FG* is parallel to *EH*. Similarly,

$$\mathbf{h} - \mathbf{g} = \tfrac{1}{2}(\mathbf{d} - \mathbf{b}) = \mathbf{e} - \mathbf{f}$$

and so *GH* is parallel to *FE*. Therefore, quadrilateral *EFGH* is a parallelogram. Adding the above four equations yields

$$\mathbf{e} + \mathbf{f} + \mathbf{g} + \mathbf{h} = \mathbf{a} + \mathbf{b} + \mathbf{c} + \mathbf{d}$$

or

$$\tfrac{1}{4}(\mathbf{e} + \mathbf{f} + \mathbf{g} + \mathbf{h}) = \tfrac{1}{4}(\mathbf{a} + \mathbf{b} + \mathbf{c} + \mathbf{d})$$

and so both quadrilaterals have the same centroid. ▲

Example 3   Prove that an angle inscribed in a semicircle is a right angle.

Figure 7

Proof. We want to show that the vector $\overrightarrow{AB}$ in Figure 7 is perpendicular to the vector $\overrightarrow{BC}$. Letting O be the center of the semicircle, we have

$$\overrightarrow{AB} \cdot \overrightarrow{BC} = (\overrightarrow{AO} + \overrightarrow{OB}) \cdot (\overrightarrow{BO} + \overrightarrow{OC})$$
$$= (\overrightarrow{OC} + \overrightarrow{OB}) \cdot (-\overrightarrow{OB} + \overrightarrow{OC})$$
$$= \overrightarrow{OC} \cdot \overrightarrow{OC} - \overrightarrow{OB} \cdot \overrightarrow{OB}$$
$$= \|\overrightarrow{OC}\|^2 - \|\overrightarrow{OB}\|^2$$
$$= 0$$

where the last equality follows from the fact that $\overrightarrow{OC}$ and $\overrightarrow{OB}$ have the same length. ▲

Example 4 Prove the Law of Cosines, that is, if a, b, and c are the lengths of the sides of a triangle and θ is the angle opposite the side of length a, then

$$a^2 = b^2 + c^2 - 2bc \cos \theta$$

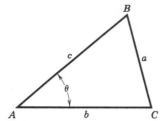

Figure 8

Proof. Letting A, B, and C be the vertices opposite the sides of lengths a, b, and c, respectively (Figure 8), we have

$$
\begin{aligned}
a^2 &= \|\overrightarrow{BC}\|^2 \\
&= \|\overrightarrow{BA} + \overrightarrow{AC}\|^2 \\
&= (\overrightarrow{BA} + \overrightarrow{AC}) \cdot (\overrightarrow{BA} + \overrightarrow{AC}) \\
&= \overrightarrow{AC} \cdot \overrightarrow{AC} + \overrightarrow{BA} \cdot \overrightarrow{BA} + 2(\overrightarrow{AC} \cdot \overrightarrow{BA}) \\
&= \overrightarrow{AC} \cdot \overrightarrow{AC} + \overrightarrow{BA} \cdot \overrightarrow{BA} - 2(\overrightarrow{AC} \cdot \overrightarrow{AB}) \\
&= \|\overrightarrow{AC}\|^2 + \|\overrightarrow{BA}\|^2 - 2\|\overrightarrow{AC}\| \, \|\overrightarrow{AB}\| \cos \theta \\
&= b^2 + c^2 - 2bc \cos \theta \quad \blacktriangle
\end{aligned}
$$

Example 5 Prove that the three altitudes of a triangle intersect at a common point. (This point is called the ***orthocenter*** of the triangle.)

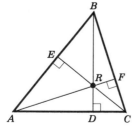

Figure 9

Proof. Let R be the point of intersection of the altitudes from vertices A and B, and let CE be the line from vertex C passing through R and meeting side AB at point E (Figure 9). We want to show that CE is perpendicular to AB.

We have that

$$\overrightarrow{CA} \cdot \overrightarrow{BR} = (\mathbf{a} - \mathbf{c}) \cdot (\mathbf{r} - \mathbf{b}) = 0$$

and

$$\overrightarrow{CB} \cdot \overrightarrow{AR} = (\mathbf{b} - \mathbf{c}) \cdot (\mathbf{r} - \mathbf{a}) = 0$$

Thus,

$$\overrightarrow{AB} \cdot \overrightarrow{CR} = (\mathbf{b} - \mathbf{a}) \cdot (\mathbf{r} - \mathbf{c})$$
$$= [(\mathbf{b} - \mathbf{c}) - (\mathbf{a} - \mathbf{c})] \cdot (\mathbf{r} - \mathbf{c})$$
$$= (\mathbf{b} - \mathbf{c}) \cdot (\mathbf{r} - \mathbf{c}) - (\mathbf{a} - \mathbf{c}) \cdot (\mathbf{r} - \mathbf{c})$$
$$= (\mathbf{b} - \mathbf{c}) \cdot (\mathbf{r} - \mathbf{a} + \mathbf{a} - \mathbf{c}) - (\mathbf{a} - \mathbf{c}) \cdot (\mathbf{r} - \mathbf{b} + \mathbf{b} - \mathbf{c})$$
$$= (\mathbf{b} - \mathbf{c}) \cdot (\mathbf{r} - \mathbf{a}) + (\mathbf{b} - \mathbf{c}) \cdot (\mathbf{a} - \mathbf{c}) - (\mathbf{a} - \mathbf{c}) \cdot (\mathbf{r} - \mathbf{b}) - (\mathbf{a} - \mathbf{c}) \cdot (\mathbf{b} - \mathbf{c})$$
$$= (\mathbf{b} - \mathbf{c}) \cdot (\mathbf{a} - \mathbf{c}) - (\mathbf{a} - \mathbf{c}) \cdot (\mathbf{b} - \mathbf{c})$$
$$= 0$$

which proves that CR, hence CE, is perpendicular to AB. ▲

Example 6   Prove that the three perpendicular bisectors of the sides of a triangle intersect at a common point. (This point is called the *circumcenter* of the triangle.)

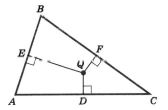

Figure 10

Proof. Let Q be the point of intersection of the perpendicular bisectors of sides AB and AC, and let F be the midpoint of side BC (Figure 10). We want to show that FQ is perpendicular to CB.

We have that

$$\overrightarrow{AB} \cdot \overrightarrow{EQ} = (\mathbf{b} - \mathbf{a}) \cdot (\mathbf{q} - \mathbf{e}) = 0$$

and

$$\overrightarrow{CA} \cdot \overrightarrow{DQ} = (\mathbf{a} - \mathbf{c}) \cdot (\mathbf{q} - \mathbf{d}) = 0$$

Thus,

$$\overrightarrow{CB} \cdot \overrightarrow{FQ} = (\mathbf{b} - \mathbf{c}) \cdot (\mathbf{q} - \mathbf{f})$$
$$= [(\mathbf{b} - \mathbf{a}) + (\mathbf{a} - \mathbf{c})] \cdot (\mathbf{q} - \mathbf{f})$$
$$= (\mathbf{b} - \mathbf{a}) \cdot (\mathbf{q} - \mathbf{f}) + (\mathbf{a} - \mathbf{c}) \cdot (\mathbf{q} - \mathbf{f})$$
$$= (\mathbf{b} - \mathbf{a}) \cdot [\mathbf{q} - \tfrac{1}{2}(\mathbf{b} + \mathbf{c})] + (\mathbf{a} - \mathbf{c}) \cdot [\mathbf{q} - \tfrac{1}{2}(\mathbf{b} + \mathbf{c})]$$
$$= (\mathbf{b} - \mathbf{a}) \cdot [\mathbf{q} - \tfrac{1}{2}(\mathbf{a} + \mathbf{b}) + \tfrac{1}{2}(\mathbf{a} - \mathbf{c})] + (\mathbf{a} - \mathbf{c}) \cdot [\mathbf{q} - \tfrac{1}{2}(\mathbf{a} + \mathbf{c}) + \tfrac{1}{2}(\mathbf{a} - \mathbf{b})]$$
$$= (\mathbf{b} - \mathbf{a}) \cdot [\mathbf{q} - \mathbf{e} + \tfrac{1}{2}(\mathbf{a} + \mathbf{c})] + (\mathbf{a} - \mathbf{c}) \cdot [\mathbf{q} - \mathbf{d} + \tfrac{1}{2}(\mathbf{a} - \mathbf{b})]$$
$$= \tfrac{1}{2}(\mathbf{b} - \mathbf{a}) \cdot (\mathbf{a} - \mathbf{c}) + \tfrac{1}{2}(\mathbf{a} - \mathbf{c}) \cdot (\mathbf{a} - \mathbf{b})$$
$$= 0$$

which proves that FQ is perpendicular to CB. ▲

EXERCISE SET 11.2

1. Let D and E be the midpoints of the sides AB and BC of a triangle ABC (Figure 11). Prove that DE is parallel to side AC and that its length is one-half that of AC.

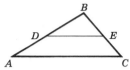

Figure 11

2. Prove that the line segments FH and EG connecting opposite midpoints of the sides of a quadrilateral $ABCD$ intersect at the centroid P of the quadrilateral (Figure 12).

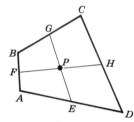

Figure 12

3. Prove that the diagonals AC and BD of a parallelogram $ABCD$ bisect each other and that the point of bisection P is the centroid of the parallelogram (Figure 13).

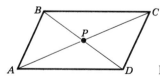

Figure 13

4. Prove that the circumcenter Q of a triangle ABC is equidistant from its three vertices, and hence is the center of the circle that circumscribes the triangle (Figure 14). (See Example 6.)

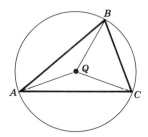

Figure 14

5. Prove that the centroid P of a triangle ABC lies on the line segment QR connecting the circumcenter Q of the triangle to its orthocenter R (Figure 15). In addition, prove that P divides this line (called the **Euler line**) in the ratio $2:1$. (See Examples 5 and 6.)

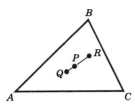

Figure 15

6. Let E and F be points on the sides AB and BC of a triangle ABC such that EF is parallel to side AC (Figure 16). Prove that if D is the point on side AC such that BD passes through the intersection G of EC and AF, then D bisects AC.

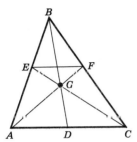

Figure 16

7. Let E be the midpoint of the side CD of a parallelogram $ABCD$ (Figure 17). Prove that the point of intersection F of BE and diagonal AC divides AC in the ratio $2:1$.

Figure 17

8. Prove that the diagonals AC and BD of a trapezoid $ABCD$ cut each other in the same ratio (Figure 18).

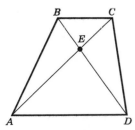

Figure 18

9. Let D, E, and F be the midpoints of the three sides of a triangle ABC (Figure 19). Prove that triangle ABC and triangle DEF have the same centroid.

10. Let D be the midpoint of side AB of a triangle ABC, and let E be the midpoint of side BC (Figure 20). Let F be an arbitrary point on side AC, and let G and H be the midpoints of AF and FC, respectively. Prove that the quadrilateral $DEHG$ is a parallelogram.

11. Let D be the midpoint of a line segment AB and also the midpoint of a line segment FC (Figure 21). Similarly, let E be the midpoint of AC and midpoint of BG. Prove that A is the midpoint of FG.

11.3 EQUILIBRIUM OF RIGID BODIES

In this section vector algebra is used to analyze the equilibrium of two-dimensional rigid bodies.

PREREQUISITE: Vector algebra in the plane

The field of mechanics has greatly stimulated the mathematical study of vectors and vector spaces. In this section we examine the use of vectors in the branch of static mechanics concerned with the equilibrium of rigid bodies. To simplify the presentation, we assume that the forces acting on the bodies all lie in a common plane, although the bodies themselves may be three dimensional.

FORCES AND MOMENTS

We are concerned here with the forces acting on rigid bodies. These forces may arise from a variety of sources, such as applied loads, gravity, and reactions of walls, floors, supports, cables, pins, or other bodies in contact with the given body. A force, whatever its nature, is specified by its magnitude, direction, and point of application. Mathematically, a force is an example of a vector. Consequently, a force can be represented graphically by an arrow or directed line segment (Figure 1) whose length is proportional to the magnitude of the force, whose tail is at

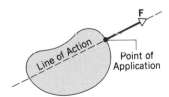

Figure 1

the point of application of the force, and whose direction is the same as that of the force. The line passing through the point of application of a force and lying along its direction is called the **line of action** of the force. For a rigid body, the following principle is of basic importance.

Principle of Transmissibility. *The effect of a force on a rigid body is not changed if the force vector is moved along its line of action.*

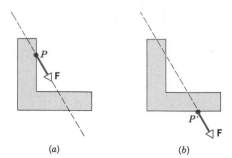

Figure 2   (a)   (b)

Thus the force **F** applied at point P of the L-shaped rigid body in Figure 2a has the same effect as the same force **F** applied at point P' in Figure 2b, as both forces are equal as vectors and have the same lines of action. It is this principle of transmissibility that distinguishes a rigid body from a deformable body. Forces that obey this principle are examples of *line vectors* or *sliding vectors*.

The following is another basic principle of mechanics.

> **Resultant of Forces.** *The effect of any number of forces applied at a single point of a rigid body is not changed if they are replaced by their vector sum applied at that same point. This vector sum is called the **resultant** of the forces.*

For example, the two forces $\mathbf{F}_1$ and $\mathbf{F}_2$ applied at a point P of the rigid body in Figure 3a can be replaced by the single force **F** in Figure 3b obtained by the usual parallelogram law of vector addition. Conversely, a single force may be decomposed vectorially into several components at its point of application without altering its effect on the body.

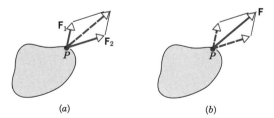

Figure 3   (a)   (b)

From Newton's first law of motion, a necessary condition for a body to be in equilibrium is that the vector sum of all forces acting on it be zero. If the forces are concurrent (i.e., if their lines of action all pass through a common point), then this condition is also sufficient to guarantee equilibrium, because then all the forces can be replaced by their zero resultant at the concurrent point. However, if the forces are not concurrent, then they may impart a rotational motion to the body even though their vector sum is zero. To determine an additional equilibrium condition to prevent rotational motion, we must examine the tendency of a force to rotate a body about a point. This leads to the concept of the *moment of a force*.

Moment of a Force. *In the plane, the* ***moment*** M_0 *of a force* **F** *about a point O is the product of the magnitude F of the force and the distance d from the point O to the line of action of the force*:

$$M_0 = Fd$$

By convention, the algebraic sign of M_0 is positive if the force **F** *tends to rotate a body counterclockwise about O, and negative if it tends to rotate a body clockwise about O (Figure 4).*

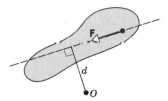

Figure 4

If there are several forces acting on a body, then the individual moments about a fixed point are added (with proper attention to their algebraic signs) to produce the total moment of the forces on the body about the point. The condition of equilibrium needed to prevent rotational motion is that the total moment about some fixed point is zero. The point about which the moments are computed is not important, as it can be shown that if the vector sum of the forces is zero and if the sum of their moments about any one point is zero, then the sum of their moments is zero about all points.

Before discussing these conditions of equilibrium in detail, we mention the following theorem of mechanics, which simplifies the calculation of moments.

Varignon's Theorem. *Let* **F** *be the resultant of the two forces* **F**$_1$ *and* **F**$_2$. *Then the moment of* **F** *about any point O is equal to the sum of the moments of* **F**$_1$ *and* **F**$_2$ *about that same point.*

For example, referring to Figure 5, Varignon's Theorem states that

$$Fd = F_1 d_1 + F_2 d_2$$

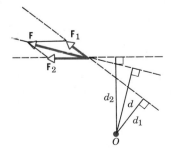

Figure 5

EQUILIBRIUM
OF RIGID BODIES

From the foregoing discussion, the following two conditions ensure that a rigid body in the plane is in equilibrium. (More precisely, these conditions serve as the definition of equilibrium.)

> **Equilibrium of Rigid Bodies in the Plane.** *A rigid body acted upon by coplanar forces is in equilibrium if*
>
> 1. *the vector sum of the forces is zero;*
> 2. *the sum of the moments of the forces about any one point in the plane is zero.*

In condition 2 any point in the plane can be used; however, a judicious choice can considerably reduce the calculations of unknown forces in most problems.

Before presenting some examples, we introduce the idea of a **free-body diagram**. This is a diagram of an isolated rigid body that graphically contains all external forces acting on it. For example, consider the 10-pound slender rod in Figure 6*a*, which is pinned to a wall at one end and whose other end leans against a smooth (frictionless) wall. Figure 6*b* is a free-body diagram of this rod. The weight of the body is indicated by a 10-pound force acting downward through its center of gravity. For a uniform rod, this is at its geometric center. Because the wall is smooth, it exerts no frictional force parallel to its surface. Thus, the reaction of the smooth wall is a force **N** of unknown magnitude whose direction is perpendicular to the wall. The reaction of the pin is a force **F** of unknown magnitude and direction.

Another example is the 5-pound rod in Figure 6*c*, which is held vertically against a rough floor by a cable. In the free-body diagram of this rod (Figure 6*d*), the weight is the 5-pound force acting downward through its geometric center. The cable can exert a force only along itself and so is replaced by a force **T**, which is along the cable. The rough floor can exert both a normal force and a frictional force parallel to its surface; consequently, the reaction of the floor is a force **R** of unknown magnitude and direction.

The general technique of applying the conditions of equilibrium to determine unknown forces is illustrated in the following four examples. In these examples the horizontal scalar component of a force is considered positive if that component is directed to the right, and negative if it is directed to the left. Similarly, vertical scalar components are considered positive if directed upward and negative if directed downward.

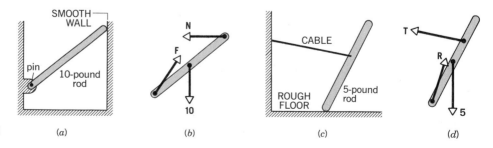

Figure 6   (*a*)   (*b*)   (*c*)   (*d*)

Example 1   A 1000-pound piano is being raised by a cable. A man guides the piano with a second cable tied to the first. Find the tension in the cable the man pulls on when the piano is in equilibrium, as in Figure 7.

Figure 7

Solution. We construct a free-body diagram of the common point of the two cables, as in Figure 8. Adding the scalar components of the forces in the horizontal direction and setting this sum equal to zero gives the equation

$$-F \sin 5° + T \cos 50° = 0$$

Similarly, equating the sum of the vertical scalar components equal to zero gives

$$F \cos 5° - 1000 - T \sin 50° = 0$$

Eliminating F between these two equations and solving for T then yields

$$T = \frac{1000}{\cot 5° \cos 50° - \sin 50°} = 152 \text{ pounds}$$

Figure 8

Example 2 A 10-foot ladder weighing 30 pounds leans against a smooth vertical wall and makes an angle of 65° with a rough floor (Figure 9). What is the magnitude of the normal force of the wall against the ladder?

Figure 9

Solution. In the free-body diagram of the ladder (Figure 10), the weight (30 pounds) acts vertically downward through the geometric center of the ladder; the reaction **N** of the smooth wall acts horizontally (perpendicular to the wall); and the force **F** at the foot of the ladder O acts in some unspecified direction. Taking moments about O and equating their sum to zero gives

$$-30(5 \cos 65°) + N(10 \sin 65°) = 0$$

or

$$N = 15 \cot 65° = 7.0 \text{ pounds}$$

Notice that by taking moments about O, the unknown force **F** does not appear in the moment equation. Once **N** has been found this way, the force **F** can be found using the equilibrium condition that the sum of all forces is zero. ▲

Figure 10

Example 3   An 8-ounce spool of thread is to be lifted over a $\frac{1}{2}$-inch step by pulling horizontally on the thread (Figure 11). What force is necessary to just start it over the step?

Figure 11

Solution. In the free-body diagram of the spool (Figure 12), the only reactive force on the spool is the force **F** at the edge O of the step. Taking moments about this point gives the equilibrium equation

$$T(\tfrac{3}{8} + \tfrac{5}{4} - \tfrac{1}{2}) - 8(1) = 0$$

from which it follows that

$$T = 7\tfrac{1}{9} \text{ ounces}$$

Notice that by taking moments about the point of application of the unknown force **F**, that unknown force does not appear in the moment equation. Once the tension force **T** is computed, **F** can be found by the equilibrium condition that the sum of all forces is zero. ▲

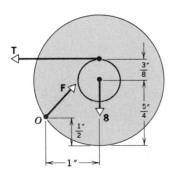

Figure 12

Example 4   A 200-pound sign is supported on a pole by two pins and a cable (Figure 13). What is the tension in the cable if the reactive force on the upper pin has no horizontal component?

Figure 13

Solution. In the free-body diagram of the sign (Figure 14), the unknown reaction of the lower pin is resolved into a horizontal component $\mathbf{F}_1$ and a vertical component $\mathbf{F}_2$. The reaction of the upper pin is the purely vertical force $\mathbf{F}_3$. The moment equation about the lower pin is

$$-(200)(33) + (T \sin 35°)(69) + (T \cos 35°)(42) = 0$$

from which it follows that

$$T = \frac{(200)(33)}{69 \sin 35° + 42 \cos 35°} = 89 \text{ pounds}$$

Notice that by taking moments about the lower pin, none of the unknown forces $\mathbf{F}_1$, $\mathbf{F}_2$, $\mathbf{F}_3$ appear in the moment equation, as their lines of action pass through the lower pin. Also notice how Varignon's Theorem was used in computing the moment of the force $\mathbf{T}$ of the cable by resolving it into a horizontal force and a vertical force and adding the individual moments. ▲

Figure 14

EXERCISE SET 11.3

1. Find the tensions in the three cables that support the 15-pound weight in Figure 15.

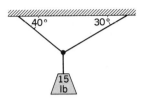

Figure 15

2. Find the tensions in all the cables supporting the two weights in Figure 16. [*Hint.* Draw a free-body diagram for each of the two points where the cables meet.]

Figure 16

3. Find the tension in the horizontal cable and the reaction of the pin that support the 8-pound rod in Figure 17. [*Hint.* First find the tension in the cable by taking moments about the pin.]

4. A 12-pound rod is pinned at one end and leans against a smooth step (Figure 18). Find the reactive forces of the pin and the step on the rod. [*Hint.* Because the step is smooth, its reaction is normal to the rod. Find this normal reaction by taking moments about the pin.]

Figure 17

Figure 18

5. Find the reactions of the three pins on the two weightless rods which support the 50-pound weight in Figure 19. [***Hint.*** Draw separate free-body diagrams of the two rods.]

Figure 19

6. Two slender 5-pound rods are pinned together and stand on a smooth surface (Figure 20). A horizontal cable connects them together at their top ends. Find the tension in the cable, the normal forces of the surface, and the reactive forces of the pin on each rod. [***Hint.*** By symmetry, the two normal forces of the surface are equal.]

Figure 20

7. Show that two parallel forces **F** and −**F** of equal magnitude and opposite direction have moment Fd or $-Fd$ where d is the distance between their lines of action, and that the moment is independent of the point about which it is computed (Figure 21). (Such a pair of forces is called a ***couple***.)

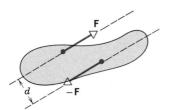

Figure 21

8. Show that if exactly two forces act on a rigid body in equilibrium, they must be of equal magnitude and opposite direction and have the same line of action.

9. (a) Show that if exactly three forces act on a rigid body in equilibrium, their lines of action must be either concurrent or parallel.
   (b) Use the result in part (a) to solve Example 2 without computing any moments.
   (c) Use the result in part (a) to solve Example 3 without computing any moments.

11.4 GRAPH THEORY

This section introduces matrix representations of relations among members of a set and uses matrix arithmetic to analyze these relationships.

PREREQUISITES: Matrix multiplication and addition

There are countless examples of sets with finitely many members in which some relation exists among members of the set. For example, the set could consist of a collection of people, animals, countries, companies, sports teams, or cities; and the relation between two members, A and B, of such a set could be that person A dominates person B, animal A feeds on animal B, country A militarily supports country B, company A sells its product to company B, sports team A consistently beats sports team B, or city A has a direct airline flight to city B.

We shall now show how the theory of *directed graphs* can be used to mathematically model relations such as those in the preceding examples.

DIRECTED GRAPHS

A *directed graph* is a finite set of elements, $\{P_1, P_2, \ldots, P_n\}$, together with a finite collection of ordered pairs (P_i, P_j) of distinct elements of this set, with no ordered pair being repeated. The elements of the set are called *vertices*, and the ordered pairs are called *directed edges*, of the directed graph. We use the notation $P_i \rightarrow P_j$ (read "P_i is connected to P_j") to indicate that the directed edge (P_i, P_j) belongs to the directed graph. Geometrically, we can visualize a directed graph (Figure 1) by representing the vertices as points in the plane, and representing the directed edge $P_i \rightarrow P_j$ by drawing a line or arc from vertex P_i to vertex P_j, with an arrow pointing from P_i to P_j. If both $P_i \rightarrow P_j$ and $P_j \rightarrow P_i$ hold (denoted $P_i \leftrightarrow P_j$), we draw a single line between P_i and P_j with two oppositely pointing arrows (as with P_2 and P_3 in the figure).

As in Figure 1, for example, a directed graph may have separate "components" of vertices that are connected only among themselves; and some vertices, such as P_5, may not be connected with any other vertex. Also, because $P_i \rightarrow P_i$ is not permitted in a directed graph, a vertex cannot be connected with itself by a single arc that does not pass through any other vertex.

Figure 1

Figure 2 presents diagrams representing three more examples of directed graphs. With a directed graph having n vertices, we may associate an $n \times n$ matrix $M = [m_{ij}]$, called the **vertex matrix** of the directed graph. Its elements are defined by

$$m_{ij} = \begin{cases} 1, & \text{if } P_i \rightarrow P_j \\ 0, & \text{otherwise} \end{cases}$$

for $i, j = 1, 2, \ldots, n$. For the three directed graphs in Figure 2, the corresponding vertex matrices are as follows:

Figure 2a:

$$M = \begin{bmatrix} 0 & 1 & 0 & 0 \\ 0 & 0 & 1 & 0 \\ 0 & 1 & 0 & 1 \\ 0 & 0 & 0 & 0 \end{bmatrix}$$

Figure 2b:

$$M = \begin{bmatrix} 0 & 1 & 0 & 0 & 1 \\ 0 & 0 & 1 & 1 & 0 \\ 0 & 0 & 0 & 1 & 0 \\ 0 & 1 & 0 & 0 & 1 \\ 0 & 1 & 1 & 0 & 0 \end{bmatrix}$$

Figure 2c:

$$M = \begin{bmatrix} 0 & 1 & 0 & 0 \\ 1 & 0 & 1 & 0 \\ 1 & 0 & 0 & 1 \\ 1 & 0 & 0 & 0 \end{bmatrix}$$

By their definition, vertex matrices have the following two properties:

(i) All entries are either 0 or 1.
(ii) All diagonal entries are 0.

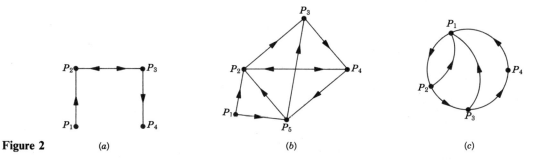

Figure 2   (a)   (b)   (c)

Conversely, any matrix with these two properties determines a unique directed graph having itself as its vertex matrix. For example, the matrix

$$M = \begin{bmatrix} 0 & 1 & 1 & 0 \\ 0 & 0 & 1 & 0 \\ 1 & 0 & 0 & 1 \\ 0 & 0 & 0 & 0 \end{bmatrix}$$

determines the directed graph in Figure 3.

Figure 3

Example 1   A certain family consists of a mother, father, daughter, and two sons. The family members have influence, or power, over each other in the following ways: the mother can influence the daugther and the oldest son; the father can influence the two sons; the daughter can influence the father; the oldest son can influence the youngest son; and the youngest son can influence the mother. We may model this family influence pattern with a directed graph whose vertices are the five family members. If family member A influences family member B, we write $A \rightarrow B$. Figure 4 is the resulting directed graph, where we have used obvious letter designations for the five family members. The vertex matrix of this directed graph is

$$\begin{array}{c} \\ M \\ F \\ D \\ OS \\ YS \end{array} \begin{array}{ccccc} M & F & D & OS & YS \\ \begin{bmatrix} 0 & 0 & 1 & 1 & 0 \\ 0 & 0 & 0 & 1 & 1 \\ 0 & 1 & 0 & 0 & 0 \\ 0 & 0 & 0 & 0 & 1 \\ 1 & 0 & 0 & 0 & 0 \end{bmatrix} \end{array}$$

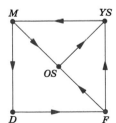

Figure 4

Example 2 In chess the knight moves in an "L"-shaped pattern about the chessboard. For the board in Figure 5 it may move horizontally two squares and then vertically one square, or it may move vertically two squares and then horizontally one square. Thus, from the center square in the figure, the knight may move to any of the eight marked shaded squares. Suppose the knight is restricted to the nine numbered squares in Figure 6. If by $i \rightarrow j$ we mean that the knight may move from square i to square j, the directed graph in Figure 7 illustrates all possible moves that the knight may make among these nine squares. In Figure 8 we have "unraveled" Figure 7 to make the pattern of possible moves clearer.

The vertex matrix of this directed graph is given by

$$
M = \begin{bmatrix}
0 & 0 & 0 & 0 & 0 & 1 & 0 & 1 & 0 \\
0 & 0 & 0 & 0 & 0 & 0 & 1 & 0 & 1 \\
0 & 0 & 0 & 1 & 0 & 0 & 0 & 1 & 0 \\
0 & 0 & 1 & 0 & 0 & 0 & 0 & 0 & 1 \\
0 & 0 & 0 & 0 & 0 & 0 & 0 & 0 & 0 \\
1 & 0 & 0 & 0 & 0 & 0 & 1 & 0 & 0 \\
0 & 1 & 0 & 0 & 0 & 1 & 0 & 0 & 0 \\
1 & 0 & 1 & 0 & 0 & 0 & 0 & 0 & 0 \\
0 & 1 & 0 & 1 & 0 & 0 & 0 & 0 & 0
\end{bmatrix}
$$

![Figure 5: a chessboard with a knight in the center and eight shaded squares marking possible moves](image)

Figure 5

1	2	3
4	5	6
7	8	9

Figure 6

Figure 7

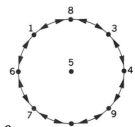

Figure 8

In Example 1 the father cannot directly influence the mother; that is, $F \to M$ is not true. But he can influence the youngest son, who can then influence the mother. We write this as $F \to YS \to M$, and call it a *2-step connection* from F to M. Analogously, we call $M \to D$ a *1-step connection*, $F \to OS \to YS \to M$ a *3-step connection*, and so forth. Let us now consider a technique for finding the number of all possible r-step connections ($r = 1, 2, \ldots$) from one vertex P_i to another vertex P_j of an arbitrary directed graph. (This will include the case when P_i and P_j are the same vertex.) The number of 1-step connections from P_i to P_j is simply m_{ij}. That is, there is either zero or one 1-step connection from P_i to P_j, depending on whether m_{ij} is zero or one. For the number of 2-step connections, we consider the square of the vertex matrix. If we let $m_{ij}^{(2)}$ be the (i, j)-th element of M^2, we have

$$m_{ij}^{(2)} = m_{i1}m_{1j} + m_{i2}m_{2j} + \cdots + m_{in}m_{nj} \tag{1}$$

Now, if $m_{i1} = m_{1j} = 1$, there is a 2-step connection $P_i \to P_1 \to P_j$ from P_i to P_j. But if either m_{i1} or m_{1j} is zero, such a 2-step connection is not possible. Thus, $P_i \to P_1 \to P_j$ is a 2-step connection if and only if $m_{i1}m_{1j} = 1$. Similarly, for any $k = 1, 2, \ldots, n$, $P_i \to P_k \to P_j$ is a 2-step connection from P_i to P_j if and only if the term $m_{ik}m_{kj}$ on the right side of (1) is one; otherwise, the term is zero. Thus, the right side of (1) is the total number of two 2-step connections from P_i to P_j.

A similar argument will work for finding the number of 3-, 4-, $\ldots$, n-step connections from P_i to P_j. In general, we have the following result.

Theorem 11.4.1. *Let M be the vertex matrix of a directed graph and let $m_{ij}^{(r)}$ be the (i, j)-th element of M^r. Then $m_{ij}^{(r)}$ is equal to the number of r-step connections from P_i to P_j.*

Example 3 Figure 9 is the route map of a small airline that services the four cities P_1, P_2, P_3, and P_4. As a directed graph, its vertex matrix is

$$M = \begin{bmatrix} 0 & 1 & 1 & 0 \\ 1 & 0 & 1 & 0 \\ 1 & 0 & 0 & 1 \\ 0 & 1 & 1 & 0 \end{bmatrix}$$

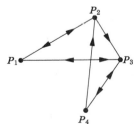

Figure 9

We have that

$$M^2 = \begin{bmatrix} 2 & 0 & 1 & 1 \\ 1 & 1 & 1 & 1 \\ 0 & 2 & 2 & 0 \\ 2 & 0 & 1 & 1 \end{bmatrix}$$

and

$$M^3 = \begin{bmatrix} 1 & 3 & 3 & 1 \\ 2 & 2 & 3 & 1 \\ 4 & 0 & 2 & 2 \\ 1 & 3 & 3 & 1 \end{bmatrix}$$

If we are interested in connections from city P_4 to city P_3, we may use Theorem 11.4.1 to find their number. Because $m_{43} = 1$, there is one 1-step connection; because $m_{43}^{(2)} = 1$, there is one 2-step connection; and because $m_{43}^{(3)} = 3$, there are three 3-step connections. To verify this, from Figure 9 we find

$$\text{1-step connections from } P_4 \text{ to } P_3: \quad P_4 \to P_3$$
$$\text{2-step connections from } P_4 \text{ to } P_3: \quad P_4 \to P_2 \to P_3$$
$$\text{3-step connections from } P_4 \text{ to } P_3: \quad P_4 \to P_3 \to P_4 \to P_3$$
$$P_4 \to P_2 \to P_1 \to P_3$$
$$P_4 \to P_3 \to P_1 \to P_3 \quad \blacktriangle$$

CLIQUES

In everyday language a "clique" is a closely knit group of people (usually three or more) that tends to communicate within itself and has no place for outsiders. In graph theory this concept is given a more precise meaning:

Definition. A subset of a directed graph is called a *clique* if it satisfies the following three conditions:

(i) The subset contains at least three vertices.
(ii) For each pair of vertices P_i and P_j in the subset, both $P_i \to P_j$ and $P_j \to P_i$ are true.
(iii) The subset is as large as possible; that is, it is not possible to add another vertex to the subset and still satisfy condition (ii).

This definition suggests that cliques are maximal subsets that are in perfect "communication" with each other. For example, if the vertices represent cities, and $P_i \to P_j$ means that there is a direct airline flight from city P_i to city P_j, then there is a direct flight between any two cities within a clique in either direction.

Example 4   The directed graph illustrated in ᵣ ˙ᵧure 10 (which might represent the route map of an airline) has two cliques:

$$\{P_1, P_2, P_3, P_4\} \qquad \text{and} \qquad \{P_3, P_4, P_6\}$$

This example shows that a directed graph may contain several cliques, and that a vertex may simultaneously belong to more than one clique. ▲

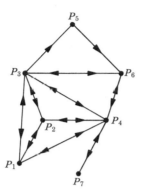

Figure 10

For simple directed graphs, cliques can be found by inspection. But for large directed graphs, it would be desirable to have a systematic procedure for detecting cliques. For this purpose it will be helpful to define a matrix $S = [s_{ij}]$ related to a given directed graph as follows:

$$s_{ij} = \begin{cases} 1, & \text{if } P_i \leftrightarrow P_j \\ 0, & \text{otherwise} \end{cases}$$

The matrix S determines a directed graph that is the same as the given directed graph, with the exception that the directed edges with only one arrow are deleted. For example, if the original directed graph is given by Figure 11a, the directed graph which has S as its vertex matrix is given in Figure 11b. The matrix S may be obtained from the vertex matrix M of the original directed graph by setting $s_{ij} = 1$ if $m_{ij} = m_{ji} = 1$ and setting $s_{ij} = 0$ otherwise.

The following theorem, which uses the matrix S, is helpful for identifying cliques.

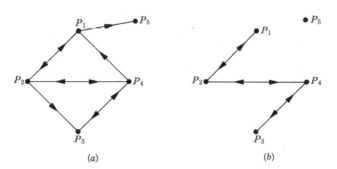

Figure 11   (a)   (b)

Theorem 11.4.2. *Let $s_{ij}^{(3)}$ be the (i, j)-th element of S^3. Then a vertex P_i belongs to some clique if and only if $s_{ii}^{(3)} \neq 0$.*

Proof. If $s_{ii}^{(3)} \neq 0$, then there is at least one 3-step connection from P_i to itself in the modified directed graph determined by S. Suppose it is $P_i \to P_j \to P_k \to P_i$. In the modified directed graph, all directed relations are two-way, so that we also have $P_i \leftrightarrow P_j \leftrightarrow P_k \leftrightarrow P_i$. But this means $\{P_i, P_j, P_k\}$ is either a clique or a subset of a clique. In either case, P_i must belong to some clique. The converse statement, if P_i belongs to a clique then $s_{ii}^{(3)} \neq 0$, follows in a similar manner. ▮

Example 5 Suppose a directed graph has as its vertex matrix

$$M = \begin{bmatrix} 0 & 1 & 1 & 1 \\ 1 & 0 & 1 & 0 \\ 0 & 1 & 0 & 1 \\ 1 & 0 & 0 & 0 \end{bmatrix}$$

Then

$$S = \begin{bmatrix} 0 & 1 & 0 & 1 \\ 1 & 0 & 1 & 0 \\ 0 & 1 & 0 & 0 \\ 1 & 0 & 0 & 0 \end{bmatrix} \quad \text{and} \quad S^3 = \begin{bmatrix} 0 & 3 & 0 & 2 \\ 3 & 0 & 2 & 0 \\ 0 & 2 & 0 & 1 \\ 2 & 0 & 1 & 0 \end{bmatrix}$$

Because all diagonal entries of S^3 are zero, it follows from Theorem 11.4.2 that the directed graph has no cliques. ▲

Example 6 Suppose a directed graph has as its vertex matrix

$$M = \begin{bmatrix} 0 & 1 & 0 & 1 & 1 \\ 1 & 0 & 0 & 1 & 0 \\ 1 & 1 & 0 & 1 & 0 \\ 1 & 1 & 0 & 0 & 0 \\ 1 & 0 & 0 & 1 & 0 \end{bmatrix}$$

Then

$$S = \begin{bmatrix} 0 & 1 & 0 & 1 & 1 \\ 1 & 0 & 0 & 1 & 0 \\ 0 & 0 & 0 & 0 & 0 \\ 1 & 1 & 0 & 0 & 0 \\ 1 & 0 & 0 & 0 & 0 \end{bmatrix} \quad \text{and} \quad S^3 = \begin{bmatrix} 2 & 4 & 0 & 4 & 3 \\ 4 & 2 & 0 & 3 & 1 \\ 0 & 0 & 0 & 0 & 0 \\ 4 & 3 & 0 & 2 & 1 \\ 3 & 1 & 0 & 1 & 0 \end{bmatrix}$$

The nonzero diagonal entries of S^3 are $s_{11}^{(3)}$, $s_{22}^{(3)}$, and $s_{44}^{(3)}$. Consequently, in the given directed graph, P_1, P_2, and P_4 belong to cliques. Because a clique must contain at least three vertices, the directed graph has only one clique, namely, $\{P_1, P_2, P_4\}$. ▲

DOMINANCE-DIRECTED GRAPHS

In many groups of individuals or animals, there is a definite "pecking order" or dominance relation between any two members of the group. That is, given any two individuals A and B, either A dominates B or B dominates A, but not both. In terms of a directed graph in which $P_i \to P_j$ means P_i dominates P_j, this means that for all distinct pairs, either $P_i \to P_j$ or $P_j \to P_i$, but not both. In general, we have the following definition.

> **Definition.** A *dominance-directed graph* is a directed graph such that for any distinct pair of vertices P_i and P_j, either $P_i \to P_j$ or $P_j \to P_i$, but not both.

An example of a directed graph satisfying this definition is a league of n sports teams that play each other exactly one time, as in one round of a round-robin tournament in which no ties are allowed. If $P_i \to P_j$ means that team P_i beat team P_j in their single match, it is easily seen that the definition of a dominance-directed group is satisfied. For this reason, dominance-directed graphs are sometimes called *tournaments*.

Figure 12 illustrates some dominance-directed graphs with three, four, and five vertices, respectively. In these three graphs, the circled vertices have the following interesting property: from each one there is either a 1-step or a 2-step connection to any other vertex in its graph. In a sports tournament these vertices would correspond to the most "powerful" teams in the sense that these teams either beat any given team or beat some other team that beat the given team. We can now state and prove a theorem that guarantees that any dominance-directed graph has at least one vertex with this property.

> **Theorem 11.4.3.** *In any dominance-directed graph there is at least one vertex from which there is a 1-step or 2-step connection to any other vertex.*

(a)

(b)

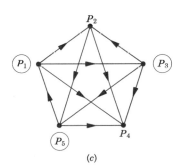

(c)

Figure 12

Proof. Consider a vertex (there may be several) with the largest total number of 1-step and 2-step connections to other vertices in the graph. By renumbering the vertices, we may assume that P_1 is such a vertex. Suppose there is some vertex P_i such that there is no 1-step or 2-step connection from P_1 to P_i. Then, in particular, $P_1 \to P_i$ is not true, so that by definition of a dominance-directed graph, it must be that $P_i \to P_1$. Next, let P_k be any vertex such that $P_1 \to P_k$ is true. Then we cannot have $P_k \to P_i$, as then $P_1 \to P_k \to P_i$ would be a 2-step connection from P_1 to P_i. Thus, it must be that $P_i \to P_k$. That is, P_i has 1-step connections to all the vertices to which P_1 has 1-step connections. The vertex P_i must then also have 2-step connections to all the vertices to which P_1 has 2-step connections. But because, in addition, we have that $P_i \to P_1$, this means that P_i has more 1-step and 2-step connections to other vertices than does P_1. However, this contradicts the way in which P_1 was chosen. Hence, there can be no vertex P_i to which P_1 has no 1-step or 2-step connection. ▮

This proof shows that a vertex with the largest total number of 1-step and 2-step connections to other vertices has the property stated in the theorem. There is a simple way of finding such vertices using the vertex matrix M and its square M^2. The sum of the entries in the ith row of M is the total number of 1-step connections from P_i to other vertices, and the sum of the entries of the ith row of M^2 is the total number of 2-step connections from P_i to other vertices. Consequently, the sum of the entries of the ith row of the matrix $A = M + M^2$ is the total number of 1-step and 2-step connections from P_i to other vertices. In other words, a row of $A = M + M^2$ with the largest row sum identifies a vertex having the property stated in Theorem 11.4.3.

Example 7 Suppose five baseball teams play each other exactly once, and the results are as indicated in the dominance-directed graph of Figure 13. The vertex matrix of the graph is

$$M = \begin{bmatrix} 0 & 0 & 1 & 1 & 0 \\ 1 & 0 & 1 & 0 & 1 \\ 0 & 0 & 0 & 1 & 0 \\ 0 & 1 & 0 & 0 & 0 \\ 1 & 0 & 1 & 1 & 0 \end{bmatrix}$$

And so

$$A = M + M^2 = \begin{bmatrix} 0 & 0 & 1 & 1 & 0 \\ 1 & 0 & 1 & 0 & 1 \\ 0 & 0 & 0 & 1 & 0 \\ 0 & 1 & 0 & 0 & 0 \\ 1 & 0 & 1 & 1 & 0 \end{bmatrix} + \begin{bmatrix} 0 & 1 & 0 & 1 & 0 \\ 1 & 0 & 2 & 3 & 0 \\ 0 & 1 & 0 & 0 & 0 \\ 1 & 0 & 1 & 0 & 1 \\ 0 & 1 & 1 & 2 & 0 \end{bmatrix} = \begin{bmatrix} 0 & 1 & 1 & 2 & 0 \\ 2 & 0 & 3 & 3 & 1 \\ 0 & 1 & 0 & 1 & 0 \\ 1 & 1 & 1 & 0 & 1 \\ 1 & 1 & 2 & 3 & 0 \end{bmatrix}$$

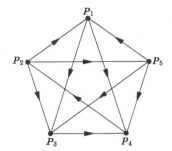

Figure 13

The row sums of A are

$$1\text{st row sum} = 4$$
$$2\text{nd row sum} = 9$$
$$3\text{rd row sum} = 2$$
$$4\text{th row sum} = 4$$
$$5\text{th row sum} = 7$$

Because the second row has the largest row sum, the vertex P_2 must have a 1-step or 2-step connection to any other vertex. This is easily verified from Figure 13.

We have informally suggested that a vertex with the largest number of 1-step and 2-step connections to other vertices is a "powerful" vertex. We can formalize this concept with the following definition.

> **Definition.** The *power* of a vertex of a dominance-directed graph is the total number of 1-step and 2-step connections from it to other vertices. Alternatively, the power of a vertex P_i is the sum of the entries of the ith row of the matrix $A = M + M^2$, where M is the vertex matrix of the directed graph.

Example 8   Let us rank the five baseball teams in Example 7 according to their powers. From the calculations for the row sums in that example we have

$$\text{Power of team } P_1 = 4$$
$$\text{Power of team } P_2 = 9$$
$$\text{Power of team } P_3 = 2$$
$$\text{Power of team } P_4 = 4$$
$$\text{Power of team } P_5 = 7$$

Hence, the ranking of the teams according to their powers would be

$$P_2 \text{ (first), } P_5 \text{ (second), } P_1 \text{ and } P_4 \text{ (tied for third), } P_3 \text{ (last)}$$

EXERCISE SET 11.4

1. Construct the vertex matrix for each of the directed graphs illustrated in Figure 14.

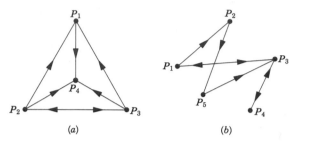

(a)   (b)   (c)   **Figure 14**

2. Draw a diagram of the directed graph corresponding to each of the following vertex matrices.

(a) $\begin{bmatrix} 0 & 1 & 1 & 0 \\ 1 & 0 & 0 & 0 \\ 0 & 0 & 0 & 1 \\ 1 & 0 & 1 & 0 \end{bmatrix}$ (b) $\begin{bmatrix} 0 & 0 & 1 & 0 & 0 \\ 1 & 0 & 0 & 0 & 1 \\ 0 & 1 & 0 & 1 & 1 \\ 0 & 0 & 0 & 0 & 0 \\ 1 & 1 & 1 & 0 & 0 \end{bmatrix}$ (c) $\begin{bmatrix} 0 & 1 & 0 & 1 & 0 & 1 \\ 1 & 0 & 0 & 0 & 1 & 0 \\ 0 & 0 & 0 & 0 & 0 & 0 \\ 1 & 1 & 0 & 0 & 1 & 0 \\ 0 & 0 & 0 & 1 & 0 & 1 \\ 0 & 1 & 0 & 0 & 1 & 0 \end{bmatrix}$

3. Let M be the following vertex matrix of a directed graph:

$$\begin{bmatrix} 0 & 1 & 1 & 1 \\ 1 & 0 & 0 & 0 \\ 0 & 1 & 0 & 1 \\ 0 & 1 & 1 & 0 \end{bmatrix}$$

(a) Draw a diagram of the directed graph.
(b) Use Theorem 11.4.1 to find the number of 1-, 2-, and 3-step connections from the vertex P_1 to the vertex P_2. Verify your answer by listing the various connections as in Example 3.
(c) Repeat part (b) for the 1-, 2-, and 3-step connections from P_1 to P_4.

4. By inspection, locate all cliques in each of the directed graphs illustrated in Figure 15.

(a)   (b)   (c)   **Figure 15**

5. For each of the following vertex matrices, use Theorem 11.4.2 to find all cliques in the corresponding directed graphs.

$$\text{(a)} \begin{bmatrix} 0 & 1 & 0 & 1 & 0 \\ 1 & 0 & 1 & 0 & 1 \\ 0 & 1 & 0 & 1 & 1 \\ 1 & 0 & 0 & 0 & 1 \\ 1 & 0 & 1 & 1 & 0 \end{bmatrix} \qquad \text{(b)} \begin{bmatrix} 0 & 1 & 0 & 1 & 1 & 0 \\ 1 & 0 & 1 & 0 & 1 & 1 \\ 0 & 1 & 0 & 1 & 0 & 1 \\ 1 & 0 & 1 & 0 & 1 & 1 \\ 0 & 1 & 0 & 1 & 0 & 0 \\ 0 & 0 & 1 & 1 & 1 & 0 \end{bmatrix}$$

6. For the dominance-directed graph illustrated in Figure 16, construct the vertex matrix and find the power of each vertex.

Figure 16

7. Five baseball teams play each other one time with the following results:

> *A* beats *B, C, D*
> *B* beats *C, E*
> *C* beats *D, E*
> *D* beats *B*
> *E* beats *A, D*

Rank the five baseball teams in accordance with the powers of the vertices they correspond to in the dominance-directed graph representing the outcomes of the games.

11.5 THEORY OF GAMES

This section discusses a general game in which two competing players choose separate strategies to reach opposing objectives. The optimal strategy of each player is found in certain cases with the use of matrix techniques.

PREREQUISITES: Matrix multiplication
Basic probability concepts

To introduce the basic concepts in the theory of games, we will consider the following carnival-type game that two people agree to play. We will call the participants in the game *player R* and *player C*. Each player has a stationary wheel with a

Figure 1

Row-wheel
of player R

Column-wheel
of player C

movable pointer on it as in Figure 1. For reasons that will become clear, we will call player R's wheel the *row-wheel* and player C's wheel the *column-wheel*. The row-wheel is divided into three sectors numbered 1, 2, and 3, and the column-wheel is divided into four sectors numbered 1, 2, 3, and 4. The fractions of the area occupied by the various sectors are indicated in the figure. To play the game each player spins the pointer of his or her wheel and lets it come to rest at random. The number of the sector in which each pointer comes to rest is called the *move* of that player. Thus, player R has three possible moves and player C has four possible moves. Depending on the move each player makes, player C then makes a payment of money to player R according to Table 1.

TABLE 1 *Payment Table from C to R*

		Player C's Move			
		1	**2**	**3**	**4**
Player R's Move	**1**	$3	$5	−$2	−$1
	2	−$2	$4	−$3	−$4
	3	$6	−$5	$0	$3

For example, if the row-wheel pointer comes to rest in sector 1 (player R makes move 1), and the column-wheel pointer comes to rest in sector 2 (player C makes move 2), then player C must pay player R the sum of $5. Some of the entries in this table are negative, indicating that player C makes a negative payment to player R. By this we mean that player R makes a positive payment to player C. For example, if the row-wheel shows 2 and the column-wheel shows 4, then player R pays player C the sum of $4, because the corresponding entry in the table is −$4. In this way the positive entries of the table are the gains of player R and the losses of player C, and the negative entries are the gains of player C and the losses of player R.

In this game the players have no control over their moves; each move is determined by chance. However, if each player can decide whether he or she wants to play, then each would want to know how much he or she can expect to win or lose over the long term if he or she chooses to play. (Later in the section we will discuss this question and also consider a more complicated situation in which the players can exercise some control over their moves by varying the sectors of their wheels.)

TWO-PERSON ZERO-SUM MATRIX GAMES

The game described above is an example of a ***two-person zero-sum matrix game***. The term "zero-sum" means that in each play of the game the positive gain of one player is equal to the negative gain (loss) of the other player. That is, the sum of the two gains is zero. The term "matrix game" is used to describe a two-person game in which each player has only a finite number of moves, so that all possible outcomes of each play, and the corresponding gains of the players, may be displayed in tabular or matrix form, as in Table 1.

In a general game of this type, let player R have m possible moves and let player C have n possible moves. In a play of the game, each player makes one of his or her possible moves, and then a *payoff* is made from player C to player R, depending on the moves. For $i = 1, 2, \ldots, m$, and $j = 1, 2, \ldots, n$, let us set

a_{ij} = payoff that player C makes to player R if player R
  makes move i and player C makes move j

This payoff need not be money; it may be any type of commodity to which we can attach a numerical value. As before, if an entry a_{ij} is negative, we mean that player C receives a payoff of $|a_{ij}|$ from player R. We arrange these mn possible payoffs in the form of an $m \times n$ matrix

$$A = \begin{bmatrix} a_{11} & a_{12} & \cdots & a_{1n} \\ a_{21} & a_{22} & \cdots & a_{2n} \\ \vdots & \vdots & & \vdots \\ a_{m1} & a_{m2} & \cdots & a_{mn} \end{bmatrix}$$

which we will call the ***payoff matrix*** of the game.

Each player is to make his or her moves on a probabilistic basis. For example, for the game discussed in the introduction, the proportion of the area of a sector to the area of the wheel would be the probability that the player makes the move corresponding to that sector. Thus, from Figure 1 we see that player R would make move 2 with probability $\frac{1}{3}$, and player C would make move 2 with probability $\frac{1}{4}$. In the general case we make the following definitions:

p_i = probability that player R makes move i $(i = 1, 2, \ldots, m)$
q_j = probability that player C makes move j $(j = 1, 2, \ldots, n)$

It follows from these definitions that

$$p_1 + p_2 + \cdots + p_m = 1$$

and

$$q_1 + q_2 + \cdots + q_n = 1$$

With the probabilities p_i and q_j we form two vectors:

$$\mathbf{p} = \begin{bmatrix} p_1 & p_2 & \cdots & p_m \end{bmatrix} \quad \text{and} \quad \mathbf{q} = \begin{bmatrix} q_1 \\ q_2 \\ \vdots \\ q_n \end{bmatrix}$$

We call the row vector $\mathbf{p}$ the ***strategy of player R*** and the column vector $\mathbf{q}$ the ***strategy of player C***. For example, from Figure 1 we have

$$\mathbf{p} = \begin{bmatrix} \frac{1}{6} & \frac{1}{3} & \frac{1}{2} \end{bmatrix} \quad \text{and} \quad \mathbf{q} = \begin{bmatrix} \frac{1}{4} \\ \frac{1}{4} \\ \frac{1}{3} \\ \frac{1}{6} \end{bmatrix}$$

for the carnival game described earlier.

From the theory of probability, if the probability that player R makes move i is p_i, and independently the probability that player C makes move j is q_j, then $p_i q_j$ is the probability that for any one play of the game, player R makes move i *and* player C makes move j. The payoff to player R for such a pair of moves is a_{ij}. Thus, $p_i q_j$ is also the probability that for any one play of the game, the payoff to player R is entry a_{ij}. If we multiply each possible payoff by its corresponding probability and sum over all possible payoffs, we obtain the expression

$$a_{11}p_1 q_1 + a_{12}p_1 q_2 + \cdots + a_{1n}p_1 q_n + a_{21}p_2 q_1 + \cdots + a_{mn}p_m q_n \tag{1}$$

Equation (1) is a weighted average of the payoffs to player R; each payoff is weighted according to the probability of its occurrence. In the theory of probability this weighted average is called the ***expected payoff*** to player R. It can be shown that if the game is played many times, the long-term average payoff per play to player R is given by this expression. We denote this expected payoff by $E(\mathbf{p}, \mathbf{q})$ to emphasize the fact that it depends on the strategies of the two players. From the definition of the payoff matrix A and the strategies $\mathbf{p}$ and $\mathbf{q}$, it can be verified that we may express the expected payoff in matrix notation as

$$E(\mathbf{p}, \mathbf{q}) = \begin{bmatrix} p_1 & p_2 & \cdots & p_m \end{bmatrix} \begin{bmatrix} a_{11} & a_{12} & \cdots & a_{1n} \\ a_{21} & a_{22} & \cdots & a_{2n} \\ \vdots & \vdots & & \vdots \\ a_{m1} & a_{m2} & \cdots & a_{mn} \end{bmatrix} \begin{bmatrix} q_1 \\ q_2 \\ \vdots \\ q_n \end{bmatrix} = \mathbf{p}A\mathbf{q} \tag{2}$$

Because $E(\mathbf{p}, \mathbf{q})$ is the expected payoff to player R, it follows that $-E(\mathbf{p}, \mathbf{q})$ is the expected payoff to player C.

Example 1 For the carnival game described earlier, we have

$$E(\mathbf{p}, \mathbf{q}) = \mathbf{p}A\mathbf{q} = \begin{bmatrix} \frac{1}{6} & \frac{1}{3} & \frac{1}{2} \end{bmatrix} \begin{bmatrix} 3 & 5 & -2 & -1 \\ -2 & 4 & -3 & -4 \\ 6 & -5 & 0 & 3 \end{bmatrix} \begin{bmatrix} \frac{1}{4} \\ \frac{1}{4} \\ \frac{1}{3} \\ \frac{1}{6} \end{bmatrix} = \frac{13}{72} = .1806\ldots$$

Thus, in the long run player R can expect to receive an average of about 18¢ from player C in each play of the game. ▲

So far we have been discussing the situation in which each player has a predetermined strategy. We will now consider the more difficult situation in which both players may change their strategies independently. For example, in the game described in the introduction, we would allow both players to alter the areas of the sectors of their wheels and thereby control the probabilities of their respective moves. This qualitatively changes the nature of the problem and puts us firmly in the field of true game theory. It is understood that neither player knows the strategy the other will choose. It is also assumed that each player will make the best possible choice of strategy and that the other player knows this. Thus, player R attempts to choose a strategy $\mathbf{p}$ such that $E(\mathbf{p}, \mathbf{q})$ is as large as possible for the best strategy $\mathbf{q}$ that player C can choose; and similarly player C attempts to choose a strategy $\mathbf{q}$ such that $E(\mathbf{p}, \mathbf{q})$ is as small as possible for the best strategy $\mathbf{p}$ that player R can choose. To see that such choices are actually possible, we shall need the following theorem, called the **Fundamental Theorem of Two-Person Zero-Sum Games**. (The general proof, which involves ideas from the theory of linear programming, will be omitted. However, later we shall prove two special cases of the theorem.)

Theorem 11.5.1. *There exist strategies* $\mathbf{p}^*$ *and* $\mathbf{q}^*$ *such that*

$$E(\mathbf{p}^*, \mathbf{q}) \geq E(\mathbf{p}^*, \mathbf{q}^*) > E(\mathbf{p}, \mathbf{q}^*) \tag{3}$$

for all strategies $\mathbf{p}$ *and* $\mathbf{q}$.

The strategies $\mathbf{p}^*$ and $\mathbf{q}^*$ in this theorem are the best possible strategies for players R and C, respectively. To see why this is so let $v = (\mathbf{p}^*, \mathbf{q}^*)$. The left-hand inequality of Equation (3) then reads

$$E(\mathbf{p}^*, \mathbf{q}) \geq v \qquad \text{for all strategies } \mathbf{q}$$

This means that if player R chooses the strategy $\mathbf{p}^*$, then no matter what strategy $\mathbf{q}$ player C chooses, the expected payoff to player R will never be below v. Moreover, it is not possible for player R to achieve an expected payoff greater than v. To see why, suppose there is some strategy $\mathbf{p}^{**}$ that player R can choose such that

$$E(\mathbf{p}^{**}, \mathbf{q}) > v \qquad \text{for all strategies } \mathbf{q}$$

Then, in particular,

$$E(\mathbf{p}^{**}, \mathbf{q}^*) > v$$

But this contradicts the right-hand inequality of Equation (3), which requires that $v \leq E(\mathbf{p}^{**}, \mathbf{q}^*)$. Consequently, the best player R can do is prevent his or her expected payoff from falling below the value v. Similarly, the best player C can do is ensure that his or her expected payoff has at least the value v, and this is achieved by using strategy $\mathbf{q}^*$.

On the basis of this discussion, we arrive at the following definitions.

Definition. If **p*** and **q*** are strategies such that

$$E(\mathbf{p}^*, \mathbf{q}) \geq E(\mathbf{p}^*, \mathbf{q}^*) \geq E(\mathbf{p}, \mathbf{q}^*) \tag{4}$$

for all strategies **p** and **q**, then

  (i) **p*** is called an *optimal strategy for player R*;
 (ii) **q*** is called an *optimal strategy for player C*;
(iii) $v = E(\mathbf{p}^*, \mathbf{q}^*)$ is called the *value* of the game.

The wording in this definition suggests that optimal strategies are not necessarily unique. This is indeed the case, and in Exercise 2 we ask the reader to show this. However, it can be proved that any two sets of optimal strategies always result in the same value v of the game. That is, if **p***, **q*** and **p****, **q**** are optimal strategies then

$$E(\mathbf{p}^*, \mathbf{q}^*) = E(\mathbf{p}^{**}, \mathbf{q}^{**}) \tag{5}$$

The value of a game is thus the expected payoff to player R when both players choose any possible optimal strategies.

To find optimal strategies we must find vectors **p*** and **q*** that satisfy Equation (4). This is generally done by using linear programming techniques. Next, we discuss special cases for which optimal strategies may be found by more elementary techniques.

STRICTLY DETERMINED GAMES

We now introduce the following definition.

Definition. An entry a_{rs} in a payoff matrix A is called a *saddle point* if

 (i) a_{rs} is the smallest entry in its row, and
(ii) a_{rs} is the largest entry in its column.

A game whose payoff matrix has a saddle point is called *strictly determined*.

For example, the shaded element in each of the following payoff matrices is a saddle point:

$$\begin{bmatrix} 3 & 1 \\ -4 & 0 \end{bmatrix} \qquad \begin{bmatrix} 30 & -50 & -5 \\ 60 & 90 & 75 \\ -10 & 60 & -30 \end{bmatrix} \qquad \begin{bmatrix} 0 & -3 & 5 & -9 \\ 15 & -8 & -2 & 10 \\ 7 & 10 & 6 & 9 \\ 6 & 11 & -3 & 2 \end{bmatrix}$$

If a matrix has a saddle point a_{rs}, it turns out that the following strategies are optimal strategies for the two players:

$$\mathbf{p}^* = [0 \quad 0 \quad \cdots \quad 1 \quad \cdots \quad 0], \qquad \mathbf{q}^* = \begin{bmatrix} 0 \\ 0 \\ \vdots \\ 1 \\ \vdots \\ 0 \end{bmatrix} \begin{matrix} \\ \\ \text{sth entry} \\ \\ \\ \end{matrix}$$

rth entry

That is, an optimal strategy for player R is to always make the rth move, and an optimal strategy for player C is to always make the sth move. Such strategies for which only one move is possible are called **pure strategies**. Strategies for which more than one move is possible are called **mixed strategies**. To show that the above pure strategies are optimal, the reader may verify the following three equations (see Exercise 6):

$$E(\mathbf{p}^*, \mathbf{q}^*) = \mathbf{p}^* A \mathbf{q}^* = a_{rs} \tag{6}$$

$$E(\mathbf{p}^*, \mathbf{q}) = \mathbf{p}^* A \mathbf{q} \geq a_{rs} \qquad \text{for any strategy } \mathbf{q} \tag{7}$$

$$E(\mathbf{p}, \mathbf{q}^*) = \mathbf{p} A \mathbf{q}^* \leq a_{rs} \qquad \text{for any strategy } \mathbf{p} \tag{8}$$

Together, these three equations imply that

$$E(\mathbf{p}^*, \mathbf{q}) \geq E(\mathbf{p}^*, \mathbf{q}^*) \geq E(\mathbf{p}, \mathbf{q}^*)$$

for all strategies $\mathbf{p}$ and $\mathbf{q}$. Because this is exactly Equation (4), it follows that $\mathbf{p}^*$ and $\mathbf{q}^*$ are optimal strategies.

From Equation (6) the value of a strictly determined game is simply the numerical value of a saddle point a_{rs}. It is possible for a payoff matrix to have several saddle points, but then the uniqueness of the value of a game guarantees that the numerical values of all saddle points are the same.

Example 2 Two competing television networks, R and C, are scheduling one-hour programs in the same time period. Network R can schedule one of three possible programs and network C can schedule one of four possible programs. Neither network knows which program the other will schedule. Both networks ask the same outside polling agency to give them an estimate of how all possible pairings of the programs will divide the viewing audience. The agency gives them each Table 2, whose (i, j)-th entry is the percentage of the viewing audience that will watch network R if network R's program i is paired against network C's program j. What program should each network schedule in order to maximize its viewing audience?

TABLE 2 *Audience Percentage for Network R*

		Network *C*'s Program			
		1	2	3	4
Network *R*'s Program	1	60	20	30	55
	2	50	75	45	60
	3	70	45	35	30

Solution. Subtract 50 from each entry in the above table to construct the following matrix:

$$\begin{bmatrix} 10 & -30 & -20 & 5 \\ 0 & 25 & -5 & 10 \\ 20 & -5 & -15 & -20 \end{bmatrix}$$

This is the payoff matrix of the two-person zero-sum game in which each network is considered to start with 50% of the audience, and the (i, j)-th entry of the matrix is the percentage of the viewing audience that network C loses to network R if programs i and j are paired against each other. It is easily seen that the entry

$$a_{23} = -5$$

is a saddle point of the payoff matrix. Hence, the optimal strategy of network R is to schedule program 2 and the optimal strategy of network C is to schedule program 3. This will result in network R's receiving 45% of the audience and network C's receiving 55% of the audience. ▲

2 × 2 MATRIX GAMES

Another case in which the optimal strategies can be found by elementary means occurs when each player has only two possible moves. In this case, the payoff matrix is a 2 × 2 matrix

$$A = \begin{bmatrix} a_{11} & a_{12} \\ a_{21} & a_{22} \end{bmatrix}$$

If the game is strictly determined, at least one of the four entries of A is a saddle point, and the techniques discussed above can then be applied to determine optimal strategies for the two players. If the game is not strictly determined, we first compute the expected payoff for arbitrary strategies **p** and **q**:

$$E(\mathbf{p}, \mathbf{q}) = \mathbf{p}A\mathbf{q} = \begin{bmatrix} p_1 & p_2 \end{bmatrix} \begin{bmatrix} a_{11} & a_{12} \\ a_{21} & a_{22} \end{bmatrix} \begin{bmatrix} q_1 \\ q_2 \end{bmatrix}$$

$$= a_{11}p_1q_1 + a_{12}p_1q_2 + a_{21}p_2q_1 + a_{22}p_2q_2 \qquad (9)$$

Because

$$p_1 + p_2 = 1 \quad \text{and} \quad q_1 + q_2 = 1 \qquad (10)$$

We may substitute $p_2 = 1 - p_1$ and $q_2 = 1 - q_1$ into (9) to obtain

$$E(\mathbf{p}, \mathbf{q}) = a_{11}p_1q_1 + a_{12}p_1(1 - q_1) + a_{21}(1 - p_1)q_1 + a_{22}(1 - p_1)(1 - q_1) \quad (11)$$

If we rearrange the terms in Equation (11), we may write

$$E(\mathbf{p}, \mathbf{q}) = [(a_{11} + a_{22} - a_{12} - a_{21})p_1 - (a_{22} - a_{21})]q_1 + (a_{12} - a_{22})p_1 + a_{22} \quad (12)$$

By examining the coefficient of the q_1 term in (12), we see that if we set

$$p_1 = p_1^* = \frac{a_{22} - a_{21}}{a_{11} + a_{22} - a_{12} - a_{21}} \quad (13)$$

then that coefficient is zero, and (12) reduces to

$$E(\mathbf{p^*}, \mathbf{q}) = \frac{a_{11}a_{22} - a_{12}a_{21}}{a_{11} + a_{22} - a_{12} - a_{21}} \quad (14)$$

Equation (14) is independent of $\mathbf{q}$; that is, if player R chooses the strategy determined by (13), player C cannot change the expected payoff by varying his or her strategy.

In a similar manner, it may be verified that if player C chooses the strategy determined by

$$q_1 = q_1^* = \frac{a_{22} - a_{12}}{a_{11} + a_{22} - a_{12} - a_{21}} \quad (15)$$

then substituting in (12) gives

$$E(\mathbf{p}, \mathbf{q^*}) = \frac{a_{11}a_{22} - a_{12}a_{21}}{a_{11} + a_{22} - a_{12} - a_{21}} \quad (16)$$

Equations (14) and (16) show that

$$E(\mathbf{p^*}, \mathbf{q}) = E(\mathbf{p^*}, \mathbf{q^*}) = E(\mathbf{p}, \mathbf{q^*}) \quad (17)$$

for all strategies $\mathbf{p}$ and $\mathbf{q}$. Thus, the strategies determined by (13), (15), and (10) are optimal strategies for players R and C, respectively, and so we have the following result.

Theorem 11.5.2. *For a 2×2 game that is not strictly determined, optimal strategies for players R and C are*

$$\mathbf{p^*} = \left[\frac{a_{22} - a_{21}}{a_{11} + a_{22} - a_{12} - a_{21}} \quad \frac{a_{11} - a_{12}}{a_{11} + a_{22} - a_{12} - a_{21}} \right]$$

and

$$\mathbf{q^*} = \begin{bmatrix} \dfrac{a_{22} - a_{12}}{a_{11} + a_{22} - a_{12} - a_{21}} \\[2mm] \dfrac{a_{11} - a_{21}}{a_{11} + a_{22} - a_{12} - a_{21}} \end{bmatrix}$$

The value of the game is

$$v = \frac{a_{11}a_{22} - a_{12}a_{21}}{a_{11} + a_{22} - a_{12} - a_{21}}$$

In order to be complete, we must show that the entries in the vectors $\mathbf{p}^*$ and $\mathbf{q}^*$ are numbers strictly between 0 and 1. In Exercise 7 we ask the reader to show that this is the case as long as the game is not strictly determined.

Equation (17) is interesting in that it implies that either player may force the expected payoff to be the value of the game by choosing his or her optimal strategy, regardless of which strategy the other player chooses. This is not true, in general, for games in which either player has more than two moves.

Example 3 The federal government desires to inoculate its citizens against a certain flu virus. The virus has two strains, and the proportions in which the two strains occur in the virus population is not known. Two vaccines have been developed. Vaccine 1 is 85% effective against strain 1 and 70% effective against strain 2. Vaccine 2 is 60% effective against strain 1 and 90% effective against strain 2. What inoculation policy should the government adopt?

Solution. We may consider this a two-person game in which player R (the government) desires to make the payoff (the fraction of citizens resistant to the virus) as large as possible, and player C (the virus) desires to make the payoff as small as possible. The payoff matrix is

$$\text{Vaccine} \begin{array}{cc} & \begin{array}{cc} \text{Strain} \\ 1 \quad\quad 2 \end{array} \\ \begin{array}{c} 1 \\ 2 \end{array} & \begin{bmatrix} .85 & .70 \\ .60 & .90 \end{bmatrix} \end{array}$$

It can be shown that this matrix has no saddle points, so that Theorem 11.5.2 is applicable. Consequently,

$$p_1^* = \frac{a_{22} - a_{21}}{a_{11} + a_{22} - a_{12} - a_{21}} = \frac{.90 - .60}{.85 + .90 - .70 - .60} = \frac{.30}{.45} = \frac{2}{3}$$

$$p_2^* = 1 - p_1^* = 1 - \frac{2}{3} = \frac{1}{3}$$

$$q_1^* = \frac{a_{22} - a_{12}}{a_{11} + a_{22} - a_{12} - a_{21}} = \frac{.90 - .70}{.85 + .90 - .70 - .60} = \frac{.20}{.45} = \frac{4}{9}$$

$$q_2^* = 1 - q_1^* = 1 - \frac{4}{9} = \frac{5}{9}$$

$$v = \frac{a_{11}a_{22} - a_{12}a_{21}}{a_{11} + a_{22} - a_{12} - a_{21}} = \frac{(.85)(.90) - (.70)(.60)}{.85 + .90 - .70 - .60} = \frac{.345}{.45} = .7666\ldots$$

Thus, the optimal strategy for the government is to inoculate $\frac{2}{3}$ of the citizens with vaccine 1 and $\frac{1}{3}$ of the citizens with vaccine 2. This will guarantee that about 76.7% of the citizens will be resistant to a virus attack regardless of the distribution of the two strains.

In contrast, a virus distribution of $\frac{4}{9}$ of strain 1 and $\frac{5}{9}$ of strain 2 will result in the same 76.7% of resistant citizens, regardless of the inoculation strategy adopted by the government. ▲

EXERCISE SET 11.5

1. Suppose a game has a payoff matrix

$$A = \begin{bmatrix} -4 & 6 & -4 & 1 \\ 5 & -7 & 3 & 8 \\ -8 & 0 & 6 & -2 \end{bmatrix}$$

(a) If players R and C use strategies

$$\mathbf{p} = \begin{bmatrix} \frac{1}{2} & 0 & \frac{1}{2} \end{bmatrix} \quad \text{and} \quad \mathbf{q} = \begin{bmatrix} \frac{1}{4} \\ \frac{1}{4} \\ \frac{1}{4} \\ \frac{1}{4} \end{bmatrix}$$

respectively, what is the expected payoff of the game?
(b) If player C keeps his strategy fixed as in part (a), what strategy should player R choose to maximize his expected payoff?
(c) If player R keeps her strategy fixed as in part (a), what strategy should player C choose to minimize the expected payoff to player R?

2. Construct a simple example to show that optimal strategies are not necessarily unique. For example, find a payoff matrix with several equal saddle points.

3. For the strictly determined games with the following payoff matrices, find optimal strategies for the two players, and find the values of the games.

(a) $\begin{bmatrix} 5 & 2 \\ 7 & 3 \end{bmatrix}$ (b) $\begin{bmatrix} -3 & -2 \\ 2 & 4 \\ -4 & 1 \end{bmatrix}$ (c) $\begin{bmatrix} 2 & -2 & 0 \\ -6 & 0 & -5 \\ 5 & 2 & 3 \end{bmatrix}$ (d) $\begin{bmatrix} -3 & 2 & -1 \\ -2 & -1 & 5 \\ -4 & 1 & 0 \\ -3 & 4 & 6 \end{bmatrix}$

4. For the 2×2 games with the following payoff matrices, find optimal strategies for the two players, and find the values of the games.

(a) $\begin{bmatrix} 6 & 3 \\ -1 & 4 \end{bmatrix}$ (b) $\begin{bmatrix} 40 & 20 \\ -10 & 30 \end{bmatrix}$ (c) $\begin{bmatrix} 3 & 7 \\ -5 & 4 \end{bmatrix}$ (d) $\begin{bmatrix} 3 & 5 \\ 5 & 2 \end{bmatrix}$ (e) $\begin{bmatrix} 7 & -3 \\ -5 & -2 \end{bmatrix}$

5. Player R has two playing cards: a black ace and a red four. Player C also has two cards: a black two and a red three. Each player secretly selects one of his or her cards. If both selected cards are the same color, player C pays player R the sum of the face values in dollars. If the cards are different colors, player R pays player C the sum of the face values. What are optimal strategies for both players, and what is the value of the game?

6. Verify Equations (6), (7), and (8).

7. Show that the entries of the optimal strategies $\mathbf{p}^*$ and $\mathbf{q}^*$ given in Theorem 11.5.2 are numbers strictly between zero and one.

11.6 THE ASSIGNMENT PROBLEM

A number of facilities are to be assigned different tasks. Each possible assignment results in a certain cost. This section describes an algorithm called the Hungarian method to find an assignment with minimum cost.

PREREQUISITE: Matrix notation

A basic problem in operations research is to assign tasks to facilities on a one-to-one basis in some optimal way. For example, the problem may be to find the best assignment of workers to jobs, sports players to field positions, equipment to construction sites, and so forth. The assignment problem requires that there be as many facilities as tasks, say n of each. In this case, there are exactly $n!$ different ways to assign the tasks to the facilities on a one-to-one basis. This follows, as there are n ways to assign the first task, $n - 1$ ways to assign the second, $n - 2$ ways to assign the third, and so on—a total of

$$n \cdot (n - 1) \cdot (n - 2) \cdot \cdots \cdot 3 \cdot 2 \cdot 1 = n!$$

possible assignments. Among these $n!$ possible assignments, we are to find one that is optimal in some sense. To define the notion of an optimal assignment precisely, we introduce the following quantities. Let

$$c_{ij} = \text{cost of assigning the } i\text{th facility the } j\text{th task}$$

for $i, j = 1, 2, \ldots, n$. The units of c_{ij} might be dollars, miles, hours—whatever is appropriate to the problem. We define the *cost matrix* to be the $n \times n$ matrix

$$C = \begin{bmatrix} c_{11} & c_{12} & \cdots & c_{1n} \\ c_{21} & c_{22} & \cdots & c_{2n} \\ \vdots & \vdots & & \vdots \\ c_{n1} & c_{n2} & \cdots & c_{nn} \end{bmatrix}$$

The requirement that each facility be assigned a unique task on a one-to-one basis is equivalent to the condition that no two of the corresponding c_{ij}'s come from the same row or column. This leads to the following definition.

Definition. Given an $n \times n$ cost matrix C, an ***assignment*** is a set of n entry positions, no two of which lie in the same row or column.

An optimal assignment is then defined as follows.

Definition. The sum of the n entries of an assignment is called its ***cost***. An assignment with the smallest possible cost is called an ***optimal assignment***.

The ***assignment problem*** is to find an optimal assignment in a given cost matrix. For example, in assigning n pieces of equipment to n construction sites, c_{ij} could be the distance in miles between the ith piece of equipment and the jth construction site. An optimal assignment is one for which the total distance traveled by the n pieces of equipment is a minimum.

Example 1 A college intends to install air-conditioning in three of its buildings during a one-week spring break. It invites three contractors to submit separate bids for the work involved in each of the three buildings. The bids it receives (in 1000-dollar units) are listed in Table 1.

TABLE 1

	Bids		
	Bldg 1	Bldg 2	Bldg 3
Contractor 1	53	96	37
Contractor 2	47	87	41
Contractor 3	60	92	36

Each contractor can install the air-conditioning for only one building during the one-week period, and so the college must assign a different contractor to each building. To which building should each contractor be assigned in order to minimize the sum of the corresponding bids?

Solution. The cost matrix for this problem is the 3×3 matrix

$$\begin{bmatrix} 53 & 96 & 37 \\ 47 & 87 & 41 \\ 60 & 92 & 36 \end{bmatrix} \tag{1}$$

Because there are only six ($= 3!$) possible assignments, we may solve this problem by computing the cost of each of them. We have shaded the entries associated with each of the six assignments and computed their sums.

$$\begin{bmatrix} 53 & 96 & 37 \\ 47 & 87 & 41 \\ 60 & 92 & 36 \end{bmatrix}$$
$$53 + 87 + 36 = 176$$
(a)

$$\begin{bmatrix} 53 & 96 & 37 \\ 47 & 87 & 41 \\ 60 & 92 & 36 \end{bmatrix}$$
$$53 + 92 + 41 = 186$$
(b)

$$\begin{bmatrix} 53 & 96 & 37 \\ 47 & 87 & 41 \\ 60 & 92 & 36 \end{bmatrix}$$
$$47 + 96 + 36 = 179$$
(c)

$$\begin{bmatrix} 53 & 96 & 37 \\ 47 & 87 & 41 \\ 60 & 92 & 36 \end{bmatrix}$$
$$47 + 92 + 37 = 176$$
(d)

$$\begin{bmatrix} 53 & 96 & 37 \\ 47 & 87 & 41 \\ 60 & 92 & 36 \end{bmatrix}$$
$$60 + 96 + 41 = 197$$
(e)

$$\begin{bmatrix} 53 & 96 & 37 \\ 47 & 87 & 41 \\ 60 & 92 & 36 \end{bmatrix}$$
$$60 + 87 + 37 = 184$$
(f)

Note that the bid totals range from a minimum of $176,000 to a maximum of $197,000. Because the minimum bid total of $176,000 is attained by both assignments (a) and (d), the college should assign the contractors to the buildings in one of the following two ways:

Contractor 1 to Building 1		Contractor 1 to Building 3
Contractor 2 to Building 2	or	Contractor 2 to Building 1
Contractor 3 to Building 3		Contractor 3 to Building 2

The method of brute force employed in this example quickly becomes impractical as the size of the cost matrix increases. For example, for a 10×10 cost matrix there are a total of 3,628,800 ($= 10!$) assignments possible. We shall now describe a practical method for solving any assignment problem. ▲

THE HUNGARIAN METHOD

Suppose a particular assignment problem has the cost matrix

$$\begin{bmatrix} 0 & 5 & 7 & 0 & 3 \\ 0 & 4 & 0 & 2 & 5 \\ 6 & 0 & 3 & 7 & 7 \\ 7 & 9 & 4 & 0 & 0 \\ 3 & 2 & 0 & 0 & 1 \end{bmatrix} \tag{2}$$

Notice that all the entires of this cost matrix are nonnegative and that it contains many zero entries. Notice also that we can find an assignment consisting entirely of zero entries, namely,

$$\begin{bmatrix} 0 & 5 & 7 & 0 & 3 \\ 0 & 4 & 0 & 2 & 5 \\ 6 & 0 & 3 & 7 & 7 \\ 7 & 9 & 4 & 0 & 0 \\ 3 & 2 & 0 & 0 & 1 \end{bmatrix} \tag{3}$$

This assignment must be optimal, as its cost is zero and it is impossible to find an assignment with a cost less than zero if all the entries of the cost matrix are nonnegative.

Very few assignment problems are as easy to solve as the preceding one. However, the following theorem leads to a method of converting an arbitrary assignment problem to one that can be solved as easily.

Theorem 11.6.1. *If a number is added to or subtracted from all of the entries of any one row or column of a cost matrix, then an optimal assignment for the resulting cost matrix is also an optimal assignment for the original cost matrix.*

To see why this theorem is true, suppose that five is added to each entry of the second row of a given cost matrix. Because each assignment contains exactly one entry from the second row, it follows that the cost of each assignment for the new matrix is exactly five more than the cost of the corresponding assignment for the original matrix. Thus, corresponding assignments preserve their ordering with respect to cost, and so an optimal assignment of either matrix corresponds to an optimal assignment of the other. A similar argument holds if a number is added to any column of the cost matrix, or if subtraction rather than addition is used.

We now introduce the **Hungarian method**, which is a procedure for applying this theorem to a given cost matrix and obtaining one with nonnegative entries that contains an assignment consisting entirely of zero entries. Such an assignment (called an *optimal assignment of zeros*) will then be an optimal assignment for the

The Hungarian Method	
Steps	**Remarks**
1. Subtract the smallest entry in each row from all the entries of its row.	After this step, each row has at least one zero entry and all other entries are nonnegative.
2. Subtract the smallest entry in each column from all the entries of its column.	After this step, each row *and* column has at least one zero entry and all other entries are nonnegative.
3. Draw lines through appropriate rows and columns so that all the zero entries of the cost matrix are covered and the *minimum* number of such lines is used.	There may be several ways to do this. The important thing is that the *minimum* number of lines be used. Algorithms suitable for computer coding are available for this; however, for small values of n, trial and error will suffice.
4. *Test for Optimality.* (*i*) If the minimum number of covering lines is n, an optimal assignment of zeros is possible and we are finished. (*ii*) If the minimum number of covering lines is less than n, an optimal assignment of zeros is not yet possible. Proceed to Step 5.	See Exercise 7 for a justification of this test. If the test is affirmative, a judicious search will produce a set of n zero entries, no two of which lie in the same row or column. Algorithms, which we will not discuss, are available to systematically find such an optimal assignment of zeros.
5. Determine the smallest entry not covered by any line. Subtract this entry from all uncovered entries and then add it to all entries covered by both a horizontal and a vertical line. Return to Step 3.	This step is equivalent to applying Theorem 11.6.1 by subtracting the smallest uncovered entry from each uncovered row and then adding it to each covered column.

Figure 1

original problem. The Hungarian method is outlined in Figure 1 for an $n \times n$ cost matrix. The first two steps use Theorem 11.6.1 to generate a cost matrix with nonnegative entries and with at least one zero entry in each row and column. The last three steps are applied iteratively as many times as necessary to generate a cost matrix that contains an optimal assignment of zeros. In Exercise 8 we ask the reader to show that each time Step 5 is applied, the sum of the entries of the new cost matrix generated is strictly less than the sum of the entries of the previous cost matrix. This guarantees that the iterative process cannot continue indefinitely.

Example 2 A construction company has four large bulldozers located at four different garages. The bulldozers are to be moved to four different construction sites. The distances in miles between the bulldozers and construction sites are given in Table 2.

TABLE 2

		Construction Site			
		1	2	3	4
Bulldozer	1	90	75	75	80
	2	35	85	55	65
	3	125	95	90	105
	4	45	110	95	115

How should the bulldozers be moved to the construction sites in order to minimize the total distance traveled?

Solution. We shall apply the Hungarian method to matrix (4), which is the cost matrix for the problem.

$$\begin{bmatrix} 90 & 75 & 75 & 80 \\ 35 & 85 & 55 & 65 \\ 125 & 95 & 90 & 105 \\ 45 & 110 & 95 & 115 \end{bmatrix} \tag{4}$$

Step 1. Subtract 75 from the first row of matrix (4), subtract 35 from its second row, subtract 90 from its third row, and subtract 45 from its fourth row to obtain matrix (5).

$$\begin{bmatrix} 15 & 0 & 0 & 5 \\ 0 & 50 & 20 & 30 \\ 35 & 5 & 0 & 15 \\ 0 & 65 & 50 & 70 \end{bmatrix} \tag{5}$$

Step 2. The first three columns of matrix (5) already contain zero entries; therefore, we need only subtract 5 from its fourth column. The result is matrix (6).

$$\begin{bmatrix} 15 & 0 & 0 & 0 \\ 0 & 50 & 20 & 25 \\ 35 & 5 & 0 & 10 \\ 0 & 65 & 50 & 65 \end{bmatrix} \tag{6}$$

Step 3. Cover the zero entries of matrix (6) with a minimum number of vertical and horizontal lines. This may be done by first trying to cover the zeros with one line, then with two, and finally with three. The indicated covering is not unique. (See Exercise 3 for an alternate covering.)

Step 4. Because the minimum number of lines used in Step 3 is three, an optimal assignment of zeros is not yet possible.

Step 5. Subtract 20, the smallest uncovered entry of matrix (6), from each of its uncovered entries and add it to the two entries covered twice with lines. The result is matrix (7).

$$\begin{bmatrix} 35 & 0 & 0 & 0 \\ 0 & 30 & 0 & 5 \\ 55 & 5 & 0 & 10 \\ 0 & 45 & 30 & 45 \end{bmatrix} \tag{7}$$

Step 6. (Step 3 of Figure 1.) Cover the zero entries of matrix (7) with a minimum number of vertical and horizontal lines.

Step 7. Because the minimum number of lines is still three, an optimal assignment of zeros is not yet possible.

Step 8. Subtract 5, the smallest uncovered entry of matrix (7), from each of its uncovered entries and add it to the two entries covered twice with lines. The result is matrix (8).

$$\begin{bmatrix} 40 & 0 & 5 & 0 \\ 0 & 25 & 0 & 0 \\ 55 & 0 & 0 & 5 \\ 0 & 40 & 30 & 40 \end{bmatrix} \tag{8}$$

Step 9. (Step 3 of Figure 1.) Cover the zero entries of matrix (8) with a minimum number of vertical and horizontal lines.

Step 10. Because the zero entries of matrix (8) cannot be covered with fewer than four lines, it must contain an optimal assignment of zeros.

By trial and error, we can find the following two optimal assignments of zeros in matrix (8):

$$\begin{bmatrix} 40 & 0 & 5 & 0 \\ 0 & 25 & 0 & 0 \\ 55 & 0 & 0 & 5 \\ 0 & 40 & 30 & 40 \end{bmatrix} \qquad \begin{bmatrix} 40 & 0 & 5 & 0 \\ 0 & 25 & 0 & 0 \\ 55 & 0 & 0 & 5 \\ 0 & 40 & 30 & 40 \end{bmatrix} \qquad (9)$$

$$\text{(a)} \qquad\qquad\qquad\qquad \text{(b)}$$

Assignment (a) leads to the following movement of bulldozers to construction sites:

Bulldozer 1 to construction site 4
Bulldozer 2 to construction site 3
Bulldozer 3 to construction site 2
Bulldozer 4 to construction site 1

From Table 2 the corresponding minimum distance traveled is

$$80 + 55 + 95 + 45 = 275 \text{ miles}$$

Similarly, assignment (b) leads to the alternative solution

Bulldozer 1 to construction site 2
Bulldozer 2 to construction site 4
Bulldozer 3 to construction site 3
Bulldozer 4 to construction site 1

with the same minimum distance traveled

$$75 + 65 + 90 + 45 = 275 \text{ miles} \quad \blacktriangle$$

Before giving another example, we should mention that the assignment problems and associated cost matrices that can be solved by the Hungarian method must satisfy the following three conditions:

1. *The cost matrix must be square.* In the next example we shall discuss a procedure for handling assignment problems for which the cost matrix is not square.
2. *The entries of the cost matrix should be integers.* For hand calculations this is more a convenience than anything else. But for machine calculations this allows for use of exact integer arithmetic and avoids roundoff error. For practical problems, noninteger entries can always be converted to integer entries by multiplying the cost matrix by a suitable power of ten.
3. *The problem must be one of minimization.* The problem of maximizing the sum of entries of a cost matrix is easily converted to one of minimizing the sum by multiplying each entry of the cost matrix by -1.

Example 3 (The Bride–Groom Problem) A marriage broker has four female clients and five male clients who desire to be married. She ranks the possible matchings between her clients on a scale of zero to ten: zero for the poorest match and ten for the best match. Her rankings are given in Table 3.

TABLE 3

		Prospective Grooms				
		Bob	Tom	Joe	Hal	Don
	Sue	7	4	7	3	10
Prospective Brides	Ann	5	9	3	8	7
	Bea	3	5	6	2	9
	Fay	6	5	0	4	8

How should she match her clients in order to maximize the sum of the rankings of the matches?

Solution. Because there is one more prospective groom than prospective bride, one of the grooms cannot be matched. Thus, the cost matrix is not square and the Hungarian method cannot be applied directly. To circumvent this problem we introduce a "dummy" prospective bride whose match with any of the five grooms has a rank of zero. The prospective groom matched with the dummy bride is then, in reality, the groom not matched. We therefore add a row of zeros to Table 3 corresponding to the dummy bride and in this way are led to the following square cost matrix:

$$\begin{bmatrix} 7 & 4 & 7 & 3 & 10 \\ 5 & 9 & 3 & 8 & 7 \\ 3 & 5 & 6 & 2 & 9 \\ 6 & 5 & 0 & 4 & 8 \\ 0 & 0 & 0 & 0 & 0 \end{bmatrix} \tag{10}$$

The problem as stated is one of maximizing a sum. We convert it to one of minimizing a sum by multiplying each entry of matrix (10) by -1 to obtain

$$\begin{bmatrix} -7 & -4 & -7 & -3 & -10 \\ -5 & -9 & -3 & -8 & -7 \\ -3 & -5 & -6 & -2 & -9 \\ -6 & -5 & 0 & -4 & -8 \\ 0 & 0 & 0 & 0 & 0 \end{bmatrix} \tag{11}$$

We now apply the Hungarian method to matrix (11) to find an optimal assignment.

Step 1. Subtract -10 from the first row of matrix (11), subtract -9 from its second row, subtract -9 from its third row, and subtract -8 from its fourth row to obtain matrix (12).

$$\begin{bmatrix} 3 & 6 & 3 & 7 & 0 \\ 4 & 0 & 6 & 1 & 2 \\ 6 & 4 & 3 & 7 & 0 \\ 2 & 3 & 8 & 4 & 0 \\ 0 & 0 & 0 & 0 & 0 \end{bmatrix} \tag{12}$$

Step 2. Because all the columns of matrix (12) contain zero entries, Step 2 in Figure 1 is not needed.

Step 3. Cover the zero entries of matrix (12) with a minimum number of vertical and horizontal lines.

Step 4. Because the minimum number of lines used in Step 3 is three, an optimal assignment of zeros is not possible.

Step 5. Subtract 2, the smallest uncovered entry of matrix (12), from each of its uncovered entries and add it to the two entries covered twice with lines. The result is matrix (13).

$$\begin{bmatrix} 1 & 4 & 1 & 5 & 0 \\ 4 & 0 & 6 & 1 & 4 \\ 4 & 2 & 1 & 5 & 0 \\ 0 & 1 & 6 & 2 & 0 \\ 0 & 0 & 0 & 0 & 2 \end{bmatrix} \tag{13}$$

Step 6. (Step 3 of Figure 1.) Cover the zero entries of matrix (13) with a minimum number of vertical and horizontal lines.

Step 7. Because the minimum number of lines is four, an optimal assignment of zeros is not yet possible.

Step 8. Subtract 1, the smallest uncovered entry of matrix (13), from each of its uncovered entries and add it to the three entries covered twice with lines. The result is matrix (14).

$$\begin{bmatrix} 0 & 3 & 0 & 4 & 0 \\ 4 & 0 & 6 & 1 & 5 \\ 3 & 1 & 0 & 4 & 0 \\ 0 & 1 & 6 & 2 & 1 \\ 0 & 0 & 0 & 0 & 3 \end{bmatrix} \tag{14}$$

Step 9.   (Step 3 of Figure 1.) Cover the zero entries of matrix (14) with a minimum number of vertical and horizontal lines.

Step 10.   Because the zero entries of matrix (14) cannot be covered with fewer than five lines, it must contain an optimal assignment of zeros. Such an assignment is given in matrix (15).

$$
\begin{bmatrix}
0 & 3 & 0 & 4 & 0 \\
4 & 0 & 6 & 1 & 5 \\
3 & 1 & 0 & 4 & 0 \\
0 & 1 & 6 & 2 & 1 \\
0 & 0 & 0 & 0 & 3
\end{bmatrix}
\tag{15}
$$

The optimal assignment indicated in (15) leads to the following bride–groom matchings:

Sue–Joe	(rank = 7)
Ann–Tom	(rank = 9)
Bea–Don	(rank = 9)
Fay–Bob	(rank = 6)
Hal (unmatched)	

and the resulting maximum ranking sum is $7 + 9 + 9 + 6 = 31$. In Exercise 4 we ask the reader to find an alternate optimal assignment of zeros in matrix (15) and to verify that it leads to the same maximum ranking sum. ▲

EXERCISE SET 11.6

1. Find an optimal assignment and corresponding cost for each of the following three cost matrices using the Hungarian method.

(a) $\begin{bmatrix} 17 & 4 & 10 \\ 15 & 5 & 8 \\ 18 & 7 & 11 \end{bmatrix}$
(b) $\begin{bmatrix} 3 & -2 & 0 & 1 \\ 5 & 3 & -3 & 4 \\ 2 & 7 & 5 & 3 \\ 5 & -2 & 0 & 1 \end{bmatrix}$
(c) $\begin{bmatrix} 12 & 9 & 7 & 7 & 10 \\ 15 & 11 & 8 & 13 & 14 \\ 9 & 6 & 5 & 12 & 12 \\ 6 & 9 & 13 & 7 & 10 \\ 8 & 13 & 12 & 9 & 13 \end{bmatrix}$

2. Solve Example 1 by the Hungarian method.

3. Find a set of three lines that covers the zero entries of matrix (6) in Example 2 different from the set given. Then use the Hungarian method and verify that it leads to the same optimal assignment of zeros given in (9).

4. Find an optimal assignment of zeros in matrix (15) of Example 3 different from the one given. Verify that the new assignment leads to the same cost as the one in the example.

5. A coin dealer is to sell four coins through a mail auction. Bids are received for each of the four coins from five bidders with instructions from each bidder that at most one of his bids is to be honored. The bids are given in Table 4.

TABLE 4

	Bids			
	Coin 1	Coin 2	Coin 3	Coin 4
Bidder 1	$150	$65	$210	$135
Bidder 2	175	75	230	155
Bidder 3	135	85	200	140
Bidder 4	140	70	190	130
Bidder 5	170	50	200	160

(a) How should the dealer assign the four coins in order to maximize the sum of the resulting bids? (Notice that a "dummy" coin must be added to produce a square cost matrix. Whichever bidder receives the dummy coin does not receive any real coin.)

(b) Suppose bidder 2 instructed the dealer that at most two of his bids are to be honored. How can the problem be modified, and what is an optimal assignment for the new problem?

6. A Little League manager has nine players to assign to the nine positions of a baseball team. The manager ranks each of the players on a scale from 0 to 25 for each of the positions, taking into account both the ability of the player in the position and the importance of the position to the game. Referring to Table 5, how should the manager assign the nine players in order to maximize the sum of the rankings?

7. Prove part (*ii*) of the test for optimality in Step 4 of the Hungarian method (Figure 1) by showing that the n zero entries of an optimal assignment of zeros cannot be covered with fewer than n lines. [**Remark.** The proof of part (*i*), which was given in 1931 by two Hungarian mathematicians, D. König and E. Egerváry, is quite difficult. For completeness, the proof is given in the solutions manual for this text.]

8. (a) Let the zero entries of an $n \times n$ nonnegative cost matrix C be covered with m lines, with $m < n$. Let $a > 0$ be the smallest uncovered entry, and let C' be the cost matrix obtained by applying Step 5 of the Hungarian method to C. Show that

$$\text{(Sum of all entries of } C) - \text{(Sum of all entries of } C') = an(n - m)$$

[**Hint.** Notice that Step 5 is equivalent to adding a to every entry of C covered by one line, adding $2a$ to every entry covered by two lines, and subtracting a from every entry.]

(b) Use the result of part (a) to show that the Hungarian method will produce a cost matrix with an optimal assignment of zeros in a finite number of steps.

TABLE 5

Position	Player								
	Sam	**Jill**	**John**	**Liz**	**Ann**	**Lois**	**Pete**	**Alex**	**Herb**
P	20	15	10	10	17	23	25	5	15
C	10	10	12	15	9	7	8	7	8
1B	12	9	9	10	10	5	7	13	9
2B	13	14	10	15	15	5	8	20	10
3B	12	13	10	15	14	5	9	20	10
SS	15	14	15	16	15	5	10	20	10
LF	7	9	12	12	7	6	7	15	12
CF	5	6	8	8	5	4	5	10	7
RF	5	6	8	8	5	4	5	10	7

11.7 MARKOV CHAINS

This section describes a general model of a system that moves from state to state and applies it to several concrete problems.

> **PREREQUISITES:** Linear systems
> Matrices
> Intuitive understanding of limits

Suppose a physical or mathematical system undergoes a process of change such that at any moment it can occupy one of a finite number of states. For example, the weather in a certain city could be in one of three possible states: sunny, cloudy, or rainy. Or an individual could be in one of four possible emotional states: happy, sad, angry, or apprehensive. Suppose such a system changes with time from one state to another and at scheduled times the state of the system is observed. If the exact state of the system at any observation cannot be determined with certainty, but the probability that a given state occurs can be specified by just knowing the state of the system at the previous observation, then the process of change is called a *Markov chain* or *Markov process*.

Definition. If a Markov chain has k possible states, which we label as $1, 2, \ldots, k$, then the probability that the system is in state i at any observation after it was in state j at the previous observation is denoted by p_{ij} and is called the **transition probability** from state j to state i. The matrix $P = [p_{ij}]$ is called the **transition matrix of the Markov chain**.

For example, in a three-state Markov chain, the transition matrix has the form

$$
\begin{array}{c}
\text{Previous State} \\
\begin{array}{ccc}
1 & 2 & 3
\end{array} \\
\begin{bmatrix}
p_{11} & p_{12} & p_{13} \\
p_{21} & p_{22} & p_{23} \\
p_{31} & p_{32} & p_{33}
\end{bmatrix}
\begin{array}{l}
1 \\
2 \quad \text{New State} \\
3
\end{array}
\end{array}
$$

In this matrix p_{32} is the probability that the system will change from state 2 to state 3, p_{11} is the probability that the system will still be in state 1 if it was previously in state 1, and so forth.

Example 1 A car rental agency has three rental locations, denoted by 1, 2, and 3. A customer may rent a car from any of the three locations and return the car to any of the three locations. The manager finds that customers return the cars to the various locations according to the following probabilities:

$$
\begin{array}{c}
\text{Rented from Location} \\
\begin{array}{ccc}
1 & 2 & 3
\end{array} \\
\begin{bmatrix}
.8 & .3 & .2 \\
.1 & .2 & .6 \\
.1 & .5 & .2
\end{bmatrix}
\begin{array}{l}
1 \quad \text{Returned} \\
2 \quad \ \ \text{to} \\
3 \quad \text{Location}
\end{array}
\end{array}
$$

This matrix is the transition matrix of the system considered as a Markov chain. From this matrix the probability is .6 that a car rented from location 3 will be returned to location 2, the probability is .8 that a car rented from location 1 will be returned to location 1, and so forth. ▲

Example 2 By reviewing its donation records, the alumni office of a college finds that 80% of its alumni who contribute to the annual fund one year will also contribute the next year, and 30% of those who do not contribute one year will contribute the next. This can be viewed as a Markov chain with two states: state 1 corresponds to an alumnus giving a donation in any one year, and state 2 corresponds to the alumnus not giving a donation in that year. The transition matrix is

$$
P = \begin{bmatrix} .8 & .3 \\ .2 & .7 \end{bmatrix} \quad ▲
$$

In the above examples, the transition matrices of the Markov chains have the property that the entries in any column sum to 1. This is not accidental. If $P = [p_{ij}]$ is the transition matrix of any Markov chain with k states, then for each j we must have

$$p_{1j} + p_{2j} + \cdots + p_{kj} = 1 \tag{1}$$

because if the system is in state j at one observation it is certain to be in one of the k possible states at the next observation.

A matrix with property (1) is called a ***stochastic matrix***, a ***probability matrix***, or a ***Markov matrix***. From the foregoing discussion, it follows that the transition matrix for a Markov chain must be a stochastic matrix.

In a Markov chain the state of the system at any observation time cannot generally be determined with certainty. The best one can usually do is specify probabilities for each of the possible states. For example, in a Markov chain with three states we might describe the possible state of the system at some observation time by a column vector

$$\mathbf{x} = \begin{bmatrix} x_1 \\ x_2 \\ x_3 \end{bmatrix}$$

in which x_1 is the probability that the system is in state 1, x_2 the probability that it is in state 2, and x_3 the probability that it is in state 3. In general we make the following definition.

> **Definition.** The ***state vector*** for an observation of a Markov chain with k states is a column vector $\mathbf{x}$ whose ith component x_i is the probability that the system is in the ith state at that time.

Observe that the entries in any state vector for a Markov chain are nonnegative and have a sum of 1 (why?). A column vector that has this property is called a ***probability vector***.

Let us suppose now that we know the state vector $\mathbf{x}^{(0)}$ for a Markov chain at some initial observation. The following theorem will enable us to determine the state vectors

$$\mathbf{x}^{(1)}, \mathbf{x}^{(2)}, \ldots, \mathbf{x}^{(n)}, \ldots$$

at the subsequent observation times.

> **Theorem 11.7.1.** *If P is the transition matrix of a Markov chain and $\mathbf{x}^{(n)}$ is the state vector at the nth observation, then $\mathbf{x}^{(n+1)} = P\mathbf{x}^{(n)}$.*

The proof of this theorem involves ideas from probability theory and will not be given here. From this theorem it follows that

$$\mathbf{x}^{(1)} = P\mathbf{x}^{(0)}$$
$$\mathbf{x}^{(2)} = P\mathbf{x}^{(1)} = P^2\mathbf{x}^{(0)}$$
$$\mathbf{x}^{(3)} = P\mathbf{x}^{(2)} = P^3\mathbf{x}^{(0)}$$
$$\vdots$$
$$\mathbf{x}^{(n)} = P\mathbf{x}^{(n-1)} = P^n\mathbf{x}^{(0)}$$

In this way, the initial state vector $\mathbf{x}^{(0)}$ and the transition matrix P determine $\mathbf{x}^{(n)}$ for $n = 1, 2, \ldots$.

Example 2 (Revisited)   The transition matrix in Example 2 was

$$P = \begin{bmatrix} .8 & .3 \\ .2 & .7 \end{bmatrix}$$

We now construct the probable future donation record of a new graduate who did not give a donation in the initial year after graduation. For such a graduate the system is initially in state 2 with certainty, and so the initial state vector is

$$\mathbf{x}^{(0)} = \begin{bmatrix} 0 \\ 1 \end{bmatrix}$$

From Theorem 11.7.1 we then have

$$\mathbf{x}^{(1)} = P\mathbf{x}^{(0)} = \begin{bmatrix} .8 & .3 \\ .2 & .7 \end{bmatrix}\begin{bmatrix} 0 \\ 1 \end{bmatrix} = \begin{bmatrix} .3 \\ .7 \end{bmatrix}$$

$$\mathbf{x}^{(2)} = P\mathbf{x}^{(1)} = \begin{bmatrix} .8 & .3 \\ .2 & .7 \end{bmatrix}\begin{bmatrix} .3 \\ .7 \end{bmatrix} = \begin{bmatrix} .45 \\ .55 \end{bmatrix}$$

$$\mathbf{x}^{(3)} = P\mathbf{x}^{(2)} = \begin{bmatrix} .8 & .3 \\ .2 & .7 \end{bmatrix}\begin{bmatrix} .45 \\ .55 \end{bmatrix} = \begin{bmatrix} .525 \\ .475 \end{bmatrix}$$

Thus, after three years the alumnus can be expected to make a donation with probability .525. Beyond three years, we find the following state vectors (to three decimal places):

$$\mathbf{x}^{(4)} = \begin{bmatrix} .563 \\ .438 \end{bmatrix}, \quad \mathbf{x}^{(5)} = \begin{bmatrix} .581 \\ .419 \end{bmatrix}, \quad \mathbf{x}^{(6)} = \begin{bmatrix} .591 \\ .409 \end{bmatrix}, \quad \mathbf{x}^{(7)} = \begin{bmatrix} .595 \\ .405 \end{bmatrix}$$

$$\mathbf{x}^{(8)} = \begin{bmatrix} .598 \\ .402 \end{bmatrix}, \quad \mathbf{x}^{(9)} = \begin{bmatrix} .599 \\ .401 \end{bmatrix}, \quad \mathbf{x}^{(10)} = \begin{bmatrix} .599 \\ .401 \end{bmatrix}, \quad \mathbf{x}^{(11)} = \begin{bmatrix} .600 \\ .400 \end{bmatrix}$$

For all n beyond eleven, we have

$$\mathbf{x}^{(n)} = \begin{bmatrix} .600 \\ .400 \end{bmatrix}$$

to three decimal places. In other words, the state vectors converge to a fixed vector as the number of observations increases. (We shall discuss this further below.) ▲

Example 1 (Revisited)   The transition matrix in Example 1 was

$$\begin{bmatrix} .8 & .3 & .2 \\ .1 & .2 & .6 \\ .1 & .5 & .2 \end{bmatrix}$$

If a car is rented initially from location 2, then the initial state vector is

$$\mathbf{x}^{(0)} = \begin{bmatrix} 0 \\ 1 \\ 0 \end{bmatrix}$$

Using this vector and Theorem 11.7.1 one obtains the later state vectors listed in Table 1.

TABLE 1

$\mathbf{x}^{(n)}$ \ n	0	1	2	3	4	5	6	7	8	9	10	11
$x_1^{(n)}$	0	.300	.400	.477	.511	.533	.544	.550	.553	.555	.556	.557
$x_2^{(n)}$	1	.200	.370	.252	.261	.240	.238	.233	.232	.231	.230	.230
$x_3^{(n)}$	0	.500	.230	.271	.228	.227	.219	.217	.215	.214	.214	.213

For all values of n greater than 11, all state vectors are equal to $\mathbf{x}^{(11)}$ to three decimal places.

Two things should be observed in this example. First, it was not necessary to know how long a customer kept the car. That is, in a Markov process the time period between observations need not be regular. Second, the state vectors approach a fixed vector as n increases, just as in the first example. ▲

Example 3 A traffic officer is assigned to control the traffic at the eight intersections indicated in Figure 1. She is instructed to remain at each intersection for an hour and then to either remain at that same intersection or move to a neighboring intersection. To avoid establishing a pattern, she is told to choose her new intersection on a random basis, with each possible choice equally likely. For example, if she is at intersection 5, her next intersection can be 2, 4, 5, or 8, each with probability $\frac{1}{4}$. Every day she starts at the location where she stopped the day before.

Figure 1

The transition matrix for this Markov chain is

Old Intersection

$$
\begin{array}{c}
\begin{array}{cccccccc} 1 & 2 & 3 & 4 & 5 & 6 & 7 & 8 \end{array} \\
\begin{bmatrix}
\frac{1}{3} & \frac{1}{3} & 0 & \frac{1}{5} & 0 & 0 & 0 & 0 \\
\frac{1}{3} & \frac{1}{3} & 0 & 0 & \frac{1}{4} & 0 & 0 & 0 \\
0 & 0 & \frac{1}{3} & \frac{1}{5} & 0 & \frac{1}{3} & 0 & 0 \\
\frac{1}{3} & 0 & \frac{1}{3} & \frac{1}{5} & \frac{1}{4} & 0 & \frac{1}{4} & 0 \\
0 & \frac{1}{3} & 0 & \frac{1}{5} & \frac{1}{4} & 0 & 0 & \frac{1}{3} \\
0 & 0 & \frac{1}{3} & 0 & 0 & \frac{1}{3} & \frac{1}{4} & 0 \\
0 & 0 & 0 & \frac{1}{5} & 0 & \frac{1}{3} & \frac{1}{4} & \frac{1}{3} \\
0 & 0 & 0 & 0 & \frac{1}{4} & 0 & \frac{1}{4} & \frac{1}{3}
\end{bmatrix}
\begin{array}{l} 1 \\ 2 \\ 3 \\ 4 \\ 5 \\ 6 \\ 7 \\ 8 \end{array}
\end{array}
\quad
\begin{array}{l} \\ \\ \\ \text{New} \\ \text{Intersection} \\ \\ \\ \end{array}
$$

If the traffic officer initially begins at intersection 5, her probable locations, hour by hour, are given by the state vectors given in Table 2.

TABLE 2

$\mathbf{x}^{(n)}$ \\ n	0	1	2	3	4	5	10	15	20	22
$x_1^{(n)}$	0	.000	.133	.116	.130	.123	.113	.109	.108	.107
$x_2^{(n)}$	0	.250	.146	.163	.140	.138	.115	.109	.108	.107
$x_3^{(n)}$	0	.000	.050	.039	.067	.073	.100	.106	.107	.107
$x_4^{(n)}$	0	.250	.113	.187	.162	.178	.178	.179	.179	.179
$x_5^{(n)}$	1	.250	.279	.190	.190	.168	.149	.144	.143	.143
$x_6^{(n)}$	0	.000	.000	.050	.056	.074	.099	.105	.107	.107
$x_7^{(n)}$	0	.000	.133	.104	.131	.125	.138	.141	.143	.143
$x_8^{(n)}$	0	.250	.146	.152	.124	.121	.108	.107	.107	.107

For all values of n greater than 22, all state vectors are equal to $\mathbf{x}^{(22)}$ to three decimal places. So, as with the first two examples, the state vectors approach a fixed vector as n increases. ▲

LIMITING BEHAVIOR OF THE STATE VECTORS

In our three examples we saw that the state vectors approached some fixed vector as the number of observations increased. We now ask whether the state vectors always approach a fixed vector in a Markov chain. A simple example shows that this is not the case.

Example 4 Let

$$
P = \begin{bmatrix} 0 & 1 \\ 1 & 0 \end{bmatrix}
$$

and

$$\mathbf{x}^{(0)} = \begin{bmatrix} 1 \\ 0 \end{bmatrix}$$

Then, because $P^2 = I$ and $P^3 = P$, we have that

$$\mathbf{x}^{(0)} = \mathbf{x}^{(2)} = \mathbf{x}^{(4)} = \cdots = \begin{bmatrix} 1 \\ 0 \end{bmatrix}$$

and

$$\mathbf{x}^{(1)} = \mathbf{x}^{(3)} = \mathbf{x}^{(6)} = \cdots = \begin{bmatrix} 0 \\ 1 \end{bmatrix}$$

This system oscillates indefinitely between the two state vectors $\begin{bmatrix} 1 \\ 0 \end{bmatrix}$ and $\begin{bmatrix} 0 \\ 1 \end{bmatrix}$, and so does not approach any fixed vector. ▲

However, if we impose a mild condition on the transition matrix, we can show that a fixed limiting state vector is approached. This condition is described by the following definition.

> **Definition.** A transition matrix is *regular* if some integer power of it has all positive entries.

Thus, for a regular transition matrix P, there is some positive integer m such that all entries of P^m are positive. This is the case with the transition matrices of Examples 1 and 2 for $m = 1$. In Example 3 it turns out that P^4 has all positive entries. Consequently, in all three examples the transition matrices are regular.

A Markov chain that is governed by a regular transition matrix is called a *regular Markov chain*. We shall see that every regular Markov chain has a fixed state vector $\mathbf{q}$ such that $P^n\mathbf{x}^{(0)}$ approaches $\mathbf{q}$ as n increases for any choice of $\mathbf{x}^{(0)}$. This result is of major importance in the theory of Markov chains. It is based on the following theorem.

> **Theorem 11.7.2.** *If P is a regular transition matrix, then as $n \to \infty$*
> $$P^n \to \begin{bmatrix} q_1 & q_1 & \cdots & q_1 \\ q_2 & q_2 & \cdots & q_2 \\ \vdots & \vdots & & \vdots \\ q_k & q_k & \cdots & q_k \end{bmatrix}$$
> *where the q_i are positive numbers such that $q_1 + q_2 + \cdots + q_k = 1$.*

We will not prove this theorem here. (The interested reader is referred to a more specialized text, such as J. Kemeny and J. Snell, *Finite Markov Chains*, New York: Springer-Verlag, 1976.)

Let us set

$$Q = \begin{bmatrix} q_1 & q_1 & \cdots & q_1 \\ q_2 & q_2 & \cdots & q_2 \\ \vdots & \vdots & & \vdots \\ q_k & q_k & \cdots & q_k \end{bmatrix} \quad \text{and} \quad \mathbf{q} = \begin{bmatrix} q_1 \\ q_2 \\ \vdots \\ q_k \end{bmatrix}$$

Thus, Q is a transition matrix, all of whose columns are equal to the probability vector $\mathbf{q}$. Q has the property that if $\mathbf{x}$ is any probability vector, then

$$Q\mathbf{x} = \begin{bmatrix} q_1 & q_1 & \cdots & q_1 \\ q_2 & q_2 & \cdots & q_2 \\ \vdots & \vdots & & \vdots \\ q_k & q_k & \cdots & q_k \end{bmatrix} \begin{bmatrix} x_1 \\ x_2 \\ \vdots \\ x_k \end{bmatrix} = \begin{bmatrix} q_1 x_1 + q_1 x_2 + \cdots + q_1 x_k \\ q_2 x_1 + q_2 x_2 + \cdots + q_2 x_k \\ \vdots \\ q_k x_1 + q_k x_2 + \cdots + q_k x_k \end{bmatrix}$$

$$= (x_1 + x_2 + \cdots + x_k) \begin{bmatrix} q_1 \\ q_2 \\ \vdots \\ q_k \end{bmatrix} = (1)\mathbf{q} = \mathbf{q}$$

That is, Q transforms any probability vector $\mathbf{x}$ into the fixed probability vector $\mathbf{q}$. This result leads to the following theorem.

Theorem 11.7.3. *If P is a regular transition matrix and $\mathbf{x}$ is any probability vector, then as $n \to \infty$*

$$P^n \mathbf{x} \to \begin{bmatrix} q_1 \\ q_2 \\ \vdots \\ q_k \end{bmatrix} = \mathbf{q}$$

where $\mathbf{q}$ is a fixed probability vector, independent of n, all of whose entries are positive.

This result holds since Theorem 11.7.2 implies that $P^n \to Q$ as $n \to \infty$. This in turn implies that $P^n\mathbf{x} \to Q\mathbf{x} = \mathbf{q}$ as $n \to \infty$. Thus, for a regular Markov chain, the system eventually approaches a fixed state vector $\mathbf{q}$. The vector $\mathbf{q}$ is called the **steady-state vector** of the regular Markov chain.

For systems with many states, usually the most efficient technique of computing the steady-state vector $\mathbf{q}$ is simply to calculate $P^n\mathbf{x}$ for some large n. Our three examples illustrate this procedure. Each is a regular Markov process, so that convergence to a steady-state vector is ensured. Another way of computing the steady-state vector is to make use of the following theorem.

Theorem 11.7.4.   *The steady-state vector* **q** *of a regular transition matrix P is the unique probability vector that satisfies the equation P***q** = **q**.

To see this, consider the matrix identity $PP^n = P^{n+1}$. By Theorem 11.7.2, both P^n and P^{n+1} approach Q as $n \to \infty$. Thus, we have $PQ = Q$. Any one column of this matrix equation gives $P\mathbf{q} = \mathbf{q}$. To show that **q** is the only probability vector that satisfies this equation, suppose **r** is another probability vector such that $P\mathbf{r} = \mathbf{r}$. Then also $P^n\mathbf{r} = \mathbf{r}$ for $n = 1, 2, \ldots$. Letting $n \to \infty$, Theorem 11.7.3 leads to $\mathbf{q} = \mathbf{r}$.

Theorem 11.7.4 can also be expressed by the statement that the homogeneous linear system

$$(I - P)\mathbf{q} = 0$$

has a unique solution vector **q** with nonnegative entries that satisfy the condition $q_1 + q_2 + \cdots + q_k = 1$. We can apply this technique to the computation of the steady-state vectors for our three examples.

Example 2   **(Revisited)**   In Example 2 the transition matrix was

$$P = \begin{bmatrix} .8 & .3 \\ .2 & .7 \end{bmatrix}$$

so that the linear system $(I - P)\mathbf{q} = 0$ is

$$\begin{bmatrix} .2 & -.3 \\ -.2 & .3 \end{bmatrix}\begin{bmatrix} q_1 \\ q_2 \end{bmatrix} = \begin{bmatrix} 0 \\ 0 \end{bmatrix} \tag{2}$$

This leads to the single independent equation

$$.2q_1 - .3q_2 = 0$$

or

$$q_1 = 1.5q_2$$

Thus, setting $q_2 = s$, any solution of (2) is of the form

$$\mathbf{q} = s\begin{bmatrix} 1.5 \\ 1 \end{bmatrix}$$

where s is an arbitrary constant. To make **q** a probability vector, we set $s = 1/(1.5 + 1) = .4$. Consequently,

$$\mathbf{q} = \begin{bmatrix} .6 \\ .4 \end{bmatrix}$$

is the steady-state vector of this regular Markov chain. This means that over the long run, 60% of the alumni will give a donation in any one year, and 40% will not. Observe that this agrees with the result obtained numerically in Example 2.

Example 1 **(Revisited)** In Example 1 the transition matrix was

$$P = \begin{bmatrix} .8 & .3 & .2 \\ .1 & .2 & .6 \\ .1 & .5 & .2 \end{bmatrix}$$

so that the linear system $(I - P)\mathbf{q} = 0$ is

$$\begin{bmatrix} .2 & -.3 & -.2 \\ -.1 & .8 & -.6 \\ -.1 & -.5 & .8 \end{bmatrix} \begin{bmatrix} q_1 \\ q_2 \\ q_3 \end{bmatrix} = \begin{bmatrix} 0 \\ 0 \\ 0 \end{bmatrix}$$

The reduced row-echelon form of the coefficient matrix is (verify)

$$\begin{bmatrix} 1 & 0 & -\frac{34}{13} \\ 0 & 1 & -\frac{14}{13} \\ 0 & 0 & 0 \end{bmatrix}$$

so that the original linear system is equivalent to the system

$$q_1 = (\tfrac{34}{13})q_3$$
$$q_2 = (\tfrac{14}{13})q_3$$

Setting $q_3 = s$, any solution of the linear system is of the form

$$\mathbf{q} = s \begin{bmatrix} \frac{34}{13} \\ \frac{14}{13} \\ 1 \end{bmatrix}$$

To make this a probability vector, we set

$$s = \frac{1}{\frac{34}{13} + \frac{14}{13} + 1} = \frac{13}{61}$$

Thus, the steady-state vector of the system is

$$\mathbf{q} = \begin{bmatrix} \frac{34}{61} \\ \frac{14}{61} \\ \frac{13}{61} \end{bmatrix} = \begin{bmatrix} .5573\ldots \\ .2295\ldots \\ .2131\ldots \end{bmatrix}$$

This agrees with the result obtained numerically in Table 1. The entries of $\mathbf{q}$ give the long-run probabilities that any one car will be returned to location 1, 2, or 3, respectively. If the car rental agency has a fleet of 1000 cars, it should design its facilities so that there are at least 558 spaces at location 1, at least 230 spaces at location 2, and at least 214 spaces at location 3. ▲

Example 3 **(Revisited)** We will not give the details of the calculations but simply state that the unique probability vector solution of the linear system $(I - P)\mathbf{q} = \mathbf{0}$ is

$$\mathbf{q} = \begin{bmatrix} \frac{3}{28} \\ \frac{3}{28} \\ \frac{3}{28} \\ \frac{5}{28} \\ \frac{4}{28} \\ \frac{3}{28} \\ \frac{4}{28} \\ \frac{3}{28} \end{bmatrix} = \begin{bmatrix} .1071\ldots \\ .1071\ldots \\ .1071\ldots \\ .1785\ldots \\ .1428\ldots \\ .1071\ldots \\ .1428\ldots \\ .1071\ldots \end{bmatrix}$$

The entries in this vector indicate the proportion of time the traffic officer spends at each intersection over the long term. Thus, if the objective is for her to spend the same proportion of time at each intersection, then the strategy of random movement with equal probabilities from one intersection to another is not a good one. (See Exercise 5.) ▲

EXERCISE SET 11.7

1. For the transition matrix $P = \begin{bmatrix} .4 & .5 \\ .6 & .5 \end{bmatrix}$

   (a) calculate $\mathbf{x}^{(n)}$ for $n = 1, 2, 3, 4, 5$, if $\mathbf{x}^{(0)} = \begin{bmatrix} 1 \\ 0 \end{bmatrix}$;

   (b) state why P is regular and find its steady-state vector.

2. Consider the transition matrix

   $$P = \begin{bmatrix} .2 & .1 & .7 \\ .6 & .4 & .2 \\ .2 & .5 & .1 \end{bmatrix}$$

   (a) Calculate $\mathbf{x}^{(1)}$, $\mathbf{x}^{(2)}$, and $\mathbf{x}^{(3)}$ to three decimal places if

   $$\mathbf{x}^{(0)} = \begin{bmatrix} 0 \\ 0 \\ 1 \end{bmatrix}$$

   (b) State why P is regular and find its steady-state vector.

3. Find the steady-state vectors of the following regular transition matrices:

   (a) $\begin{bmatrix} \frac{1}{3} & \frac{3}{4} \\ \frac{2}{3} & \frac{1}{4} \end{bmatrix}$ (b) $\begin{bmatrix} .81 & .26 \\ .19 & .74 \end{bmatrix}$ (c) $\begin{bmatrix} \frac{1}{3} & \frac{1}{2} & 0 \\ \frac{1}{3} & 0 & \frac{1}{4} \\ \frac{1}{3} & \frac{1}{2} & \frac{3}{4} \end{bmatrix}$

4. Let P be the transition matrix

$$\begin{bmatrix} \frac{1}{2} & 0 \\ \frac{1}{2} & 1 \end{bmatrix}$$

(a) Show that P is not regular.

(b) Show that as n increases, $P^n \mathbf{x}^{(0)}$ approaches $\begin{bmatrix} 0 \\ 1 \end{bmatrix}$ for any initial state vector $\mathbf{x}^{(0)}$.

(c) What conclusion of Theorem 11.7.3 is not valid for the steady-state of this transition matrix?

5. Verify that if P is a $k \times k$ regular transition matrix all of whose row sums are equal to one, then the entries of its steady-state vector are all equal to $1/k$.

6. Show that the transition matrix

$$P = \begin{bmatrix} 0 & \frac{1}{2} & \frac{1}{2} \\ \frac{1}{2} & \frac{1}{2} & 0 \\ \frac{1}{2} & 0 & \frac{1}{2} \end{bmatrix}$$

is regular, and use Exercise 5 to find its steady-state vector.

7. John is either happy or sad. If he is happy one day, he is happy the next day four times out of five. If he is sad one day, he is sad the next day one time out of three. Over the long term, what are the chances that John is happy on any given day?

8. A country is divided into three demographic regions. It is found that each year 5% of the residents of region 1 move to region 2, and 5% move to region 3. Of the residents of region 2, 15% move to region 1 and 10% move to region 3. And of the residents of region 3, 10% move to region 1 and 5% move to region 2. What percentage of the population resides in each of the three regions after a long period of time?

11.8 LEONTIEF ECONOMIC MODELS

In this section two linear models for economic systems are discussed. Some results about nonnegative matrices are applied to determine equilibrium price structures and outputs necessary to satisfy demand.

PREREQUISITES: Linear systems
Matrices

Matrix theory has been very successful in describing the interrelations among prices, outputs, and demands in economic systems. In this section we discuss some simple models based on the ideas of Nobel-laureate Wassily Leontief. We examine two different but related models: the closed or input-output model, and the open or production model. In each, we are given certain economic parameters that describe the interrelations between the "industries" in the economy under consideration. Using matrix theory, we then evaluate certain other parameters, such as prices or output levels, in order to satisfy a desired economic objective. We begin with the closed model.

**LEONTIEF
CLOSED
(INPUT-OUTPUT)
MODEL**

First we present a simple example, then we proceed to the general theory of the model.

Example 1   Three homeowners—a carpenter, an electrician, and a plumber—agree to make repairs in their three homes. They agree to work a total of 10 days each according to the following schedule:

	Work Performed by		
	Carpenter	Electrician	Plumber
Days of Work in Home of Carpenter	2	1	6
Days of Work in Home of Electrician	4	5	1
Days of Work in Home of Plumber	4	4	3

For tax purposes they must report and pay each other a reasonable daily wage, even for the work each does on his or her own home. Their normal daily wages are about $100, but they agree to adjust their respective daily wages so that each homeowner will come out even, that is, so that the total amount paid out by each is the same as the total amount each receives. We can set

$$p_1 = \text{daily wage of carpenter}$$
$$p_2 = \text{daily wage of electrician}$$
$$p_3 = \text{daily wage of plumber}$$

To satisfy the "equilibrium" condition that each homeowner comes out even, we require that

$$\text{total expenditures} = \text{total income}$$

for each of the homeowners for the 10-day period. For example, the carpenter pays a total of $2p_1 + p_2 + 6p_3$ for the repairs in his own home, and receives a total income of $10p_1$ for the repairs that he performs on all three homes. Equating these two expressions then gives the first of the following three equations:

$$2p_1 + \ p_2 + 6p_3 = 10p_1$$
$$4p_1 + 5p_2 + \ p_3 = 10p_2$$
$$4p_1 + 4p_2 + 3p_3 = 10p_3$$

The remaining two equations are the equilibrium equations for the electrician and the plumber. Dividing these equations by 10 and rewriting them in matrix form yields

$$
\begin{bmatrix} .2 & .1 & .6 \\ .4 & .5 & .1 \\ .4 & .4 & .3 \end{bmatrix} \begin{bmatrix} p_1 \\ p_2 \\ p_3 \end{bmatrix} = \begin{bmatrix} p_1 \\ p_2 \\ p_3 \end{bmatrix} \tag{1}
$$

Equation (1) can be rewritten as a homogeneous system by subtracting the left side from the right side to obtain

$$
\begin{bmatrix} .8 & -.1 & -.6 \\ -.4 & .5 & -.1 \\ -.4 & -.4 & .7 \end{bmatrix} \begin{bmatrix} p_1 \\ p_2 \\ p_3 \end{bmatrix} = \begin{bmatrix} 0 \\ 0 \\ 0 \end{bmatrix}
$$

The solution of this homogeneous system is found to be (verify)

$$
\begin{bmatrix} p_1 \\ p_2 \\ p_3 \end{bmatrix} = s \begin{bmatrix} 31 \\ 32 \\ 36 \end{bmatrix}
$$

where s is an arbitrary constant. This constant is a scale factor, which the home-owners may choose for their convenience. For example, they may set $s = 3$ so that the corresponding daily wages—\$93, \$96, and \$108—are about \$100. ▲

This example illustrates the salient features of the Leontief input-output model of a closed economy. In the basic Equation (1) each column sum of the coefficient matrix is 1, corresponding to the fact that each of the homeowners' "output" of labor is completely distributed among these same homeowners in the proportions given by the entries in the column. Our problem is to determine suitable "prices" for these outputs so as to put the system in equilibrium, that is, so that each home-owner's total expenditures equal his total income.

In the general model we have an economic system consisting of a finite num-ber of "industries," which we number as industries $1, 2, \ldots, k$. Over some fixed period of time, each industry produces an "output" of some good or service that is completely utilized in a predetermined manner by the k industries. An important problem is to find suitable "prices" to be charged for these k outputs so that for each industry, total expenditures equal total income. Such a price structure repre-sents an equilibrium position for the economy.

For the fixed time period in question, let us set

p_i = price charged by the ith industry for its total output

e_{ij} = fraction of the total output of the jth industry purchased by the ith industry

for $i, j = 1, 2, \ldots, k$. By definition, we have

(i) $p_i \geq 0, \qquad i = 1, 2, \ldots, k$

(ii) $e_{ij} \geq 0, \qquad i, j = 1, 2, \ldots, k$

(iii) $e_{1j} + e_{2j} + \cdots + e_{kj} = 1, \qquad j = 1, 2, \ldots, k$

With these quantities, we form the *price vector*

$$\mathbf{p} = \begin{bmatrix} p_1 \\ p_2 \\ \vdots \\ p_k \end{bmatrix}$$

and the *exchange* or *input-output matrix*

$$E = \begin{bmatrix} e_{11} & e_{12} & \cdots & e_{1k} \\ e_{21} & e_{22} & \cdots & e_{2k} \\ \vdots & \vdots & & \vdots \\ e_{k1} & e_{k2} & \cdots & e_{kk} \end{bmatrix}$$

Condition (iii) expresses the fact that all the column sums of the exchange matrix are 1.

As in the example, in order that the expenditures of each industry be equal to its income, the following matrix equation must be satisfied [see (1)]:

$$E\mathbf{p} = \mathbf{p} \tag{2}$$

or

$$(I - E)\mathbf{p} = \mathbf{0} \tag{3}$$

Equation (3) is a homogeneous linear system for the price vector $\mathbf{p}$. It will have a nontrivial solution if and only if the determinant of its coefficient matrix $I - E$ is zero. In Exercise 6 we ask the reader to show that this is the case for any exchange matrix E. Thus, (3) always has nontrivial solutions for the price vector $\mathbf{p}$.

Actually, for our economic model to make sense we need more than just the fact that (3) has nontrivial solutions for $\mathbf{p}$. We also need the prices p_i of the k outputs to be nonnegative numbers. We express this condition as $\mathbf{p} \geq 0$. (In general, if A is any vector or matrix, the notation $A \geq 0$ means that every entry of A is nonnegative, and the notation $A > 0$ means that every entry of A is positive. Similarly, $A \geq B$ means $A - B \geq 0$, and $A > B$ means $A - B > 0$.) To show that (3) has a nontrivial solution for which $\mathbf{p} \geq 0$ is a bit more difficult than showing merely that some nontrivial solution exists. But it is true, and we state this fact without proof in the following theorem.

Theorem 11.8.1. *If E is an exchange matrix, then E$\mathbf{p}$ = $\mathbf{p}$ always has a nontrivial solution $\mathbf{p}$ whose entries are nonnegative.*

Let us consider a few simple examples of this theorem.

Example 2   Let

$$E = \begin{bmatrix} \frac{1}{2} & 0 \\ \frac{1}{2} & 1 \end{bmatrix}$$

Then $(I - E)\mathbf{p} = \mathbf{0}$ is

$$\begin{bmatrix} \frac{1}{2} & 0 \\ -\frac{1}{2} & 0 \end{bmatrix} \begin{bmatrix} p_1 \\ p_2 \end{bmatrix} = \begin{bmatrix} 0 \\ 0 \end{bmatrix}$$

which has the general solution

$$\mathbf{p} = s \begin{bmatrix} 0 \\ 1 \end{bmatrix}$$

where s is an arbitrary constant. We then have nontrivial solutions $\mathbf{p} > \mathbf{0}$ for any $s > 0$. ▲

Example 3　Let

$$E = \begin{bmatrix} 1 & 0 \\ 0 & 1 \end{bmatrix}$$

Then $(I - E)\mathbf{p} = \mathbf{0}$ has the general solution

$$\mathbf{p} = s \begin{bmatrix} 1 \\ 0 \end{bmatrix} + t \begin{bmatrix} 0 \\ 1 \end{bmatrix}$$

where s and t are independent arbitrary constants. Nontrivial solutions $\mathbf{p} \geq \mathbf{0}$ then result from any $s \geq 0$ and $t \geq 0$, not both zero. ▲

Example 2 indicates that in some situations one of the prices must be zero in order to satisfy the equilibrium condition. Example 3 indicates that there may be several linearly independent price structures available. Neither of these situations describes a truly interdependent economic structure. The following theorem gives sufficient conditions for both cases to be excluded.

Theorem 11.8.2. *Let E be an exchange matrix such that for some positive integer m all the entries of E^m are positive. Then there is exactly one linearly independent solution of $(I - E)\mathbf{p} = \mathbf{0}$, and it may be chosen so that all its entries are positive.*

We will not give a proof of this theorem. The reader who has read Section 11.7 on Markov chains may observe that this theorem is essentially the same as Theorem 11.7.4. What we are calling exchange matrices in this chapter were called stochastic or Markov matrices in Section 11.7.

Example 4　The exchange matrix in Example 1 was

$$E = \begin{bmatrix} .2 & .1 & .6 \\ .4 & .5 & .1 \\ .4 & .4 & .3 \end{bmatrix}$$

Because $E > 0$, the condition $E^m > 0$ in Theorem 11.8.2 is satisfied for $m = 1$. Consequently, we are guaranteed that there is exactly one linearly independent solution of $(I - E)\mathbf{p} = \mathbf{0}$, and it can be chosen so that $\mathbf{p} > 0$. In that example, we found that

$$\mathbf{p} = \begin{bmatrix} 31 \\ 32 \\ 36 \end{bmatrix}$$

is such a solution. ▲

LEONTIEF OPEN (PRODUCTION) MODEL

In contrast with the closed model, in which the outputs of k industries are distributed only among themselves, the open model attempts to satisfy an outside demand for the outputs. Portions of these outputs may still be distributed among the industries themselves, to keep them operating, but there is to be some excess, some net production, with which to satisfy the outside demand. In the closed model the outputs of the industries are fixed, and our objective is to determine prices for these outputs so that the equilibrium condition, that expenditures equal incomes, is satisfied. In the open model it is the prices that are fixed, and our objective is to determine levels of the outputs of the industries needed to satisfy the outside demand. We will measure the levels of the outputs in terms of their economic values using the fixed prices. To be precise, over some fixed period of time, let

x_i = monetary value of the total output of the ith industry

d_i = monetary value of the output of the ith industry needed to satisfy the outside demand

c_{ij} = monetary value of the output of the ith industry needed by the jth industry to produce one unit of monetary value of its own output

With these quantities, we define the *production vector*

$$\mathbf{x} = \begin{bmatrix} x_1 \\ x_2 \\ \vdots \\ x_k \end{bmatrix}$$

the *demand vector*

$$\mathbf{d} = \begin{bmatrix} d_1 \\ d_2 \\ \vdots \\ d_k \end{bmatrix}$$

and the *consumption matrix*

$$C = \begin{bmatrix} c_{11} & c_{12} & \cdots & c_{1k} \\ c_{21} & c_{22} & \cdots & c_{2k} \\ \vdots & \vdots & & \vdots \\ c_{k1} & c_{k2} & \cdots & c_{kk} \end{bmatrix}$$

By their nature, we have that

$$\mathbf{x} \geq 0, \qquad \mathbf{d} \geq 0, \qquad \text{and} \qquad C \geq 0$$

From the definition of c_{ij} and x_j, it can be seen that the quantity

$$c_{i1}x_1 + c_{i2}x_2 + \cdots + c_{ik}x_k$$

is the value of the output of the ith industry needed by all k industries to produce a total output specified by the production vector $\mathbf{x}$. Because this quantity is simply the ith entry of the column vector $C\mathbf{x}$, we can say further that the ith entry of the column vector

$$\mathbf{x} - C\mathbf{x}$$

is the value of the excess output of the ith industry available to satisfy the outside demand. The value of the outside demand for the output of the ith industry is the ith entry of the demand vector $\mathbf{d}$. Consequently, we are led to the following equation

$$\mathbf{x} - C\mathbf{x} = \mathbf{d}$$

or

$$(I - C)\mathbf{x} = \mathbf{d} \qquad\qquad (4)$$

for the demand to be exactly met, without any surpluses or shortages. Thus, given C and $\mathbf{d}$, our objective is to find a production vector $\mathbf{x} \geq 0$ that satisfies Equation (4).

Example 5   A town has three main industries: a coal-mining operation, an electric power generating plant, and a local railroad. To mine $1 of coal, the mining operation must purchase $.25 of electricity to run its equipment and $.25 of transportation for its shipping needs. To produce $1 of electricity, the generating plant requires $.65 of coal for fuel, $.05 of its own electricity to run auxilliary equipment, and $.05 of transportation. To provide $1 of transportation, the railroad requires $.55 of coal for fuel and $.10 of electricity for its auxilliary equipment. In a certain week the coal-mining operation receives orders for $50,000 of coal from outside the town and the generating plant receives orders for $25,000 of electricity from outside. There is no outside demand for the local railroad. How much must each of the three industries produce in that week to exactly satisfy their own demand and the outside demand?

Solution.   For the one-week period let

$$x_1 = \text{value of total output of coal-mining operation}$$
$$x_2 = \text{value of total output of power generating plant}$$
$$x_3 = \text{value of total output of local railroad}$$

From the information supplied the consumption matrix of the system is

$$C = \begin{bmatrix} 0 & .65 & .55 \\ .25 & .05 & .10 \\ .25 & .05 & 0 \end{bmatrix}$$

The linear system $(I - C)\mathbf{x} = \mathbf{d}$ is then

$$\begin{bmatrix} 1.00 & -.65 & -.55 \\ -.25 & .95 & -.10 \\ -.25 & -.05 & 1.00 \end{bmatrix} \begin{bmatrix} x_1 \\ x_2 \\ x_3 \end{bmatrix} = \begin{bmatrix} 50{,}000 \\ 25{,}000 \\ 0 \end{bmatrix}$$

The coefficient matrix on the left is invertible, and the solution is given by

$$\mathbf{x} = (I - C)^{-1}\mathbf{d} = \frac{1}{503} \begin{bmatrix} 756 & 542 & 470 \\ 220 & 690 & 190 \\ 200 & 170 & 630 \end{bmatrix} \begin{bmatrix} 50{,}000 \\ 25{,}000 \\ 0 \end{bmatrix} = \begin{bmatrix} 102{,}087 \\ 56{,}163 \\ 28{,}330 \end{bmatrix}$$

Thus, the total output of the coal-mining operation should be $102,087, the total output of the power generating plant should be $56,163, and the total output of the railroad should be $28,330. ▲

Let us reconsider Equation (4):

$$(I - C)\mathbf{x} = \mathbf{d}$$

If the square matrix $I - C$ is invertible, we can write

$$\mathbf{x} = (I - C)^{-1}\mathbf{d} \tag{5}$$

In addition, if the matrix $(I - C)^{-1}$ has only nonnegative entries, then we are guaranteed that for any $\mathbf{d} \geq 0$, Equation (5) has a unique nonnegative solution for $\mathbf{x}$. This is a particularly desirable situation, as it means that any outside demand can be met. The terminology used to describe this case is given in the following definition.

Definition. A consumption matrix C is said to be **productive** if $(I - C)^{-1}$ exists and $(I - C)^{-1} \geq 0$.

We will now consider some simple criteria that guarantee that a consumption matrix is productive. The first is given in the following theorem.

Theorem 11.8.3. *A consumption matrix C is productive if and only if there is some production vector $\mathbf{x} \geq 0$ such that $\mathbf{x} > C\mathbf{x}$.*

(The proof is outlined in Exercise 8.) The condition $\mathbf{x} > C\mathbf{x}$ means that there is some production schedule possible such that each industry produces more than it consumes.

Theorem 11.8.3 has two interesting corollaries. Suppose that all the row sums of C are less than one. If

$$\mathbf{x} = \begin{bmatrix} 1 \\ 1 \\ \vdots \\ 1 \end{bmatrix}$$

then $C\mathbf{x}$ is a column vector whose entries are these row sums. Therefore, $\mathbf{x} > C\mathbf{x}$, and the condition of Theorem 11.8.3 is satisfied. Thus, we arrive at the following corollary:

> **Corollary 11.8.4.** A consumption matrix is productive if each of its row sums is less than 1.

As we ask the reader to show in Exercise 7, this corollary leads to the following:

> **Corollary 11.8.5.** A consumption matrix is productive if each of its column sums is less than 1.

Recalling the definition of the entries of the consumption matrix C, we see that the jth column sum of C is the total value of the outputs of all k industries needed to produce one unit of value of output of the jth industry. The jth industry is thus said to be *profitable* if that jth column sum is less than 1. In other words, Corollary 11.8.5 says that a consumption matrix is productive if all k industries in the economic system are profitable.

Example 6   The consumption matrix in Example 5 was

$$C = \begin{bmatrix} 0 & .65 & .55 \\ .25 & .05 & .10 \\ .25 & .05 & 0 \end{bmatrix}$$

All three column sums in this matrix are less than 1 and so all three industries are profitable. Consequently, by Corollary 11.8.5 the consumption matrix C is productive. This can also be seen in the calculations in Example 5, as $(I - C)^{-1}$ is nonnegative. ▲

EXERCISE SET 11.8

1. For the following exchange matrices, find nonnegative price vectors that satisfy the equilibrium condition (3).

(a) $\begin{bmatrix} \frac{1}{2} & \frac{1}{3} \\ \frac{1}{2} & \frac{2}{3} \end{bmatrix}$
  (b) $\begin{bmatrix} \frac{1}{2} & 0 & \frac{1}{2} \\ \frac{1}{3} & 0 & \frac{1}{2} \\ \frac{1}{6} & 1 & 0 \end{bmatrix}$
  (c) $\begin{bmatrix} .35 & .50 & .30 \\ .25 & .20 & .30 \\ .40 & .30 & .40 \end{bmatrix}$

2. Using Theorem 11.8.3 and its corollaries, show that each of the following consumption matrices is productive.

(a) $\begin{bmatrix} .8 & .1 \\ .3 & .6 \end{bmatrix}$
  (b) $\begin{bmatrix} .70 & .30 & .25 \\ .20 & .40 & .25 \\ .05 & .15 & .25 \end{bmatrix}$
  (c) $\begin{bmatrix} .7 & .3 & .2 \\ .1 & .4 & .3 \\ .2 & .4 & .1 \end{bmatrix}$

3. Using Theorem 11.8.2, show that there is only one linearly independent price vector for the closed economic system with exchange matrix

$$E = \begin{bmatrix} 0 & .2 & .5 \\ 1 & .2 & .5 \\ 0 & .6 & 0 \end{bmatrix}$$

4. Three neighbors have backyard vegetable gardens. Neighbor A grows tomatoes, neighbor B grows corn, and neighbor C grows lettuce. They agree to divide their crops among themselves as follows: A gets $\frac{1}{2}$ of the tomatoes, $\frac{1}{3}$ of the corn, and $\frac{1}{4}$ of the lettuce. B gets $\frac{1}{3}$ of the tomatoes, $\frac{1}{3}$ of the corn, and $\frac{1}{4}$ of the lettuce. C gets $\frac{1}{6}$ of the tomatoes, $\frac{1}{3}$ of the corn, and $\frac{1}{2}$ of the lettuce. What prices should the neighbors assign to their respective crops if the equilibrium condition of a closed economy is to be satisfied, and if the lowest-priced crop is to have a price of $100?

5. Three engineers—a civil engineer (CE), an electrical engineer (EE), and a mechanical engineer (ME)—each have a consulting firm. The consulting they do is of a multidisciplinary nature, and so they buy a portion of each others' services. For each $1 of consulting the CE does, she buys $.10 of the EE's services and $.30 of the ME's services. For each $1 of consulting the EE does, she buys $.20 of the CE's services and $.40 of the ME's services. And for each $1 of consulting the ME does, she buys $.30 of the CE's services and $.40 of the EE's services. In a certain week the CE receives outside consulting orders of $500, the EE receives outside consulting orders of $700, and the ME receives outside consulting orders of $600. What dollar amount of consulting does each engineer perform in that week?

6. Using the fact that the column sums of an exchange matrix E are all 1, show that the column sums of $I - E$ are zero. From this, show that $I - E$ has zero determinant, and so $(I - E)\mathbf{p} = \mathbf{0}$ has nontrivial solutions for $\mathbf{p}$.

7. Show that Corollary 11.8.5 follows from Corollary 11.8.4. [HINT. Use the fact that $(A^t)^{-1} = (A^{-1})^t$ for any invertible matrix A.]

8. **(For readers who have studied calculus.)** Prove Theorem 11.8.3 as follows:
   (a) Prove the "only if" part of the theorem; that is, show that if C is a productive consumption matrix, then there is a vector $\mathbf{x} \geq 0$ such that $\mathbf{x} > C\mathbf{x}$.
   (b) Prove the "if" part of the theorem as follows:

   Step 1. Show that if there is a vector $\mathbf{x}^* \geq 0$ such that $C\mathbf{x}^* < \mathbf{x}^*$, then $\mathbf{x}^* > 0$.
   Step 2. Show that there is a number λ such that $0 < \lambda < 1$ and $C\mathbf{x}^* < \lambda\mathbf{x}^*$.
   Step 3. Show that $C^n\mathbf{x}^* < \lambda^n\mathbf{x}^*$ for $n = 1, 2, \ldots$.
   Step 4. Show that $C^n \to 0$ as $n \to \infty$.
   Step 5. By multiplying out, show that

   $$(I - C)(I + C + C^2 + \cdots + C^{n-1}) = I - C^n$$

   for $n = 1, 2, \ldots$.
   Step 6. By letting $n \to \infty$ in Step 5, show that the matrix infinite sum

   $$S = I + C + C^2 + \cdots$$

   exists and that $(I - C)S = I$.
   Step 7. Show that $S \geq 0$ and that $S = (I - C)^{-1}$.
   Step 8. Show that C is a productive consumption matrix.

11.9 FOREST MANAGEMENT

This section discusses a matrix model for the management of a forest whose trees are grouped into height classes. The optimal sustainable yield of a periodic harvest is calculated when the trees of different height classes can have different economic values.

PREREQUISITE: Matrix operations

Our objective is to introduce a simplified model for the sustainable harvesting of a forest whose trees are classified by height. The height of a tree is assumed to determine its economic value when it is cut down and sold. Initially, there is a distribution of trees of various heights. The forest is then allowed to grow for a certain period of time, after which some of the trees of various heights are harvested. The trees left unharvested are to be of the same height configuration as the original forest, so that the harvest is sustainable. As we will see, there are many such sustainable harvesting procedures. We want to find one for which the total economic value of all the trees removed is as large as possible. This determines the *optimal sustainable yield* of the forest and is the largest yield that can be attained continually without depleting the forest.

THE MODEL

Suppose a harvester has a forest of Douglas fir trees that are to be sold as Christmas trees year after year. Every December the harvester cuts down some of the trees to be sold. For each tree cut down a seedling is planted in its place. In this way the total number of trees in the forest is always the same. (In this simplified model, we will not take into account trees that die between harvests. We assume that every seedling planted survives and grows until it is harvested.)

In the marketplace, trees of different heights have different economic values. Suppose there are n different price classes corresponding to certain height intervals, as shown in Table 1 and Figure 1.

TABLE 1

Class	Value (dollars)	Height Interval
1 (seedlings)	None	$[0, h_1)$
2	p_2	$[h_1, h_2)$
3	p_3	$[h_2, h_3)$
$\vdots$	$\vdots$	$\vdots$
$n - 1$	p_{n-1}	$[h_{n-2}, h_{n-1})$
n	p_n	$[h_{n-1}, \infty)$

Figure 1

The first class consists of seedlings with heights in the interval $[0, h_1)$, and these seedlings are of no economic value. The nth class consists of trees with heights greater than or equal to h_{n-1}.

Let x_i $(i = 1, 2, \ldots, n)$ be the number of trees within the ith class that remain after each harvest. We form a column vector with the numbers and call it the **nonharvest vector**:

$$\mathbf{x} = \begin{bmatrix} x_1 \\ x_2 \\ \vdots \\ x_n \end{bmatrix}$$

For a sustainable harvesting policy, the forest is to be returned after each harvest to the fixed configuration given by the nonharvest vector $\mathbf{x}$. Part of our problem is to find those nonharvest vectors $\mathbf{x}$ for which sustainable harvesting is possible.

Because the total number of trees in the forest is fixed, we can set

$$x_1 + x_2 + \cdots + x_n = s \tag{1}$$

where s is predetermined by the amount of land available and the amount of space each tree requires. Referring to Figure 2, we have the following situation. The for-

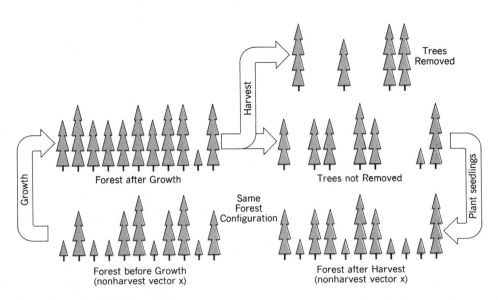

Figure 2

est configuration is given by the vector **x** after each harvest. Between harvests the trees grow and produce a new forest configuration before each harvest. A certain number of trees are removed from each class at the harvest. Finally, a seedling is planted in place of each tree removed to return the forest again to the configuration **x**.

Consider first the growth of the forest between harvests. During this period a tree in the ith class may grow and move up to a higher height class. Or its growth may be retarded for some reason, and it will remain in the same class. We consequently define the following growth parameters g_i for $i = 1, 2, \ldots, n - 1$:

g_i = the fraction of trees in the ith class that grow into
the $(i + 1)$-st class during a growth period

For simplicity we assume that a tree can move at most one height class upward in one growth period. With this assumption we have

$1 - g_i$ = the fraction of trees in the ith class that remain in
the ith class during a growth period

With these $n - 1$ growth parameters we form the following $n \times n$ **growth matrix**:

$$G = \begin{bmatrix} 1 - g_1 & 0 & 0 & \cdots & 0 \\ g_1 & 1 - g_2 & 0 & \cdots & 0 \\ 0 & g_2 & 1 - g_3 & \cdots & 0 \\ \vdots & \vdots & \vdots & & \vdots \\ 0 & 0 & 0 & \cdots & 1 - g_{n-1} & 0 \\ 0 & 0 & 0 & \cdots & g_{n-1} & 1 \end{bmatrix} \quad (2)$$

Because the entries of the vector **x** are the numbers of trees in the n classes before the growth period, the reader can verify that the entries of the vector

$$G\mathbf{x} = \begin{bmatrix} (1 - g_1)x_1 \\ g_1 x_1 + (1 - g_2)x_2 \\ g_2 x_2 + (1 - g_3)x_3 \\ \vdots \\ g_{n-2}x_{n-2} + (1 - g_{n-1})x_{n-1} \\ g_{n-1}x_{n-1} + x_n \end{bmatrix} \quad (3)$$

are the numbers of trees in the n classes after the growth period.

Suppose during the harvest we remove y_i ($i = 1, 2, \ldots, n$) trees from the ith class. We will call the column vector

$$\mathbf{y} = \begin{bmatrix} y_1 \\ y_2 \\ \vdots \\ y_n \end{bmatrix}$$

the **harvest vector**. Thus, a total of

$$y_1 + y_2 + \cdots + y_n$$

trees are removed at each harvest. This is also the total number of trees added to the first class (the new seedlings) after each harvest. If we define the following $n \times n$ **replacement matrix**

$$R = \begin{bmatrix} 1 & 1 & 1 & \cdots & 1 & 1 \\ 0 & 0 & 0 & \cdots & 0 & 0 \\ \vdots & \vdots & \vdots & & \vdots & \vdots \\ 0 & 0 & 0 & \cdots & 0 & 0 \end{bmatrix} \tag{4}$$

then the column vector

$$R\mathbf{y} = \begin{bmatrix} y_1 + y_2 + \cdots + y_n \\ 0 \\ 0 \\ \vdots \\ 0 \end{bmatrix} \tag{5}$$

specifies the configuration of trees planted after each harvest.

At this point we are ready to write the following equation, which characterizes a sustainable harvesting policy:

$$\begin{bmatrix} \text{configuration} \\ \text{at end of} \\ \text{growth period} \end{bmatrix} - [\text{harvest}] + \begin{bmatrix} \text{new seedling} \\ \text{replacement} \end{bmatrix} = \begin{bmatrix} \text{configuration} \\ \text{at beginning of} \\ \text{growth period} \end{bmatrix}$$

or, mathematically,

$$G\mathbf{x} - \mathbf{y} + R\mathbf{y} = \mathbf{x}$$

This equation can be rewritten as

$$(I - R)\mathbf{y} = (G - I)\mathbf{x} \tag{6}$$

or, more fully,

$$\begin{bmatrix} 0 & -1 & -1 & \cdots & -1 & -1 \\ 0 & 1 & 0 & \cdots & 0 & 0 \\ 0 & 0 & 1 & \cdots & 0 & 0 \\ \vdots & \vdots & \vdots & & \vdots & \vdots \\ 0 & 0 & 0 & \cdots & 1 & 0 \\ 0 & 0 & 0 & \cdots & 0 & 1 \end{bmatrix} \begin{bmatrix} y_1 \\ y_2 \\ y_3 \\ \vdots \\ y_{n-1} \\ y_n \end{bmatrix}$$

$$= \begin{bmatrix} -g_1 & 0 & 0 & \cdots & 0 & 0 \\ g_1 & -g_2 & 0 & \cdots & 0 & 0 \\ 0 & g_2 & -g_3 & \cdots & 0 & 0 \\ \vdots & \vdots & \vdots & & \vdots & \vdots \\ 0 & 0 & 0 & \cdots & -g_{n-1} & 0 \\ 0 & 0 & 0 & \cdots & g_{n-1} & 0 \end{bmatrix} \begin{bmatrix} x_1 \\ x_2 \\ x_3 \\ \vdots \\ x_{n-1} \\ x_n \end{bmatrix}$$

We will refer to Equation (6) as the **sustainable harvesting condition**. Any vectors

x and **y** with nonnegative entries, and such that $x_1 + x_2 + \cdots + x_n = s$, which satisfy this matrix equation, determine a sustainable harvesting policy for the forest. Note that if $y_1 > 0$, then the harvester is removing seedlings of no economic value and replacing them with new seedlings. As there is no point in doing this, we assume

$$y_1 = 0 \tag{7}$$

With this assumption, it can be verified that (6) is the matrix form of the following set of equations:

$$
\begin{aligned}
y_2 + y_3 + \cdots + y_n &= g_1 x_1 \\
y_2 &= g_1 x_1 - g_2 x_2 \\
y_3 &= g_2 x_2 - g_3 x_3 \\
&\;\;\vdots \\
y_{n-1} &= g_{n-2} x_{n-2} - g_{n-1} x_{n-1} \\
y_n &= g_{n-1} x_{n-1}
\end{aligned}
\tag{8}
$$

Notice that the first equation in (8) is the sum of the remaining $n - 1$ equations. Because we must have $y_i \geq 0$ for $i = 2, 3, \ldots, n$, Equations (8) require that

$$g_1 x_1 \geq g_2 x_2 \geq \cdots \geq g_{n-1} x_{n-1} \geq 0 \tag{9}$$

Conversely, if **x** is a column vector with nonnegative entries that satisfy Equation (9), then (7) and (8) define a column vector **y** with nonnegative entries. Furthermore, **x** and **y** then satisfy the sustainable harvesting condition (6). In other words, a necessary and sufficient condition that a nonnegative column vector **x** determine a forest configuration that is capable of sustainable harvesting is that its entries satisfy (9).

OPTIMAL SUSTAINABLE YIELD

Because we remove y_i trees from the ith class ($i = 2, 3, \ldots, n$) and each tree in the ith class has an economic value of p_i, the total yield of the harvest, Yld, is given by

$$Yld = p_2 y_2 + p_3 y_3 + \cdots + p_n y_n \tag{10}$$

Using (8) we may substitute for the y_i's in (10) to obtain

$$Yld = p_2 g_1 x_1 + (p_3 - p_2) g_2 x_2 + \cdots + (p_n - p_{n-1}) g_{n-1} x_{n-1} \tag{11}$$

Combining (11), (1), and (9), we can now state the problem of maximizing the yield of the forest over all possible sustainable harvesting policies as follows:

Problem. Find nonnegative numbers $x_1, x_2, \ldots, x_n$ that maximize

$$Yld = p_2 g_1 x_1 + (p_3 - p_2) g_2 x_2 + \cdots + (p_n - p_{n-1}) g_{n-1} x_{n-1}$$

subject to

$$x_1 + x_2 + \cdots + x_n = s$$

and

$$g_1 x_1 \geq g_2 x_2 \geq \cdots \geq g_{n-1} x_{n-1} \geq 0$$

As formulated above, this problem belongs to the field of linear programming. However, we shall prove the following result below, without linear programming theory, by actually exhibiting a sustainable harvesting policy.

> **Theorem 11.9.1.** *The optimal sustainable yield is achieved by harvesting all the trees from one particular height class and none of the trees from any other height class.*

Let us first set

$$Yld_k = \text{yield obtained by harvesting all of the } k\text{th}$$
$$\text{class and none of the other classes}$$

The largest value of Yld_k for $k = 2, 3, \ldots, n$ will then be the optimal sustainable yield, and the corresponding value of k will be the class that should be completely harvested to attain the optimal sustainable yield. Because no class but the kth is harvested, we have

$$y_2 = y_3 = \cdots = y_{k-1} = y_{k+1} = \cdots = y_n = 0 \tag{12}$$

In addition, because all of the kth class is harvested, no trees are left unharvested in the kth class, and no trees are ever present in the height clases above the kth class. Thus,

$$x_k = x_{k+1} = \cdots = x_n = 0 \tag{13}$$

Substituting (12) and (13) into the sustainable harvesting condition (8) gives

$$
\begin{aligned}
y_k &= g_1 x_1 \\
0 &= g_1 x_1 - g_2 x_2 \\
0 &= g_2 x_2 - g_3 x_3 \\
&\ \ \vdots \\
0 &= g_{k-2} x_{k-2} - g_{k-1} x_{k-1} \\
y_k &= g_{k-1} x_{k-1}
\end{aligned}
\tag{14}
$$

Equations (14) can also be written as

$$y_k = g_1 x_1 = g_2 x_2 = \cdots = g_{k-1} x_{k-1} \tag{15}$$

from which it follows that

$$
\begin{aligned}
x_2 &= g_1 x_1 / g_2 \\
x_3 &= g_1 x_1 / g_3 \\
&\ \ \vdots \\
x_{k-1} &= g_1 x_1 / g_{k-1}
\end{aligned}
\tag{16}
$$

If we substitute Equations (13) and (16) into

$$x_1 + x_2 + \cdots + x_n = s$$

[which is Equation (1)], we can solve for x_1 and obtain

$$x_1 = \frac{s}{1 + \dfrac{g_1}{g_2} + \dfrac{g_1}{g_3} + \cdots + \dfrac{g_1}{g_{k-1}}} \tag{17}$$

For the yield Yld_k, we combine (10), (12), (15), and (17) to obtain

$$
\begin{aligned}
Yld_k &= p_2 y_2 + p_3 y_3 + \cdots + p_n y_n \\
&= p_k y_k \\
&= p_k g_1 x_1 \\
&= \frac{p_k s}{\dfrac{1}{g_1} + \dfrac{1}{g_2} + \cdots + \dfrac{1}{g_{k-1}}}
\end{aligned}
\tag{18}
$$

Equation (18) determines Yld_k in terms of the known growth and economic parameters for any $k = 2, 3, \ldots, n$. Thus, the optimal sustainable yield is found as follows.

Theorem 11.9.2. *The optimal sustainable yield is the largest value of*

$$\frac{p_k s}{\dfrac{1}{g_1} + \dfrac{1}{g_2} + \cdots + \dfrac{1}{g_{k-1}}}$$

for $k = 2, 3, \ldots, n$. The corresponding value of k is the number of the class that is completely harvested.

In Exercise 4 we ask the reader to show that the nonharvest vector **x** for the optimal sustainable yield is

$$
\mathbf{x} = \frac{s}{\dfrac{1}{g_1} + \dfrac{1}{g_2} + \cdots + \dfrac{1}{g_{k-1}}}
\begin{bmatrix}
1/g_1 \\
1/g_2 \\
\vdots \\
1/g_{k-1} \\
0 \\
0 \\
\vdots \\
0
\end{bmatrix}
\tag{19}
$$

Theorem 11.9.2 implies that it is not necessarily the highest-priced class of trees that should be totally cropped. The growth parameters g_i must also be taken into account to determine the optimal sustainable yield.

Example 1   For a Scots Pine forest in Scotland with a growth period of six years, the following growth matrix was found (see M. B. Usher, "A Matrix Approach to the Management of Renewable Resources, with Special Reference to Selection

Forests," *Journal of Applied Ecology*, vol. 3, 1966, pp. 355–367):

$$G = \begin{bmatrix} .72 & 0 & 0 & 0 & 0 & 0 \\ .28 & .69 & 0 & 0 & 0 & 0 \\ 0 & .31 & .75 & 0 & 0 & 0 \\ 0 & 0 & .25 & .77 & 0 & 0 \\ 0 & 0 & 0 & .23 & .63 & 0 \\ 0 & 0 & 0 & 0 & .37 & 1.00 \end{bmatrix}$$

Suppose the prices of trees in the five tallest height classes are

$$p_2 = \$50, \qquad p_3 = \$100, \qquad p_4 = \$150, \qquad p_5 = \$200, \qquad p_6 = \$250$$

Which class should be completely harvested to obtain the optimal sustainable yield, and what is the yield?

Solution. From the matrix G we have that

$$g_1 = .28, \qquad g_2 = .31, \qquad g_3 = .25, \qquad g_4 = .23, \qquad g_5 = .37$$

Equation (18) then gives

$$\begin{aligned} Yld_2 &= 50s/(.28^{-1}) = 14.0s \\ Yld_3 &= 100s/(.28^{-1} + .31^{-1}) = 14.7s \\ Yld_4 &= 150s/(.28^{-1} + .31^{-1} + .25^{-1}) = 13.9s \\ Yld_5 &= 200s/(.28^{-1} + .31^{-1} + .25^{-1} + .23^{-1}) = 13.2s \\ Yld_6 &= 250s/(.28^{-1} + .31^{-1} + .25^{-1} + .23^{-1} + .37^{-1}) = 14.0s \end{aligned}$$

We see that Yld_3 is the largest of these five quantities, so from Theorem 11.9.2 the third class should be completely harvested every six years to maximize the sustainable yield. The corresponding optimal sustainable yield is $\$14.7s$, where s is the total number of trees in the forest. ▲

EXERCISE SET 11.9

1. A certain forest is divided into three height classes and has a growth matrix between harvests given by

$$G = \begin{bmatrix} \frac{1}{2} & 0 & 0 \\ \frac{1}{2} & \frac{1}{3} & 0 \\ 0 & \frac{2}{3} & 1 \end{bmatrix}$$

If the price of trees in the second class is $30 and the price of trees in the third class is $50, which class should be completely harvested to attain the optimal sustainable yield? What is the optimal yield of there are 1000 trees in the forest?

2. In Example 1, to what level must the price of trees in the fifth class rise so that class is the one to completely harvest in order to attain the optimal sustainable yield?

3. In Example 1, what must the ratio of the prices $p_2:p_3:p_4:p_5:p_6$ be in order that the yields Yld_k, $k = 2, 3, 4, 5, 6$, all be the same? (In this case, any sustainable harvesting policy will produce the same optimal sustainable yield.)

4. Derive Equation (19) for the nonharvest vector **x** corresponding to the optimal sustainable harvesting policy described in Theorem 11.9.2.

5. For the optimal sustainable harvesting policy described in Theorem 11.9.2, how many trees are removed from the forest during each harvest?

6. If all the growth parameters $g_1, g_2, \ldots, g_{n-1}$ in the growth matrix G are equal, what should the ratio of the prices $p_2 : p_3 : \ldots : p_n$ be in order that any sustainable harvesting policy be an optimal sustainable harvesting policy? (See Exercise 3.)

11.10 COMPUTER GRAPHICS

In this section we assume that a view of a three-dimensional object is displayed on a video screen and show how matrix algebra can be used to obtain new views of the object by rotation, translation, and scaling.

PREREQUISITES: Analytic geometry
Matrix algebra

Suppose we want to visualize a three-dimensional object by displaying various views of it on a video screen. The object we have in mind to display is to be determined by a finite number of straight line segments. As an example, consider the truncated right pyramid with hexagonal base illustrated in Figure 1. We first introduce an *xyz*-coordinate system in which to embed the object. As in Figure 1, we orient the coordinate system so that its origin is at the center of the video screen and the *xy*-plane coincides with the plane of the screen. Consequently, an observer will see only the projection of the view of the three-dimensional object onto the two-dimensional *xy*-plane.

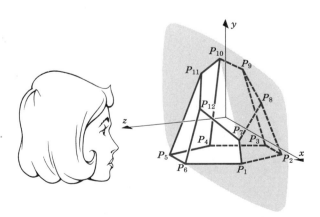

Figure 1

In the *xyz*-coordinate system, the endpoints $P_1, P_2, \ldots, P_n$ of the straight line segments that determine the view of the object will have certain coordinates, say,

$$(x_1, y_1, z_1), \quad (x_2, y_2, z_2), \quad \ldots, \quad (x_n, y_n, z_n)$$

These coordinates, together with a specification of which pairs are to be connected by straight line segments, are to be stored in the memory of the video display system. For example, assume that the 12 vertices of the truncated pyramid in Figure 1 have the following coordinates (the screen is 4 units wide by 3 units high):

$$P_1: (1.000, -.800, .000), \qquad P_2: (.500, -.800, -.866),$$
$$P_3: (-.500, -.800, -.866), \qquad P_4: (-1.000, -.800, .000),$$
$$P_5: (-.500, -.800, .866), \qquad P_6: (.500, .800, .866),$$
$$P_7: (.840, -.400, .000), \qquad P_8: (.315, .125, -.546),$$
$$P_9: (-.210, .650, -.364) \qquad P_{10}: (-.360, .800, .000),$$
$$P_{11}: (-.210, .650, .364), \qquad P_{12}: (.315, .125, .546)$$

These 12 vertices are connected pairwise by 18 straight line segments as follows, where $P_i \leftrightarrow P_j$ denotes that point P_i is connected to point P_j:

$$P_1 \leftrightarrow P_2, \quad P_2 \leftrightarrow P_3, \quad P_3 \leftrightarrow P_4, \quad P_4 \leftrightarrow P_5, \quad P_5 \leftrightarrow P_6, \quad P_6 \leftrightarrow P_1,$$
$$P_7 \leftrightarrow P_8, \quad P_8 \leftrightarrow P_9, \quad P_9 \leftrightarrow P_{10}, \quad P_{10} \leftrightarrow P_{11}, \quad P_{11} \leftrightarrow P_{12}, \quad P_{12} \leftrightarrow P_7,$$
$$P_1 \leftrightarrow P_7, \quad P_2 \leftrightarrow P_8, \quad P_3 \leftrightarrow P_9, \quad P_4 \leftrightarrow P_{10}, \quad P_5 \leftrightarrow P_{11}, \quad P_6 \leftrightarrow P_{12}$$

In View 1 these 18 straight line segments are shown as they would appear on the video screen. It should be noticed that only the *x*- and *y*-coordinates of the vertices are needed by the video display system to draw the view, as only the projection of the object onto the *xy*-plane is displayed. However, we must keep track of the *z*-coordinates to carry out certain transformations discussed later.

We now show how to form new views of the object that are produced by scaling, translating, or rotating the initial view. We first construct a $3 \times n$ matrix P, referred to as the *coordinate matrix of the view*, whose columns are the coordinates of the n points of a view:

$$P = \begin{bmatrix} x_1 & x_2 & \cdots & x_n \\ y_1 & y_2 & \cdots & y_n \\ z_1 & z_2 & \cdots & z_n \end{bmatrix}$$

View 1

For example, the coordinate matrix P corresponding to View 1 is the 3×12 matrix

$$\begin{bmatrix} 1.000 & .500 & -.500 & -1.000 & -.500 & .500 & .840 & .315 & -.210 & -.360 & -.210 & .315 \\ -.800 & -.800 & -.800 & -.800 & -.800 & -.800 & -.400 & .125 & .650 & .800 & .650 & .125 \\ .000 & -.866 & -.866 & .000 & .866 & .866 & .000 & -.546 & -.364 & .000 & .364 & .546 \end{bmatrix}$$

We will show below how to transform the coordinate matrix P of a view to a new coordinate matrix P' corresponding to a new view of the object. The straight line segments connecting the various points move with the points as they are transformed. In this way each view is uniquely determined by its coordinate matrix once we have specified which pairs of points in the original view are to be connected by straight lines.

SCALING

The first type of transformation we consider consists of scaling a view along the x, y, and z directions by factors of α, β, and γ, respectively. By this we mean that if a point P_i has coordinates (x_i, y_i, z_i) in the original view, it is to move to a new point P'_i with coordinates $(\alpha x_i, \beta y_i, \gamma z_i)$ in the new view. This has the effect of transforming a unit cube in the original view to a rectangular parallelepiped of dimensions $\alpha \times \beta \times \gamma$ (Figure 2). Mathematically, this may be accomplished with matrix multiplication as follows. Define a 3×3 diagonal matrix

$$S = \begin{bmatrix} \alpha & 0 & 0 \\ 0 & \beta & 0 \\ 0 & 0 & \gamma \end{bmatrix}$$

Then, if a point P_i in the original view is represented by the column vector

$$\begin{bmatrix} x_i \\ y_i \\ z_i \end{bmatrix}$$

the transformed point P'_i is represented by the column vector

$$\begin{bmatrix} x'_i \\ y'_i \\ z'_i \end{bmatrix} = \begin{bmatrix} \alpha & 0 & 0 \\ 0 & \beta & 0 \\ 0 & 0 & \gamma \end{bmatrix} \begin{bmatrix} x_i \\ y_i \\ z_i \end{bmatrix}$$

Using the coordinate matrix P, which contains the coordinates of all n points of the original view as its columns, these n points can be transformed simultaneously to produce the coordinate matrix P' of the scaled view, as follows:

$$SP = \begin{bmatrix} \alpha & 0 & 0 \\ 0 & \beta & 0 \\ 0 & 0 & \gamma \end{bmatrix} \begin{bmatrix} x_1 & x_2 & \cdots & x_n \\ y_1 & y_2 & \cdots & y_n \\ z_1 & z_2 & \cdots & z_n \end{bmatrix}$$

$$= \begin{bmatrix} \alpha x_1 & \alpha x_2 & \cdots & \alpha x_n \\ \beta y_1 & \beta y_2 & \cdots & \beta y_n \\ \gamma z_1 & \gamma z_2 & \cdots & \gamma z_n \end{bmatrix} = P'$$

The new coordinate matrix can then be entered into the video display system to produce the new view of the object. As an example, View 2 is View 1 scaled by setting $\alpha = 1.8$, $\beta = 0.5$, and $\gamma = 3.0$. Notice that the scaling $\gamma = 3.0$ along the z-axis is not visible in View 2, as we see only the projection of the object onto the xy-plane.

View 2   View 1 scaled by $\alpha = 1.8$, $\beta = 0.5$, $\gamma = 3.0$.

TRANSLATION

We next consider the transformation of translating or displacing an object to a new position on the screen. Referring to Figure 3, suppose we desire to change an existing view so that each point P_i with coordinates (x_i, y_i, z_i) moves to a new point P_i' with coordinates $(x_i + x_0, y_i + y_0, z_i + z_0)$. The vector

$$\begin{bmatrix} x_0 \\ y_0 \\ z_0 \end{bmatrix}$$

Figure 3

is called the ***translation vector*** of the transformation. By defining a $3 \times n$ matrix T as

$$T = \begin{bmatrix} x_0 & x_0 & \cdots & x_0 \\ y_0 & y_0 & \cdots & y_0 \\ z_0 & z_0 & \cdots & z_0 \end{bmatrix}$$

all n points of the view determined by the coordinate matrix P can be translated by matrix addition by means of the equation

$$P' = P + T$$

The coordinate matrix P' then specifies the new coordinates of the n points. For example, if we wish to translate View 1 according to the translation vector

$$\begin{bmatrix} 1.2 \\ 0.4 \\ 1.7 \end{bmatrix}$$

the result is View 3. Notice, again, that the translation $z_0 = 1.7$ along the z-axis does not show up explicitly in View 3.

In Exercise 7 a technique of performing translations by matrix multiplication rather than by matrix addition is explained.

ROTATION

A more complicated type of transformation is a rotation of a view about one of the three coordinate axes. We begin with a rotation about the z-axis (the axis perpendicular to the screen) through an angle θ. Given a point P_i in the original view with coordinates (x_i, y_i, z_i), we wish to compute the new coordinates (x_i', y_i', z_i') of the rotated point P_i'. Referring to Figure 4 and using a little trigonometry, the reader should be able to derive the following:

$$x_i' = \rho \cos(\phi + \theta) = \rho \cos \phi \cos \theta - \rho \sin \phi \sin \theta = x_i \cos \theta - y_i \sin \theta$$
$$y_i' = \rho \sin(\phi + \theta) = \rho \cos \phi \sin \theta + \rho \sin \phi \cos \theta = x_i \sin \theta + y_i \cos \theta$$
$$z_i' = z_i$$

View 3   View 1 translated by $x_0 = 1.2$, $y_0 = 0.4$, $z_0 = 1.7$.

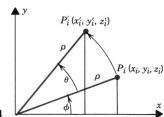

Figure 4

These equations can be written in matrix form as

$$\begin{bmatrix} x_i' \\ y_i' \\ z_i' \end{bmatrix} = \begin{bmatrix} \cos\theta & -\sin\theta & 0 \\ \sin\theta & \cos\theta & 0 \\ 0 & 0 & 1 \end{bmatrix} \begin{bmatrix} x_i \\ y_i \\ z_i \end{bmatrix}$$

If we let R denote the 3×3 matrix in this equation, all n points can be rotated by the matrix product

$$P' = RP$$

to yield the coordinate matrix P' of the rotated view.

Rotations about the x- and y-axes can be accomplished analogously, and the resulting rotation matrices are given with Views 4, 5, and 6. These three new views of the truncated pyramid correspond to rotations of View 1 about the x-, y-, and z-axes, respectively, each through an angle of 90°.

Rotation about the x-axis

$$\begin{bmatrix} 1 & 0 & 0 \\ 0 & \cos\theta & -\sin\theta \\ 0 & \sin\theta & \cos\theta \end{bmatrix}$$

View 4　View 1 rotated 90° about x-axis.

Rotation about the y-axis

$$\begin{bmatrix} \cos\theta & 0 & \sin\theta \\ 0 & 1 & 0 \\ -\sin\theta & 0 & \cos\theta \end{bmatrix}$$

View 5　View 1 rotated 90° about y-axis.

Rotation about the z-axis

$$\begin{bmatrix} \cos\theta & -\sin\theta & 0 \\ \sin\theta & \cos\theta & 0 \\ 0 & 0 & 1 \end{bmatrix}$$

View 6　View 1 rotated 90° about z-axis.

Rotations about the three coordinate axes may be combined to give oblique views of an object. For example, View 7 is View 1 rotated first about the x-axis through $30°$, then about the y-axis through $-70°$, and finally about the z-axis through $-27°$. Mathematically, these three successive rotations can be embodied in the single transformation equation $P' = RP$, where R is the product of three individual rotation matrices:

$$R_1 = \begin{bmatrix} 1 & 0 & 0 \\ 0 & \cos(30°) & -\sin(30°) \\ 0 & \sin(30°) & \cos(30°) \end{bmatrix}$$

$$R_2 = \begin{bmatrix} \cos(-70°) & 0 & \sin(-70°) \\ 0 & 1 & 0 \\ -\sin(-70°) & 0 & \cos(-70°) \end{bmatrix}$$

$$R_3 = \begin{bmatrix} \cos(-27°) & -\sin(-27°) & 0 \\ \sin(-27°) & \cos(-27°) & 0 \\ 0 & 0 & 1 \end{bmatrix}$$

in the order

$$R = R_3 R_2 R_1 = \begin{bmatrix} .305 & -.025 & -.952 \\ -.155 & .985 & -.076 \\ .940 & .171 & .296 \end{bmatrix}$$

As a final illustration, in View 8 we have two separate views of the truncated pyramid, which constitute a stereoscopic pair. They were produced by first rotating View 7 about the y-axis through an angle of $-3°$ and translating it to the right, then rotating the same View 7 about the y-axis through an angle of $+3°$ and translating it to the left. The translation distances were chosen so that the stereoscopic views are about $2\frac{1}{2}$ inches apart—the approximate distance between a pair of eyes.

View 7 Oblique view of truncated pyramid.

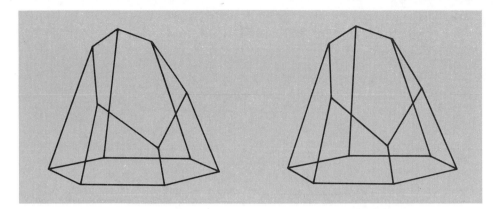

View 8   Stereoscopic figure of truncated pyramid. The three-dimensionality of the diagram can be seen by holding the book about one foot away and focusing on a distant object. Then by shifting gaze to View 8 without refocusing, the two views of the steroscopic pair can be made to merge together and produce the desired effect.

EXERCISE SET 11.10

1. View 9 is a view of a square with vertices $(0, 0, 0)$, $(1, 0, 0)$, $(1, 1, 0)$, and $(0, 1, 0)$.
   (a) What is the coordinate matrix of View 9?
   (b) What is the coordinate matrix of View 9 after it is scaled by a factor $1\frac{1}{2}$ in the x-direction and $\frac{1}{2}$ in the y-direction? Draw a sketch of the scaled view.
   (c) What is the coordinate matrix of View 9 after it is translated by the vector

$$\begin{bmatrix} -2 \\ -1 \\ 3 \end{bmatrix}$$

   Draw a sketch of the translated view.
   (d) What is the coordinate matrix of View 9 after it is rotated through an angle of $-30°$ about the z-axis? Draw a sketch of the rotated view.

View 9   Square with vertices $(0, 0, 0)$, $(1, 0, 0)$, $(1, 1, 0)$, and $(0, 1, 0)$ (Exercises 1 and 2).

2. (a) If the coordinate matrix of View 9 is multiplied by the matrix

$$\begin{bmatrix} 1 & \frac{1}{2} & 0 \\ 0 & 1 & 0 \\ 0 & 0 & 1 \end{bmatrix}$$

the result is the coordinate matrix of View 10. Such a transformation is called a *shear in the x-direction with factor $\frac{1}{2}$ with respect to the y-coordinate*. Show that under such a transformation, a point with coordinates (x_i, y_i, z_i) has new coordinates $(x_i + \frac{1}{2}y_i, z_i)$.

View 10   View 9 sheared along x-axis by $\frac{1}{2}$ with respect to the y-coordinate (Exercise 2).

(b) What are the coordinates of the four vertices of the shear square in View 10?

(c) The matrix

$$\begin{bmatrix} 1 & 0 & 0 \\ 6 & 1 & 0 \\ 0 & 0 & 1 \end{bmatrix}$$

determines a *shear in the y-direction with factor .6 with respect to the x-coordinate* (e.g., View 11). Sketch a view of the square in View 9 after such a shearing transformation, and find the new coordinates of its four vertices.

View 11   View 1 sheared along y-axis by 6 with respect to the x-coordinate (Exercise 2).

3. (a) The *reflection about the xz-plane* is defined as the transformation that takes a point (x_i, y_i, z_i) to the point $(z_i, -y_i, z_i)$ (e.g., View 12). If P and P' are the coordinate matrices of a view and its reflection about the xz-plane, respectively, find a matrix M such that $P' = MP$.

(b) Analogous to part (a), define the *reflection about the yz-plane* and construct the corresponding transformation matrix. Draw a sketch of View 1 reflected about the yz-plane.

(c) Analogous to part (a), define the *reflection about the xy-plane* and construct the corresponding transformation matrix. Draw a sketch of View 1 reflected about the xy-plane.

View 12   View 1 reflected about the *xz*-plane (Exercise 3).

4. (a) View 13 is View 1 subject to the following five transformations:
   1. Scale by a factor of $\frac{1}{2}$ in the *x*-direction, 2 in the *y*-direction, and $\frac{1}{3}$ in the *z*-direction.
   2. Translate $\frac{1}{2}$ unit in the *x*-direction.
   3. Rotate $20°$ about the *x*-axis.
   4. Rotate $-45°$ about the *y*-axis.
   5. Rotate $90°$ about the *z*-axis.
   Construct the five matrices M_1, M_2, M_3, M_4, and M_5 associated with these five transformations.
   (b) If P is the coordinate matrix of View 1 and P' is the coordinate matrix of View 13, express P' in terms of M_1, M_2, M_3, M_4, M_5, and P.

View 13   View 1 scaled, translated, and rotated (Exercise 4).

5. (a) View 14 is View 1 subject to the following seven transformations:
   1. Scale by a factor of .3 in the *x*-direction and .5 in the *y*-direction.
   2. Rotate $45°$ about the *x*-axis.
   3. Translate 1 unit in the *x*-direction.
   4. Rotate $35°$ about the *y*-axis.
   5. Rotate $-45°$ about the *z*-axis.
   6. Translate 1 unit in the *z*-direction.
   7. Scale by a factor of 2 in the *x*-direction.
   Construct the seven matrices M_1, M_2, ..., M_7 associated with these seven transformations.
   (b) If P is the coordinate matrix of View 1 and P' is the coordinate matrix of View 14, express P' in terms of M_1, M_2, ..., M_7, and P.

View 14   View 1 scaled, translated, and rotated (Exercise 5).

6. Suppose a view with coordinate matrix P is to be rotated through an angle θ about an axis through the origin and specified by two angles α and β (Figure 5). If P' is the coordinate matrix of the rotated view, find rotation matrices R_1, R_2, R_3, R_4, and R_5 such that

$$P' = R_5 R_4 R_3 R_2 R_1 P$$

[**Hint.** The desired rotation can be accomplished in the following five steps:
1. Rotate through an angle of β about the y-axis.
2. Rotate through an angle of α about the z-axis.
3. Rotate through an angle of θ about the y-axis.
4. Rotate through an angle of $-\alpha$ about the z-axis.
5. Rotate through an angle of $-\beta$ about the y-axis.]

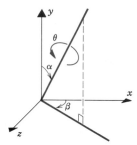

Figure 5

7. This exercise illustrates a technique for translating a point with coordinates (x_i, y_i, z_i) to a point with coordinates $(x_i + x_0, y_i + y_0, z_i + z_0)$ by matrix multiplication rather than matrix addition.

(a) Let the point (x_i, y_i, z_i) be associated with the column vector

$$\mathbf{v}_i = \begin{bmatrix} x_i \\ y_i \\ z_i \\ 1 \end{bmatrix}$$

and let the point $(x_i + x_0, y_i + y_0, z_i + z_0)$ be associated with the column vector

$$\mathbf{v}'_i = \begin{bmatrix} x_i + x_0 \\ y_i + y_0 \\ z_i + z_0 \\ 1 \end{bmatrix}$$

Find a 4×4 matrix M such that $\mathbf{v}'_i = M\mathbf{v}_i$.

(b) Find the specific 4×4 matrix of the above form that will effect the translation of the point $(4, -2, 3)$ to the point $(-1, 7, 0)$.

8. For the three rotation matrices given with Views 4, 5, and 6, show that

$$R^{-1} = R^t$$

(A matrix with this property is called an *orthogonal matrix*.)

11.11 EQUILIBRIUM TEMPERATURE DISTRIBUTIONS

In this section the equilibrium temperature distribution within a trapezoidal plate is found when the temperature around the edges of the plate is specified. The problem is reduced to solving a system of linear equations, and an iterative technique for solving the system is described. This section also discusses a "random walk" approach to the problem.

PREREQUISITES: Linear systems
Matrices
Intuitive understanding of limits

Suppose we are given a thin trapezoidal plate (Figure 1*a*) whose two faces are insulated from heat. Suppose we are also given the temperature along the four edges of the plate. For example, let the temperature be constant on each edge with values of 0°, 0°, 1°, and 2°, as in the figure. After a period of time the temperature inside the plate will stabilize. Our objective in this section is to determine this equilibrium temperature distribution at the points inside the plate. As we will see, the interior equilibrium temperature is completely determined by the **boundary data**, that is, the temperature along the edges of the plate.

The equilibrium temperature distribution can be visualized by the use of curves that connect points of equal temperature. Such curves are called **isotherms** of the temperature distribution. In Figure 1*b* we have sketched a few isotherms using information we derive later in the chapter.

Although all our calculations will be for the trapezoidal plate illustrated, our techniques generalize easily to a plate of any practical shape. They also generalize to the problem of finding the temperature within a three-dimensional body. In fact, our "plate" could be the cross section of some solid object if the flow of heat

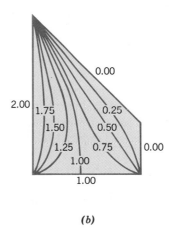

Figure 1   *(a)*   *(b)*

perpendicular to the cross section is negligible. For example, Figure 1 could represent the cross section of a long dam. The dam is exposed to three different temperatures: the temperature of the ground at its base, the temperature of the water on one side, and the temperature of the air on the other side. A knowledge of the temperature distribution inside the dam is necessary to determine the thermal stresses to which it is subjected.

Next, we shall consider a certain thermodynamic principle that characterizes the temperature distribution we are seeking.

THE MEAN-VALUE PROPERTY

There are many different ways to obtain a mathematical model for our problem. The approach we use is based on the following property of equilibrium temperature distributions.

> **Theorem 11.11.1 (*The Mean-Value Property*).** *Let a plate be in thermal equilibrium and let P be a point inside the plate. Then if C is any circle with center at P that is completely contained in the plate, the temperature at P is the average value of the temperature on the circle* (Figure 2).

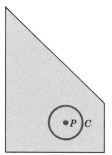

Figure 2

This property is a consequence of certain basic laws of molecular motion, and we will not attempt to derive it. Basically, this property states that in equilibrium, thermal energy tends to distribute itself as evenly as possible consistent with the boundary conditions. It can be shown that the mean-value property uniquely determines the equilibrium temperature distribution of a plate.

Unfortunately, determining the equilibrium temperature distribution from the mean-value property is not an easy matter. However, if we restrict ourselves to finding the temperature only at a finite set of points within the plate, the problem can be reduced to solving a linear system. We pursue this idea next.

DISCRETE FORMULATION OF THE PROBLEM

We can overlay our trapezoidal plate with a succession of finer and finer square nets or meshes (Figure 3). In (*a*) we have a rather coarse net; in (*b*) we have a net with half the spacing as in (*a*); and in (*c*) we have a net with the spacing again reduced by half. The points of intersection of the net lines are called *mesh points*. We classify them as ***boundary mesh points*** if they fall on the boundary of the plate,

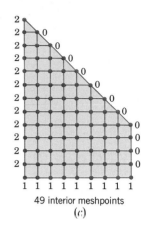

1 interior meshpoint
(a)

9 interior meshpoints
(b)

49 interior meshpoints
(c)

Figure 3

or *interior mesh points* if they lie in the interior of the plate. For the three net spacings we have chosen, there are 1, 9, and 49 interior mesh points, respectively.

In the discrete formulation of our problem, we try to find the temperature only at the interior mesh points of some particular net. For a rather fine net, as in (c), this will provide an excellent picture of the temperature distribution throughout the entire plate.

At the boundary mesh points the temperature is given by the boundary data. (In Figure 3 we have labeled all the boundary mesh points with their corresponding (temperatures.) At the interior mesh points, we shall apply the following discrete version of the mean-value property.

> **Theorem 11.11.2 (*Discrete Mean-Value Property*).**   *At each interior mesh point, the temperature is the average of the temperatures at the four neighboring mesh points.*

This discrete version is a reasonable approximation to the true mean-value property. But because it is only an approximation, it will provide only an approximation to the true temperatures at the interior mesh points. However, the approximations will get better as the mesh spacing decreases. In fact, as the mesh spacing approaches zero, the approximations approach the exact temperature distribution, a fact proved in advanced courses in numerical analysis. We will illustrate this convergence by computing the approximate temperatures at the mesh points for the three mesh spacings given in Figure 3.

Case (*a*) of Figure 3 is simple, as there is only one interior mesh point. If we let t_0 be the temperature at this mesh point, the discrete mean-value property immediately gives

$$t_0 = \tfrac{1}{4}(2 + 1 + 0 + 0) = .75$$

In case (*b*) we can label the temperatures at the nine interior mesh points $t_1, t_2, \ldots, t_9$, as in Figure 3b. (The particular ordering is not important.) By applying the discrete mean-value property successively to each of these nine mesh points, we obtain the following nine equations:

$$
\begin{aligned}
t_1 &= \tfrac{1}{4}(t_2 + 2 + 0 + 0) \\
t_2 &= \tfrac{1}{4}(t_1 + t_3 + t_4 + 2) \\
t_3 &= \tfrac{1}{4}(t_2 + t_5 + 0 + 0) \\
t_4 &= \tfrac{1}{4}(t_2 + t_5 + t_7 + 2) \\
t_5 &= \tfrac{1}{4}(t_3 + t_4 + t_6 + t_8) \\
t_6 &= \tfrac{1}{4}(t_5 + t_9 + 0 + 0) \\
t_7 &= \tfrac{1}{4}(t_4 + t_8 + 1 + 2) \\
t_8 &= \tfrac{1}{4}(t_5 + t_7 + t_9 + 1) \\
t_9 &= \tfrac{1}{4}(t_6 + t_8 + 1 + 0)
\end{aligned}
\tag{1}
$$

This is a system of nine linear equations in nine unknowns. We can rewrite it in matrix form as

$$\mathbf{t} = M\mathbf{t} + \mathbf{b} \tag{2}$$

where

$$
\mathbf{t} = \begin{bmatrix} t_1 \\ t_2 \\ t_3 \\ t_4 \\ t_5 \\ t_6 \\ t_7 \\ t_8 \\ t_9 \end{bmatrix}, \quad
M = \begin{bmatrix}
0 & \tfrac{1}{4} & 0 & 0 & 0 & 0 & 0 & 0 & 0 \\
\tfrac{1}{4} & 0 & \tfrac{1}{4} & \tfrac{1}{4} & 0 & 0 & 0 & 0 & 0 \\
0 & \tfrac{1}{4} & 0 & 0 & \tfrac{1}{4} & 0 & 0 & 0 & 0 \\
0 & \tfrac{1}{4} & 0 & 0 & \tfrac{1}{4} & 0 & \tfrac{1}{4} & 0 & 0 \\
0 & 0 & \tfrac{1}{4} & \tfrac{1}{4} & 0 & \tfrac{1}{4} & 0 & \tfrac{1}{4} & 0 \\
0 & 0 & 0 & 0 & \tfrac{1}{4} & 0 & 0 & 0 & \tfrac{1}{4} \\
0 & 0 & 0 & \tfrac{1}{4} & 0 & 0 & 0 & \tfrac{1}{4} & 0 \\
0 & 0 & 0 & 0 & \tfrac{1}{4} & 0 & \tfrac{1}{4} & 0 & \tfrac{1}{4} \\
0 & 0 & 0 & 0 & 0 & \tfrac{1}{4} & 0 & \tfrac{1}{4} & 0
\end{bmatrix}, \quad
\mathbf{b} = \begin{bmatrix} \tfrac{1}{2} \\ \tfrac{1}{2} \\ 0 \\ \tfrac{1}{2} \\ 0 \\ 0 \\ \tfrac{3}{4} \\ \tfrac{1}{4} \\ \tfrac{1}{4} \end{bmatrix}
$$

To solve Equation (2) we write it as

$$(I - M)\mathbf{t} = \mathbf{b}$$

The solution for **t** is thus

$$\mathbf{t} = (I - M)^{-1}\mathbf{b} \tag{3}$$

as long as the matrix $(I - M)$ is invertible. This is indeed the case, and the solution for **t** as calculated by (3) is

$$
\mathbf{t} = \begin{bmatrix}
0.7846 \\
1.1383 \\
0.4719 \\
1.2967 \\
0.7491 \\
0.3265 \\
1.2995 \\
0.9014 \\
0.5570
\end{bmatrix}
\tag{4}
$$

Figure 4 is a diagram of the plate with the nine interior mesh points labeled with their temperatures as given by this solution.

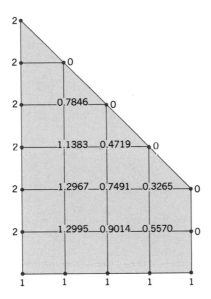

Figure 4

For case (*c*) of Figure 3, we repeat this same procedure. We label the temperatures at the 49 interior mesh points as $t_1, t_2, \ldots, t_{49}$ in some manner. For example, we may begin at the top of the plate and proceed from left to right along each row of mesh points. Applying the discrete mean-value property to each mesh point gives a system of 49 linear equations in 49 unknowns:

$$t_1 = \tfrac{1}{4}(t_2 + 2 + 0 + 0)$$
$$t_2 = \tfrac{1}{4}(t_1 + t_3 + t_4 + 2)$$
$$\vdots \tag{5}$$
$$t_{48} = \tfrac{1}{4}(t_{41} + t_{47} + t_{49} + 1)$$
$$t_{49} = \tfrac{1}{4}(t_{42} + t_{48} + 0 + 1)$$

In matrix form, Equations (5) are

$$\mathbf{t} = M\mathbf{t} + \mathbf{b}$$

where **t** and **b** are column vectors with 49 entries, and M is a 49 × 49 matrix. As in (3) the solution for **t** is

$$\mathbf{t} = (I - M)^{-1}\mathbf{b} \tag{6}$$

In Figure 5 we display the temperatures at the 49 mesh points found by Equation (6). The nine unshaded temperatures in this figure fall on the mesh points of Figure 4. In Table 1 we compare the temperatures at these nine common mesh points for the three different mesh spacings used.

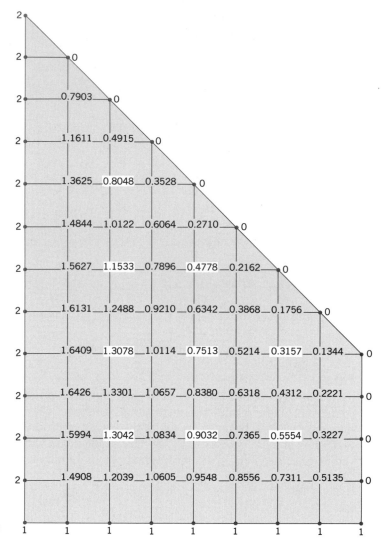

Figure 5

Knowing that the temperatures of the discrete problem approach the exact temperatures as the mesh spacing decreases, we may surmise that the nine temperatures obtained in case (c) are closer to the exact values than those in case (b).

A NUMERICAL TECHNIQUE

To obtain the 49 temperatures in case (c) of Figure 3 it was necessary to solve a linear system with 49 unknowns. A finer net might involve a linear system with hundreds or even thousands of unknowns. Exact algorithms for the solutions of such large systems are impractical, and for this reason we now discuss a numerical technique for the practical solution of these systems.

To describe this technique, we look again at Equation (2):

$$\mathbf{t} = M\mathbf{t} + \mathbf{b} \tag{7}$$

TABLE 1

	Temperatures at Common Mesh Points		
	Case (*a*)	**Case (*b*)**	**Case (*c*)**
t_1	—	0.7846	0.8048
t_2	—	1.1383	1.1533
t_3	—	0.4719	0.4778
t_4	—	1.2967	1.3078
t_5	0.7500	0.7491	0.7513
t_6	—	0.3265	0.3157
t_7	—	1.2995	1.3042
t_8	—	0.9014	0.9032
t_9	—	0.5570	0.5554

The vector **t** we are seeking appears on both sides of this equation. We consider a way of generating better and better approximations to the vector solution **t**. For the initial approximation $\mathbf{t}^{(0)}$ we may take $\mathbf{t}^{(0)} = \mathbf{0}$ if no better choice is available. If we substitute $\mathbf{t}^{(0)}$ into the right side of (7) and label the resulting left side as $\mathbf{t}^{(1)}$, we have

$$\mathbf{t}^{(1)} = M\mathbf{t}^{(0)} + \mathbf{b} \tag{8}$$

Usually $\mathbf{t}^{(1)}$ is a better approximation to the solution than is $\mathbf{t}^{(0)}$. If we substitute $\mathbf{t}^{(1)}$ into the right side of (7) we generate another approximation, which we label $\mathbf{t}^{(2)}$:

$$\mathbf{t}^{(2)} = M\mathbf{t}^{(1)} + \mathbf{b} \tag{9}$$

Continuing in this way, we generate a sequence of approximations as follows:

$$\mathbf{t}^{(1)} = M\mathbf{t}^{(0)} + \mathbf{b}$$
$$\mathbf{t}^{(2)} = M\mathbf{t}^{(1)} + \mathbf{b}$$
$$\mathbf{t}^{(3)} = M\mathbf{t}^{(2)} + \mathbf{b}$$
$$\vdots \tag{10}$$
$$\mathbf{t}^{(n)} = M\mathbf{t}^{(n-1)} + \mathbf{b}$$
$$\vdots$$

One would hope that this sequence of approximations $\mathbf{t}^{(0)}, \mathbf{t}^{(1)}, \mathbf{t}^{(2)}, \ldots$ converges to the exact solution of (7). We do not have the space here to go into the theoretical considerations necessary to show this. However, suffice it to say that for the particular problem we are considering, the sequence converges to the exact solution for any mesh size and for any initial approximation $\mathbf{t}^{(0)}$.

This technique of generating successive approximations to the solution of (7) is a variation of a technique called *Jacobi iteration* (see Section 9.3); the approximations themselves are called *iterates*. As a numerical example, let us apply Jacobi iteration to the calculation of the nine mesh point temperatures of case (*b*). Setting $\mathbf{t}^{(0)} = \mathbf{0}$, we have from Equation (2)

$$\mathbf{t}^{(1)} = M\mathbf{t}^{(0)} + \mathbf{b} = M\mathbf{0} + \mathbf{b} = \mathbf{b} = \begin{bmatrix} .5000 \\ .5000 \\ .0000 \\ .5000 \\ .0000 \\ .0000 \\ .7500 \\ .2500 \\ .2500 \end{bmatrix}$$

$$\mathbf{t}^{(2)} = M\mathbf{t}^{(1)} + \mathbf{b}$$

$$= \begin{bmatrix} 0 & \frac{1}{4} & 0 & 0 & 0 & 0 & 0 & 0 & 0 \\ \frac{1}{4} & 0 & \frac{1}{4} & \frac{1}{4} & 0 & 0 & 0 & 0 & 0 \\ 0 & \frac{1}{4} & 0 & 0 & \frac{1}{4} & 0 & 0 & 0 & 0 \\ 0 & \frac{1}{4} & 0 & 0 & \frac{1}{4} & 0 & \frac{1}{4} & 0 & 0 \\ 0 & 0 & \frac{1}{4} & \frac{1}{4} & 0 & \frac{1}{4} & 0 & \frac{1}{4} & 0 \\ 0 & 0 & 0 & 0 & \frac{1}{4} & 0 & 0 & 0 & \frac{1}{4} \\ 0 & 0 & 0 & \frac{1}{4} & 0 & 0 & 0 & \frac{1}{4} & 0 \\ 0 & 0 & 0 & 0 & \frac{1}{4} & 0 & \frac{1}{4} & 0 & \frac{1}{4} \\ 0 & 0 & 0 & 0 & 0 & \frac{1}{4} & 0 & \frac{1}{4} & 0 \end{bmatrix} \begin{bmatrix} .5000 \\ .5000 \\ .0000 \\ .5000 \\ .0000 \\ .0000 \\ .7500 \\ .2500 \\ .2500 \end{bmatrix} + \begin{bmatrix} .5000 \\ .5000 \\ .0000 \\ .5000 \\ .0000 \\ .0000 \\ .7500 \\ .2500 \\ .2500 \end{bmatrix} = \begin{bmatrix} .6250 \\ .7500 \\ .1250 \\ .8125 \\ .1875 \\ .0625 \\ .9375 \\ .5000 \\ .3125 \end{bmatrix}$$

Some additional iterates are

$$\mathbf{t}^{(3)} = \begin{bmatrix} 0.6875 \\ 0.8906 \\ 0.2344 \\ 0.9688 \\ 0.3750 \\ 0.1250 \\ 1.0781 \\ 0.6094 \\ 0.3906 \end{bmatrix}, \quad \mathbf{t}^{(10)} = \begin{bmatrix} 0.7791 \\ 1.1230 \\ 0.4573 \\ 1.2770 \\ 0.7236 \\ 0.3131 \\ 1.2848 \\ 0.8827 \\ 0.5446 \end{bmatrix}, \quad \mathbf{t}^{(20)} = \begin{bmatrix} 0.7845 \\ 1.1380 \\ 0.4716 \\ 1.2963 \\ 0.7486 \\ 0.3263 \\ 1.2992 \\ 0.9010 \\ 0.5567 \end{bmatrix}, \quad \mathbf{t}^{(30)} = \begin{bmatrix} 0.7846 \\ 1.1383 \\ 0.4719 \\ 1.2967 \\ 0.7491 \\ 0.3265 \\ 1.2995 \\ 0.9014 \\ 0.5570 \end{bmatrix}$$

All iterates beginning with the thirtieth are equal to $\mathbf{t}^{(30)}$ to four decimal places. Consequently, $\mathbf{t}^{(30)}$ is the exact solution to four decimal places. This agrees with our previous result given in Equation (4).

The Jacobi iteration scheme applied to the linear system (5) with 49 unknowns produces iterates that begin repeating to four decimal places after 119 iterations. Thus, $\mathbf{t}^{(119)}$ would provide the 49 temperatures of case (*c*) correct to four decimal places.

**A MONTE
CARLO
TECHNIQUE**

In this section we describe a so-called ***Monte Carlo technique*** for computing the temperature at a single interior mesh point of the discrete problem without having to compute the temperatures at the remaining interior mesh points. First we define a ***discrete random walk*** along the net. By this we mean a directed path along the net lines (Figure 6) which joins a succession of mesh points such that the direction of departure from each mesh point is chosen at random. Each of the four possible directions of departure from each mesh point along the path is to be equally probable.

Figure 6

By the use of random walks, we can compute the temperature at a specified interior mesh point on the basis of the following property.

Theorem 11.11.3 (*Random Walk Property*). *Let $W_1, W_2, \ldots, W_n$ be a succession of random walks, all of which begin at a specified interior mesh point. Let $t_1^*, t_2^*, \ldots, t_n^*$ be the temperatures at the boundary mesh points first encountered along each of these random walks. Then the average value $\frac{1}{n}(t_1^* + t_2^* + \cdots + t_n^*)$ of these boundary temperatures approaches the temperature at the specified interior mesh point as the number of random walks n increases without bound.*

This property is a consequence of the discrete mean-value property which the mesh point temperatures satisfy. The proof of the random walk property involves elementary concepts from probability theory and we will not give it here.

In Table 2 we display the results of a large number of computer-generated random walks for the evaluation of the temperature t_5 of the nine-point mesh of case (*b*) in Figure 6. The first column lists the number n of the random walk. The second column lists the temperature t_n^* of the boundary point first encountered along the corresponding random walk. The last column contains the cumulative average of the boundary temperatures encountered along the n random walks.

TABLE 2

n	t_n^*	$(t_1^* + \cdots + t_n^*)/n$	n	t_n^*	$(t_1^* + \cdots + t_n^*)/n$
1	1	1.0000	20	1	0.9500
2	2	1.5000	30	0	0.8000
3	1	1.3333	40	0	0.8250
4	0	1.0000	50	2	0.8400
5	2	1.2000	100	0	0.8300
6	0	1.0000	150	1	0.8000
7	2	1.1429	200	0	0.8050
8	0	1.0000	250	1	0.8240
9	2	1.1111	500	1	0.7860
10	0	1.0000	1000	0	0.7550

Thus, after 1000 random walks we have the approximation $t_5 \simeq .7550$. This compares with the exact value $t_5 = .7491$ we had previously evaluated. As can be seen, the convergence to the exact value is not too rapid.

EXERCISE SET 11.11

1. A plate in the form of a circular disk has boundary temperatures of $0°$ on the left half of its circumference and $1°$ on the right half of its circumference. A net with four interior mesh points is overlaid on the disk (Figure 7).
   (a) Using the discrete mean-value property, write the 4×4 linear system $\mathbf{t} = M\mathbf{t} + \mathbf{b}$ that determines the approximate temperatures at the four interior mesh points.
   (b) Solve the linear system in part (a).
   (c) Use the Jacobi iteration scheme with $\mathbf{t}^{(0)} = \mathbf{0}$ to generate the iterates $\mathbf{t}^{(1)}$, $\mathbf{t}^{(2)}$, $\mathbf{t}^{(3)}$, $\mathbf{t}^{(4)}$, and $\mathbf{t}^{(5)}$ for the linear system in part (a). What is the "error vector" $\mathbf{t}^{(5)} - \mathbf{t}$ where $\mathbf{t}$ is the solution found in part (b)?
   (d) By certain advanced methods, it can be determined that the exact temperatures at the four mesh points are $t_1 = t_3 = .2371$ and $t_2 = t_4 = .7629$. What are the percentage errors in the values found in part (b)?

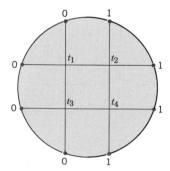

Figure 7

2. Use Theorem 11.11.1 to find the exact equilibrium temperature at the center of the disk in Exercise 1.

3. Calculate the first two iterates $t^{(1)}$ and $t^{(2)}$ for case (b) with nine interior mesh points [Equation (2)] when the initial iterate is chosen as

$$t^{(0)} = [1 \quad 1 \quad 1 \quad 1 \quad 1 \quad 1 \quad 1 \quad 1 \quad 1]^t$$

4. The random walk illustrated in Figure 8 can be described by six arrows

$$\leftarrow \downarrow \rightarrow \rightarrow \uparrow \rightarrow$$

that specify the directions of departure from the successive mesh points along the path. Figure 9 is an array of 100 computer-generated, randomly oriented arrows arranged in a 10×10 array. Use these arrows to determine random walks to approximate the temperature t_5, as in Table 2. Proceed as follows:

(1) Take the last two digits of your telephone number. Use the last digit to specify a row and the other to specify a column.

(2) Go to the arrow in the array with that row and column number.

(3) Using this arrow as a starting point, move through the array of arrows as you would read a book (left to right and top to bottom). Beginning at the point labeled t_5 in Figure 8 and using this sequence of arrows to specify a sequence of directions, move from mesh point to mesh point until you reach a boundary mesh point. This completes your first random walk. Record the temperature at the boundary mesh point. (If you reach the end of the arrow array, continue with the arrow in the upper left corner.)

(4) Return to the interior mesh point labeled t_5 and begin where you left off in the arrow array; generate your next random walk. Repeat this process until you have completed 10 random walks and have recorded 10 boundary temperatures.

(5) Calculate the average of the 10 boundary temperatures recorded. (The exact value is $t_5 = .7491$.)

Figure 8

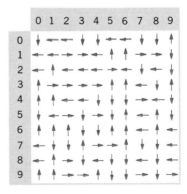

Figure 9

11.12   SOME APPLICATIONS IN GENETICS

This section investigates the propagation of an inherited trait in successive generations by computing powers of a matrix.

> **PREREQUISITES:** Eigenvalues and eigenvectors
> Diagonalization of a matrix
> Intuitive understanding of limits

In this section we examine the inheritance of traits in animals or plants. The inherited trait under consideration is assumed to be governed by a set of two genes, which we designate by *A* and *a*. Under ***autosomal inheritance*** each individual in the population of either gender possesses two of these genes, the possible pairings being designated *AA*, *Aa*, and *aa*. This pair of genes is called the individual's ***genotype***, and it determines how the trait controlled by the genes is manifested in the individual. For example, in snapdragons a set of two genes determines the color of the flower. Genotype *AA* produces red flowers, genotype *Aa* produces pink flowers, and genotype *aa* produces white flowers. In humans, eye coloration is controlled through autosomal inheritance. Genotypes *AA* and *Aa* have brown eyes, and genotype *aa* has blue eyes. In this case we say that gene *A* ***dominates*** gene *a*, or that gene *a* is ***recessive*** to gene *A*, because genotype *Aa* has the same outward trait as genotype *AA*.

In addition to autosomal inheritance we will also discuss ***X-linked inheritance***. In this type of inheritance the male of the species possesses only one of the two possible genes (*A* or *a*), and the female possesses a pair of the two genes (*AA*, *Aa*, or *aa*). In humans, color blindness, hereditary baldness, hemophilia, and muscular dystrophy, to name a few, are traits controlled by X-linked inheritance.

Below we explain the manner in which the genes of the parents are passed on to their offspring for the two types of inheritance. We construct matrix models that give the probable genotypes of the offspring in terms of the genotypes of the parents, and we use these matrix models to follow the genotype distribution of a population through successive generations.

AUTOSOMAL INHERITANCE

In autosomal inheritance an individual inherits one gene from each of its parents' pairs of genes to form its own particular pair. As far as we know, it is a matter of chance which of the two genes a parent passes on to the offspring. Thus, if one parent is of genotype *Aa*, it is equally likely that the offspring will inherit the *A* gene or the *a* gene from that parent. If one parent is of genotype *aa* and the other parent is of genotype *Aa*, the offspring will always receive an *a* gene from the *aa* parent, and will receive either an *A* gene or an *a* gene, with equal probability, from the *Aa* parent. Consequently, each of the offspring has equal probability of being genotype *aa* or *Aa*. In Table 1 we list the probabilities of the possible genotypes of the offspring for all possible combinations of the genotypes of the parents.

TABLE 1

Genotype of Offspring	Genotypes of Parents					
	$AA-AA$	$AA-Aa$	$AA-aa$	$Aa-Aa$	$Aa-aa$	$aa-aa$
AA	1	$\frac{1}{2}$	0	$\frac{1}{4}$	0	0
Aa	0	$\frac{1}{2}$	1	$\frac{1}{2}$	$\frac{1}{2}$	0
aa	0	0	0	$\frac{1}{4}$	$\frac{1}{2}$	1

Example 1 Suppose a farmer has a large population of plants consisting of some distribution of all three possible genotypes AA, Aa, and aa. The farmer desires to undertake a breeding program in which each plant in the population is always fertilized with a plant of genotype AA. We want to derive an expression for the distribution of the three possible genotypes in the population after any number of generations.

For $n = 0, 1, 2, \ldots$, let us set

a_n = fraction of plants of genotype AA in nth generation

b_n = fraction of plants of genotype Aa in nth generation

c_n = fraction of plants of genotype aa in nth generation

Thus, a_0, b_0, and c_0 specify the initial distribution of the genotypes. We also have that

$$a_n + b_n + c_n = 1 \qquad \text{for } n = 0, 1, 2, \ldots$$

From Table 1 we can determine the genotype distribution of each generation from the genotype distribution of the preceding generation by the following equations:

$$
\begin{aligned}
a_n &= a_{n-1} + \tfrac{1}{2}b_{n-1} \\
b_n &= c_{n-1} + \tfrac{1}{2}b_{n-1} \qquad n = 1, 2, \ldots \\
c_n &= 0
\end{aligned}
\tag{1}
$$

For example, the first of these three equations states that all the offspring of a plant of genotype AA will be of genotype AA under this breeding program, and half of the offspring of a plant of genotype Aa will be of genotype AA.

Equations (1) can be written in matrix notation as

$$\mathbf{x}^{(n)} = M\mathbf{x}^{(n-1)} \qquad n = 1, 2, \ldots \tag{2}$$

where

$$
\mathbf{x}^{(n)} = \begin{bmatrix} a_n \\ b_n \\ c_n \end{bmatrix}, \qquad
\mathbf{x}^{(n-1)} = \begin{bmatrix} a_{n-1} \\ b_{n-1} \\ c_{n-1} \end{bmatrix}, \qquad \text{and} \qquad
M = \begin{bmatrix} 1 & \tfrac{1}{2} & 0 \\ 0 & \tfrac{1}{2} & 1 \\ 0 & 0 & 0 \end{bmatrix}
$$

Notice that the three columns of the matrix M are the same as the first three columns of Table 1.

From Equation (2) it follows that

$$\mathbf{x}^{(n)} = M\mathbf{x}^{(n-1)} = M^2\mathbf{x}^{(n-2)} = \cdots = M^n\mathbf{x}^{(0)} \tag{3}$$

Consequently, if we can find an explicit expression for M^n, we can use (3) to obtain an explicit expression for $\mathbf{x}^{(n)}$. To find an explicit expression for M^n, we first diagonalize M. That is, we find an invertible matrix P and a diagonal matrix D such that

$$M = PDP^{-1} \tag{4}$$

With such a diagonalization, we then have (see Exercise 1)

$$M^n = PD^nP^{-1} \qquad \text{for } n = 1, 2, \ldots$$

where

$$D^n = \begin{bmatrix} \lambda_1 & 0 & 0 & \cdots & 0 \\ 0 & \lambda_2 & 0 & \cdots & 0 \\ \vdots & \vdots & \vdots & & \vdots \\ 0 & 0 & 0 & \cdots & \lambda_k \end{bmatrix}^n = \begin{bmatrix} \lambda_1^n & 0 & 0 & \cdots & 0 \\ 0 & \lambda_2^n & 0 & \cdots & 0 \\ \vdots & \vdots & \vdots & & \vdots \\ 0 & 0 & 0 & \cdots & \lambda_k^n \end{bmatrix}$$

The diagonalization of M is accomplished by finding its eigenvalues and corresponding eigenvectors. These are as follows (verify):

Eigenvalues: $\qquad \lambda_1 = 1, \qquad \lambda_2 = \tfrac{1}{2}, \qquad \lambda_3 = 0$

Corresponding eigenvectors: $\quad \mathbf{v}_1 = \begin{bmatrix} 1 \\ 0 \\ 0 \end{bmatrix}, \qquad \mathbf{v}_2 = \begin{bmatrix} 1 \\ -1 \\ 0 \end{bmatrix}, \qquad \mathbf{v}_3 = \begin{bmatrix} 1 \\ -2 \\ 1 \end{bmatrix}$

Thus, in Equation (4) we have

$$D = \begin{bmatrix} \lambda_1 & 0 & 0 \\ 0 & \lambda_2 & 0 \\ 0 & 0 & \lambda_2 \end{bmatrix} = \begin{bmatrix} 1 & 0 & 0 \\ 0 & \tfrac{1}{2} & 0 \\ 0 & 0 & 0 \end{bmatrix}$$

and

$$P = [\mathbf{v}_1 \mid \mathbf{v}_2 \mid \mathbf{v}_3] = \begin{bmatrix} 1 & 1 & 1 \\ 0 & -1 & -2 \\ 0 & 0 & 1 \end{bmatrix}$$

Therefore,

$$\mathbf{x}^{(n)} = PD^nP^{-1}\mathbf{x}^{(0)} = \begin{bmatrix} 1 & 1 & 1 \\ 0 & -1 & -2 \\ 0 & 0 & 1 \end{bmatrix} \begin{bmatrix} 1 & 0 & 0 \\ 0 & (\tfrac{1}{2})^n & 0 \\ 0 & 0 & 0 \end{bmatrix} \begin{bmatrix} 1 & 1 & 1 \\ 0 & -1 & -2 \\ 0 & 0 & 1 \end{bmatrix} \begin{bmatrix} a_0 \\ b_0 \\ c_0 \end{bmatrix}$$

or

$$\mathbf{x}^{(n)} = \begin{bmatrix} a_n \\ b_n \\ c_n \end{bmatrix} = \begin{bmatrix} 1 & 1 - (\frac{1}{2})^n & 1 - (\frac{1}{2})^{n-1} \\ 0 & (\frac{1}{2})^n & (\frac{1}{2})^{n-1} \\ 0 & 0 & 0 \end{bmatrix} \begin{bmatrix} a_0 \\ b_0 \\ c_0 \end{bmatrix}$$

$$= \begin{bmatrix} a_0 + b_0 + c_0 - (\frac{1}{2})^n b_0 - (\frac{1}{2})^{n-1} c_0 \\ (\frac{1}{2})^n b_0 + (\frac{1}{2})^{n-1} c_0 \\ 0 \end{bmatrix}$$

Using the fact that $a_0 + b_0 + c_0 = 1$, we thus have

$$\begin{aligned} a_n &= 1 - (\tfrac{1}{2})^n b_0 - (\tfrac{1}{2})^{n-1} c_0 \\ b_n &= (\tfrac{1}{2})^n b_0 + (\tfrac{1}{2})^{n-1} c_0 \qquad n = 1, 2, \dots \\ c_n &= 0 \end{aligned} \qquad (5)$$

These are explicit formulas for the fractions of the three genotypes in the nth generation of plants in terms of the initial genotype fractions.

Because $(\frac{1}{2})^n$ tends to zero as n approaches infinity, it follows from these equations that

$$\begin{aligned} a_n &\to 1 \\ b_n &\to 0 \\ c_n &= 0 \end{aligned}$$

as n approaches infinity. That is, in the limit all plants in the population will be genotype AA. ▲

Example 2   We can modify Example 1 so that instead of fertilizing each plant with one of genotype AA, each plant is fertilized with a plant of its own genotype. Using the same notation as in Example 1, we then find

$$\mathbf{x}^{(n)} = M^n \mathbf{x}^{(0)}$$

where

$$M = \begin{bmatrix} 1 & \frac{1}{4} & 0 \\ 0 & \frac{1}{2} & 0 \\ 0 & \frac{1}{4} & 1 \end{bmatrix}$$

The columns of this new matrix M are the same as the columns of Table 1 corresponding to parents with genotypes AA–AA, Aa–Aa, and aa–aa.

The eigenvalues of M are (verify)

$$\lambda_1 = 1, \qquad \lambda_2 = 1, \qquad \lambda_3 = \tfrac{1}{2}$$

The eigenvalue $\lambda_1 = 1$ has multiplicity two and its corresponding eigenspace is two-dimensional. Picking two linearly independently eigenvectors $\mathbf{v}_1$ and $\mathbf{v}_2$ in that eigenspace, and a single eigenvector $\mathbf{v}_3$ for the simple eigenvalue $\lambda_3 = \frac{1}{2}$, we have (verify)

$$\mathbf{v}_1 = \begin{bmatrix} 1 \\ 0 \\ 0 \end{bmatrix}, \qquad \mathbf{v}_2 = \begin{bmatrix} 0 \\ 0 \\ 1 \end{bmatrix}, \qquad \mathbf{v}_3 = \begin{bmatrix} 1 \\ -2 \\ 1 \end{bmatrix}$$

The calculations for $\mathbf{x}^{(n)}$ are then

$$\mathbf{x}^{(n)} = M^n \mathbf{x}^{(0)} = PD^nP^{-1}\mathbf{x}^{(0)}$$

$$= \begin{bmatrix} 1 & 0 & 1 \\ 0 & 0 & -2 \\ 0 & 1 & 1 \end{bmatrix} \begin{bmatrix} 1 & 0 & 0 \\ 0 & 1 & 0 \\ 0 & 0 & (\frac{1}{2})^n \end{bmatrix} \begin{bmatrix} 1 & \frac{1}{2} & 0 \\ 0 & \frac{1}{2} & 1 \\ 0 & -\frac{1}{2} & 0 \end{bmatrix} \begin{bmatrix} a_0 \\ b_0 \\ c_0 \end{bmatrix}$$

$$= \begin{bmatrix} 1 & \frac{1}{2} - (\frac{1}{2})^{n+1} & 0 \\ 0 & (\frac{1}{2})^n & 0 \\ 0 & \frac{1}{2} - (\frac{1}{2})^{n+1} & 1 \end{bmatrix} \begin{bmatrix} a_0 \\ b_0 \\ c_0 \end{bmatrix}$$

Thus,

$$\begin{aligned} a_n &= a_0 + [\tfrac{1}{2} - (\tfrac{1}{2})^{n+1}]b_0 \\ b_n &= (\tfrac{1}{2})^n b_0 \qquad\qquad\qquad n = 1, 2, \ldots \\ c_n &= c_0 + [\tfrac{1}{2} - (\tfrac{1}{2})^{n+1}]b_0 \end{aligned} \qquad (6)$$

In the limit, as n tends to infinity, $(\frac{1}{2})^n \to 0$ and $(\frac{1}{2})^{n+1} \to 0$ so that

$$\begin{aligned} a_n &\to a_0 + \tfrac{1}{2}b_0 \\ b_n &\to 0 \\ c_n &\to c_0 + \tfrac{1}{2}b_0 \end{aligned}$$

Thus, fertilization of each plant with one of its own genotype produces a population that in the limit contains only genotypes AA and aa. ▲

AUTOSOMAL RECESSIVE DISEASES

There are many genetic diseases governed by autosomal inheritance in which a normal gene A dominates an abnormal gene a. Genotype AA is a normal individual; genotype Aa is a carrier of the disease but is not afflicted with the disease; and genotype aa is afflicted with the disease. In humans such genetic diseases are often associated with a particular racial group; for instance, cystic fibrosis (predominant among Caucasians), sickle-cell anemia (predominant among blacks), Cooley's anemia (predominant among people of Mediterranean origin), and Tay-Sachs disease (predominant among Eastern European Jews).

Suppose an animal breeder has a population of animals that carry an autosomal recessive disease. Suppose, further, that those animals afflicted with the disease do not survive to maturity. One possible way to control such a disease is for the breeder to always mate a female, regardless of her genotype, with a normal male. In this way, all future offspring will either have a normal father and a normal mother (AA–AA matings) or a normal father and a carrier mother (AA–Aa matings). There can be no AA–aa matings since animals of genotype aa do not survive to maturity. Under this type of mating program no future offspring will be afflicted with the disease, although there will still be carriers in future generations. Let us now determine the fraction of carriers in future generations. We set

$$\mathbf{x}^{(n)} = \begin{bmatrix} a_n \\ b_n \end{bmatrix}, \qquad n = 1, 2, \ldots$$

where

a_n = fraction of population of genotype AA in nth generation

b_n = fraction of population of genotype Aa (carriers) in nth generation

Because each offspring has at least one normal parent, we may consider the controlled mating program as one of continual mating with genotype AA, as in Example 1. Thus, the transition of genotype distributions from one generation to the next is governed by the equation

$$\mathbf{x}^{(n)} = M\mathbf{x}^{(n-1)}, \qquad n = 1, 2, \ldots$$

where

$$M = \begin{bmatrix} 1 & \frac{1}{2} \\ 0 & \frac{1}{2} \end{bmatrix}$$

Knowing the initial distribution $\mathbf{x}^{(0)}$, the distribution of genotypes in the nth generation is thus given by

$$\mathbf{x}^{(0)} = M^n\mathbf{x}^{(0)}, \qquad n = 1, 2, \ldots$$

The diagonalization of M is easily carried out (see Exercise 4) and leads to

$$\mathbf{x}^{(n)} = PD^nP^{-1}\mathbf{x}^{(0)} = \begin{bmatrix} 1 & 1 \\ 0 & -1 \end{bmatrix}\begin{bmatrix} 1 & 0 \\ 0 & (\frac{1}{2})^n \end{bmatrix}\begin{bmatrix} 1 & 1 \\ 0 & -1 \end{bmatrix}\begin{bmatrix} a_0 \\ b_0 \end{bmatrix}$$

$$= \begin{bmatrix} 1 & 1 - (\frac{1}{2})^n \\ 0 & (\frac{1}{2})^n \end{bmatrix}\begin{bmatrix} a_0 \\ b_0 \end{bmatrix} = \begin{bmatrix} a_0 + b_0 - (\frac{1}{2})^n b_0 \\ (\frac{1}{2})^n b_0 \end{bmatrix}$$

Because $a_0 + b_0 = 1$, we have

$$\begin{aligned} a_n &= 1 - (\tfrac{1}{2})^n b_0 \\ b_n &= (\tfrac{1}{2})^n b_0 \end{aligned} \qquad n = 1, 2, \ldots \tag{7}$$

Thus, as n tends to infinity we have

$$\begin{aligned} a_n &\to 1 \\ b_n &\to 0 \end{aligned}$$

so that in the limit there will be no carriers in the population.

From (7) we see that

$$b_n = \tfrac{1}{2}b_{n-1}, \qquad n = 1, 2, \ldots \tag{8}$$

That is, the fraction of carriers in each generation is one-half the fraction of carriers in the preceding generation. It would be of interest to also investigate the propagation of carriers under random mating, when two animals mate without regard to their genotypes. Unfortunately, such random mating leads to nonlinear equations and the techniques of this chapter are not applicable. However, by other techniques it can be shown that under random mating Equation (8) is replaced by

$$b_n = \frac{b_{n-1}}{1 + \frac{1}{2}b_{n-1}}, \qquad n = 1, 2, \ldots \tag{9}$$

As a numerical example, suppose the breeder starts with a population in which 10% of the animals are carriers. Under the controlled mating program governed by Equation (8), the percentage of carriers can be reduced to 5% in one generation. But under random mating, Equation (9) predicts that 9.5% of the population will be carriers after one generation (i.e., $b_n = .095$ if $b_{n-1} = .10$). In addition, under controlled mating no offspring will ever be afflicted with the disease, but with random mating it can be shown that about 1 in 400 offspring will be born with the disease when 10% of the population are carriers.

X-LINKED INHERITANCE

As mentioned in the introduction, in X-linked inheritance the male possesses one gene (*A* or *a*) and the female possesses two genes (*AA*, *Aa*, or *aa*). The term "X-linked" is used because such genes are found on the X-chromosome, of which the male has one and the female has two. The inheritance of such genes is as follows: A male offspring receives one of his mother's two genes with equal probability, and a female offspring receives the one gene of her father and one of her mother's two genes with equal probability. Readers familiar with basic probability can verify that this type of inheritance leads to the genotype probabilities in Table 2.

TABLE 2

			Genotypes of Parents (Father, Mother)					
			(*A, AA*)	(*A, Aa*)	(*A, aa*)	(*a, AA*)	(*a, Aa*)	(*a, aa*)
Offspring	Male	*A*	1	$\frac{1}{2}$	0	1	$\frac{1}{2}$	0
		a	0	$\frac{1}{2}$	1	0	$\frac{1}{2}$	1
	Female	*AA*	1	$\frac{1}{2}$	0	0	0	0
		Aa	0	$\frac{1}{2}$	1	1	$\frac{1}{2}$	0
		aa	0	0	0	0	$\frac{1}{2}$	1

We will discuss a program of inbreeding in connection with X-linked inheritance. We begin initially with a male and female; select two of their offspring at random, one of each gender, and mate them; select two of the resulting offspring and mate them; and so forth. Such inbreeding is commonly performed with animals. (Among humans, such brother-sister marriages were used by the rulers of ancient Egypt to keep the royal line pure.)

The original male-female pair can be one of the six types, corresponding to the six columns of Table 2:

$$(A, AA), \quad (A, Aa), \quad (A, aa), \quad (a, AA), \quad (a, Aa), \quad (a, aa)$$

The sibling-pairs mated in each successive generation have certain probabilities of being one of these six types. To compute these probabilities, for $n = 0, 1, 2, \ldots$, let us set

a_n = probability sibling-pair mated in nth generation is type (A, AA)
b_n = probability sibling-pair mated in nth generation is type (A, Aa)
c_n = probability sibling-pair mated in nth generation is type (A, aa)
d_n = probability sibling-pair mated in nth generation is type (a, AA)
e_n = probability sibling-pair mated in nth generation is type (a, AA)
f_n = probability sibling-pair mated in nth generation is type (a, aa)

With these probabilities we form a column vector

$$\mathbf{x}^{(n)} = \begin{bmatrix} a_n \\ b_n \\ c_n \\ d_n \\ e_n \\ f_n \end{bmatrix} \qquad n = 0, 1, 2, \ldots$$

From Table 2 it follows that

$$\mathbf{x}^{(n)} = M\mathbf{x}^{(n-1)}, \qquad n = 1, 2, \ldots \tag{10}$$

where

$$
M = \begin{array}{c} \\ \end{array}
\begin{array}{cccccc}
(A, AA) & (A, Aa) & (A, aa) & (a, AA) & (a, Aa) & (a, aa) \\
\end{array}
$$

	(A, AA)	(A, Aa)	(A, aa)	(a, AA)	(a, Aa)	(a, aa)	
	1	$\frac{1}{4}$	0	0	0	0	(A, AA)
	0	$\frac{1}{4}$	0	1	$\frac{1}{4}$	0	(A, Aa)
$M =$	0	0	0	0	$\frac{1}{4}$	0	(A, aa)
	0	$\frac{1}{4}$	0	0	0	0	(a, AA)
	0	$\frac{1}{4}$	1	0	$\frac{1}{4}$	0	(a, Aa)
	0	0	0	0	$\frac{1}{4}$	1	(a, aa)

For example, suppose in the $(n-1)$-st generation the sibling-pair mated is type (A, Aa). Then their male offspring will be either genotype A or a with equal probability, and their female offspring will be either genotype AA or Aa with equal probability. Because one of the male offspring and one of the female offspring are chosen at random for mating, the next sibling-pair will be one of type (A, AA), (A, Aa), (a, AA), (a, Aa), with equal probability. Thus, the second column of M contains "$\frac{1}{4}$" in each of the four rows corresponding to these four sibling-pairs. (See Exercise 9 for the remaining columns.)

As in our previous examples, it follows from (10) that

$$\mathbf{x}^{(n)} = M^n \mathbf{x}^{(0)}, \qquad n = 1, 2, \ldots \tag{11}$$

After lengthy calculations, the eigenvalues and eigenvectors of M turn out to be

$$\lambda_1 = 1, \quad \lambda_2 = 1, \quad \lambda_3 = \tfrac{1}{2}, \quad \lambda_4 = -\tfrac{1}{2}, \quad \lambda_5 = \tfrac{1}{4}(1 + \sqrt{5}), \quad \lambda_6 = \tfrac{1}{4}(1 - \sqrt{5})$$

$$
\mathbf{v}_1 = \begin{bmatrix} 1 \\ 0 \\ 0 \\ 0 \\ 0 \\ 0 \end{bmatrix}, \qquad
\mathbf{v}_2 = \begin{bmatrix} 0 \\ 0 \\ 0 \\ 0 \\ 0 \\ 1 \end{bmatrix}, \qquad
\mathbf{v}_3 = \begin{bmatrix} -1 \\ 2 \\ -1 \\ 1 \\ -2 \\ 1 \end{bmatrix}, \qquad
\mathbf{v}_4 = \begin{bmatrix} 1 \\ -6 \\ -3 \\ 3 \\ 6 \\ -1 \end{bmatrix},
$$

$$
\mathbf{v}_5 = \begin{bmatrix} \tfrac{1}{4}(-3 - \sqrt{5}) \\ 1 \\ \tfrac{1}{4}(-1 + \sqrt{5}) \\ \tfrac{1}{4}(-1 + \sqrt{5}) \\ 1 \\ \tfrac{1}{4}(-3 - \sqrt{5}) \end{bmatrix}, \qquad
\mathbf{v}_6 = \begin{bmatrix} \tfrac{1}{4}(-3 + \sqrt{5}) \\ 1 \\ \tfrac{1}{4}(-1 - \sqrt{5}) \\ \tfrac{1}{4}(-1 - \sqrt{5}) \\ 1 \\ \tfrac{1}{4}(-3 + \sqrt{5}) \end{bmatrix}
$$

The diagonalization of M then leads to

$$\mathbf{x}^{(n)} = PD^nP^{-1}\mathbf{x}^{(0)}, \qquad n = 1, 2, \ldots \tag{12}$$

where

$$
P = \begin{bmatrix}
1 & 0 & -1 & 1 & \tfrac{1}{4}(-3 - \sqrt{5}) & \tfrac{1}{4}(-3 + \sqrt{5}) \\
0 & 0 & 2 & -6 & 1 & 1 \\
0 & 0 & -1 & -3 & \tfrac{1}{4}(-1 + \sqrt{5}) & \tfrac{1}{4}(-1 - \sqrt{5}) \\
0 & 0 & 1 & 3 & \tfrac{1}{4}(-1 + \sqrt{5}) & \tfrac{1}{4}(-1 - \sqrt{5}) \\
0 & 0 & -2 & 6 & 1 & 1 \\
0 & 1 & 1 & -1 & \tfrac{1}{4}(-3 - \sqrt{5}) & \tfrac{1}{4}(-3 + \sqrt{5})
\end{bmatrix}
$$

$$
D^n = \begin{bmatrix}
1 & 0 & 0 & 0 & 0 & 0 \\
0 & 1 & 0 & 0 & 0 & 0 \\
0 & 0 & (\tfrac{1}{2})^n & 0 & 0 & 0 \\
0 & 0 & 0 & (-\tfrac{1}{2})^n & 0 & 0 \\
0 & 0 & 0 & 0 & [\tfrac{1}{4}(1 + \sqrt{5})]^n & 0 \\
0 & 0 & 0 & 0 & 0 & [\tfrac{1}{4}(1 - \sqrt{5})]^n
\end{bmatrix}
$$

$$
P^{-1} = \begin{bmatrix}
1 & \tfrac{2}{3} & \tfrac{1}{3} & \tfrac{2}{3} & \tfrac{1}{3} & 0 \\
0 & \tfrac{1}{3} & \tfrac{2}{3} & \tfrac{1}{3} & \tfrac{2}{3} & 1 \\
0 & \tfrac{1}{8} & -\tfrac{1}{4} & \tfrac{1}{4} & -\tfrac{1}{8} & 0 \\
0 & -\tfrac{1}{24} & -\tfrac{1}{12} & \tfrac{1}{12} & \tfrac{1}{24} & 0 \\
0 & \tfrac{1}{20}(5 + \sqrt{5}) & \tfrac{1}{5}\sqrt{5} & \tfrac{1}{5}\sqrt{5} & \tfrac{1}{20}(5 + \sqrt{5}) & 0 \\
0 & \tfrac{1}{20}(5 + \sqrt{5}) & -\tfrac{1}{5}\sqrt{5} & -\tfrac{1}{5}\sqrt{5} & \tfrac{1}{20}(5 - \sqrt{5}) & 0
\end{bmatrix}
$$

We will not write out the matrix product in (12), as it is rather unwieldy. However, if a specific vector $\mathbf{x}^{(0)}$ is given the calculation for $\mathbf{x}^{(n)}$ is not too cumbersome (see Exercise 6.)

Because the absolute values of the last four diagonal entries of D are less than 1, we see that as n tends to infinity

$$
D^n \to
\begin{bmatrix}
1 & 0 & 0 & 0 & 0 & 0 \\
0 & 1 & 0 & 0 & 0 & 0 \\
0 & 0 & 0 & 0 & 0 & 0 \\
0 & 0 & 0 & 0 & 0 & 0 \\
0 & 0 & 0 & 0 & 0 & 0 \\
0 & 0 & 0 & 0 & 0 & 0
\end{bmatrix}
$$

And so from Equation (12)

$$
\mathbf{x}^{(n)} \to P
\begin{bmatrix}
1 & 0 & 0 & 0 & 0 & 0 \\
0 & 1 & 0 & 0 & 0 & 0 \\
0 & 0 & 0 & 0 & 0 & 0 \\
0 & 0 & 0 & 0 & 0 & 0 \\
0 & 0 & 0 & 0 & 0 & 0 \\
0 & 0 & 0 & 0 & 0 & 0
\end{bmatrix}
P^{-1}\mathbf{x}^{(0)}
$$

Performing the matrix multiplication on the right we obtain (verify)

$$
\mathbf{x}^{(n)} \to
\begin{bmatrix}
a_0 + \frac{2}{3}b_0 + \frac{1}{3}c_0 + \frac{2}{3}d_0 + \frac{1}{3}e_0 \\
0 \\
0 \\
0 \\
0 \\
f_0 + \frac{1}{3}b_0 + \frac{2}{3}c_0 + \frac{1}{3}d_0 + \frac{2}{3}e_0
\end{bmatrix}
\tag{13}
$$

That is, in the limit all sibling-pairs will be either type (A, AA) or type (a, aa). For example, if the initial parents are type (A, Aa) (i.e., $b_0 = 1$ and $a_0 = c_0 = d_0 = e_0 = f_0 = 0$), then as n tends to infinity

$$
\mathbf{x}^{(n)} \to
\begin{bmatrix}
\frac{2}{3} \\
0 \\
0 \\
0 \\
0 \\
\frac{1}{3}
\end{bmatrix}
$$

Thus, in the limit there is probability $\frac{2}{3}$ that the sibling-pairs will all be (A, AA), and probability $\frac{1}{3}$ that they will all be (a, aa).

EXERCISE SET 11.12

1. Show that if $M = PDP^{-1}$, then $M^n = PD^nP^{-1}$ for $n = 1, 2, \ldots$.

2. In Example 1 suppose the plants are always fertilized with a plant of genotype Aa rather than one of genotype AA. Derive formulas for the fractions of the plants of genotypes AA, Aa, and aa in the nth generation. Also, find the limiting genotype distribution as n tends to infinity.

3. In Example 1 suppose the initial plants are fertilized with genotype AA, the first generation is fertilized with genotype Aa, the second generation is fertilized with genotype AA, and this alternating pattern of fertilization is kept up. Find formulas for the fractions of the plants of genotypes AA, Aa, and aa in the nth generation.

4. In the section on autosomal recessive diseases, find the eigenvalues and eigenvectors of the matrix M and verify Equation (7).

5. Suppose a breeder has an animal population in which 25% of the population are carriers of an autosomal recessive disease. If the breeder allows the animals to mate irrespective of their genotype, use Equation (9) to calculate the number of generations required for the percentage of carriers to fall from 25% to 10%. If the breeder instead implements the controlled mating program determined by Equation (8), what will the percentage of carriers be after the same number of generations?

6. In the section on X-linked inheritance, suppose the initial parents are equally likely to be of any of the six possible genotype parents; that is,

$$\mathbf{x}^{(0)} = \begin{bmatrix} \frac{1}{6} \\ \frac{1}{6} \\ \frac{1}{6} \\ \frac{1}{6} \\ \frac{1}{6} \\ \frac{1}{6} \end{bmatrix}$$

Using Equation (12) calculate $\mathbf{x}^{(n)}$ and also calculate the limit of $\mathbf{x}^{(n)}$ as n tends to infinity.

7. From (13) show that under X-linked inheritance with inbreeding, the probability that the limiting sibling-pairs will be of type (A, AA) is the same as the proportion of A genes in the initial population.

8. In X-linked inheritance suppose none of the females of genotype Aa survive to maturity. Under inbreeding the possible sibling-pairs are then

$$(A, AA), \quad (A, aa), \quad (a, AA), \quad \text{and} \quad (a, aa)$$

Find the transition matrix that describes how the genotype distribution changes in one generation.

9. Derive the matrix M in Equation (10) from Table 2.

11.13   AGE-SPECIFIC POPULATION GROWTH

Using the Leslie matrix model, this section investigates the growth in time of a female population divided into age classes. The limiting age distribution and growth rate of the population are determined.

PREREQUISITES:	Eigenvalues and eigenvectors
	Diagonalization of a matrix
	Intuitive understanding of limits

One of the most common models of population growth used by demographers is the so-called Leslie model, developed in the 1940s. This model describes the growth of the female portion of a human or animal population. In this model the females are divided into age classes of equal duration. To be specific, suppose the maximum age attained by any female in the population is L years (or some other time unit) and we divide the population into n age classes. Then each class is L/n years in duration. We label the age classes according to Table 1.

TABLE 1

Age Class	Age Interval
1	$[0, L/n)$
2	$[L/n, 2L/n)$
3	$[2L/n, 3L/n)$
$\vdots$	$\vdots$
$n - 1$	$[(n - 2)L/n, (n - 1)L/n]$
n	$[(n - 1)L/n, L]$

Suppose we know the number of females in each of the n classes at time $t = 0$. In particular, let there be $x_1^{(0)}$ females in the first class, $x_2^{(0)}$ females in the second class, and so forth. With these n numbers we form a column vector:

$$\mathbf{x}^{(0)} = \begin{bmatrix} x_1^{(0)} \\ x_2^{(0)} \\ \vdots \\ x_4^{(0)} \end{bmatrix}$$

We call this vector the ***initial age distribution vector***.

As time progresses, the number of females within each of the n classes changes because of three biological processes: birth, death, and aging. By describing these three processes quantitatively, we will see how to project the initial age distribution vector into the future.

The easiest way to study the aging process is to observe the population at discrete times, say, $t_0, t_1, t_2, \ldots, t_k, \ldots$. The Leslie model requires that the duration between any two successive observation times be the same as the duration of the age intervals. Therefore, we set

$$t_0 = 0$$
$$t_1 = L/n$$
$$t_2 = 2L/n$$
$$\vdots$$
$$t_k = kL/n$$
$$\vdots$$

With this assumption, all females in the $(i + 1)$-st class at time t_{k+1} were in the ith class at time t_k.

The birth and death processes between two successive observation times can be described by means of the following demographic parameters:

a_i $(i = 1, 2, \ldots, n)$	The average number of daughters born to a single female during the time she is in the ith age class.
b_i $(i = 1, 2, \ldots, n - 1)$	The fraction of females in the ith age class that can be expected to survive and pass into the $(i + 1)$-st age class.

By their definitions, we have that

(i) $a_i \geq 0$ for $i = 1, 2, \ldots, n$
(ii) $0 < b_i \leq 1$ for $i = 1, 2, \ldots, n - 1$

Notice that we do not allow any b_i to equal zero, as then no females will survive beyond the ith age class. We also assume that at least one a_i is positive so that some births occur. Any age class for which the corresponding value of a_i is positive is called a ***fertile age class***.

We next define the age distribution vector $\mathbf{x}^{(k)}$ at time t_k by

$$\mathbf{x}^{(k)} = \begin{bmatrix} x_1^{(k)} \\ x_2^{(k)} \\ \vdots \\ x_n^{(k)} \end{bmatrix}$$

where $x_i^{(k)}$ is the number of females in the ith age class at time t_k. Now, at time t_k, the females in the first age class are just those daughters born between times t_{k-1} and t_k. Thus, we can write

$$\left\{\begin{array}{c}\text{number of}\\\text{females}\\\text{in class 1}\\\text{at time } t_k\end{array}\right\} = \left\{\begin{array}{c}\text{number of}\\\text{daughters}\\\text{born to}\\\text{females in}\\\text{class 1}\\\text{between times}\\t_{k-1}\text{ and }t_k\end{array}\right\} + \left\{\begin{array}{c}\text{number of}\\\text{daughters}\\\text{born to}\\\text{females in}\\\text{class 2}\\\text{between times}\\t_{k-1}\text{ and }t_k\end{array}\right\} + \cdots + \left\{\begin{array}{c}\text{number of}\\\text{daughters}\\\text{born to}\\\text{females in}\\\text{class } n\\\text{between times}\\t_{k-1}\text{ and }t_k\end{array}\right\}$$

or, mathematically,

$$x_1^{(k)} = a_1 x_1^{(k-1)} + a_2 x_2^{(k-1)} + \cdots + a_n x_n^{(k-1)} \tag{1}$$

The females in the $(i+1)$-st age class $(i = 1, 2, \ldots, n-1)$ at time t_k are those females in the ith class at time t_{k-1} who are still alive at time t_k. Thus,

$$\left\{\begin{array}{c}\text{number of}\\\text{females in}\\\text{class } i + 1\\\text{at time } t_k\end{array}\right\} = \left\{\begin{array}{c}\text{fraction of}\\\text{females in}\\\text{class } i\\\text{who survive}\\\text{and pass into}\\\text{class } i + 1\end{array}\right\}\left\{\begin{array}{c}\text{number of}\\\text{females in}\\\text{class } i\\\text{at time } t_{k-1}\end{array}\right\}$$

or, mathematically,

$$x_{i+1}^{(k)} = b_i x_i^{(k-1)}, \qquad i = 1, 2, \ldots, n - 1 \tag{2}$$

Using matrix notation, Equations (1) and (2) can be written

$$\begin{bmatrix} x_1^{(k)} \\ x_2^{(k)} \\ x_3^{(k)} \\ \vdots \\ x_n^{(k)} \end{bmatrix} = \begin{bmatrix} a_1 & a_2 & a_3 & \cdots & a_{n-1} & a_n \\ b_1 & 0 & 0 & \cdots & 0 & 0 \\ 0 & b_2 & 0 & \cdots & 0 & 0 \\ \vdots & \vdots & \vdots & & \vdots & \vdots \\ 0 & 0 & 0 & \cdots & b_{n-1} & 0 \end{bmatrix} \begin{bmatrix} x_1^{(k-1)} \\ x_2^{(k-1)} \\ x_3^{(k-1)} \\ \vdots \\ x_n^{(k-1)} \end{bmatrix}$$

or, more compactly,

$$\mathbf{x}^{(k)} = L\mathbf{x}^{(k-1)}, \qquad k = 1, 2, \ldots \tag{3}$$

where L is the **Leslie matrix**

$$L = \begin{bmatrix} a_1 & a_2 & a_3 & \cdots & a_{n-1} & a_n \\ b_1 & 0 & 0 & \cdots & 0 & 0 \\ 0 & b_2 & 0 & \cdots & 0 & 0 \\ \vdots & \vdots & \vdots & & \vdots & \vdots \\ 0 & 0 & 0 & \cdots & b_{n-1} & 0 \end{bmatrix} \tag{4}$$

From Equation (3) it follows that

$$\mathbf{x}^{(1)} = L\mathbf{x}^{(0)}$$
$$\mathbf{x}^{(2)} = L\mathbf{x}^{(1)} = L^2\mathbf{x}^{(0)}$$
$$\mathbf{x}^{(3)} = L\mathbf{x}^{(2)} = L^3\mathbf{x}^{(0)}$$
$$\vdots$$
$$\mathbf{x}^{(k)} = L\mathbf{x}^{(k-1)} = L^k\mathbf{x}^{(0)} \tag{5}$$

Thus, if we know the initial age distribution $\mathbf{x}^{(0)}$ and the Leslie matrix L, we can determine the female age distribution at any later time.

Example 1 Suppose the oldest age attained by the females in a certain animal population is 15 years and we divide the population into three age classes with equal durations of five years. Let the Leslie matrix for this population be

$$L = \begin{bmatrix} 0 & 4 & 3 \\ \frac{1}{2} & 0 & 0 \\ 0 & \frac{1}{4} & 0 \end{bmatrix}$$

If there are initially 1000 females in each of the three age classes, then from Equation (3) we have

$$\mathbf{x}^{(0)} = \begin{bmatrix} 1,000 \\ 1,000 \\ 1,000 \end{bmatrix}$$

$$\mathbf{x}^{(1)} = L\mathbf{x}^{(0)} = \begin{bmatrix} 0 & 4 & 3 \\ \frac{1}{2} & 0 & 0 \\ 0 & \frac{1}{4} & 0 \end{bmatrix} \begin{bmatrix} 1,000 \\ 1,000 \\ 1,000 \end{bmatrix} = \begin{bmatrix} 7,000 \\ 500 \\ 250 \end{bmatrix}$$

$$\mathbf{x}^{(2)} = L\mathbf{x}^{(1)} = \begin{bmatrix} 0 & 4 & 3 \\ \frac{1}{2} & 0 & 0 \\ 0 & \frac{1}{4} & 0 \end{bmatrix} \begin{bmatrix} 7,000 \\ 500 \\ 250 \end{bmatrix} = \begin{bmatrix} 2,750 \\ 3,500 \\ 125 \end{bmatrix}$$

$$\mathbf{x}^{(3)} = L\mathbf{x}^{(2)} = \begin{bmatrix} 0 & 4 & 3 \\ \frac{1}{2} & 0 & 0 \\ 0 & \frac{1}{4} & 0 \end{bmatrix} \begin{bmatrix} 2,750 \\ 3,500 \\ 125 \end{bmatrix} = \begin{bmatrix} 14,375 \\ 1,375 \\ 875 \end{bmatrix}$$

Thus, after 15 years there are 14,375 females between 0 and 5 years of age, 1375 females between 5 and 10 years of age, and 875 females between 10 and 15 years of age. ▲

LIMITING BEHAVIOR

Although Equation (5) gives the age distribution of the population at any time, it does not immediately give a general picture of the dynamics of the growth process. For this we need to investigate the eigenvalues and eigenvectors of the Leslie matrix. The eigenvalues of L are the roots of its characteristic polynomial. As we ask the reader to verify in Exercise 2, this characteristic polynomial is

$$p(\lambda) = |\lambda I - L|$$
$$= \lambda^n - a_1\lambda^{n-1} - a_2b_1\lambda^{n-2} - a_3b_1b_2\lambda^{n-3} \cdots - a_nb_1b_2 \cdots b_{n-1}$$

To analyze the roots of this polynomial, it will be convenient to introduce the function

$$q(\lambda) = \frac{a_1}{\lambda} + \frac{a_2b_1}{\lambda^2} + \frac{a_3b_1b_2}{\lambda^3} + \cdots + \frac{a_nb_1b_2 \cdots b_{n-1}}{\lambda^n} \tag{6}$$

Using this function, the characteristic equation $p(\lambda) = 0$ can be written (verify)

$$q(\lambda) = 1 \qquad \text{for } \lambda \neq 0 \tag{7}$$

Because all the a_i and b_i are nonnegative, we see that $q(\lambda)$ is monotonically decreasing for λ greater than zero. Furthermore, $q(\lambda)$ has a vertical asymptote at $\lambda = 0$ and approaches zero as $\lambda \to \infty$. Consequently, as Figure 1 indicates, there

Figure 1

is a unique λ, say $\lambda = \lambda_1$, such that $q(\lambda_1) = 1$. That is, the matrix L has a unique positive eigenvalue. It can also be shown (see Exercise 3) that λ_1 has multiplicity 1; that is, λ_1 is not a repeated root of the characteristic equation. Although we omit the computational details, the reader can verify that an eigenvector corresponding to λ_1 is

$$\mathbf{x}_1 = \begin{bmatrix} 1 \\ b_1/\lambda_1 \\ b_1b_2/\lambda_1^2 \\ b_1b_2b_3/\lambda_1^3 \\ \vdots \\ b_1b_2 \cdots b_{n-1}/\lambda_1^{n-1} \end{bmatrix} \tag{8}$$

Because λ_1 has multiplicity 1, its corresponding eigenspace has dimension 1 (Exercise 3), and so any eigenvector corresponding to it is some multiple of $\mathbf{x}_1$. We can summarize these results in the following theorem.

Theorem 11.13.1. *A Leslie matrix L has a unique positive eigenvalue λ_1. This eigenvalue has multiplicity 1 and an eigenvector $\mathbf{x}_1$ all of whose entries are positive.*

We will now show that the long-term behavior of the age distribution of the population is determined by the positive eigenvalue λ_1 and its eigenvector $\mathbf{x}_1$.

In Exercise 9 we ask the reader to prove the following result.

Theorem 11.13.2. *If λ_1 is the unique positive eigenvalue of a Leslie matrix L and λ_k is any other real or complex eigenvalue of L, then $|\lambda_k| \le \lambda_1$.*

For our purposes the conclusion in Theorem 11.13.2 is not strong enough; we need λ_1 to satisfy $|\lambda_k| < \lambda_1$. In this case λ_1 would be called the **dominant eigenvalue** of L. However, as the following example shows, not all Leslie matrices satisfy this condition.

Example 2   Let

$$L = \begin{bmatrix} 0 & 0 & 6 \\ \frac{1}{2} & 0 & 0 \\ 0 & \frac{1}{3} & 0 \end{bmatrix}$$

Then the characteristic polynomial of L is

$$p(\lambda) = |\lambda I - L| = \lambda^3 - 1$$

The eigenvalues of L are thus the solutions of $\lambda^3 = 1$; namely,

$$\lambda = 1, \qquad -\frac{1}{2} + \frac{\sqrt{3}}{2}\,i, \qquad -\frac{1}{2} - \frac{\sqrt{3}}{2}\,i$$

All three eigenvalues have absolute value 1, and so the unique positive eigenvalue $\lambda_1 = 1$ is not dominant. Note that this matrix has the property that $L^3 = I$. This means that for any choice of the initial age distribution $\mathbf{x}^{(0)}$, we have

$$\mathbf{x}^{(0)} = \mathbf{x}^{(3)} = \mathbf{x}^{(6)} = \cdots = \mathbf{x}^{(3k)} = \cdots$$

The age distribution vector thus oscillates with a period of three time units. Such oscillations (or **population waves**, as they are called) could not occur if λ_1 were dominant, as we will see below. ▲

It is beyond the scope of this book to discuss necessary and sufficient conditions for λ_1 to be a dominant eigenvalue. However, we will state the following sufficient condition without proof.

Theorem 11.13.3. *If two successive entries a_i and a_{i+1} in the first row of a Leslie matrix L are nonzero, then the positive eigenvalue of L is dominant.*

Thus, if the female population has two successive fertile age classes, then its Leslie matrix has a dominant eigenvalue. This is always the case for realistic populations if the duration of the age classes is sufficiently small. Notice that in Example 2 there is only one fertile age class (the third), and so the condition of Theorem 11.13.3 is not satisfied. In what follows, we always assume that the condition of Theorem 11.13.3 is satisfied.

Let us assume that L is diagonalizable. This is not really necessary for the conclusions we will draw, but it does simplify the arguments. In this case, L has n eigenvalues, $\lambda_1, \lambda_2, \ldots, \lambda_n$, not necessarily distinct, and n linearly independent eigenvectors, $\mathbf{x}_1, \mathbf{x}_2, \ldots, \mathbf{x}_n$, corresponding to them. In this listing we place the dominant eigenvalue λ_1 first. We construct a matrix P whose columns are the eigenvectors of L:

$$P = [\mathbf{x}_1 \mid \mathbf{x}_2 \mid \mathbf{x}_3 \mid \cdots \mid \mathbf{x}_n]$$

The diagonalization of L is then given by the equation

$$L = P \begin{bmatrix} \lambda_1 & 0 & 0 & \cdots & 0 \\ 0 & \lambda_2 & 0 & \cdots & 0 \\ \vdots & \vdots & \vdots & & \vdots \\ 0 & 0 & 0 & \cdots & \lambda_n \end{bmatrix} P^{-1}$$

From this it follows that

$$L^k = P \begin{bmatrix} \lambda_1^k & 0 & 0 & \cdots & 0 \\ 0 & \lambda_2^k & 0 & \cdots & 0 \\ \vdots & \vdots & \vdots & & \vdots \\ 0 & 0 & 0 & \cdots & \lambda_n^k \end{bmatrix} P^{-1}$$

for $k = 1, 2, \ldots$. For any initial age distribution vector $\mathbf{x}^{(0)}$ we then have

$$L^k \mathbf{x}^{(0)} = P \begin{bmatrix} \lambda_1^k & 0 & 0 & \cdots & 0 \\ 0 & \lambda_2^k & 0 & \cdots & 0 \\ \vdots & \vdots & \vdots & & \vdots \\ 0 & 0 & 0 & \cdots & \lambda_n^k \end{bmatrix} P^{-1} \mathbf{x}^{(0)}$$

for $k = 1, 2, \ldots$. Dividing both sides of this equation by λ_1^k and using the fact that $\mathbf{x}^{(k)} = L^k \mathbf{x}^{(0)}$, we have

$$\frac{1}{\lambda_1^k} \mathbf{x}^{(k)} = P \begin{bmatrix} 1 & 0 & 0 & \cdots & 0 \\ 0 & \left(\dfrac{\lambda_2}{\lambda_1}\right)^k & 0 & \cdots & 0 \\ \vdots & \vdots & \vdots & & \vdots \\ 0 & 0 & 0 & \cdots & \left(\dfrac{\lambda_n}{\lambda_1}\right)^k \end{bmatrix} P^{-1} \mathbf{x}^{(0)} \tag{9}$$

Because λ_1 is the dominant eigenvalue, we have $|\lambda_i/\lambda_1| < 1$ for $i = 2, 3, \ldots, n$. It follows that

$$(\lambda_i/\lambda_1)^k \to 0 \text{ as } k \to \infty \qquad \text{for } i = 2, 3, \ldots, n$$

Using this fact, we may take the limit of both sides of (9) to obtain

$$\lim_{k \to \infty} \left\{ \frac{1}{\lambda_1^k} \mathbf{x}^{(k)} \right\} = P \begin{bmatrix} 1 & 0 & 0 & \cdots & 0 \\ 0 & 0 & 0 & \cdots & 0 \\ \vdots & \vdots & \vdots & & \vdots \\ 0 & 0 & 0 & \cdots & 0 \end{bmatrix} P^{-1} \mathbf{x}^{(0)} \tag{10}$$

Let us denote the first entry of the column vector $P^{-1}\mathbf{x}^{(0)}$ by the constant c. As we ask the reader to show in Exercise 4, the right side of (10) can be written as $c\mathbf{x}_1$, where c is a positive constant that depends only on the initial age distribution vector $\mathbf{x}^{(0)}$. Thus, (10) becomes

$$\lim_{k \to \infty} \left\{ \frac{1}{\lambda_1^k} \mathbf{x}^{(k)} \right\} = c\mathbf{x}_1 \tag{11}$$

Equation (11) gives us the approximation

$$\mathbf{x}^{(k)} \simeq c\lambda_1^k \mathbf{x}_1 \tag{12}$$

for large values of k. From (12) we also have

$$\mathbf{x}^{(k-1)} \simeq c\lambda_1^{k-1} \mathbf{x}_1 \tag{13}$$

Comparing Equations (12) and (13) we see that

$$\mathbf{x}^{(k)} \simeq \lambda_1 \mathbf{x}^{(k-1)} \tag{14}$$

for large values of k. This means that for large values of time, each age distribution vector is a scalar multiple of the preceding age distribution vector, the scalar being the positive eigenvalue of the Leslie matrix. Consequently, the *proportion* of females in each of the age classes becomes constant. As we will see in the following example, these limiting proportions can be determined from the eigenvector $\mathbf{x}_1$.

Example 1   (Revisited)   The Leslie matrix in Example 1 was

$$L = \begin{bmatrix} 0 & 4 & 3 \\ \frac{1}{2} & 0 & 0 \\ 0 & \frac{1}{4} & 0 \end{bmatrix}$$

Its characteristic polynomial is $p(\lambda) = \lambda^3 - 2\lambda - \frac{3}{8}$, and the reader can verify that the positive eigenvalue is $\lambda_1 = \frac{3}{2}$. From (8) the corresponding eigenvector $\mathbf{x}_1$ is

$$\mathbf{x}_1 = \begin{bmatrix} 1 \\ b_1/\lambda_1 \\ b_1 b_2/\lambda_1^2 \end{bmatrix} = \begin{bmatrix} 1 \\ \frac{1}{2} \\ \frac{3}{2} \\ \frac{(\frac{1}{2})(\frac{1}{4})}{(\frac{3}{2})^2} \end{bmatrix} = \begin{bmatrix} 1 \\ \frac{1}{3} \\ \frac{1}{18} \end{bmatrix}$$

From (14) we have

$$\mathbf{x}^{(k)} \simeq \tfrac{3}{2}\mathbf{x}^{(k-1)}$$

for large values of k. Hence, every five years the number of females in each of the three classes will increase by about 50%, as will the total number of females in the population.

From (12) we have

$$\mathbf{x}^{(k)} \simeq c(\tfrac{3}{2})^k \begin{bmatrix} 1 \\ \tfrac{1}{3} \\ \tfrac{1}{18} \end{bmatrix}$$

Consequently, eventually the females will be distributed among the three age classes in the ratios $1:\tfrac{1}{3}:\tfrac{1}{18}$. This corresponds to a distribution of 72% of the females in the first age class, 24% of the females in the second age class, and 4% of the females in the third age class. ▲

Example 3 In this example we use birth and death parameters from the year 1965 for Canadian females. Because few women over 50 years of age bear children, we restrict ourselves to the portion of the female population between 0 and 50 years of age. The data are for 5-year age classes, so there are a total of 10 age classes. Rather than write out the 10×10 Leslie matrix in full, we list the birth and death parameters as follows:

Age Interval	a_i	b_i
$[0, 5)$	0.00000	0.99651
$[5, 10)$	0.00024	0.99820
$[10, 15)$	0.05861	0.99802
$[15, 20)$	0.28608	0.99729
$[20, 25)$	0.44791	0.99694
$[25, 30)$	0.36399	0.99621
$[30, 35)$	0.22259	0.99460
$[35, 40)$	0.10457	0.99184
$[40, 45)$	0.02826	0.98700
$[45, 50)$	0.00240	—

Using numerical techniques, the positive eigenvalue and corresponding eigenvector can be approximated by

$$\lambda_1 = 1.07622 \quad \text{and} \quad \mathbf{x}_1 = \begin{bmatrix} 1.00000 \\ 0.92594 \\ 0.85881 \\ 0.79641 \\ 0.73800 \\ 0.68364 \\ 0.63281 \\ 0.58482 \\ 0.53897 \\ 0.49429 \end{bmatrix}$$

Thus, if Canadian women continued to reproduce and die as they did in 1965, eventually every 5 years their numbers would increase by 7.622%. From the eigenvector x_1, we see that, in the limit, for every 100,000 females between 0 and 5 years of age, there will be 92,594 females between 5 and 10 years of age, 85,881 females between 10 and 15 years of age, and so forth. ▲

Let us look again at Equation (12), which gives the age distribution vector of the population for large times:

$$x^{(k)} \simeq c\lambda_1^k x_1 \tag{15}$$

Three cases arise according to the value of the positive eigenvalue λ_1:

    (i) The population is eventually increasing if $\lambda_1 > 1$

    (ii) The population is eventually decreasing if $\lambda_1 < 1$

    (iii) The population eventually stabilizes if $\lambda_1 = 1$

The case $\lambda_1 = 1$ is particularly interesting as it determines a population that has **zero population growth**. For any initial age distribution, the population approaches a limiting age distribution that is some multiple of the eigenvector x_1. From Equations (6) and (7) we see that $\lambda_1 = 1$ is an eigenvalue if and only if

$$a_1 + a_2 b_1 + a_3 b_1 b_2 + \cdots + a_n b_1 b_2 \cdots b_{n-1} = 1 \tag{16}$$

The expression

$$R = a_1 + a_2 b_1 + a_3 b_1 b_2 + \cdots + a_n b_1 b_2 \cdots b_{n-1} \tag{17}$$

is called the **net reproduction rate** of the population. (See Exercise 5 for a demographic interpretation of R.) Thus, we can say that a population has zero population growth if and only if its net reproduction rate is 1.

EXERCISE SET 11.13

1. Suppose a certain animal population is divided into two age classes and has a Leslie matrix

$$L = \begin{bmatrix} 1 & \frac{2}{3} \\ \frac{1}{2} & 0 \end{bmatrix}$$

(a) Calculate the positive eigenvalue λ_1 of L and the corresponding eigenvector x_1.

(b) Beginning with the initial age distribution vector

$$x^{(0)} = \begin{bmatrix} 100 \\ 0 \end{bmatrix}$$

    calculate $x^{(1)}, x^{(2)}, x^{(3)}, x^{(4)}$, and $x^{(5)}$, rounding off to the nearest integer when necessary.

(c) Calculate $x^{(6)}$ using the exact formula $x^{(6)} = Lx^{(5)}$ and the approximate formula $x^{(6)} \simeq \lambda_1 x^{(5)}$.

2. Find the characteristic polynomial of a general Leslie matrix given by Equation (4).

3. (a) Show that the positive eigenvalue λ_1 of a Leslie matrix is always simple. Recall that a root λ_0 of a polynomial $q(\lambda)$ is simple if and only if $q'(\lambda_0) \neq 0$.
   (b) Show that the eigenspace corresponding to λ_1 has dimension 1.

4. Show that the right side of Equation (10) is $c\mathbf{x}_1$ where c is the first entry of the column vector $P^{-1}\mathbf{x}^{(0)}$.

5. Show that the net reproduction rate R, defined by (17), can be interpreted as the average number of daughters born to a single female during her expected lifetime.

6. Show that a population is eventually decreasing if and only if its net reproduction rate is less than one. Similarly, show that a population is eventually increasing if and only if its net reproduction rate is greater than 1.

7. Calculate the net reproduction rate of the animal population in Example 1.

8. **(For readers with a hand calculator.)** Calculate the net reproduction rate of the Canadian female population in Example 3.

9. **(For readers who have read Sections 10.1–10.3.)** Prove Theorem 11.13.2. [**Hint.** Write $\lambda_k = re^{i\theta}$, substitute into (7), take the real parts of both sides, and show that $r \leq \lambda_1$.]

11.14 HARVESTING OF ANIMAL POPULATIONS

This section employs the Leslie matrix model of population growth to model the sustainable harvesting of an animal population and investigates the effect of harvesting different fractions of different age groups.

> **PREREQUISITE:** Section 11.13: Age-Specific Population Growth

Section 11.13 described the Leslie matrix model for the growth of a female population divided into discrete age classes. In this section, we investigate the effects of harvesting an animal population growing according to such a model. By *harvesting* we mean the removal of animals from the population. (The word "harvesting" is not necessarily a euphemism for "slaughtering"; the animals may be removed from the population for other purposes.)

We restrict ourselves to *sustainable harvesting policies*. By this we mean the following:

> **Definition.** A harvesting policy in which an animal population is periodically harvested is said to be **sustainable** if the yield of each harvest is the same and the age distribution of the population remaining after each harvest is the same.

Thus, the animal population is not depleted by a sustainable harvesting policy; only the excess growth is removed.

As in Section 11.13 we will discuss only the females of the population. If the number of males in each age class is equal to the number of females—a reasonable assumption for many populations—then our harvesting policies will also apply to the male portion of the population.

THE HARVESTING MODEL

Figure 1 illustrates the basic idea of the model. We begin with a population having a particular age distribution. It undergoes a growth period that will be described by the Leslie matrix. At the end of the growth period, a certain fraction of each age class is harvested in such a way that the unharvested population has the same age distribution as the original population. This cycle repeats after each harvest so that the yield is sustainable. The duration of the harvest is assumed to be short in comparison with the growth period so that any growth or change in the population during the harvest period can be neglected.

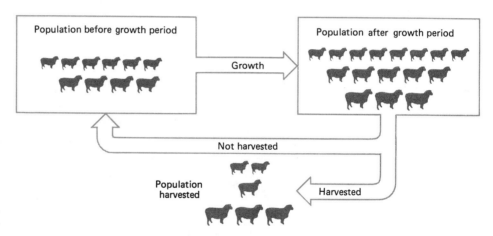

Figure 1

To describe this harvesting model mathematically, let

$$\mathbf{x} = \begin{bmatrix} x_1 \\ x_2 \\ \vdots \\ x_n \end{bmatrix}$$

be the age distribution vector of the population at the beginning of the growth period. Thus, x_i is the number of females in the ith class left unharvested. As in Section 11.13 we require that the duration of each age class be identical with the duration of the growth period. For example, if the population is harvested once a year, then the population is divided into 1-year age classes.

If L is the Leslie matrix describing the growth of the population, then the vector $L\mathbf{x}$ is the age distribution vector of the population at the end of the growth period, immediately before the periodic harvest. Let h_i, for $i = 1, 2, \ldots, n$, be the fraction of females from the ith class which is harvested. We use these n numbers to form an $n \times n$ diagonal matrix

$$H = \begin{bmatrix} h_1 & 0 & 0 & \cdots & 0 \\ 0 & h_2 & 0 & \cdots & 0 \\ 0 & 0 & h_3 & \cdots & 0 \\ \vdots & \vdots & \vdots & & \vdots \\ 0 & 0 & 0 & \cdots & h_n \end{bmatrix}$$

which we shall call the **harvesting matrix**. By definition, we have

$$0 \le h_i \le 1 \qquad (i - 1, 2, \ldots, n)$$

That is, we may harvest none ($h_i = 0$), all ($h_i = 1$), or some proper fraction ($0 < h_i < 1$) of each of the n classes. Because the number of females in the ith class immediately before each harvest is the ith entry $(L\mathbf{x})_i$ of the vector $L\mathbf{x}$, it can be seen that the ith entry of the column vector

$$HL\mathbf{x} = \begin{bmatrix} h_1(L\mathbf{x})_1 \\ h_2(L\mathbf{x})_2 \\ \vdots \\ h_n(L\mathbf{x})_n \end{bmatrix}$$

is the number of females harvested from the ith class.

From the definition of a sustainable harvesting policy, we have

$$\begin{bmatrix} \text{age distribution} \\ \text{at end of} \\ \text{growth period} \end{bmatrix} - [\text{harvest}] = \begin{bmatrix} \text{age distribution} \\ \text{at beginning of} \\ \text{growth period} \end{bmatrix}$$

or, mathematically,

$$L\mathbf{x} - HL\mathbf{x} = \mathbf{x} \tag{1}$$

If we write Equation (1) in the form

$$(I - H)L\mathbf{x} = \mathbf{x} \tag{2}$$

we see that $\mathbf{x}$ must be an eigenvector of the matrix $(I - H)L$ corresponding to the eigenvalue 1. As we will now show, this places certain restrictions on the values of h_i and $\mathbf{x}$.

Suppose the Leslie matrix of the population is

$$L = \begin{bmatrix} a_1 & a_2 & a_3 & \cdots & a_{n-1} & a_n \\ b_1 & 0 & 0 & \cdots & 0 & 0 \\ 0 & b_2 & 0 & \cdots & 0 & 0 \\ \vdots & \vdots & \vdots & & \vdots & \vdots \\ 0 & 0 & 0 & \cdots & b_{n-1} & 0 \end{bmatrix} \tag{3}$$

Then the matrix $(I - H)L$ is (verify)

$$(I - H)L = \begin{bmatrix} (1-h_1)a_1 & (1-h_1)a_2 & (1-h_1)a_3 & \cdots & (1-h_1)a_{n-1} & (1-h_1)a_n \\ (1-h_2)b_1 & 0 & 0 & \cdots & 0 & 0 \\ 0 & (1-h_3)b_2 & 0 & \cdots & 0 & 0 \\ \vdots & \vdots & \vdots & & \vdots & \vdots \\ 0 & 0 & 0 & \cdots & (1-h_n)b_{n-1} & 0 \end{bmatrix}$$

Thus, we see that $(I - H)L$ is a matrix with the same mathematical form as a Leslie matrix. In Section 11.13 we showed that a necessary and sufficient condition for a Leslie matrix to have 1 as an eigenvalue is that its net reproduction rate also be 1 [see Eq. (16) of Section 11.13]. Calculating the net reproduction rate of $(I - H)L$ and setting it equal to 1, we obtain (verify)

$$(1 - h_1)[a_1 + a_2 b_1(1 - h_2) + a_3 b_1 b_2(1 - h_2)(1 - h_3) + \cdots$$
$$+ a_n b_1 b_2 \cdots b_{n-1}(1 - h_2)(1 - h_3) \cdots (1 - h_n)] = 1 \qquad (4)$$

This equation places a restriction on the allowable harvesting fractions. Only those values of $h_1, h_2, \ldots, h_n$ which satisfy (4) and which lie in the interval $[0, 1]$ can produce a sustainable yield.

If $h_1, h_2, \ldots, h_n$ do satisfy (4), then the matrix $(I - H)L$ has the desired eigenvalue $\lambda_1 = 1$; and, furthermore, this eigenvalue has multiplicity 1, as the positive eigenvalue of a Leslie matrix always has multiplicity 1 (Theorem 11.13.1). This means that there is only one linearly independent eigenvector $\mathbf{x}$ satisfying Equation (2). [See Exercise 3(b) of Section 11.13.] One possible choice for $\mathbf{x}$ is the following normalized eigenvector:

$$\mathbf{x}_1 = \begin{bmatrix} 1 \\ b_1(1 - h_2) \\ b_1 b_2(1 - h_2)(1 - h_3) \\ b_1 b_2 b_3(1 - h_2)(1 - h_3)(1 - h_4) \\ \vdots \\ b_1 b_2 b_3 \cdots b_{n-1}(1 - h_2)(1 - h_3) \cdots (1 - h_n) \end{bmatrix} \qquad (5)$$

Any other solution $\mathbf{x}$ of (2) is a multiple of $\mathbf{x}_1$. Thus, the vector $\mathbf{x}_1$ determines the proportion of females within each of the n classes after a harvest under a sustainable harvesting policy. But there is an ambiguity in the total number of females in the population after each harvest. This can be determined by some auxilliary condition, such as an ecological or economic constraint. For example, for a population economically supported by the harvester, the largest population the harvester can afford to raise between harvests would determine the particular constant that $\mathbf{x}_1$ is multiplied by to produce the appropriate vector $\mathbf{x}$ in Equation (2). For a wild population the natural habitat of the population would determine how large the total population could be between harvests.

Summarizing our results so far, we see that there is a wide choice in the values of $h_1, h_2, \ldots, h_n$ that will produce a sustainable yield. But once these values are selected, the proportional age distribution of the population after each harvest is uniquely determined by the normalized eigenvector $\mathbf{x}_1$ defined by Equation (5). We now consider a few particular harvesting strategies of this type.

UNIFORM HARVESTING

With many populations it is difficult to distinguish or catch animals of specific ages. If animals are caught at random, we can reasonably assume the same fraction of each age class is harvested. We therefore set

$$h = h_1 = h_2 = \cdots = h_n$$

Equation (2) then reduces to (verify)

$$L\mathbf{x} = \left(\frac{1}{1-h}\right)\mathbf{x}$$

Hence, $1/(1-h)$ must be the unique positive eigenvalue λ_1 of the Leslie growth matrix L. That is,

$$\lambda_1 = \frac{1}{1-h}$$

Solving for the harvesting fraction h, we obtain

$$h = 1 - (1/\lambda_1) \tag{6}$$

The vector $\mathbf{x}_1$, in this case, is the same as the eigenvector of L corresponding to the eigenvalue λ_1. From Equation (8) of Section 11.13, this is

$$\mathbf{x}_1 = \begin{bmatrix} 1 \\ b_1/\lambda_1 \\ b_1 b_2/\lambda_1^2 \\ b_1 b_2 b_3/\lambda_1^3 \\ \vdots \\ b_1 b_2 \cdots b_{n-1}/\lambda_1^{n-1} \end{bmatrix} \tag{7}$$

From (6) we can see that the larger λ_1 is, the larger the fraction of animals we can harvest without depleting the population. Notice that we need $\lambda_1 > 1$ in order that the harvesting fraction h lie in the interval $(0, 1)$. This is to be expected, as $\lambda_1 > 1$ is the condition that the population be increasing.

Example 1 For a certain species of domestic sheep in New Zealand with a growth period of one year, the following Leslie matrix was found (see G. Caughley, "Parameters for Seasonally Breeding Populations," *Ecology*, vol. 48, 1967, pp. 834–839):

$$
L = \begin{bmatrix}
.000 & .045 & .391 & .472 & .484 & .546 & .543 & .502 & .468 & .459 & .433 & .421 \\
.845 & 0 & 0 & 0 & 0 & 0 & 0 & 0 & 0 & 0 & 0 & 0 \\
0 & .975 & 0 & 0 & 0 & 0 & 0 & 0 & 0 & 0 & 0 & 0 \\
0 & 0 & .965 & 0 & 0 & 0 & 0 & 0 & 0 & 0 & 0 & 0 \\
0 & 0 & 0 & .950 & 0 & 0 & 0 & 0 & 0 & 0 & 0 & 0 \\
0 & 0 & 0 & 0 & .926 & 0 & 0 & 0 & 0 & 0 & 0 & 0 \\
0 & 0 & 0 & 0 & 0 & .895 & 0 & 0 & 0 & 0 & 0 & 0 \\
0 & 0 & 0 & 0 & 0 & 0 & .850 & 0 & 0 & 0 & 0 & 0 \\
0 & 0 & 0 & 0 & 0 & 0 & 0 & .786 & 0 & 0 & 0 & 0 \\
0 & 0 & 0 & 0 & 0 & 0 & 0 & 0 & .691 & 0 & 0 & 0 \\
0 & 0 & 0 & 0 & 0 & 0 & 0 & 0 & 0 & .561 & 0 & 0 \\
0 & 0 & 0 & 0 & 0 & 0 & 0 & 0 & 0 & 0 & .370 & 0
\end{bmatrix}
$$

The sheep have a lifespan of 12 years so that they are divided into 12 age classes of duration 1 year each. By the use of numerical techniques, the unique positive eigenvalue of L can be found to be

$$\lambda_1 = 1.221$$

From Equation (6) the harvesting fraction h is

$$h = 1 - (1/\lambda_1) = 1 - (1/1.221) = 0.181$$

Thus, the uniform harvesting policy is one in which 18.1% of the sheep from each of the 12 age classes is harvested every year. From (7) the age distribution vector of the sheep after each harvest is proportional to

$$
\mathbf{x}_1 = \begin{bmatrix}
1.000 \\
0.692 \\
0.552 \\
0.436 \\
0.339 \\
0.257 \\
0.189 \\
0.131 \\
0.084 \\
0.048 \\
0.022 \\
0.007
\end{bmatrix}
\tag{8}
$$

From (8) we see that for every 1000 sheep between 0 and 1 year of age which are not harvested, there are 692 sheep between 1 and 2 years of age, 552 sheep between 2 and 3 years of age, and so forth. ▲

HARVESTING ONLY THE YOUNGEST AGE CLASS

In some populations only the youngest females are of any economic value, and so the harvester seeks to harvest only the females from the youngest age class. Accordingly, let us set

$$h_1 = h$$
$$h_2 = h_3 = \cdots = h_n = 0$$

Equation (4) then reduces to

$$(1 - h)(a_1 + a_2 b_1 + a_3 b_1 b_2 + \cdots + a_n b_1 b_2 \cdots b_{n-1}) = 1$$

or

$$(1 - h)R = 1$$

where R is the net reproduction rate of the population. [See Equation (17) of Section 11.13.] Solving for h, we obtain

$$h = 1 - (1/R) \tag{9}$$

Notice from this equation that a sustainable harvesting policy is possible only if $R > 1$. This is reasonable, as only if $R > 1$ is the population increasing. From Equation (5) the age distribution vector after each harvest is proportional to the vector

$$\mathbf{x}_1 = \begin{bmatrix} 1 \\ b_1 \\ b_1 b_2 \\ b_1 b_2 b_3 \\ \vdots \\ b_1 b_2 b_3 \cdots b_{n-1} \end{bmatrix} \tag{10}$$

Example 2 Let us apply this type of sustainable harvesting policy to the sheep population in Example 1. For the net reproduction rate of the population we find

$$R = a_1 + a_2 b_1 + a_3 b_1 b_2 + \cdots + a_n b_1 b_2 \cdots b_{n-1}$$
$$= (.000) + (.045)(.845) + \cdots + (.421)(.845)(.975) \cdots (.370)$$
$$= 2.513$$

From Equation (9) the fraction of the first age class harvested is

$$h = 1 - (1/R) = 1 - (1/2.513) = .602$$

From Equation (10) the age distribution of the sheep population after the harvest is proportional to the vector

$$
\mathbf{x}_1 = \begin{bmatrix} 1.000 \\ 0.845 \\ (.845)(.975) \\ (.845)(.975)(.965) \\ \vdots \\ (.845)(.975) \cdots (.370) \end{bmatrix} = \begin{bmatrix} 1.000 \\ 0.845 \\ 0.824 \\ 0.795 \\ 0.755 \\ 0.699 \\ 0.626 \\ 0.532 \\ 0.418 \\ 0.289 \\ 0.162 \\ 0.060 \end{bmatrix} \tag{11}
$$

A direct calculation gives us the following (see also Exercise 3):

$$
L\mathbf{x}_1 = \begin{bmatrix} 2.513 \\ 0.845 \\ 0.824 \\ 0.795 \\ 0.755 \\ 0.699 \\ 0.626 \\ 0.532 \\ 0.418 \\ 0.289 \\ 0.162 \\ 0.060 \end{bmatrix} \tag{12}
$$

The vector $L\mathbf{x}_1$ is the age distribution vector immediately before the harvest. The total of all entries in $L\mathbf{x}_1$ is 8.518, so that the first entry 2.513 is 29.5% of the total. This means that immediately before each harvest, 29.5% of the population is in the youngest age class. Since 60.2% of this class is harvested, it follows that 17.8% ($=60.2\%$ of 29.5%) of the entire sheep population is harvested each year. This can be compared with the uniform harvesting policy of Example 1, in which 18.1% of the sheep population is harvested each year. ▲

OPTIMAL SUSTAINABLE YIELD

We saw in Example 1 that a sustainable harvesting policy in which the same fraction of each age class is harvested produces a yield of 18.1% of the sheep population. In Example 2 we saw that if only the youngest age class is harvested, the resulting yield is 17.8% of the population. There are many other possible sustainable harvesting policies and each will generally provide a different yield. It would be of interest to find a sustainable harvesting policy that produces the largest possible yield. Such a policy is called an *optimal sustainable harvesting policy*, and the resulting yield is called the *optimal sustainable yield*. However, determining the optimal sustainable yield requires linear programming theory, which we will not discuss here. We refer the reader to the following result, which appears in J. R. Beddington and D. B. Taylor, "Optimum Age Specific Harvesting of a Population," *Biometrics*, vol. 29, 1973, pp. 801–809.

> **Theorem 11.14.1 (*Optimal Sustainable Yield*).** *An optimal sustainable harvesting policy is one in which either one or two age classes are harvested. If two age classes are harvested, then the older age class is completely harvested.*

As an illustration, it can be shown that the optimal sustainable yield of the sheep population is attained when

$$h_1 = 0.522$$
$$h_9 = 1.000$$

$$(13)$$

and all other values of h_i are zero. Thus, 52.2% of the sheep between 0 and 1 year of age and all the sheep between 8 and 9 years of age are harvested. As we ask the reader to show in Exercise 2, the resulting optimal sustainable yield is 19.9% of the population.

EXERCISE SET 11.14

1. Let a certain animal population be divided into three 1-year age classes and have as its Leslie matrix

$$L = \begin{bmatrix} 0 & 4 & 3 \\ \frac{1}{2} & 0 & 0 \\ 0 & \frac{1}{4} & 0 \end{bmatrix}$$

   (a) Find the yield and the age distribution vector after each harvest if the same fraction of each of the three age classes is harvested every year.
   (b) Find the yield and the age distribution vector after each harvest if only the youngest age class is harvested every year. Also, find the fraction of the youngest age class that is harvested.

2. For the optimal sustainable harvesting policy described by Equations (13), find the vector x_1 that specifies the age distribution of the population after each harvest. Also calculate the vector Lx_1 and verify that the optimal sustainable yield is 19.9% of the population.

3. If only the first age class of an animal population is harvested, use Equation (10) to show that

$$L\mathbf{x}_1 - \mathbf{x}_1 = \begin{bmatrix} R - 1 \\ 0 \\ 0 \\ \vdots \\ 0 \end{bmatrix}$$

   where R is the net reproduction rate of the population.

4. If only the Ith class of an animal population is to be periodically harvested ($I = 1, 2, \ldots, n$), find the corresponding harvesting fraction h_I.

5. Suppose all of the Jth class and a certain fraction h_I of the Ith class of an animal population is to be periodically harvested ($1 \leq I < J \leq n$). Calculate h_I.

11.15   A LEAST SQUARES MODEL FOR HUMAN HEARING

In this section the method of least squares approximation, motivated by energy considerations, is applied to a model for human hearing.

> **PREREQUISITES:**   Inner product spaces
> Orthogonal projection
> Fourier series (Section 8.4)

We begin with a brief discussion of the nature of sound and human hearing. Figure 1 is a schematic diagram of the ear showing its three main components: the outer ear, middle ear, and inner ear. Sound waves enter the outer ear where they are channeled to the eardrum, causing it to vibrate. Three tiny bones in the middle ear mechanically link the eardrum with the snail-shaped cochlea within the inner ear. These bones pass on the vibrations of the eardrum to a fluid within the cochlea. The cochlea contains thousands of minute hairs that oscillate with the fluid. Those near the entrance of the cochlea are stimulated by high frequencies and those near the tip are stimulated by low frequencies. The movements of these hairs activate nerve cells that send signals along various neural pathways to the brain, where the signals are interpreted as sound.

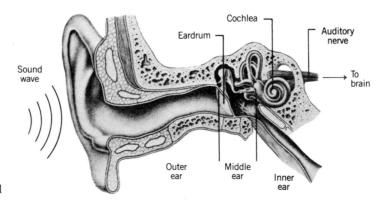

Figure 1

The sound waves themselves are variations in time of the air pressure. For the auditory system, the most elementary type of sound wave is a sinusoidal variation in the air pressure. This type of sound wave stimulates the hairs within the cochlea in such a way that a nerve impulse along a single neural pathway is produced (Figure 2). A sinusoidal sound wave can be described by a function of time

$$q(t) = A_0 + A \sin \omega(t - \delta) \tag{1}$$

where $q(t)$ is the atmospheric pressure at the eardrum, A_0 is the normal atmospheric pressure, A is the maximum deviation of the pressure from the normal atmospheric pressure, $\omega/2\pi$ is the frequency of the wave in cycles per second, and δ is the phase angle of the wave. To be perceived as sound, such sinusoidal waves must have frequencies within a certain range. For humans this range is roughly 20 cycles per second (cps) to 20,000 cps. Frequencies outside this range will not stimulate the hairs within the cochlea enough to produce nerve signals.

Figure 2

To a reasonable degree of accuracy, the ear is a linear system. This means that if a complex sound wave is a finite sum of sinusoidal components of different amplitudes, frequencies, and phase angles, say,

$$q(t) = A_0 + A_1 \sin \omega_1(t - \delta_1) + A_2 \sin \omega_2(t - \delta_2) + \cdots + A_n \sin \omega_n(t - \delta_n) \tag{2}$$

then the response of the ear consists of nerve impulses along the same neural pathways that would be stimulated by the individual components (Figure 3).

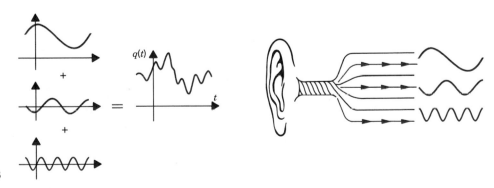

Figure 3

Let us now consider some periodic sound wave $p(t)$ with period T [i.e., $p(t) \equiv p(t + T)$] which is *not* a finite sum of sinusoidal waves. If we examine the response of the ear to such a periodic wave, we find that it is the same as the response to some wave that is the sum of sinusoidal waves. That is, there is some sound wave $q(t)$ as given by Equation (2) that produces the same response as $p(t)$, even though $p(t)$ and $q(t)$ are different functions of time.

We now want to determine the frequencies, amplitudes, and phase angles of the sinusoidal components of $q(t)$. Because $q(t)$ produces the same response as the periodic wave $p(t)$, it is reasonable to expect that $q(t)$ has the same period T as $p(t)$. This requires that each sinusoidal term in $q(t)$ have period T. Consequently, the frequencies of the sinusoidal components must be integer multiples of the basic frequency $1/T$ of the function $p(t)$. Thus, the ω_k in Equation (2) must be of the form

$$\omega_k = 2k\pi/T, \qquad k = 1, 2, \ldots$$

But because the ear cannot perceive sinusoidal waves with frequencies greater than 20,000 cps, we may omit those values of k for which $\omega_k/2\pi = k/T$ is greater than 20,000. Thus, $q(t)$ is of the form

$$q(t) = A_0 + A_1 \sin \frac{2\pi}{T} (t - \delta_1) + \cdots + A_n \sin \frac{2n\pi}{T} (t - \delta_n) \qquad (3)$$

where n is the largest integer such that n/T is not greater than 20,000.

We now turn our attention to the values of the amplitudes $A_0, A_1, \ldots, A_n$ and the phase angles $\delta_1, \delta_2, \ldots, \delta_n$ that appear in Equation (3). There is some criterion by which the auditory system "picks" these values so that $q(t)$ produces the same response as $p(t)$. To examine this criterion, let us set

$$e(t) = p(t) - q(t)$$

If we consider $q(t)$ as an approximation to $p(t)$, then $e(t)$ is the error in this approximation, an error that the ear cannot perceive. In terms of $e(t)$, the criterion for the determination of the amplitudes and the phase angles is that the quantity

$$\int_0^T [e(t)]^2 \, dt = \int_0^T [p(t) - q(t)]^2 \, dt \qquad (4)$$

be as small as possible. We cannot go into the physiological reasons for this, but we note that this expression is proportional to the *acoustic energy* of the error wave $e(t)$ over one period. In other words, it is the energy of the difference between the two sound waves $p(t)$ and $q(t)$ that determines whether the ear perceives any difference between them. If this energy is as small as possible, then the two waves produce the same sensation of sound. Mathematically, the function $q(t)$ in (4) is the least squares approximation to $p(t)$ from the vector space $C[0, T]$ of continuous functions on the interval $[0, T]$. (See Section 8.4.)

Least squares approximations by continuous functions arise in a wide variety of engineering and scientific approximation problems. Apart from the acoustics problem just discussed, some other examples are as follows.

1. Let $S(x)$ be the axial strain distribution in a uniform rod lying along the x-axis from $x = 0$ to $x = l$ (Figure 4). The strain energy in the rod is proportional to the integral

$$\int_0^l [S(x)]^2 \, dx$$

The closeness of an approximation $q(x)$ to $S(x)$ can be judged according to the strain energy of the difference of the two strain distributions. That energy is proportional to

$$\int_0^l [S(x) - q(x)]^2 \, dx$$

which is a least squares criterion.

Figure 4

2. Let $E(t)$ be a periodic voltage across a resistor in an electrical circuit (Figure 5). The electrical energy transferred to the resistor during one period T is proportional to

$$\int_0^T [E(t)]^2 \, dt$$

If $q(t)$ has the same period as $E(t)$ and is to be an approximation to $E(t)$, then the criterion of closeness might be taken as the energy of the difference voltage. This is proportional to

$$\int_0^T [E(t) - q(t)]^2 \, dt$$

which is again a least squares criterion.

Figure 5

3. Let $y(x)$ be the vertical displacement of a uniform flexible string whose equilibrium position is along the x-axis from $x = 0$ to $x = l$ (Figure 6). The elastic

Figure 6

potential energy of the string is proportional to

$$\int_0^l [y(x)]^2 \, dx$$

If $q(x)$ is to be an approximation to the displacement, then as before, the energy integral

$$\int_0^l [y(x) - q(x)]^2 \, dx$$

determines a least squares criterion for the closeness of the approximation.

Least squares approximation is also used in situations where there is no a priori justification for its use, such as for approximating business cycles, population growth curves, sales curves, and so forth. It is used in these cases because of its mathematical simplicity. In general, if no other error criterion is immediately apparent for an approximation problem, the least squares criterion is the one most often chosen.

The following result was obtained in Section 8.4.

Theorem 11.15.1. *If $f(t)$ is continuous on $[0, 2\pi]$, the trigonometric function $g(t)$ of the form*

$$g(t) = \tfrac{1}{2}a_0 + a_1 \cos t + \cdots + a_n \cos nt + b_1 \sin t + \cdots + b_n \sin nt$$

which minimizes the mean square error

$$\int_0^{2\pi} [f(t) - g(t)]^2 \, dt$$

has coefficients

$$a_k = \frac{1}{\pi} \int_0^{2\pi} f(t) \cos kt \, dt, \qquad k = 0, 1, 2, \ldots, n$$

$$b_k = \frac{1}{\pi} \int_0^{2\pi} f(t) \sin kt \, dt, \qquad k = 1, 2, \ldots, n$$

If the original function $f(t)$ is defined over the interval $[0, T]$ instead of $[0, 2\pi]$, a change of scale will yield the following result (see Exercise 8):

Theorem 11.15.2. *If $f(t)$ is continuous on $[0, T]$, the trigonometric function $g(t)$ of the form*

$$g(t) = \frac{1}{2}a_0 + a_1 \cos \frac{2\pi}{T} t + \cdots + a_n \cos \frac{2n\pi}{T} t$$

$$+ b_1 \sin \frac{2\pi}{T} t + \cdots + b_n \sin \frac{2n\pi}{T} t$$

which minimizes the mean square error

$$\int_0^T [f(t) - g(t)]^2 \, dt$$

has coefficients

$$a_k = \frac{2}{T} \int_0^T f(t) \cos \frac{2k\pi t}{T} \, dt, \qquad k = 0, 1, 2, \ldots, n$$

$$b_k = \frac{2}{T} \int_0^T f(t) \sin \frac{2k\pi t}{T} \, dt, \qquad k = 1, 2, \ldots, n$$

Example 1   Let a sound wave $p(t)$ have a saw-tooth pattern with a basic frequency of 5000 cps (Figure 7). Assume units are chosen so that the normal atmospheric pressure is at the zero level and the maximum amplitude of the wave is A. The basic period of the wave is $T = 1/5000 = .0002$ second. From $t = 0$ to $t = T$, the function $p(t)$ has the equation

$$p(t) = \frac{2A}{T} \left(\frac{T}{2} - t \right)$$

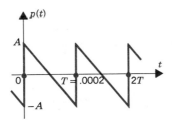

Figure 7

Theorem 11.15.2 then yields the following (verify):

$$a_0 = \frac{2}{T} \int_0^T p(t) \, dt = \frac{2}{T} \int_0^T \frac{2A}{T} \left(\frac{T}{2} - t \right) dt = 0$$

$$a_k = \frac{2}{T} \int_0^T p(t) \cos \frac{2k\pi t}{T} \, dt = \frac{2}{T} \int_0^T \frac{2A}{T} \left(\frac{T}{2} - t \right) \cos \frac{2k\pi t}{T} \, dt = 0, \quad k = 1, 2, \ldots$$

$$b_k = \frac{2}{T} \int_0^T p(t) \sin \frac{2k\pi t}{T} \, dt = \frac{2}{T} \int_0^T \frac{2A}{T} \left(\frac{T}{2} - t \right) \sin \frac{2k\pi t}{T} \, dt = \frac{2A}{k\pi}, \quad k = 1, 2, \ldots$$

We can now investigate how the sound wave $p(t)$ is perceived by the human ear. We notice that $4/T = 20{,}000$ cps, so that we need only go up to $k = 4$ in the above formulas. The least squares approximation to $p(t)$ is then

$$q(t) = \frac{2A}{\pi}\left[\sin\frac{2\pi}{T}t + \frac{1}{2}\sin\frac{4\pi}{T}t + \frac{1}{3}\sin\frac{6\pi}{T}t + \frac{1}{4}\sin\frac{8\pi}{T}t\right]$$

The four sinusoidal terms have frequencies of 5,000, 10,000, 15,000, and 20,000 cps, respectively. In Figure 8 we have plotted $p(t)$ and $q(t)$ over one period. Although $q(t)$ is not a very good point-by-point approximation to $p(t)$, to the ear, both $p(t)$ and $q(t)$ produce the same sensation of sound. ▲

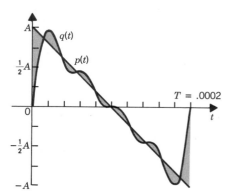

Figure 8

As discussed in Section 8.4, the least squares approximation becomes better as the number of terms in the approximating trigonometric polynomial becomes larger. More precisely,

$$\int_0^{2\pi}\left[f(t) - \frac{1}{2}a_0 - \sum_{k=1}^n (a_k\cos kt + b_k\sin kt)\right]^2 dt$$

tends to zero as n approaches infinity. We denote this by writing

$$f(t) = \frac{1}{2}a_0 + \sum_{k=1}^\infty (a_k\cos kt + b_k\sin kt)$$

where the right side of this equation is the Fourier series of $f(t)$. Whether the Fourier series of $f(t)$ converges to $f(t)$ for each t is another question, and a more difficult one. For most continuous functions encountered in applications, the Fourier series does indeed converge to its corresponding function for each value of t.

EXERCISE SET 11.15

1. Find the trigonometric polynomial of order three that is the least squares approximation to the function $f(t) = (t - \pi)^2$ over the interval $[0, 2\pi]$.

2. Find the trigonometric polynomial of order four that is the least squares approximation to the function $f(t) = t^2$ over the interval $[0, T]$.

3. Find the trigonometric polynomial of order four that is the least squares approximation to the function $f(t)$ over the interval $[0, 2\pi]$, where

$$f(t) = \begin{cases} \sin t & 0 \le t \le \pi \\ 0 & \pi < t \le 2\pi \end{cases}$$

4. Find the trigonometric polynomial of arbitrary order n that is the least squares approximation to the function $f(t) = \sin \frac{1}{2}t$ over the interval $[0, 2\pi]$.

5. Find the trigonometric polynomial of arbitrary order n that is the least squares approximation to the function $f(t)$ over the interval $[0, T]$, where

$$f(t) = \begin{cases} t & 0 \le t \le \frac{1}{2}T \\ T - t & \frac{1}{2}T < t \le T \end{cases}$$

6. Show that the $2n + 1$ functions

$$1, \cos t, \cos 2t, \ldots, \cos nt, \sin t, \sin 2t, \ldots, \sin nt$$

are orthogonal over the interval $[0, 2\pi]$ relative to the inner product $\langle \mathbf{u}, \mathbf{v} \rangle$ defined by Equation (8).

7. For the inner product

$$\langle \mathbf{u}, \mathbf{v} \rangle = \int_0^{2\pi} u(t)v(t)\, dt$$

show that
   (a) $\|1\| = \sqrt{2\pi}$ (b) $\|\cos kt\| = \sqrt{\pi}$ for $k = 1, 2, \ldots$ (c) $\|\sin kt\| = \sqrt{\pi}$ for $k = 1, 2, \ldots$

8. If $f(t)$ is defined and continuous on the interval $[0, T]$, show that $f(T\tau/2\pi)$ is defined and continuous for τ in the interval $[0, 2\pi]$. Use this fact to show how Theorem 11.15.2 follows from Theorem 11.15.1.

11.16 CUBIC SPLINE INTERPOLATION

In this section an artist's drafting aid is used as a physical model for the mathematical problem of finding a curve that passes through specified points in the plane. The parameters of the curve are determined by solving a linear system.

PREREQUISITES: Linear systems
Matrix algebra
Differential calculus

Figure 1

Figure 2

Fitting a curve through specified points in a plane is a common problem encountered in analyzing experimental data, in ascertaining the relations among variables, and in design work. In Figure 1 seven points in the xy-plane are displayed, and in Figure 2 a smooth curve has been drawn that passes through them. A curve that passes through a set of points in the plane is said to **interpolate** those points, and the curve is called an **interpolating curve** for those points. The interpolating curve in Figure 2 was drawn with the aid of a *drafting spline* (Figure 3). This drafting aid consists of a thin, flexible strip of wood or other material that is bent to pass through the points to be interpolated. Attached weights hold the spline in position while the artist draws the interpolating curve. The drafting spline will serve as the physical model for a mathematical theory of interpolation that we will discuss in this section.

Figure 3

STATEMENT OF THE PROBLEM

Suppose we are given n points in the xy-plane,

$$(x_1, y_1), (x_2, y_2), \ldots, (x_n, y_n)$$

which we wish to interpolate with a "well-behaved" curve (Figure 4). For convenience, we take the points to be equally spaced in the x-direction, although our results can easily be extended to the case of unequally spaced points. If we let the common distance between the x-coordinates of the points be h, then we have

$$x_2 - x_1 = x_3 - x_2 = \cdots = x_n - x_{n-1} = h$$

Let $y = S(x)$, $x_1 \leq x \leq x_n$ denote the interpolating curve that we seek. We assume that this curve describes the displacement of a drafting spline that interpolates the

Figure 4

n points when the weights holding down the spline are situated precisely at the n points. It is known from linear beam theory that the fourth derivative of the displacement of a beam is zero along any interval of the x-axis that contains no external forces acting on the beam. If we treat our drafting spline as a thin beam and realize that the only external forces acting on it arise from the weights at the n specified points, then it follows that

$$S^{(iv)}(x) \equiv 0 \tag{1}$$

for values of x lying in the $n - 1$ open intervals

$$(x_1, x_2), (x_2, x_3), \ldots, (x_{n-1}, x_n)$$

between the n points.

We also need the result from linear beam theory which states that for a beam acted upon only by external forces the displacement must have two continuous derivatives. In the case of the interpolating curve $y = S(x)$ constructed by the drafting spline, this means that $S(x)$, $S'(x)$, and $S''(x)$ must be continuous for $x_1 \leq x \leq x_n$.

The condition that $S''(x)$ be continuous is what causes a drafting spline to produce a pleasing curve, as it results in continuous *curvature*. The eye can perceive sudden changes in curvature—that is, discontinuities in $S''(x)$—but sudden changes in higher derivatives are not discernible. Thus, the condition that $S''(x)$ be continuous is the minimal prerequisite for the interpolating curve to be perceptible as a single smooth curve, rather than as a series of separate curves pieced together.

To determine the mathematical form of the function $S(x)$, we observe that because $S^{(iv)}(x) \equiv 0$ in the intervals between the n specified points, it follows by integrating this equation four times that $S(x)$ must be a *cubic polynomial* in x in each such interval. In general, however, $S(x)$ will be a different cubic polynomial in each interval, and so $S(x)$ must have the form

$$S(x) = \begin{cases} S_1(x), & x_1 \leq x \leq x_2 \\ S_2(x), & x_2 \leq x \leq x_3 \\ \vdots \\ S_{n-1}(x), & x_{n-1} \leq x \leq x_n \end{cases} \tag{2}$$

where $S_1(x), S_2(x), \ldots, S_{n-1}(x)$ are cubic polynomials. For convenience, we will write these in the form

$$\begin{aligned} S_1(x) &= a_1(x - x_1)^3 + b_1(x - x_1)^2 + c_1(x - x_1) + d_1, & x_1 \leq x \leq x_2 \\ S_2(x) &= a_2(x - x_2)^3 + b_2(x - x_2)^2 + c_2(x - x_2) + d_2, & x_2 \leq x \leq x_3 \\ &\vdots \\ S_{n-1}(x) &= a_{n-1}(x - x_{n-1})^3 + b_{n-1}(x - x_{n-1})^2 \\ & \qquad + c_{n-1}(x - x_{n-1}) + d_{n-1}, & x_{n-1} \leq x \leq x_n \end{aligned} \tag{3}$$

The a_i's, b_i's, c_i's, and d_i's constitute a total of $4n - 4$ coefficients that we must determine to completely specify $S(x)$. If we choose these coefficients so that $S(x)$ interpolates the n specified points in the plane and $S(x)$, $S'(x)$, and $S''(x)$ are continuous, then the resulting interpolating curve is called a ***cubic spline***.

DERIVATION OF THE FORMULA OF A CUBIC SPLINE

From Equations (2) and (3) we have

$$S(x) = S_1(x) = a_1(x - x_1)^3 + b_1(x - x_1)^2 + c_1(x - x_1) + d_1, \qquad x_1 \le x \le x_2$$
$$S(x) = S_2(x) = a_2(x - x_2)^3 + b_2(x - x_2)^2 + c_2(x - x_2) + d_2, \qquad x_2 \le x \le x_3$$
$$\vdots \qquad \vdots \tag{4}$$
$$S(x) = S_{n-1}(x) = a_{n-1}(x - x_{n-1})^3 + b_{n-1}(x - x_{n-1})^2$$
$$+ c_{n-1}(x - x_{n-1}) + d_{n-1}, \qquad x_{n-1} \le x \le x_n$$

and so

$$S'(x) = S_1'(x) = 3a_1(x - x_1)^2 + 2b_1(x - x_1) + c_1, \qquad x_1 \le x \le x_2$$
$$S'(x) = S_2'(x) = 3a_2(x - x_2)^3 + 2b_2(x - x_2) + c_2, \qquad x_2 \le x \le x_3$$
$$\vdots \qquad \vdots$$
$$S'(x) = S_{n-1}'(x) = 3a_{n-1}(x - x_{n-1})^2 + 2b_{n-1}(x - x_{n-1}) + c_{n-1}, \qquad x_{n-1} \le x \le x_n$$
$$\tag{5}$$

and

$$S''(x) = S_1''(x) = 6a_1(x - x_1) + 2b_1, \qquad x_1 \le x \le x_2$$
$$S''(x) = S_2''(x) = 6a_2(x - x_2) + 2b_2, \qquad x_2 \le x \le x_3$$
$$\vdots \qquad \vdots \tag{6}$$
$$S''(x) = S_{n-1}''(x) = 6a_{n-1}(x - x_{n-1}) + 2b_{n-1}, \qquad x_{n-1} \le x \le x_n$$

We will now use these equations and the four properties of cubic splines stated below to express the unknown coefficients $a_i, b_i, c_i, d_i, i = 1, 2, \ldots, n - 1$, in terms of the known coordinates $y_1, y_2, \ldots, y_n$.

1. $S(x)$ *interpolates the points* $(x_i, y_i), i = 1, 2, \ldots, n.$

Because $S(x)$ interpolates the points $(x_i, y_i), i = 1, 2, \ldots, n$, we have

$$S(x_1) = y_1, \ S(x_2) = y_2, \ldots, S(x_n) = y_n \tag{7}$$

From the first $n - 1$ of these equations and (4) we obtain

$$d_1 = y_1$$
$$d_2 = y_2$$
$$\vdots \tag{8}$$
$$d_{n-1} = y_{n-1}$$

From the last equation in (7), the last equation in (4), and the fact that $x_n - x_{n-1} = h$, we obtain

$$a_{n-1}h^3 + b_{n-1}h^2 + c_{n-1}h + d_{n-1} = y_n \tag{9}$$

2. $S(x)$ *is continuous on* $[x_1, x_n].$

Because $S(x)$ is continuous for $x_1 \le x \le x_n$, it follows that at each point x_i in the set $x_2, x_3, \ldots, x_{n-1}$ we must have

$$S_{i-1}(x_i) = S_i(x_i), \qquad i = 2, 3, \ldots, n - 1 \tag{10}$$

Otherwise, the graphs of $S_{i-1}(x)$ and $S_i(x)$ would not join together to form a continuous curve at x_i. Using the interpolating property $S_i(x_i) = y_i$, it follows from (10) that $S_{i-1}(x_i) = y_i$, $i = 2, 3, \ldots, n-1$, or from (4) that

$$
\begin{aligned}
a_1 h^3 + b_1 h^2 + c_1 h + d_1 &= y_2 \\
a_2 h^3 + b_2 h^2 + c_2 h + d_2 &= y_3 \\
&\;\;\vdots \\
a_{n-2} h^3 + b_{n-2} h^2 + c_{n-2} h + d_{n-2} &= y_{n-1}
\end{aligned}
\tag{11}
$$

3. $S'(x)$ *is continuous on* $[x_1, x_n]$.

Because $S'(x)$ is continuous for $x_1 \le x \le x_n$, it follows that

$$
S_{i-1}''(x_i) = S_i''(x_i), \qquad i = 2, 3, \ldots, n-1
$$

or from (5)

$$
\begin{aligned}
3a_1 h^2 + 2b_1 h + c_1 &= c_2 \\
3a_2 h^2 + 2b_2 h + c_2 &= c_3 \\
&\;\;\vdots \\
3a_{n-2} h^2 + 2b_{n-2} h + c_{n-2} &= c_{n-1}
\end{aligned}
\tag{12}
$$

4. $S''(x)$ *is continuous on* $[x_1, x_2]$.

Because $S''(x)$ is continuous for $x_1 \le x \le x_n$, it follows that

$$
S_{i-1}'''(x_i) = S_i''(x_i), \qquad i = 2, 3, \ldots, n-1
$$

or from (6)

$$
\begin{aligned}
6a_1 h + 2b_1 &= 2b_2 \\
6a_2 h + 2b_2 &= 2b_3 \\
&\;\;\vdots \\
6a_{n-2} h + 2b_{n-2} &= 2b_{n-1}
\end{aligned}
\tag{13}
$$

Equations (8), (9), (11), (12), and (13) constitute a system of $4n - 6$ linear equations in the $4n - 4$ unknown coefficients $a_i, b_i, c_i, d_i, i = 1, 2, \ldots, n-1$. Consequently, we need two more equations to determine these coefficients uniquely. Before obtaining these additional equations, however, we can simplify our existing system by expressing the unknowns a_i, b_i, c_i, and d_i in terms of new unknown quantities

$$
M_1 = S''(x_1), M_2 = S''(x_2), \ldots, M_n = S''(x_n)
$$

and the known quantities

$$
y_1, y_2, \ldots, y_n
$$

For example, from (6) it follows that

$$
\begin{aligned}
M_1 &= 2b_1 \\
M_2 &= 2b_2 \\
&\;\;\vdots \\
M_{n-1} &= 2b_{n-1}
\end{aligned}
$$

and so

$$b_1 = \tfrac{1}{2}M_1, b_2 = \tfrac{1}{2}M_2, \ldots, b_{n-1} = \tfrac{1}{2}M_{n-1}$$

Moreover, we already known from (8) that

$$d_1 = y_1, d_2 = y_2, \ldots, d_{n-1} = y_{n-1}$$

We leave it as an exercise for the reader to derive the expressions for the a_i's and c_i's in terms of the M_i's and y_i's. The final result is as follows:

Theorem 11.16.1. *Given n points $(x_1, y_1), (x_2, y_2), \ldots, (x_n, y_n)$ with $x_{i+1} - x_i = h$, $i = 1, 2, \ldots, n-1$, the cubic spline*

$$S(x) = \begin{cases} a_1(x - x_1)^3 + b_1(x - x_1)^2 + c_1(x - x_1) + d_1, & x_1 \leq x \leq x_2 \\ a_2(x - x_2)^3 + b_2(x - x_2)^2 + c_2(x - x_2) + d_2, & x_2 \leq x \leq x_3 \\ \vdots \quad\quad \vdots \\ a_{n-1}(x - x_{n-1})^3 + b_{n-1}(x - x_{n-1})^2 \\ \quad\quad + c_{n-1}(x - x_{n-1}) + d_{n-1}, & x_{n-1} \leq x \leq x_n \end{cases}$$

that interpolates these points has coefficients given by

$$\begin{aligned} a_i &= (M_{i+1} - M_i)/6h \\ b_i &= M_i/2 \\ c_i &= (y_{i+1} - y_i)/h - [(M_{i+1} + 2M_i)h/6] \\ d_i &= y_i \end{aligned} \quad\quad (14)$$

for $i = 1, 2, \ldots, n-1$, where $M_i = S''(x_i)$, $i = 1, 2, \ldots, n$.

From this result, we see that the quantities $M_1, M_2, \ldots, M_n$ uniquely determine the cubic spline. To find these quantities we substitute the expressions for a_i, b_i, and c_i given in (14) into (12). After some algebraic simplification, we obtain

$$\begin{aligned} M_1 + 4M_2 + M_3 &= 6(y_1 - 2y_2 + y_3)/h^2 \\ M_2 + 4M_3 + M_4 &= 6(y_2 - 2y_3 + y_4)/h^2 \\ &\vdots \\ M_{n-2} + 4M_{n-1} + M_n &= 6(y_{n-2} - 2y_{n-1} + y_n)/h^2 \end{aligned} \quad\quad (15)$$

or, in matrix form,

$$\begin{bmatrix} 1 & 4 & 1 & 0 & \cdots & 0 & 0 & 0 & 0 \\ 0 & 1 & 4 & 1 & \cdots & 0 & 0 & 0 & 0 \\ 0 & 0 & 1 & 4 & \cdots & 0 & 0 & 0 & 0 \\ \vdots & \vdots & \vdots & \vdots & & \vdots & \vdots & \vdots & \vdots \\ 0 & 0 & 0 & 0 & \cdots & 4 & 1 & 0 & 0 \\ 0 & 0 & 0 & 0 & \cdots & 1 & 4 & 1 & 0 \\ 0 & 0 & 0 & 0 & \cdots & 0 & 1 & 4 & 1 \end{bmatrix} \begin{bmatrix} M_1 \\ M_2 \\ M_3 \\ M_4 \\ \vdots \\ M_{n-3} \\ M_{n-2} \\ M_{n-1} \\ M_n \end{bmatrix} = \frac{6}{h^2} \begin{bmatrix} y_1 - 2y_2 + y_3 \\ y_2 - 2y_3 + y_4 \\ y_3 - 2y_4 + y_5 \\ \vdots \\ y_{n-4} - 2y_{n-3} + y_{n-2} \\ y_{n-3} - 2y_{n-2} + y_{n-1} \\ y_{n-2} - 2y_{n-1} + y_n \end{bmatrix}$$

This is a linear system of $n - 2$ equations for the n unknowns $M_1, M_2, \ldots, M_n$. Thus, we still need two additional equations to determine $M_1, M_2, \ldots, M_n$ uniquely. The reason for this is that there are infinitely many cubic splines that interpolate the given points, so that we simply do not have enough conditions to determine a unique cubic spline passing through the points. We discuss below three possible ways of specifying the two additional conditions required to obtain a unique cubic spline through the points. (The exercises present two more.) They are summarized in Table 1 shown on the following page.

THE NATURAL SPLINE

The two simplest mathematical conditions we can impose are

$$M_1 = M_n = 0$$

These conditions together with (15) result in an $n \times n$ linear system for $M_1, M_2, \ldots, M_n$, which can be written in matrix form as

$$
\begin{bmatrix}
1 & 0 & 0 & 0 & \cdots & 0 & 0 & 0 \\
1 & 4 & 1 & 0 & \cdots & 0 & 0 & 0 \\
0 & 1 & 4 & 1 & \cdots & 0 & 0 & 0 \\
\vdots & \vdots & \vdots & \vdots & & \vdots & \vdots & \vdots \\
0 & 0 & 0 & 0 & \cdots & 1 & 4 & 1 \\
0 & 0 & 0 & 0 & \cdots & 0 & 0 & 1
\end{bmatrix}
\begin{bmatrix}
M_1 \\ M_2 \\ M_3 \\ \vdots \\ M_{n-1} \\ M_n
\end{bmatrix}
= \frac{6}{h^2}
\begin{bmatrix}
0 \\
y_1 - 2y_2 + y_3 \\
y_2 - 2y_3 + y_4 \\
\vdots \\
y_{n-2} - 2y_{n-1} + y_n \\
0
\end{bmatrix}
$$

For numerical calculations it is more convenient to eliminate M_1 and M_n from this system and write

$$
\begin{bmatrix}
4 & 1 & 0 & 0 & \cdots & 0 & 0 & 0 \\
1 & 4 & 1 & 0 & \cdots & 0 & 0 & 0 \\
0 & 1 & 4 & 1 & \cdots & 0 & 0 & 0 \\
\vdots & \vdots & \vdots & \vdots & & \vdots & \vdots & \vdots \\
0 & 0 & 0 & 0 & \cdots & 1 & 4 & 1 \\
0 & 0 & 0 & 0 & \cdots & 0 & 1 & 4
\end{bmatrix}
\begin{bmatrix}
M_2 \\ M_3 \\ M_4 \\ \vdots \\ M_{n-2} \\ M_{n-1}
\end{bmatrix}
= \frac{6}{h^2}
\begin{bmatrix}
y_1 - 2y_2 + y_3 \\
y_2 - 2y_3 + y_4 \\
y_3 - 2y_4 + y_5 \\
\vdots \\
y_{n-3} - 2y_{n-2} + y_{n-1} \\
y_{n-2} - 2y_{n-1} + y_n
\end{bmatrix}
\tag{16}
$$

together with

$$M_1 = 0 \tag{17}$$
$$M_n = 0 \tag{18}$$

Thus, the $(n - 2) \times (n - 2)$ linear system can be solved for the $n - 2$ coefficients $M_2, M_3, \ldots, M_{n-1}$, and M_1 and M_n are determined by (17) and (18).

Physically, the natural spline results when the ends of a drafting spline extend freely beyond the interpolating points without constraint. The end portions of the spline outside the interpolating points will fall on straight line paths, causing $S''(x)$ to vanish at the end points x_1 and x_n and resulting in the mathematical conditions $M_1 = M_n = 0$.

The natural spline tends to flatten the interpolating curve at the endpoints, which may be undesirable. Of course, if it is required that $S''(x)$ vanish at the endpoints, then the natural spline must be used.

TABLE 1

NATURAL SPLINE	The second derivative of the spline is zero at the endpoints.	$M_1 = 0$ $M_n = 0$	$$\begin{bmatrix} 4 & 1 & 0 & \cdots & 0 & 0 & 0 \\ 1 & 4 & 1 & \cdots & 0 & 0 & 0 \\ \vdots & & & \cdots & & & \vdots \\ 0 & 0 & 0 & \cdots & 1 & 4 & 1 \\ 0 & 0 & 0 & \cdots & 0 & 1 & 4 \end{bmatrix} \begin{bmatrix} M_2 \\ M_3 \\ \vdots \\ M_{n-2} \\ M_{n-1} \end{bmatrix} = \frac{6}{h^2} \begin{bmatrix} y_1 - 2y_2 + y_3 \\ y_2 - 2y_3 + y_4 \\ \vdots \\ y_{n-2} - 2y_{n-1} + y_n \end{bmatrix}$$
PARABOLIC RUNOUT SPLINE	The spline reduces to a parabolic curve on the first and last intervals.	$M_1 = M_2$ $M_n = M_{n-1}$	$$\begin{bmatrix} 5 & 1 & 0 & \cdots & 0 & 0 & 0 \\ 1 & 4 & 1 & \cdots & 0 & 0 & 0 \\ \vdots & & & \cdots & & & \vdots \\ 0 & 0 & 0 & \cdots & 1 & 4 & 1 \\ 0 & 0 & 0 & \cdots & 0 & 1 & 5 \end{bmatrix} \begin{bmatrix} M_2 \\ M_3 \\ \vdots \\ M_{n-2} \\ M_{n-1} \end{bmatrix} = \frac{6}{h^2} \begin{bmatrix} y_1 - 2y_2 + y_3 \\ y_2 - 2y_3 + y_4 \\ \vdots \\ y_{n-2} - 2y_{n-1} + y_n \end{bmatrix}$$
CUBIC RUNOUT SPLINE	The spline is a single cubic curve on the first two and last two intervals.	$M_1 = 2M_2 - M_3$ $M_n = 2M_{n-1} - M_{n-2}$	$$\begin{bmatrix} 6 & 0 & 0 & \cdots & 0 & 0 & 0 \\ 1 & 4 & 1 & \cdots & 0 & 0 & 0 \\ \vdots & & & \cdots & & & \vdots \\ 0 & 0 & 0 & \cdots & 1 & 4 & 1 \\ 0 & 0 & 0 & \cdots & 0 & 0 & 6 \end{bmatrix} \begin{bmatrix} M_2 \\ M_3 \\ \vdots \\ M_{n-2} \\ M_{n-1} \end{bmatrix} = \frac{6}{h^2} \begin{bmatrix} y_1 - 2y_2 + y_3 \\ y_2 - 2y_3 + y_4 \\ \vdots \\ y_{n-2} - 2y_{n-1} + y_n \end{bmatrix}$$

THE PARABOLIC RUNOUT SPLINE

The two additional constraints imposed for this type of spline are

$$M_1 = M_2 \tag{19}$$

$$M_n = M_{n-1} \tag{20}$$

If we use these two equations to eliminate M_1 and M_n from (15), we obtain the $(n-2) \times (n-2)$ linear system

$$\begin{bmatrix} 5 & 1 & 0 & 0 & \cdots & 0 & 0 & 0 \\ 1 & 4 & 1 & 0 & \cdots & 0 & 0 & 0 \\ 0 & 1 & 4 & 1 & \cdots & 0 & 0 & 0 \\ \vdots & \vdots & \vdots & \vdots & & \vdots & \vdots & \vdots \\ 0 & 0 & 0 & 0 & \cdots & 1 & 4 & 1 \\ 0 & 0 & 0 & 0 & \cdots & 0 & 1 & 5 \end{bmatrix} \begin{bmatrix} M_2 \\ M_3 \\ M_4 \\ \vdots \\ M_{n-2} \\ M_{n-1} \end{bmatrix} = \frac{6}{h^2} \begin{bmatrix} y_1 - 2y_2 + y_3 \\ y_2 - 2y_3 + y_4 \\ y_3 - 2y_4 + y_5 \\ \vdots \\ y_{n-3} - 2y_{n-2} + y_{n-1} \\ y_{n-2} - 2y_{n-1} + y_n \end{bmatrix} \tag{21}$$

for $M_2, M_3, \ldots, M_n$. Once these $n-2$ values have been determined, then M_1 and M_n are determined from (19) and (20).

From (14) we see that $M_1 = M_2$ implies that $a_1 = 0$, and $M_n = M_{n-1}$ implies that $a_{n-1} = 0$. Thus, from (3) there are no cubic terms in the formula for the spline over the end intervals $[x_1, x_2]$ and $[x_{n-1}, x_n]$. Hence, as the name suggests, the parabolic runout spline reduces to a parabolic curve over these end intervals.

THE CUBIC RUNOUT SPLINE

For this type of spline we impose the two additional conditions

$$M_1 = 2M_2 - M_3 \tag{22}$$

$$M_n = 2M_{n-1} - M_{n\ 2} \tag{23}$$

Using these two equations to eliminate M_1 and M_n from (15) results in the following $(n-2) \times (n-2)$ linear system for $M_2, M_3, \ldots, M_{n-1}$:

$$\begin{bmatrix} 6 & 0 & 0 & 0 & \cdots & 0 & 0 & 0 \\ 1 & 4 & 1 & 0 & \cdots & 0 & 0 & 0 \\ 0 & 1 & 4 & 1 & \cdots & 0 & 0 & 0 \\ \vdots & \vdots & \vdots & \vdots & & \vdots & \vdots & \vdots \\ 0 & 0 & 0 & 0 & \cdots & 1 & 4 & 1 \\ 0 & 0 & 0 & 0 & \cdots & 0 & 0 & 6 \end{bmatrix} \begin{bmatrix} M_2 \\ M_3 \\ M_4 \\ \vdots \\ M_{n-2} \\ M_{n-1} \end{bmatrix} = \frac{6}{h^2} \begin{bmatrix} y_1 - 2y_2 + y_3 \\ y_2 - 2y_3 + y_4 \\ y_3 - 2y_4 + y_5 \\ \vdots \\ y_{n-3} - 2y_{n-2} + y_{n-1} \\ y_{n-2} - 2y_{n-1} + y_n \end{bmatrix} \tag{24}$$

After we solve this linear system for $M_2, M_3, \ldots, M_{n-1}$, we can use (22) and (23) to determine M_1 and M_n.

If we rewrite (22) as

$$M_2 - M_1 = M_3 - M_2$$

it then follows from (14) that $a_1 = a_2$. Because $S'''(x) = 6a_1$ on $[x_1, x_2]$ and $S'''(x) = 6a_2$ on $[x_2, x_3]$, we see that $S'''(x)$ is constant over the entire interval $[x_1, x_3]$. Consequently, $S(x)$ consists of a *single* cubic curve over the interval $[x_1, x_3]$ rather than two different cubic curves pieced together at x_2. [To see this, integrate $S'''(x)$ three times.] A similar analysis shows that $S(x)$ consists of a single cubic curve over the last two intervals.

Whereas the natural spline tends to produce an interpolating curve that is flat at the endpoints, the cubic runout spline has the opposite tendency: it produces a curve with pronounced curvature at the endpoints. If neither behavior is desired, the parabolic runout spline is a reasonable compromise.

Example 1 The density of water is well known to reach a maximum at a temperature slightly above freezing. Table 2, from the *Handbook of Chemistry and Physics* (Cleveland, Ohio: Chemical Rubber Publishing Company), gives the density of water in grams per cubic centimeter for five equally spaced temperatures from $-10°C$ to $30°C$. We will interpolate these five temperature-density measurements with a parabolic runout spline and attempt to find the maximum density of water in this range by finding the maximum value on this cubic spline. In the exercises we ask the reader to perform similar calculations using a natural spline and a cubic runout spline to interpolate the data points.

TABLE 2

Temperature (°C)	Density (gm/cm³)
−10	.99815
0	.99987
10	.99973
20	.99823
30	.99567

Set

$$x_1 = -10, \qquad y_1 = .99815$$
$$x_2 = 0, \qquad y_2 = .99987$$
$$x_3 = 10, \qquad y_3 = .99973$$
$$x_4 = 20, \qquad y_4 = .99823$$
$$x_5 = 30, \qquad y_5 = .99567$$

Then

$$6[y_1 - 2y_2 + y_3]/h^2 = -.0001116$$
$$6[y_2 - 2y_3 + y_4]/h^2 = -.0000816$$
$$6[y_3 - 2y_4 + y_5]/h^2 = -.0000636$$

and the linear system (21) for the parabolic runout spline becomes

$$\begin{bmatrix} 5 & 1 & 0 \\ 1 & 4 & 1 \\ 0 & 1 & 5 \end{bmatrix} \begin{bmatrix} M_2 \\ M_3 \\ M_4 \end{bmatrix} = \begin{bmatrix} -.0001116 \\ -.0000816 \\ -.0000636 \end{bmatrix}$$

Solving this system yields

$$M_2 = -.00001973$$
$$M_3 = -.00001293$$
$$M_4 = -.00001013$$

From (19) and (20) we have

$$M_1 = M_2 = -.00001973$$
$$M_5 = M_4 = -.00001013$$

Solving for the a_i's, b_i's, c_i's, and d_i's in (14), we obtain the following expression for the interpolating parabolic runout spline:

$$S(x) = \begin{cases} & -.0000098\,7(x + 10)^2 + .0002707(x + 10) + .99815, & -10 \leq x \leq 0 \\ .000000113(x - 0)^3 & -.00000987(x - 0)^2 + .0000733(x - 0) + .99987, & 0 \leq x \leq 10 \\ .000000047(x - 10)^3 & -.00000647(x - 10)^2 - .0000900(x - 10) + .99973, & 10 \leq x \leq 20 \\ & -.00000507(x - 20)^2 - .0002054(x - 20) + .99823, & 20 \leq x \leq 30 \end{cases}$$

This spline is plotted in Figure 5. From that figure we see that the maximum is attained in the interval $[0, 10]$. To find this maximum, we set $S'(x)$ equal to zero in the interval $[0, 10]$:

$$S'(x) = .000000339x^2 - .0000197x + .0000733 = 0$$

To three significant digits the root of this quadratic in the interval $[0, 10]$ is $x = 3.99$, and for this value of x, $S(3.99) = 1.00001$. Thus, according to our interpolated estimate, the maximum density of water is 1.00001 gm/cm^3 attained at $3.99°$C. This agrees well with the experimental maximum density of 1.00000 gm/cm^3 attained at $3.98°$C. (In the original metric system, the gram was *defined* as the mass of one cubic centimeter of water at its maximum density.) ▲

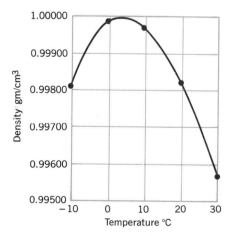

Figure 5

CLOSING REMARKS

In addition to producing excellent interpolating curves, cubic splines and their generalizations are useful for numerical integration and differentiation, for the numerical solution of differential and integral equations, and in optimization theory. Readers desiring to pursue this topic further may find the following references useful.

1. J. H. Ahlberg, E. N. Nilson, and J. L. Walsh, *The Theory of Splines and Their Applications* (New York: Academic Press, 1967).
2. C. de Boor, *A Practical Guide to Splines* (New York: Springer-Verlag, 1978).
3. J. R. Rice, *Numerical Methods, Software, and Analysis* (New York: McGraw-Hill, 1983).

EXERCISE SET 11.16

1. Derive the expressions for a_i and c_i in Equations (14) of Theorem 11.16.1.

2. The six points

$$(0, .00000)$$
$$(.2, .19867)$$
$$(.4, .38942)$$
$$(.6, .56464)$$
$$(.8, .71736)$$
$$(1.0, .84147)$$

lie on the graph of $y = \sin x$, where x is in radians.
(a) Find the parabolic runout spline that interpolates these six points for $.4 \le x \le .6$. Maintain an accuracy of five decimal places in your calculations.
(b) Calculate $S(.5)$ for the spline you found in part (a). What is the percentage error of $S(.5)$ with respect to the "exact" value of $\sin(.5) = .47943$?

3. The following five points

$$(0, 1)$$
$$(1, 7)$$
$$(2, 27)$$
$$(3, 79)$$
$$(4, 181)$$

lie on a single cubic curve.
(a) Which of the three types of cubic splines (natural, parabolic runout, or cubic runout) would agree exactly with the single cubic curve on which the five points lie?
(b) Determine the cubic spline you chose in part (a), and verify that it is a single cubic curve that interpolates the five points.

4. Repeat the calculations in Example 1 using a natural spline to interpolate the five data points.

5. Repeat the calculations in the example in this section using a cubic runout spline to interpolate the five data points.

6. **(The Periodic Spline.)** If it is known or if it is desired that the n points $(x_1, y_1), (x_2, y_2), \ldots,$ (x_n, y_n) to be interpolated lie on a single cycle of a periodic curve with period $x_n - x_1$, then an interpolating cubic spline $S(x)$ must satisfy

$$S(x_1) = S(x_n)$$
$$S'(x_1) = S'(x_n)$$
$$S''(x_1) = S''(x_n)$$

(a) Show that these three periodicity conditions require that

$$y_1 = y_n$$
$$M_1 = M_n$$
$$4M_1 + M_2 + M_{n-1} = 6(y_{n-1} - 2y_1 + y_2)/h^2$$

(b) Using the three equations in part (a) and Equations (15), construct an $(n-1) \times (n-1)$ linear system for $M_1, M_2, \ldots, M_{n-1}$ in matrix form.

7. **(The Clamped Spline.)** Suppose, in addition to the n points to be interpolated, we are given specific values y_1' and y_n' for the slopes $S'(x_1)$ and $S'(x_n)$ of the interpolating cubic spline at the endpoints x_1 and x_n.
(a) Show that

$$2M_1 + M_2 = 6(y_2 - y_1 - hy_1')/h^2$$
$$2M_n + M_{n-1} = 6(y_{n-1} - y_n + hy_n')/h^2$$

(b) Using the equations in part (a) and Equations (15), construct an $n \times n$ linear system for $M_1, M_2, \ldots, M_n$ in matrix form. [**Remark.** The clamped spline described in this exercise is the most accurate type of spline for interpolation work if the slopes at the endpoints are known or can be estimated.]

11.17 CRYPTOGRAPHY

This section presents a method of encoding and decoding messages using matrix transformations and modular arithmetic and shows how Gaussian elimination can sometimes be used to break an opponent's code.

PREREQUISITES: Matrices
Gaussian elimination
Linear independence
Linear transformations (Sections 7.1 and 7.2)

The study of encoding and decoding secret messages is called **cryptography**. Although secret codes date to the earliest days of written communication, there has been a recent surge of interest in the subject because of the need to maintain the privacy of information transmitted over public communications lines.

In the language of cryptography, codes are called **ciphers**, uncoded messages are called **plaintext**, and coded messages are called **ciphertext**. The process of converting from plaintext to ciphertext is called **enciphering**, and the reverse process of converting from ciphertext to plaintext is called **deciphering**.

The simplest ciphers, called **substitution ciphers**, are those which replace each letter of the alphabet by a different letter. For example, in the substitution cipher

Plain	A B C D E F G H I J K L M N O P Q R S T U V W X Y Z
Cipher	D E F G H I J K L M N O P Q R S T U V W X Y Z A B C

the plaintext letter A is replaced by D, the plaintext letter B by E, and so forth. With this cipher the plaintext message

$$ROME\ WAS\ NOT\ BUILT\ IN\ A\ DAY$$

becomes

$$URPH\ ZDV\ QRW\ EXLOW\ LQ\ D\ GDB$$

HILL CIPHERS

A disadvantage of substitution ciphers is that they preserve the frequencies of individual letters, making it relatively easy to break the code by statistical methods. One way to overcome this problem is to divide the plaintext into groups of letters and encipher the plaintext group by group, rather than one letter at a time. A system of cryptography in which the plaintext is divided into sets of n letters, each of which is replaced by a set of n cipher letters, is called a **polygraphic system**. In this section we will study a class of polygraphic systems based on matrix transformations. (The ciphers that we will discuss are called **Hill ciphers** after Lester S. Hill, who introduced them in two papers: "Cryptography in an Algebraic Alphabet," *American Mathematical Monthly*, *36*, June–July 1929, pp. 306–312; and "Concerning Certain Linear Transformation Apparatus of Cryptography," *American Mathematical Monthly*, *38*, March 1931, pp. 135–154.)

In the discussion to follow we assume that each plaintext and ciphertext letter except Z is assigned the numerical value that specifies its position in the standard alphabet (Table 1). For reasons that will become clear later, Z is assigned a value of zero.

TABLE 1

A	B	C	D	E	F	G	H	I	J	K	L	M	N	O	P	Q	R	S	T	U	V	W	X	Y	Z
1	2	3	4	5	6	7	8	9	10	11	12	13	14	15	16	17	18	19	20	21	22	23	24	25	0

In the simplest Hill ciphers, successive *pairs* of plaintext are transformed into ciphertext by the following procedure:

Step 1.   Choose a 2 × 2 matrix with integer entries

$$A = \begin{bmatrix} a_{11} & a_{12} \\ a_{21} & a_{22} \end{bmatrix}$$

to perform the encoding. Certain additional conditions on A will be imposed later.

Step 2.   Group successive plaintext letters into pairs, adding an arbitrary "dummy" letter to fill out the last pair if the plaintext has an odd number of letters, and replace each plaintext letter by its numerical value.

Step 3.   Successively convert each plaintext pair $p_1 p_2$ into a column vector

$$\mathbf{p} = \begin{bmatrix} p_1 \\ p_2 \end{bmatrix}$$

and form the product $A\mathbf{p}$. We will call $\mathbf{p}$ a *plaintext vector* and $A\mathbf{p}$ the corresponding *ciphertext vector*.

Step 4.   Convert each ciphertext vector into its alphabetic equivalent.

Example 1   Use the matrix

$$\begin{bmatrix} 1 & 2 \\ 0 & 3 \end{bmatrix}$$

to obtain the Hill cipher for the plaintext message

I AM HIDING

Solution. If we group the plaintext into pairs and add the dummy letter *G* to fill out the last pair, we obtain

IA   MH   ID   IN   GG

or equivalently from Table 1,

9 1   13 8   9 4   9 14   7 7

To encipher the pair *IA*, we form the matrix product

$$\begin{bmatrix} 1 & 2 \\ 0 & 3 \end{bmatrix}\begin{bmatrix} 9 \\ 1 \end{bmatrix} = \begin{bmatrix} 11 \\ 3 \end{bmatrix}$$

which, from Table 1, yields the ciphertext *KC*.

To encipher the pair *MH* we form the product

$$\begin{bmatrix} 1 & 2 \\ 0 & 3 \end{bmatrix} \begin{bmatrix} 13 \\ 8 \end{bmatrix} = \begin{bmatrix} 29 \\ 24 \end{bmatrix} \qquad (1)$$

However, there is a problem here, because the number 29 has no alphabet equivalent (Table 1). To resolve this problem we make the following agreement:

> *Whenever an integer greater than 25 occurs, it will be replaced by the remainder that results when this integer is divided by 26.*

Because the remainder after division by 26 is one of the integers 0, 1, 2, . . . , 25, this procedure will always yield an integer with an alphabet equivalent.

Thus, in (1) we replace 29 by 3, which is the remainder after dividing 29 by 26. It now follows from Table 1 that the ciphertext for the pair *MH* is *CX*.

The computations for the remaining ciphertext vectors are

$$\begin{bmatrix} 1 & 2 \\ 0 & 3 \end{bmatrix} \begin{bmatrix} 9 \\ 4 \end{bmatrix} = \begin{bmatrix} 17 \\ 12 \end{bmatrix}$$

$$\begin{bmatrix} 1 & 2 \\ 0 & 3 \end{bmatrix} \begin{bmatrix} 9 \\ 14 \end{bmatrix} = \begin{bmatrix} 37 \\ 42 \end{bmatrix} \quad \text{or} \quad \begin{bmatrix} 11 \\ 16 \end{bmatrix}$$

$$\begin{bmatrix} 1 & 2 \\ 0 & 3 \end{bmatrix} \begin{bmatrix} 7 \\ 7 \end{bmatrix} = \begin{bmatrix} 21 \\ 21 \end{bmatrix}$$

These correspond to the ciphertext pairs *QL*, *KP*, and *UU*, respectively. In summary, the entire ciphertext message is

$$KC \qquad CX \qquad QL \qquad KP \qquad UU$$

which would usually be transmitted as a single string without spaces:

$$KCCXQLKPUU \quad \blacktriangle$$

Because the plaintext was grouped in pairs and enciphered by a 2×2 matrix, the Hill cipher in Example 1 is referred to as a ***Hill 2-cipher***. It is obviously also possible to group the plaintext in triples and encipher by a 3×3 matrix with integer entries; this is called a ***Hill 3-cipher***. In general, for a ***Hill n-cipher***, plaintext is grouped into sets of *n* letters and enciphered by an $n \times n$ matrix with integer entries.

MODULAR ARITHMETIC

In Example 1 integers greater than 25 were replaced by their remainders after division by 26. This technique of working with remainders is at the core of a body of mathematics called *modular arithmetic*. Because of its importance in cryptography, we will digress for a moment to touch on some of the main ideas in this area.

In modular arithmetic we are given a positive integer *m*, called the ***modulus***, and any two integers whose difference is an integer multiple of the modulus are regarded as "equal" or "equivalent" with respect to the modulus. More precisely, we make the following definition.

Definition. If m is a positive integer and a and b are any integers, then we say that a is ***equivalent*** to b modulo m, written

$$a = b \quad (\text{mod } m)$$

if $a - b$ is an integer multiple of m.

Example 2

$$7 = 2 \quad (\text{mod } 5)$$
$$19 = 3 \quad (\text{mod } 2)$$
$$-1 = 25 \quad (\text{mod } 26)$$
$$12 = 0 \quad (\text{mod } 4)$$

For any modulus m it can be proved that every integer a is equivalent, modulo m, to exactly one of the integers

$$0, 1, 2, \ldots, m - 1$$

We call this integer the ***residue*** of a modulo m, and we write

$$Z_m = \{0, 1, 2, \ldots, m - 1\}$$

to denote the set of residues modulo m.

If a is a *nonnegative* integer, then its residue modulo m is simply the remainder that results when a is divided by m. For an arbitrary integer a, the residue can be found using the following theorem.

Theorem 11.17.1. *For any integer a and modulus m, let*

$$R = remainder \ of \ \frac{|a|}{m}$$

Then the residue r of a modulo m is given by

$$r = \begin{cases} R & \textit{if } a \geq 0 \\ m - R & \textit{if } a < 0 \quad \textit{and} \quad R \neq 0 \\ 0 & \textit{if } a < 0 \quad \textit{and} \quad R = 0 \end{cases}$$

Example 3   Find the residue modulo 26 of
   (a) 87   (b) -38 (c) -26

Solution (a). Dividing $|87| = 87$ by 26 yields a remainder of $R = 9$, so $r = 9$. Thus,

$$87 = 9 \quad (\text{mod } 26)$$

Solution (b). Dividing $|-38| = 38$ by 26 yields a remainder of $R = 12$, so $r = 26 - 12 = 14$. Thus,

$$-38 = 14 \quad (\text{mod } 26)$$

Solution (c). Dividing $|-26| = 26$ by 26 yields a remainder of $R = 0$. Thus,

$$-26 = 0 \quad (\text{mod } 26) \quad \blacktriangle$$

In ordinary arithmetic every nonzero number a has a *reciprocal* or *multiplicative inverse*, denoted by a^{-1}, such that

$$aa^{-1} = a^{-1}a = 1$$

In modular arithmetic we have the following corresponding concept:

Definition. If a is a number in Z_m, then a number a^{-1} in Z_m is called a *reciprocal* or **multiplicative inverse** of a modulo m if $aa^{-1} = a^{-1}a = 1 \ (\text{mod } m)$.

It can be proved that if a and m have no common prime factors, then a has a unique reciprocal modulo m; conversely, if a and m have a common prime factor, then a has no reciprocal modulo m.

Example 4   The number 3 has a reciprocal modulo 26 because 3 and 26 have no common prime factors. This reciprocal can be obtained by finding the number x in Z_{26} that satisfies the modular equation

$$3x = 1 \quad (\text{mod } 26)$$

Although there are general methods for solving such modular equations, it would take us too far afield to study them. However, because 26 is relatively small, this equation can be solved by trying the possible solutions, 0 to 25, one at a time. With this approach we find that $x = 9$ is the solution, as

$$3 \cdot 9 = 27 = 1 \quad (\text{mod } 26)$$

Thus,

$$3^{-1} = 9 \quad (\text{mod } 26) \quad \blacktriangle$$

Example 5   The number 4 has no reciprocal modulo 26 because 4 and 26 have 2 as a common prime factor (see Exercise 8). $\blacktriangle$

For future reference, we provide the following table of reciprocals modulo 26:

TABLE 2 *Reciprocals Modulo 26*

a	1	3	5	7	9	11	15	17	19	21	23	25
a^{-1}	1	9	21	15	3	19	7	23	11	5	17	25

DECIPHERING   Every useful cipher must have a procedure for decipherment. In the case of a Hill cipher, decipherment uses the inverse (mod 26) of the enciphering matrix. To be precise, if m is a positive integer, then a square matrix A with entries in Z_m is said to be ***invertible modulo m*** if there is a matrix B with entries in Z_m such that

$$AB = BA = I \quad (\text{mod } m)$$

Suppose now that

$$A = \begin{bmatrix} a_{11} & a_{12} \\ a_{21} & a_{22} \end{bmatrix}$$

is invertible modulo 26 and this matrix is used in a Hill 2-cipher. If

$$\mathbf{p} - \begin{bmatrix} p_1 \\ p_2 \end{bmatrix}$$

is a plaintext vector, then

$$\mathbf{c} = A\mathbf{p}$$

is the corresponding ciphertext vector and

$$\mathbf{p} = A^{-1}\mathbf{c}$$

Thus, each plaintext vector can be recovered from the corresponding ciphertext vector by multiplying it on the left by A^{-1} (mod m).

In cryptography it is important to know which matrices are invertible modulo 26 and how to obtain their inverses. We now investigate these questions.

In ordinary arithmetic, a square matrix A is invertible if and only if $\det(A) \neq 0$; or equivalently, if and only if $\det(A)$ has a reciprocal. The following theorem is the analog of this result in modular arithmetic.

Theorem 11.17.2.   *A square matrix A with entries in Z_m is invertible modulo m if and only if the residue of $\det(A)$ modulo m has a reciprocal modulo m.*

Because the residue of $\det(A)$ modulo m will have a reciprocal modulo m if and only if this residue and m have no common prime factors, we have the following corollary.

Corollary 11.17.3.   *A square matrix A with entries in Z_m is invertible modulo m if and only if m and the residue of $\det(A)$ modulo m have no common prime factors.*

Because the only prime factors of $m = 26$ are 2 and 13, we have the following corollary, which is useful in cryptography.

Corollary 11.17.4.   *A square matrix A with entries in Z_{26} is invertible modulo 26 if and only if the residue of $\det(A)$ modulo 26 is not divisible by 2 or 13.*

We leave it for the reader to verify that if

$$A = \begin{bmatrix} a & b \\ c & d \end{bmatrix}$$

has entries in Z_{26} and the residue of $\det(A) = ad - bc$ modulo 26 is not divisible by 2 or 13, then the inverse of A (mod 26) is given by

$$A^{-1} = (ad - bc)^{-1} \begin{bmatrix} d & -b \\ -c & a \end{bmatrix} \quad \text{(mod 26)} \tag{2}$$

where $(ad - bc)^{-1}$ is the reciprocal of the residue of $ad - bc$ (mod 26).

Example 6 Find the inverse of

$$A = \begin{bmatrix} 5 & 6 \\ 2 & 3 \end{bmatrix}$$

modulo 26.

Solution.

$$\det(A) = ad - bc = 5 \cdot 3 - 6 \cdot 2 = 3$$

so, from Table 2

$$(ad - bc)^{-1} = 3^{-1} = 9 \quad \text{(mod 26)}$$

Thus, from (2)

$$A^{-1} = 9 \begin{bmatrix} 3 & -6 \\ -2 & 5 \end{bmatrix} = \begin{bmatrix} 27 & -54 \\ -18 & 45 \end{bmatrix} = \begin{bmatrix} 1 & 24 \\ 8 & 19 \end{bmatrix} \quad \text{(mod 26)}$$

As a check,

$$AA^{-1} = \begin{bmatrix} 5 & 6 \\ 2 & 3 \end{bmatrix}\begin{bmatrix} 1 & 24 \\ 8 & 19 \end{bmatrix} = \begin{bmatrix} 53 & 234 \\ 26 & 105 \end{bmatrix} = \begin{bmatrix} 1 & 0 \\ 0 & 1 \end{bmatrix} \quad \text{(mod 26)}$$

Similarly, $A^{-1}A = I$. ▲

Example 7 Decode the following Hill 2-cipher, which was enciphered by the matrix in Example 6:

$$GTNKGKDUSK$$

Solution. From Table 1 the numerical equivalent of this ciphertext is

$$7 \quad 20 \qquad 14 \quad 11 \qquad 7 \quad 11 \qquad 4 \quad 21 \qquad 19 \quad 11$$

To obtain the plaintext pairs, we multiply each ciphertext vector by the inverse of A (obtained in Example 5):

$$\begin{bmatrix} 1 & 24 \\ 8 & 19 \end{bmatrix} \begin{bmatrix} 7 \\ 20 \end{bmatrix} = \begin{bmatrix} 487 \\ 436 \end{bmatrix} = \begin{bmatrix} 19 \\ 20 \end{bmatrix} \quad (\text{mod } 26)$$

$$\begin{bmatrix} 1 & 24 \\ 8 & 19 \end{bmatrix} \begin{bmatrix} 14 \\ 11 \end{bmatrix} = \begin{bmatrix} 278 \\ 321 \end{bmatrix} = \begin{bmatrix} 18 \\ 9 \end{bmatrix} \quad (\text{mod } 26)$$

$$\begin{bmatrix} 1 & 24 \\ 8 & 19 \end{bmatrix} \begin{bmatrix} 7 \\ 11 \end{bmatrix} = \begin{bmatrix} 271 \\ 265 \end{bmatrix} = \begin{bmatrix} 11 \\ 5 \end{bmatrix} \quad (\text{mod } 26)$$

$$\begin{bmatrix} 1 & 24 \\ 8 & 19 \end{bmatrix} \begin{bmatrix} 4 \\ 21 \end{bmatrix} = \begin{bmatrix} 508 \\ 431 \end{bmatrix} = \begin{bmatrix} 14 \\ 15 \end{bmatrix} \quad (\text{mod } 26)$$

$$\begin{bmatrix} 1 & 24 \\ 8 & 19 \end{bmatrix} \begin{bmatrix} 19 \\ 11 \end{bmatrix} = \begin{bmatrix} 283 \\ 361 \end{bmatrix} = \begin{bmatrix} 23 \\ 23 \end{bmatrix} \quad (\text{mod } 26)$$

From Table 1 the alphabet equivalents of these vectors are

$$ST \quad RI \quad KE \quad NO \quad WW$$

which yields the message

$$STRIKE\ NOW\ \blacktriangle$$

BREAKING A HILL CIPHER

Because the purpose of enciphering messages and information is to prevent "opponents" from learning their contents, cryptographers are concerned with the *security* of their ciphers—that is, how readily they can be broken (deciphered by their opponents). We will conclude this section by discussing one technique for breaking Hill ciphers.

Suppose that you are able to obtain some corresponding plaintext and ciphertext from an opponent's message. For example, on examining some intercepted ciphertext, you may be able to deduce that the message is a letter that begins *DEAR SIR*. We will show that with a small amount of such data it may be possible to determine the deciphering matrix of a Hill code and consequently obtain access to the rest of the message.

It is a basic result in linear algebra that a linear transformation is completely determined by its values at a basis. This principle suggests that if we have a Hill n-cipher, and if

$$\mathbf{p}_1, \mathbf{p}_2, \dots, \mathbf{p}_n$$

are linearly independent plaintext vectors whose corresponding ciphertext vectors

$$A\mathbf{p}_1, A\mathbf{p}_2, \dots, A\mathbf{p}_n$$

are known, then there is enough information available to determine the matrix A, hence A^{-1} (mod m).

The following theorem, whose proof is discussed in the exercises, provides a way to do this.

> **Theorem 11.17.5.** *Let* $\mathbf{p}_1, \mathbf{p}_2, \ldots, \mathbf{p}_n$ *be linearly independent plaintext vectors, and let* $\mathbf{c}_1, \mathbf{c}_2, \ldots, \mathbf{c}_n$ *be the corresponding ciphertext vectors in a Hill n-cipher. If*
>
> $$P = \begin{bmatrix} \mathbf{p}_1^t \\ \mathbf{p}_2^t \\ \vdots \\ \mathbf{p}_n^t \end{bmatrix}$$
>
> *is the* $n \times n$ *matrix with row vectors* $\mathbf{p}_1^t, \mathbf{p}_2^t, \ldots, \mathbf{p}_n^t$ *and if*
>
> $$C = \begin{bmatrix} \mathbf{c}_1^t \\ \mathbf{c}_2^t \\ \vdots \\ \mathbf{c}_n^t \end{bmatrix}$$
>
> *is the* $n \times n$ *matrix with row vectors* $\mathbf{c}_1^t, \mathbf{c}_2^t, \ldots, \mathbf{c}_n^t,$ *then the sequence of elementary row operations that reduces C to I transforms P to* $(A^{-1})^t.$

This theorem tells us that to find the transpose of the deciphering matrix A^{-1}, we must find a sequence of row operations that reduces C to I and then perform this same sequence of operations on P. The following example illustrates a simple algorithm for doing this.

Example 8 The following Hill 2-cipher is intercepted:

$$IOSBTGXESPXHOPDE$$

Decipher the message, given that it starts with the word *DEAR*.

Solution. From Table 1, the numerical equivalent of the known plaintext is

$$
\begin{array}{cc}
DE & AR \\
4\ 5 & 1\ 18
\end{array}
$$

and the numerical equivalent of the corresponding ciphertext is

$$
\begin{array}{cc}
IO & SB \\
9\ 15 & 19\ 2
\end{array}
$$

so the corresponding plaintext and ciphertext vectors are

$$\mathbf{p}_1 = \begin{bmatrix} 4 \\ 5 \end{bmatrix} \leftrightarrow \mathbf{c}_1 = \begin{bmatrix} 9 \\ 15 \end{bmatrix}$$

$$\mathbf{p}_2 = \begin{bmatrix} 1 \\ 18 \end{bmatrix} \leftrightarrow \mathbf{c}_2 = \begin{bmatrix} 19 \\ 2 \end{bmatrix}$$

We want to reduce

$$C = \begin{bmatrix} \mathbf{c}_1^t \\ \mathbf{c}_2^t \end{bmatrix} = \begin{bmatrix} 9 & 15 \\ 19 & 2 \end{bmatrix}$$

to *I* by elementary row operations and simultaneously apply these operations to

$$P = \begin{bmatrix} \mathbf{p}_1^t \\ \mathbf{p}_2^t \end{bmatrix} = \begin{bmatrix} 4 & 5 \\ 1 & 18 \end{bmatrix}$$

to obtain $(A^{-1})^t$ (the transpose of the deciphering matrix). This can be accomplished by adjoining *P* to the right of *C* and applying row operations to the resulting matrix $[C \mid P]$ until the left side is reduced to *I*. The final matrix will then have the form $[I \mid (A^{-1})^t]$. The computations can be carried out as follows:

$$\begin{bmatrix} 9 & 15 & \vdots & 4 & 5 \\ 19 & 2 & \vdots & 1 & 18 \end{bmatrix}$$
We formed the matrix $[C \mid P]$.

$$\begin{bmatrix} 1 & 45 & \vdots & 12 & 15 \\ 19 & 2 & \vdots & 1 & 18 \end{bmatrix}$$
We multiplied the first row by $9^{-1} = 3$.

$$\begin{bmatrix} 1 & 19 & \vdots & 12 & 15 \\ 19 & 2 & \vdots & 1 & 18 \end{bmatrix}$$
We replaced 45 by its residue modulo 26.

$$\begin{bmatrix} 1 & 19 & \vdots & 12 & 15 \\ 0 & -359 & \vdots & -227 & -267 \end{bmatrix}$$
We added −19 times the first row to the second.

$$\begin{bmatrix} 1 & 19 & \vdots & 12 & 15 \\ 0 & 5 & \vdots & 7 & 19 \end{bmatrix}$$
We replaced the entries in the second row by their residues modulo 26.

$$\begin{bmatrix} 1 & 19 & \vdots & 12 & 15 \\ 0 & 1 & \vdots & 147 & 399 \end{bmatrix}$$
We multiplied the second row by $5^{-1} = 21$.

$$\begin{bmatrix} 1 & 19 & \vdots & 12 & 15 \\ 0 & 1 & \vdots & 17 & 9 \end{bmatrix}$$
We replaced the entries in the second row by their residues modulo 26.

$$\begin{bmatrix} 1 & 0 & \vdots & -311 & -156 \\ 0 & 1 & \vdots & 17 & 9 \end{bmatrix}$$
We added −19 times the second row to the first.

$$\begin{bmatrix} 1 & 0 & \vdots & 1 & 0 \\ 0 & 1 & \vdots & 17 & 9 \end{bmatrix}$$
We replaced the entries in the first row by their residues modulo 26.

Thus,

$$(A^{-1})^t = \begin{bmatrix} 1 & 0 \\ 17 & 9 \end{bmatrix}$$

so the deciphering matrix is

$$A^{-1} = \begin{bmatrix} 1 & 17 \\ 0 & 9 \end{bmatrix}$$

To decipher the message, we first group the ciphertext into pairs and find the numerical equivalent of each letter:

IO	SB	TG	XE	SP	XH	OP	DE
9 15	19 2	20 7	24 5	19 16	24 8	15 16	4 5

Next, we multiply successive ciphertext vectors on the left by A^{-1} and find the alphabet equivalents of the resulting plaintext pairs:

$$\begin{bmatrix} 1 & 17 \\ 0 & 9 \end{bmatrix}\begin{bmatrix} 9 \\ 15 \end{bmatrix} = \begin{bmatrix} 4 \\ 5 \end{bmatrix} \quad \begin{matrix} D \\ E \end{matrix}$$

$$\begin{bmatrix} 1 & 17 \\ 0 & 9 \end{bmatrix}\begin{bmatrix} 19 \\ 2 \end{bmatrix} = \begin{bmatrix} 1 \\ 18 \end{bmatrix} \quad \begin{matrix} A \\ R \end{matrix}$$

$$\begin{bmatrix} 1 & 17 \\ 0 & 9 \end{bmatrix}\begin{bmatrix} 20 \\ 7 \end{bmatrix} = \begin{bmatrix} 9 \\ 11 \end{bmatrix} \quad \begin{matrix} I \\ K \end{matrix}$$

$$\begin{bmatrix} 1 & 17 \\ 0 & 9 \end{bmatrix}\begin{bmatrix} 24 \\ 5 \end{bmatrix} = \begin{bmatrix} 5 \\ 19 \end{bmatrix} \quad \begin{matrix} E \\ S \end{matrix}$$

$$\begin{bmatrix} 1 & 17 \\ 0 & 9 \end{bmatrix}\begin{bmatrix} 19 \\ 16 \end{bmatrix} = \begin{bmatrix} 5 \\ 14 \end{bmatrix} \quad \begin{matrix} E \\ N \end{matrix} \qquad \text{(mod 26)}$$

$$\begin{bmatrix} 1 & 17 \\ 0 & 9 \end{bmatrix}\begin{bmatrix} 24 \\ 8 \end{bmatrix} = \begin{bmatrix} 4 \\ 20 \end{bmatrix} \quad \begin{matrix} D \\ T \end{matrix}$$

$$\begin{bmatrix} 1 & 17 \\ 0 & 9 \end{bmatrix}\begin{bmatrix} 15 \\ 16 \end{bmatrix} = \begin{bmatrix} 1 \\ 14 \end{bmatrix} \quad \begin{matrix} A \\ N \end{matrix}$$

$$\begin{bmatrix} 1 & 17 \\ 0 & 9 \end{bmatrix}\begin{bmatrix} 4 \\ 5 \end{bmatrix} = \begin{bmatrix} 11 \\ 9 \end{bmatrix} \quad \begin{matrix} K \\ S \end{matrix}$$

Finally, we construct the message from the plaintext pairs:

DE	AR	IK	ES	EN	DT	AN	KS

DEAR IKE SEND TANKS ▲

FURTHER READINGS

Readers interested in learning more about mathematical cryptography are referred to the following books, the first of which is elementary and the second more advanced:

1. Abraham Sinkov, *Elementary Cryptanalysis, a Mathematical Approach* (Mathematical Association of America, Mathematical Library, 1966).
2. Alan G. Konheim, *Cryptography, a Primer* (New York: Wiley-Interscience, 1981).

EXERCISE SET 11.17

1. Obtain the Hill cipher of the message

   DARK NIGHT

   for each of the following enciphering matrices:

   (a) $\begin{bmatrix} 1 & 3 \\ 2 & 1 \end{bmatrix}$ (b) $\begin{bmatrix} 4 & 3 \\ 1 & 2 \end{bmatrix}$

2. In each part determine whether the matrix is invertible modulo 26. If so, find its inverse modulo 26 and check your work by verifying that $AA^{-1} = A^{-1}A = I$ (mod 26).

   (a) $A = \begin{bmatrix} 9 & 1 \\ 7 & 2 \end{bmatrix}$ (b) $A = \begin{bmatrix} 3 & 1 \\ 5 & 3 \end{bmatrix}$ (c) $A = \begin{bmatrix} 8 & 11 \\ 1 & 9 \end{bmatrix}$

   (d) $A = \begin{bmatrix} 2 & 1 \\ 1 & 7 \end{bmatrix}$ (e) $A = \begin{bmatrix} 3 & 1 \\ 6 & 2 \end{bmatrix}$ (f) $A = \begin{bmatrix} 1 & 8 \\ 1 & 3 \end{bmatrix}$

3. Decode the message

   SAKNOXAOJX

   given that it is a Hill cipher with enciphering matrix

   $\begin{bmatrix} 4 & 1 \\ 3 & 2 \end{bmatrix}$

4. A Hill 2-cipher is intercepted that starts with the pairs

   SL   HK

   Find the deciphering and enciphering matrices, given that the plaintext is known to start with the word *ARMY*.

5. Decode the following Hill 2-cipher if the last four plaintext letters are known to be *ATOM*.

   LNGIHGYBVRENJYQO

6. Decode the following Hill 3-cipher if the first nine plaintext letters are *IHAVECOME*:

   HPAFQGGDUGDDHPGODYNOR

7. If, in addition to the standard alphabet, a period, comma, and question mark are allowed, then 29 plaintext and ciphertext symbols are available and all matrix arithmetic would be done modulo 29. Under what conditions would a matrix with entries in Z_{29} be invertible modulo 29?

8. Show that the modular equation $4x = 1$ (mod 26) has no solution in Z_{26} by successively substituting the values $x = 0, 1, 2, \ldots, 25$.

9. (a) Let P and C be the matrices in Theorem 11.17.5. Show that $P = C(A^{-1})^t$.
   (b) To prove Theorem 11.17.5, let $E_1, E_2, \ldots, E_n$ be the elementary matrices that correspond to the row operations that reduce C to I, so

$$E_n \cdots E_2 E_1 C = I$$

Show that

$$E_n \cdots E_2 E_1 P = (A^{-1})^t$$

from which it follows that the same sequence of row operations that reduces C to I converts P to $(A^{-1})^t$.

10. (a) If A is the enciphering matrix of a Hill n-cipher, show that

$$A^{-1} = (C^{-1}P)^t \quad (\text{mod } 26)$$

where C and P are the matrices defined in Theorem 11.17.5.
   (b) Instead of using Theorem 11.17.5 as in the text, find the deciphering matrix A^{-1} of Example 8 by using the result in part (a) and Equation (2) to compute C^{-1}. [*Note.* Although this method is practical for Hill 2-ciphers, Theorem 11.17.5 is more efficient for Hill n-ciphers with $n > 2$.]

11.18 COMPUTED TOMOGRAPHY

Constructing a cross-sectional view of a human body by analyzing X-ray scans leads to an inconsistent linear system for the digital representation of the view. This section presents an iteration technique that provides an "approximate solution" of the linear system.

PREREQUISITES: Linear systems
Natural logarithms
Euclidean space R^n

The basic problem of computed tomography is to construct an image of a cross section of the human body using data collected from many individual beams of X rays that are passed through the cross section. These data are processed by a computer, and the computed cross section is displayed on a video monitor. Figure 1 is a diagram of General Electric's CT/T system showing a patient prepared to have a cross section of his head scanned by X-ray beams. Such a system is also known as a *CAT scanner*, for *C*omputer-*A*ided *T*omography scanner. Figure 2 displays a typical cross section of a human head produced by the system.

The first commercial system of computed tomography for medical use was developed in 1971 by G. N. Hounsfield of EMI, Ltd., in England. In 1979, Hounsfield and A. M. Cormack were awarded the Nobel prize for their pioneering work in the field. As we will see in this section, the construction of a cross section or tomograph requires the solution of a large linear system of equations. Certain

Figure 1

algorithms, called algebraic reconstruction techniques (ARTs), can be used to solve these linear systems, whose solutions yield the cross sections in digital form.

**SCANNING
MODES**

Unlike conventional X-ray pictures that are formed by X rays that are projected *perpendicular* to the plane of the picture, tomographs are constructed from thousands of individual, hairline-thin X-ray beams that *lie in the plane* of the cross section. After they pass through the cross section, the intensities of the X-ray beams are measured by an X-ray detector, and these measurements are relayed to a computer where they are processed. Figures 3 and 4 illustrate two possible modes of scanning the cross section: the *parallel mode* and the *fan-beam mode*. In the parallel mode a single X-ray source and X-ray detector pair are translated across the field of view containing the cross section, and many measurements of the parallel beams are recorded. Then the source and detector pair are rotated through a small angle, and another set of measurements is taken. This is repeated until the desired number of beam measurements is completed. For example, in the original 1971 machine, 160 parallel measurements were taken through 180 angles spaced 1° apart: a total of $160 \times 180 = 28{,}800$ beam measurements. Each such scan took approximately $5\frac{1}{2}$ minutes.

Figure 2

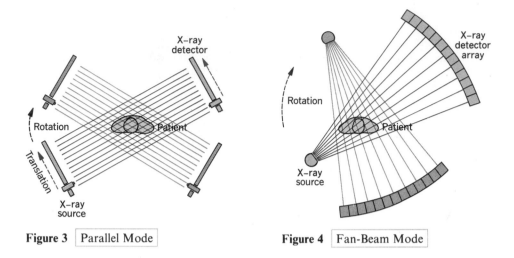

Figure 3 | Parallel Mode

Figure 4 | Fan-Beam Mode

In the fan-beam mode of scanning, a single X-ray tube generates a fan of collimated beams whose intensities are measured simultaneously by an array of detectors on the other side of the field of view. The X-ray tube and detector array are rotated through many angles, and a set of measurements is taken at each angle until the scan is completed. In the General Electric CT/T system, which uses the fan-beam mode, each scan takes 4.6 seconds, and a total of 184,320 beam measurements are recorded.

DERIVATION OF EQUATIONS

To see how the cross section is reconstructed from the many individual beam measurements, refer to Figure 5. Here the field of view in which the cross section is situated has been divided into many square *pixels* (picture elements) numbered 1 through N as indicated. It is our desire to determine the X-ray density of each

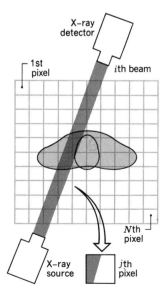

Figure 5

pixel. In the EMI system, 6400 pixels were used, arranged in a square 80 × 80 array. The G.E. CT/T system uses 102,400 pixels in a 320 × 320 array, each pixel being about 1 mm on a side. After the densities of the pixels are determined by the method we will describe, they are reproduced on a video monitor, with each pixel shaded a level of gray proportional to its X-ray density. Because different tissues within the human body have different X-ray densities, the video display clearly distinguishes the various tissues and organs within the cross section.

Figure 6 shows a single pixel with an X-ray beam of roughly the same width as the pixel passing squarely through it. The photons constituting the X-ray beam

Figure 6   Photons entering *j*th pixel   Photons leaving *j*th pixel

are absorbed by the tissue within the pixel at a rate proportional to the X-ray density of the tissue. Quantitatively, the X-ray density of the *j*th pixel is denoted by x_j and is defined by

$$x_j = \ln\left(\frac{\text{number of photons entering the jth pixel}}{\text{number of photons leaving the jth pixel}}\right)$$

where "ln" denotes the natural logarithmic function. Using the logarithm property $\ln(a/b) = -\ln(b/a)$, we also have

$$x_j = -\ln\left(\begin{array}{c}\text{fraction of photons that pass through}\\ \text{the jth pixel without being absorbed}\end{array}\right)$$

If the X-ray beam passes through an entire row of pixels (Figure 7), then the number of photons leaving one pixel is equal to the number of photons entering the next pixel in the row. If the pixels are numbered $1, 2, \ldots, n$, then the additive property of the logarithmic function gives

$$x_1 + x_2 + \cdots + x_n = \ln\left(\frac{\text{number of photons entering the first pixel}}{\text{number of photons leaving the nth pixel}}\right)$$

$$= -\ln\left(\begin{array}{c}\text{fraction of photons that pass}\\ \text{through the row of n pixels}\\ \text{without being absorbed}\end{array}\right) \tag{1}$$

Thus, to determine the total X-ray density of a row of pixels, we simply sum the individual pixel densities.

Figure 7   Photons entering first pixel   First pixel   Second pixel   Third pixel   *n*th pixel   Photons leaving *n*th pixel

Next, consider the X-ray beam in Figure 5. By the **beam density** of the ith beam of a scan, denoted by b_i, we mean

$$b_i = \ln\left(\frac{\begin{array}{c}\textit{number of photons of the ith beam entering the detector}\\ \textit{without the cross section in the field of view}\end{array}}{\begin{array}{c}\textit{number of photons of the ith beam entering the detector}\\ \textit{with the cross section in the field of view}\end{array}}\right)$$

$$= -\ln\left(\begin{array}{c}\textit{fraction of photons of the ith beam that}\\ \textit{pass through the cross section without}\\ \textit{being absorbed}\end{array}\right) \qquad (2)$$

The numerator in the first expression for b_i is obtained by performing a calibration scan without the cross section in the field of view. The resulting detector measurements are stored within the computer's memory. Then a clinical scan is performed with the cross section in the field of view, the b_i's of all the beams constituting the scan are computed, and the values are stored for further processing.

For each beam that passes squarely through a row of pixels, we must have

$$\left(\begin{array}{c}\textit{fraction of photons of the}\\ \textit{beam that pass through the}\\ \textit{row of pixels without being}\\ \textit{absorbed}\end{array}\right) = \left(\begin{array}{c}\textit{fraction of photons of the}\\ \textit{beam that pass through the}\\ \textit{cross section without being}\\ \textit{absorbed}\end{array}\right)$$

Thus, if the ith beam passes squarely through a row of n pixels, then it follows from Equations (1) and (2) that

$$x_1 + x_2 + \cdots + x_n = b_i$$

In this equation, b_i is known from the clinical and calibration measurements, and $x_1, x_2, \ldots, x_n$ are unknown pixel densities that must be determined.

More generally, if the ith beam passes squarely through a row (or column) of pixels with numbers $j_1, j_2, \ldots, j_i$, then we have

$$x_{j_1} + x_{j_2} + \cdots + x_{j_i} = b_i$$

If we set

$$a_{ij} = \begin{cases} 1 & \textit{if } j = j_1, j_2, \ldots, j_i \\ 0 & \textit{otherwise} \end{cases}$$

then we may write this equation as

$$a_{i1}x_1 + a_{i2}x_2 + \cdots + a_{iN}x_N = b_i \qquad (3)$$

We shall refer to Equation (3) as the *ith beam equation.*

Referring to Figure 5, however, we see that the beams of a scan do not necessarily pass through a row or column of pixels squarely. Instead, a typical beam passes diagonally through each pixel in its path. There are many ways to take this into account. In Figure 8 we outline three methods of defining the quantities a_{ij} that appear in Equation (3), each of which reduces to our previous definition when the beam passes squarely through a row or column of pixels. Reading down the figure, each method is more exact than its predecessor, but with successively more computational difficulty.

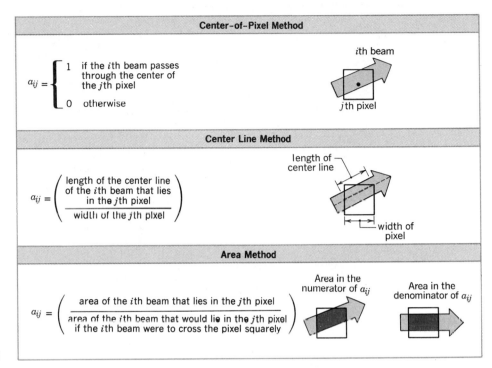

Figure 8

Using any one of the three methods to define the a_{ij}'s in the ith beam equation, we can write the set of M beam equations in a complete scan as

$$
\begin{aligned}
a_{11}x_1 + a_{12}x_2 + \cdots + a_{1N}x_N &= b_1 \\
a_{21}x_1 + a_{22}x_2 + \cdots + a_{2N}x_N &= b_2 \\
&\ \ \vdots \\
a_{M1}x_1 + a_{M2}x_2 + \cdots + a_{MN}x_N &= b_M
\end{aligned}
\tag{4}
$$

In this way we have a linear system of M equations (the M beam equations) in N unknowns (the N pixel densities).

Depending on the number of beams and pixels used, we may have $M > N$, $M = N$, or $M < N$. We will consider only the case $M > N$, the so-called *overdetermined case*, in which there are more beams in the scan than pixels in the field of view. Because of inherent modeling and experimental errors in the problem, we should not expect our linear system to have an exact mathematical solution for the pixel densities. In the next section we attempt to find an "approximate" solution to this linear system.

ALGEBRAIC RECONSTRUCTION TECHNIQUES

There have been many mathematical algorithms devised to treat the overdetermined linear system (4). The one we will describe belongs to the class of so-called *Algebraic Reconstruction Techniques* (*ART*). This method, which can be traced to an iterative technique originally introduced by S. Kaczmarz in 1937, was the one used in the first commercial machine. To introduce this technique, consider the following system of three equations in two unknowns:

$$L_1: \quad x_1 + x_2 = 2$$
$$L_2: \quad x_1 - 2x_2 = -2 \tag{5}$$
$$L_3: \quad 3x_1 - x_2 = 3$$

The lines L_1, L_2, L_3 determined by these three equations are plotted in the x_1x_2-plane. As shown in Figure 9a, the three lines do not have a common intersection, and so the three equations do not have an exact solution. However, the points (x_1, x_2) on the shaded triangle formed by the three lines are all situated "near" these three lines and can be thought of as constituting "approximate" solutions to our system. The following iterative procedure describes a geometric construction for generating points on the boundary of that triangular region (Figure 9b):

Algorithm 1

Step 0. Choose an arbitrary starting point $\mathbf{x}_0$ in the x_1x_2-plane.

Step 1. Project $\mathbf{x}_0$ orthogonally onto the first line L_1 and call the projection $\mathbf{x}_1^{(1)}$. The superscript (1) indicates that this is the first of several cycles through the steps.

Step 2. Project $\mathbf{x}_1^{(1)}$ orthogonally onto the second line L_2 and call the projection $\mathbf{x}_2^{(1)}$.

Step 3. Project $\mathbf{x}_2^{(1)}$ orthogonally onto the third line L_3 and call the projection $\mathbf{x}_3^{(1)}$.

Step 4. Take $\mathbf{x}_3^{(1)}$ as the new value of $\mathbf{x}_0$ and cycle through Steps 1 through 3 again. In the second cycle, label the projected points $\mathbf{x}_1^{(2)}$, $\mathbf{x}_2^{(2)}$, $\mathbf{x}_3^{(2)}$; in the third cycle label the projected points $\mathbf{x}_1^{(3)}$, $\mathbf{x}_2^{(3)}$, $\mathbf{x}_3^{(3)}$; and so forth.

(a)

(b)

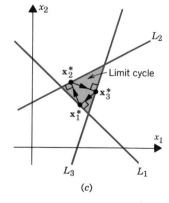

(c)

Figure 9

This algorithm generates three sequences of points

$$L_1: \quad \mathbf{x}_1^{(1)}, \mathbf{x}_1^{(2)}, \mathbf{x}_1^{(3)}, \ldots$$
$$L_2: \quad \mathbf{x}_2^{(1)}, \mathbf{x}_2^{(2)}, \mathbf{x}_2^{(3)}, \ldots$$
$$L_3: \quad \mathbf{x}_3^{(1)}, \mathbf{x}_3^{(2)}, \mathbf{x}_3^{(3)}, \ldots$$

that lie on the three lines L_1, L_2, and L_3, respectively. It can be shown that as long as the three lines are not all parallel, then the first sequence converges to a point $\mathbf{x}_1^*$ on L_1, the second sequence converges to a point $\mathbf{x}_2^*$ on L_2, and the third sequence converges to a point $\mathbf{x}_3^*$ on L_3 (Figure 9c). These three limit points form what is called the **limit cycle** of the iterative process. It can be shown that the limit cycle is independent of the starting point $\mathbf{x}_0$.

We next discuss the specific formulas needed to effect the orthogonal projections in Algorithm 1. First, because the equation of a line in x_1x_2-space is

$$a_1 x_1 + a_2 x_2 = b$$

we can express it in vector form as

$$\mathbf{a}^t \mathbf{x} = b$$

where

$$\mathbf{a} = \begin{bmatrix} a_1 \\ a_2 \end{bmatrix} \quad \text{and} \quad \mathbf{x} = \begin{bmatrix} x_1 \\ x_2 \end{bmatrix}$$

The following theorem gives the necessary projection formula (Exercise 5).

Theorem 11.18.1. *Let L be a line in R^2 with equation $\mathbf{a}^t\mathbf{x} = b$, and let $\mathbf{x}^*$ be any point in R^2 (Figure 10). Then the orthogonal projection, $\mathbf{x}_p$, of $\mathbf{x}^*$ onto L is given by*

$$\mathbf{x}_p = \mathbf{x}^* + \frac{(b - \mathbf{a}^t\mathbf{x}^*)}{\mathbf{a}^t\mathbf{a}} \mathbf{a}$$

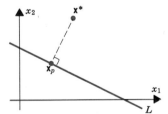

Figure 10

Example 1 We can use Algorithm 1 to find an approximate solution of the linear system given in (5) and illustrated in Figure 10. If we write the equations of the three lines as

$$L_1: \quad \mathbf{a}_1^t \mathbf{x} = b_1$$
$$L_2: \quad \mathbf{a}_2^t \mathbf{x} = b_2$$
$$L_3: \quad \mathbf{a}_3^t \mathbf{x} = b_3$$

where

$$\mathbf{x} = \begin{bmatrix} x_1 \\ x_2 \end{bmatrix}, \qquad \mathbf{a}_1 = \begin{bmatrix} 1 \\ 1 \end{bmatrix}, \qquad \mathbf{a}_2 = \begin{bmatrix} 1 \\ -2 \end{bmatrix}, \qquad \mathbf{a}_3 = \begin{bmatrix} 3 \\ -1 \end{bmatrix},$$

$$b_1 = 2, \qquad b_2 = -2, \qquad b_3 = 3$$

then, using Theorem 11.18.1, the iteration scheme in Algorithm 1 can be expressed as

$$\mathbf{x}_k^{(p)} = \mathbf{x}_{k-1}^{(p)} + \frac{b_k - \mathbf{a}_k^t \mathbf{x}_{k-1}^{(p)}}{\mathbf{a}_k^t \mathbf{a}_k} \mathbf{a}_k, \qquad k = 1, 2, 3$$

where $p = 1$ for the first cycle of iterates, $p = 2$ for the second cycle of iterates, and so forth. After each cycle of iterates (i.e., after $\mathbf{x}_3^{(p)}$ is computed), the next cycle of iterates is begun with $\mathbf{x}_0$ set equal to $\mathbf{x}_3^{(p)}$.

Table 1 gives the numerical results of six cycles of iterations starting with the initial point $\mathbf{x}_0 = (1, 3)$.

TABLE 1

	x_1	x_2
$\mathbf{x}_0$	1.00000	3.00000
$\mathbf{x}_1^{(1)}$	.00000	2.00000
$\mathbf{x}_2^{(1)}$	.40000	1.20000
$\mathbf{x}_3^{(1)}$	1.30000	.90000
$\mathbf{x}_1^{(2)}$	1.20000	.80000
$\mathbf{x}_2^{(2)}$	.88000	1.44000
$\mathbf{x}_3^{(2)}$	1.42000	1.26000
$\mathbf{x}_1^{(3)}$	1.08000	.92000
$\mathbf{x}_2^{(3)}$	.83200	1.41600
$\mathbf{x}_3^{(3)}$	1.40800	1.22400
$\mathbf{x}_1^{(4)}$	1.09200	.90800
$\mathbf{x}_2^{(4)}$	.83680	1.41840
$\mathbf{x}_3^{(4)}$	1.40920	1.22760
$\mathbf{x}_1^{(5)}$	1.09080	.90920
$\mathbf{x}_2^{(5)}$	.83632	1.41816
$\mathbf{x}_3^{(5)}$	1.40908	1.22724
$\mathbf{x}_1^{(6)}$	1.09092	.90908
$\mathbf{x}_2^{(6)}$	.83637	1.41818
$\mathbf{x}_3^{(6)}$	1.40909	1.22728

Using certain techniques that are impractical for large linear systems, the exact values of the points of the limit cycle in this example can be shown to be

$$\mathbf{x}_1^* = (\tfrac{12}{11}, \tfrac{10}{11}) = (1.09090\ldots, .90909\ldots)$$
$$\mathbf{x}_2^* = (\tfrac{46}{55}, \tfrac{78}{55}) = (.83636\ldots, 1.41818\ldots)$$
$$\mathbf{x}_3^* = (\tfrac{31}{22}, \tfrac{27}{22}) = (1.40909\ldots, 1.22727\ldots)$$

It can be seen that the sixth cycle of iterates provides an excellent approximation to the limit cycle. Any one of the three iterates $\mathbf{x}_1^{(6)}$, $\mathbf{x}_2^{(6)}$, $\mathbf{x}_3^{(6)}$ can be used as an approximate solution of the linear system. (The large discrepancies in the values of $\mathbf{x}_1^{(6)}$, $\mathbf{x}_2^{(6)}$, and $\mathbf{x}_3^{(6)}$ are due to the artificial nature of this illustrative example. In practical problems these discrepancies would be much smaller.) ▲

To generalize Algorithm 1 so it applies to an overdetermined system of M equations in N unknowns,

$$a_{11}x_1 + a_{12}x_2 + \cdots + a_{1N}x_N = b_1$$
$$a_{21}x_1 + a_{22}x_2 + \cdots + a_{2N}x_N = b_2$$
$$\vdots \qquad\qquad\qquad\qquad \vdots$$
$$a_{M1}x_1 + a_{M2}x_2 + \cdots + a_{MN}x_N = b_M$$

we introduce column vectors $\mathbf{x}$ and $\mathbf{a}_i$ as follows:

$$\mathbf{x} = \begin{bmatrix} x_1 \\ x_2 \\ \vdots \\ x_N \end{bmatrix}, \qquad \mathbf{a}_i = \begin{bmatrix} a_{i1} \\ a_{i2} \\ \vdots \\ a_{iN} \end{bmatrix}, \qquad i = 1, 2, \ldots, M$$

With these vectors, the M equations constituting our linear system (6) can be written in vector form as

$$\mathbf{a}_i^t\mathbf{x} = b_i, \qquad i = 1, 2, \ldots, M$$

Each of these M equations defines what is called a **hyperplane** in the N-dimensional Euclidean space R^N. In general these M hyperplanes have no common intersection, and so we seek instead some point in R^N that is reasonably "close" to all of them. Such a point will constitute an approximate solution of the linear system, and its N entries will determine approximate pixel densities with which to form the desired cross section.

As in the two-dimensional case, we will introduce an iterative process that generates cycles of successive orthogonal projections onto the M hyperplanes beginning with some arbitrary initial point in R^N. Our notation for these successive iterates is

$$\mathbf{x}_k^{(p)} = \begin{pmatrix} \textit{the iterate lying on the kth hyperplane} \\ \textit{generated during the pth cycle of iterations} \end{pmatrix}$$

The algorithm is as follows:

Algorithm 2

Step 0. Choose any point in R^N and label it $\mathbf{x}_0$.

Step 1. For the first cycle of iterates, set $p = 1$.

Step 2. For $k = 1, 2, \ldots, M$, compute

$$\mathbf{x}_k^{(p)} = \mathbf{x}_{k-1}^{(p)} + \frac{(b_k - \mathbf{a}_k^t \mathbf{x}_{k-1}^{(p)})}{\mathbf{a}_k^t \mathbf{a}_k} \mathbf{a}_k$$

Step 3. Set $\mathbf{x}_0 = \mathbf{x}_M^{(p)}$.

Step 4. Increase the cycle number p by 1 and return to Step 2.

In Step 2 the iterate $\mathbf{x}_k^{(p)}$ is called the ***orthogonal projection*** of $\mathbf{x}_{k-1}^{(p)}$ onto the hyperplane $\mathbf{a}_k^t \mathbf{x} = b_k$. Consequently, as in the two-dimensional case, this algorithm determines a sequence of orthogonal projections from one hyperplane onto the next in which we cycle back to the first hyperplane after each projection onto the last hyperplane.

It can be shown that if the vectors $\mathbf{a}_1, \mathbf{a}_2, \ldots, \mathbf{a}_M$ span R^N, then the iterates $\mathbf{x}_M^{(1)}, \mathbf{x}_M^{(2)}, \mathbf{x}_M^{(3)}, \ldots$ lying on the Mth hyperplane will converge to a point $\mathbf{x}_M^*$ on that hyperplane which does not depend on the choice of the initial point $\mathbf{x}_0$. In computed tomography one of the iterates $\mathbf{x}_M^{(p)}$ for p sufficiently large is taken as an approximate solution of the linear system for the pixel densities.

Notice that for the center-of-pixel method, the scalar quantity $\mathbf{a}_k^t \mathbf{a}_k$ appearing in the equation in Step 2 of the algorithm is simply the number of pixels in which the kth beam passes through the center. Similarly, notice that the scalar quantity

$$b_k - \mathbf{a}_k^t \mathbf{x}_{k-1}^{(p)}$$

in that same equation can be interpreted as the *excess kth beam density* that results if the pixel densities are set equal to the entries of $\mathbf{x}_{k-1}^{(p)}$. This provides the following interpretation of our ART iteration scheme for the center-of-pixel method: *Generate the pixel densities of each iterate by distributing the excess beam density of successive beams in the scan evenly among those pixels in which the beam passes through the center. When the last beam in the scan has been reached, return to the first beam and continue.*

Example 2 We can use Algorithm 2 to find the unknown pixel densities of the 9 pixels arranged in the 3×3 array illustrated in Figure 11. These 9 pixels are scanned using the parallel mode with 12 beams whose measured beam densities are indicated in the figure. We choose the center-of-pixel method to set up the 12 beam equations. (In Exercises 7 and 8 the reader is asked to set up the beam equations using the center line and area methods.) As the reader can verify, the

Figure 11

beam equations are

$$x_7 + x_8 + x_9 = 13.00 \qquad x_3 + x_6 + x_9 = 18.00$$
$$x_4 + x_5 + x_6 = 15.00 \qquad x_2 + x_5 + x_8 = 12.00$$
$$x_1 + x_2 + x_3 = 8.00 \qquad x_1 + x_4 + x_7 = 6.00$$
$$x_6 + x_8 + x_9 = 14.79 \qquad x_2 + x_3 + x_6 = 10.51$$
$$x_3 + x_5 + x_7 = 14.31 \qquad x_1 + x_5 + x_9 = 16.13$$
$$x_1 + x_2 + x_4 = 3.81 \qquad x_4 + x_7 + x_8 = 7.04$$

Table 2 illustrates the results of the iteration scheme starting with an initial iterate $x_0 = 0$. The table gives the values of each of the first cycle of iterates, $x_1^{(1)}$ through $x_{12}^{(1)}$, but thereafter gives the iterates $x_{12}^{(p)}$ only for various values of p. The iterates $x_{12}^{(p)}$ start repeating to 2 decimal places for $p \geq 45$, and so we take the entries of $x_{12}^{(45)}$ as approximate values of the 9 pixel densities. ▲

We close this section by noting that the field of computed tomography is presently a very active research area. In fact, the ART scheme discussed here has been replaced in commercial systems by more sophisticated techniques that are faster and provide a more accurate view of the cross section. However, all the new techniques address the same basic mathematical problem: finding a good approximate solution of a large overdetermined inconsistent linear system of equations.

Readers interested in pursuing computed tomography further might find the following references useful:

1. Richard Gordon, Gabor T. Herman, and Steven A. Johnson, "Image Reconstruction from Projections," *Scientific American*, October 1975, pp. 56–68.
2. Gabor T. Herman, *Image Reconstruction from Projections—The Fundamentals of Computerized Tomography* (New York: Academic Press, 1980).
3. L. A. Shepp and J. B. Kruskal, "Computerized Tomography: The New Medical X-Ray Technology," *American Mathematical Monthly*, vol. 85, June–July 1978, pp. 420–439.
4. Richard Gordon, "A Tutorial on ART," *IEEE Transactions on Nuclear Science*, Vol. NS-21, June 1974, pp. 78–93.

TABLE 2

	Pixel Densities								
	x_1	x_2	x_3	x_4	x_5	x_6	x_7	x_8	x_9
$\mathbf{x}_0$	.00	.00	.00	.00	.00	.00	.00	.00	.00
$\mathbf{x}_1^{(1)}$	.00	.00	.00	.00	.00	.00	4.33	4.33	4.33
$\mathbf{x}_2^{(1)}$	.00	.00	.00	5.00	5.00	5.00	4.33	4.33	4.33
$\mathbf{x}_3^{(1)}$	2.67	2.67	2.67	5.00	5.00	5.00	4.33	4.33	4.33
$\mathbf{x}_4^{(1)}$	2.67	2.67	2.67	5.00	5.00	5.37	4.33	4.71	4.71
$\mathbf{x}_5^{(1)}$	2.67	2.67	3.44	5.00	5.77	5.37	5.10	4.71	4.71
$\mathbf{x}_6^{(1)}$	.49	.49	3.44	2.83	5.77	5.37	5.10	4.71	4.71
$\mathbf{x}_7^{(1)}$	.49	.49	4.93	2.83	5.77	6.87	5.10	4.71	6.20
$\mathbf{x}_8^{(1)}$	.49	.84	4.93	2.83	6.11	6.87	5.10	5.05	6.20
$\mathbf{x}_9^{(1)}$	−.31	.84	4.93	2.02	6.11	6.87	4.30	5.05	6.20
$\mathbf{x}_{10}^{(1)}$	−.31	.13	4.22	2.02	6.11	6.16	4.30	5.05	6.20
$\mathbf{x}_{11}^{(1)}$	1.06	.13	4.22	2.02	7.49	6.16	4.30	5.05	7.58
$\mathbf{x}_{12}^{(1)}$	1.06	.13	4.22	.58	7.49	6.16	2.85	3.61	7.58
$\mathbf{x}_{12}^{(2)}$	2.03	.69	4.42	1.34	7.49	5.39	2.65	3.04	6.61
$\mathbf{x}_{12}^{(3)}$	1.78	.51	4.52	1.26	7.49	5.48	2.56	3.22	6.86
$\mathbf{x}_{12}^{(4)}$	1.82	.52	4.62	1.37	7.49	5.37	2.45	3.22	6.82
$\mathbf{x}_{12}^{(5)}$	1.79	.49	4.71	1.43	7.49	5.31	2.37	3.25	6.85
$\mathbf{x}_{12}^{(10)}$	1.68	.44	5.03	1.70	7.49	5.03	2.04	3.29	6.96
$\mathbf{x}_{12}^{(20)}$	1.49	.48	5.29	2.00	7.49	4.73	1.79	3.25	7.15
$\mathbf{x}_{12}^{(30)}$	1.38	.55	5.34	2.11	7.49	4.62	1.74	3.19	7.26
$\mathbf{x}_{12}^{(40)}$	1.33	.59	5.33	2.14	7.49	4.59	1.75	3.15	7.31
$\mathbf{x}_{12}^{(45)}$	1.32	.60	5.32	2.15	7.49	4.59	1.76	3.14	7.32

←——————— First Cycle of Iterates ———————→

EXERCISE SET 11.18

1. (a) Setting $\mathbf{x}_k^{(p)} = (x_{k1}^{(p)}, x_{k2}^{(p)})$, show that the three projection equations

$$\mathbf{x}_k^{(p)} = \mathbf{x}_{k-1}^{(p)} + \frac{(b_k - \mathbf{a}_k^t \mathbf{x}_{k-1}^{(p)})}{\mathbf{a}_k^t \mathbf{a}_k}\, \mathbf{a}_k, \qquad k = 1, 2, 3$$

for the three lines in Equation (5) can be written as

$$k = 1: \quad \begin{aligned} x_{11}^{(p)} &= \tfrac{1}{2}[2 + x_{01}^{(p)} - x_{02}^{(p)}] \\ x_{12}^{(p)} &= \tfrac{1}{2}[2 - x_{01}^{(p)} + x_{02}^{(p)}] \end{aligned}$$

$$k = 2: \quad \begin{aligned} x_{21}^{(p)} &= \tfrac{1}{5}[-2 + 4x_{11}^{(p)} + 2x_{12}^{(p)}] \\ x_{22}^{(p)} &= \tfrac{1}{5}[4 + 2x_{11}^{(p)} + x_{12}^{(p)}] \end{aligned}$$

$$k = 3: \quad \begin{aligned} x_{31}^{(p)} &= \tfrac{1}{10}[9 + x_{21}^{(p)} + 3x_{22}^{(p)}] \\ x_{32}^{(p)} &= \tfrac{1}{10}[-3 + 3x_{21}^{(p)} + 9x_{22}^{(p)}] \end{aligned}$$

where $(x_{01}^{(p+1)}, x_{02}^{(p+1)}) = (x_{31}^{(p)}, x_{32}^{(p)})$ for $p = 1, 2, \ldots$.

   (b) Show that the three pairs of equations in part (a) can be combined to produce

$$\begin{aligned} x_{31}^{(p)} &= \tfrac{1}{20}[28 + x_{31}^{(p-1)} - x_{32}^{(p-1)}] \\ x_{32}^{(p)} &= \tfrac{1}{20}[24 + 3x_{31}^{(p-1)} - 3x_{32}^{(p-1)}] \end{aligned} \qquad p = 1, 2, \ldots$$

where $(x_{31}^{(0)}, x_{32}^{(0)}) = (x_{01}^{(1)}, x_{02}^{(1)}) = \mathbf{x}_0^{(1)}$. [**Note.** Using this pair of equations, one complete cycle of three orthogonal projections can be performed in a single step.]

   (c) Because $\mathbf{x}_3^{(p)}$ tends to the limit point $\mathbf{x}_3^*$ as $p \to \infty$, the equations in part (b) become

$$\begin{aligned} x_{31}^* &= \tfrac{1}{20}[28 + x_{31}^* - x_{32}^*] \\ x_{32}^* &= \tfrac{1}{20}[24 + 3x_{31}^* - 3x_{32}^*] \end{aligned}$$

as $p \to \infty$. Solve this linear system for $\mathbf{x}_3^* - (x_{31}^*, x_{32}^*)$. [**Note.** The simplifications of the ART formulas described in this exercise are impractical for the large linear systems that arise in realistic computed tomography problems.]

2. Use the result of Exercise 1(b) to find $\mathbf{x}_3^{(1)}, \mathbf{x}_3^{(2)}, \ldots, x_3^{(6)}$ to five decimal places in Example 2 using the following initial points:
   (a) $\mathbf{x}_0 = (0, 0)$ (b) $\mathbf{x}_0 = (1, 1)$ (c) $\mathbf{x}_0 = (148, -15)$

3. (a) Show directly that the points of the limit cycle in Example 1,

$$\mathbf{x}_1^* = (\tfrac{12}{11}, \tfrac{10}{11}), \qquad \mathbf{x}_2^* = (\tfrac{46}{55}, \tfrac{78}{55}), \qquad \mathbf{x}_3^* = (\tfrac{31}{22}, \tfrac{27}{22})$$

form a triangle whose vertices lie on the lines $L_1, L_2,$ and L_3 and whose sides are perpendicular to these lines (Figure 9c).

   (b) Using the equations derived in Exercise 1(a), show that if $\mathbf{x}_0^{(1)} = \mathbf{x}_3^* = (\tfrac{31}{22}, \tfrac{27}{22})$, then

$$\mathbf{x}_1^{(1)} = \mathbf{x}_1^* = (\tfrac{12}{11}, \tfrac{10}{11})$$
$$\mathbf{x}_2^{(1)} = \mathbf{x}_2^* = (\tfrac{46}{55}, \tfrac{78}{55})$$
$$\mathbf{x}_3^{(1)} = \mathbf{x}_3^* = (\tfrac{31}{22}, \tfrac{27}{22})$$

[**Note.** Either part of this exercise shows that successive orthogonal projections of any point on the limit cycle will move around the limit cycle indefinitely.]

4. The following three lines in the $x_1 x_2$-plane,

$$L_1: \qquad x_2 = 1$$
$$L_2: \quad x_1 - x_2 = 2$$
$$L_3: \quad x_1 - x_2 = 0$$

do not have a common intersection. Draw an accurate sketch of the three lines and graphically perform several cycles of the orthogonal projections described in Algorithm 1, beginning with the initial point $\mathbf{x}_0 = (0, 0)$. On the basis of your sketch, determine the three points of the limit cycle.

5. Prove Theorem 11.18.1 by verifying that
   (a) the point $\mathbf{x}_p$ as defined in the theorem lies on the line $\mathbf{a}^t \mathbf{x} = b$ (i.e., $\mathbf{a}^t \mathbf{x}_p = b$);
   (b) the vector $\mathbf{x}_p - \mathbf{x}^*$ is orthogonal to the line $\mathbf{a}^t \mathbf{x} = b$ (i.e., $\mathbf{x}_p - \mathbf{x}^*$ is parallel to $\mathbf{a}$).

6. As stated in the text, the iterates $\mathbf{x}_M^{(1)}, \mathbf{x}_M^{(2)}, \mathbf{x}_M^{(3)}, \ldots$ defined in Algorithm 2 will converge to a unique limit point $\mathbf{x}_M^*$ if the vectors $\mathbf{a}_1, \mathbf{a}_2, \ldots, \mathbf{a}_M$ span R^N. Show that if this is the case and if the center-of-pixel method is used, then the center of each of the N pixels in the field of view is crossed by at least one of the M beams in the scan.

7. Construct the 12 beam equations in Example 2 using the center line method. Assume that the distance between the center lines of adjacent beams is equal to the width of a single pixel.

8. Construct the 12 beam equations in Example 2 using the area method. Assume that the width of each beam is equal to the width of a single pixel and that the distance between the center lines of adjacent beams is also equal to the width of a single pixel.

11.19 ELECTRICAL NETWORKS

Basic laws of electrical circuits are discussed, and it is shown how these laws can be used to obtain systems of linear equations whose solutions yield the currents flowing in an electrical circuit.

PREREQUISITE: Linear systems

The simplest electrical circuits consist of two basic components:

electrical sources   denoted by

resistors   denoted by

Electrical sources, such as a batteries, create currents in an electrical circuit. Resistors, such as light bulbs, limit the magnitudes of the currents.

There are three basic quantities associated with electrical circuits: *electrical potential* (E), *resistance* (R), and *current* (I). These are commonly measured in the following units:

$$E \quad \text{in} \quad \text{volts} \quad \text{(V)}$$
$$R \quad \text{in} \quad \text{ohms} \quad (\Omega)$$
$$I \quad \text{in} \quad \text{amperes} \quad \text{(A)}$$

Electrical potential is associated with two points in an electrical circuit and is measured in practice by connecting those points to a device called a *voltmeter*. For example, a common AA battery is rated at 1.5 volts, which means that this is the electrical potential across its positive and negative terminals (Figure 1).

Figure 1

In an electrical circuit the electrical potential between two points is called the *voltage drop* between those points. As we shall see, currents and voltage drops can be either positive or negative. (A negative voltage drop is called a *voltage rise*.)

The flow of current in an electrical circuit is governed by three basic principles:

1. ***Ohm's Law*** The voltage drop across a resistor is the product of the current passing through it and its resistance; that is, $E = IR$.
2. ***Kirchhoff's Current Law*** The sum of the currents flowing into any point equals the sum of the currents flowing out from the point.
3. ***Kirchhoff's Voltage Law*** Around any closed loop the sum of the voltage drops is equal to the sum of the voltage rises.

Example 1 Find the unknown currents I_1, I_2, and I_3 in the circuit shown in Figure 2.

Solution. The flow directions for the currents I_1, I_2, and I_3 (marked by the arrowheads) were picked arbitrarily. Any of these currents that turn out to be negative actually flow opposite to the direction selected.

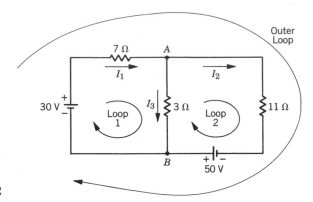

Figure 2

Applying Kirchhoff's current law to points A and B yields

$$I_1 = I_2 + I_3 \quad \text{(Point } A\text{)}$$
$$I_3 + I_2 = I_1 \qquad \text{(Point } B\text{)}$$

Since these equations both simplify to the same linear equation

$$I_1 - I_2 - I_3 = 0 \tag{1}$$

we still need two more equations to determine I_1, I_2, and I_3 uniquely. We will obtain them using Kirchhoff's voltage law.

To apply Kirchhoff's voltage law to a loop, select a positive direction around the loop (say clockwise) and make the following sign conventions:

- A current passing through a resistor produces a voltage drop if it flows in the positive direction of the loop and a voltage rise if it flows in the negative direction of the loop.

- A current passing through an electrical source in the positive direction of the loop produces a voltage drop if it flows from $+$ to $-$ and a voltage rise if it flows from $-$ to $+$. A current passing through an electrical source in the negative direction of the loop produces a voltage rise if it flows from $+$ to $-$ and a voltage drop if it flows from $-$ to $+$.

Applying Kirchhoff's voltage law and Ohm's law to Loop 1 in Figure 1 yields

$$7I_1 + 3I_3 = 30 \tag{2}$$

and to the Loop 2 yields

$$11I_2 = 3I_3 + 50 \tag{3}$$

Combining (1), (2), and (3) yields the linear system

$$
\begin{aligned}
I_1 - \quad I_2 - \ I_3 &= \ 0 \\
7I_1 \qquad\quad + 3I_3 &= 30 \\
11I_2 - 3I_3 &= 50
\end{aligned}
$$

Solving this linear system yields the following values for the currents:

$$I_1 = \tfrac{570}{131}(A), \qquad I_2 = \tfrac{590}{131}(A), \qquad I_3 = -\tfrac{20}{131}(A)$$

Note that I_3 is negative, which means that this current flows opposite to the direction indicated in Figure 2. Also note that we could have applied Kirchhoff's voltage law to the outer loop of the circuit. However, this produces a redundant equation (try it). ▲

EXERCISE SET 11.19

In Exercises 1–4 find the currents in the circuits.

1.

2.

3.

4.

5. Show that if the current I_5 in the circuit below is zero, then $R_4 = R_3R_2/R_1$.

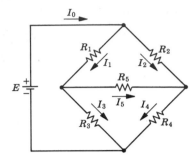

[***Remark.*** This circuit, called a Wheatstone bridge circuit, is used for the precise measurement of resistance. Here, R_4 is an unknown resistance and $R_1, R_2,$ and R_3 are adjustable calibrated resistors. R_5 represents a galvanometer—a device for measuring current. After varying the resistances $R_1, R_2,$ and R_3 until the galvanometer reading is zero, the formula $R_4 = R_3R_2/R_1$ determines the unknown resistance R_4.]

6. Show that if the two currents labeled I in the circuits below are equal, then

$$R = \frac{1}{\dfrac{1}{R_1} + \dfrac{1}{R_2}}.$$

11.20 FRACTALS

In this section certain classes of linear transformations are used to describe and generate intricate sets in the Euclidean plane called fractals. Such sets are currently the focus of much mathematical and scientific research.

PREREQUISITES: Linear operators on R^2 (Section 8.2)
Euclidean space R^n
Natural logarithms
Intuitive understanding of limits

At the end of the nineteenth century and beginning of the twentieth century, various bizarre and wild sets of points in the Euclidean plane began appearing in mathematics. Although they were initially mathematical curiosities, these sets, called *fractals*, are rapidly growing in importance. It is now recognized that they reveal a regularity in physical and biological phenomena previously dismissed as "random," "noisy," or "chaotic." For example, fractals are around us in the shapes of clouds, mountains, coastlines, trees, and ferns.

In this section we give a brief description of certain types of fractals in the Euclidean plane R^2. Much of this description is due to two mathematicians, Benoit B. Mandelbrot and Michael Barnsley, who are both active researchers in the field.

SELF-SIMILAR SETS

To start our study of fractals, we need to introduce some terminology about sets in R^2. We shall call a set in R^2 **bounded** if it can be enclosed by a large enough circle (Figure 1) and **closed** if it contains all of its boundary points (Figure 2). Two sets in R^2 will be called **congruent** if they can be made to coincide exactly by translating and rotating them appropriately within R^2 (Figure 3). We will also rely on the reader's intuitive concept of **overlapping** and **nonoverlapping sets**, as illustrated in Figure 4.

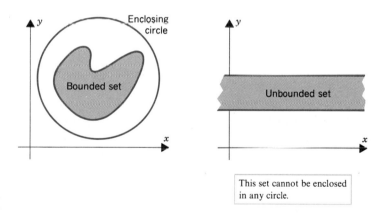

Figure 1

This set cannot be enclosed in any circle.

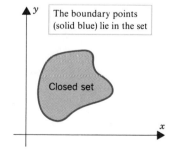

Figure 2

The boundary points (solid blue) lie in the set

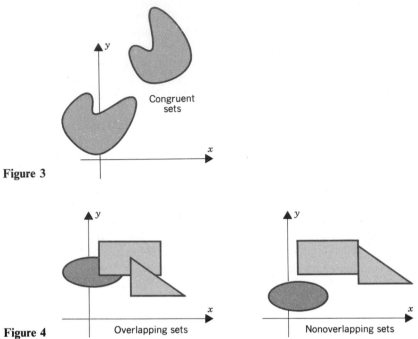

Figure 3

Figure 4

If $T:R^2 \rightarrow R^2$ is the linear operator that scales by a factor of s (see Example 7 of Section 8.2), and if Q is a set in R^2, then the set $T(Q)$ (the set of images of points in Q under T) is called an ***expansion*** of the set Q if $s > 1$ and a ***contraction*** of Q if $0 < s < 1$ (Figure 5). In either case we say that $T(Q)$ is the set Q **scaled by the factor s.**

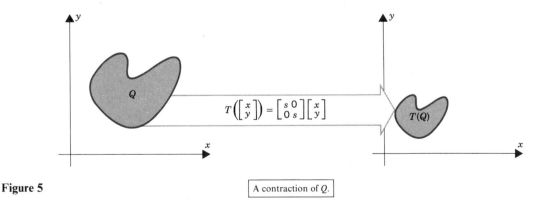

Figure 5

A contraction of Q.

The types of fractals we shall consider first are called *self-similar*. In general, we define a self-similar set in R^2 as follows:

Definition. A closed and bounded subset of the Euclidean plane R^2 is said to be ***self-similar*** if it can be expressed in the form

$$S = S_1 \cup S_2 \cup S_3 \cup \cdots \cup S_k \tag{1}$$

where $S_1, S_2, S_3, \ldots, S_k$ are nonoverlapping sets, each of which is congruent to S scaled by the same factor s ($0 < s < 1$).

If S is a self-similar set, then (1) is sometimes called a ***decomposition*** of S into nonoverlapping congruent sets.

Example 1   (Line Segment)   A line segment in R^2 (Figure 6a) can be expressed as the union of two nonoverlapping congruent line segments (Figure 6b). In Figure 6b we have separated the two line segments slightly so that they can be seen more easily. Each of these two smaller line segments is congruent to the original line segment scaled by a factor of $\frac{1}{2}$. Hence, a line segment is a self-similar set with $k = 2$ and $s = \frac{1}{2}$. ▲

Figure 6   (a)   (b)

Example 2   (Square)   A square (Figure 7a) can be expressed as the union of four nonoverlapping congruent squares (Figure 7b), where we have again separated the smaller squares slightly. Each of the four smaller squares is congruent to the original square scaled by a factor of $\frac{1}{2}$. Hence, a square is a self-similar set with $k = 4$ and $s = \frac{1}{2}$. ▲

Figure 7   (a)   (b)

Example 3 (Sierpinski Carpet) The set suggested by Figure 8*a* was first described by the Polish mathematician Waclaw Sierpinski (1882–1969). It can be expressed as the union of eight nonoverlapping congruent subsets (Figure 8*b*), each of which is congruent to the original set scaled by a factor of $\frac{1}{3}$. Hence, it is a self-similar set with $k = 8$ and $s = \frac{1}{3}$. Notice that the intricate square-within-a-square pattern continues forever on a smaller and smaller scale (although this can only be suggested in a figure such as the one shown). ▲

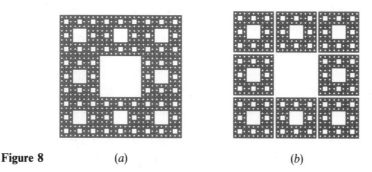

Figure 8   (*a*)   (*b*)

Example 4 (Sierpinski Triangle) Figure 9*a* illustrates another set due to Sierpinski. It is a self-similar set with $k = 3$ and $s = \frac{1}{2}$ (Figure 9*b*). As with the Sierpinski carpet, the intricate triangle-within-a-triangle pattern continues forever on a smaller and smaller scale. ▲

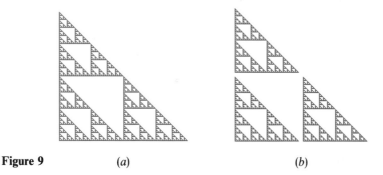

Figure 9   (*a*)   (*b*)

The Sierpinski carpet and triangle have a more intricate structure than the line segment and the square in that they exhibit a pattern that is repeated indefinitely. This difference will be explored more fully later in this section.

TOPOLOGICAL DIMENSION OF A SET

In Section 4.5 we defined the dimension of a *subspace* of a vector space to be the number of vectors in a basis, and we found that definition to coincide with our intuitive sense of dimension. For example, the origin of R^2 is zero-dimensional, lines through the origin are one-dimensional, and R^2 itself is two-dimensional.

This definition of dimension is a special case of a more general concept, called **topological dimension**, which is applicable to sets in R^n that are not necessarily subspaces. A precise definition of this concept is studied in a branch of mathematics called *topology*. Although that definition is beyond the scope of this text, we can state informally that

- a point in R^2 has topological dimension zero;
- a curve in R^2 has topological dimension one;
- a region in R^2 has topological dimension two.

It can be proved that the topological dimension of a set in R^n must be an integer between 0 and n, inclusive. In this text we shall denote the topological dimension of a set S by $d_T(S)$.

Example 5 The following table gives the topological dimensions of the sets studied in our earlier examples.

TABLE 1

Set S	$d_T(S)$
Line segment	1
Square	2
Sierpinski carpet	1
Sierpinski triangle	1

The first two results in this table are intuitively obvious; however, the last two are not. Informally stated, both the Sierpinski carpet and triangle contain so many "holes" that those sets resemble web-like networks of lines rather than regions. Hence they have topological dimension one. The proofs are quite difficult. ▲

HAUSDORFF DIMENSION OF A SELF-SIMILAR SET

In 1919 the German mathematician Felix Hausdorff (1868–1942) gave an alternative definition for the dimension of an arbitrary set in R^n. His definition is quite complicated, but for a self-similar set it reduces to something rather simple:

> **Definition.** The **Hausdorff dimension** of a self-similar set S of form (1) is denoted by $d_H(S)$ and is defined by
> $$d_H(S) = \frac{\ln k}{\ln(1/s)} \tag{2}$$

In this definition *ln* denotes the natural logarithm function. Equation (2) can also be expressed as

$$s^{d_H(S)} = \frac{1}{k} \tag{3}$$

in which the Hausdorff dimension $d_H(S)$ appears as an exponent. Formula (3) is more helpful for interpreting the concept of Hausdorff dimension; it states, for example, that if you scale a self-similar set by a factor of $s = \frac{1}{2}$, then its area (or, more properly, its *measure*) decreases by a factor of $(\frac{1}{2})^{d_H(S)}$. Thus, scaling a line segment by a factor of $\frac{1}{2}$ reduces its measure (length) by a factor of $(\frac{1}{2})^1 = \frac{1}{2}$, and scaling a square region by a factor of $\frac{1}{2}$ reduces its measure (area) by a factor of $(\frac{1}{2})^2 = \frac{1}{4}$.

Before proceeding to some examples, a few facts about the Hausdorff dimension of a set are in order:

- The topological dimension and Hausdorff dimension of a set need not be the same.

- The Hausdorff dimension of a set need not be an integer.

- The topological dimension of a set is less than or equal to its Hausdorff dimension, that is $d_T(S) \le d_H(S)$.

Example 6 The following table lists the Hausdorff dimensions of the sets studied in our earlier examples.

TABLE 2

Set S	s	k	$d_H(S) = \dfrac{\ln k}{\ln(1/s)}$
Line segment	$\frac{1}{2}$	2	$\ln 2/\ln 2 = 1$
Square	$\frac{1}{2}$	4	$\ln 4/\ln 2 = 2$
Sierpinski carpet	$\frac{1}{3}$	8	$\ln 8/\ln 3 = 1.893\ldots$
Sierpinski triangle	$\frac{1}{2}$	3	$\ln 3/\ln 2 = 1.585\ldots$

FRACTALS

Comparing Tables 1 and 2, we see that the Hausdorff and topological dimensions are equal for both the line segment and square, but unequal for the Sierpinski carpet and triangle. In 1977 Benoit B. Mandelbrot suggested that sets for which the topological and Hausdorff dimensions differ must be quite complicated (as Hausdorff had earlier suggested in 1919). Mandelbrot proposed calling such sets *fractals* and he offered the following definition:

> **Definition.** A *fractal* is a subset of a Euclidean space whose Hausdorff dimension and topological dimension are not equal.

Mandelbrot has suggested that this definition is rather limited and probably will be replaced in the future. But in the meantime, it remains the formal definition of a fractal. According to this definition the Sierpinski carpet and Sierpinski triangle are fractals, whereas the line segment and square are not.

It follows from the foregoing definition that a set whose Hausdorff dimension is not an integer must be a fractal (why?). However, we will see later that the converse is not true; that is, it is possible for a fractal to have an integer Hausdorff dimension.

SIMILITUDES

We shall now show how some techniques from linear algebra can be used to generate fractals. This linear algebra approach also leads to algorithms that can be exploited to draw fractals on a computer. We begin with a definition.

> **Definition.** A *similitude* with scale factor s is a mapping of R^2 into R^2 of the form
>
> $$T\left(\begin{bmatrix} x \\ y \end{bmatrix}\right) = s\begin{bmatrix} \cos\theta & -\sin\theta \\ \sin\theta & \cos\theta \end{bmatrix}\begin{bmatrix} x \\ y \end{bmatrix} + \begin{bmatrix} e \\ f \end{bmatrix}$$
>
> where s, θ, e, and f are scalars.

Geometrically, a similitude is a composition of three simpler mappings: a scaling by a factor of s, a rotation about the origin through an angle θ, and a translation (e units in the x-direction and f units in the y-direction). Figure 10 illustrates the effect of a similitude on the unit square U.

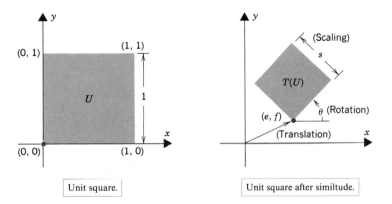

Figure 10

| Unit square. | Unit square after similtude. |

For our application to fractals we shall only need similitudes that are *contractions*, by which we mean that the scale factor s is restricted to the range $0 < s < 1$. Consequently, when we refer to similitudes we shall always mean similitudes subject to this restriction.

Similitudes are important in the study of fractals because of the following fact:

> If $T:R^2 \rightarrow R^2$ is a similitude with scale factor s and S is a closed and bounded set in R^2, then the image $T(S)$ of the set S under T is congruent to S scaled by s.

Recall from the definition of a self-similar set in R^2 that a closed and bounded set S in R^2 is self-similar if it can be expressed in the form

$$S = S_1 \cup S_2 \cup S_3 \cup \cdots \cup S_k$$

where $S_1, S_2, S_3, \ldots, S_k$ are nonoverlapping sets each of which is congruent to S scaled by the same factor s $(0 < s < 1)$ [see (1)]. In the following examples we will find similitudes that produce the sets $S_1, S_2, S_3, \ldots, S_k$ from S for the line segment, square, Sierpinski carpet, and Sierpinski triangle.

Example 7 **(Line Segment)** We shall take as our line segment the line S connecting the points $(0, 0)$ and $(1, 0)$ in the xy-plane (Figure 11). Consider the two similitudes

$$T_1\left(\begin{bmatrix} x \\ y \end{bmatrix}\right) = \frac{1}{2}\begin{bmatrix} 1 & 0 \\ 0 & 1 \end{bmatrix}\begin{bmatrix} x \\ y \end{bmatrix}$$

$$T_2\left(\begin{bmatrix} x \\ y \end{bmatrix}\right) = \frac{1}{2}\begin{bmatrix} 1 & 0 \\ 0 & 1 \end{bmatrix}\begin{bmatrix} x \\ y \end{bmatrix} + \begin{bmatrix} \frac{1}{2} \\ 0 \end{bmatrix} \tag{4}$$

both of which have $s = \frac{1}{2}$ and $\theta = 0$. In Figure 11 we show how these two similitudes map the unit square U. The similitude T_1 maps U onto the smaller square $T_1(U)$, and the similitude T_2 maps U onto the smaller square $T_2(U)$. At the same time, T_1 maps the line segment S onto the smaller line segment $T_1(S)$, and T_2 maps S onto the smaller nonoverlapping line segment $T_2(S)$. The union of these two smaller nonoverlapping line segments is precisely the original line segment S; that is,

$$S = T_1(S) \cup T_2(S) \quad \blacktriangle \tag{5}$$

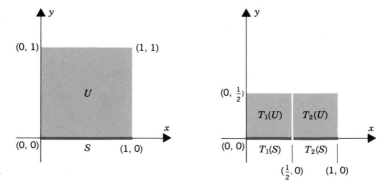

Figure 11

Example 8   (Square)   Let us consider the unit square U in the xy-plane and the following four similitudes, all having $s = \frac{1}{2}$ and $\theta = 0$:

$$T_1\left(\begin{bmatrix} x \\ y \end{bmatrix}\right) = \frac{1}{2}\begin{bmatrix} 1 & 0 \\ 0 & 1 \end{bmatrix}\begin{bmatrix} x \\ y \end{bmatrix}$$

$$T_2\left(\begin{bmatrix} x \\ y \end{bmatrix}\right) = \frac{1}{2}\begin{bmatrix} 1 & 0 \\ 0 & 1 \end{bmatrix}\begin{bmatrix} x \\ y \end{bmatrix} + \begin{bmatrix} \frac{1}{2} \\ 0 \end{bmatrix}$$

$$T_3\left(\begin{bmatrix} x \\ y \end{bmatrix}\right) = \frac{1}{2}\begin{bmatrix} 1 & 0 \\ 0 & 1 \end{bmatrix}\begin{bmatrix} x \\ y \end{bmatrix} + \begin{bmatrix} 0 \\ \frac{1}{2} \end{bmatrix}$$

$$T_4\left(\begin{bmatrix} x \\ y \end{bmatrix}\right) = \frac{1}{2}\begin{bmatrix} 1 & 0 \\ 0 & 1 \end{bmatrix}\begin{bmatrix} x \\ y \end{bmatrix} + \begin{bmatrix} \frac{1}{2} \\ \frac{1}{2} \end{bmatrix}$$

(6)

The images of the unit square U under these four similitudes are the four squares shown in Figure 12. Thus,

$$U = T_1(U) \cup T_2(U) \cup T_3(U) \cup T_4(U) \tag{7}$$

is a decomposition of U into four nonoverlapping squares that are congruent to U with the same scale factor ($s = \frac{1}{2}$). ▲

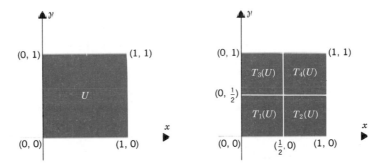

Figure 12

Example 9   (Sierpinski Carpet)   Let us consider a Sierpinski carpet S over the unit square U of the xy-plane (Figure 13) and the following eight similitudes, all having $s = \frac{1}{3}$ and $\theta = 0$:

$$T_i\left(\begin{bmatrix} x \\ y \end{bmatrix}\right) = \frac{1}{3}\begin{bmatrix} 1 & 0 \\ 0 & 1 \end{bmatrix}\begin{bmatrix} x \\ y \end{bmatrix} + \begin{bmatrix} e_i \\ f_i \end{bmatrix}, \qquad i = 1, 2, 3, \ldots, 8 \tag{8}$$

where the eight values of $\begin{bmatrix} e_i \\ f_i \end{bmatrix}$ are

$$\begin{bmatrix} 0 \\ 0 \end{bmatrix}, \ \begin{bmatrix} \frac{1}{3} \\ 0 \end{bmatrix}, \ \begin{bmatrix} \frac{2}{3} \\ 0 \end{bmatrix}, \ \begin{bmatrix} 0 \\ \frac{1}{3} \end{bmatrix}, \ \begin{bmatrix} \frac{2}{3} \\ \frac{1}{3} \end{bmatrix}, \ \begin{bmatrix} 0 \\ \frac{2}{3} \end{bmatrix}, \ \begin{bmatrix} \frac{1}{3} \\ \frac{2}{3} \end{bmatrix}, \ \begin{bmatrix} \frac{2}{3} \\ \frac{2}{3} \end{bmatrix}$$

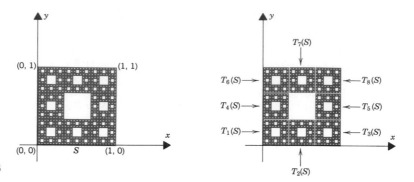

Figure 13

The images of S under these eight similitudes are the eight sets shown in Figure 13. Thus,

$$S = T_1(S) \cup T_2(S) \cup T_3(S) \cup \cdots \cup T_8(S) \tag{9}$$

is a decomposition of S into eight nonoverlapping sets that are congruent to S with the same scale factor ($s = \frac{1}{3}$). ▲

Example 10   (Sierpinski Triangle)   Let us consider a Sierpinski triangle S fitted inside the unit square U of the xy-plane, as shown in Figure 14, and the following three similitudes, all having $s = \frac{1}{2}$ and $\theta = 0$:

$$T_1\left(\begin{bmatrix} x \\ y \end{bmatrix}\right) = \frac{1}{2}\begin{bmatrix} 1 & 0 \\ 0 & 1 \end{bmatrix}\begin{bmatrix} x \\ y \end{bmatrix}$$

$$T_2\left(\begin{bmatrix} x \\ y \end{bmatrix}\right) = \frac{1}{2}\begin{bmatrix} 1 & 0 \\ 0 & 1 \end{bmatrix}\begin{bmatrix} x \\ y \end{bmatrix} + \begin{bmatrix} \frac{1}{2} \\ 0 \end{bmatrix} \tag{10}$$

$$T_3\left(\begin{bmatrix} x \\ y \end{bmatrix}\right) = \frac{1}{2}\begin{bmatrix} 1 & 0 \\ 0 & 1 \end{bmatrix}\begin{bmatrix} x \\ y \end{bmatrix} + \begin{bmatrix} 0 \\ \frac{1}{2} \end{bmatrix}$$

The images of S under these three similitudes are the three sets in Figure 14. Thus,

$$S = T_1(S) \cup T_2(S) \cup T_3(S) \tag{11}$$

is a decomposition of S into three nonoverlapping sets that are congruent to S with the same scale factor ($s = \frac{1}{2}$). ▲

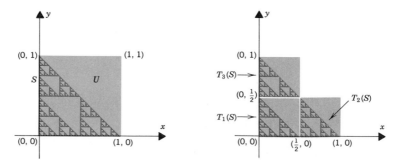

Figure 14

In the preceding examples we started with a specific set S and showed that it was self-similar by finding similitudes $T_1, T_2, T_3, \ldots, T_k$ with the same scale factor such that $T_1(S), T_2(S), T_3(S), \ldots, T_k(S)$ were nonoverlapping sets and such that

$$S = T_1(S) \cup T_2(S) \cup T_3(S) \cup \cdots \cup T_k(S) \tag{12}$$

The following theorem addresses the converse problem of determining a self-similar set from a collection of similitudes.

Theorem 11.20.1. *If $T_1, T_2, T_3, \ldots, T_k$ are contracting similitudes with the same scale factor, then there is a unique nonempty closed and bounded set S such that*

$$S = T_1(S) \cup T_2(S) \cup T_3(S) \cup \cdots \cup T_k(S)$$

Furthermore, if the sets $T_1(S), T_2(S), T_3(S), \ldots, T_k(S)$ are nonoverlapping, then S is self-similar.

ALGORITHMS FOR GENERATING FRACTALS

In general, there is no simple way to obtain the set S in the foregoing theorem. We now describe an iterative procedure that will determine S from the similitudes that define it. We first give an example of the procedure and then give an algorithm for the general case.

Example 11   (Sierpinski Carpet)   Figure 15 shows the unit square region S_0 in the xy-plane, which will serve as an "initial" set for an iterative procedure for the construction of the Sierpinski carpet. The set S_1 in the figure is the result of mapping S_0 with each of the eight similitudes $T_i\ (i = 1, 2, \ldots, 8)$ in (8) that deter-

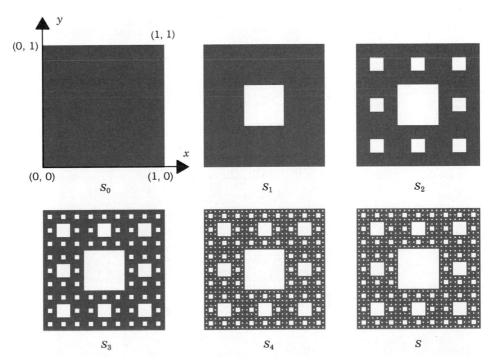

Figure 15

mine the Sierpinski carpet. It consists of eight square regions, each of side length $\frac{1}{3}$, surrounding an empty middle square. We next apply the eight similitudes to S_1 and arrive at the set S_2. Similarly, applying the eight similitudes to S_2 results in the set S_3. If we continue this process indefinitely, the sequence of sets S_1, S_2, S_3, ... will "converge" to a set S, which is the Sierpinski carpet. ▲

REMARK. Although we should properly give a definition of what it means for a sequence of sets to "converge" to a given set, an intuitive interpretation will suffice in this introductory treatment.

Although we started in Figure 15 with the unit square region to arrive at the Sierpinski carpet, we could have started with any nonempty set S_0. The only restriction is that the set S_0 be closed and bounded. For example, if we start with the particular set S_0 shown in Figure 16, then S_1 is the set obtained by applying each of the eight similitudes in (8). Applying the eight similitudes to S_1 results in the set S_2. As before, applying the eight similitudes indefinitely yields the Sierpinski carpet S as the limiting set.

The general algorithm illustrated in the preceding example is as follows: Let T_1, T_2, T_3, ..., T_k be contracting similitudes with the same scale factor, and for an arbitrary set Q in R^2 define the set $\mathcal{T}(Q)$ by

$$\mathcal{T}(Q) = T_1(Q) \cup T_2(Q) \cup T_3(Q) \cup \cdots \cup T_k(Q)$$

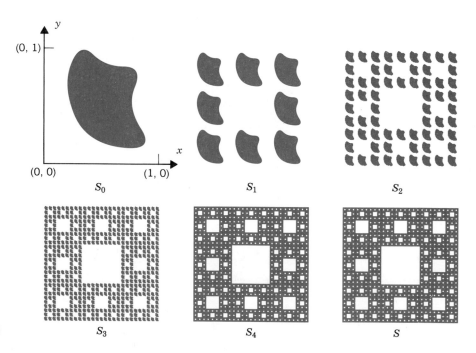

Figure 16

The following algorithm generates a sequence of sets $S_0, S_1, \ldots, S_n, \ldots$ that converges to the set S in Theorem 11.20.1.

Algorithm 1

Step 0. Choose an arbitrary nonempty closed and bounded set S_0 in R^2.

Step 1. Compute $S_1 = \mathcal{T}(S_0)$.

Step 2. Compute $S_2 = \mathcal{T}(S_1)$.

Step 3. Compute $S_3 = \mathcal{T}(S_2)$.

$\vdots$

Step n. Compute $S_n = \mathcal{T}(S_{n-1})$.

$\vdots$

Example 12   (Sierpinski Triangle)   Let us construct the Sierpinski triangle determined by the three similitudes given in (10). The corresponding set mapping is $\mathcal{T}(Q) = T_1(Q) \cup T_2(Q) \cup T_3(Q)$. Figure 17 shows an arbitrary closed and bounded set S_0, the first four iterates S_1, S_2, S_3, S_4, and the limiting set S (the Sierpinski triangle). ▲

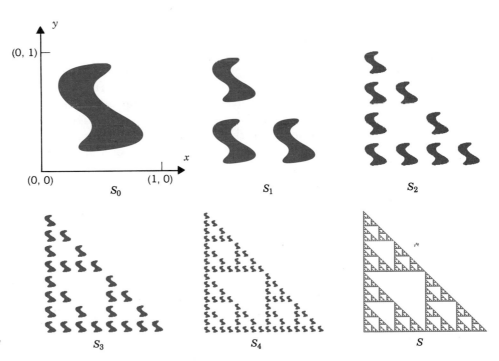

Figure 17

Example 13 Consider the following two similitudes:

$$T_1\left(\begin{bmatrix} x \\ y \end{bmatrix}\right) = \frac{1}{2}\begin{bmatrix} 1 & 0 \\ 0 & 1 \end{bmatrix}$$

$$T_2\left(\begin{bmatrix} x \\ y \end{bmatrix}\right) = \frac{1}{2}\begin{bmatrix} \cos\theta & -\sin\theta \\ \sin\theta & \cos\theta \end{bmatrix}\begin{bmatrix} x \\ y \end{bmatrix} + \begin{bmatrix} .3 \\ .3 \end{bmatrix}$$

The actions of these two similitudes on the unit square U are illustrated in Figure 18.

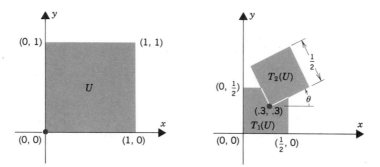

Figure 18

Here, the rotation angle θ is a parameter that we shall vary to generate different self-similar sets. The self-similar sets determined by these two similitudes are shown in Figure 19 for various values of θ. For simplicity, we have not drawn the xy axes, but in each case the origin is the lower left point of the set. These sets were generated on a computer using Algorithm 1 for the various values of θ. Because $k = 2$ and $s = \frac{1}{2}$, it follows from (2) that the Hausdorff dimension of these sets for any value of θ is 1. It can be shown that the topological dimension of these sets is 1 for $\theta = 0$ and 0 for all other values of θ. It follows that the self-similar set for $\theta = 0$ is not a fractal [it is the straight line segment from $(0, 0)$ to $(.6, .6)$], while the self-similar sets for all other values of θ are fractals. In particular, they are examples of fractals with integer Hausdorff dimension. ▲

Figure 19 $\theta = 60°$ $\theta = 50°$ $\theta = 40°$ $\theta = 30°$ $\theta = 20°$ $\theta = 10°$ $\theta = 0°$

A MONTE CARLO APPROACH

The set-mapping approach of constructing self-similar sets described in Algorithm 1 is rather time consuming on a computer because the similitudes involved must be applied to each of the many computer screen pixels in the successive iterated sets. In 1985 Michael Barnsley described an alternative, more practical, method of generating a self-similar set defined through its similitudes. It is a so-called *Monte Carlo method* that takes advantage of probability theory. Barnsley refers to it as the *Random Iteration Algorithm*.

Let $T_1, T_2, T_3, \ldots, T_k$ be contracting similitudes with the same scale factor. The following algorithm generates a sequence of points

$$\begin{bmatrix} x_0 \\ y_0 \end{bmatrix}, \begin{bmatrix} x_1 \\ y_1 \end{bmatrix}, \ldots, \begin{bmatrix} x_n \\ y_n \end{bmatrix}, \ldots$$

that collectively converge to the set S in Theorem 11.20.1.

Algorithm 2

Step 0. Choose an arbitrary point $\begin{bmatrix} x_0 \\ y_0 \end{bmatrix}$ in S.

Step 1. Choose one of the k similitudes at random, say T_{k_1}, and compute

$$\begin{bmatrix} x_1 \\ y_1 \end{bmatrix} = T_{k_1}\left(\begin{bmatrix} x_0 \\ y_0 \end{bmatrix}\right)$$

Step 2. Choose one of the k similitudes at random, say T_{k_2}, and compute

$$\begin{bmatrix} x_2 \\ y_2 \end{bmatrix} = T_{k_2}\left(\begin{bmatrix} x_1 \\ y_1 \end{bmatrix}\right)$$

$\vdots$

Step n. Choose one of the k similitudes at random, say T_{k_n}, and compute

$$\begin{bmatrix} x_n \\ y_n \end{bmatrix} = T_{k_n}\left(\begin{bmatrix} x_{n-1} \\ y_{n-1} \end{bmatrix}\right)$$

$\vdots$

On a computer screen the pixels corresponding to the points generated by this algorithm will fill out the pixel representation of the limiting set S.

Figure 20 shows four stages of the Random Iteration Algorithm that generate the Sierpinski carpet, starting with the initial point $\begin{bmatrix} 0 \\ 0 \end{bmatrix}$.

REMARK. Although Step 0 in the preceding algorithm requires the selection of an initial point in the set S, which may not be known in advance, this is not a serious problem. In practice, one can usually start with any point in R^2 and after a few iterations (say ten or so) the point generated will be sufficiently close to S that the algorithm will work correctly from that point on.

| 5000 iterations | 15,000 iterations | 45,000 iterations | 100,000 iterations |

Figure 20

MORE GENERAL FRACTALS

So far, we have discussed fractals that are self-similar sets according to the definition of a self-similar set in R^2. However, Theorem 11.20.1 remains true if the similitudes $T_1, T_2, \ldots, T_k$ are replaced by more general transformations, called *contracting affine transformations*. An affine transformation is defined as follows:

Definition. An ***affine transformation*** is a mapping of R^2 into R^2 of the form

$$T\left(\begin{bmatrix} x \\ y \end{bmatrix}\right) = \begin{bmatrix} a & b \\ c & d \end{bmatrix}\begin{bmatrix} x \\ y \end{bmatrix} + \begin{bmatrix} e \\ f \end{bmatrix}$$

where $a, b, c, d, e,$ and f are scalars.

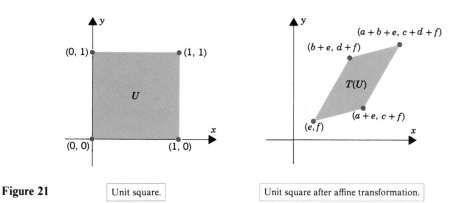

Figure 21

| Unit square. | Unit square after affine transformation. |

Figure 21 shows how an affine transformation maps the unit square U onto a parallelogram $T(U)$. An affine transformation is said to be ***contracting*** if the Euclidean distance between any two points in the plane is strictly decreased after the two points are mapped by the transformation. It can be shown that any k contracting affine transformations $T_1, T_2, \ldots, T_k$ determine a unique closed and bounded set S satisfying the equation

$$S = T_1(S) \cup T_2(S) \cup T_3(S) \cup \cdots \cup T_k(S) \tag{13}$$

Equation (13) has the same form as Equation (12), which we used to find self-similar sets. Although Equation (13), which uses contracting affine transformations, does not determine a self-similar set S, the set it does determine has many of the features of self-similar sets. For example, Figure 22 shows how a set in the plane resembling a fern (an example made famous by Barnsley) can be generated through four contracting affine transformations. Notice how the middle fern is the slightly overlapping union of the four smaller affine-image ferns surrounding it. Notice also how T_3, because the determinant of its matrix part is zero, maps the entire fern onto the small straight-line segment between the points (.50, 0) and (.50, .16). Figure 22 contains a wealth of information and should be studied carefully.

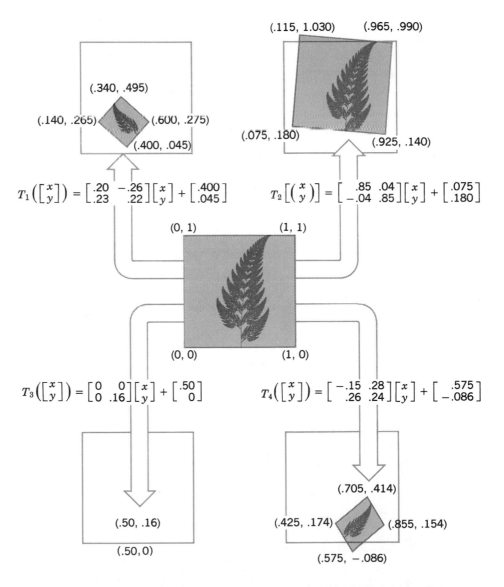

Figure 22

Michael Barnsley is actively pursuing an application of the above theory to the field of data compression and transmission. The fern, for example, is completely determined by the four affine transformations T_1, T_2, T_3, T_4. These four transformations, in turn, are determined by the 24 numbers given in Figure 22 defining their corresponding values of a, b, c, d, e, and f. In other words, these 24 numbers completely *encode* the picture of the fern. Storing these 24 numbers in a computer requires considerably less memory space than storing a pixel-by-pixel description of the fern. In principle, any picture represented by a pixel-map on a computer screen can be described through a finite number of affine transformations, although it is not easy to determine which transformations to use. Nevertheless, once encoded, the affine transformations generally require several orders of magnitude less computer memory than a pixel-by-pixel description of the pixel map.

FURTHER READINGS

Readers interested in learning more about fractals are referred to the following books, the first of which elaborates on the linear transformation approach of this section.

1. Michael Barnsley, *Fractals Everywhere* (New York: Academic Press, 1988).
2. Benoit B. Mandelbrot, *The Fractal Geometry of Nature* (New York: W. H. Freeman, 1982).
3. Heinz-Otto Peitgen and P. H. Richter, *The Beauty of Fractals* (New York: Springer-Verlag, 1986).
4. Heinz-Otto Peitgen and Dietmar Saupe, *The Science of Fractal Images* (New York: Springer-Verlag, 1988).

EXERCISE SET 11.20

1. The self-similar set in Figure 23 has the sizes indicated. Given that its lower left corner is situated at the origin of the xy-plane, find the similitudes that determine the set. What is its Hausdorff dimension? Is it a fractal?

Figure 23

2. Find the Hausdorff dimension of the self-similar set shown in Figure 24. Use a ruler to measure the figure and determine an approximate value of the scale factor s. What are the rotation angles of the similitudes determining this set?

Figure 24

3. For each of the self-similar sets in Figure 25 find: (i) the scale factor s of the similitudes describing the set; (ii) the rotation angles θ of all similitudes describing the set (all rotation angles are multiples of 90°); and (iii) the Hausdorff dimension of the set. Which of the sets are fractals and why?

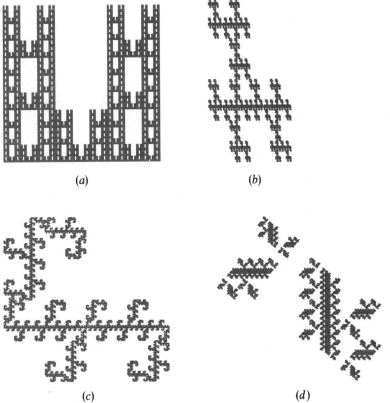

(a)

(b)

(c)

(d)

Figure 25

4. Of the four affine transformations shown in Figure 22, only the transformation T_2 is a similitude. Determine its scale factor s and rotation angle θ.

5. Find the coordinates of the tip of the fern in Figure 22. [**Hint.** The transformation T_2 maps the tip of the fern to itself.]

6. The square in Figure 7a was expressed as the union of four nonoverlapping squares as in Figure 7b. Suppose instead it were expressed as the union of 16 nonoverlapping squares. Verify that its Hausdorff dimension is still two, as determined by Equation (2).

7. Show that the four similitudes

$$T_1\left(\begin{bmatrix} x \\ y \end{bmatrix}\right) = \frac{3}{4}\begin{bmatrix} 1 & 0 \\ 0 & 1 \end{bmatrix}\begin{bmatrix} x \\ y \end{bmatrix}$$

$$T_2\left(\begin{bmatrix} x \\ y \end{bmatrix}\right) = \frac{3}{4}\begin{bmatrix} 1 & 0 \\ 0 & 1 \end{bmatrix}\begin{bmatrix} x \\ y \end{bmatrix} + \begin{bmatrix} \frac{1}{4} \\ 0 \end{bmatrix}$$

$$T_3\left(\begin{bmatrix} x \\ y \end{bmatrix}\right) = \frac{3}{4}\begin{bmatrix} 1 & 0 \\ 0 & 1 \end{bmatrix}\begin{bmatrix} x \\ y \end{bmatrix} + \begin{bmatrix} 0 \\ \frac{1}{4} \end{bmatrix}$$

$$T_4\left(\begin{bmatrix} x \\ y \end{bmatrix}\right) = \frac{3}{4}\begin{bmatrix} 1 & 0 \\ 0 & 1 \end{bmatrix}\begin{bmatrix} x \\ y \end{bmatrix} + \begin{bmatrix} \frac{1}{4} \\ \frac{1}{4} \end{bmatrix}$$

express the unit square as the union of four *overlapping* squares. Evaluate the right-hand side of Equation (2) for the values of k and s determined by these similitudes, and show that the result is not the correct value of the Hausdorff dimension of the unit square. [*Note*: This exercise shows the necessity of the nonoverlapping condition in the definition of a self-similar set and its Hausdorff dimension.]

8. All of the results in this section can be extended to R^n. Compute the Hausdorff dimension of the unit cube in R^3 (Figure 26). Given that the topological dimension of the unit cube is 3, determine whether it is a fractal. [*Hint*: Express the unit cube as the union of 8 smaller congruent nonoverlapping cubes.]

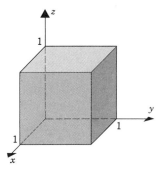

Figure 26

9. The set in Figure 27 in R^3 is called the *Menger sponge*. It is a self-similar set obtained by drilling out certain square holes from the unit cube. Notice that each face of the Menger sponge is a Sierpinski carpet and that the holes in the Sierpinski carpet now run all the way through the Menger sponge. Determine the values of k and s for the Menger sponge and find its Hausdorff dimension. Is the Menger sponge a fractal?

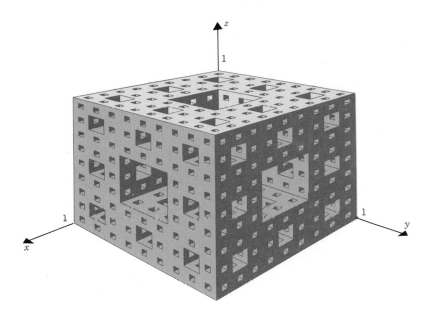

Figure 27

10. The two similitudes

$$T_1\left(\begin{bmatrix} x \\ y \end{bmatrix}\right) = \frac{1}{3}\begin{bmatrix} 1 & 0 \\ 0 & 1 \end{bmatrix}\begin{bmatrix} x \\ y \end{bmatrix} \quad \text{and} \quad T_2\left(\begin{bmatrix} x \\ y \end{bmatrix}\right) = \frac{1}{3}\begin{bmatrix} 1 & 0 \\ 0 & 1 \end{bmatrix}\begin{bmatrix} x \\ y \end{bmatrix} + \begin{bmatrix} \frac{2}{3} \\ 0 \end{bmatrix}$$

determine a fractal known as the **Cantor set**. Starting with the unit square region U as an initial set, sketch the first four sets that Algorithm 1 determines. Also, find the Hausdorff dimension of the Cantor set. (This famous set was the first example that Hausdorff gave in his 1919 paper of a set whose Hausdorff dimension is not equal to its topological dimension.)

11. Compute the areas of the sets S_0, S_1, S_2, S_3, and S_4 in Figure 15.

ANSWERS TO EXERCISES

EXERCISE SET 1.1 (page 6)

1. (a), (c), (f)   2. (a), (b), (c)

3. (a) $x = \frac{3}{7} + \frac{4}{7}t$ (b) $x_1 = \frac{4}{3}s - \frac{4}{3}t + \frac{7}{3}$ $x_1 = \frac{1}{4}r - \frac{5}{8}s + \frac{3}{4}t - \frac{1}{8}$ $v = \frac{8}{3}q - \frac{2}{3}r + \frac{1}{3}s - \frac{4}{3}t$

   $y = t$ $x_2 = s$ $x_2 = r$ $w = q$

   $x_3 = t$ $x_3 = s$ $x = r$

   $x_4 = t$ $y = s$

   $z = t$

4. (a) $\begin{bmatrix} 3 & -2 & -1 \\ 4 & 5 & 3 \\ 7 & 3 & 2 \end{bmatrix}$ (b) $\begin{bmatrix} 2 & 0 & 2 & 1 \\ 3 & -1 & 4 & 7 \\ 6 & 1 & -1 & 0 \end{bmatrix}$

   (c) $\begin{bmatrix} 1 & 2 & 0 & -1 & 1 & 1 \\ 0 & 3 & 1 & 0 & -1 & 2 \\ 0 & 0 & 1 & 7 & 0 & 1 \end{bmatrix}$ (d) $\begin{bmatrix} 1 & 0 & 0 & 1 \\ 0 & 1 & 0 & 2 \\ 0 & 0 & 1 & 3 \end{bmatrix}$

5. (a) $2x_1 \qquad\qquad = 0$ (b) $3x_1 \qquad\quad - 2x_3 = 5$ (c) $7x_1 + 2x_2 + x_3 - 3x_4 = 5$

   $3x_1 - 4x_2 = 0$ $7x_1 + x_2 + 4x_3 = -3$ $x_1 + 2x_2 + 4x_3 \qquad = 1$

   $x_2 = 1$ $-2x_2 + x_3 = 7$

   (d) $x_1 \qquad\qquad = 7$

   $x_2 \qquad\quad = -2$

   $x_3 \qquad = 3$

   $x_4 = 4$

6. (a) $x - 2y = 5$ (b) Let $x = t$; then $t - 2y = 5$. Solving for y yields $y = \frac{1}{2}t - \frac{5}{2}$.

8. $k = 6$: infinitely many solutions

   $k \neq 6$: no solutions

   No value of k yields one solution.

9. (a) The lines have no common point of intersection.   (b) The lines intersect in exactly one point.
   (c) The three lines coincide.

EXERCISE SET 1.2 (page 17)

1. (a), (b), (c), (d), (h), (i)   2. (a), (b), (d)   3. (a) Both   (b) Neither   (c) Both
   (d) Row-echelon   (e) Neither   (f) Both

4. (a) $x_1 = -3, x_2 = 0, x_3 = 7$
   (b) $x_1 = 7t + 8, x_2 = -3t + 2, x_3 = -t - 5, x_4 = t$
   (c) $x_1 = 6s - 3t - 2, x_2 = s, x_3 = -4t + 7, x_4 = -5t + 8, x_5 = t$
   (d) Inconsistent

5. (a) $x_1 = -37, x_2 = -8, x_3 = 5$
   (b) $x_1 = 13t - 10, x_2 = 13t - 5, x_3 = -t + 2, x_4 = t$
   (c) $x_1 = -7s + 2t - 11, x_2 = s, x_3 = -3t - 4, x_4 = -3t + 9, x_5 = t$
   (d) Inconsistent

6. (a) $x_1 = 3, x_2 = 1, x_3 = 2$ (b) $x_1 = -\frac{1}{2}, x_2 = \frac{1}{2}, x_3 = 0$
   (c) $x = s - 1, y = 2r, z = r, w = s$ (d) Inconsistent

8. (a) Inconsistent   (b) $x_1 = -4, x_2 = 2, x_3 = 7$ (c) $x_1 = 3 + 2t, x_2 = t$
   (d) $x = \frac{8}{3} - \frac{2}{3}t - \frac{2}{3}s, y = \frac{1}{10} + \frac{2}{3}t - \frac{1}{10}s, z = t, w = s$

10. (a) $x_1 = 2 + 12t, x_2 = 5 + 27t, x_3 = t$ (b) Inconsistent
    (c) $u = -2s - 3t - 6, v = s, w = -t - 2, x = t + 3, y = t$

12. $I_1 = -1, I_2 = 0, I_3 = 1, I_4 = 2$

13. (a) $x = \frac{2}{3}a - \frac{1}{5}b, y = -\frac{1}{3}a + \frac{2}{5}b$ (b) $x_1 = a - \frac{1}{3}c, x_2 = a - \frac{1}{2}b, x_3 = -a + \frac{1}{2}b + \frac{1}{3}c$

14. $a = -4$, none; $a \neq \pm 4$, exactly one; $a = 4$, infinitely many   16. $\begin{bmatrix} 1 & 3 \\ 0 & 1 \end{bmatrix}$ and $\begin{bmatrix} 1 & 0 \\ 0 & 1 \end{bmatrix}$ are possible answers.

17. $\alpha = \dfrac{\pi}{2}, \beta = \pi, \gamma = 0$ 18. $x = \pm 1, y = \pm \sqrt{3}, z = \pm \sqrt{2}$ 19. $a = 1, b = -6, c = 2, d = 10$

EXERCISE SET 1.3 (page 23)

1. (a), (c), (d)   2. (a) $x_1 = 0, x_2 = 0, x_3 = 0$
   (b) $x_1 = -s, x_2 = -t - s, x_3 = 4s, x_4 = t$
   (c) $w = t, x = -t, y = t, z = 0$

3. (a) Only the trivial solution   (b) $u = 7s - 5t, v = -6s + 4t, w = 2s, x = 2t$
   (c) Only the trivial solution

4. $Z_1 = s, Z_2 = -t - s, Z_3 = -t, Z_4 = 0, Z_5 = t$ 6. $\lambda = 4, \lambda = 2$ 8. $x^2 + y^2 - 2x - 4y - 29 = 0$

12. One possible answer is $x + y + z = 0$
    $$x + y + z = 1$$

EXERCISE SET 1.4 (page 32)

1. (a) Undefined   (b) 4×2 (c) Undefined   (d) Undefined   2. $a = 5, b = -3, c = 4, d = 1$
   (e) 5×5 (f) 5×2 (g) Undefined   (h) 5×2

3. (a) $\begin{bmatrix} 7 & 6 & 5 \\ -2 & 1 & 3 \\ 7 & 3 & 7 \end{bmatrix}$ (b) $\begin{bmatrix} -5 & 4 & -1 \\ 0 & -1 & -1 \\ -1 & 1 & 1 \end{bmatrix}$ (c) $\begin{bmatrix} 15 & 0 \\ -5 & 10 \\ 5 & 5 \end{bmatrix}$ (d) $\begin{bmatrix} -7 & -28 & -14 \\ -21 & -7 & -35 \end{bmatrix}$

   (e) Undefined   (f) $\begin{bmatrix} 22 & -6 & 8 \\ -2 & 4 & 6 \\ 10 & 0 & 4 \end{bmatrix}$ (g) $\begin{bmatrix} -39 & -21 & -24 \\ 9 & -6 & -15 \\ -33 & -12 & -30 \end{bmatrix}$ (h) $\begin{bmatrix} 0 & 0 \\ 0 & 0 \\ 0 & 0 \end{bmatrix}$

   (i) 5   (j) -25 (k) 168   (l) Undefined

4. (a) $\begin{bmatrix} 7 & 2 & 4 \\ 3 & 5 & 7 \end{bmatrix}$ (b) $\begin{bmatrix} -5 & 0 & -1 \\ 4 & -1 & 1 \\ -1 & -1 & 1 \end{bmatrix}$ (c) $\begin{bmatrix} -5 & 0 & -1 \\ 4 & -1 & 0 \\ -1 & -1 & 1 \end{bmatrix}$ (d) Undefined

(e) $\begin{bmatrix} -\frac{1}{4} & \frac{3}{2} \\ \frac{9}{4} & 0 \\ \frac{3}{4} & \frac{9}{4} \end{bmatrix}$ (f) $\begin{bmatrix} 0 & -1 \\ 1 & 0 \end{bmatrix}$ (g) $\begin{bmatrix} 9 & 1 & -1 \\ -13 & 2 & -4 \\ 0 & 1 & -6 \end{bmatrix}$ (h) $\begin{bmatrix} 9 & -13 & 0 \\ 1 & 2 & 1 \\ -1 & -4 & -6 \end{bmatrix}$

5. (a) $\begin{bmatrix} 12 & -3 \\ -4 & 5 \\ 4 & 1 \end{bmatrix}$ (b) Undefined (c) $\begin{bmatrix} 42 & 108 & 75 \\ 12 & -3 & 21 \\ 36 & 78 & 63 \end{bmatrix}$ (d) $\begin{bmatrix} 3 & 45 & 9 \\ 11 & -11 & 17 \\ 7 & 17 & 13 \end{bmatrix}$

(e) $\begin{bmatrix} 3 & 45 & 9 \\ 11 & -11 & 17 \\ 7 & 17 & 13 \end{bmatrix}$ (f) $\begin{bmatrix} 21 & 17 \\ 17 & 35 \end{bmatrix}$ (g) $\begin{bmatrix} 0 & -2 & 11 \\ 12 & 1 & 8 \end{bmatrix}$ (h) $\begin{bmatrix} 12 & 6 & 9 \\ 48 & -20 & 14 \\ 24 & 8 & 16 \end{bmatrix}$

(i) 61 (j) 35 (k) 28

6. (a) $\begin{bmatrix} -6 & -3 \\ 36 & 0 \\ 4 & 7 \end{bmatrix}$ (b) Undefined (c) $\begin{bmatrix} 2 & -10 & 11 \\ 13 & 2 & 5 \\ 4 & -3 & 13 \end{bmatrix}$ (d) $\begin{bmatrix} 10 & -6 \\ -14 & 2 \\ -1 & -8 \end{bmatrix}$

(e) $\begin{bmatrix} 40 & 72 \\ 26 & 42 \end{bmatrix}$ (f) $\begin{bmatrix} 0 & 0 & 0 \\ 0 & 0 & 0 \\ 0 & 0 & 0 \end{bmatrix}$

7. (a) $[67 \quad 41 \quad 41]$ (b) $[63 \quad 67 \quad 57]$ (c) $\begin{bmatrix} 41 \\ 21 \\ 67 \end{bmatrix}$ (d) $\begin{bmatrix} 6 \\ 6 \\ 63 \end{bmatrix}$ (e) $[24 \quad 56 \quad 97]$ (f) $\begin{bmatrix} 76 \\ 98 \\ 97 \end{bmatrix}$

10. (a) $A = \begin{bmatrix} 2 & -3 & 5 \\ 9 & -1 & 1 \\ 1 & 5 & 4 \end{bmatrix}, X = \begin{bmatrix} x_1 \\ x_2 \\ x_3 \end{bmatrix}, B = \begin{bmatrix} 7 \\ -1 \\ 0 \end{bmatrix}$ (b) $A = \begin{bmatrix} 4 & 0 & -3 & 1 \\ 5 & 1 & 0 & -8 \\ 2 & -5 & 9 & -1 \\ 0 & 3 & -1 & 7 \end{bmatrix}, X = \begin{bmatrix} x_1 \\ x_2 \\ x_3 \\ x_4 \end{bmatrix}, B = \begin{bmatrix} 1 \\ 3 \\ 0 \\ 2 \end{bmatrix}$

11. (a)
$$3x_1 - x_2 + 2x_3 = 2$$
$$4x_1 + 3x_2 + 7x_3 = -1$$
$$-2x_1 + x_2 + 5x_3 = 4$$
(b)
$$3w - 2x + z = 0$$
$$5w + 2y - 2z = 0$$
$$3w + x + 4y + 7z = 0$$
$$-2w + 5x + y + 6z = 0$$

12. (a) $DA = \begin{bmatrix} d_1 a_{11} & d_1 a_{12} & \cdots & d_1 a_{1n} \\ d_2 a_{21} & d_2 a_{22} & \cdots & d_2 a_{2n} \\ \vdots & \vdots & & \vdots \\ d_m a_{m1} & d_m a_{m2} & \cdots & d_m a_{mn} \end{bmatrix}, AE = \begin{bmatrix} a_{11} e_1 & a_{12} e_2 & \cdots & a_{1m} e_m \\ a_{21} e_1 & a_{22} e_2 & \cdots & a_{2m} e_m \\ \vdots & \vdots & & \vdots \\ a_{m1} e_1 & a_{m2} e_2 & \cdots & a_{mm} e_m \end{bmatrix}$

(b) i. To multiply a matrix A on the left by a diagonal matrix D, multiply every element of the ith row of A by the ith diagonal entry of D to obtain the ith row of the product.

ii. To multiply a matrix A on the right by a diagonal matrix E, multiply every element of the jth column of A by the jth diagonal entry of E to obtain the jth column of the product.

(c) $AB = \begin{bmatrix} -8 & 1 & -15 \\ 12 & 0 & -6 \\ -28 & 1 & -15 \end{bmatrix}, BA = \begin{bmatrix} -8 & 4 & 20 \\ 3 & 0 & 2 \\ 21 & 3 & -15 \end{bmatrix}$

15. It will appear in the jth row and the ith column.

18. (a) $\begin{bmatrix} a_{11} & 0 & 0 & 0 & 0 & 0 \\ 0 & a_{22} & 0 & 0 & 0 & 0 \\ 0 & 0 & a_{33} & 0 & 0 & 0 \\ 0 & 0 & 0 & a_{44} & 0 & 0 \\ 0 & 0 & 0 & 0 & a_{55} & 0 \\ 0 & 0 & 0 & 0 & 0 & a_{66} \end{bmatrix}$ (b) $\begin{bmatrix} a_{11} & a_{12} & a_{13} & a_{14} & a_{15} & a_{16} \\ 0 & a_{22} & a_{23} & a_{24} & a_{25} & a_{26} \\ 0 & 0 & a_{33} & a_{34} & a_{35} & a_{36} \\ 0 & 0 & 0 & a_{44} & a_{45} & a_{46} \\ 0 & 0 & 0 & 0 & a_{55} & a_{56} \\ 0 & 0 & 0 & 0 & 0 & a_{66} \end{bmatrix}$

(c) $\begin{bmatrix} a_{11} & 0 & 0 & 0 & 0 & 0 \\ a_{21} & a_{22} & 0 & 0 & 0 & 0 \\ a_{31} & a_{32} & a_{33} & 0 & 0 & 0 \\ a_{41} & a_{42} & a_{43} & a_{44} & 0 & 0 \\ a_{51} & a_{52} & a_{53} & a_{54} & a_{55} & 0 \\ a_{61} & a_{62} & a_{63} & a_{64} & a_{65} & a_{66} \end{bmatrix}$ (d) $\begin{bmatrix} a_{11} & a_{12} & 0 & 0 & 0 & 0 \\ a_{21} & a_{22} & a_{23} & 0 & 0 & 0 \\ 0 & a_{32} & a_{33} & a_{34} & 0 & 0 \\ 0 & 0 & a_{43} & a_{44} & a_{45} & 0 \\ 0 & 0 & 0 & a_{54} & a_{55} & a_{56} \\ 0 & 0 & 0 & 0 & a_{65} & a_{66} \end{bmatrix}$

EXERCISE SET 1.5 (page 44)

4. $A^{-1} = \begin{bmatrix} 2 & -1 \\ -5 & 3 \end{bmatrix}$, $B^{-1} = \begin{bmatrix} \frac{1}{5} & \frac{3}{20} \\ -\frac{1}{5} & \frac{1}{10} \end{bmatrix}$, $C^{-1} = \begin{bmatrix} \frac{1}{2} & 0 \\ 0 & \frac{1}{3} \end{bmatrix}$ 6. No   7. $A = \begin{bmatrix} \frac{5}{13} & \frac{1}{13} \\ -\frac{3}{13} & \frac{2}{13} \end{bmatrix}$ 8. $A = \begin{bmatrix} \frac{2}{7} & 1 \\ \frac{1}{7} & \frac{3}{7} \end{bmatrix}$

9. $A^3 = \begin{bmatrix} 8 & 0 \\ 28 & 1 \end{bmatrix}$, $A^{-3} = \begin{bmatrix} \frac{1}{8} & 0 \\ -\frac{7}{2} & 1 \end{bmatrix}$, $A^2 - 2A + I = \begin{bmatrix} 1 & 0 \\ 4 & 0 \end{bmatrix}$ 10. $A^{-1} = X = \begin{bmatrix} \frac{1}{2} & \frac{1}{2} & -\frac{1}{2} \\ -\frac{1}{2} & \frac{1}{2} & \frac{1}{2} \\ \frac{1}{2} & -\frac{1}{2} & \frac{1}{2} \end{bmatrix}$

11. $\begin{bmatrix} \cos\theta & -\sin\theta \\ \sin\theta & \cos\theta \end{bmatrix}$ 12. (c) $(A+B)^2 = A^2 + AB + BA + B^2$ 13. $A^{-1} = \begin{bmatrix} \frac{1}{a_{11}} & 0 & \cdots & 0 \\ 0 & \frac{1}{a_{22}} & \cdots & 0 \\ \vdots & \vdots & & \vdots \\ 0 & 0 & \cdots & \frac{1}{a_{nn}} \end{bmatrix}$

18. $0A$ and $A0$ may not have the same size.   19. $\begin{bmatrix} \pm 1 & 0 & 0 \\ 0 & \pm 1 & 0 \\ 0 & 0 & \pm 1 \end{bmatrix}$

EXERCISE SET 1.6 (page 53)

1. (c), (d), (f)   2. (a) Add three times the first row to the second row.
         (b) Multiply the third row by $\frac{1}{3}$.
         (c) Interchange the first row and the fourth row.
         (d) Multiply the third row by $\frac{1}{7}$. Then add this result to the first row.

3. (a) $\begin{bmatrix} 0 & 0 & 1 \\ 0 & 1 & 0 \\ 1 & 0 & 0 \end{bmatrix}$ (b) $\begin{bmatrix} 0 & 0 & 1 \\ 0 & 1 & 0 \\ 1 & 0 & 0 \end{bmatrix}$ (c) $\begin{bmatrix} 1 & 0 & 0 \\ 0 & 1 & 0 \\ -2 & 0 & 1 \end{bmatrix}$ (d) $\begin{bmatrix} 1 & 0 & 0 \\ 0 & 1 & 0 \\ 2 & 0 & 1 \end{bmatrix}$

4. No, since C cannot be obtained by performing a single row operation on B.

5. (a) $\begin{bmatrix} -7 & 4 \\ 2 & 1 \end{bmatrix}$ (b) $\begin{bmatrix} -\frac{5}{39} & \frac{2}{13} \\ \frac{4}{39} & \frac{1}{13} \end{bmatrix}$ (c) Not invertible

6. (a) $\begin{bmatrix} \frac{3}{2} & -\frac{11}{10} & -\frac{6}{5} \\ -1 & 1 & 1 \\ -\frac{1}{2} & \frac{7}{10} & \frac{2}{5} \end{bmatrix}$ **(b)** Not invertible **(c)** $\begin{bmatrix} \frac{1}{2} & -\frac{1}{2} & \frac{1}{2} \\ -\frac{1}{2} & \frac{1}{2} & \frac{1}{2} \\ \frac{1}{2} & \frac{1}{2} & -\frac{1}{2} \end{bmatrix}$

(d) $\begin{bmatrix} \frac{7}{2} & 0 & -3 \\ -1 & 1 & 0 \\ 0 & -1 & 1 \end{bmatrix}$ **(e)** $\begin{bmatrix} \frac{1}{2} & -\frac{1}{2} & \frac{1}{2} \\ 0 & 0 & 1 \\ \frac{1}{2} & \frac{1}{2} & -\frac{1}{2} \end{bmatrix}$

7. (a) $\begin{bmatrix} 1 & 3 & 1 \\ 0 & 1 & -1 \\ -2 & 2 & 0 \end{bmatrix}$ **(b)** $\begin{bmatrix} \frac{\sqrt{2}}{26} & \frac{-3\sqrt{2}}{26} & 0 \\ \frac{4\sqrt{2}}{26} & \frac{\sqrt{2}}{26} & 0 \\ 0 & 0 & 1 \end{bmatrix}$ **(c)** $\begin{bmatrix} 1 & 0 & 0 & 0 \\ -\frac{1}{3} & \frac{1}{3} & 0 & 0 \\ 0 & -\frac{1}{5} & \frac{1}{5} & 0 \\ 0 & 0 & -\frac{1}{7} & \frac{1}{7} \end{bmatrix}$ **(d)** Not invertible

8. (a) $\begin{bmatrix} \frac{1}{k_1} & 0 & 0 & 0 \\ 0 & \frac{1}{k_2} & 0 & 0 \\ 0 & 0 & \frac{1}{k_3} & 0 \\ 0 & 0 & 0 & \frac{1}{k_4} \end{bmatrix}$ **(b)** $\begin{bmatrix} 0 & 0 & 0 & \frac{1}{k_4} \\ 0 & 0 & \frac{1}{k_3} & 0 \\ 0 & \frac{1}{k_2} & 0 & 0 \\ \frac{1}{k_1} & 0 & 0 & 0 \end{bmatrix}$ **(c)** $\begin{bmatrix} \frac{1}{k} & 0 & 0 & 0 \\ -\frac{1}{k^2} & \frac{1}{k} & 0 & 0 \\ \frac{1}{k^3} & -\frac{1}{k^2} & \frac{1}{k} & 0 \\ -\frac{1}{k^4} & \frac{1}{k^3} & -\frac{1}{k^2} & \frac{1}{k} \end{bmatrix}$

9. (a) $E_1 = \begin{bmatrix} 1 & 0 \\ 5 & 1 \end{bmatrix}$, $E_2 = \begin{bmatrix} 1 & 0 \\ 0 & \frac{1}{2} \end{bmatrix}$ **(b)** $A^{-1} = E_2 E_1$ **(c)** $A = E_1^{-1} E_2^{-1}$

11. $\begin{bmatrix} 1 & 0 & 0 \\ 0 & 1 & 0 \\ 0 & -2 & 1 \end{bmatrix} \begin{bmatrix} 1 & 0 & 0 \\ 0 & 1 & 0 \\ 1 & 0 & 1 \end{bmatrix} \begin{bmatrix} 0 & 1 & 0 \\ 1 & 0 & 0 \\ 0 & 0 & 1 \end{bmatrix} \begin{bmatrix} 1 & 3 & 3 & 8 \\ 0 & 1 & 7 & 8 \\ 0 & 0 & 0 & 0 \end{bmatrix}$

EXERCISE SET 1.7 (page 61)

1. $x_1 = 3, x_2 = -1$ **2.** $x_1 = -3, x_2 = -3$ **3.** $x_1 = -1, x_2 = 4, x_3 = -7$ **4.** $x_1 = 1, x_2 = -11, x_3 = 16$

5. $x_1 = 1, x_2 = 5, x_3 = -1$ **6.** $x = -6, y = 1, z = 10, w = -7$ **7.** $x_1 = 2b_1 - 5b_2, x_2 = -b_1 + 3b_2$

8. $x_1 = -\frac{15}{2}b_1 + \frac{1}{2}b_2 + \frac{5}{2}b_3, x_2 = \frac{1}{2}b_1 + \frac{1}{2}b_2 - \frac{1}{2}b_3, x_3 = \frac{5}{2}b_1 - \frac{1}{2}b_2 - \frac{1}{2}b_3$

9. (a) $x_1 = \frac{16}{3}, x_2 = -\frac{4}{3}, x_3 = -\frac{11}{3}$ **(b)** $x_1 = -\frac{5}{3}, x_2 = \frac{5}{3}, x_3 = \frac{10}{3}$ **(c)** $x_1 = 3, x_2 = 0, x_3 = -4$

11. (a) $x_1 = \frac{22}{17}, x_2 = \frac{1}{17}$ **12. (a)** $x_1 = -18, x_2 = -1, x_3 = -14$

(b) $x_1 = \frac{21}{17}, x_2 = \frac{11}{17}$ **(b)** $x_1 = -\frac{421}{2}, x_2 = -\frac{25}{2}, x_3 = -\frac{327}{2}$

13. (a) $x_1 = \frac{7}{15}, x_2 = \frac{14}{15}$ **14. (a)** $x_1 = 18, x_2 = -19, x_3 = 2$ **15. (a)** $x_1 = -12 - 3t, x_2 = -5 - t$

(b) $x_1 = \frac{34}{15}, x_2 = \frac{28}{15}$ **(b)** $x_1 = -23, x_2 = 11, x_3 = -2$ **(b)** $x_1 = 7 - 3t, x_2 = 3 - t$

(c) $x_1 = \frac{19}{15}, x_2 = \frac{13}{15}$ **(c)** $x_1 = 5, x_2 = -2, x_3 = 0$

(d) $x_1 = -\frac{1}{5}, x_2 = \frac{3}{5}$

16. $b_1 = 2b_2$ **17.** $b_1 = b_2 + b_3$ **18.** No restrictions **19.** $b_1 = b_3 + b_4, b_2 = 2b_3 + b_4$

21. $X = \begin{bmatrix} 11 & 12 & -3 & 27 & 26 \\ -6 & -8 & 1 & -18 & -17 \\ -15 & -21 & 9 & -38 & -35 \end{bmatrix}$ **22. (a)** Only the trivial solution $x_1 = x_2 = x_3 = x_4 = 0$; invertible

(b) Infinitely many solutions; not invertible

CHAPTER 1 SUPPLEMENTARY EXERCISES (page 63)

1. $x' = \frac{3}{5}x + \frac{4}{5}y, \; y' = -\frac{4}{5}x + \frac{3}{5}y$ 2. $x' = x\cos\theta + y\sin\theta, \; y' = -x\sin\theta + y\cos\theta$

3. One possible answer is
$$x_1 - 2x_2 - x_3 - \; x_4 = 0$$
$$x_1 + 5x_2 \qquad + 2x_4 = 0$$

4. 3 pennies, 4 nickels, 6 dimes

5. $x = 4, \; y = 2, \; z = 3$ 6. Infinitely many if $a = 2$ or $a = -\frac{3}{2}$; none otherwise

7. (a) $a \neq 0, b \neq 2$ (b) $a \neq 0, b = 2$ 8. $x = \frac{5}{3}, y = 9, z = \frac{1}{3}$ 9. $K = \begin{bmatrix} 0 & 2 \\ 1 & 1 \end{bmatrix}$
   (c) $a = 0, b = 2$ (d) $a = 0, b \neq 2$

10. $a = 2, b = -1, c = 1$ 11. (a) $X = \begin{bmatrix} -1 & 3 & -1 \\ 6 & 0 & 1 \end{bmatrix}$ (b) $X = \begin{bmatrix} 1 & -2 \\ 3 & 1 \end{bmatrix}$ (c) $X = \begin{bmatrix} -\frac{113}{37} & -\frac{160}{37} \\ -\frac{20}{37} & -\frac{46}{37} \end{bmatrix}$

12. (a) $Z = \begin{bmatrix} -1 & -7 & 11 \\ 14 & 10 & -26 \end{bmatrix} X$ (b) $z_1 = -x_1 - 7x_2 + 11x_3$
$$z_2 = 14x_1 + 10x_2 - 26x_3$$

13. *mpn* multiplications and $mp(n-1)$ additions   15. $a = 1, b = -2, c = 3$

16. $a = 1, b = -4, c = -5$ 26. $A = -\frac{7}{3}, B = \frac{4}{3}, C = \frac{2}{3}$

EXERCISE SET 2.1 (page 73)

1. (a) 5   (b) 9   (c) 6   (d) 10   (e) 0   (f) 2

2. (a) Odd   (b) Odd   (c) Even   (d) Even   (e) Even   (f) Even

3. 22   4. 0   5. 52   6. $-3\sqrt{6}$ 7. $a^2 - 5a + 21$ 8. 0   9. -65 10. -4 11. -123

12. $-c^4 + c^3 - 16c^2 + 8c - 2$ 13. (a) $\lambda = 1, \lambda = -3$ (b) $\lambda = -2, \lambda = 3, \lambda = 4$

16. 275   17. (a) $= -120$ (b) $= -120$ 18. $x = \dfrac{3 \pm \sqrt{33}}{4}$

EXERCISE SET 2.2 (page 78)

1. (a) -30 (b) -2 (c) 0   (d) 0   2. 30   3. 5   4. -17 5. 33   6. 39   7. 6

8. $-\frac{1}{6}$ 9. -2 10. (a) -6 (b) 72   (c) -6 (d) 18

EXERCISE SET 2.3 (page 84)

1. (a) $\det(A) = -11 = \det(A^t)$ (b) $\det(A) = -5 = \det(A^t)$ 2. $\det AB = -170 = (\det A)(\det B)$

4. (a) Invertible   (b) Not invertible   5. (a) -189 (b) $-\frac{8}{7}$
   (c) Not invertible   (d) Not invertible   (c) $-\frac{1}{56}$ (d) 7

6. If $x = 0$, the first and third rows are proportional.   12. (a) $k = \dfrac{5 \pm \sqrt{17}}{2}$ (b) $k = -1$
   If $x = 2$, the first and second rows are proportional.

EXERCISE SET 2.4 (page 94)

1. (a) $M_{11} = 29, M_{12} = 21, M_{13} = 27, M_{21} = -11, M_{22} = 13, M_{23} = -5, M_{31} = -19, M_{32} = -19, M_{33} = 19$
   (b) $C_{11} = 29, C_{12} = -21, C_{13} = 27, C_{21} = 11, C_{22} = 13, C_{23} = 5, C_{31} = -19, C_{32} = 19, C_{33} = 19$

2. (a) $M_{13} = 0, C_{13} = 0$ **(b)** $M_{23} = -96, C_{23} = 96$ **3.** 152
(c) $M_{22} = -48, C_{22} = -48$ **(d)** $M_{21} = 72, C_{21} = -72$

4. (a) $\text{adj}(A) = \begin{bmatrix} 29 & -11 & -19 \\ -21 & 13 & 19 \\ 27 & -5 & 19 \end{bmatrix}$ **(b)** $A^{-1} = \begin{bmatrix} \frac{29}{152} & -\frac{11}{152} & -\frac{19}{152} \\ -\frac{21}{152} & \frac{13}{152} & \frac{19}{152} \\ \frac{27}{152} & \frac{5}{152} & \frac{19}{152} \end{bmatrix}$

5. -40 **6.** -66 **7.** 0   **8.** $k^3 - 8k^2 - 10k + 95$ **9.** -240 **10.** 0

11. $A^{-1} = \begin{bmatrix} 3 & -5 & -5 \\ -3 & 4 & 5 \\ 2 & -2 & -3 \end{bmatrix}$ **12.** $A^{-1} = \begin{bmatrix} 2 & 0 & \frac{3}{2} \\ \frac{2}{3} & \frac{1}{3} & \frac{2}{3} \\ -1 & 0 & -1 \end{bmatrix}$ **13.** $A^{-1} = \begin{bmatrix} \frac{1}{2} & \frac{3}{2} & 1 \\ 0 & 1 & \frac{3}{2} \\ 0 & 0 & \frac{1}{2} \end{bmatrix}$

14. $A^{-1} = \begin{bmatrix} \frac{1}{2} & 0 & 0 \\ -4 & 1 & 0 \\ \frac{29}{12} & -\frac{1}{2} & \frac{1}{6} \end{bmatrix}$ **15.** $A^{-1} = \begin{bmatrix} -4 & 3 & 0 & -1 \\ 2 & -1 & 0 & 0 \\ -7 & 0 & -1 & 8 \\ 6 & 0 & 1 & -7 \end{bmatrix}$ **16.** $x_1 = 1, x_2 = 2$

17. $x = \frac{3}{11}, y = \frac{2}{11}, z = -\frac{1}{11}$ **18.** $x = -\frac{144}{55}, y = -\frac{61}{55}, z = \frac{46}{11}$

19. $x_1 = -\frac{30}{11}, x_2 = -\frac{38}{11}, x_3 = -\frac{40}{11}$ **20.** $x_1 - 5, x_2 = 8, x_3 = 3, x_4 = -1$

21. Cramer's Rule does not apply.   **22.** $A^{-1} = \begin{bmatrix} \cos\theta & -\sin\theta & 0 \\ \sin\theta & \cos\theta & 0 \\ 0 & 0 & 1 \end{bmatrix}$ **23. (a)** $x = 1, y = 0, z = 2, w = 0$

CHAPTER 2 SUPPLEMENTARY EXERCISES (page 97)

1. $x' = \frac{2}{3}x + \frac{4}{3}y, y' = -\frac{4}{3}x + \frac{2}{3}y$ **2.** $x' = x\cos\theta + y\sin\theta, y' = -x\sin\theta + y\cos\theta$ **4.** 2

5. $\cos\beta = \dfrac{c^2 + a^2 - b^2}{2ac}, \cos\gamma = \dfrac{a^2 + b^2 - c^2}{2ab}$ **10. (b)** $\frac{19}{2}$ **12.** $\det(B) = -1^{n(n-1)/2}\det(A)$

13. (a) The ith and jth columns will be interchanged.
(b) The ith column will be divided by c.
(c) $-c$ times the jth column will be added to the ith column.

15. (a) $\lambda^3 + (-a_{11} - a_{22} - a_{33})\lambda^2 + (a_{11}a_{22} + a_{11}a_{33} + a_{22}a_{33} - a_{12}a_{21} - a_{13}a_{31} - a_{23}a_{32})\lambda +$
$(a_{11}a_{23}a_{32} + a_{12}a_{21}a_{33} + a_{13}a_{22}a_{31} - a_{11}a_{22}a_{33} - a_{12}a_{23}a_{31} - a_{13}a_{21}a_{32})$

EXERCISE SET 3.1 (page 109)

3. (a) $\overrightarrow{P_1P_2} = (-1, -1)$ **(b)** $\overrightarrow{P_1P_2} = (-7, -2)$ **(c)** $\overrightarrow{P_1P_2} = (2, 1)$ **(d)** $\overrightarrow{P_1P_2} = (a, b)$
(e) $\overrightarrow{P_1P_2} = (-5, 12, -6)$ **(f)** $\overrightarrow{P_1P_2} = (1, -1, -2)$ **(g)** $\overrightarrow{P_1P_2} = (-a, -b, -c)$ **(h)** $\overrightarrow{P_1P_2} = (a, b, c)$

4. (a) $\mathbf{u} = (5, 10, -8)$ **5. (a)** $\mathbf{u} = (-1, 2, -4)$ is one possible answer.
(b) $\mathbf{u} = (-5, -10, 8)$ **(b)** $\mathbf{u} = (7, -2, -6)$ is one possible answer.

6. (a) $(-2, 1, -4)$ **(b)** $(-10, 6, 4)$ **(c)** $(-7, 1, 10)$ **7.** $\mathbf{x} = (-\frac{8}{3}, \frac{1}{2}, \frac{8}{3})$
(d) $(80, -20, -80)$ **(e)** $(132, -24, -72)$ **(f)** $(-77, 8, 94)$

8. $c_1 = 2, c_2 = -1, c_3 = 2$ **10.** $c_1 = c_2 = c_3 = 0$ **11. (a)** $(\frac{9}{2}, -\frac{1}{2}, -\frac{1}{2})$ **12. (a)** $x' = 5, y' = 8$
(b) $(\frac{23}{4}, -\frac{9}{4}, \frac{1}{4})$ **(b)** $x = -1, y = 3$

14. $\mathbf{u} = \left(\dfrac{\sqrt{3}}{2}, \dfrac{1}{2}\right), \mathbf{v} = \left(-\dfrac{1}{2}, -\dfrac{\sqrt{3}}{2}\right), \mathbf{u} + \mathbf{v} = \left(\dfrac{\sqrt{3}-1}{2}, \dfrac{1-\sqrt{3}}{2}\right), \mathbf{u} - \mathbf{v} = \left(\dfrac{\sqrt{3}+1}{2}, \dfrac{\sqrt{3}+1}{2}\right)$

EXERCISE SET 3.2 (page 113)

1. (a) 5   (b) $\sqrt{13}$ (c) 5   (d) $2\sqrt{3}$ (e) $3\sqrt{6}$ (f) 6

2. (a) $\sqrt{13}$ (b) $2\sqrt{26}$ (c) $\sqrt{209}$ (d) $3\sqrt{2}$

3. (a) $\sqrt{83}$ (b) $\sqrt{17}+\sqrt{26}$ (c) $4\sqrt{17}$ (d) $\sqrt{466}$ (e) $\left(\dfrac{3}{\sqrt{61}},\dfrac{6}{\sqrt{61}},-\dfrac{4}{\sqrt{61}}\right)$ (f) 1

4. $k=\pm\dfrac{4}{\sqrt{30}}$ 8. A sphere of radius 1 centered at (x_0,y_0,z_0)

EXERCISE SET 3.3 (page 123)

1. (a) -11 (b) -24 (c) 0   (d) 0   2. (a) $-\dfrac{11}{\sqrt{13}\sqrt{74}}$ (b) $-\dfrac{3}{\sqrt{10}}$ (c) 0   (d) 0

3. (a) Orthogonal   (b) Obtuse   (c) Acute   (d) Obtuse

4. (a) $(0,0)$ (b) $(\frac{8}{13},-\frac{12}{13})$ (c) $(-\frac{16}{13},0,-\frac{80}{13})$ (d) $(\frac{16}{89},\frac{12}{89},\frac{32}{89})$

5. (a) $(6,2)$ (b) $(-\frac{21}{13},-\frac{14}{13})$ (c) $(\frac{55}{13},1,-\frac{11}{13})$ (d) $(\frac{78}{89},-\frac{12}{89},-\frac{32}{89})$

6. (a) $\frac{2}{3}$ (b) $\dfrac{4\sqrt{5}}{5}$ (c) $\dfrac{18}{\sqrt{22}}$ (d) $\dfrac{43}{\sqrt{54}}$ 9. (a) 102   (b) $125\sqrt{2}$ (c) 170   (d) 170

11. $\cos\theta_1=\dfrac{\sqrt{10}}{10}$, $\cos\theta_2=\dfrac{3\sqrt{10}}{10}$, $\cos\theta_3=0$ 12. The right angle is at B.

13. No. The equation will hold if b and c are perpendicular to a, and $b\neq c$.

15. (a) 1   (b) $\dfrac{1}{\sqrt{17}}$ (c) $\dfrac{6}{\sqrt{10}}$ 19. (b) $\cos\beta=\dfrac{b}{\|v\|}$, $\cos\gamma=\dfrac{c}{\|v\|}$ 20. $\theta_1\approx71.0°$, $\theta_2\approx60.9°$, $\theta_3\approx35.8°$

EXERCISE SET 3.4 (page 134)

1. (a) $(32,-6,-4)$ (b) $(-14,-20,-82)$ (c) $(27,40,-42)$
   (d) $(0,176,-264)$ (e) $(-44,55,-22)$ (f) $(-8,-3,-8)$

2. (a) $(18,36,-18)$ (b) $(-3,9,-3)$ 3. (a) $\dfrac{\sqrt{374}}{2}$ (b) $\sqrt{285}$ 4. (a) $\sqrt{59}$ (b) $\sqrt{101}$

8. (a) -10 (b) -110 9. (a) -3 (b) 3   (c) 3   (d) -3 (e) -3 (f) 0   10. (a) 16   (b) 45

11. (a) No   (b) Yes   (c) No   12. $\left(0,-\dfrac{2}{\sqrt{5}},-\dfrac{1}{\sqrt{5}}\right)$ and any unit vector parallel to this such as $\left(0,\dfrac{2}{\sqrt{5}},\dfrac{1}{\sqrt{5}}\right)$

13. $\left(\dfrac{6}{\sqrt{61}},-\dfrac{3}{\sqrt{61}},\dfrac{4}{\sqrt{61}}\right)$, $\left(-\dfrac{6}{\sqrt{61}},\dfrac{3}{\sqrt{61}},-\dfrac{4}{\sqrt{61}}\right)$ 15. $2(v\times u)$ 16. $\dfrac{12\sqrt{13}}{49}$

17. (a) $\dfrac{\sqrt{26}}{2}$ (b) $\dfrac{\sqrt{26}}{3}$ 19. (a) $\dfrac{2\sqrt{141}}{\sqrt{29}}$ (b) $\dfrac{10\sqrt{2}}{3}$ 21. (a) $\sqrt{122}$ (b) $\theta\approx40°19''$

23. (a) $m=(0,1,0)$ and $n=(1,0,0)$ (b) $(-1,0,0)$ (c) $(0,0,-1)$ 27. $(-8,0,-8)$ 30. (a) $\frac{2}{3}$ (b) $\frac{1}{2}$

EXERCISE SET 3.5 (page 144)

1. (a) $-2(x+1)+(y-3)-(z+2)=0$
   (b) $(x-1)+9(y-1)+8(z-4)=0$
   (c) $2z=0$
   (d) $x+2y+3z=0$

2. (a) $-2x+y-z-7=0$
   (b) $x+9y+8z-42=0$
   (c) $2z=0$
   (d) $x+2y+3y=0$

3. (a) $(0,0,5)$ is a point in the plane and $\mathbf{n}=(-3,7,2)$ is a normal vector so that $-3(x-0)+7(y-0)+2(z-5)=0$ is a point-normal form; other points and normals yield other correct answers.
   (b) $1(x-0)-4(z-0)=0$ is one possible answer.

4. (a) $2y-z+1=0$ (b) $x+9y-5z-26=0$ 5. (a) Not parallel   (b) Parallel   (c) Parallel

6. (a) Parallel   (b) Not parallel   7. (a) Not perpendicular   (b) Perpendicular

8. (a) Perpendicular   (b) Not perpendicular   9. (a) $x=3+2t,\ y=-1+t,\ z=2+3t$
   (b) $x=-2+6t,\ y=3-6t,\ z=-3-2t$
   (c) $x=2,\ y=2+t,\ z=6$
   (d) $x=t,\ y=-2t,\ z=3t$

10. (a) $\dfrac{x-3}{2}=y+1=\dfrac{z-2}{3}$ (b) $\dfrac{x+2}{6}=-\dfrac{y-3}{6}=-\dfrac{z+3}{2}$

11. (a) $x=5+2t,\ y=-2+2t,\ z=4-4t$
    (b) $x=2t,\ y=-t,\ z=-3t$

12. (a) $x=-12-7t,\ y=-41+23t,\ z=t$ 13. (a) $x-2y-17=0$ and $x+4z-27=0$
    (b) $x=\tfrac{5}{2}t,\ y=0,\ z=t$ (b) $x-2y=0$ and $-7y+2z=0$

15. $2x+3y-5z+36=0$ 16. (a) $z=0$ (b) $y=0$ (c) $x=0$

17. (a) $z-z_0=0$ (b) $x-x_0=0$ (c) $y-y_0=0$ 18. $7x+4y-2z=0$ 19. $5x-2y+z-34=0$

20. $(-\tfrac{173}{3},-\tfrac{43}{3},\tfrac{49}{3})$ 21. $y+2z-9=0$ 22. $x-y-4z-2=0$ 24. $x=\tfrac{4}{3}t-2,\ y=-\tfrac{2}{3}t+5,\ z=t$

25. $x+5y+3z-18=0$ 26. $(x-2)+(y+1)-3(z-4)=0$ 27. $4x+13y-z-17=0$

28. $3x+10y+4z-53=0$ 29. $3x-y-z-2=0$ 30. $5x-3y+2z-5=0$

31. $2x+4y+8z+13=0$ 34. $x-4y+4z+9=0$ 35. (a) $x=\tfrac{11}{23}+\tfrac{7}{23}t,\ y=\tfrac{41}{23}-\tfrac{2}{23}t,\ z=t$
                                                    (b) $x=-\tfrac{2}{3}t,\ y=0,\ z=t$

37. (a) $\tfrac{5}{3}$ (b) $\dfrac{1}{\sqrt{29}}$ (c) $\dfrac{4}{\sqrt{3}}$ 38. (a) $\dfrac{1}{2\sqrt{26}}$ (b) 0 (c) $\dfrac{2}{\sqrt{6}}$

39. (a) $\theta\approx 35.3°$ (b) $\theta\approx 79.0°$ 40. $\theta\approx 75.2°$

EXERCISE SET 4.1 (page 154)

1. (a) $(-1,9,-11,1)$ (b) $(22,53,-19,14)$ (c) $(-13,13,-36,-2)$
   (d) $(-90,-114,60,-36)$ (e) $(-9,-5,-5,-3)$ (f) $(27,29,-27,9)$

2. $(\tfrac{6}{3},\tfrac{2}{3},\tfrac{2}{3},\tfrac{2}{3})$ 3. $c_1=1,\ c_2=1,\ c_3=-1,\ c_4=1$ 5. (a) $\sqrt{29}$ (b) 3 (c) 13 (d) $\sqrt{31}$

6. (a) $\sqrt{133}$ (b) $\sqrt{30}+\sqrt{77}$ (c) $4\sqrt{30}$ (d) $\sqrt{1811}$ (e) $\left(\dfrac{1}{\sqrt{2}},\dfrac{1}{3\sqrt{2}},\dfrac{2}{3\sqrt{2}},\dfrac{2}{3\sqrt{2}}\right)$ (f) 1

8. $k=\pm\tfrac{5}{7}$ 9. (a) 7 (b) 14 (c) 7 (d) 11 10. (a) $\left(\dfrac{1}{\sqrt{10}},\dfrac{3}{\sqrt{10}}\right),\ \left(-\dfrac{1}{\sqrt{10}},-\dfrac{3}{\sqrt{10}}\right)$

11. (a) $\sqrt{10}$ (b) $2\sqrt{14}$ (c) $\sqrt{59}$ (d) 10   20. (a) Euclidean measure of "box" in R^n is $a_1 a_2 \cdots a_n$

                                              (b) Length of diagonal is $\sqrt{a_1^2 + a_2^2 + \cdots + a_n^2}$

EXERCISE SET 4.2 (page 161)

1. Not a vector space. Axiom 8 fails.

2. Not a vector space. Axiom 10 fails.

3. Not a vector space. Axioms 9 and 10 fail.

4. The set is a vector space under the given operations.

5. The set is a vector space under the given operations.

6. Not a vector space. Axioms 5 and 6 fail.

7. The set is a vector space under the given operations.

8. Not a vector space. Axioms 7 and 8 fail.

9. The set is a vector space under the given operations.

10. Not a vector space. Axioms 1, 4, 5, and 6 fail.

11. The set is a vector space under the given operations.

12. The set is a vector space under the given operations.

13. The set is a vector space under the given operations.

14. The set is a vector space under the given operations.

EXERCISE SET 4.3 (page 171)

1. (a), (c)   2. (b), (c)   3. (a), (b), (d)   4. (b), (d), (e)   5. (a), (b), (d)

6. (a) $(-9, -7, -15) = -2\mathbf{u} + \mathbf{v} - 2\mathbf{w}$ 7. (a) $-9 - 7x - 15x^2 = -2\mathbf{p}_1 + \mathbf{p}_2 - 2\mathbf{p}_3$
     (b) $(6, 11, 6) = 4\mathbf{u} - 5\mathbf{v} + \mathbf{w}$ (b) $6 + 11x + x^2 = 4\mathbf{p}_1 - 5\mathbf{p}_2 + \mathbf{p}_3$
     (c) $(0, 0, 0) = 0\mathbf{u} + 0\mathbf{v} + 0\mathbf{w}$ (c) $0 = 0\mathbf{p}_1 + 0\mathbf{p}_2 + 0\mathbf{p}_3$
     (d) $(7, 8, 9) = 0\mathbf{u} - 2\mathbf{v} + 3\mathbf{w}$ (d) $7 + 8x + 9x^2 = 0\mathbf{p}_1 - 2\mathbf{p}_2 + 3\mathbf{p}_3$

9. (a) The vectors span.   (b) The vectors do not span.   10. (a), (c), (e)   11. No
   (c) The vectors do not span.   (d) The vectors span.

12. (a), (b), (d)   13. $y = z$ 14. $x = 3t, y = -2t, z = 5t$, where $-\infty < t < +\infty$

EXERCISE SET 4.4 (page 178)

1. (a) $\mathbf{u}_2$ is a scalar multiple of $\mathbf{u}_1$.   (b) The vectors are linearly dependent by Theorem 4.4.3.
   (c) $\mathbf{p}_2$ is a scalar multiple of $\mathbf{p}_1$.   (d) B is a scalar multiple of A.

2. (d)   3. None   4. (d)   5. (a), (d), (e), (f)

6. (a) They do not lie in a plane.   7. (a) They do not lie on the same line.
   (b) They do lie in a plane.   (b) They do not lie on the same line.
                                   (c) They do lie on the same line.

8. (b) $\mathbf{v}_1 = \tfrac{2}{7}\mathbf{v}_2 - \tfrac{3}{7}\mathbf{v}_3$, $\mathbf{v}_2 = \tfrac{7}{2}\mathbf{v}_1 + \tfrac{3}{2}\mathbf{v}_3$, $\mathbf{v}_3 = -\tfrac{7}{3}\mathbf{v}_1 + \tfrac{2}{3}\mathbf{v}_2$ 9. $\lambda = -\tfrac{1}{2}, \lambda = 1$

20. If and only if the vector is not zero

21. (a) They are linearly independent since $\mathbf{v}_1, \mathbf{v}_2$, and $\mathbf{v}_3$ do not lie in the same plane when they are placed with their initial points at the origin.
   (b) They are not linearly independent since $\mathbf{v}_1, \mathbf{v}_2$, and $\mathbf{v}_3$ lie in the same plane when they are placed with their initial points at the origin.

EXERCISE SET 4.5 (page 190)

1. (a) A basis for R^2 has two linearly independent vectors.   (b) A basis for R^3 has three linearly independent vectors.

   (c) A basis for P_2 has three linearly independent vectors.   (d) A basis for M_{22} has four linearly independent vectors.

2. (a), (b)   3. (a), (b)   4. (c), (d)   6. (b) Any two of the vectors $\mathbf{v}_1, \mathbf{v}_2, \mathbf{v}_3$

7. (a) $(\mathbf{w})_S = (3, -7)$ (b) $(\mathbf{w})_S = (\frac{5}{28}, \frac{3}{14})$ (c) $(\mathbf{w})_S = \left(a, \dfrac{b-a}{2}\right)$

8. (a) $(\mathbf{v})_S = (3, -2, 1)$ (b) $(\mathbf{v})_S = (-2, 0, 1)$ 9. (a) $(\mathbf{p})_S = (4, -3, 1)$ (b) $(\mathbf{p})_S = (0, 2, -1)$

10. $(A)_S = (-1, 1, -1, 3)$ 11. Basis: $(1, 0, 1)$; dimension $= 1$

12. Basis: $(-\frac{1}{4}, -\frac{1}{4}, 1, 0)$, $(0, -1, 0, 1)$; dimension $= 2$

13. Basis: $(4, 1, 0, 0)$, $(-3, 0, 1, 0)$, $(1, 0, 0, 1)$; dimension $= 3$

14. Basis: $(3, 1, 0)$, $(-1, 0, 1)$; dimension $= 2$ 15. No basis; dimension $= 0$

16. Basis: $(4, -5, 1)$; dimension $= 1$ 17. (a) $(\frac{2}{3}, 1, 0)$, $(-\frac{4}{3}, 0, 1)$ (b) $(1, 1, 0)$, $(0, 0, 1)$
                                                      (c) $(2, -1, 4)$ (d) $(1, 1, 0)$, $(0, 1, 1)$

18. (a) 3-dimensional   (b) 2-dimensional   (c) 1-dimensional   19. 3-dimensional

27. (a) $S = \{1, x, x^2\}$; $(\mathbf{v}_1)_S = (-1, 1, 2)$, $(\mathbf{v}_2)_S = (3, 3, 6)$, $(\mathbf{v}_3)_S = (9, 0, 0)$
    (b) Any three of the four vectors $(\mathbf{v}_1)_S = (1, 1, 0)$, $(\mathbf{v}_2)_S = (0, 0, 1)$, $(\mathbf{v}_3)_S = (-2, 0, 2)$, $(\mathbf{v}_4)_S = (0, -3, 0)$ are a basis.
    (c) $(\mathbf{v}_1)_S = (1, 1, -3)$ is a basis.

28. (a) $(0, \sqrt{2})$ (b) $(1, 0)$ (c) $(-1, \sqrt{2})$ (d) $(a - b, \sqrt{2}b)$

29. (a) $(2, 0)$ (b) $\left(\dfrac{2}{\sqrt{3}}, -\dfrac{1}{\sqrt{3}}\right)$ (c) $(0, 1)$ (d) $\left(\dfrac{2}{\sqrt{3}}\,a, b - \dfrac{a}{\sqrt{3}}\right)$

EXERCISE SET 4.6 (page 202)

1. $\mathbf{r}_1 = (2, -1, 0, 1)$, $\mathbf{r}_2 = (3, 5, 7, -1)$, $\mathbf{r}_3 = (1, 4, 2, 7)$; $\mathbf{c}_1 = \begin{bmatrix} 2 \\ 3 \\ 1 \end{bmatrix}$, $\mathbf{c}_2 = \begin{bmatrix} -1 \\ 5 \\ 4 \end{bmatrix}$, $\mathbf{c}_3 = \begin{bmatrix} 0 \\ 7 \\ 2 \end{bmatrix}$, $\mathbf{c}_4 = \begin{bmatrix} 1 \\ -1 \\ 7 \end{bmatrix}$

2. (a) $1\begin{bmatrix} 2 \\ -1 \end{bmatrix} + 2\begin{bmatrix} 3 \\ 4 \end{bmatrix} = \begin{bmatrix} 8 \\ 7 \end{bmatrix}$ (b) $-2\begin{bmatrix} 4 \\ 3 \\ 0 \end{bmatrix} + 3\begin{bmatrix} 0 \\ 6 \\ -1 \end{bmatrix} + 5\begin{bmatrix} -1 \\ 2 \\ 4 \end{bmatrix} = \begin{bmatrix} -13 \\ 22 \\ 17 \end{bmatrix}$

   (c) $-1\begin{bmatrix} -3 \\ 5 \\ 2 \\ 1 \end{bmatrix} + 2\begin{bmatrix} 6 \\ -4 \\ 3 \\ 8 \end{bmatrix} + 5\begin{bmatrix} 2 \\ 0 \\ -1 \\ 3 \end{bmatrix} = \begin{bmatrix} 25 \\ -13 \\ -1 \\ 30 \end{bmatrix}$ (d) $3\begin{bmatrix} 2 \\ 6 \end{bmatrix} + 0\begin{bmatrix} 1 \\ 3 \end{bmatrix} + (-5)\begin{bmatrix} 5 \\ -8 \end{bmatrix} = \begin{bmatrix} -19 \\ 58 \end{bmatrix}$

3. (a) $\begin{bmatrix} 16 \\ 19 \\ 1 \end{bmatrix}$ (b) $\begin{bmatrix} 1 \\ 0 \\ 2 \end{bmatrix}, \begin{bmatrix} 0 \\ 1 \\ 0 \end{bmatrix}$ (c) $\begin{bmatrix} -1 \\ -1 \\ 1 \\ 0 \end{bmatrix}, \begin{bmatrix} 2 \\ -4 \\ 0 \\ 7 \end{bmatrix}$

   (d) $\begin{bmatrix} -1 \\ -1 \\ 1 \\ 0 \\ 0 \end{bmatrix}, \begin{bmatrix} -2 \\ -1 \\ 0 \\ 1 \\ 0 \end{bmatrix}, \begin{bmatrix} -1 \\ -2 \\ 0 \\ 0 \\ 1 \end{bmatrix}$ (e) $\begin{bmatrix} -2 \\ 0 \\ 0 \\ 1 \\ 0 \end{bmatrix}, \begin{bmatrix} -16 \\ 2 \\ 5 \\ 0 \\ 12 \end{bmatrix}$

4. (a) $(1, -1, 3)$, $(0, 1, -19)$ (b) $(1, 0, -\frac{1}{2})$
   (c) $(1, 4, 5, 2)$, $(0, 1, 1, \frac{4}{7})$ (d) $(1, 4, 5, 6, 9)$, $(0, 1, 1, 1, 2)$
   (e) $(1, -3, 2, 2, 1)$, $(0, 1, 2, 0, -1)$, $(0, 0, 1, 0, -\frac{5}{12})$

5. (a) $\begin{bmatrix} 1 \\ 5 \\ 7 \end{bmatrix}, \begin{bmatrix} -1 \\ -4 \\ -6 \end{bmatrix}$ (b) $\begin{bmatrix} 2 \\ 4 \\ 0 \end{bmatrix}$ (c) $\begin{bmatrix} 1 \\ 2 \\ -1 \end{bmatrix}, \begin{bmatrix} 4 \\ 1 \\ 3 \end{bmatrix}$ (d) $\begin{bmatrix} 1 \\ 3 \\ -1 \\ 2 \end{bmatrix}, \begin{bmatrix} 4 \\ -2 \\ 0 \\ 3 \end{bmatrix}$ (e) $\begin{bmatrix} 1 \\ 0 \\ 2 \\ 3 \\ -2 \end{bmatrix}, \begin{bmatrix} -3 \\ 3 \\ -3 \\ -6 \\ 9 \end{bmatrix}, \begin{bmatrix} 2 \\ 6 \\ -2 \\ 0 \\ 2 \end{bmatrix}$

6. (a) $(1, -1, 3), (5, -4, -4)$ (b) $(2, 0, -1)$ (c) $(1, 4, 5, 2), (2, 1, 3, 0)$
(d) $(1, 4, 5, 6, 9), (3, -2, 1, 4, -1)$ (e) $(1, -3, 2, 2, 1), (0, 3, 6, 0, -3), (2, -3, -2, 4, 4)$

7. (a) $(1, 1, -4, -3), (0, 1, -5, -2), (0, 0, 1, -\frac{1}{2})$ **8.** (a) $\{v_1, v_2\}; v_3 = 2v_1 + v_2, v_4 = -2v_1 + v_2$
(b) $(1, -1, 2, 0), (0, 1, 0, 0), (0, 0, 1, -\frac{1}{6})$ (b) $\{v_1, v_3\}; v_2 = 2v_1, v_4 = v_1 + v_3$
(c) $(1, 1, 0, 0), (0, 1, 1, 1), (0, 0, 1, 1), (0, 0, 0, 1)$ (c) $\{v_1, v_2, v_4\}; v_3 = 2v_1 - v_2, v_5 = -v_1 + 3v_2 + 2v_4$

9. (b) $\begin{bmatrix} 0 & 0 & 0 \\ 0 & 1 & 0 \\ 0 & 0 & 1 \end{bmatrix}$

EXERCISE SET 4.7 (page 211)

1. (a) Nullity $= 1$, rank $= 2$; $n = 3$ (b) Nullity $= 2$, rank $= 1$; $n = 3$
(c) Nullity $= 2$, rank $= 2$; $n = 4$ (d) Nullity $= 3$, rank $= 2$; $n = 5$
(e) Nullity $= 2$, rank $= 3$; $n = 5$

2. (a) $\begin{bmatrix} -2 \\ 10 \end{bmatrix} = \begin{bmatrix} 1 \\ 4 \end{bmatrix} - \begin{bmatrix} 3 \\ -6 \end{bmatrix}$ (b) **b** is not in the column space of A. (c) $\begin{bmatrix} 1 \\ 9 \\ 1 \end{bmatrix} - 3\begin{bmatrix} -1 \\ 3 \\ 1 \end{bmatrix} + \begin{bmatrix} 1 \\ 1 \\ 1 \end{bmatrix} = \begin{bmatrix} 5 \\ 1 \\ -1 \end{bmatrix}$

(d) $\begin{bmatrix} 2 \\ 0 \\ 0 \end{bmatrix} = \begin{bmatrix} 1 \\ 1 \\ -1 \end{bmatrix} + (t-1)\begin{bmatrix} -1 \\ 1 \\ -1 \end{bmatrix} + t\begin{bmatrix} 1 \\ -1 \\ 1 \end{bmatrix}$ (e) $\begin{bmatrix} 4 \\ 3 \\ 5 \\ 7 \end{bmatrix} = -26\begin{bmatrix} 1 \\ 0 \\ 1 \\ 0 \end{bmatrix} + 13\begin{bmatrix} 2 \\ 1 \\ 2 \\ 1 \end{bmatrix} - 7\begin{bmatrix} 0 \\ 2 \\ 1 \\ 2 \end{bmatrix} + 4\begin{bmatrix} 1 \\ 1 \\ 3 \\ 2 \end{bmatrix}$

3. (a) $\mathbf{x} = r\begin{bmatrix} -3 \\ 1 \\ 1 \\ 0 \end{bmatrix} + s\begin{bmatrix} 4 \\ -1 \\ 0 \\ 1 \end{bmatrix}$ (b) $\begin{bmatrix} -1 \\ 2 \\ 4 \\ -3 \end{bmatrix} + r\begin{bmatrix} -3 \\ 1 \\ 1 \\ 0 \end{bmatrix} + s\begin{bmatrix} 4 \\ -1 \\ 0 \\ 1 \end{bmatrix}$

4. (a) $\begin{bmatrix} 1 \\ 0 \end{bmatrix} + t\begin{bmatrix} 3 \\ 1 \end{bmatrix}; t\begin{bmatrix} 3 \\ 1 \end{bmatrix}$ (b) $\begin{bmatrix} -2 \\ 7 \\ 0 \end{bmatrix} + t\begin{bmatrix} -1 \\ -1 \\ 1 \end{bmatrix}; t\begin{bmatrix} -1 \\ -1 \\ 1 \end{bmatrix}$

(c) $\begin{bmatrix} -1 \\ 0 \\ 0 \\ 0 \end{bmatrix} + r\begin{bmatrix} 2 \\ 1 \\ 0 \\ 0 \end{bmatrix} + s\begin{bmatrix} -1 \\ 0 \\ 1 \\ 0 \end{bmatrix} + t\begin{bmatrix} -2 \\ 0 \\ 0 \\ 1 \end{bmatrix}; r\begin{bmatrix} 2 \\ 1 \\ 0 \\ 0 \end{bmatrix} + s\begin{bmatrix} -1 \\ 0 \\ 1 \\ 0 \end{bmatrix} + t\begin{bmatrix} -2 \\ 0 \\ 0 \\ 1 \end{bmatrix}$

(d) $\begin{bmatrix} \frac{6}{5} \\ \frac{7}{5} \\ 0 \\ 0 \end{bmatrix} + s\begin{bmatrix} \frac{7}{5} \\ \frac{4}{5} \\ 1 \\ 0 \end{bmatrix} + t\begin{bmatrix} \frac{1}{5} \\ -\frac{3}{5} \\ 0 \\ 1 \end{bmatrix}; s\begin{bmatrix} \frac{7}{5} \\ \frac{4}{5} \\ 1 \\ 0 \end{bmatrix} + t\begin{bmatrix} \frac{1}{5} \\ -\frac{3}{5} \\ 0 \\ 1 \end{bmatrix}$

5. (a) Rank $= 4$, nullity $= 0$ (b) Rank $= 3$, nullity $= 2$ (c) Rank $= 3$, nullity $= 2$

6. Rank $= \min(m, n)$, nullity $= n - \min(m, n)$

7. (a) Yes, 0
   (b) No
   (c) Yes, 2
   (d) Yes, 7
   (e) No
   (f) Yes, 4
   (g) Yes, 4

8. (a) Nullity $= 0$, number of parameters $= 0$
   (b) Nullity $= 1$, number of parameters $= 1$
   (c) Nullity $= 2$, number of parameters $= 2$
   (d) Nullity $= 7$, number of parameters $= 7$
   (e) Nullity $= 7$, number of parameters $= 7$
   (f) Nullity $= 4$, number of parameters $= 4$
   (g) Nullity $= 0$, number of parameters $= 0$

CHAPTER 4 SUPPLEMENTARY EXERCISES (page 213)

1. (a) All of R^3
   (b) Plane: $2x - 3y + z = 0$
   (c) Line: $x = 2t, y = t, z = 0$
   (d) The origin: $(0, 0, 0)$

2. (a) $a(4, 1, 1) + b(0, -1, 2)$
   (b) $(a + c)(3, -1, 2) + b(1, 4, 1)$
   (c) $a(2, 3, 0) + b(-1, 0, 4) + c(4, -1, 1)$

4. (a) $\mathbf{v} = (-1 + r)\mathbf{v}_1 + (\tfrac{2}{3} - r)\mathbf{v}_2 + r\mathbf{v}_3$; r arbitrary   5. A must be invertible.   6. No

7. (a) Rank $= 2$, nullity $= 1$
   (b) Rank $= 2$, nullity $= 2$
   (c) Rank $= 2$, nullity $= n - 2$

8. (a) Rank $= 2$, nullity $= 1$
   (b) Rank $= 3$, nullity $= 2$
   (c) Rank $= n + 1$, nullity $= n$

10. $\{1, x^2, x^3, x^4, x^5, x^6, \ldots, x^n\}$ 11. (a) 2   (b) 1   (c) 2   (d) 3   12. 0, 1, or 2

EXERCISE SET 5.1 (page 224)

1. (a) 2   (b) 11   (c) -13 (d) -8 (e) 0   2. (a) -2 (b) 62   (c) -74 (d) 8   (e) 0

3. (a) 3   (b) 56   4. (a) -29 (b) -15 5. (b) 29   6. (b) -42

7. (a) $\begin{bmatrix} \sqrt{3} & 0 \\ 0 & \sqrt{5} \end{bmatrix}$ (b) $\begin{bmatrix} 2 & 0 \\ 0 & \sqrt{6} \end{bmatrix}$

9. (a) No. Axiom 4 fails.   (b) No. Axioms 2 and 3 fail.   (c) Yes   (d) No. Axiom 4 fails.

10. (a) $\sqrt{10}$ (b) $\sqrt{21}$ (c) $5\sqrt{5}$ 11. (a) $3\sqrt{2}$ (b) $3\sqrt{5}$ (c) $3\sqrt{13}$ 12. (a) $\sqrt{17}$ (b) 5

13. (a) $\sqrt{74}$ (b) 0   14. $3\sqrt{2}$ 15. (a) $\sqrt{105}$ (b) $\sqrt{47}$ 16. (a) 8   (b) -33 (c) -40
   (d) 3   (e) $2\sqrt{13}$ (f) $\sqrt{241}$

17. $2\sqrt{\tfrac{2}{3}}$ 18. (a) $\dfrac{x^2}{4} + \dfrac{y^2}{16} = 1$

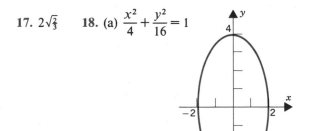

(b) $\dfrac{x^2}{\frac{1}{2}} + \dfrac{y^2}{1} = 1$

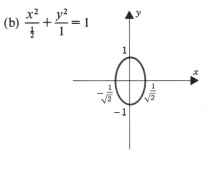

19. $\langle \mathbf{u}, \mathbf{v} \rangle = \tfrac{1}{9} u_1 v_1 + u_2 v_2$ 22. No. Axiom 4 fails.   27. (a) $-\tfrac{28}{13}$ (b) 0   28. (a) 0   (b) 1

EXERCISE SET 5.2 (page 232)

1. (a) Yes (b) No (c) Yes (d) No (e) No (f) Yes 2. No

3. (a) $-\dfrac{1}{\sqrt{2}}$ (b) $-\dfrac{3}{\sqrt{73}}$ (c) 0 4. (a) 0 (b) 0 6. (a) $\dfrac{19}{10\sqrt{7}}$ (b) 0

  (d) $-\dfrac{20}{9\sqrt{10}}$ (e) $-\dfrac{1}{\sqrt{2}}$ (f) $\dfrac{2}{\sqrt{55}}$

7. (a) Orthogonal (b) Orthogonal 8. (a) $k=-3$ (b) $k=-2, k=-3$
  (c) Orthogonal (d) Not orthogonal

9. $\pm\frac{1}{57}(-34, 44, -6, 11)$ 26. $\langle \mathbf{u}, \mathbf{v}\rangle = \frac{1}{2}u_1v_1 + \frac{1}{6}u_2v_2$

EXERCISE SET 5.3 (page 245)

1. (a), (b), (d) 2. (b) 3. (b), (d) 4. (b), (d) 5. (a) 6. (a)

7. (a) $\left(-\dfrac{1}{\sqrt{5}}, \dfrac{2}{\sqrt{5}}\right), \left(\dfrac{2}{\sqrt{5}}, \dfrac{1}{\sqrt{5}}\right)$ (b) $\left(\dfrac{1}{\sqrt{2}}, 0, -\dfrac{1}{\sqrt{2}}\right), \left(\dfrac{1}{\sqrt{2}}, 0, \dfrac{1}{\sqrt{2}}\right), (0, 1, 0)$

  (c) $\left(\dfrac{1}{\sqrt{3}}, \dfrac{1}{\sqrt{3}}, \dfrac{1}{\sqrt{3}}\right), \left(-\dfrac{1}{\sqrt{2}}, \dfrac{1}{\sqrt{2}}, 0\right), \left(\dfrac{1}{\sqrt{6}}, \dfrac{1}{\sqrt{6}}, -\dfrac{2}{\sqrt{6}}\right)$

9. (a) $-\frac{7}{3}\mathbf{v}_1 + \frac{1}{3}\mathbf{v}_2 + 2\mathbf{v}_3$ (b) $-\frac{37}{3}\mathbf{v}_1 - \frac{9}{3}\mathbf{v}_2 + 4\mathbf{v}_3$ (c) $-\frac{3}{7}\mathbf{v}_1 - \frac{1}{7}\mathbf{v}_2 + \frac{5}{7}\mathbf{v}_3$

10. (a) $\frac{1}{7}\mathbf{v}_1 + \frac{5}{21}\mathbf{v}_2 + \frac{1}{3}\mathbf{v}_3 + \mathbf{v}_4$ (b) $\dfrac{15\sqrt{2}}{7}\mathbf{v}_1 + \dfrac{5\sqrt{2}}{21}\mathbf{v}_2 - \dfrac{2\sqrt{2}}{3}\mathbf{v}_3 + \frac{1}{2}\mathbf{v}_4$ (c) $-\frac{3}{7}\mathbf{v}_1 + \frac{11}{63}\mathbf{v}_2 - \frac{1}{18}\mathbf{v}_3 + \frac{1}{2}\mathbf{v}_4$

11. (a) $(\mathbf{w})_S = (-2\sqrt{2}, 5\sqrt{2}),\ [\mathbf{w}]_S = \begin{bmatrix} -2\sqrt{2} \\ 5\sqrt{2} \end{bmatrix}$ (b) $(\mathbf{w})_S = (0, -2, 1),\ [\mathbf{w}]_S = \begin{bmatrix} 0 \\ -2 \\ 1 \end{bmatrix}$

12. (a) $\mathbf{u} = (\frac{7}{5}, -\frac{1}{5}),\ \mathbf{v} = (\frac{13}{5}, \frac{16}{5})$ (b) $\|\mathbf{u}\| = \sqrt{2},\ d(\mathbf{u}, \mathbf{v}) = \sqrt{13},\ \langle \mathbf{u}, \mathbf{v}\rangle = 3$

13. (a) $\mathbf{u} = (1, \frac{14}{5}, -\frac{2}{5}),\ \mathbf{v} = (0, -\frac{17}{5}, \frac{6}{5}),\ \mathbf{w} = (-4, -\frac{11}{5}, \frac{23}{5})$ (b) $\|\mathbf{v}\| = \sqrt{13},\ d(\mathbf{u}, \mathbf{v}) = 5\sqrt{3},\ \langle \mathbf{w}, \mathbf{v}\rangle = 13$

14. (a) $\|\mathbf{u}\| = \sqrt{15},\ \|\mathbf{v} - \mathbf{w}\| = 5,\ \|\mathbf{v} + \mathbf{w}\| = \sqrt{105},\ \langle \mathbf{v}, \mathbf{w}\rangle = 20$

  (b) $\|\mathbf{u}\| = \sqrt{2},\ \|\mathbf{v} - \mathbf{w}\| = \sqrt{34},\ \|\mathbf{v} + \mathbf{w}\| = \sqrt{118},\ \langle \mathbf{v}, \mathbf{w}\rangle = 21$ 15. (b) $\mathbf{u} = -\frac{4}{5}\mathbf{v}_1 - \frac{11}{10}\mathbf{v}_2 + 0\mathbf{v}_3 + \frac{1}{2}\mathbf{v}_4$

16. (a) $\left(\dfrac{1}{\sqrt{10}}, -\dfrac{3}{\sqrt{10}}\right), \left(\dfrac{3}{\sqrt{10}}, \dfrac{1}{\sqrt{10}}\right)$ (b) $(1, 0), (0, -1)$

17. (a) $\left(\dfrac{1}{\sqrt{3}}, \dfrac{1}{\sqrt{3}}, \dfrac{1}{\sqrt{3}}\right), \left(-\dfrac{1}{\sqrt{2}}, \dfrac{1}{\sqrt{2}}, 0\right), \left(\dfrac{1}{\sqrt{6}}, \dfrac{1}{\sqrt{6}}, -\dfrac{2}{\sqrt{6}}\right)$ (b) $(1, 0, 0), \left(0, \dfrac{7}{\sqrt{53}}, -\dfrac{2}{\sqrt{53}}\right), \left(0, \dfrac{2}{\sqrt{53}}, \dfrac{7}{53}\right)$

18. $\left(0, \dfrac{2}{\sqrt{5}}, \dfrac{1}{\sqrt{5}}, 0\right), \left(\dfrac{5}{\sqrt{30}}, -\dfrac{1}{\sqrt{30}}, \dfrac{2}{\sqrt{30}}, 0\right), \left(\dfrac{1}{\sqrt{10}}, \dfrac{1}{\sqrt{10}}, -\dfrac{2}{\sqrt{10}}, -\dfrac{2}{\sqrt{10}}\right), \left(\dfrac{1}{\sqrt{15}}, \dfrac{1}{\sqrt{15}}, -\dfrac{2}{\sqrt{15}}, \dfrac{3}{\sqrt{15}}\right)$

19. $\left(0, \dfrac{1}{\sqrt{5}}, \dfrac{2}{\sqrt{5}}\right), \left(-\dfrac{\sqrt{5}}{\sqrt{6}}, -\dfrac{2}{\sqrt{30}}, \dfrac{1}{\sqrt{30}}\right)$ 20. $\left(\dfrac{1}{\sqrt{6}}, \dfrac{1}{\sqrt{6}}, \dfrac{1}{\sqrt{6}}\right), \left(\dfrac{1}{\sqrt{6}}, \dfrac{1}{\sqrt{6}}, -\dfrac{1}{\sqrt{6}}\right), \left(\dfrac{2}{\sqrt{6}}, -\dfrac{1}{\sqrt{6}}, 0\right)$

21. $\mathbf{w}_1 = (-\frac{4}{5}, 2, \frac{2}{5}),\ \mathbf{w}_2 = (\frac{9}{5}, 0, \frac{18}{5})$ 22. $\mathbf{w}_1 = (\frac{39}{42}, \frac{93}{42}, \frac{120}{42}),\ \mathbf{w}_2 = (\frac{3}{42}, -\frac{9}{42}, \frac{6}{42})$

23. $\mathbf{w}_1 = (-\frac{5}{4}, -\frac{1}{4}, \frac{5}{4}, \frac{9}{4}),\ \mathbf{w}_2 = (\frac{1}{4}, \frac{9}{4}, \frac{19}{4}, -\frac{9}{4})$ 27. $v_1 = \dfrac{1}{\sqrt{2}},\ v_2 = \sqrt{\dfrac{3}{2}}\,x,\ v_3 = \dfrac{\sqrt{5}}{2\sqrt{2}}(3x^2 - 1)$

28. (a) $1 + x + 4x^2 = \frac{7}{3}\sqrt{2}\,v_1 + \frac{1}{3}\sqrt{6}\,v_2 + \frac{8}{15}\sqrt{5}\,v_3$ (b) $2 - 7x^2 = -\frac{\sqrt{2}}{3}\,v_1 - \frac{28}{15}\sqrt{\frac{5}{2}}\,v_3$ (c) $4 + 3x = 4\sqrt{2}\,v_1 + \sqrt{6}\,v_2$

29. $v_1 = 1$, $v_2 = \sqrt{3}(2x - 1)$, $v_3 = \sqrt{5}(6x^2 - 6x + 1)$ **30.** $Q(-\frac{8}{7}, -\frac{4}{7}, \frac{25}{7}); \frac{3}{7}\sqrt{35}$ **31.** $Q(-\frac{8}{7}, \frac{4}{7}, -\frac{16}{7})$

EXERCISE SET 5.4 (page 261)

1. (a) $(\mathbf{w})_S = (3, -7)$, $[\mathbf{w}]_S = \begin{bmatrix} 3 \\ -7 \end{bmatrix}$ (b) $(\mathbf{w})_S = (\frac{5}{28}, \frac{3}{14})$, $[\mathbf{w}]_S = \begin{bmatrix} \frac{5}{28} \\ \frac{3}{14} \end{bmatrix}$

(c) $(\mathbf{w})_S = \left(a, \dfrac{b-a}{2}\right)$, $[\mathbf{w}]_S = \begin{bmatrix} a \\ \dfrac{b-a}{2} \end{bmatrix}$

2. (a) $(\mathbf{v})_S = (3, -2, 1)$, $[\mathbf{v}]_S = \begin{bmatrix} 3 \\ -2 \\ 1 \end{bmatrix}$ (b) $(\mathbf{v})_S = (-2, 0, 1)$, $[\mathbf{v}]_S = \begin{bmatrix} -2 \\ 0 \\ 1 \end{bmatrix}$

3. (a) $(\mathbf{p})_S = (4, -3, 1)$, $[\mathbf{p}]_S = \begin{bmatrix} 4 \\ -3 \\ 1 \end{bmatrix}$ (b) $(\mathbf{p})_S = (0, 2, -1)$, $[\mathbf{p}]_S = \begin{bmatrix} 0 \\ 2 \\ -1 \end{bmatrix}$

4. $(A)_S = (-1, 1, -1, 3)$, $[A]_S = \begin{bmatrix} -1 \\ 1 \\ -1 \\ 3 \end{bmatrix}$ **5.** (a) $\mathbf{w} = (16, 10, 12)$ (b) $\mathbf{q} = 3 + 4x^2$ (c) $B = \begin{bmatrix} 15 & -1 \\ 6 & 3 \end{bmatrix}$

6. (a) $\begin{bmatrix} 2 & -3 \\ 1 & 4 \end{bmatrix}$ (b) $\begin{bmatrix} \frac{4}{11} & \frac{3}{11} \\ -\frac{1}{11} & \frac{2}{11} \end{bmatrix}$ (c) $[\mathbf{w}]_B = \begin{bmatrix} 3 \\ -5 \end{bmatrix}$, $[\mathbf{w}]_{B'} = \begin{bmatrix} -\frac{3}{11} \\ -\frac{13}{11} \end{bmatrix}$

7. (a) $\begin{bmatrix} \frac{13}{10} & -\frac{1}{2} \\ -\frac{2}{5} & 0 \end{bmatrix}$ (b) $\begin{bmatrix} 0 & -\frac{5}{2} \\ -2 & -\frac{13}{2} \end{bmatrix}$ (c) $[\mathbf{w}]_B = \begin{bmatrix} -\frac{17}{10} \\ \frac{8}{5} \end{bmatrix}$, $[\mathbf{w}]_{B'} = \begin{bmatrix} -4 \\ -7 \end{bmatrix}$

8. (a) $\begin{bmatrix} \frac{3}{4} & \frac{3}{4} & \frac{1}{12} \\ -\frac{3}{4} & -\frac{17}{12} & -\frac{17}{12} \\ 0 & \frac{2}{3} & \frac{2}{3} \end{bmatrix}$ (b) $\begin{bmatrix} \frac{19}{12} \\ -\frac{43}{12} \\ \frac{4}{3} \end{bmatrix}$ **9.** (a) $\begin{bmatrix} 3 & 2 & \frac{5}{2} \\ -2 & -3 & -\frac{1}{2} \\ 5 & 1 & 6 \end{bmatrix}$ (b) $\begin{bmatrix} -\frac{7}{2} \\ \frac{23}{2} \\ 6 \end{bmatrix}$

10. (a) $\begin{bmatrix} -\frac{2}{9} & \frac{7}{9} \\ \frac{1}{3} & -\frac{1}{6} \end{bmatrix}$ (b) $\begin{bmatrix} \frac{3}{4} & \frac{7}{4} \\ \frac{3}{2} & 1 \end{bmatrix}$ (c) $[\mathbf{p}]_B = \begin{bmatrix} 1 \\ -1 \end{bmatrix}$ (d) $[\mathbf{p}]_{B'} = \begin{bmatrix} -\frac{11}{4} \\ \frac{1}{2} \end{bmatrix}$

11. (b) $\begin{bmatrix} 2 & 0 \\ 1 & 3 \end{bmatrix}$ (c) $\begin{bmatrix} \frac{1}{2} & 0 \\ -\frac{1}{6} & \frac{1}{3} \end{bmatrix}$ (d) $[\mathbf{h}]_B = \begin{bmatrix} 2 \\ -5 \end{bmatrix}$, $[\mathbf{h}]_{B'} = \begin{bmatrix} 1 \\ -2 \end{bmatrix}$

12. (a) $(4\sqrt{2}, -2\sqrt{2})$ (b) $(-\frac{7}{2}\sqrt{2}, \frac{3}{2}\sqrt{2})$ **13.** (a) $(-1 + 3\sqrt{3}, 3 + \sqrt{3})$ (b) $(\frac{5}{2} - \sqrt{3}, \frac{5}{2}\sqrt{3} + 1)$

14. (a) $(\frac{1}{2}\sqrt{2}, \frac{3}{2}\sqrt{2}, 5)$ (b) $(-\frac{5}{2}\sqrt{2}, \frac{7}{2}\sqrt{2}, -3)$

15. (a) $(-\frac{1}{2} - \frac{5}{2}\sqrt{3}, 2, \frac{5}{2} - \frac{1}{2}\sqrt{3})$ (b) $(\frac{1}{2} - \frac{3}{2}\sqrt{3}, 6, -\frac{3}{2} - \frac{1}{2}\sqrt{3})$

16. (a) $(-1, \frac{3}{2}\sqrt{2}, -\frac{7}{2}\sqrt{2})$ (b) $(1, -\frac{3}{2}\sqrt{2}, \frac{9}{2}\sqrt{3})$ **17.** (a), (b), (d), (e)

18. (a) $\begin{bmatrix} 1 & 0 \\ 0 & 1 \end{bmatrix}$ (b) $\begin{bmatrix} \dfrac{1}{\sqrt{2}} & \dfrac{1}{\sqrt{2}} \\ -\dfrac{1}{\sqrt{2}} & \dfrac{1}{\sqrt{2}} \end{bmatrix}$ (d) $\begin{bmatrix} -\dfrac{1}{\sqrt{2}} & 0 & \dfrac{1}{\sqrt{2}} \\ \dfrac{1}{\sqrt{6}} & -\dfrac{2}{\sqrt{6}} & \dfrac{1}{\sqrt{6}} \\ \dfrac{1}{\sqrt{3}} & \dfrac{1}{\sqrt{3}} & \dfrac{1}{\sqrt{3}} \end{bmatrix}$ (e) $\begin{bmatrix} \frac{1}{2} & \frac{1}{2} & \frac{1}{2} & \frac{1}{2} \\ \frac{1}{2} & -\frac{5}{6} & \frac{1}{6} & \frac{1}{6} \\ \frac{1}{2} & \frac{1}{6} & \frac{1}{6} & -\frac{5}{6} \\ \frac{1}{2} & \frac{1}{6} & -\frac{5}{6} & \frac{1}{6} \end{bmatrix}$

20. (a) $\begin{bmatrix} \cos\theta & \sin\theta \\ -\sin\theta & \cos\theta \end{bmatrix}$ **(b)** $\begin{bmatrix} \cos\theta & \sin\theta & 0 \\ -\sin\theta & \cos\theta & 0 \\ 0 & 0 & 1 \end{bmatrix}$

21. (a) $(-2, -1)$ **(b)** $(-\frac{4}{3}, -\frac{22}{3})$ **(c)** $(-\frac{4}{3}, \frac{22}{3})$ **(d)** $(0, 0)$ **23. (a), (c)**

25. (a) $(\frac{12}{5}, -\frac{9}{5}, -7)$ **(b)** $(2, 1, 6)$ **(c)** $(-\frac{42}{5}, \frac{19}{5}, -3)$ **(d)** $(0, 0, 0)$ **27. (b)**

29. (a) $A = \begin{bmatrix} \cos\theta & 0 & -\sin\theta \\ 0 & 1 & 0 \\ \sin\theta & 0 & \cos\theta \end{bmatrix}$ **(b)** $A = \begin{bmatrix} 1 & 0 & 0 \\ 0 & \cos\theta & \sin\theta \\ 0 & -\sin\theta & \cos\theta \end{bmatrix}$ **30.** $\begin{bmatrix} \frac{\sqrt{2}}{4} & \frac{\sqrt{6}}{4} & -\frac{\sqrt{2}}{2} \\ -\frac{\sqrt{3}}{2} & \frac{1}{2} & 0 \\ \frac{\sqrt{2}}{4} & \frac{\sqrt{6}}{4} & \frac{\sqrt{2}}{2} \end{bmatrix}$

CHAPTER 5 SUPPLEMENTARY EXERCISES (page 264)

1. (a) $(0, a, a, 0)$ with $a \neq 0$ **(b)** $\pm\left(0, \frac{2}{\sqrt{5}}, \frac{1}{\sqrt{5}}, 0\right)$ **6.** $\pm\left(\frac{1}{\sqrt{2}}, 0, \frac{1}{\sqrt{2}}\right)$

7. $w_k = \frac{1}{k}, k = 1, 2, \ldots, n$ **8. No** **11. (b)** θ approaches $\frac{\pi}{2}$

12. (b) The diagonals of a parallelogram are perpendicular if and only if its sides have the same length.

EXERCISE SET 6.1 (page 274)

1. (a) $\lambda^2 - 2\lambda - 3 = 0$ **(b)** $\lambda^2 - 8\lambda + 16 = 0$ **(c)** $\lambda^2 - 12 = 0$
(d) $\lambda^2 + 3 = 0$ **(e)** $\lambda^2 = 0$ **(f)** $\lambda^2 - 2\lambda + 1 = 0$

2. (a) $\lambda = 3, \lambda = -1$ **(b)** $\lambda = 4$ **(c)** $\lambda = \sqrt{12}, \lambda = -\sqrt{12}$
(d) No real eigenvalues **(e)** $\lambda = 0$ **(f)** $\lambda = 1$

3. (a) Basis for eigenspace corresponding to $\lambda = 3$: $\begin{bmatrix} \frac{1}{2} \\ 1 \end{bmatrix}$; basis for eigenspace corresponding to $\lambda = -1$: $\begin{bmatrix} 0 \\ 1 \end{bmatrix}$

(b) Basis for eigenspace corresponding to $\lambda = 4$: $\begin{bmatrix} \frac{3}{2} \\ 1 \end{bmatrix}$

(c) Basis for eigenspace corresponding to $\lambda = \sqrt{12}$: $\begin{bmatrix} 3 \\ \sqrt{12} \\ 1 \end{bmatrix}$;

   basis for eigenspace corresponding to $\lambda = -\sqrt{12}$: $\begin{bmatrix} -3 \\ \sqrt{12} \\ 1 \end{bmatrix}$

(d) There are no eigenspaces. **(e)** Basis for eigenspace corresponding to $\lambda = 0$: $\begin{bmatrix} 1 \\ 0 \end{bmatrix}, \begin{bmatrix} 0 \\ 1 \end{bmatrix}$

(f) Basis for eigenspace corresponding to $\lambda = 1$: $\begin{bmatrix} 1 \\ 0 \end{bmatrix} \begin{bmatrix} 0 \\ 1 \end{bmatrix}$

4. (a) $\lambda^3 - 6\lambda^2 + 11\lambda - 6 = 0$ **(b)** $\lambda^3 - 2\lambda = 0$
(c) $\lambda^3 + 8\lambda^2 + \lambda + 8 = 0$ **(d)** $\lambda^3 - \lambda^2 - \lambda - 2 = 0$
(e) $\lambda^3 - 6\lambda^2 + 12\lambda - 8 = 0$ **(f)** $\lambda^3 - 2\lambda^2 - 15\lambda + 36 = 0$

5. (a) $\lambda = 1, \lambda = 2, \lambda = 3$ (b) $\lambda = 0, \lambda = \sqrt{2}, \lambda = -\sqrt{2}$ (c) $\lambda = -8$
   (d) $\lambda = 2$ (e) $\lambda = 2$ (f) $\lambda = -4, \lambda = 3$

6. (a) $\lambda = 1$: basis $\begin{bmatrix} 0 \\ 1 \\ 0 \end{bmatrix}$; $\lambda = 2$: basis $\begin{bmatrix} -\frac{1}{2} \\ 1 \\ 1 \end{bmatrix}$; $\lambda = 3$: basis $\begin{bmatrix} -1 \\ 1 \\ 1 \end{bmatrix}$

   (b) $\lambda = 0$: basis $\begin{bmatrix} \frac{5}{3} \\ \frac{1}{3} \\ 1 \end{bmatrix}$; $\lambda = \sqrt{2}$: basis $\begin{bmatrix} \frac{1}{7}(15 + 5\sqrt{2}) \\ \frac{1}{7}(-1 + 2\sqrt{2}) \\ 1 \end{bmatrix}$; $\lambda = -\sqrt{2}$: basis $\begin{bmatrix} \frac{1}{7}(15 - 5\sqrt{2}) \\ \frac{1}{7}(-1 - 2\sqrt{2}) \\ 1 \end{bmatrix}$

   (c) $\lambda = -8$: basis $\begin{bmatrix} -\frac{1}{6} \\ -\frac{1}{6} \\ 1 \end{bmatrix}$ (d) $\lambda = 2$: basis $\begin{bmatrix} \frac{1}{3} \\ \frac{1}{3} \\ 1 \end{bmatrix}$ (e) $\lambda = 2$: basis $\begin{bmatrix} -\frac{1}{3} \\ -\frac{1}{3} \\ 1 \end{bmatrix}$

   (f) $\lambda = -4$: basis $\begin{bmatrix} -2 \\ \frac{8}{3} \\ 1 \end{bmatrix}$; $\lambda = 3$: basis $\begin{bmatrix} 5 \\ -2 \\ 1 \end{bmatrix}$

7. (a) $(\lambda - 1)^2(\lambda + 2)(\lambda + 1) = 0$ (b) $(\lambda - 4)^2(\lambda^2 + 3) = 0$ 8. (a) $\lambda = 1, \lambda = -2, \lambda = -1$ (b) $\lambda = 4$

9. (a) $\lambda = 1$: basis $\begin{bmatrix} 0 \\ 0 \\ 0 \\ 1 \end{bmatrix}$ and $\begin{bmatrix} 2 \\ 3 \\ 1 \\ 0 \end{bmatrix}$; $\lambda = -2$: basis $\begin{bmatrix} -1 \\ 0 \\ 1 \\ 0 \end{bmatrix}$; $\lambda = -1$: basis $\begin{bmatrix} -2 \\ 1 \\ 1 \\ 0 \end{bmatrix}$ (b) $\lambda = 4$: basis $\begin{bmatrix} \frac{3}{2} \\ 1 \\ 0 \\ 0 \end{bmatrix}$

10. (a) $\lambda = -1, \lambda = 5$ (b) $\lambda = 3, \lambda = 7, \lambda = 1$ (c) $\lambda = -\frac{1}{3}, \lambda = 1, \lambda = \frac{1}{4}$ 11. $\lambda = 1, \lambda = \frac{1}{512}, \lambda = 512, \lambda = 0$

12. For A^{25}, $\lambda = 1, -1$; basis for $\lambda = 1$: $\begin{bmatrix} -1 \\ 1 \\ 0 \end{bmatrix}, \begin{bmatrix} -1 \\ 0 \\ 1 \end{bmatrix}$; basis for $\lambda = -1$: $\begin{bmatrix} 2 \\ -1 \\ 1 \end{bmatrix}$

13. (a) $y = x$ and $y = 2x$ (b) No lines (c) $y = 0$

EXERCISE SET 6.2 (page 283)

5. $P = \begin{bmatrix} \frac{4}{3} & \frac{3}{4} \\ 1 & 1 \end{bmatrix}$; $P^{-1}AP = \begin{bmatrix} 1 & 0 \\ 0 & 2 \end{bmatrix}$ 6. $P = \begin{bmatrix} \frac{1}{3} & 0 \\ 1 & 1 \end{bmatrix}$; $P^{-1}AP = \begin{bmatrix} 1 & 0 \\ 0 & -1 \end{bmatrix}$

7. $P = \begin{bmatrix} 0 & 1 & 0 \\ 1 & 0 & 1 \\ -1 & 0 & 1 \end{bmatrix}$; $P^{-1}AP = \begin{bmatrix} 0 & 0 & 0 \\ 0 & 1 & 0 \\ 0 & 0 & 2 \end{bmatrix}$ 8. $P = \begin{bmatrix} -2 & 0 & 1 \\ 0 & 1 & 0 \\ 1 & 0 & 0 \end{bmatrix}$; $P^{-1}AP = \begin{bmatrix} 3 & 0 & 0 \\ 0 & 3 & 0 \\ 0 & 0 & 2 \end{bmatrix}$

9. Not diagonalizable 10. $P = \begin{bmatrix} 1 & 2 & 1 \\ 1 & 3 & 3 \\ 1 & 3 & 4 \end{bmatrix}$; $P^{-1}AP = \begin{bmatrix} 1 & 0 & 0 \\ 0 & 2 & 0 \\ 0 & 0 & 3 \end{bmatrix}$ 11. Not diagonalizable

12. $P = \begin{bmatrix} -\frac{1}{3} & 0 & 0 \\ 0 & 1 & 0 \\ 1 & 0 & 1 \end{bmatrix}$; $P^{-1}AP = \begin{bmatrix} 0 & 0 & 0 \\ 0 & 0 & 0 \\ 0 & 0 & 1 \end{bmatrix}$ 13. Not diagonalizable 14. $P = \begin{bmatrix} 1 & 1 & 0 & 0 \\ 0 & 1 & 1 & 0 \\ 0 & 0 & 1 & 1 \\ 0 & 0 & 0 & 1 \end{bmatrix}$;

$P^{-1}AP = \begin{bmatrix} -2 & 0 & 0 & 0 \\ 0 & -2 & 0 & 0 \\ 0 & 0 & 3 & 0 \\ 0 & 0 & 0 & 3 \end{bmatrix}$ 15. $\begin{bmatrix} 1 & 0 \\ -1023 & 1024 \end{bmatrix}$ 16. $\begin{bmatrix} -1 & 10237 & -2047 \\ 0 & 1 & 0 \\ 0 & 10245 & -2048 \end{bmatrix}$

17. (a) $\begin{bmatrix} 1 & 0 & 0 \\ 0 & 1 & 0 \\ 0 & 0 & 1 \end{bmatrix}$ **(b)** $\begin{bmatrix} 1 & 0 & 0 \\ 0 & 1 & 0 \\ 0 & 0 & 1 \end{bmatrix}$ **(c)** $\begin{bmatrix} 1 & -2 & 8 \\ 0 & -1 & 0 \\ 0 & 0 & -1 \end{bmatrix}$ **(d)** $\begin{bmatrix} 1 & -2 & 8 \\ 0 & -1 & 0 \\ 0 & 0 & -1 \end{bmatrix}$

18. $A^n = PD^nP^{-1} = \begin{bmatrix} 1 & 1 & 1 \\ 2 & 0 & -1 \\ 1 & -1 & 1 \end{bmatrix} \begin{bmatrix} 1^n & 0 & 0 \\ 0 & 3^n & 0 \\ 0 & 0 & 4^n \end{bmatrix} \begin{bmatrix} \frac{1}{6} & \frac{1}{3} & \frac{1}{6} \\ \frac{1}{2} & 0 & \frac{1}{2} \\ \frac{1}{3} & -\frac{1}{3} & \frac{1}{3} \end{bmatrix}$

EXERCISE SET 6.3 (page 291)

1. (a) $\lambda = 0$: 1-dimensional; $\lambda = 5$: 1-dimensional **(b)** $\lambda = 6$: 1-dimensional; $\lambda = -3$: 2-dimensional
  (c) $\lambda = 3$: 1-dimensional; $\lambda = 0$: 2-dimensional **(d)** $\lambda = 2$: 2-dimensional; $\lambda = 8$: 1-dimensional
  (e) $\lambda = 0$: 3-dimensional; $\lambda = 8$: 1-dimensional **(f)** $\lambda = 1$: 2-dimensional; $\lambda = 3$: 2-dimensional

2. $P = \begin{bmatrix} \frac{1}{\sqrt{2}} & -\frac{1}{\sqrt{2}} \\ \frac{1}{\sqrt{2}} & \frac{1}{\sqrt{2}} \end{bmatrix}$; $P^{-1}AP = \begin{bmatrix} 4 & 0 \\ 0 & 2 \end{bmatrix}$ **3.** $P = \begin{bmatrix} -\frac{2}{\sqrt{7}} & \frac{\sqrt{3}}{\sqrt{7}} \\ \frac{\sqrt{3}}{\sqrt{7}} & \frac{2}{\sqrt{7}} \end{bmatrix}$; $P^{-1}AP = \begin{bmatrix} 3 & 0 \\ 0 & 10 \end{bmatrix}$

4. $P = \begin{bmatrix} -\frac{2}{\sqrt{5}} & \frac{1}{\sqrt{5}} \\ \frac{1}{\sqrt{5}} & \frac{2}{\sqrt{5}} \end{bmatrix}$; $P^{-1}AP = \begin{bmatrix} 7 & 0 \\ 0 & 2 \end{bmatrix}$ **5.** $P = \begin{bmatrix} -\frac{4}{5} & 0 & \frac{3}{5} \\ 0 & 1 & 0 \\ \frac{3}{5} & 0 & \frac{4}{5} \end{bmatrix}$; $P^{-1}AP = \begin{bmatrix} 25 & 0 & 0 \\ 0 & -3 & 0 \\ 0 & 0 & -50 \end{bmatrix}$

6. $\begin{bmatrix} \frac{1}{\sqrt{2}} & \frac{1}{\sqrt{2}} & 0 \\ \frac{1}{\sqrt{2}} & \frac{1}{\sqrt{2}} & 0 \\ 0 & 0 & 1 \end{bmatrix}$ **7.** $\begin{bmatrix} \frac{1}{\sqrt{3}} & \frac{1}{\sqrt{6}} & \frac{1}{\sqrt{2}} \\ \frac{1}{\sqrt{3}} & -\frac{2}{\sqrt{6}} & 0 \\ \frac{1}{\sqrt{3}} & \frac{1}{\sqrt{6}} & -\frac{1}{\sqrt{2}} \end{bmatrix}$ **8.** $\begin{bmatrix} 0 & 0 & \frac{1}{\sqrt{2}} & \frac{1}{\sqrt{2}} \\ 0 & 0 & \frac{1}{\sqrt{2}} & -\frac{1}{\sqrt{2}} \\ 1 & 0 & 0 & 0 \\ 0 & 1 & 0 & 0 \end{bmatrix}$

9. $P = \begin{bmatrix} -\frac{4}{5} & \frac{3}{5} & 0 & 0 \\ \frac{3}{5} & \frac{4}{5} & 0 & 0 \\ 0 & 0 & -\frac{4}{5} & \frac{3}{5} \\ 0 & 0 & \frac{3}{5} & \frac{4}{5} \end{bmatrix}$; $P^{-1}AP = \begin{bmatrix} 25 & 0 & 0 & 0 \\ 0 & 25 & 0 & 0 \\ 0 & 0 & -25 & 0 \\ 0 & 0 & 0 & 25 \end{bmatrix}$ **10.** $\begin{bmatrix} \frac{1}{\sqrt{2}} & -\frac{1}{\sqrt{2}} \\ \frac{1}{\sqrt{2}} & \frac{1}{\sqrt{2}} \end{bmatrix}$

CHAPTER 6 SUPPLEMENTARY EXERCISES (page 292)

1. (b) The transformation rotates vectors through the angle θ; therefore, if $0 < \theta < \pi$, then no nonzero vector is transformed into a vector in the same or opposite direction.

2. $\lambda = k$ with multiplicity 3. **3. (c)** $\begin{bmatrix} 1 & 1 & 0 \\ 0 & 2 & 1 \\ 0 & 0 & 3 \end{bmatrix}$ **9.** $A^2 = \begin{bmatrix} 15 & 30 \\ 5 & 10 \end{bmatrix}$, $A^3 = \begin{bmatrix} 75 & 150 \\ 25 & 50 \end{bmatrix}$,

$A^4 = \begin{bmatrix} 375 & 750 \\ 125 & 250 \end{bmatrix}$, $A^5 = \begin{bmatrix} 1875 & 3750 \\ 625 & 1250 \end{bmatrix}$ **10.** $A^3 = \begin{bmatrix} 1 & -3 & 3 \\ 3 & -8 & 6 \\ 6 & -15 & 10 \end{bmatrix}$, $A^4 = \begin{bmatrix} 3 & -8 & 6 \\ 6 & -15 & 10 \\ 10 & -24 & 15 \end{bmatrix}$

12. (b) $\begin{bmatrix} 0 & 0 & 0 & -1 \\ 1 & 0 & 0 & 2 \\ 0 & 1 & 0 & -1 \\ 0 & 0 & 1 & -3 \end{bmatrix}$ **17. (a)** $\lambda_1 = 1: \begin{bmatrix} 1 \\ 0 \\ 1 \end{bmatrix}; \lambda_2 = \frac{1}{2}: \begin{bmatrix} 1 \\ \frac{1}{2} \\ 0 \end{bmatrix}; \lambda_3 = \frac{1}{3}: \begin{bmatrix} 1 \\ 1 \\ 1 \end{bmatrix}$

(b) $\lambda_1 = -2: \begin{bmatrix} 1 \\ 0 \\ 1 \end{bmatrix}; \lambda_2 = -1: \begin{bmatrix} 1 \\ \frac{1}{2} \\ 0 \end{bmatrix}; \lambda_3 = 0: \begin{bmatrix} 1 \\ 1 \\ 1 \end{bmatrix}$ **(c)** $\lambda_1 = 3: \begin{bmatrix} 1 \\ 0 \\ 1 \end{bmatrix}; \lambda_2 = 4: \begin{bmatrix} 1 \\ \frac{1}{2} \\ 0 \end{bmatrix}; \lambda_3 = 5: \begin{bmatrix} 1 \\ 1 \\ 1 \end{bmatrix}$

EXERCISE SET 7.1 (page 307)

1. (a) Linear   **(b)** Nonlinear   **(c)** Linear   **(d)** Linear

2. (a) Linear   **(b)** Nonlinear   **(c)** Linear   **(d)** Nonlinear   **3. (a)** Linear   **(b)** Nonlinear

4. (a) Linear   **(b)** Linear   **5. (a)** Linear   **(b)** Nonlinear   **6. (a)** Linear   **(b)** Nonlinear

7. (a) Linear   **(b)** Linear   **8. (a)** Linear   **(b)** Nonlinear   **9.** $F(x, y) = (-x, y)$

11. (a) $\begin{bmatrix} 42 \\ -55 \end{bmatrix}$ **(b)** $\begin{bmatrix} x + 3y + 4z \\ x - 7z \end{bmatrix}$ **(c)** $\begin{bmatrix} 1 & 3 & 4 \\ 1 & 0 & -7 \end{bmatrix}$ **12. (a)** $T(x, y, z) = (x, 0, z)$ **(b)** $(2, 0, -1)$

13. (a) $T(x, y, z) = \frac{1}{3}(2x - y - z, -x + 2y - z, -x - y + 2z)$ **(b)** $T(3, 8, 4) = (-2, 3, -1)$

14. (a) $T(-1, 2) = \left(-\frac{3}{\sqrt{2}}, \frac{1}{\sqrt{2}}\right); T(x, y) = \left(\frac{x}{\sqrt{2}} - \frac{y}{\sqrt{2}}, \frac{x}{\sqrt{2}} + \frac{y}{\sqrt{2}}\right)$ **(b)** $T(-1, 2) = (1, -2); T(x, y) = (-x, -y)$

(c) $T(-1, 2) = \left(-\frac{\sqrt{3}}{2} - 1, \sqrt{3} - \frac{1}{2}\right); T(x, y) = \left(\frac{\sqrt{3}}{2} x - \frac{1}{2} y, \frac{1}{2} x + \frac{\sqrt{3}}{2} y\right)$

(d) $T(-1, 2) = \left(-\frac{1}{2} + \sqrt{3}, \frac{\sqrt{3}}{2} + 1\right); T(x, y) = \left(\frac{1}{2} x + \frac{\sqrt{3}}{2} y, -\frac{\sqrt{3}}{2} x + \frac{1}{2} y\right)$

15. (a) $(\frac{5}{3}, -\frac{17}{9}, \frac{37}{9})$ **(b)** $(\frac{5}{3}, \frac{10}{9}, \frac{10}{9})$ **(c)** $\left(\frac{x + 2y + 2z}{9}, \frac{4x + 17y - z}{18}, \frac{4x - y + 17z}{18}\right)$

17. (a) Domain: R^2; image space: R^2; $(T_2 \circ T_1)(x, y) = (2x - 3y, 2x + 3y)$
(b) Domain: R^2; image space: the line $y = \frac{3}{4}x$; $(T_2 \circ T_1)(x, y) = (4x - 12y, 3x - 9y)$

18. (a) Domain: R^2; image space: R^2; $(T_2 \circ T_1)(x, y) = (2x + 3y, x - 2y)$
(b) Domain: R^2; image space: the line $x = 0$; $(T_2 \circ T_1)(x, y) = (0, 2x)$

19. (a) Domain: R^2; image space: R^2; $(T_3 \circ T_2 \circ T_1)(x, y) = (3x - 2y, x)$
(b) Domain: R^2; image space: the line $y = \frac{3}{2}x$; $(T_3 \circ T_2 \circ T_1)(x, y) = (4y, 6y)$

21. $(T_2 \circ T_1)(a_0 + a_1 x + a_2 x^2) = (a_0 + a_1 + a_2)x + (a_1 + 2a_2)x^2 + a_2 x^3$

22. (a) $T_1(\mathbf{v}) = \begin{bmatrix} \frac{1 + \sqrt{3}}{2} \\ \frac{\sqrt{3} - 1}{2} \end{bmatrix}; T_2(\mathbf{v}) = \begin{bmatrix} 1 \\ 1 \end{bmatrix}; (T_2 \circ T_1)(\mathbf{v}) = \begin{bmatrix} \frac{1 - \sqrt{3}}{2} \\ \frac{1 + \sqrt{3}}{2} \end{bmatrix}$

(b) $T_1(\mathbf{v}) = \begin{bmatrix} -1 - \frac{3\sqrt{3}}{2} \\ -\sqrt{3} + \frac{3}{2} \end{bmatrix}; T_2(\mathbf{v}) = \begin{bmatrix} -3 \\ -2 \end{bmatrix}; (T_2 \circ T_1)(\mathbf{v}) = \begin{bmatrix} \sqrt{3} - \frac{3}{2} \\ -1 - \frac{3\sqrt{3}}{2} \end{bmatrix}$

(c) $T_1(\mathbf{v}) = \begin{bmatrix} \dfrac{x - \sqrt{3}y}{2} \\ \dfrac{\sqrt{3}x + y}{2} \end{bmatrix}$; $T_2(\mathbf{v}) = \begin{bmatrix} -y \\ x \end{bmatrix}$; $(T_2 \circ T_1)(\mathbf{v}) = \begin{bmatrix} -\dfrac{\sqrt{3}x - y}{2} \\ \dfrac{x - \sqrt{3}y}{2} \end{bmatrix}$

23. $T(2\mathbf{v}_1 - 3\mathbf{v}_2 + 4\mathbf{v}_3) = (-10, -7, 6)$ **27.** (b) $(3T)(x_1, x_2) = (6x_1 - 3x_2, 3x_2 + 3x_1)$

28. (b) $(T_1 + T_2)(x, y) = (3y, 4x)$; $(T_2 - T_1)(x, y) = (y, 2x)$ **30.** (b) Nonlinear

31. (a) 4   (b) .8415   (c) 1

EXERCISE SET 7.2 (page 315)

1. *a, c*   **2.** *a*   **3.** *a, b, c*   **4.** *a*   **5.** *b*   **6.** *a*   **7.** (a) $\begin{bmatrix} \frac{1}{2} \\ 1 \end{bmatrix}$ (b) $\begin{bmatrix} \frac{3}{2} \\ -4 \\ 1 \\ 0 \end{bmatrix}$ (c) No basis exists.

8. (a) $\begin{bmatrix} 1 \\ -4 \end{bmatrix}$ (b) $\begin{bmatrix} 4 \\ 2 \\ 6 \end{bmatrix}, \begin{bmatrix} 1 \\ 1 \\ 0 \end{bmatrix}, \begin{bmatrix} -3 \\ -4 \\ 9 \end{bmatrix}$ (c) $\begin{bmatrix} x \\ x^2 \\ x^3 \end{bmatrix}$

10. (a) $\begin{bmatrix} 1 \\ 5 \\ 7 \end{bmatrix}, \begin{bmatrix} 0 \\ 1 \\ 1 \end{bmatrix}$ (b) $\begin{bmatrix} -\frac{14}{11} \\ \frac{19}{11} \\ 1 \end{bmatrix}$ (c) Rank $(T) = 2$, nullity $(T) = 1$

11. (a) $\begin{bmatrix} 1 \\ 2 \\ 0 \end{bmatrix}$ (b) $\begin{bmatrix} \frac{1}{2} \\ 0 \\ 1 \end{bmatrix}, \begin{bmatrix} 0 \\ 1 \\ 0 \end{bmatrix}$ (c) Rank $(T) = 1$, nullity $(T) = 2$

12. (a) $\begin{bmatrix} 1 \\ \frac{1}{4} \end{bmatrix}, \begin{bmatrix} 0 \\ 1 \end{bmatrix}$ (b) $\begin{bmatrix} -1 \\ -1 \\ 1 \\ 0 \end{bmatrix}, \begin{bmatrix} -\frac{4}{7} \\ \frac{2}{7} \\ 0 \\ 1 \end{bmatrix}$ (c) Rank $(T) = 2$, nullity $(T) = 2$

13. (a) $\begin{bmatrix} 1 \\ 3 \\ -1 \\ 2 \end{bmatrix}, \begin{bmatrix} 0 \\ 1 \\ -\frac{2}{7} \\ \frac{5}{14} \end{bmatrix}, \begin{bmatrix} 0 \\ 0 \\ 0 \\ 1 \end{bmatrix}$ (b) $\begin{bmatrix} -1 \\ -1 \\ 1 \\ 0 \\ 0 \end{bmatrix}, \begin{bmatrix} -1 \\ -2 \\ 0 \\ 0 \\ 1 \end{bmatrix}$ (c) Rank $(T) = 3$, nullity $(T) = 2$

14. (a) Range: xz-plane; nullspace: y-axis   **15.** $\ker(T) = \{\mathbf{0}\}$; $R(T) = V$
(b) Range: yz-plane; nullspace: x-axis
(c) Range: plane $y = x$; nullspace: the line $x = -t, y = t, z = 0$

16. (a) Nullity $(T) = 2$ (b) Nullity $(T) = 4$ **17.** Nullity $(T) = 0$, rank $(T) = 6$
(c) Nullity $(T) = 3$ (d) Nullity $(T) = 1$

18. (a) Dimension = nullity $(T) = 3$
(b) No. In order for $A\mathbf{x} = \mathbf{b}$ to be consistent for all $\mathbf{b}$ in R^5, we must have $R(T) = R^5$. But $R(T) \neq R^5$, since rank $(T) = \dim R(T) = 4$.

21. (a) $x = -t, y = -t, z = t, -\infty < t < +\infty$ (b) $14x - 8y - 5z = 0$

25. $\ker(D)$ consists of all constant polynomials.   **26.** $\ker(J)$ consists of all polynomials of the form kx.

EXERCISE SET 7.3 (page 323)

1. (a) $\ker(T) = \{0\}$; T is one-to-one. (b) $\ker(T) = \left\{ k \begin{bmatrix} -\frac{3}{2} \\ 1 \end{bmatrix} \right\}$; T is not one-to-one.

(c) $\ker(T) = \{0\}$; T is one-to-one. (d) $\ker(T) = \{0\}$; T is one-to-one.

(e) $\ker(T) = \left\{ k \begin{bmatrix} 1 \\ 1 \end{bmatrix} \right\}$; T is not one-to-one. (f) $\ker(T) = \left\{ k \begin{bmatrix} 0 \\ 1 \\ 1 \end{bmatrix} \right\}$; T is not one-to-one.

2. (a) $T^{-1} \begin{bmatrix} x_1 \\ x_2 \end{bmatrix} = \begin{bmatrix} x_1 - 2x_2 \\ -2x_1 + 5x_2 \end{bmatrix}$ (b) T has no inverse. (c) $T^{-1} \begin{bmatrix} x_1 \\ x_2 \end{bmatrix} = \begin{bmatrix} \frac{3}{19}x_1 - \frac{7}{19}x_2 \\ \frac{1}{19}x_1 + \frac{4}{19}x_2 \end{bmatrix}$

3. (a) T has no inverse. (b) $T^{-1} \begin{bmatrix} x_1 \\ x_2 \\ x_3 \end{bmatrix} = \begin{bmatrix} \frac{1}{4}x_1 + \frac{1}{8}x_2 - \frac{3}{4}x_3 \\ \frac{1}{4}x_1 + \frac{1}{8}x_2 + \frac{1}{4}x_3 \\ -\frac{3}{8}x_1 + \frac{3}{8}x_2 + \frac{1}{4}x_3 \end{bmatrix}$ (c) $T^{-1} \begin{bmatrix} x_1 \\ x_2 \\ x_3 \end{bmatrix} = \begin{bmatrix} \frac{1}{2}x_1 - \frac{1}{2}x_2 + \frac{1}{2}x_3 \\ -\frac{1}{2}x_1 + \frac{1}{2}x_2 + \frac{1}{2}x_3 \\ \frac{1}{2}x_1 + \frac{1}{2}x_2 - \frac{1}{2}x_3 \end{bmatrix}$

(d) $T^{-1} \begin{bmatrix} x_1 \\ x_2 \\ x_3 \end{bmatrix} = \begin{bmatrix} 3x_1 + 3x_2 - x_3 \\ -2x_1 - 2x_2 + x_3 \\ -4x_1 - 5x_2 + 2x_3 \end{bmatrix}$ 4. (a) Not one-to-one (b) Not one-to-one (c) One-to-one

5. (a) $\ker(T) = \left\{ k \begin{bmatrix} -1 \\ 1 \end{bmatrix} \right\}$ (b) T is not one-to-one since $\ker(T) \neq \{0\}$.

6. (a) $\ker(T) = \{0\}$ (b) T is one-to-one from Theorem 7.3.2.

7. (a) T is one-to-one. (b) T is not one-to-one. (c) T is not one-to-one. (d) T is one-to-one.

8. (a) T is one-to-one. (b) T is one-to-one. 9. No. A is not invertible.

10. (a) T is not one-to-one. (b) $T^{-1}(x_1, x_2, x_3, \ldots, x_n) = (x_n, x_{n-1}, x_{n-2}, \ldots, x_1)$
(c) $T^{-1}(x_1, x_2, x_3, \ldots, x_n) = (x_n, x_1, x_2, \ldots, x_{n-1})$

11. (a) $a_i \neq 0$ for $i = 1, 2, 3, \ldots, n$ (b) $T^{-1}(x_1, x_2, x_3, \ldots, x_n) = \left(\frac{1}{a_1}x_1, \frac{1}{a_2}x_2, \frac{1}{a_3}x_3, \ldots, \frac{1}{a_n}x_n \right)$

14. (a) T is not one-to-one. (b) T is one-to-one. $T^{-1} \begin{bmatrix} a & b \\ c & d \end{bmatrix} = \begin{bmatrix} a & c \\ b & d \end{bmatrix}$

(c) T is one-to-one. $T^{-1} \begin{bmatrix} a & b \\ c & d \end{bmatrix} = \begin{bmatrix} d & -b \\ -c & a \end{bmatrix}$

17. D is not one-to-one since $D(a + bx) = D(c + bx)$, where $a \neq c$.

18. J is not one-to-one since $J(x) = J(x^3)$.

EXERCISE SET 7.4 (page 332)

1. $T(x_1, x_2) = (-x_2, x_1 + 2x_2, -\frac{5}{2}x_1 + \frac{13}{2}x_2)$; $T(3, -6) = (6, -9, -\frac{93}{2})$

2. $T(x, y, z) = (30x - 10y - 3z, -9x + 3y + z)$, $T(1, 1, 1) = (17, -5)$

3. $T(a_0 + a_1x + a_2x^2) = (a_0 + 3a_1 + 4a_2) + (a_0 + 2a_2)x + (-a_1 - 3a_2)x^2$
$T(2 - 2x + 3x^2) = 8 + 8x - 7x^2$

4. $T(x_1, x_2, x_3) = (x_2 + x_3, x_1 + 2x_2 + x_3, x_1 + x_2 + x_3)$; $T(2, 1, -3) = (-2, 1, 0)$

5. (a) $\begin{bmatrix} 2 & -1 \\ 1 & 1 \end{bmatrix}$ (b) $\begin{bmatrix} 1 & 0 \\ 0 & 1 \end{bmatrix}$ (c) $\begin{bmatrix} 1 & 2 & 1 \\ 1 & 5 & 0 \\ 0 & 0 & 1 \end{bmatrix}$ (d) $\begin{bmatrix} 4 & 0 & 0 \\ 0 & 7 & 0 \\ 0 & 0 & -8 \end{bmatrix}$

6. (a) $\begin{bmatrix} 0 & 1 \\ -1 & 0 \\ 1 & 3 \\ 1 & -1 \end{bmatrix}$ **(b)** $\begin{bmatrix} 7 & 2 & -1 & 1 \\ 0 & 1 & 1 & 0 \\ -1 & 0 & 0 & 0 \end{bmatrix}$ **(c)** $\begin{bmatrix} 0 & 0 & 0 \\ 0 & 0 & 0 \\ 0 & 0 & 0 \\ 0 & 0 & 0 \\ 0 & 0 & 0 \end{bmatrix}$ **(d)** $\begin{bmatrix} 0 & 0 & 0 & 1 \\ 1 & 0 & 0 & 0 \\ 0 & 0 & 1 & 0 \\ 0 & 1 & 0 & 0 \\ 1 & 0 & -1 & 0 \end{bmatrix}$

7. (a) $\begin{bmatrix} -1 \\ 1 \end{bmatrix}$ **(b)** $\begin{bmatrix} 3 \\ 13 \end{bmatrix}$ **(c)** $\begin{bmatrix} -2x_1 + x_2 + 4x_3 \\ 3x_1 + 5x_2 + 7x_3 \\ 6x_1 - x_3 \end{bmatrix}$ **(d)** $\begin{bmatrix} -x_1 + x_2 \\ 2x_1 + 4x_2 \\ 7x_1 + 8x_2 \end{bmatrix}$

8. (a) $T(-1, 4) = (5, 4)$
(b) $T(2, 1, -3) = (0, -2, 0)$

9. (a) Yes. $T^{-1}(x_1, x_2) = (\frac{1}{3}x_1 - \frac{2}{3}x_2, \frac{1}{3}x_1 + \frac{1}{3}x_2)$
(b) No
(c) Yes. $T^{-1}(x_1, x_2) = (-x_2, -x_1)$
(d) No

10. (a) Yes. $T^{-1}(x_1, x_2, x_3) = (x_1 + 2x_2 - 4x_3, -x_1 - x_2 + 3x_3, -x_1 - 2x_2 + 5x_3)$ **(b)** No
(c) Yes. $T^{-1}(x_1, x_2, x_3) = (-\frac{3}{2}x_1 - \frac{3}{2}x_2 + \frac{11}{2}x_3, \frac{1}{2}x_1 + \frac{1}{2}x_2 - \frac{3}{2}x_3, -\frac{1}{2}x_1 + \frac{1}{2}x_2 - \frac{1}{2}x_3)$ **(d)** No

11. (a) $\begin{bmatrix} 1 & -2 & 1 & 3 \\ 2 & 1 & 0 & 4 \\ -1 & -3 & 1 & -1 \\ 3 & 3 & 1 & 7 \end{bmatrix}$, rank $(T) = 3$, nullity $(T) = 1$

(b) $\begin{bmatrix} 1 & 4 & 5 & 0 & 9 \\ 3 & -2 & 1 & 0 & -1 \\ -1 & 0 & -1 & 0 & -1 \\ 2 & 3 & 5 & 1 & 8 \end{bmatrix}$, rank $(T) = 3$, nullity $(T) = 2$

16. (a) $(10, 1, -14, -2)$
(b) No. $(1, -2, -1)$ is not a linear combination of the vectors $(1, 0, 0)$ and $(-1, 0, 1)$.

17. (a) $\begin{bmatrix} 0 & -c_0 & b_0 \\ c_0 & 0 & -a_0 \\ -b_0 & a_0 & 0 \end{bmatrix}$ **(b)** $T(x, y, z) = (b_0 z - c_0 y, c_0 x - a_0 z, a_0 y - b_0 x)$

EXERCISE SET 7.5 (page 345)

1. (a) $\begin{bmatrix} 0 & 0 & 0 \\ 1 & 0 & 0 \\ 0 & 1 & 0 \\ 0 & 0 & 1 \end{bmatrix}$ **2. (a)** $\begin{bmatrix} 1 & 1 & 0 \\ 0 & -2 & -3 \end{bmatrix}$ **3. (a)** $\begin{bmatrix} 1 & -1 & 1 \\ 0 & 1 & -2 \\ 0 & 0 & 1 \end{bmatrix}$ **4. (a)** $\begin{bmatrix} 2 & -1 \\ 2 & 0 \end{bmatrix}$

5. (a) $\begin{bmatrix} 0 & 0 \\ -\frac{1}{2} & 1 \\ \frac{8}{3} & \frac{4}{3} \end{bmatrix}$ **6. (a)** $\begin{bmatrix} 1 & -\frac{3}{2} & \frac{1}{2} \\ -1 & \frac{1}{2} & \frac{1}{2} \\ 0 & \frac{1}{2} & -\frac{1}{2} \end{bmatrix}$ **7. (a)** $\begin{bmatrix} 1 & 1 & 1 \\ 0 & 2 & 4 \\ 0 & 0 & 4 \end{bmatrix}$ **(b)** $3 + 10x + 16x^2$

8. (a) $\begin{bmatrix} 0 & 0 & 0 \\ 1 & -3 & 9 \\ 0 & 1 & -6 \\ 0 & 0 & 1 \end{bmatrix}$ **(b)** $-11x + 7x^2 - x^3$

9. (a) $[T(\mathbf{v}_1)]_B = \begin{bmatrix} 1 \\ -2 \end{bmatrix}$, $[T(\mathbf{v}_2)]_B = \begin{bmatrix} 3 \\ 5 \end{bmatrix}$ **(b)** $T(\mathbf{v}_1) = \begin{bmatrix} 3 \\ -5 \end{bmatrix}$, $T(\mathbf{v}_2) = \begin{bmatrix} -2 \\ 29 \end{bmatrix}$

(c) $T\left(\begin{bmatrix} x_1 \\ x_2 \end{bmatrix}\right) = \begin{bmatrix} \frac{18}{7} & \frac{1}{7} \\ -\frac{107}{7} & \frac{24}{7} \end{bmatrix}\begin{bmatrix} x_1 \\ x_2 \end{bmatrix}$ **(d)** $\begin{bmatrix} \frac{19}{7} \\ -\frac{83}{7} \end{bmatrix}$

10. (a) $[T(\mathbf{v}_1)]_{B'} = \begin{bmatrix} 3 \\ 1 \\ -3 \end{bmatrix}$, $[T(\mathbf{v}_2)]_{B'} = \begin{bmatrix} -2 \\ 6 \\ 0 \end{bmatrix}$, $[T(\mathbf{v}_3)]_{B'} = \begin{bmatrix} 1 \\ 2 \\ 7 \end{bmatrix}$, $[T(\mathbf{v}_4)]_{B'} = \begin{bmatrix} 0 \\ 1 \\ 1 \end{bmatrix}$

(b) $T(\mathbf{v}_1) = \begin{bmatrix} 11 \\ 5 \\ 22 \end{bmatrix}$, $T(\mathbf{v}_2) = \begin{bmatrix} -42 \\ 32 \\ -10 \end{bmatrix}$, $T(\mathbf{v}_3) = \begin{bmatrix} -56 \\ 87 \\ 17 \end{bmatrix}$, $T(\mathbf{v}_4) = \begin{bmatrix} -13 \\ 17 \\ 2 \end{bmatrix}$

(c) $T\left(\begin{bmatrix} x_1 \\ x_2 \\ x_3 \\ x_4 \end{bmatrix}\right) = \begin{bmatrix} -\frac{253}{10} & \frac{49}{5} & \frac{241}{10} & -\frac{229}{10} \\ \frac{115}{2} & -39 & -\frac{65}{2} & \frac{153}{2} \\ 66 & -60 & -9 & 91 \end{bmatrix} \begin{bmatrix} x_1 \\ x_2 \\ x_3 \\ x_4 \end{bmatrix}$ (d) $\begin{bmatrix} -31 \\ 37 \\ 12 \end{bmatrix}$

11. (a) $[T(\mathbf{v}_1)]_B = \begin{bmatrix} 1 \\ 2 \\ 6 \end{bmatrix}$, $[T(\mathbf{v}_2)]_B = \begin{bmatrix} 3 \\ 0 \\ -2 \end{bmatrix}$, $[T(\mathbf{v}_3)]_B = \begin{bmatrix} -1 \\ 5 \\ 4 \end{bmatrix}$

(b) $T(\mathbf{v}_1) = 16 + 51x + 19x^2$, $T(\mathbf{v}_2) = -6 - 5x + 5x^2$, $T(\mathbf{v}_3) = 7 + 40x + 15x^2$

(c) $T(a_0 + a_1x + a_2x^2) = \dfrac{239a_0 - 161a_1 + 289a_2}{24} + \dfrac{201a_0 - 111a_1 + 247a_2}{8}x + \dfrac{61a_0 - 31a_1 + 107a_2}{12}x^2$

(d) $T(Hx^2) = 22 + 56x + 14x^2$

12. (a) $[T_2 \circ T_1]_{B',B} = \begin{bmatrix} 1 & 1 \\ 2 & 4 \\ 0 & 4 \end{bmatrix}$, $[T_2]_{B'} = \begin{bmatrix} 1 & 1 & 1 \\ 0 & 2 & 4 \\ 0 & 0 & 4 \end{bmatrix}$, $[T_1]_{B',B} = \begin{bmatrix} 0 & 0 \\ 1 & 0 \\ 0 & 1 \end{bmatrix}$ (b) $[T_2 \circ T_1]_{B',B} = [T_2]_{B'}[T_1]_{B',B}$

13. (a) $[T_2 \circ T_1]_{B',B} = \begin{bmatrix} 0 & 0 \\ 6 & 0 \\ 0 & 0 \\ 0 & -9 \end{bmatrix}$, $[T_2]_{B',B''} = \begin{bmatrix} 0 & 0 & 0 \\ 3 & 0 & 0 \\ 0 & 3 & 0 \\ 0 & 0 & 3 \end{bmatrix}$, $[T_1]_{B'',B} = \begin{bmatrix} 2 & 0 \\ 0 & 0 \\ 0 & -3 \end{bmatrix}$

(b) $[T_2 \circ T_1]_{B',B} = [T_2]_{B',B''}[T_1]_{B'',B}$ **16.** $\begin{bmatrix} 0 & 0 & 0 & 1 \\ 1 & 0 & 0 & 0 \\ 0 & 1 & 0 & 0 \\ 0 & 0 & 1 & 0 \end{bmatrix}$

17. (a) $\begin{bmatrix} 0 & 1 & 0 \\ 0 & 0 & 2 \\ 0 & 0 & 0 \end{bmatrix}$ (b) $\begin{bmatrix} 0 & -\frac{3}{2} & \frac{23}{6} \\ 0 & 0 & -\frac{16}{3} \\ 0 & 0 & 0 \end{bmatrix}$ (c) $-6 + 48x$

18. (a) $\begin{bmatrix} 0 & 0 & 0 \\ 0 & 0 & -1 \\ 0 & 1 & 0 \end{bmatrix}$ (b) $\begin{bmatrix} 0 & 0 & 0 \\ 0 & 1 & 0 \\ 0 & 0 & 2 \end{bmatrix}$ (c) $\begin{bmatrix} 2 & 1 & 0 \\ 0 & 2 & 2 \\ 0 & 0 & 2 \end{bmatrix}$

EXERCISE SET 7.6 (page 360)

1. $[T]_B = \begin{bmatrix} 1 & -2 \\ 0 & -1 \end{bmatrix}$, $[T]_{B'} = \begin{bmatrix} -\frac{3}{11} & -\frac{56}{11} \\ -\frac{2}{11} & \frac{3}{11} \end{bmatrix}$ **2.** $[T]_B = \begin{bmatrix} \frac{4}{5} & \frac{61}{10} \\ \frac{18}{5} & -\frac{19}{5} \end{bmatrix}$, $[T]_{B'} = \begin{bmatrix} -\frac{31}{2} & \frac{9}{2} \\ -\frac{75}{2} & \frac{25}{2} \end{bmatrix}$

3. $[T]_B = \begin{bmatrix} \frac{1}{\sqrt{2}} & -\frac{1}{\sqrt{2}} \\ \frac{1}{\sqrt{2}} & \frac{1}{\sqrt{2}} \end{bmatrix}$, $[T]_{B'} = \begin{bmatrix} \frac{13}{11\sqrt{2}} & -\frac{25}{11\sqrt{2}} \\ \frac{5}{11\sqrt{2}} & \frac{9}{11\sqrt{2}} \end{bmatrix}$

4. $[T]_B = \begin{bmatrix} 1 & 2 & -1 \\ 0 & -1 & 0 \\ 1 & 0 & 7 \end{bmatrix}$, $[T]_{B'} = \begin{bmatrix} 1 & 4 & 3 \\ -1 & -2 & -9 \\ 1 & 1 & 8 \end{bmatrix}$ **5.** $[T]_B = \begin{bmatrix} 1 & 0 & 0 \\ 0 & 1 & 0 \\ 0 & 0 & 0 \end{bmatrix}$, $[T]_{B'} = \begin{bmatrix} 1 & 0 & 0 \\ 0 & 1 & 1 \\ 0 & 0 & 0 \end{bmatrix}$

6. $[T]_B = \begin{bmatrix} 5 & 0 \\ 0 & 5 \end{bmatrix}$, $[T]_{B'} = \begin{bmatrix} 5 & 0 \\ 0 & 5 \end{bmatrix}$ **7.** $[T]_B = \begin{bmatrix} \frac{2}{3} & -\frac{2}{3} \\ \frac{1}{2} & \frac{4}{3} \end{bmatrix}$, $[T]_{B'} = \begin{bmatrix} 1 & 1 \\ 0 & 1 \end{bmatrix}$

8. (a) $\det(T) = 17$ **10.** (a) $[T]_B = \begin{bmatrix} 1 & 1 & 1 & 1 & 1 \\ 0 & 2 & 4 & 6 & 8 \\ 0 & 0 & 4 & 12 & 24 \\ 0 & 0 & 0 & 8 & 32 \\ 0 & 0 & 0 & 0 & 16 \end{bmatrix}$, where B is the standard   (b) T is one-to-one.
(b) $\det(T) = 0$ basis for P_4; rank $(T) = 5$
(c) $\det(T) = 1$ and nullity $(T) = 0$.

11. (a) $u'_1 = \begin{bmatrix} -1 \\ 1 \end{bmatrix}$, $u'_2 = \begin{bmatrix} 1 \\ -2 \end{bmatrix}$ (b) $u'_1 = \begin{bmatrix} 1 \\ \dfrac{3 - \sqrt{21}}{2} \end{bmatrix}$, $u'_2 = \begin{bmatrix} 1 \\ \dfrac{3 + \sqrt{21}}{2} \end{bmatrix}$

12. (a) $u'_1 = \begin{bmatrix} -1 \\ 1 \\ 0 \end{bmatrix}$, $u'_2 = \begin{bmatrix} 1 \\ 0 \\ 1 \end{bmatrix}$, $u'_3 = \begin{bmatrix} -1 \\ -1 \\ 1 \end{bmatrix}$ (b) $u'_1 = \begin{bmatrix} -1 \\ 1 \\ 0 \end{bmatrix}$, $u'_2 = \begin{bmatrix} 1 \\ 0 \\ 1 \end{bmatrix}$, $u'_3 = \begin{bmatrix} -1 \\ -1 \\ 1 \end{bmatrix}$

(c) $u'_1 = \begin{bmatrix} 1 \\ 2 \\ 1 \end{bmatrix}$, $u'_2 = \begin{bmatrix} 0 \\ 1 \\ 0 \end{bmatrix}$, $u'_3 = \begin{bmatrix} -1 \\ 0 \\ 1 \end{bmatrix}$

13. (a) $\lambda = -4$, $\lambda = 3$ (b) Basis for eigenspace corresponding to $\lambda = -4$: $-2 + \frac{8}{3}x + x^2$;
basis for eigenspace corresponding to $\lambda = 3$: $5 - 2x + x^2$

14. (a) $\lambda = 1$, $\lambda = -2$, $\lambda = -1$

(b) Basis for eigenspace corresponding to $\lambda = 1$: $\begin{bmatrix} 0 & 0 \\ 0 & 1 \end{bmatrix}$ and $\begin{bmatrix} 2 & 3 \\ 1 & 0 \end{bmatrix}$;

basis for eigenspace corresponding to $\lambda = -2$: $\begin{bmatrix} -1 & 0 \\ 1 & 0 \end{bmatrix}$;

basis for eigenspace corresponding to $\lambda = -1$: $\begin{bmatrix} -2 & 1 \\ 1 & 0 \end{bmatrix}$

18. (b) $\left(\dfrac{3 + 5\sqrt{3}}{4}, \dfrac{\sqrt{3} + 5}{4} \right)$

CHAPTER 7 SUPPLEMENTARY EXERCISES (page 362)

1. No. $T(x_1 + x_2) = A(x_1 + x_2) + B \neq (Ax_1 + B) + (Ax_2 + B) = T(x_1) + T(x_2)$, and if $c \neq 1$, then
$T(cx) = cAx + B \neq c(Ax + B) = cT(x)$.

2. (b) $A^n = \begin{bmatrix} \cos n\theta & -\sin n\theta \\ \sin n\theta & \cos n\theta \end{bmatrix}$

5. (a) $T(e_3)$ and any two of $T(e_1)$, $T(e_2)$, and $T(e_4)$ form bases for the range; $(-1, 1, 0, 1)$ is a basis for the kernel.
(b) Rank $= 3$, nullity $= 1$

6. (a) 2   (b) 0   **7.** (a) Rank $(T) = 2$ and nullity $(T) = 2$ (b) T is not one-to-one.

11. Rank $= 3$, nullity $= 1$ **13.** $\begin{bmatrix} 1 & 0 & 0 & 0 \\ 0 & 0 & 1 & 0 \\ 0 & 1 & 0 & 0 \\ 0 & 0 & 0 & 1 \end{bmatrix}$

14. (a) $v_1 = 2u_1 + u_2$, $v_2 = -u_1 + u_2 + u_3$, $v_3 = 3u_1 + 4u_2 + 2u_3$
  (b) $u_1 = -2v_1 - 2v_2 + v_3$, $u_2 = 5v_1 + 4v_2 - 2v_3$, $u_3 = -7v_1 - 5v_2 + 3v_3$

15. $[T]_{B'} = \begin{bmatrix} -4 & 0 & 9 \\ 1 & 0 & -2 \\ 0 & 1 & 1 \end{bmatrix}$ **17.** $[T]_B = \begin{bmatrix} 1 & -1 & 1 \\ 0 & 1 & 0 \\ 1 & 0 & -1 \end{bmatrix}$ **19.** (b) $(D \circ J)(p(x)) = p(x)$

20. (a) $\begin{bmatrix} 2 \\ 6 \\ 12 \end{bmatrix}$ (d) $-3x^2 + 3$ (e)

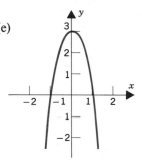

24. $\begin{bmatrix} 0 & 1 & 0 & 0 & \cdots & 0 \\ 0 & 0 & 1 & 0 & \cdots & 0 \\ 0 & 0 & 0 & 1 & \cdots & 0 \\ \vdots & \vdots & \vdots & \vdots & & \vdots \\ 0 & 0 & 0 & 0 & \cdots & 1 \\ 0 & 0 & 0 & 0 & \cdots & 0 \end{bmatrix}$ **25.** $\begin{bmatrix} 0 & 0 & 0 & \cdots & 0 \\ 1 & 0 & 0 & \cdots & 0 \\ 0 & \frac{1}{2} & 0 & \cdots & 0 \\ 0 & 0 & \frac{1}{3} & \cdots & 0 \\ \vdots & \vdots & \vdots & & \vdots \\ 0 & 0 & 0 & \cdots & \frac{1}{n+1} \end{bmatrix}$

EXERCISE SET 8.1 (page 372)

1. (a) $y_1 = c_1 e^{5x} - 2c_2 e^{-x}$ (b) $y_1 = 0$ **2.** (a) $y_1 = c_1 e^{7x} - 3c_2 e^{-x}$ (b) $y_1 = -\frac{1}{40} e^{7x} + \frac{81}{40} e^{-x}$
       $y_2 = c_1 e^{5x} + c_2 e^{-x}$ $y_2 = 0$ $y_2 = 2c_1 e^{7x} + 2c_2 e^{-x}$ $y_2 = -\frac{1}{20} e^{7x} - \frac{27}{20} e^{-x}$

3. (a) $y_1 = -c_2 e^{2x} + c_3 e^{3x}$ (b) $y_1 = e^{2x} - 2e^{3x}$ **4.** $y_1 = (c_1 + c_2)e^{2x} + c_3 e^{8x}$
       $y_2 = c_1 e^x + 2c_2 e^{2x} - c_3 e^{3x}$ $y_2 = e^x - 2e^{2x} + 2e^{3x}$ $y_2 = -c_2 e^{2x} + c_3 e^{8x}$
       $y_3 = 2c_2 e^{2x} - c_3 e^{3x}$ $y_3 = -2e^{2x} + 2e^{3x}$ $y_3 = -c_1 e^{2x} + c_3 e^{8x}$

5. $y = c_1 e^{3x} + c_2 e^{-2x}$ **6.** $y = c_1 e^x + c_2 e^{2x} + c_3 e^{3x}$

EXERCISE SET 8.2 (page 384)

1. (a) $\begin{bmatrix} 0 & -1 \\ -1 & 0 \end{bmatrix}$ (b) $\begin{bmatrix} -1 & 0 \\ 0 & -1 \end{bmatrix}$ (c) $\begin{bmatrix} 1 & 0 \\ 0 & 0 \end{bmatrix}$ (d) $\begin{bmatrix} 0 & 0 \\ 0 & 1 \end{bmatrix}$

2. (a) $(-1, -2)$ (b) $(-2, -1)$ (c) $(2, 0)$ (d) $(0, 1)$

3. (a) $\begin{bmatrix} 1 & 0 & 0 \\ 0 & 1 & 0 \\ 0 & 0 & -1 \end{bmatrix}$ (b) $\begin{bmatrix} 1 & 0 & 0 \\ 0 & -1 & 0 \\ 0 & 0 & 1 \end{bmatrix}$ (c) $\begin{bmatrix} -1 & 0 & 0 \\ 0 & 1 & 0 \\ 0 & 0 & 1 \end{bmatrix}$

4. (a) $(1, 1, -1)$ (b) $(1, -1, 1)$ (c) $(-1, 1, 1)$

5. (a) $\begin{bmatrix} 0 & -1 & 0 \\ 1 & 0 & 0 \\ 0 & 0 & 1 \end{bmatrix}$ (b) $\begin{bmatrix} 1 & 0 & 0 \\ 0 & 0 & -1 \\ 0 & 1 & 0 \end{bmatrix}$ (c) $\begin{bmatrix} 0 & 0 & 1 \\ 0 & 1 & 0 \\ -1 & 0 & 0 \end{bmatrix}$

6. (a) Rectangle with vertices at $(0, 0)$, $(1, 0)$, $(1, -2)$, $(0, -2)$
   (b) Rectangle with vertices at $(0, 0)$, $(-1, 0)$, $(-1, 2)$, $(0, 2)$
   (c) Rectangle with vertices at $(0, 0)$, $(1, 0)$, $(1, \frac{1}{2})$, $(0, \frac{1}{2})$
   (d) Square with vertices at $(0, 0)$, $(2, 0)$, $(2, 2)$, $(0, 2)$
   (e) Parallelogram with vertices at $(0, 0)$, $(1, 0)$, $(7, 2)$, $(6, 2)$
   (f) Parallelogram with vertices at $(0, 0)$, $(1, -2)$, $(1, 0)$, $(0, 2)$

7. Rectangle with vertices at $(0, 0)$, $(-3, 0)$, $(0, 1)$, $(-3, 1)$

8. (a) $\begin{bmatrix} \dfrac{1}{\sqrt{2}} & -\dfrac{1}{\sqrt{2}} \\ \dfrac{1}{\sqrt{2}} & \dfrac{1}{\sqrt{2}} \end{bmatrix}$ (b) $\begin{bmatrix} 0 & -1 \\ 1 & 0 \end{bmatrix}$ (c) $\begin{bmatrix} -1 & 0 \\ 0 & -1 \end{bmatrix}$ (d) $\begin{bmatrix} 0 & 1 \\ -1 & 0 \end{bmatrix}$ (e) $\begin{bmatrix} \dfrac{\sqrt{3}}{2} & \dfrac{1}{2} \\ -\dfrac{1}{2} & \dfrac{\sqrt{3}}{2} \end{bmatrix}$

9. (a) $\begin{bmatrix} 1 & 0 \\ 4 & 1 \end{bmatrix}$ (b) $\begin{bmatrix} 1 & -2 \\ 0 & 1 \end{bmatrix}$ **10.** (a) $\begin{bmatrix} 1 & 0 \\ 0 & \frac{1}{3} \end{bmatrix}$ (b) $\begin{bmatrix} 6 & 0 \\ 0 & 1 \end{bmatrix}$

11. (a) Expansion by a factor of 3 in the x-direction   (b) Expansion by a factor of -5 in the y-direction
   (c) Shear by a factor of 4 in the x-direction

12. (a) $\begin{bmatrix} 2 & 0 \\ 0 & 1 \end{bmatrix}\begin{bmatrix} 3 & 0 \\ 0 & 1 \end{bmatrix}$; expansion in the y-direction by a factor of 3, then expansion in the x-direction by a factor of 2

   (b) $\begin{bmatrix} 1 & 0 \\ 2 & 1 \end{bmatrix}\begin{bmatrix} 1 & 4 \\ 0 & 1 \end{bmatrix}$; shear in the x-direction by a factor of 4, then shear in the y-direction by a factor of 2

   (c) $\begin{bmatrix} 0 & 1 \\ 1 & 0 \end{bmatrix}\begin{bmatrix} 4 & 0 \\ 0 & 1 \end{bmatrix}\begin{bmatrix} 1 & 0 \\ 0 & -2 \end{bmatrix}$; expansion in the y-direction by a factor of -2, then expansion in the x-direction by a factor of 4, then reflection about $y = x$

   (d) $\begin{bmatrix} 1 & 0 \\ 4 & 1 \end{bmatrix}\begin{bmatrix} 1 & 0 \\ 1 & 18 \end{bmatrix}\begin{bmatrix} 1 & -3 \\ 0 & 1 \end{bmatrix}$; shear in the x-direction by a factor of -3, then expansion in the y-direction by a factor of 18, then shear in the y-direction by a factor of 4

13. (a) $\begin{bmatrix} \frac{1}{2} & 0 \\ 0 & 5 \end{bmatrix}$ (b) $\begin{bmatrix} 1 & 0 \\ 2 & 5 \end{bmatrix}$ (c) $\begin{bmatrix} 0 & 1 \\ -1 & 0 \end{bmatrix}$ **14.** (a) $\begin{bmatrix} 0 & 1 \\ -5 & 0 \end{bmatrix}$ (b) $\dfrac{1}{2}\begin{bmatrix} \sqrt{3} & -1 \\ -6\sqrt{3}+3 & 6+3\sqrt{3} \end{bmatrix}$

16. $16y - 11x - 3 = 0$ **17.** (a) $y = \frac{2}{3}x$ (b) $y = x$ (c) $y = \frac{1}{2}x$ (d) $y = -2x$ **18.** $\begin{bmatrix} 1 & -2 \\ 0 & 1 \end{bmatrix}$

19. (b) No. A is not invertible.   **22.** (a) $\begin{bmatrix} 1 & 0 & 0 \\ 0 & 0 & 1 \\ 0 & 1 & 0 \end{bmatrix}$ (b) $\begin{bmatrix} 0 & 0 & 1 \\ 0 & 1 & 0 \\ 1 & 0 & 0 \end{bmatrix}$ (c) $\begin{bmatrix} 0 & 1 & 0 \\ 1 & 0 & 0 \\ 0 & 0 & 1 \end{bmatrix}$

23. (a) $\begin{bmatrix} 1 & 0 & k \\ 0 & 1 & k \\ 0 & 0 & 1 \end{bmatrix}$ (b) xz-direction: $\begin{bmatrix} 1 & k & 0 \\ 0 & 1 & 0 \\ 0 & k & 1 \end{bmatrix}$; yz-direction: $\begin{bmatrix} 1 & 0 & 0 \\ k & 1 & 0 \\ k & 0 & 1 \end{bmatrix}$

24. (a) $\lambda_1 = 1$: $\begin{bmatrix} 1 \\ 0 \end{bmatrix}$; $\lambda_2 = -1$: $\begin{bmatrix} 0 \\ 1 \end{bmatrix}$ (b) $\lambda_1 = 1$: $\begin{bmatrix} 0 \\ 1 \end{bmatrix}$; $\lambda_2 = -1$: $\begin{bmatrix} 1 \\ 0 \end{bmatrix}$ (c) $\lambda_1 = 1$: $\begin{bmatrix} 1 \\ 1 \end{bmatrix}$; $\lambda_2 = -1$: $\begin{bmatrix} -1 \\ 1 \end{bmatrix}$

   (d) $\lambda = 1$: $\begin{bmatrix} 1 \\ 0 \end{bmatrix}$ (e) $\lambda = 1$: $\begin{bmatrix} 0 \\ 1 \end{bmatrix}$ (f) (θ an odd integer multiple of π)$\lambda = -1$: $\begin{bmatrix} 1 \\ 0 \end{bmatrix}$

   (θ an even integer with multiple of π)$\lambda = 1$: $\begin{bmatrix} 1 \\ 0 \end{bmatrix}, \begin{bmatrix} 0 \\ 1 \end{bmatrix}$
   (θ not an integer multiple of π) no real eigenvalues

EXERCISE SET 8.3 (page 395)

1. $y = -\dfrac{1}{2} + \dfrac{7}{2}x$ 2. $y = \dfrac{2}{3} + \dfrac{1}{6}x$ 3. $y = 2 + 5x - 3x^2$ 4. $y = -5 + 3x - 4x^2 + 2x^3$

9. $y = 4 - .2x + .2x^2$; if $x = 12$, then $y = 30.4$ ($\$30.4$ thousand)

EXERCISE SET 8.4 (page 402)

1. (a) $(1 + \pi) - 2\sin x - \sin 2x$ (b) $(1 + \pi) - 2\left[\sin x + \dfrac{\sin 2x}{2} + \dfrac{\sin 3x}{3} + \cdots + \dfrac{\sin nx}{n}\right]$

2. (a) $\tfrac{4}{3}\pi^2 + 4\cos x + \cos 2x + \tfrac{4}{9}\cos 3x - 4\pi\sin x - 2\pi\sin 2x - \dfrac{4\pi}{3}\sin 3x$

(b) $\tfrac{4}{3}\pi^2 + 4\sum_{k=1}^{n}\dfrac{\cos kx}{k^2} - 4\pi\sum_{k=1}^{n}\dfrac{\sin kx}{k}$ 3. (a) $-\dfrac{1}{2} + \dfrac{1}{e-1}e^x$ (b) $\dfrac{1}{12} - \dfrac{3-e}{2e-2}$

4. (a) $(4e - 10) + (18 - 6e)x$ (b) $\dfrac{(3-e)(7e-19)}{2}$ 5. (a) $\dfrac{3}{\pi}x$ (b) $1 - \dfrac{6}{\pi^2}$ 8. $\sum_{k=1}^{\infty}\dfrac{2}{k}\sin(kx)$

EXERCISE SET 8.5 (page 409)

1. (a), (c), (e), (g), (h)

2. (a) $A = \begin{bmatrix} 3 & 0 \\ 0 & 7 \end{bmatrix}$ (b) $A = \begin{bmatrix} 4 & -3 \\ -3 & -9 \end{bmatrix}$ (c) $A = \begin{bmatrix} 5 & \frac{3}{2} \\ \frac{3}{2} & 0 \end{bmatrix}$ (d) $A = \begin{bmatrix} 0 & -\frac{7}{2} \\ -\frac{7}{2} & 0 \end{bmatrix}$

3. (a) $A = \begin{bmatrix} 9 & 3 & -4 \\ 3 & -1 & \frac{1}{2} \\ -4 & \frac{1}{2} & 4 \end{bmatrix}$ (b) $\begin{bmatrix} 1 & -\frac{5}{2} & \frac{9}{2} \\ -\frac{5}{2} & 1 & 0 \\ \frac{9}{2} & 0 & -3 \end{bmatrix}$ (c) $A = \begin{bmatrix} 0 & \frac{1}{2} & \frac{1}{2} \\ \frac{1}{2} & 0 & \frac{1}{2} \\ \frac{1}{2} & \frac{1}{2} & 0 \end{bmatrix}$

(d) $A = \begin{bmatrix} \sqrt{2} & \sqrt{2} & -4\sqrt{3} \\ \sqrt{2} & 0 & 0 \\ -4\sqrt{3} & 0 & -\sqrt{3} \end{bmatrix}$ (e) $A = \begin{bmatrix} 1 & 1 & 0 & -5 \\ 1 & 1 & 0 & 0 \\ 0 & 0 & -1 & 2 \\ -5 & 0 & 2 & -1 \end{bmatrix}$

4. (a) $2x^2 + 5y^2 - 6xy$ (b) $7x_1^2 + 5x_1x_2$ (c) $x^2 - 3y^2 + 5z^2$
(d) $-2x_1^2 + 3x_3^2 + 7x_1x_2 + x_1x_3 + 12x_2x_3$ (e) $2x_1x_2 + 2x_1x_3 + 2x_1x_4 + 2x_2x_3 + 2x_2x_4 + 2x_3x_4$

5. max value $= 5$ at $\pm(1, 0)$; min value $= -1$ at $\pm(0, 1)$

(b) max value $= \dfrac{11 + \sqrt{10}}{2}$ at $\pm\left(\dfrac{-1}{\sqrt{20 + 6\sqrt{10}}}, \dfrac{1}{\sqrt{20 + 6\sqrt{10}}}\right)$;

min value $= \dfrac{11 - \sqrt{10}}{2}$ at $\pm\left(\dfrac{-1}{\sqrt{20 + 6\sqrt{10}}}, \dfrac{1}{\sqrt{20 + 6\sqrt{10}}}\right)$

(c) max value $= \dfrac{7 + \sqrt{10}}{2}$ at $\pm\left(\dfrac{3 + \sqrt{10}}{\sqrt{20 + 6\sqrt{10}}}, \dfrac{-1}{\sqrt{20 + 6\sqrt{10}}}\right)$;

min value $= \dfrac{7 - \sqrt{10}}{2}$ at $\pm\left(\dfrac{1}{\sqrt{20 + 6\sqrt{10}}}, \dfrac{3 + \sqrt{10}}{\sqrt{20 + 6\sqrt{10}}}\right)$

(d) max value $= \dfrac{7}{2}$ at $\pm\left(\dfrac{1}{\sqrt{2}}, \dfrac{1}{\sqrt{2}}\right)$; min value $= \dfrac{1}{2}$ at $\pm\left(\dfrac{1}{\sqrt{2}}, -\dfrac{1}{\sqrt{2}}\right)$

6. (a) max value $= 4$ at $\pm\left(\dfrac{1}{\sqrt{6}}, \dfrac{1}{\sqrt{6}}, \dfrac{2}{\sqrt{6}}\right)$; min value $= -2$ at $\pm\left(-\dfrac{1}{\sqrt{3}}, -\dfrac{1}{\sqrt{3}}, \dfrac{1}{\sqrt{3}}\right)$

(b) max value $= 3$ at $\left(\dfrac{2}{\sqrt{6}}, \dfrac{1}{\sqrt{6}}, \dfrac{1}{\sqrt{6}}\right)$; min value $= 0$ at $\left(\dfrac{1}{\sqrt{3}}, -\dfrac{1}{\sqrt{3}}, \dfrac{-1}{\sqrt{3}}\right)$

(c) max value $= 4$ at $\pm\left(\dfrac{1}{\sqrt{2}}, 0, \dfrac{1}{\sqrt{2}}\right)$; min value $= 2$ at $\pm\left(\dfrac{1}{\sqrt{2}}, 0, -\dfrac{1}{\sqrt{2}}\right)$ and $\pm(0, 1, 0)$ **7.** (b)   **9.** (a)

11. (a) Positive definite   (b) Negative definite   (c) Positive semidefinite
(d) Negative semidefinite   (e) Indefinite   (f) Indefinite

12. (a) Indefinite   (b) Indefinite   (c) Positive definite   (d) Indefinite
(e) Positive and negative semidefinite   (f) Positive definite

13. (b) No. $T(k\mathbf{x}) \neq kT(\mathbf{x})$, unless $k = 1$.

14. (a) $k > 4$ (b) $k > 2$ (c) $-\tfrac{1}{3}\sqrt{15} < k < \tfrac{1}{3}\sqrt{15}$

15. $A = \begin{bmatrix} c_1^2 & c_1 c_2 & c_1 c_3 & \cdots & c_1 c_n \\ c_1 c_2 & c_2^2 & c_2 c_3 & \cdots & c_2 c_n \\ \vdots & \vdots & \vdots & & \vdots \\ c_1 c_n & c_2 c_n & c_3 c_n & \cdots & c_n^2 \end{bmatrix}$

16. (a) $A = \begin{bmatrix} \dfrac{1}{n} & \dfrac{-1}{n(n-1)} & \dfrac{-1}{n(n-1)} & \cdots & \dfrac{-1}{n(n-1)} \\ \dfrac{-1}{n(n-1)} & \dfrac{1}{n} & \dfrac{-1}{n(n-1)} & \cdots & \dfrac{-1}{n(n-1)} \\ \vdots & \vdots & \vdots & & \vdots \\ \dfrac{-1}{n(n-1)} & \dfrac{-1}{n(n-1)} & \dfrac{-1}{n(n-1)} & \cdots & \dfrac{1}{n} \end{bmatrix}$ (b) Positive semidefinite

EXERCISE SET 8.6 (page 421)

1. (a) $\begin{bmatrix} x_1 \\ x_2 \end{bmatrix} = \begin{bmatrix} \dfrac{1}{\sqrt{2}} & \dfrac{1}{\sqrt{2}} \\ \dfrac{1}{\sqrt{2}} & -\dfrac{1}{\sqrt{2}} \end{bmatrix} \begin{bmatrix} y_1 \\ y_2 \end{bmatrix}$; $y_1^2 + 3y_2^2$ (b) $\begin{bmatrix} x_1 \\ x_2 \end{bmatrix} = \begin{bmatrix} \dfrac{1}{\sqrt{5}} & \dfrac{2}{\sqrt{5}} \\ -\dfrac{2}{\sqrt{5}} & \dfrac{1}{\sqrt{5}} \end{bmatrix} \begin{bmatrix} y_1 \\ y_2 \end{bmatrix}$; $y_1^2 + 6y_2^2$

(c) $\begin{bmatrix} x_1 \\ x_2 \end{bmatrix} = \begin{bmatrix} \dfrac{1}{\sqrt{2}} & \dfrac{1}{\sqrt{2}} \\ \dfrac{1}{\sqrt{2}} & -\dfrac{1}{\sqrt{2}} \end{bmatrix} \begin{bmatrix} y_1 \\ y_2 \end{bmatrix}$; $y_1^2 - y_2^2$

(d) $\begin{bmatrix} x_1 \\ x_2 \end{bmatrix} = \begin{bmatrix} \dfrac{\sqrt{17}-4}{\sqrt{34-8\sqrt{17}}} & \dfrac{\sqrt{17}-4}{\sqrt{34+8\sqrt{17}}} \\ \dfrac{1}{\sqrt{34-8\sqrt{17}}} & \dfrac{-1}{\sqrt{34+8\sqrt{17}}} \end{bmatrix} \begin{bmatrix} y_1 \\ y_2 \end{bmatrix}$; $(1+\sqrt{17})y_1^2 + (1-\sqrt{17})y_2^2$

2. (a) $\begin{bmatrix} x_1 \\ x_2 \\ x_3 \end{bmatrix} = \begin{bmatrix} \dfrac{1}{\sqrt{2}} & \dfrac{-4}{\sqrt{66+2\sqrt{33}}} & \dfrac{-4}{\sqrt{66-2\sqrt{33}}} \\ 0 & \dfrac{-1-\sqrt{33}}{\sqrt{66+2\sqrt{33}}} & \dfrac{1-\sqrt{33}}{\sqrt{66-2\sqrt{33}}} \\ \dfrac{1}{\sqrt{2}} & \dfrac{4}{\sqrt{66+2\sqrt{33}}} & \dfrac{4}{\sqrt{66-2\sqrt{33}}} \end{bmatrix} \begin{bmatrix} y_1 \\ y_2 \\ y_3 \end{bmatrix}$; $3y_1^2 + \dfrac{7\sqrt{33}}{2}y_2^2 + \dfrac{7-\sqrt{33}}{2}y_3^2$

(b) $\begin{bmatrix} x_1 \\ x_2 \\ x_3 \end{bmatrix} = \begin{bmatrix} \frac{1}{3} & \frac{2}{3} & \frac{2}{3} \\ \frac{2}{3} & \frac{1}{3} & -\frac{2}{3} \\ -\frac{2}{3} & \frac{2}{3} & -\frac{1}{3} \end{bmatrix} \begin{bmatrix} y_1 \\ y_2 \\ y_3 \end{bmatrix}$; $7y_1^2 + 4y_2^2 + y_3^2$

(c) $\begin{bmatrix} x_1 \\ x_2 \\ x_3 \end{bmatrix} = \begin{bmatrix} \dfrac{1}{\sqrt{14}} & \dfrac{1}{\sqrt{6}} & -\dfrac{4}{\sqrt{21}} \\ -\dfrac{2}{\sqrt{14}} & \dfrac{2}{\sqrt{6}} & \dfrac{1}{\sqrt{21}} \\ \dfrac{3}{\sqrt{14}} & \dfrac{1}{\sqrt{6}} & \dfrac{2}{\sqrt{21}} \end{bmatrix} \begin{bmatrix} y_1 \\ y_2 \\ y_3 \end{bmatrix}$; $2y_2^2 - 7y_3^2$

(d) $\begin{bmatrix} x_1 \\ x_2 \\ x_3 \end{bmatrix} = \begin{bmatrix} \dfrac{3}{\sqrt{10}} & \dfrac{1}{\sqrt{20}} & \dfrac{1}{\sqrt{20}} \\ -\dfrac{1}{\sqrt{10}} & \dfrac{3}{\sqrt{20}} & \dfrac{3}{\sqrt{20}} \\ 0 & \dfrac{1}{\sqrt{2}} & -\dfrac{1}{\sqrt{2}} \end{bmatrix} \begin{bmatrix} y_1 \\ y_2 \\ y_3 \end{bmatrix}$; $\sqrt{10}y_2^2 - \sqrt{10}y_3^2$

3. (a) $2x^2 - 3xy + 4y^2$ (b) $x^2 - xy$ (c) $5xy$ (d) $4x^2 - 2y^2$ (e) y^2

4. (a) $\begin{bmatrix} 2 & -\frac{3}{2} \\ -\frac{3}{2} & 4 \end{bmatrix}$ (b) $\begin{bmatrix} 1 & -\frac{1}{2} \\ -\frac{1}{2} & 0 \end{bmatrix}$ (c) $\begin{bmatrix} 0 & \frac{5}{2} \\ \frac{5}{2} & 0 \end{bmatrix}$ (d) $\begin{bmatrix} 4 & 0 \\ 0 & -2 \end{bmatrix}$ (e) $\begin{bmatrix} 0 & 0 \\ 0 & 1 \end{bmatrix}$

5. (a) $[x\ y]\begin{bmatrix} 2 & -\frac{3}{2} \\ -\frac{3}{2} & 4 \end{bmatrix}\begin{bmatrix} x \\ y \end{bmatrix} + [-7\ 2]\begin{bmatrix} x \\ y \end{bmatrix} + 7 = 0$ (b) $[x\ y]\begin{bmatrix} 1 & -\frac{1}{2} \\ -\frac{1}{2} & 0 \end{bmatrix}\begin{bmatrix} x \\ y \end{bmatrix} + [5\ 8]\begin{bmatrix} x \\ y \end{bmatrix} - 3 = 0$

(c) $[x\ y]\begin{bmatrix} 0 & \frac{5}{2} \\ \frac{5}{2} & 0 \end{bmatrix}\begin{bmatrix} x \\ y \end{bmatrix} - 8 = 0$ (d) $[x\ y]\begin{bmatrix} 4 & 0 \\ 0 & -2 \end{bmatrix}\begin{bmatrix} x \\ y \end{bmatrix} - 7 = 0$

(e) $[x\ y]\begin{bmatrix} 0 & 0 \\ 0 & 1 \end{bmatrix}\begin{bmatrix} x \\ y \end{bmatrix} + [7\ -8]\begin{bmatrix} x \\ y \end{bmatrix} - 5 = 0$

6. (a) Ellipse (b) Ellipse (c) Hyperbola (d) Hyperbola (e) Circle
(f) Parabola (g) Parabola (h) Parabola (i) Parabola (j) Circle

7. (a) $9x'^2 + 4y'^2 = 36$, ellipse (b) $x'^2 - 16y'^2 = 16$, hyperbola
(c) $y'^2 = 8x'$, parabola (d) $x'^2 + y'^2 = 16$, circle
(e) $18y'^2 - 12x'^2 = 419$, hyperbola (f) $y' = -\frac{1}{4}x'^2$, parabola

8. (a) Hyperbola; possible equations are
$3x'^2 - 2y'^2 + 8 = 0$, $-2x'^2 + 3y'^2 + 8 = 0$

(b) Parabola; possible equations are
$2\sqrt{2}x'^2 + 9x' - 7y' = 0$, $2\sqrt{2}y'^2 + 7x' + 9y' = 0$
$2\sqrt{2}y'^2 - 7x' - 9y' = 0$, $2\sqrt{2}x'^2 - 9x' + 7y' = 0$

(c) Ellipse; possible equations are
$7x'^2 + 3y'^2 = 9$, $3x'^2 + 7y'^2 = 9$

(d) Hyperbola; possible equations are
$4x'^2 - y'^2 = 3$, $4y'^2 - x'^2 = 3$

9. $2x''^2 + y''^2 = 6$, ellipse   **10.** $13y''^2 - 4x''^2 = 81$, hyperbola   **11.** $2x''^2 - 3y''^2 = 24$, hyperbola

12. $6x''^2 + 11y''^2 = 66$, ellipse   **13.** $4y''^2 - x''^2 = 0$, hyperbola   **14.** $\sqrt{29}x'^2 - 3y'^2 = 0$, parabola

15. (a) Two intersecting lines, $y = x$ and $y = -x$ (b) No graph
   (c) The graph is the single point $(0, 0)$.   (d) The graph is the line $y = x$.

   (e) The graph consists of two parallel lines $\dfrac{3}{\sqrt{13}} x + \dfrac{2}{\sqrt{13}} y = \pm 2$.   (f) The graph is the single point $(1, 2)$.

EXERCISE SET 8.7 (page 427)

1. (a) $x^2 + 2y^2 - z^2 + 4xy - 5yz$ (b) $3x^2 + 7z^2 + 2xy - 3xz + 4yz$
   (c) $xy + xz + yz$ (d) $x^2 + y^2 - z^2$
   (e) $3z^2 + 3xz$ (f) $2z^2 + 2xz + y^2$

2. (a) $\begin{bmatrix} 1 & 2 & 0 \\ 2 & 2 & -\frac{5}{2} \\ 0 & -\frac{5}{2} & -1 \end{bmatrix}$ (b) $\begin{bmatrix} 3 & 1 & -\frac{3}{2} \\ 1 & 0 & 2 \\ -\frac{3}{2} & 2 & 7 \end{bmatrix}$ (c) $\begin{bmatrix} 0 & \frac{1}{2} & \frac{1}{2} \\ \frac{1}{2} & 0 & \frac{1}{2} \\ \frac{1}{2} & \frac{1}{2} & 0 \end{bmatrix}$ (d) $\begin{bmatrix} 1 & 0 & 0 \\ 0 & 1 & 0 \\ 0 & 0 & -1 \end{bmatrix}$

   (e) $\begin{bmatrix} 0 & 0 & \frac{3}{2} \\ 0 & 0 & 0 \\ \frac{3}{2} & 0 & 3 \end{bmatrix}$ (f) $\begin{bmatrix} 0 & 0 & 1 \\ 0 & 1 & 0 \\ 1 & 0 & 2 \end{bmatrix}$

3. (a) $[x \; y \; z] \begin{bmatrix} 1 & 2 & 0 \\ 2 & 2 & -\frac{5}{2} \\ 0 & -\frac{5}{2} & -1 \end{bmatrix} \begin{bmatrix} x \\ y \\ z \end{bmatrix} + [7 \; 0 \; 2] \begin{bmatrix} x \\ y \\ z \end{bmatrix} - 3 = 0$

   (b) $[x \; y \; z] \begin{bmatrix} 3 & 1 & -\frac{3}{2} \\ 1 & 0 & 2 \\ -\frac{3}{2} & 2 & 7 \end{bmatrix} \begin{bmatrix} x \\ y \\ z \end{bmatrix} + [-3 \; 0 \; 0] \begin{bmatrix} x \\ y \\ z \end{bmatrix} - 4 = 0$

   (c) $[x \; y \; z] \begin{bmatrix} 0 & \frac{1}{2} & \frac{1}{2} \\ \frac{1}{2} & 0 & \frac{1}{2} \\ \frac{1}{2} & \frac{1}{2} & 0 \end{bmatrix} \begin{bmatrix} x \\ y \\ z \end{bmatrix} - 1 = 0$ (d) $[x \; y \; z] \begin{bmatrix} 1 & 0 & 0 \\ 0 & 1 & 0 \\ 0 & 0 & -1 \end{bmatrix} \begin{bmatrix} x \\ y \\ z \end{bmatrix} - 7 = 0$

   (e) $[x \; y \; z] \begin{bmatrix} 0 & 0 & \frac{3}{2} \\ 0 & 0 & 0 \\ \frac{3}{2} & 0 & 3 \end{bmatrix} \begin{bmatrix} x \\ y \\ z \end{bmatrix} + [0 \; -14 \; 0] \begin{bmatrix} x \\ y \\ z \end{bmatrix} + 9 = 0$

   (f) $[x \; y \; z] \begin{bmatrix} 0 & 0 & 1 \\ 0 & 1 & 0 \\ 1 & 0 & 2 \end{bmatrix} \begin{bmatrix} x \\ y \\ z \end{bmatrix} + [2 \; -1 \; 3] \begin{bmatrix} x \\ y \\ z \end{bmatrix} = 0$

4. (a) Ellipsoid   **5.** (a) $9x'^2 + 36y'^2 + 4z'^2 = 36$, ellipsoid
   (b) Hyperboloid of one sheet   (b) $6x'^2 + 3y'^2 - 2z'^2 = 18$, hyperboloid of one sheet
   (c) Hyperboloid of two sheets   (c) $3x'^2 - 3y'^2 - z'^2 = 3$, hyperboloid of two sheets
   (d) Elliptic cone   (d) $4x'^2 + 9y'^2 - z'^2 = 0$, elliptic cone
   (e) Elliptic paraboloid   (e) $x'^2 + 16y'^2 - 16z' = 0$, elliptic paraboloid
   (f) Hyperbolic paraboloid   (f) $7x'^2 - 3y'^2 + z' = 0$, hyperbolic paraboloid
   (g) Sphere   (g) $x'^2 + y'^2 + z'^2 = 25$, sphere

6. (a) $25x'^2 - 3y'^2 - 50z'^2 - 150 = 0$, hyperboloid of two sheets
   (b) $2x'^2 + 2y'^2 + 8z'^2 - 5 = 0$, ellipsoid
   (c) $9x'^2 + 4y'^2 - 36z = 0$, elliptic paraboloid
   (d) $x'^2 - y'^2 + z' = 0$, hyperbolic paraboloid

7. $x''^2 + y''^2 - 2z''^2 = -1$, hyperboloid of two sheets   8. $x''^2 + y''^2 + 2z''^2 = 4$, ellipsoid

9. $x''^2 - y''^2 + z'' = 0$, hyperbolic paraboloid   10. $6x''^2 + 3y''^2 - 8\sqrt{2}z'' = 0$, elliptic paraboloid

EXERCISE SET 9.1 (page 437)

1. Multiplications: mpn; additions: $mp(n-1)$ 2. Multiplications: $(k-1)n^3$; additions: $(k-1)(n^3 - n^2)$

3.

	$n = 5$	$n = 10$	$n = 100$	$n = 1000$
Solve $A\mathbf{x} = \mathbf{b}$ by Gauss-Jordan elimination	+: 50 ×: 65	+: 375 ×: 430	+: 383,250 ×: 343,300	+: 333,832,500 ×: 334,333,000
Solve $A\mathbf{x} = \mathbf{b}$ by Gaussian elimination	+: 50 ×: 65	+: 375 ×: 430	+: 383,250 ×: 343,300	+: 333,832,500 ×: 334,333,000
Find A^{-1} by reducing $[A|I]$ to $[I|A^{-1}]$	+: 80 ×: 125	+: 810 ×: 1000	+: 980,100 ×: 1,000,000	+: 998,001,000 ×: 1,000,000,000
Solve $A\mathbf{x} = \mathbf{b}$ as $\mathbf{x} = A^{-1}\mathbf{b}$	+: 100 ×: 150	+: 900 ×: 1100	+: 990,000 ×: 1,010,000	+: 999,000,000 ×: 1,001,000,000
Find $\det(A)$ by row reduction	+: 30 ×: 44	+: 285 ×: 339	+: 328,350 ×: 333,399	+: 332,833,500 ×: 333,333,999
Solve $A\mathbf{x} = \mathbf{b}$ by Cramer's Rule	+: 180 ×: 264	+: 3135 ×: 3729	+: 33,163,350 ×: 33,673,299	+: 33,299,933 × 10^4 ×: 33,366,733 × 10^4

4.

	$n = 5$	$n = 10$	$n = 100$	$n = 1000$
	Execution Time (sec)	Execution Time (sec)	Execution Time (sec)	Execution Time (sec)
Solve $A\mathbf{x} = \mathbf{b}$ by Gauss-Jordan elimination	1.55×10^{-4}	1.05×10^{-3}	.878	836
Solve $A\mathbf{x} = \mathbf{b}$ by Gaussian elimination	1.55×10^{-4}	1.05×10^{-3}	.878	836
Find A^{-1} by reducing $[A|I]$ to $[I|A^{-1}]$	2.84×10^{-4}	2.41×10^{-3}	2.49	2499
Solve $A\mathbf{x} = \mathbf{b}$ as $\mathbf{x} = A^{-1}\mathbf{b}$	3.50×10^{-4}	2.65×10^{-3}	2.52	2502
Find $\det(A)$ by row reduction	1.03×10^{-4}	8.21×10^{-4}	.831	833
Solve $A\mathbf{x} = \mathbf{b}$ by Cramer's Rule	6.18×10^{-4}	90.3×10^{-4}	83.9	834×10^3

EXERCISE SET 9.2 (page 447)

1. $x_1 = 2, x_2 = 1$ 2. $x_1 = -2, x_2 = 1, x_3 = -3$

3. $x_1 = 3, x_2 = -1$ 4. $x_1 = 4, x_2 = -1$

5. $x_1 = -1, x_2 = 1, x_3 = 0$ 6. $x_1 = 1, x_2 = -2, x_3 = 1$

7. $x_1 = -1, x_2 = 1, x_3 = 0$

8. $x_1 = -1, x_2 = 1, x_3 = 1$

9. $x_1 = -3, x_2 = 1, x_3 = 2, x_4 = 1$

10. $x_1 = 2, x_2 = -1, x_3 = 0, x_4 = 0$

11. (a) $A = Lu = \begin{bmatrix} 2 & 0 & 0 \\ -2 & 1 & 0 \\ 2 & 1 & 1 \end{bmatrix} \begin{bmatrix} 1 & \frac{1}{2} & -\frac{1}{2} \\ 0 & 0 & 1 \\ 0 & 0 & 0 \end{bmatrix}$

13. (b) $\begin{bmatrix} a & b \\ b & d \end{bmatrix} = \begin{bmatrix} 1 & 0 \\ \frac{c}{a} & 1 \end{bmatrix} \begin{bmatrix} a & b \\ 0 & \frac{ad-bc}{a} \end{bmatrix}$

(b) $A = L_1 DU = \begin{bmatrix} 1 & 0 & 0 \\ -1 & 1 & 0 \\ 1 & 1 & 1 \end{bmatrix} \begin{bmatrix} 2 & 0 & 0 \\ 0 & 1 & 0 \\ 0 & 0 & 1 \end{bmatrix} \begin{bmatrix} 1 & \frac{1}{2} & -\frac{1}{2} \\ 0 & 0 & 1 \\ 0 & 0 & 0 \end{bmatrix}$

(c) $A = L_2 U_2 = \begin{bmatrix} 1 & 0 & 0 \\ -1 & 1 & 0 \\ 1 & 1 & 1 \end{bmatrix} \begin{bmatrix} 2 & 1 & -1 \\ 0 & 0 & 1 \\ 0 & 0 & 0 \end{bmatrix}$

14. Additions: $\dfrac{n^3}{3} + \dfrac{n^2}{2} - \dfrac{5n}{6}$; multiplications: $\dfrac{n^3}{3} + n^2 - \dfrac{n}{3}$

18. $A = PLU = \begin{bmatrix} 1 & 0 & 0 \\ 0 & 0 & 1 \\ 0 & 1 & 0 \end{bmatrix} \begin{bmatrix} 3 & 0 & 0 \\ 0 & 2 & 0 \\ 3 & 0 & 1 \end{bmatrix} \begin{bmatrix} 1 & -\frac{1}{3} & 0 \\ 0 & 1 & \frac{1}{2} \\ 0 & 0 & 1 \end{bmatrix}$

EXERCISE SET 9.3 (page 454)

1. $x_1 \approx 2.81, x_2 \approx .940$; exact solution is $x_1 = 3, x_2 = 1$

2. $x_1 \approx .954, x_2 \approx -.90$; exact solution is $x_1 = 1, x_2 = -2$

3. $x_1 \approx -2.99, x_2 \approx -.999$; exact solution is $x_1 = -3, x_2 = -1$

4. $x_1 \approx 0.00, x_2 \approx 2.00$; exact solution is $x_1 = 0, x_2 = 2$

5. $x_1 \approx 3.03, x_2 \approx 1.02$; exact solution is $x_1 = 3, x_2 = 1$

6. $x_1 \approx 1.03, x_2 \approx -2.02$; exact solution is $x_1 = 1, x_2 = -2$

7. $x_1 \approx -3.00, x_2 \approx -1.00$; exact solution is $x_1 = -3, x_2 = -1$

8. $x_1 \approx .005, x_2 \approx 2.00$; exact solution is $x_1 = 0, x_2 = 2$

9. $x_1 \approx .492, x_2 \approx .006, x_3 \approx -.996$; exact solution is $x_1 = \frac{1}{2}, x_2 = 0, x_3 = -1$

10. $x_1 \approx 1.00, x_2 \approx .998, x_3 = 1.00$; exact solution is $x_1 = 1, x_2 = 1, x_3 = 1$

11. $x_1 \approx .499, x_2 \approx .0004, x_3 \approx -1.00$; exact solution is $x_1 = \frac{1}{2}, x_2 = 0, x_3 = -1$

12. $x_1 \approx 1.00, x_2 \approx 1.00, x_3 \approx 1.00$; exact solution is $x_1 = 1, x_2 = 1, x_3 = 1$

13. (a), (d), (e)

EXERCISE SET 9.4 (page 461)

1. (a) $.28 \times 10^1$ (b) $.3452 \times 10^4$ (c) $.3879 \times 10^{-5}$
(d) $-.135 \times 10^0$ (e) $.17921 \times 10^2$ (f) $-.863 \times 10^{-1}$

2. (a) $.280 \times 10^1$ (b) $.345 \times 10^4$ (c) $.388 \times 10^{-5}$
(d) $-.135 \times 10^0$ (e) $.179 \times 10^2$ (f) $-.863 \times 10^{-1}$

3. (a) $.28 \times 10^1$ (b) $.35 \times 10^4$ (c) $.39 \times 10^{-5}$
(d) $-.14 \times 10^0$ (e) $.18 \times 10^2$ (f) $-.86 \times 10^{-1}$

4. $x_1 = -3, x_2 = 7$ **5.** $x_1 = \frac{13}{8}, x_2 = \frac{14}{8}, x_3 = \frac{21}{8}$ **6.** $x_1 = 1, x_2 = 2, x_3 = 3$

7. $x_1 = 0, x_2 = 0, x_3 = 1, x_4 = -1$ **8.** $x_1 = .997, x_2 = 1.00$ **9.** $x_1 = -2, x_2 = 0, x_3 = 1$

10. $x_1 = 0, x_2 = 1$ (without pivoting); $x_1 = 1, x_2 = 1$ (with pivoting)

EXERCISE SET 9.5 (page 469)

1. (a) $\lambda = -3$ (b) No dominant eigenvalue   (c) $\lambda = 6$ (d) $\lambda = 3$

2. (a) $\begin{bmatrix} 1.00 \\ .503 \end{bmatrix}$ (b) 5.02   (c) The dominant eigenvector is $\begin{bmatrix} 1 \\ \frac{1}{2} \end{bmatrix}$; the dominant eigenvalue is 5.

(d) The percentage error is .4%.

3. (a) $\begin{bmatrix} 1.00 \\ .750 \end{bmatrix}$ (b) 8.01   (c) The dominant eigenvector is $\begin{bmatrix} 1 \\ \frac{3}{4} \end{bmatrix}$; the dominant eigenvalue is 8.

(d) The percentage error is .125%.

4. $\begin{bmatrix} 1.00 \\ -.560 \end{bmatrix}$ (b) -4.00 (c) The dominant eigenvector is $\begin{bmatrix} 1 \\ -\frac{1}{2} \end{bmatrix}$; the dominant eigenvalue is -4.

(d) The percentage error is 0%.

5. (a) At the end of two iterations the dominant eigenvalue and eigenvector are approximately $\lambda_1 \approx 20.1$ and

$$\mathbf{x} \approx \begin{bmatrix} 1 \\ .119 \end{bmatrix}.$$

(b) The exact values of the dominant eigenvalue and eigenvector are $\lambda_1 = 20$ and $\mathbf{x} = \begin{bmatrix} 1 \\ \frac{2}{17} \end{bmatrix}$.

6. (a) At the end of three iterations, the dominant eigenvalue and eigenvector are approximately $\lambda_1 \approx -9.95$ and

$$\mathbf{x} \approx \begin{bmatrix} -.978 \\ 1 \end{bmatrix}.$$

(b) The exact values of the dominant eigenvalue and eigenvector are $\lambda_1 = -10$ and $\mathbf{x} = \begin{bmatrix} -1 \\ 1 \end{bmatrix}$.

7. (a) $\begin{bmatrix} .027 \\ .027 \\ 1 \end{bmatrix}$ (b) 10.0   (c) The dominant eigenvector is $\begin{bmatrix} 0 \\ 0 \\ 1 \end{bmatrix}$; the dominant eigenvalue is 10.

EXERCISE SET 9.6 (page 474)

1. (a) $\begin{bmatrix} 1 \\ .509 \end{bmatrix}$ (b) 7.00   (c) $\lambda_2 \approx 2.00$, $\mathbf{v}_2 \approx \begin{bmatrix} -.51 \\ 1 \end{bmatrix}$

(d) Exact eigenvalues 7, 2; exact eigenvectors $\mathbf{v}_1 = \begin{bmatrix} 1 \\ \frac{1}{2} \end{bmatrix}$, $\mathbf{v}_2 = \begin{bmatrix} -\frac{1}{2} \\ 1 \end{bmatrix}$

2. (a) $\begin{bmatrix} 1 \\ .503 \end{bmatrix}$ (b) 12.0   (c) $\lambda_2 \approx 2.02$, $\mathbf{v}_2 \approx \begin{bmatrix} -.532 \\ 1 \end{bmatrix}$

(d) Exact eigenvalues 12, 2; exact eigenvectors $\mathbf{v}_1 = \begin{bmatrix} 1 \\ \frac{1}{2} \end{bmatrix}$, $\mathbf{v}_2 = \begin{bmatrix} -\frac{1}{2} \\ 1 \end{bmatrix}$

3. $\lambda_{\max} \approx -1.37$, $\mathbf{v}_{\max} \approx \begin{bmatrix} -.448 \\ 1.00 \end{bmatrix}$; $\lambda_{\min} \approx \frac{1}{.482} \approx 2.07$, $\mathbf{v}_{\min} \approx \begin{bmatrix} .125 \\ 1.00 \end{bmatrix}$

4. $\lambda_{max} \approx -3.99$, $\mathbf{v}_{max} \approx \begin{bmatrix} -1.00 \\ .560 \end{bmatrix}$; $\lambda_{min} \approx \dfrac{1}{1.00} = 1.00$, $\mathbf{v}_{min} \approx \begin{bmatrix} .494 \\ 1.00 \end{bmatrix}$

EXERCISE SET 10.1 (page 483)

1. (a–d)

2. (a) $(2, 3)$ (b) $(-4, 0)$ (c) $(-3, -2)$ (d) $(0, -5)$ 3. (a) $x = -2, y = -3$ (b) $x = 2, y = 1$

4. (a) $5 + 3i$ (b) $-3 - 7i$ (c) $4 - 8i$ (d) $-4 - 5i$ (e) $19 + 14i$ (f) $-\frac{11}{2} - \frac{17}{2}i$

5. (a) $2 + 3i$ (b) $-1 - 2i$ (c) $-2 + 9i$

6. (a)

(b)

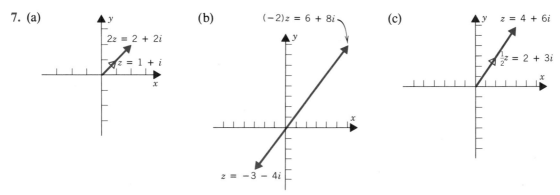

7. (a)

(b)

(c)

8. (a) $k_1 = -5, k_2 = 3$ 9. (a) $z_1 z_2 = 3 + 3i$, $z_1^2 = -9$, $z_2^2 = -2i$
   (b) $k_1 = 3, k_2 = 1$ (b) $z_1 z_2 = 26$, $z_1^2 = -20 + 48i$, $z_2^2 = -5 - 12i$
                                  (c) $z_1 z_2 = \frac{11}{3} - i$, $z_1^2 = \frac{4}{9}(-3 + 4i)$, $z_2^2 = -6 - \frac{5}{2}i$

10. (a) $9 - 8i$ (b) $-63 + 16i$ (c) $-32 - 24i$ (d) $22 + 19i$ **11.** $76 - 88i$ **12.** $26 - 18i$

13. $-26 + 18i$ **14.** $-1 - 11i$ **15.** $-\frac{63}{16} + i$ **16.** $(2 + \sqrt{2}) + i(1 - \sqrt{2})$ **17.** 0 **18.** $-24i$

19. (a) $\begin{bmatrix} 1 + 6i & -3 + 7i \\ 3 + 8i & 3 + 12i \end{bmatrix}$ (b) $\begin{bmatrix} 3 - 2i & 6 + 5i \\ 3 - 5i & 13 + 3i \end{bmatrix}$ (c) $\begin{bmatrix} 3 + 3i & 2 + 5i \\ 9 - 5i & 13 - 2i \end{bmatrix}$ (d) $\begin{bmatrix} 9 + i & 12 + 2i \\ 18 - 2i & 13 + i \end{bmatrix}$

20. (a) $\begin{bmatrix} 13 + 13i & -8 + 12i & -33 - 22i \\ 1 + i & 0 & i \\ 7 + 9i & -6 + 6i & -16 - 16i \end{bmatrix}$ (b) $\begin{bmatrix} 6 + 2i & -11 + 19i \\ -1 + 6i & -9 - 5i \end{bmatrix}$ (c) $\begin{bmatrix} 6i & 1 + i \\ -6 - i & 5 - 9i \end{bmatrix}$

(d) $\begin{bmatrix} 22 - 7i & 2 + 10i \\ -5 - 4i & 6 - 8i \\ 9 - i & -1 - i \end{bmatrix}$ **22.** (a) $z = -1 \pm i$ (b) $z = \frac{1}{2} \pm \frac{\sqrt{3}}{2} i$ **23.** (b) i

EXERCISE SET 10.2 (page 490)

1. (a) $2 - 7i$ (b) $-3 + 5i$ (c) $-5i$ **2.** (a) 1 (b) 7 (c) 5
   (d) i (e) -9 (f) 0 (d) $\sqrt{2}$ (c) 8 (f) 0

4. (a) $-\frac{17}{25} - \frac{19}{25}i$ (b) $\frac{23}{23} + \frac{11}{23}i$ (c) $\frac{23}{23} - \frac{11}{23}i$ (d) $-\frac{17}{23} + \frac{19}{23}i$ (e) $\frac{1}{3} - i$ (f) $\dfrac{\sqrt{26}}{5}$

5. (a) $-i$ (b) $\frac{1}{26} + \frac{5}{26}i$ (c) $7i$ **6.** (a) $\frac{6}{5} + \frac{2}{3}i$ (b) $-\frac{2}{3} + \frac{1}{3}i$ (c) $\frac{3}{5} + \frac{11}{5}i$ (d) $\frac{3}{5} + \frac{1}{5}i$ **7.** $\frac{1}{2} + \frac{1}{2}i$

8. $\frac{2}{5} + \frac{1}{5}i$ **9.** $-\frac{7}{625} - \frac{24}{625}i$ **10.** $-\frac{11}{25} + \frac{2}{25}i$ **11.** $\dfrac{1 - \sqrt{3}}{4} + \dfrac{1 + \sqrt{3}}{4} i$ **12.** $-\frac{1}{26} - \frac{5}{26}i$ **13.** $-\frac{1}{10} + \frac{1}{10}i$

14. $-\frac{2}{5}$ **15.** (a) $-1 - 2i$ (b) $-\frac{3}{25} - \frac{4}{25}i$

17. (a)   (b)   (c)   (d)

18. (a)   (b)   (c)   (d)

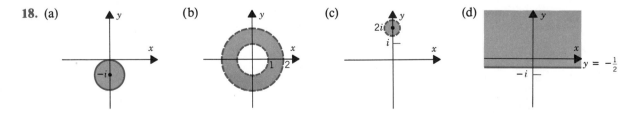

19. (a) $-y$ (b) $-x$ (c) y (d) x **20.** (b) $-i$ **23.** (a) $\dfrac{x_1 x_2 + y_1 y_2}{x_2^2 + y_2^2}$ (b) $\dfrac{x_2 y_1 - x_1 y_2}{x_2^2 + y_2^2}$

27. (c) Yes, if $z \neq 0$.   **28.** $x_1 = i, x_2 = -i$ **29.** $x_1 = 1 + i, x_2 = 1 - i$ **30.** $x_1 = \frac{1}{2} + i, x_2 = 2, x_3 = \frac{1}{2} - i$

31. $x_1 = i, x_2 = 0, x_3 = -i$ **32.** $x_1 = -(1 + i)t, x_2 = t$ **33.** $x_1 = (1 + i)t, x_2 = 2t$

34. $x_1 = -(1-i)t,\ x_2 = -it,\ x_3 = t$ **35.** (a) $\begin{bmatrix} i & 2 \\ -1 & i \end{bmatrix}$ (b) $\begin{bmatrix} 0 & 1 \\ -i & 2i \end{bmatrix}$

38. (a) $\begin{bmatrix} -i & -2-2i & -1+i \\ 1 & 2 & -i \\ i & i & 1 \end{bmatrix}$ (b) $\begin{bmatrix} 1+i & -i & 1 \\ -7+6i & 5-i & 1+4i \\ 1+2i & -i & 1 \end{bmatrix}$

EXERCISE SET 10.3 (page 500)

1. (a) 0   (b) $\pi/2$ (c) $-\pi/2$ (d) $\pi/4$ (e) $2\pi/3$ (f) $-\pi/4$ **2.** (a) $5\pi/3$ (b) $-\pi/3$ (c) $5\pi/3$

3. (a) $2\left[\cos\left(\dfrac{\pi}{2}\right) + i\sin\left(\dfrac{\pi}{2}\right)\right]$ (b) $4[\cos\pi + i\sin\pi]$

    (c) $5\sqrt{2}\left[\cos\left(\dfrac{\pi}{4}\right) + i\sin\left(\dfrac{\pi}{4}\right)\right]$ (d) $12\left[\cos\left(\dfrac{2\pi}{3}\right) + i\sin\left(\dfrac{2\pi}{3}\right)\right]$

    (e) $3\sqrt{2}\left[\cos\left(-\dfrac{3\pi}{4}\right) + i\sin\left(-\dfrac{3\pi}{4}\right)\right]$ (f) $4\left[\cos\left(-\dfrac{\pi}{6}\right) + i\sin\left(-\dfrac{\pi}{6}\right)\right]$

4. (a) $6\left[\cos\left(\dfrac{5\pi}{12}\right) + i\sin\left(\dfrac{5\pi}{12}\right)\right]$ (b) $\dfrac{2}{3}\left[\cos\left(\dfrac{\pi}{12}\right) + i\sin\left(\dfrac{\pi}{12}\right)\right]$

    (c) $\dfrac{3}{2}\left[\cos\left(-\dfrac{\pi}{12}\right) + i\sin\left(-\dfrac{\pi}{12}\right)\right]$ (d) $\dfrac{32}{9}\left[\cos\left(\dfrac{11\pi}{12}\right) + i\sin\left(\dfrac{11\pi}{12}\right)\right]$

5. 1   **6.** (a) -64 (b) $-i$ (c) $-64\sqrt{3} - 64i$ (d) $-\dfrac{1 + \sqrt{3}i}{2048}$

7. (a)

(b)

(c)

(d)

(e)

(f)

8.

9.

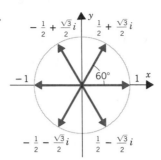

10. $\sqrt[4]{2}\left[\cos\left(\dfrac{\pi}{8}\right)+i\sin\left(\dfrac{\pi}{8}\right)\right], \sqrt[4]{2}\left[\cos\left(\dfrac{9\pi}{8}\right)+i\sin\left(\dfrac{9\pi}{8}\right)\right]$ **11.** (a) $\pm 2, \pm 2i$ (b) $\pm(2+2i), \pm(2-2i)$

12. The roots are $\pm(2^{1/4}+2^{1/4}i), \pm(2^{1/4}-2^{1/4}i)$ and the factorization is
$z^4+8=(z^2-2^{5/4}z+2^{3/2})\cdot(z^2+2^{5/4}z+2^{3/2})$.

13. Rotates z clockwise by $90°$.   **14.** (a) 16 (b) $\dfrac{i}{4^9}$

15. (a) $\mathrm{Re}(z)=-3, \mathrm{Im}(z)=0$ (b) $\mathrm{Re}(z)=-3, \mathrm{Im}(z)=0$
   (c) $\mathrm{Re}(z)=0, \mathrm{Im}(z)=-\sqrt{2}$ (d) $\mathrm{Re}(z)=-3, \mathrm{Im}(z)=0$

EXERCISE SET 10.4 (page 505)

1. (a) $(3i, -i, -2-i, 4)$ (b) $(3+2i, -1-2i, -3+5i, -i)$ **2.** $(2+i, 0, -3+i, -4i)$
   (c) $(-1-2i, 2i, 2-i, -1)$ (d) $(-3+9i, 3-3i, -3-6i, 12+3i)$
   (e) $(-3+2i, 3, -3-3i, i)$ (f) $(-1-5i, 3i, 4, -5)$

3. $c_1=-2-i, c_2=0, c_3=2-i$ **5.** (a) $\sqrt{2}$ (b) $2\sqrt{3}$ (c) $\sqrt{10}$ (d) $\sqrt{37}$

6. (a) $\sqrt{43}$ (b) $\sqrt{10}+\sqrt{29}$ (c) $\sqrt{10}+\sqrt{10}i$ (d) $\sqrt{699}$ (e) $\left(\dfrac{1+i}{\sqrt{6}}, \dfrac{2i}{\sqrt{6}}, 0\right)$ (f) 1

8. All k such that $|k|=\frac{1}{3}$ **9.** (a) 3 (b) $2-27i$ (c) $-5-10i$

10. The set is a vector space under the given operations.

11. Not a vector space. Axiom 6 fails, that is, the set is not closed under scalar multiplication. (Multiply by i, for example.)

12. No, R^n is not closed under scalar multiplication. (Multiply a nonzero vector in R^n by i.)

13. (a)   **14.** (b)   **15.** (a), (d)   **16.** (a), (b), (d)

17. (a) $(3-2i)\mathbf{u}+(3-i)\mathbf{v}+(1+2i)\mathbf{w}$ (b) $(2+i)\mathbf{u}+(-1+i)\mathbf{v}+(-1-i)\mathbf{w}$
   (c) $0\mathbf{u}+0\mathbf{v}+0\mathbf{w}$ (d) $(-5-4i)\mathbf{u}+(5-2i)\mathbf{v}+(2+4i)\mathbf{w}$

18. (a) Yes   (b) No   (c) Yes   (d) No   **19.** (a), (b), (c)

20. (a) $\mathbf{u}_2=i\mathbf{u}_1$ (b) Three vectors in a two-dimensional space   (c) A is a scalar multiple of B.

21. (b), (c)   **22.** $\mathbf{f}-3\mathbf{g}-3\mathbf{h}=\mathbf{0}$ **23.** (a) Three vectors in a two-dimensional space   **24.** (a), (b)
   (b) Two vectors in a three-dimensional space

25. (a), (b), (c), (d)   **26.** $(-1-i, 1)$; dimension $=1$ **27.** $(1, 1-i)$; dimension $=1$

28. $(3+6i, -3i, 1)$; dimension $=1$ **29.** $(\frac{5}{2}i, -\frac{1}{2}, 1, 0), (-\frac{1}{4}, \frac{3}{4}i, 0, 1)$; dimension $=2$

EXERCISE SET 10.5 (page 512)

2. (a) -12 (b) 0   (c) $2i$ (d) 37   4. (a) $-4+5i$ (b) 0   (c) $4-4i$ (d) 42

5. (a) Axiom 4 fails.   (b) Axiom 4 fails.   (c) Axioms 2 and 3 fail.   (d) Axioms 1 and 4 fail.
    (e) This is an inner product.   6. $-9-5i$ 7. No. Axioms 1 and 4 fail.

9. (a) $\sqrt{21}$ (b) $\sqrt{10}$ (c) $\sqrt{10}$ (d) 0   10. (a) $\sqrt{10}$ (b) 2   (c) $\sqrt{5}$ (d) 0

11. (a) $\sqrt{2}$ (b) $2\sqrt{3}$ (c) 5   (d) 0   12. (a) $3\sqrt{10}$ (b) $\sqrt{14}$ 13. (a) $\sqrt{10}$ (b) $2\sqrt{5}$

14. (a) 2   (b) $2\sqrt{2}$ 15. (a) $2\sqrt{3}$ (b) $2\sqrt{2}$ 16. (a) $7\sqrt{2}$ (b) $2\sqrt{3}$ 17. (a) $-\frac{8}{3}i$ (b) None

18. (a), (b), (c)   20. (b)   21. (b), (c)

23. $\left(\dfrac{i}{\sqrt{2}},0,0,\dfrac{i}{\sqrt{2}}\right),\left(-\dfrac{i}{\sqrt{6}},0,\dfrac{2i}{\sqrt{6}},\dfrac{i}{\sqrt{6}}\right),\left(\dfrac{2i}{\sqrt{21}},\dfrac{3i}{\sqrt{21}},\dfrac{2i}{\sqrt{21}},\dfrac{-2i}{\sqrt{21}}\right),\left(-\dfrac{i}{\sqrt{7}},\dfrac{2i}{\sqrt{7}},-\dfrac{i}{\sqrt{7}},\dfrac{i}{\sqrt{7}}\right)$

24. (a) $\mathbf{v}_1=\left(\dfrac{i}{\sqrt{10}},-\dfrac{3i}{\sqrt{10}}\right),\mathbf{v}_2=\left(\dfrac{3i}{\sqrt{10}},\dfrac{i}{\sqrt{10}}\right)$ (b) $\mathbf{v}_1=(i,0),\mathbf{v}_2=(0,-i)$

25. (a) $\mathbf{v}_1=\left(\dfrac{i}{\sqrt{3}},\dfrac{i}{\sqrt{3}},\dfrac{i}{\sqrt{3}}\right),\mathbf{v}_2=\left(-\dfrac{i}{\sqrt{2}},\dfrac{i}{\sqrt{2}},0\right),\mathbf{v}_3=\left(\dfrac{i}{\sqrt{6}},\dfrac{i}{\sqrt{6}},-\dfrac{2i}{\sqrt{6}}\right)$

   (b) $\mathbf{v}_1=(i,0,0),\mathbf{v}_2=\left(0,\dfrac{7i}{\sqrt{53}},\dfrac{-2i}{\sqrt{53}}\right),\mathbf{v}_3=\left(0,\dfrac{2i}{\sqrt{53}},\dfrac{7i}{\sqrt{53}}\right)$

26. $\left(0,\dfrac{2i}{\sqrt{5}},\dfrac{i}{\sqrt{5}},0\right),\left(\dfrac{5i}{\sqrt{30}},-\dfrac{i}{\sqrt{30}},\dfrac{2i}{\sqrt{30}},0\right),\left(\dfrac{i}{\sqrt{10}},\dfrac{i}{\sqrt{10}},-\dfrac{2i}{\sqrt{10}},-\dfrac{2i}{\sqrt{10}}\right),\left(\dfrac{i}{\sqrt{15}},\dfrac{i}{\sqrt{15}},-\dfrac{2i}{\sqrt{15}},\dfrac{3i}{\sqrt{15}}\right)$

27. $\mathbf{v}_1=\left(0,\dfrac{i}{\sqrt{3}},\dfrac{1-i}{\sqrt{3}}\right),\mathbf{v}_2=\left(-\dfrac{3i}{\sqrt{15}},\dfrac{2}{\sqrt{15}},\dfrac{1+i}{\sqrt{15}}\right)$

28. $\mathbf{w}_1=\left(-\dfrac{5i}{4},-\dfrac{i}{4},\dfrac{5i}{4},\dfrac{9i}{4}\right),\mathbf{w}_2=\left(\dfrac{i}{4},\dfrac{9i}{4},\dfrac{19i}{4},-\dfrac{9i}{4}\right)$ 36. $\mathbf{u}=-\sqrt{3}i\mathbf{v}_1+\dfrac{3}{\sqrt{6}}\mathbf{v}_2-\dfrac{1}{\sqrt{2}}\mathbf{v}_3$

EXERCISE SET 10.6 (page 523)

1. (a) $\begin{bmatrix}-2i & 4 & 5-i \\ 1+i & 3-i & 0\end{bmatrix}$ (b) $\begin{bmatrix}-2i & 4 & -i \\ 1+i & 5+7i & 3 \\ -1-i & i & 1\end{bmatrix}$ (c) $\begin{bmatrix}-7i \\ 0 \\ 3i\end{bmatrix}$ (d) $\begin{bmatrix}\bar{a}_{11} & \bar{a}_{21} \\ \bar{a}_{12} & \bar{a}_{22} \\ \bar{a}_{13} & \bar{a}_{23}\end{bmatrix}$

2. (b), (d), (e)   3. $k=3+5i,\, l=i,\, m=2-4i$ 4. (a), (b)

5. (a) $A^{-1}=\begin{bmatrix}\dfrac{3}{5} & -\dfrac{4}{5} \\ -\dfrac{4}{5}i & -\dfrac{3}{5}i\end{bmatrix}$ (b) $A^{-1}=\begin{bmatrix}\dfrac{1}{\sqrt{2}} & \dfrac{-1+i}{2} \\ \dfrac{1}{\sqrt{2}} & \dfrac{1-i}{2}\end{bmatrix}$ (c) $A^{-1}=\begin{bmatrix}\frac{1}{4}(\sqrt{3}-i) & \frac{1}{4}(1-i\sqrt{3}) \\ \frac{1}{4}(1+\sqrt{3}i) & \frac{1}{4}(1+\sqrt{3}i)\end{bmatrix}$

(d) $A^{-1}=\begin{bmatrix}\dfrac{1-i}{2} & -\dfrac{i}{\sqrt{3}} & \dfrac{3-i}{2\sqrt{15}} \\ -\dfrac{1}{2} & \dfrac{1}{\sqrt{3}} & \dfrac{4-3i}{2\sqrt{15}} \\ \dfrac{1}{2} & \dfrac{i}{\sqrt{3}} & -\dfrac{5i}{2\sqrt{15}}\end{bmatrix}$ 7. $P=\begin{bmatrix}\dfrac{-1+i}{\sqrt{3}} & \dfrac{1-i}{\sqrt{6}} \\ \dfrac{1}{\sqrt{3}} & \dfrac{2}{\sqrt{6}}\end{bmatrix}$; $P^{-1}AP=\begin{bmatrix}3 & 0 \\ 0 & 6\end{bmatrix}$

8. $P = \begin{bmatrix} -i/\sqrt{2} & i/\sqrt{2} \\ 1/\sqrt{2} & 1/\sqrt{2} \end{bmatrix}$; $P^{-1}AP = \begin{bmatrix} 4 & 0 \\ 0 & 2 \end{bmatrix}$

9. $P = \begin{bmatrix} -\dfrac{1+i}{\sqrt{6}} & \dfrac{1+i}{\sqrt{3}} \\ \dfrac{2}{\sqrt{6}} & \dfrac{1}{\sqrt{3}} \end{bmatrix}$; $P^{-1}AP = \begin{bmatrix} 2 & 0 \\ 0 & 8 \end{bmatrix}$

10. $P = \begin{bmatrix} -\dfrac{2}{\sqrt{14}} & \dfrac{5}{\sqrt{35}} \\ \dfrac{3-i}{\sqrt{14}} & \dfrac{3-i}{\sqrt{35}} \end{bmatrix}$; $P^{-1}AP = \begin{bmatrix} -5 & 0 \\ 0 & 2 \end{bmatrix}$

11. $P = \begin{bmatrix} 0 & 1 & 0 \\ -\dfrac{1-i}{\sqrt{6}} & 0 & \dfrac{1-i}{\sqrt{3}} \\ \dfrac{2}{\sqrt{6}} & 0 & \dfrac{1}{\sqrt{3}} \end{bmatrix}$; $P^{-1}AP = \begin{bmatrix} 1 & 0 & 0 \\ 0 & 5 & 0 \\ 0 & 0 & -2 \end{bmatrix}$

12. $P = \begin{bmatrix} \dfrac{i}{\sqrt{2}} & 0 & -\dfrac{i}{\sqrt{2}} \\ -\dfrac{1}{2} & \dfrac{1}{\sqrt{2}} & -\dfrac{1}{2} \\ \dfrac{1}{2} & \dfrac{1}{\sqrt{2}} & \dfrac{1}{2} \end{bmatrix}$; $P^{-1}AP = \begin{bmatrix} 1 & 0 & 0 \\ 0 & 2 & 0 \\ 0 & 0 & 3 \end{bmatrix}$

13. $\lambda = 2 \pm i\sqrt{15}$; no, since A has complex entries.

14. $\begin{bmatrix} 0 & i \\ -i & 0 \end{bmatrix}$ is one possibility.

CHAPTER 10 SUPPLEMENTARY EXERCISES (page 525)

3. $\begin{bmatrix} -i \\ 1 \\ 0 \end{bmatrix}, \begin{bmatrix} 1 \\ 0 \\ 1 \end{bmatrix}$ is one possibility.   **5.** $\lambda = 1, \omega, \omega^2(=\overline{\omega})$ **10.** (b) Dimension $= 2$

EXERCISE SET 11.1 (page 532)

1. (a) $y = 3x - 4$ **2.** (a) $x^2 + y^2 - 4x - 6y + 4 = 0$ or $(x - 2)^2 + (y - 3)^2 = 9$
   (b) $y = -2x + 1$ (b) $x^2 + y^2 + 2x - 4y - 20 = 0$ or $(x + 1)^2 + (y - 2)^2 = 25$

3. $4x^2 + 8xy + 4y^2 - 8x + 4y = 0$ (a parabola)   **4.** (a) $x + 2y + z = 0$ (b) $-x + y - 2z + 1 = 0$

5. (a) $x^2 + y^2 + z^2 - 2x - 4y - 2z = -2$ or $(x - 1)^2 + (y - 2)^2 + (z - 1)^2 = 4$
   (b) $x^2 + y^2 + z^2 - 2x - 2y = 3$ or $(x - 1)^2 + (y - 1)^2 + z^2 = 5$ **9.** $\begin{vmatrix} y & x^2 & x & 1 \\ y_1 & x_1^2 & x_1 & 1 \\ y_2 & x_2^2 & x_2 & 1 \\ y_3 & x_3^3 & x_3 & 1 \end{vmatrix} = 0$

EXERCISE SET 11.3 (page 551)

1.

2.

3.

4.

5.

6.

EXERCISE SET 11.4 (page 564)

1. (a) $\begin{bmatrix} 0 & 0 & 0 & 1 \\ 1 & 0 & 1 & 1 \\ 1 & 1 & 0 & 1 \\ 0 & 0 & 0 & 0 \end{bmatrix}$ (b) $\begin{bmatrix} 0 & 1 & 1 & 0 & 0 \\ 0 & 0 & 0 & 0 & 1 \\ 1 & 0 & 0 & 1 & 0 \\ 0 & 0 & 1 & 0 & 0 \\ 0 & 0 & 1 & 0 & 0 \end{bmatrix}$ (c) $\begin{bmatrix} 0 & 1 & 0 & 1 & 0 & 0 \\ 1 & 0 & 0 & 0 & 0 & 0 \\ 0 & 1 & 0 & 1 & 1 & 1 \\ 0 & 0 & 0 & 0 & 0 & 1 \\ 0 & 0 & 0 & 0 & 0 & 1 \\ 0 & 0 & 1 & 0 & 1 & 0 \end{bmatrix}$

2. (a)   (b)   (c)

3. (a)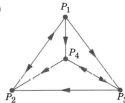

(b) 1-step: $P_1 \rightarrow P_2$

2-step: $P_1 \rightarrow P_4 \rightarrow P_2$
$P_1 \rightarrow P_3 \rightarrow P_2$

3-step: $P_1 \rightarrow P_2 \rightarrow P_1 \rightarrow P_2$
$P_1 \rightarrow P_3 \rightarrow P_4 \rightarrow P_2$
$P_1 \rightarrow P_4 \rightarrow P_3 \rightarrow P_2$

(c) 1-step: $P_1 \rightarrow P_4$

2-step: $P_1 \rightarrow P_3 \rightarrow P_4$

3-step: $P_1 \rightarrow P_2 \rightarrow P_1 \rightarrow P_4$
$P_1 \rightarrow P_4 \rightarrow P_3 \rightarrow P_4$

4. (a) $\{P_1, P_2, P_3\}$ (b) $\{P_3, P_4, P_5\}$ (c) $\{P_2, P_4, P_6, P_8\}$ and $\{P_4, P_5, P_6\}$ 5. (a) None   (b) $\{P_3, P_4, P_6\}$

6. $\begin{bmatrix} 0 & 0 & 1 & 1 \\ 1 & 0 & 0 & 0 \\ 0 & 1 & 0 & 1 \\ 0 & 1 & 0 & 0 \end{bmatrix}$ Power of $P_1 = 5$
Power of $P_2 = 3$
Power of $P_3 = 4$
Power of $P_4 = 2$

7. First, A; second, B and E (tie); fourth, C; fifth, D

EXERCISE SET 11.5 (page 575)

1. (a) $-5/8$ (b) $[0 \ 1 \ 0]$ (c) $[1 \ 0 \ 0 \ 0]^t$ 2. Let $A = \begin{bmatrix} 1 & 1 \\ 1 & 1 \end{bmatrix}$, for example.

3. (a) $\mathbf{p}^* = [0 \ 1]$, $\mathbf{q}^* = \begin{bmatrix} 0 \\ 1 \end{bmatrix}$, $v = 3$ (b) $\mathbf{p}^* = [0 \ 1 \ 0]$, $\mathbf{q}^* = \begin{bmatrix} 1 \\ 0 \end{bmatrix}$, $v = 2$

(c) $\mathbf{p}^* = [0 \ 0 \ 1]$, $\mathbf{q}^* = \begin{bmatrix} 0 \\ 1 \\ 0 \end{bmatrix}$, $v = 2$ (d) $\mathbf{p}^* = [0 \ 1 \ 0 \ 0]$, $\mathbf{q}^* = \begin{bmatrix} 1 \\ 0 \\ 0 \end{bmatrix}$, $v = -2$

4. (a) $\mathbf{p}^* = [\frac{5}{8} \ \frac{3}{8}]$, $\mathbf{q}^* = \begin{bmatrix} \frac{1}{8} \\ \frac{7}{8} \end{bmatrix}$, $v = \frac{27}{8}$ (b) $\mathbf{p}^* = [\frac{2}{3} \ \frac{1}{3}]$, $\mathbf{q}^* = \begin{bmatrix} \frac{1}{6} \\ \frac{5}{6} \end{bmatrix}$, $v = \frac{70}{3}$

(c) $\mathbf{p}^* = [1 \ 0]$, $\mathbf{q}^* = \begin{bmatrix} 1 \\ 0 \end{bmatrix}$, $v = 3$ (d) $\mathbf{p}^* = [\frac{1}{3} \ \frac{2}{3}]$, $\mathbf{q}^* = \begin{bmatrix} \frac{3}{5} \\ \frac{2}{5} \end{bmatrix}$, $v = \frac{19}{5}$

(e) $\mathbf{p}^* = [\frac{3}{13} \ \frac{10}{13}]$, $\mathbf{q}^* = \begin{bmatrix} \frac{1}{13} \\ \frac{12}{13} \end{bmatrix}$, $v = -\frac{29}{13}$ 5. $\mathbf{p}^* = [\frac{13}{20} \ \frac{7}{20}]$, $\mathbf{q}^* = \begin{bmatrix} \frac{11}{20} \\ \frac{9}{20} \end{bmatrix}$, $v = -\frac{3}{20}$

EXERCISE SET 11.6 (page 585)

1. (a) $\begin{bmatrix} 17 & 4 & 10 \\ 15 & 5 & 8 \\ 18 & 7 & 11 \end{bmatrix}$ or $\begin{bmatrix} 17 & 4 & 10 \\ 15 & 5 & 8 \\ 18 & 7 & 11 \end{bmatrix}$ (b) $\begin{bmatrix} 3 & -2 & 0 & 1 \\ 5 & 3 & -3 & 4 \\ 2 & 7 & 5 & 3 \\ 5 & -2 & 0 & 1 \end{bmatrix}$ or $\begin{bmatrix} 3 & -2 & 0 & 1 \\ 5 & 3 & -3 & 4 \\ 2 & 7 & 5 & 3 \\ 5 & -2 & 0 & 1 \end{bmatrix}$

cost $= 30$ cost $= 30$ cost $= -2$ cost $= -2$

(c) $\begin{bmatrix} 12 & 9 & 7 & 7 & \boxed{10} \\ 15 & 11 & \boxed{8} & 13 & 14 \\ 9 & \boxed{6} & 5 & 12 & 12 \\ 6 & 9 & 13 & \boxed{7} & 10 \\ \boxed{8} & 13 & 12 & 9 & 13 \end{bmatrix}$ or $\begin{bmatrix} 12 & 9 & 7 & \boxed{7} & 10 \\ 15 & 11 & \boxed{8} & 13 & 14 \\ 9 & \boxed{6} & 5 & 12 & 12 \\ 6 & 9 & 13 & 7 & \boxed{10} \\ \boxed{8} & 13 & 12 & 9 & 13 \end{bmatrix}$ or $\begin{bmatrix} 12 & 9 & 7 & 7 & \boxed{10} \\ 15 & 11 & \boxed{8} & 13 & 14 \\ 9 & \boxed{6} & 5 & 12 & 12 \\ \boxed{6} & 9 & 13 & 7 & 10 \\ 8 & 13 & 12 & \boxed{9} & 13 \end{bmatrix}$

$\quad$ cost $= 39$ $\qquad\qquad$ cost $= 39$ $\qquad\qquad$ cost $= 39$

3. $\begin{bmatrix} -15 & 0 & 0 & 0 \\ 0 & 50 & 20 & 25 \\ 35 & 5 & 0 & 10 \\ 0 & 65 & 50 & 65 \end{bmatrix}$ $\qquad$ 4. $\begin{bmatrix} 0 & 3 & 0 & 4 & 0 \\ 4 & 0 & 6 & 1 & 5 \\ 3 & 1 & 0 & 4 & 0 \\ 0 & 1 & 6 & 2 & 1 \\ 0 & 0 & 0 & 0 & 3 \end{bmatrix}$

5. (a) Bidder 1 receives Coin 3 ($210).
   Bidder 2 receives Coin 1 ($175).
   Bidder 3 receives Coin 2 ($85).
   Bidder 4 receives nothing.
   Bidder 5 receives Coin 4 ($160).

   Sum of bids = $630

   (b) Repeat bidder 2's row twice in the cost matrix; add two dummy coins to end up with a 6×6 cost matrix. Optimal assignment is

   Bidder 1 receives nothing.
   Bidder 2 receives Coins 1 and 3 ($175 + $230).
   Bidder 3 receives Coin 2 ($85).
   Bidder 4 receives nothing.
   Bidder 5 receives Coin 4 ($160).

   Sum of bids = $650

6. P—Pete $\qquad$ 2B—Jill or Ann $\qquad$ LF—Herb
   C—Liz $\qquad$ 3B—Alex $\qquad$ CF—John or Lois
   1B—Sam $\qquad$ SS—Ann or Jill $\qquad$ RF—Lois or John

EXERCISE SET 11.7 (page 597)

1. (a) $\mathbf{x}^{(1)} = \begin{bmatrix} .4 \\ .6 \end{bmatrix}$, $\quad \mathbf{x}^{(2)} = \begin{bmatrix} .46 \\ .54 \end{bmatrix}$, $\quad \mathbf{x}^{(3)} = \begin{bmatrix} .454 \\ .546 \end{bmatrix}$, $\quad \mathbf{x}^{(4)} = \begin{bmatrix} .4546 \\ .5454 \end{bmatrix}$, $\quad \mathbf{x}^{(5)} = \begin{bmatrix} .45454 \\ .54546 \end{bmatrix}$

   (b) P is regular since all entries of P are positive; $\mathbf{q} = \begin{bmatrix} \frac{5}{11} \\ \frac{6}{11} \end{bmatrix}$

2. (a) $\mathbf{x}^{(1)} = \begin{bmatrix} .7 \\ .2 \\ .1 \end{bmatrix}$, $\quad \mathbf{x}^{(2)} = \begin{bmatrix} .23 \\ .52 \\ .25 \end{bmatrix}$, $\quad \mathbf{x}^{(3)} = \begin{bmatrix} .273 \\ .396 \\ .331 \end{bmatrix}$

   (b) P is regular, since all entries of P are positive: $\mathbf{q} = \begin{bmatrix} \frac{22}{72} \\ \frac{29}{72} \\ \frac{21}{72} \end{bmatrix}$ $\qquad$ 3. (a) $\begin{bmatrix} \frac{9}{17} \\ \frac{8}{17} \end{bmatrix}$ $\quad$ (b) $\begin{bmatrix} \frac{26}{45} \\ \frac{19}{45} \end{bmatrix}$ $\quad$ (c) $\begin{bmatrix} \frac{3}{19} \\ \frac{4}{19} \\ \frac{12}{19} \end{bmatrix}$

4. (a) $P^n = \begin{bmatrix} (\frac{1}{2})^n & 0 \\ 1 - (\frac{1}{2})^n & 1 \end{bmatrix}$, $\quad n = 1, 2, \ldots$ by induction

   Thus, no integer power of P has all positive entries.

   (b) $P^n \to \begin{bmatrix} 0 & 0 \\ 1 & 1 \end{bmatrix}$ as n increases, and so $P^n \mathbf{x}^{(0)} \to \begin{bmatrix} 0 \\ 1 \end{bmatrix}$ for any $\mathbf{x}^{(0)}$ as n increases.

   (c) The entries of the limiting vector $\begin{bmatrix} 0 \\ 1 \end{bmatrix}$ are not all positive.

6. $P^2 = \begin{bmatrix} \frac{1}{2} & \frac{1}{4} & \frac{1}{4} \\ \frac{1}{4} & \frac{1}{2} & \frac{1}{4} \\ \frac{1}{4} & \frac{1}{4} & \frac{1}{2} \end{bmatrix}$ has all positive entries; $\mathbf{q} = \begin{bmatrix} \frac{1}{3} \\ \frac{1}{3} \\ \frac{1}{3} \end{bmatrix}$ 7. $\frac{10}{13}$

8. $54\frac{1}{6}$ percent in region 1, $16\frac{2}{3}$ percent in region 2, and $29\frac{1}{6}$ percent in region 3

EXERCISE SET 11.8 (page 606)

1. (a) $\begin{bmatrix} 2 \\ 3 \end{bmatrix}$ (b) $\begin{bmatrix} 6 \\ 5 \\ 6 \end{bmatrix}$ (c) $\begin{bmatrix} 78 \\ 54 \\ 79 \end{bmatrix}$

2. (a) Use Corollary 11.8.4; all row sums are less than one.
   (b) Use Corollary 11.8.5; all column sums are less than one.
   (c) Use Theorem 11.8.3, with $\mathbf{x} = \begin{bmatrix} 2 \\ 1 \\ 1 \end{bmatrix} > C\mathbf{x} = \begin{bmatrix} 1.9 \\ .9 \\ .9 \end{bmatrix}$.

3. E^2 has all positive entries.   4. Price of tomatoes, $120.00; price of corn, $100.00; price of lettuce, $106.67

5. $1256 for the CE, $1448 for the EE, $1556 for the ME

EXERCISE SET 11.9 (page 615)

1. The second class; $15,000   2. $223   3. $1:1.90:3.02:4.24:5.00$

4. $s/(g_1^{-1} + g_2^{-1} + \cdots + g_{k-1}^{-1})$ 5. $1:2:3\cdots:n-1$

EXERCISE SET 11.10 (page 623)

1. (a) $\begin{bmatrix} 0 & 1 & 1 & 0 \\ 0 & 0 & 1 & 1 \\ 0 & 0 & 0 & 0 \end{bmatrix}$ (b) $\begin{bmatrix} 0 & 1\frac{1}{2} & 1\frac{1}{2} & 0 \\ 0 & 0 & \frac{1}{2} & \frac{1}{2} \\ 0 & 0 & 0 & 0 \end{bmatrix}$

(c) $\begin{bmatrix} -2 & -1 & -1 & -2 \\ -1 & -1 & 0 & 0 \\ 3 & 3 & 3 & 3 \end{bmatrix}$

(d) $\begin{bmatrix} 0 & .866 & 1.366 & .500 \\ 0 & -.500 & .366 & .866 \\ 0 & 0 & 0 & 0 \end{bmatrix}$

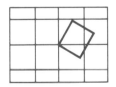

2. (b) $(0, 0, 0)$, $(1, 0, 0)$, $(1\frac{1}{2}, 1, 0)$, and $(\frac{1}{2}, 1, 0)$
   (c) $(0, 0, 0)$, $(1, .6, 0)(1, 1.6, 0)$, $(0, 1, 0)$

3. (a) $\begin{bmatrix} 1 & 0 & 0 \\ 0 & -1 & 0 \\ 0 & 0 & 1 \end{bmatrix}$ (b) $\begin{bmatrix} -1 & 0 & 0 \\ 0 & 1 & 0 \\ 0 & 0 & 1 \end{bmatrix}$

(c) $\begin{bmatrix} 1 & 0 & 0 \\ 0 & 1 & 0 \\ 0 & 0 & -1 \end{bmatrix}$

4. (a) $M_1 = \begin{bmatrix} \frac{1}{2} & 0 & 0 \\ 0 & 2 & 0 \\ 0 & 0 & \frac{1}{3} \end{bmatrix}$, $M_2 = \begin{bmatrix} \frac{1}{2} & \frac{1}{2} & \cdots & \frac{1}{2} \\ 0 & 0 & \cdots & 0 \\ 0 & 0 & \cdots & 0 \end{bmatrix}$, $M_3 = \begin{bmatrix} 1 & 0 & 0 \\ 0 & \cos 20° & -\sin 20° \\ 0 & \sin 20° & \cos 20° \end{bmatrix}$,

$M_4 = \begin{bmatrix} \cos(-45°) & 0 & \sin(-45°) \\ 0 & 1 & 0 \\ -\sin(-45°) & 0 & \cos(-45°) \end{bmatrix}$, $M_5 = \begin{bmatrix} 0 & -1 & 0 \\ 1 & 0 & 0 \\ 0 & 0 & 1 \end{bmatrix}$ (b) $P' = M_5 M_4 M_3 (M_1 P + M_2)$

5. (a) $M_1 = \begin{bmatrix} .3 & 0 & 0 \\ 0 & .5 & 0 \\ 0 & 0 & 1 \end{bmatrix}$, $M_2 = \begin{bmatrix} 1 & 0 & 0 \\ 0 & \cos 45° & -\sin 45° \\ 0 & \sin 45° & \cos 45° \end{bmatrix}$, $M_3 = \begin{bmatrix} 1 & 1 & \cdots & 1 \\ 0 & 0 & \cdots & 0 \\ 0 & 0 & \cdots & 0 \end{bmatrix}$,

$M_4 = \begin{bmatrix} \cos 35° & 0 & \sin 35° \\ 0 & 1 & 0 \\ -\sin 35° & 0 & \cos 35° \end{bmatrix}$, $M_5 = \begin{bmatrix} \cos(-45°) & -\sin(-45°) & 0 \\ \sin(-45°) & \cos(-45°) & 0 \\ 0 & 0 & 1 \end{bmatrix}$,

$M_6 = \begin{bmatrix} 0 & 0 & \cdots & 0 \\ 0 & 0 & \cdots & 0 \\ 1 & 1 & \cdots & 1 \end{bmatrix}$, $M_7 = \begin{bmatrix} 2 & 0 & 0 \\ 0 & 1 & 0 \\ 0 & 0 & 1 \end{bmatrix}$ (b) $P' = M_7 (M_5 M_4 (M_2 M_1 P + M_3) + M_6)$

6. $R_1 = \begin{bmatrix} \cos \beta & 0 & \sin \beta \\ 0 & 1 & 0 \\ -\sin \beta & 0 & \cos \beta \end{bmatrix}$, $R_2 = \begin{bmatrix} \cos \alpha & -\sin \alpha & 0 \\ \sin \alpha & \cos \alpha & 0 \\ 0 & 0 & 1 \end{bmatrix}$, $R_3 = \begin{bmatrix} \cos \theta & 0 & \sin \theta \\ 0 & 1 & 0 \\ -\sin \theta & 0 & \cos \theta \end{bmatrix}$,

$R_4 = \begin{bmatrix} \cos \alpha & \sin \alpha & 0 \\ -\sin \alpha & \cos \alpha & 0 \\ 0 & 0 & 1 \end{bmatrix}$, $R_5 = \begin{bmatrix} \cos \beta & 0 & -\sin \beta \\ 0 & 1 & 0 \\ \sin \beta & 0 & \cos \beta \end{bmatrix}$

7. (a) $M = \begin{bmatrix} 1 & 0 & 0 & x_0 \\ 0 & 1 & 0 & y_0 \\ 0 & 0 & 1 & z_0 \\ 0 & 0 & 0 & 1 \end{bmatrix}$ (b) $\begin{bmatrix} 1 & 0 & 0 & -5 \\ 0 & 1 & 0 & 9 \\ 0 & 0 & 1 & -3 \\ 0 & 0 & 0 & 1 \end{bmatrix}$

EXERCISE SET 11.11 (page 636)

1. (a) $\begin{bmatrix} t_1 \\ t_2 \\ t_3 \\ t_4 \end{bmatrix} = \begin{bmatrix} 0 & \frac{1}{4} & \frac{1}{4} & 0 \\ \frac{1}{4} & 0 & 0 & \frac{1}{4} \\ \frac{1}{4} & 0 & 0 & \frac{1}{4} \\ 0 & \frac{1}{4} & \frac{1}{4} & 0 \end{bmatrix} \begin{bmatrix} t_1 \\ t_2 \\ t_3 \\ t_4 \end{bmatrix} + \begin{bmatrix} 0 \\ \frac{1}{2} \\ 0 \\ \frac{1}{2} \end{bmatrix}$ (b) $\mathbf{t} = \begin{bmatrix} \frac{1}{4} \\ \frac{3}{4} \\ \frac{1}{4} \\ \frac{3}{4} \end{bmatrix}$

(c) $\mathbf{t}^{(1)} = \begin{bmatrix} 0 \\ \frac{1}{2} \\ 0 \\ \frac{5}{2} \end{bmatrix}$, $\quad \mathbf{t}^{(2)} = \begin{bmatrix} \frac{1}{8} \\ \frac{5}{8} \\ \frac{1}{8} \\ \frac{5}{8} \end{bmatrix}$, $\quad \mathbf{t}^{(3)} = \begin{bmatrix} \frac{3}{16} \\ \frac{11}{16} \\ \frac{3}{16} \\ \frac{11}{16} \end{bmatrix}$, $\quad \mathbf{t}^{(4)} = \begin{bmatrix} \frac{7}{32} \\ \frac{23}{32} \\ \frac{7}{32} \\ \frac{23}{32} \end{bmatrix}$, $\quad \mathbf{t}^{(5)} = \begin{bmatrix} \frac{15}{64} \\ \frac{47}{64} \\ \frac{15}{64} \\ \frac{47}{64} \end{bmatrix}$, $\quad \mathbf{t}^{(5)} - \mathbf{t} = \begin{bmatrix} -\frac{1}{64} \\ -\frac{1}{64} \\ -\frac{1}{64} \\ -\frac{1}{64} \end{bmatrix}$

(d) for t_1, 5.4 percent; for t_2, -1.7 percent   2. $\frac{1}{2}$ 3. $\mathbf{t}^{(1)} = [\frac{3}{4} \quad \frac{5}{4} \quad \frac{2}{4} \quad \frac{5}{4} \quad \frac{4}{4} \quad \frac{2}{4} \quad \frac{5}{4} \quad \frac{4}{4} \quad \frac{3}{4}]^t$

$\mathbf{t}^{(2)} = [\frac{13}{16} \quad \frac{18}{16} \quad \frac{9}{16} \quad \frac{22}{16} \quad \frac{13}{16} \quad \frac{7}{16} \quad \frac{21}{16} \quad \frac{16}{16} \quad \frac{10}{18}]^t$

EXERCISE SET 11.12 (page 648)

2. $\begin{cases} a_n = \frac{1}{4} + (\frac{1}{2})^{n+1}(a_0 - c_0) \\ b_n = \frac{1}{2} \\ c_n = \frac{1}{4} - (\frac{1}{2})^{n+1}(a_0 - c_0) \end{cases}$ $\quad n = 1, 2, \ldots$ $\quad \begin{cases} a_n \to \frac{1}{4} \\ b_n = \frac{1}{2} \\ c_n \to \frac{1}{4} \end{cases}$ as $n \to \infty$

3. $\begin{cases} a_{2n+1} = \dfrac{2}{3} + \dfrac{1}{6(4)^n}(2a_0 - b_0 - 4c_0) \\ b_{2n+1} = \dfrac{1}{3} - \dfrac{1}{6(4)^n}(2a_0 - b_0 - 4c_0) \quad n = 0, 1, 2, \ldots \\ c_{2n+1} = 0 \end{cases}$

$\begin{cases} a_{2n} = \dfrac{5}{12} + \dfrac{1}{6(4)^n}(2a_0 - b_0 - 4c_0) \\ b_{2n} = \dfrac{1}{2} \quad\quad\quad\quad\quad\quad n = 1, 2, \ldots \\ c_{2n} = \dfrac{1}{12} - \dfrac{1}{6(4)^n}(2a_0 - b_0 - 4c_0) \end{cases}$

4. Eigenvalues: $\lambda_1 = 1$, $\lambda_2 = \frac{1}{2}$; eigenvectors: $\mathbf{e}_1 = \begin{bmatrix} 1 \\ 0 \end{bmatrix}$, $\mathbf{e}_2 = \begin{bmatrix} 1 \\ -1 \end{bmatrix}$

5. 12 generations; from 25 percent to .006 percent

6. $\mathbf{x}^{(n)} = \begin{bmatrix} \frac{1}{2} + \frac{1}{2^{2n+3}}[(-3 - \sqrt{5})(1 + \sqrt{5})^{n+1} + (-3 + \sqrt{5})(1 - \sqrt{5})^{n+1}] \\ \frac{1}{2^{2n+1}}[(1 + \sqrt{5})^{n+1} + (1 - \sqrt{5})^{n+1}] \\ \frac{1}{2^{2n+1}}[(1 + \sqrt{5})^n + (1 - \sqrt{5})^n] \\ \frac{1}{2^{2n+1}}[(1 + \sqrt{5})^n + (1 - \sqrt{5})^n] \\ \frac{1}{2^{2n+1}}[(1 + \sqrt{5})^{n+1} + (1 - \sqrt{5})^{n+1}] \\ \frac{1}{2} + \frac{1}{2^{2n+3}}[(-3 - \sqrt{5})(1 + \sqrt{5})^{n+1} + (-3 + \sqrt{5})(1 - \sqrt{5})^{n+1}] \end{bmatrix}$; $\mathbf{x}^{(n)} \to \begin{bmatrix} \frac{1}{2} \\ 0 \\ 0 \\ 0 \\ 0 \\ \frac{1}{2} \end{bmatrix}$ as $n \to \infty$

8. $\begin{bmatrix} 1 & 0 & 0 & 0 \\ 0 & 0 & 0 & 0 \\ 0 & 0 & 0 & 0 \\ 0 & 0 & 0 & 1 \end{bmatrix}$

EXERCISE SET 11.13 (page 658)

1. (a) $\lambda_1 = \frac{3}{2}$, $\quad x_1 = \begin{bmatrix} 1 \\ \frac{1}{3} \end{bmatrix}$

(b) $x^{(1)} = \begin{bmatrix} 100 \\ 50 \end{bmatrix}$, $\quad x^{(2)} = \begin{bmatrix} 175 \\ 50 \end{bmatrix}$, $\quad x^{(3)} = \begin{bmatrix} 250 \\ 88 \end{bmatrix}$, $\quad x^{(4)} = \begin{bmatrix} 382 \\ 125 \end{bmatrix}$, $\quad x^{(5)} = \begin{bmatrix} 570 \\ 191 \end{bmatrix}$

(c) $x^{(6)} = Lx^{(5)} = \begin{bmatrix} 857 \\ 285 \end{bmatrix}$, $\quad x^{(6)} \approx \lambda_1 x^{(5)} = \begin{bmatrix} 855 \\ 287 \end{bmatrix}$ **7.** 2.375   **8.** 1.49611

EXERCISE SET 11.14 (page 667)

1. (a) Yield $= 33\frac{1}{3}$ percent of population; $x_1 = \begin{bmatrix} 1 \\ \frac{1}{3} \\ \frac{1}{18} \end{bmatrix}$

(b) Yield $= 45.8$ percent of population; $x_1 = \begin{bmatrix} 1 \\ \frac{1}{2} \\ \frac{1}{8} \end{bmatrix}$; harvest 57.9 percent of youngest age class

2. $x_1 = \begin{bmatrix} 1.000 \\ .845 \\ .824 \\ .795 \\ .755 \\ .699 \\ .626 \\ .532 \\ 0 \\ 0 \\ 0 \\ 0 \end{bmatrix}$, $\quad Lx_1 = \begin{bmatrix} 2.090 \\ .845 \\ .824 \\ .795 \\ .755 \\ .699 \\ .626 \\ .532 \\ .418 \\ 0 \\ 0 \\ 0 \end{bmatrix}$, $\quad \dfrac{1.090 + .418}{7.584} = .199$

4. $h_I = (R - 1)/(a_I b_1 b_2 \cdots b_{I-1} + \cdots + a_n b_1 b_2 \cdots b_{n-1})$

5. $h_I = \dfrac{a_1 + a_2 b_1 + \cdots + (a_{J-1} b_1 b_2 \cdots b_{J-2}) - 1}{a_I b_1 b_2 \cdots b_{I-1} + \cdots + a_{J-1} b_1 b_2 \cdots b_{J-2}}$

EXERCISE SET 11.15 (page 675)

1. $\dfrac{\pi^2}{3} + 4 \cos t + \cos 2t + \dfrac{4}{9} \cos 3t$

2. $\dfrac{T^2}{3} + \dfrac{T^2}{\pi^2} \left(\cos \dfrac{2\pi}{T} t + \dfrac{1}{2^2} \cos \dfrac{4\pi}{T} t + \dfrac{1}{3^2} \cos \dfrac{6\pi}{T} t + \dfrac{1}{4^2} \cos \dfrac{8\pi}{T} t \right)$

$$- \dfrac{T^2}{\pi} \left(\sin \dfrac{2\pi}{T} t + \dfrac{1}{2} \sin \dfrac{4\pi}{T} t + \dfrac{1}{3} \sin \dfrac{6\pi}{T} t + \dfrac{1}{4} \sin \dfrac{8\pi}{T} t \right)$$

3. $\dfrac{1}{\pi} + \dfrac{1}{2} \sin t - \dfrac{2}{3\pi} \cos 2t - \dfrac{2}{15\pi} \cos 4t$

4. $\dfrac{4}{\pi}\left(\dfrac{1}{2} - \dfrac{1}{1 \cdot 3}\cos t - \dfrac{1}{3 \cdot 5}\cos 2t - \dfrac{1}{5 \cdot 7}\cos 3t - \cdots - \dfrac{1}{(2n-1)(2n+1)}\cos nt\right)$

5. $\dfrac{T}{4} - \dfrac{8T}{\pi^2}\left(\dfrac{1}{2^2}\cos\dfrac{2\pi t}{T} + \dfrac{1}{6^2}\cos\dfrac{6\pi t}{T} + \dfrac{1}{10^2}\cos\dfrac{10\pi t}{T} + \cdots + \dfrac{1}{(2n)^2}\cos\dfrac{2n\pi t}{T}\right)$

EXERCISE SET 11.16 (page 686)

2. (a) $S(x) = -.12643(x - .4)^3 - .20211(x - .4)^2 + .92158(x - .4) + .38942$
   (b) $S(.5) = .47943$; error $= 0$ percent   3. (a) The cubic runout spline   (b) $S(x) = 3x^3 - 2x^2 + 5x + 1$

4. $S(x) = \begin{cases} -.00000042(x + 10)^3 & +.000214(x + 10) + .99815, & -10 \le x \le 0 \\ \ .00000024(x)^3 & -.0000126(x)^2 & +.000088(x) & +.99987, & 0 \le x \le 10 \\ -.00000004(x - 10)^3 - .0000054(x - 10)^2 - .000092(x - 10) + .99973, & 10 \le x \le 20 \\ \ .00000022(x - 20)^3 - .0000066(x - 20)^2 - .000212(x - 20) + .99823, & 20 \le x \le 30 \end{cases}$

Maximum at $(x, S(x)) = (3.93, 1.00004)$

5. $S(x) = \begin{cases} .00000009(x + 10)^3 - .0000121(x + 10)^2 + .000282(x + 10) + .99815, & -10 \le x \le 0 \\ .00000009(x)^3 & -.0000093(x)^2 & +.000070(x) & +.99987, & 0 \le x \le 10 \\ .00000004(x + 10)^3 - .0000066(x - 10)^2 - .000087(x - 10) + .99973, & 10 \le x \le 20 \\ .00000004(x - 20)^3 - .0000053(x - 20)^2 - .000207(x - 20) + .99823, & 20 < x < 30 \end{cases}$

Maximum at $(x, S(x)) = (4.00, 1.00001)$

6. (b) $\begin{bmatrix} 4 & 1 & 0 & 0 & \cdots & 0 & 0 & 0 & 1 \\ 1 & 4 & 1 & 0 & \cdots & 0 & 0 & 0 & 0 \\ 0 & 1 & 4 & 1 & \cdots & 0 & 0 & 0 & 0 \\ \vdots & \vdots & \vdots & \vdots & & \vdots & \vdots & \vdots & \vdots \\ 0 & 0 & 0 & 0 & \cdots & 0 & 1 & 4 & 1 \\ 1 & 0 & 0 & 0 & \cdots & 0 & 0 & 1 & 4 \end{bmatrix} \begin{bmatrix} M_1 \\ M_2 \\ M_3 \\ \vdots \\ M_{n-2} \\ M_{n-1} \end{bmatrix} = \dfrac{6}{h^2} \begin{bmatrix} y_{n-1} - 2y_1 & +y_2 \\ y_1 & -2y_2 & +y_3 \\ y_2 & -2y_3 & +y_4 \\ & \vdots & \\ y_{n-3} - 2y_{n-2} + y_{n-1} \\ y_{n-2} - 2y_{n-1} + y_1 \end{bmatrix}$

7. (b) $\begin{bmatrix} 2 & 1 & 0 & 0 & \cdots & 0 & 0 & 0 & 0 \\ 1 & 4 & 1 & 0 & \cdots & 0 & 0 & 0 & 0 \\ 0 & 1 & 4 & 1 & \cdots & 0 & 0 & 0 & 0 \\ \vdots & \vdots & \vdots & \vdots & & \vdots & \vdots & \vdots & \vdots \\ 0 & 0 & 0 & 0 & \cdots & 0 & 1 & 4 & 1 \\ 0 & 0 & 0 & 0 & \cdots & 0 & 0 & 1 & 2 \end{bmatrix} \begin{bmatrix} M_1 \\ M_2 \\ M_3 \\ \vdots \\ M_{n-1} \\ M_n \end{bmatrix} = \dfrac{6}{h^2} \begin{bmatrix} -hy_1' & -y_1 & +y_2 \\ y_1 & -2y_2 & +y_3 \\ y_2 & -2y_3 & +y_4 \\ & \vdots & \\ y_{n-2} - 2y_{n-1} + y_n \\ y_{n-1} - y_n & +hy_n' \end{bmatrix}$

EXERCISE SET 11.17 (page 699)

1. (a) *GIYUOKEVBH*   (b) *SFANEFZWJH*

2. (a) $A^{-1} = \begin{bmatrix} 12 & 7 \\ 23 & 15 \end{bmatrix}$ (b) Not invertible   (c) $A^{-1} = \begin{bmatrix} 1 & 19 \\ 23 & 24 \end{bmatrix}$ (d) Not invertible

   (e) Not invertible   (f) $A^{-1} = \begin{bmatrix} 15 & 12 \\ 21 & 5 \end{bmatrix}$ 3. *WE LOVE MATH*

4. Deciphering matrix $= \begin{bmatrix} 7 & 15 \\ 6 & 5 \end{bmatrix}$; enciphering matrix $= \begin{bmatrix} 7 & 5 \\ 2 & 15 \end{bmatrix}$ 5. *THEY SPLIT THE ATOM*

6. *I HAVE COME TO BURY CAESAR*   7. A is invertible modulo 29 if and only if $\det(A) \ne 0 \pmod{29}$

EXERCISE SET 11.18 (page 713)

1. (c) $x_3^* = (\frac{31}{22}, \frac{27}{22})$ 2. (a) $\mathbf{x}_3^{(1)} = (1.40000, 1.20000)$ (b) Same as part (a)   (c) $\mathbf{x}_3^{(1)} = (9.55000, 25.65000)$

$\mathbf{x}_3^{(2)} = (1.41000, 1.23000)$ $\qquad\qquad\qquad\qquad$ $\mathbf{x}_3^{(2)} = (.59500, -1.21500)$

$\mathbf{x}_3^{(3)} = (1.40900, 1.22700)$ $\qquad\qquad\qquad\qquad$ $\mathbf{x}_3^{(3)} = (1.49050, 1.47150)$

$\mathbf{x}_3^{(4)} = (1.40910, 1.22730)$ $\qquad\qquad\qquad\qquad$ $\mathbf{x}_3^{(4)} = (1.40095, 1.20285)$

$\mathbf{x}_3^{(5)} = (1.40909, 1.22727)$ $\qquad\qquad\qquad\qquad$ $\mathbf{x}_3^{(5)} = (1.40991, 1.22972)$

$\mathbf{x}_3^{(6)} = (1.40909, 1.22727)$ $\qquad\qquad\qquad\qquad$ $\mathbf{x}_3^{(6)} = (1.40901, 1.22703)$

4. $\mathbf{x}_1^* = (1, 1)$, $\mathbf{x}_2^* = (2, 0)$, $\mathbf{x}_3^* = (1, 1)$ 7.

$$x_7 + x_8 + x_9 = 13.00$$
$$x_4 + x_5 + x_6 = 15.00$$
$$x_1 + x_2 + x_3 = 8.00$$
$$.82843(x_6 + x_8) + .58579x_9 = 14.79$$
$$1.41421(x_3 + x_5 + x_7) = 14.31$$
$$.82843(x_2 + x_4) + .58579x_1 = 3.81$$
$$x_3 + x_6 + x_9 = 18.00$$
$$x_2 + x_5 + x_8 = 12.00$$
$$x_1 + x_4 + x_7 = 6.00$$
$$.82843(x_2 + x_6) + .58579x_3 = 10.51$$
$$1.41421(x_1 + x_5 + x_9) = 16.13$$
$$.82843(x_4 + x_8) + .58579x_7 = 7.04$$

8. (a)

$$x_7 + x_8 + x_9 = 13.00$$
$$x_4 + x_5 + x_6 = 15.00$$
$$x_1 + x_2 + x_3 = 8.00$$
$$.04289(x_3 + x_5 + x_7) + .75000(x_6 + x_8) + .61396x_9 = 14.79$$
$$.91421(x_3 + x_5 + x_7) + .25000(x_2 + x_4 + x_6 + x_8) = 14.31$$
$$.04289(x_3 + x_5 + x_7) + .75000(x_2 + x_4) + .61396x_1 = 3.81$$
$$x_3 + x_6 + x_9 = 18.00$$
$$x_2 + x_5 + x_8 = 12.00$$
$$x_1 + x_4 + x_7 = 6.00$$
$$.04289(x_1 + x_5 + x_9) + .75000(x_2 + x_6) + .61396x_3 = 10.51$$
$$.91421(x_1 + x_5 + x_9) + .25000(x_2 + x_4 + x_6 + x_8) = 16.13$$
$$.04289(x_1 + x_5 + x_9) + .75000(x_4 + x_8) + .61396x_7 = 7.04$$

EXERCISE SET 11.19 (page 717)

1. $I_1 = \frac{255}{317}, I_2 = \frac{97}{317}, I_3 = \frac{158}{317}$ 2. $I_1 = \frac{3}{2}, I_2 = -\frac{3}{2}, I_3 = 1$ 3. $I_1 = \frac{13}{5}, I_2 = -\frac{2}{5}, I_3 = \frac{11}{5}$

4. $I_1 = \frac{1}{2}, I_2 = 0, I_3 = 0, I_4 = \frac{1}{2}, I_5 = \frac{1}{2}, I_6 = \frac{1}{2}$

EXERCISE SET 11.20 (page 736)

1. $T_i\left(\begin{bmatrix} x \\ y \end{bmatrix}\right) = \frac{12}{25}\begin{bmatrix} 1 & 0 \\ 0 & 1 \end{bmatrix}\begin{bmatrix} x \\ y \end{bmatrix} + \begin{bmatrix} e_i \\ f_i \end{bmatrix}$, $i = 1, 2, 3, 4$, where the four values of $\begin{bmatrix} e_i \\ f_i \end{bmatrix}$ are $\begin{bmatrix} 0 \\ 0 \end{bmatrix}$, $\begin{bmatrix} \frac{13}{23} \\ 0 \end{bmatrix}$, $\begin{bmatrix} 0 \\ \frac{13}{23} \end{bmatrix}$,

and $\begin{bmatrix} \frac{13}{23} \\ \frac{13}{23} \end{bmatrix}$. $d_H(S) = \ln(4)/\ln(\frac{25}{12}) = 1.889 \ldots$

2. $s \approx .47$. $d_H(S) \approx \ln(4)/\ln(1/.47) = 1.8. \ldots$ Rotation angles: $0°$ (upper left); $90°$ (upper right); $180°$ (lower left); $180°$ (lower right)

3. (a) (i) $s = \frac{1}{3}$; (ii) all rotation angles are 0°; (iii) $d_H(S) = \ln(7)/\ln(3) = 1.771. \ldots$ This set is a fractal.
   (b) (i) $s = \frac{1}{2}$; (ii) all rotation angles are 180°; (iii) $d_H(S) = \ln(3)/\ln(2) = 1.584. \ldots$ This set is a fractal.
   (c) (i) $s = \frac{1}{2}$; (ii) rotation angles: 90° (top); 180° (lower left); 180° (lower right);
   (iii) $d_H(S) = \ln(3)/\ln(2) = 1.584. \ldots$ This set is a fractal.
   (d) (i) $s = \frac{1}{2}$; (ii) rotation angles: 90° (upper left); 180° (upper right); 180° (lower right);
   (iii) $d_H(S) = \ln(3)/\ln(2) = 1.584. \ldots$ This set is a fractal.

4. $s = .8509 \ldots, \theta = -2.69° \ldots$ 5. $(0.766, 0.996)$ rounded to three decimal places

6. $d_H(S) = \ln(16)/\ln(4) = 2$ 7. $\ln(4)/\ln(\frac{4}{3}) = 4.818 \ldots$

8. $d_H(S) = \ln(8)/\ln(2) = 3$; the cube is not a fractal.

9. $k = 20$; $s = \frac{1}{3}$; $d_H(S) = \ln(20)/\ln(3) = 2.726 \ldots$; the set is a fractal.

10.

Initial set

First iterate

Second iterate

Third iterate

Fourth iterate

$d_H(S) = \ln(2)/\ln(3) = 0.6309 \ldots$

11. Area of $S_0 = 1$; area of $S_1 = \frac{8}{9} = 0.888 \ldots$; area of $S_2 = (\frac{8}{9})^2 = 0.790 \ldots$; area of $S_3 = (\frac{8}{9})^3 = 0.702 \ldots$;
    area of $S_4 = (\frac{8}{9})^4 = 0.624 \ldots$

INDEX